Lecture Notes in Mathematics

Edited by A. Dold and B. Eckmann

499

$\models$ ISILC Logic Conference

Proceedings of the International Summer Institute and Logic Colloquium, Kiel 1974

Edited by G. H. Müller, A. Oberschelp, and K. Potthoff

Springer-Verlag
Berlin · Heidelberg · New York 1975

Editors
Prof. Gert H. Müller
Mathematisches Institut
der Universität Heidelberg
Im Neuenheimer Feld 288
69 Heidelberg 1/BRD

Prof. Arnold Oberschelp
Philosophisches Seminar
der Universität Kiel
Ohlshausenstr. 40–60
23 Kiel/BRD

Dr. Klaus Potthoff
Philosophisches Seminar
der Universität Kiel
Ohlshausenstr. 40–60
23 Kiel/BRD

Library of Congress Cataloging in Publication Data

International Summer Institute and Logic Colloquium, University of Kiel, 1974.
ISILC Logic Conference.

(Lecture notes in mathematics ; 499)
English and French.
Chiefly comprised of selected course lecture notes from the institute, July 17-31, 1974, and selected papers from the conference, Aug. 1-3, 1974.
1. Logic, Symbolic and mathematical--Congresses. 2. Set theory--Congresses. I. Müller, Gert Heinz, 1923- II. Oberschelp, Arnold. III. Potthoff, Karl. IV. Title. V. Title: Logic Conference.

VI. Series: Lecture notes in mathematics (Berlin) ; 499.
QA3.L28 no. 499 [QA9.A1] 510'.8s [511'.3]
75-40481

AMS Subject Classifications (1970): 02-XX, 04-XX, 68-XX

ISBN 3-540-07534-8 Springer-Verlag Berlin · Heidelberg · New York
ISBN 0-387-07534-8 Springer-Verlag New York · Heidelberg · Berlin

Printed in Germany
Offsetdruck: Julius Beltz, Hemsbach/Bergstr.

PREFACE

An International Summer Institute and Logic Colloquium (ISILC) was held in Kiel, Federal Republic of Germany from Wednesday, 17th July to Saturday, 3rd August 1974. The conference was organized by the Deutsche Vereinigung für mathematische Logik und Grundlagenforschung der exakten Wissenschaften and was sponsored by the Association for Symbolic Logic as the European Summer meeting of the ASL. The congress received financial support from: The International Union for History and Philosophy of Science/Division for Logic, Methodology and Philosophy of Science, from the Deutsche Forschungsgemeinschaft, from the Land Schleswig-Holstein, from the City of Kiel, and from the firms Siemens AG and AEG-Telefunken. A part of the conference was financed as "Tagung über Modelltheorie" by the Stiftung Volkswagenwerk. The University of Kiel provided the lecture hall and supported the conference in other helpful ways.

182 people from 17 countries took part in the conference. The Organizing Committee consisted of the undersigned from Heidelberg (G. H. M.) and Kiel (A. O. and K. P.).

The Summer Institute (17th July to 31st July) was mainly devoted to series of lectures, the Logic Colloquium (1st August to 3rd August) was devoted to invited lectures and contributed papers. S. Kochen dedicated his course to the memory of A. Robinson. The following courses were given during the Summer Institute (the first six courses were 8 hours each, the last two were 4-hour courses):

W. Boos (Iowa City): An outline of the theory of large cardinals.

J. Flum (Freiburg): First order logic and its extensions.

S. Kochen (Princeton): The model theory of local fields.

D. Prawitz (Oslo): General proof theory of first and higher order systems: Normalization of proofs.

W. Richter (Minneapolis): Inductive definitions.

D. Scott (Oxford): Lambda calculus: Models and applications.

J. E. Fenstad (Oslo): Computation theories: An axiomatic approach to recursion on general structures.

R. B. Jensen (Bonn): The solution of the gap-2 case of the two cardinal problems.

During the Summer Institute and the Logic Colloquium there were invited lectures by W. W. Boone (Urbana), W. Boos (Iowa City), F. Drake (Leeds), K. Gloede (Heidelberg), S. Grigorieff (Paris), C. Imbert (Paris), P. Krauss (New Paltz), F. von Kutschera (Regensburg), A. Mostowski (Warszawa), H. Rasiowa (Warszawa), and E. Specker (Zürich).

On Friday, 2nd August, there was a symposium on proof theory organized by J. Diller (Münster) to honour Kurt Schütte on the occasion of his 65th birthday with lectures by J. Diller (Münster), H. Luckhardt (Frankfurt), D. Prawitz (Oslo) and G. Takeuti (Urbana). In addition 28 papers were contributed to the congress, some of them are included in this volume (Adamowicz, Börger, Flannagan, Fittler).

These proceedings contain lecture notes of courses and invited and contributed papers as they were made available by the authors. Jensen and Richter decided to substitute another paper for the one actually read at the conference. The paper of Devlin is included since he originally agreed to give an invited lecture but finally was not able to attend the congress.

Papers on proof theory dedicated to Kurt Schütte are published as a separate volume of these proceedings edited by J. Diller and G. H. Müller.

G. H. Müller

A. Oberschelp

K. Potthoff

The first invited lecture of the Logic Colloquium was given by Professor Mostowski, president of the Division for Logic, Methodology and Philosophy of Science. When the editorial work on this volume was almost finished, the editors were dismayed to hear of the sudden and untimely death of Professor Mostowski. We want to express how deeply we admired and respected Professor Mostowski, both as a man and as a scholar who stimulated the science of logic for decades and who made so many deep contributions to it.

TABLE OF CONTENTS

An observation on the product of Silver's forcing

Zofia Adamowicz, Warszawa

It can be shown that if $\mathbb{P}$ is the Sacks notion of forcing, M is a countable standard model of ZF + V = L and G is generic over $\mathbb{P} \times \mathbb{P}$ and M, then

$$M[G] \models (x)(x \in L[G_1] \vee x \in L[G_2] \vee G \in L[x])$$

It is well known that if G is generic over $\mathbb{P}$ and M then

$$M[G] \models (x)(x \in L \vee G \in L[x])$$

i.e., there is a minimal degree of constructibility in M[G].

So forcing with the product $\mathbb{P} \times \mathbb{P}$ gives the following image of the degrees of constructibility:

(note that G_1, G_2 are generic over $\mathbb{P}$).

A natural question that arises is whether for any $\mathbb{P}$ giving a minimal degree of constructibility it is true that $\mathbb{P} \times \mathbb{P}$ gives the above image.

This paper gives a partial solution of this problem. It is shown that the product of Silver's forcing does not give the above image even if only the degrees of constructibility of reals are considered.

But Silver's forcing does not satisfy the whole assumption; it is known only that it gives a minimal degree in the sense of reals, i.e.,

$$M[G] \models (x)_{2^\omega} (x \in L \vee G \in L[x])$$

for a generic G.

Thus the hypothesis in question, weakened to the case of real numbers, is not true.

Let $\mathbb{P}$ be Silver's notion of forcing, i.e.,

$$p \in \mathbb{P} \leftrightarrow \operatorname{dom} p \subseteq \omega \ \& \ \omega - \operatorname{dom} p \text{ is infinite}$$
$$\& \ p : \operatorname{dom} p \to 2$$

$$p \leq q \leftrightarrow q \subseteq p$$

We shall define an element $\underline{x}$ of $M^{\overline{\mathbb{P}\times\mathbb{P}}}$ (the Scott boolean model where the boolean completion $\overline{\mathbb{P}\times\mathbb{P}}$ of $\mathbb{P}\times\mathbb{P}$ is taken as the algebra) such that for any G generic over M and $\mathbb{P}\times\mathbb{P}$ the following holds in M[G]:

(*) $\quad i_G(\underline{x}) \notin L[G_1] \ \& \ i_G(\underline{x}) \notin L[G_2] \ \& \ G \notin L[i_G(\underline{x})]$

To show (*) it is enough to show

$$i_G(\underline{x}) \notin L[G_1] \ \& \ G_2 \notin L[i_G(\underline{x})]$$

because if we had $i_G(\underline{x}) \in L[G_2]$, then we would obtain $G_2 \in L[i_G(\underline{x})]$ by the properties of Silver's forcing.

In order to define $\underline{x}$, let us introduce the following definitions:

Def. 1.

We shall define a subset A_n of $2^n \times 2^n$. Let $\langle s, t\rangle \in A_n \longleftrightarrow$

1) $\sum_{m\in n} s(m)$ is even

or 2) $t(n-1) = 1$.

Def. 2.

Let $s \in 2^n$. Then $s \in \mathbb{P}$.

Let $p \in \mathbb{P}$. Let $p_s = p \wedge s$, where $\wedge$ denotes the greatest lower bound in $\mathbb{P}$. Let $\check{\alpha}$, $\underline{G}$, $\underline{G}_1$, $\underline{G}_2$, $\check{n}$ be element of $M^{\overline{\mathbb{P}\times\mathbb{P}}}$ with the usual meaning.

Let $\mathbb{O}$ be the least element in $\overline{\mathbb{P}\times\mathbb{P}}$. Sometimes it will denote a least element added to $\mathbb{P}$.

Def. 3.

Let $\underline{x} \in M^{\overline{\mathbb{P}\times\mathbb{P}}}$ be defined as follows:

$$\operatorname{dom} \underline{x} = \{ \check{n} : n \in \omega \}$$

$$\underline{x}(\check{n}) = \sum_{\langle s,t\rangle \in A_{n+1}} \langle s,t\rangle$$

where $\sum$ denotes the boolean union in $\overline{\mathbb{P}\times\mathbb{P}}$ (assume $\mathbb{P}\times\mathbb{P} \subseteq \overline{\mathbb{P}\times\mathbb{P}}$).

Remark.

Let G be generic over M, $\mathbb{P}\times\mathbb{P}$. Assume that G is an ultrafilter in $\overline{\mathbb{P}\times\mathbb{P}}$, according to the boolean symbolism.

Let

$$t_{G_1} = \{ n \in \omega : (Ep)_{G_1} (p(n) = 1) \}$$

$$t_{G_2} = \{ n \in \omega : (Eq)_{G_2} (q(n) = 1) \}.$$

Consider $i_G(\underline{x})$

$$i_G(\underline{x}) = \{ n \in \omega : (E\langle s,t\rangle)_{A_{n+1}} (\langle s,t\rangle \in G) \}.$$

Notice that:

$\langle s,t\rangle \in G$ and $\langle s,t\rangle \in A_{n+1}$

$\longleftrightarrow$ ($\sum_{m \in n+1} s(m)$ is even or $t(n) = 1$) and $\langle s,t\rangle \in G$.

$\longleftrightarrow$ the number of m's belonging to n+1 such that $s(m) = 1$ is even or $t(n) = 1$ and $\langle s,t\rangle \in G$

$\longleftrightarrow$ the number of m's belonging to n+1 such that $s(m) = 1$ is even or $t(n) = 1$ and $s \in G_1$ and $t \in G_2$

$\longleftrightarrow$ the number of m's less than n+1 belonging to t_{G_1} is even or n belongs to t_{G_2}.

So $i_G(\underline{x}) = t_{G_2} \cup \{ n : \overline{\overline{(n+1) \cap t_{G_1}}} \text{ is even} \}$

$\supseteq t_{G_2}$ plus "every second element of t_{G_1}".

Now we shall prove two lemmas, leading to the proof that

$i_G(\underline{x}) \notin L[G_1]$.

Lemma 1.

For any $\langle p,q\rangle$ in $\mathbb{P} \times \mathbb{P}$ there are such $n \in \omega$ and s', t_1, $t_2 \in 2^{n+1}$ that p_s, q_{t_1}, $q_{t_2} \neq \mathbb{O}$ and $t_1 \upharpoonright n = t_2 \upharpoonright n$ and that $\langle s',t_1\rangle \in A_{n+1} \longleftrightarrow \langle s',t_2\rangle \notin A_{n+1}$.

Proof.

Let m, n be such that $m \notin$ domp and $n \notin$ domq and $m \in n+1$. Let s, t be such that $s \in 2^{n+1-\{m\}}$, $t \in 2^n$, $p \wedge s \neq \mathbb{O}$, $q_t \neq \mathbb{O}$. Then

$$S_0 = \sum_{\substack{m' \in n+1 \\ m' \neq m}} s(m') \text{ is even} \longleftrightarrow S_1 = \sum_{\substack{m' \in n+1 \\ m' \neq m}} s(m') + 1 \text{ is odd}$$

Take the value which is odd. Let it be S_0. Now take $s' \in 2^{n+1}$ such that, for $m' \neq m$, $s'(m') = s(m')$ and $s'(m) = 0$. Then $\sum_{m \in n+1} s'(m)$ is odd.

Consider t_1, t_2 such that $t \subseteq t_1 \cap t_2$ and $t_1(n) = 0$ and $t_2(n) = 1$.

Then $\langle s',t_1\rangle \in A_{n+1} \longleftrightarrow \langle s',t_2\rangle \notin A_{n+1}$ and p_s, q_{t_1}, $q_{t_2} \neq \mathbb{O}$,

Q.e.d.

Lemma 2.

Let $\underline{y} \in M^{\overline{\mathbb{P}\times\mathbb{P}}}$ be such that

$$M^{\overline{\mathbb{P}\times\mathbb{P}}} \models y = F_{\check{\xi}}[\underline{G_1}]$$

which means that y satisfies in $M^{\overline{\mathbb{P}\times\mathbb{P}}}$ the formula "to be $F_z[w]$" with the parameters $\check{\xi}$ and $\underline{G_1}$ substituted for z and w respectively, where F is Gödel's enumeration of constructible stes. Then for any

p, q_1, q_2 in $\mathbb{P}$ and for any n

$$\langle p,q_1\rangle \Vdash \check{n} \in y \leftrightarrow \langle p,q_2\rangle \Vdash \check{n} \in y \quad \text{and}$$

$$\langle p,q_1\rangle \Vdash \check{n} \notin y \leftrightarrow \langle p,q_2\rangle \Vdash \check{n} \notin y$$

Proof.

Assume that $\langle p,q_1\rangle \Vdash \check{n} \in \underline{y}$ and take an arbitrary G generic over $\mathbb{P}\times\mathbb{P}$ such that $\langle p,q_2\rangle \in G$. Take another generic G' such that $G_1' = G_1$ and $\langle p,q_1\rangle \in G'$. Then

$$M[G'] \models n \in F_\xi[G_1']$$

So $n \in F_\xi[G_1]$. Thus we have

$$M[G] \models n \in F_\xi[G] \qquad \text{i.e.,}$$

$$M[G] \models n \in i_G(y).$$

To prove the second equivalence and the inverse implication use the same argument.

Lemma 3.

For any generic G

$$M[G] \models i_G(\underline{x}) \notin L[G_1]$$

Proof.

We shall show that the set

$$D_\xi = \{\langle p,q\rangle : (En)\omega\,((\langle p,q\rangle \Vdash (\check{n}\in \underline{y} \Leftrightarrow \check{n}\notin \underline{x}))\}$$

is dense for any ξ in $On \cap M$ where $\underline{y}$ is the same as in Lemma 2. Let us take an arbitrary $\langle p,q\rangle$. Now take n, s', t_1, t_2 as in Lemma 1. Then we have

$$\langle p_{s'}, q_{t_1}\rangle \leqslant \langle p,q\rangle \;\&\; \langle p_{s'}, q_{t_2}\rangle \leqslant \langle p,q\rangle$$
$$\&\; \langle p_{s'}, q_{t_1}\rangle \Vdash \check{n}\in\underline{x} \;\&\; \langle p_{s'}, q_{t_2}\rangle \Vdash \check{n}\notin\underline{x}$$

or conversely. Assume that the above statement is true. Then the following is possible:

(1) $\langle p_{s'}, q_{t_1}\rangle \Vdash \check{n} \in y$ & $\langle p_{s'}, q_{t_2}\rangle \Vdash \check{n} \in y$

or (2) $\langle p_{s'}, q_{t_1}\rangle \Vdash \check{n} \notin y$ & $\langle p_{s'}, q_{t_2}\rangle \Vdash \check{n} \notin y$

or (3) $\langle p_{s'}, q_{t_1}\rangle \nVdash \check{n} \in y$ v $\langle p_{s'}, q_{t_2}\rangle \nVdash \check{n} \in y$

In each case there is an $r \leqslant \langle p_{s'} q_{t_1}\rangle$ such that $r \Vdash \check{n} \in \underline{y}$ or there is an $r \leqslant \langle p_{s'}, q_{t_2}\rangle$ such that $r \Vdash \check{n} \in \underline{y}$. This r is stronger than $\langle p,q\rangle$ and belongs to D_ξ. So far any ξ $\{\langle p,q\rangle : \langle p,q\rangle \Vdash \underline{x} \neq F_{\check{\xi}} [\underline{G}_1]\}$ is dense. Thus

$$M[G] \models i_G(\underline{x}) \neq F_\xi [G_1]$$

for any ξ. Q.e.d.

Now she shall prove two analogous lemmas leading to the proof that $G_2 \notin L[i_G(\underline{x})]$.

Lemma 4.

For any $\langle p,q\rangle$ there are n, s', t_1, $t_2 \in 2^{n+1}$ such that p_s, q_{t_1}, $q_{t_2} \neq \mathbb{O}$ and $\langle s',t_1\rangle \in A_{n+1}$ and $\langle s',t_2\rangle \in A_{n+1}$ and $t_1 \restriction n = t_2 \restriction n$ and $t_1(n) \neq t_2(n)$.

Proof.

We take m, n, s, t in the same way as in Lemma 1. Let S_0, S_1 mean the same as in Lemma 1. Now we take the value which is even. Let is be S_1.

Let us now take $s' \in 2^{n+1}$ such that, for $m' \neq m$, $s'(m') = s(m')$ and $s'(m) = 1$.

Then $\sum_{m' \in n+1} s'(m')$ is even.

Consider t_1, t_2 such that $t \subseteq t_1 \cap t_2$ and $t_1(n) = 0$ and $t_2(n) = 1$; then

$$\langle s,t_1\rangle \in A_{n+1} \quad \text{and} \quad \langle s,t_2\rangle \in A_{n+1} .$$

Lemma 5.

Let $\underline{y} \in M^{\overline{\mathbb{P}\times\mathbb{P}}}$ be such that

$$M^{\overline{\mathbb{P}\times\mathbb{P}}} \models y = F_{\check{\xi}}[\underline{x}]$$

Then the following holds:

for any $\langle p,q\rangle$, if s', t_1, t_2 are as in Lemma 4, then for any m

$$\langle p_{s'}, q_{t_1}\rangle \Vdash \check{m} \in y \leftrightarrow \langle p_{s'}, q_{t_2}\rangle \Vdash \check{m} \in y$$
$$\langle p_{s'}, q_{t_1}\rangle \Vdash \check{m} \notin y \leftrightarrow \langle p_{s'}, q_{t_2}\rangle \Vdash \check{m} \notin y$$

Proof.

Assume $\langle p_s, q_{t_1}\rangle \Vdash \check{m} \in \underline{y}$.

Take an arbitrary generic G such that

$$\langle p_{s'}, q_{t_2}\rangle \in G .$$

Now consider an automorphism φ of $\mathbb{P}\times\mathbb{P}$ defined as follows:

$$\varphi(\langle p, q_n\rangle) = \langle p, q_n'\rangle$$

where $\mathrm{dom}\,q' = \mathrm{dom}\,q$ and for $m \in \mathrm{dom}\,q$, $m \neq n$,

$$q'(m) = q(m)$$

and $\qquad q'(n) = 1-q(n)$ if $n \in \mathrm{dom}\,q$. Then

$$\varphi(\langle p_{s'}, q_{t_1}\rangle) = \langle p_{s'}, q_{t_2}\rangle$$

Taken an ultrafilter G' in $\overline{\mathbb{P}\times\mathbb{P}}$ containing $\varphi''(G \wedge \mathbb{P}\times\mathbb{P})$. Then G' is generic.

We shall show that $i_{G'}(\underline{x}) = i_G(\underline{x})$. Indeed

$$\underline{x}(\check{m}) \in G' \leftrightarrow \langle s,t\rangle \in G'$$

for some $\langle s,t\rangle$ belonging to A_{m+1}.

But $\langle p_{s'}, q_{t_1}\rangle \in G'$, and so $\underline{x}(\check{m}) \in G' \leftrightarrow \langle s,t\rangle \in G'$, $\langle s,t\rangle$ belonging to A_{m+1} and such that $s \subseteq s'$ or $s' \subseteq s$ and $t \subseteq t_1$ or $t_1 \subseteq t$ (otherwise $\langle p_{s'}, q_{t_1}\rangle \wedge \langle s,t\rangle = 0$).

But $\langle s,t\rangle \in G' \leftrightarrow \varphi^{-1}(\langle s,t\rangle) = \langle s,t'\rangle \in G$. Now notice that t and t' differ at most at n, and $s \restriction n+1 = s'$, and $\sum_{m' \in n+1} s'(m')$ is even, and, for any m, $\sum_{m' \in m} s(m')$ is the same for the two pairs $\langle s,t\rangle$ and $\langle s,t'\rangle$.

So we have

$$\langle s,t\rangle \in A_{m+1} \leftrightarrow \langle s,t'\rangle \in A_{m+1}.$$

Thus

$$\underline{x}(\check{m}) \in G' \leftrightarrow \underline{x}(\check{m}) \in G.$$

Hence $i_{G'}(\underline{x}) = i_G(\underline{x})$.

From the assumption $\langle p_{s'}, q_{t_1}\rangle \Vdash \check{m} \in \underline{y}$ we infer that $m \in i_{G'}(\underline{y})$.

So $m \in F_{\xi}[i_G(\underline{x})]$ for an arbitrary G containing $\langle p_{s'}, q_{t_2}\rangle$.

Thus $\langle p_{s'}, q_{t_2}\rangle \Vdash \check{m} \in \underline{y}$.

The case where $\langle p_{s'}, q_{t_1}\rangle \Vdash \check{m} \notin y$ and the inverse implication is treated similarly. Q.e.d.

Lemma 6.

For any generic G,

$$M[G] \models G_2 \notin L[i_G(\underline{x})].$$

Proof.

We shall proceed as in Lemma 3. Let

$$D_{\xi} = \{\langle p,q_n\rangle : (En)\omega\, (\langle p,q_n\rangle \Vdash (\check{n} \in y \leftrightarrow \check{n} \notin \underline{t_{G_2}}))\}$$

where $\underline{y}$ is such that $M^{\overline{P\times P}} \models \underline{y} = F_{\check{\xi}}[\underline{x}]$ and tG_2 is the "name" for the real tG_2. We shall show that D_{ξ} is dense.

Let $\langle p,q\rangle \in P \times P$

Take n, s', t_1, t_2 as in Lemma 4.

Then we have

$$\langle p_{s'}, q_{n t_1}\rangle \leq \langle p,q_n\rangle, \quad \langle p_{s'}, q_{n t_2}\rangle \leq \langle p,q_n\rangle \;\&$$

$$\langle p_{s'}, q_{n t_1}\rangle \Vdash \check{n} \notin \underline{t_{G_2}} \;\&\; \langle p_{s'}, q_{n t_2}\rangle \Vdash \check{n} \in \underline{t_{G_2}}$$

The following is possible:

1) $\langle p_{s'}, q_{t_1} \rangle \Vdash \check{n} \in \underline{y}$ and $\langle p_{s'}, q_{t_2} \rangle \Vdash \check{n} \in \underline{y}$

or

2) $\langle p_{s'}, q_{t_1} \rangle \Vdash \check{n} \notin \underline{y}$ and $\langle p_{s'}, q_{t_2} \rangle \Vdash \check{n} \notin \underline{y}$

or

3) $\langle p_{s'}, q_{t_1} \rangle \nVdash \check{n} \in \underline{y}$ and $\langle p_{s'}, q_{t_2} \rangle \nVdash \check{n} \in \underline{y}$.

In each case there is an $r \leqslant \langle p_{s'}, q_{t_1} \rangle$ such that $r \Vdash \check{n} \in \underline{y}$ or there is an $r \leqslant \langle p_{s'}, q_{t_2} \rangle$ such that $r \Vdash \check{n} \notin \underline{y}$. Consequently, $r \leqslant \langle p, q \rangle$ and $r \in D_\xi$. So for any ξ

$$\{\langle p,q\rangle : \langle p,q\rangle \Vdash t_{G_2} \neq F_{\xi}^{y} [\underline{x}]\}$$

is dense. Thus $M[G] \models G_2 \notin L[i_G(\underline{x})]$. Q.e.d.

RECURSIVELY UNSOLVABLE ALGORITHMIC PROBLEMS AND RELATED QUESTIONS REEXAMINED

Egon Börger
Istituto di Scienze dell'Informazione
Università di Salerno
Salerno / ITALY

Introduction. This paper starts by defining inductively a particular kind of formal systems M (see Rödding [1968: 200-202]) and then shows how these systems represent frequently used combinatorial systems such as register machines (as introduced by Minsky [1961] and Shepherdson and Sturgis [1962]), Semi-Thue, Thue, Post normal systems and Markov algorithms (see Malcew [1974]) when interpreted in a natural and straight-forward way avoiding any complicated gödelizations or simulation tricks as still used in the literature on this subject. The intention is that the proofs should be elementary, short and at the same time complete, and it seems that without proving really anything at all which would not be trivial or immedeately clear (when adequately formulated), one gets from only basic facts about the class of partial recursive functions a) the equivalence between partial recursive and register machine computable functions, Kleenes normal form theorem and some characterizations of the class of partial recursive functions by a simple basis and closure Operations of the type proposed in Eilenberg and Elgot [1970] , b) Minsky's [1961] theorem about universal 2-register machines, c) the creativity of the general (and the special) word and halting problem for (some) 2-register machines, Semi-Thue and Thue systems, Markov algorithms (without concluding rules) and Post normal systems and finally the creativity of (the class of all formulas derivable in) classical first order logic with equality (using only a small portion of some correct axiomatization of it.) In the meaning of the author the method gives strong evidence of the fruitfulness of inductively defined computing devices with only a few closure operations such as concatenation, case and while operations, at least with regard to theoretical questions about computability.

§1. Register machines and recursive functions

Let us start by defining, following Rödding [1968: 200-202], [1972] n-register operators inductively by a_j resp. s_j (+1 resp. $\dot{-}1$ in register j), $M_1 M_2$ (apply first M_1 and then M_2 to the result of M_1) and iteration $(M)_j$ of M until the register j has become empty, where $1 \leq j \leq n$. For technical reasons we follow the approach advocated in Eilenberg and Elgot [1970] and take the n-register operators as partial functions from n-tuples of numbers into n-tuples of numbers, i.e. strictly speaking the n-register operators form the smallest class of functions which contains the initial functions $a_j := \lambda x_1 \ldots x_n (x_1, \ldots, x_j+1, \ldots, x_n)$, $s_j := \lambda x_1 \ldots x_n (x_1, \ldots, x_j \dot{-} 1, \ldots, x_n)$ for $1 \leq j \leq n$ and is closed with respect to the substitution gf of a function g into another f and to the iteration of a function until the first component becomes 0 . But we recommend the reader to think about these operators as computing devices which transform the content of their registers (i.e. their arguments) in the indicated way (into their values). The intuitive background of (and the jargon about) computing machines makes it extremly easy to grasp the simple ideas underlying the following constructions. (In particular we make free use of the usual terminology about register machines.)

<u>Lemma 1.</u> One can construct to every n-register operator M a 3-register operator $\bar{M}$ "simulating" it by

$$(x_1, \ldots, x_n) \vdash_M (y_1, \ldots, y_n) \quad \text{iff} \quad \langle x_1, \ldots, x_n \rangle, 0, 0 \vdash_{\bar{M}} \langle y_1, \ldots, y_n \rangle, 0, 0$$

and the stop condition:

$$(x_1, \ldots, x_n) \vdash_M \text{Stop} \qquad \text{iff} \quad \langle x_1, \ldots, x_n \rangle, 0, 0 \vdash_{\bar{M}} \text{Stop}$$

for all numbers $x_1, \ldots, x_n, y_1, \ldots, y_n$.

($\langle x_1, \ldots, x_n \rangle$ denotes the codification of the n-tuple $x_1, \ldots, x_n$ into a number with $(\langle x_1, \ldots, x_n \rangle)_i = x_{i+1}$ for $0 \leq i < n$. Let us take for ex. the usual prime number codification.) For the proof of Lemma 1 one defines for every number k the 3-register operators:

$$\text{Mult}[k] := (s_1 a_2^k)_1 (s_2 a_1)_2 \qquad (\text{"multiply the first register by } k\text{"})$$

$$\text{Div}[k] := (s_1^k a_2)_1 (s_2 a_1)_2 \qquad (\text{"divide the first register by } k\text{"})$$

$$\text{Test } [k] \quad \text{with: } x,0,0 \vdash \begin{cases} x,0,1 & \text{if } k \text{ divides } x \\ x,0,0 & \text{otherwise} \end{cases}$$

Now one defines $\bar{M}$ by induction on M with

$\bar{a}_j := \text{Mult}\,[p_{j-1}]$, $\bar{s}_j := \text{Test}\,[p_{j-1}]\,(s_3\ \text{Div}\,[p_{j-1}]\,)_3$, $\overline{M_1\, M_2} := \overline{M_1}\,\overline{M_2}$,

$\overline{(M)_j} := \text{Test}\,[p_{j-1}]\,(s_3\ \bar{M}\ \text{Test}\,[p_{j-1}]\,)_3$. The following 3-register operator Test $[k]$ is due to my student G.d'Amico and improves an own construction with 4 registers. The idea is to calculate the difference d between the number x in the first register and the smallest number bigger than or equal to x and divisible by k: $x\ 0\ 0 \Rightarrow 0\ x\ x \Rightarrow x{+}d\ 0\ x \Rightarrow d\ x\ 0 \Rightarrow d\ x\ 1 \Rightarrow \begin{cases} 0\ x\ 1 & \text{if } k \mid x \\ 0\ x\ 0 & \text{else} \end{cases}$

$$\text{Test}\,[k] := (s_1\ a_2\ a_3)_1\ (s_2^k\ a_1^k)_2\ (s_1\ s_3\ a_2)_3\ a_3\ (s_3\ s_1)_1\ (s_2\ a_1)_2$$

(In the appendix is shown that reduction to 2-register operators is no more possible.)

Corollary 1. The class of register operator computable functions coincides with the class of partial recursive functions.

Proof. To show the register operator computability of all partial recursive functions is just an easy exercise. For the other direction it is convenient to use the inductively defined structure of register operators (see Cohors-Fresenborg [1973]). By lemma 1 it is sufficient to define for every 3-register operator M a partial recursive function f_M such that $M(x,y,z) = (a,b,c)$ iff $f_M(\langle x,y,z\rangle) = \langle a,b,c\rangle$ for all x,y,z,a,b,c , because then every function g computable by a register operator M in the sense of $x_1,\ldots,x_n,0,\ldots,0 \vdash_M g(x_1,\ldots,x_n),x_1,\ldots,x_n,0,\ldots,0$ admits the partial recursive description

$$g(x_1,\ldots,x_n) = ((f_{\bar{M}}(\langle\langle x_1,\ldots,x_n\rangle\rangle))_o)_o \quad .$$

Observe that every 3-register operator can be obtained from $a = a_1$, $s = s_1$ and $p := \lambda xyz(y,z,x)$ by composition and iteration $(M) = (M)_1$ restricted to the first register. Define therefore inductively the partial recursive function f_M by:

$$f_a(u) := \langle (u)_o + 1, (u)_1, (u)_2 \rangle$$

$$f_s(u) := \langle (u)_o \dot{-} 1, (u)_1, (u)_2 \rangle$$

$$(1) \qquad f_p(u) := \langle (u)_1, (u)_2, (u)_o \rangle$$

$$f_{MN}(u) := f_N(f_M(u))$$

$$f_{(M)}(u) := \begin{cases} u & \text{if } (u)_o = 0 \\ f_{(M)}(f_M(u)) & \text{otherwise} \end{cases}$$

(The last equation could be described more explicitely as substitution of the μ-operator into the iteration f'_M of f_M defined by $f'_M(u,0):=u$, $f'_M(u,n+1):= f_M(f'_M(u,n))$ and $f_{(M)}(u) = f'_M(u,\mu y(f'_M(u,y))_o=0)$.)

From Kleene's first recursion theorem it is clear that the equations (1) define a universal partial recursive function $f = \lambda Mu\, f_M(u)$ of two arguments so that one has obtained a normal form theorem. But this can be shown from more elementary facts about partial recursive functions without using the recursion theorem. Since every 3-register operator M can be decomposed uniquely into a finite sequence $M = e_o \ldots e_n$ of symbols (elementary operators) from (,),a, s,p , gödelize $\underline{M}:= \langle \underline{e}_o,\ldots,\underline{e}_n\rangle$ with $\underline{(}:= 2$, $\underline{)}:= 3$, $\underline{a}:= 5$, $\underline{s}:= 7$, $\underline{p}:= 11$ and define two primitive recursive functions f and e such that for all 3-register operators M and all u,y one has:

$f(\underline{M},u,y)$ = (codification of the) result of y steps of computation by M started with input (codificated by) u

$e(\underline{M},u,y)$ = (Gödel number of) that elementary operator in M which will be executed in step y of the computation by M started with input u

One can define f and e by a straightforward simultaneous recursion, without any need to analyse register operators as register machines as defined originally in Minsky [1961] by program tables of elementary instructions (see Rödding [1972a]):

$$f(k,u,0) = u \quad , \quad e(k,u,0) = (k)_o$$

$$f(k,u,y+1) = \begin{cases} q'(f(k,u,y)) & \text{if } e(k,u,y) = \underline{q} \in \{\underline{a},\underline{s},\underline{p}\} \\ f(k,u,y) & \text{else} \end{cases}$$

with a primitive recursive q' such that $q'(\langle a,b,c\rangle) = \langle q(a,b,c)\rangle$. For the description of the correct transfer from the zero test by a left parenthesis "("to the corresponding right one")" and vice versa in case the first register is resp. is not empty, one needs to look only at the next corresponding parenthesis to the right resp. to the left such that between these two, the number of occurences of "("equals that of")". So introduce the primitive recursive auxiliary functions $L(\underline{M},j,z)$ = number of occurences of "("in M between $(\underline{M})_j$ and $(\underline{M})_z$ (if $j\leq z$) = $\Big(\prod_{\substack{i=j \\ (\underline{M})_i=2}}^{z} (\underline{M})_i\Big)_o$ and the same for $R(\underline{M},j,z)$ with right parenthesis and define:

$$e(k,u,y+1)=\begin{cases}(k)_{r+1} & \text{if } e(k,u,y)=(k)_j=\underline{(}, \ (f(k,u,y))_o=0 \text{ and } r=\mu z_{j\le z\le k}(L(k,j,z)=R(k,j,z))\\ (k)_{l+1} & \text{if } e(k,u,y)=(k)_j=\underline{)}, (f(k,u,y))_o\ne 0 \text{ and } l=\max z_{z\le j}(L(k,z,j)=R(k,z,j))\\ (k)_{j+1} & \text{if } e(k,u,y)=(k)_j \quad \text{else}\end{cases}$$

Therefore every partial recursive g admits the description $g(x) = (f(\underline{M},x, \mu y\, e(\underline{M},x,y)=0))_o$ for a 3-register operator M computing it, so we have proved the existence of a universal 3-register operator (computing f) in the form of:

Corollary 1'. Kleenes normal form theorem.

(For another interesting way of transforming equations (1) into a partial recursive scheme to obtain Kleenes normal form theorem see Ottmann [1974]. See also Cohors-Fresenborg [1974: 75-77].)

By lemma 1, every partial recursive function $f^{(n)}$ is computable by a n+2 -register operator M which transforms an arbitrary input $(x_1,\dots,x_n 0,0)$ into the output $(f(x_1,\dots,x_n),0,\dots)$ - if any - by making intermediate calculations using its auxiliary registers R_{n+1} and R_{n+2}. Obviously this could be analysed in the algebraic fashion as introduced by Eilenberg and Elgot [1970]: the function **0** without arguments and constant value zero describes the introduction of a new (auxiliary) empty register, whereas to apply the left projection function L with $L(x,y) = x$ corresponds to the inverse operation of throwing away one register (with arbitrary context x, but used essentially only for x=0); the fact that an elementary register operator operates only locally, that is to say on one or only a few (adjacent) registers is reflected by considering operations of "left cylindrification" ${}^cf = id\times f$ and "right cylindrification" $f^c = f\times id$ of a given function f. For example the 6-register operator $a_4={}^{ccc}a^{cc}$ or the 7-register operator $s_2={}^cs^{ccccc}$ with a and s here and in the following taken as 1-register operators. In presence of the operations of cylindrification we use p for the permutation $p(x,y) = (y,x)$ and Perm to denote an arbitrary permutation function (which one could always be reconstructed from the context but this isn't of any interest, as it is of no interest to indicate always the correct number and types of cylindrifications.)

Now Lemma 1 exprèsses clearly some characterization of the class of all partial recursive functions by means of a small (finite) basis and a few closure operations including Minsky's [1961: Theorem II.a] ; using the language introduced above Lemma 1 reads now:

Corollary 2. The class of all partial recursive functions is characterized by the basis a,s,p and the closure against substitution, cylindrification and iteration ( ) restricted to the first argument.

Indeed one has to check only that coding and decoding can be realized with 3-register operators, just another trivial exercise in programming: for every k, the 3-register operator $E[k]$ does $y,0,x \Rightarrow y\cdot(k^x),0,0$ with $E[k] := (s_3 \text{ Mult}[k])_3$, and $D[k]$ computes $k^x,0,0 \Rightarrow 0,0,x$ by $D[k] := s_1\ (a_1 \text{ Div}[k]\ a_3\ s_1)_1$. The rest is Perm.

As one sees here, treatment in terms of machines may cut down long chains of algebraically expressed complicated equations to a few simple programming tasks. For another such immediate application of lemma 1 consider the following variants of the "repetition operation" ( ) : $(M)'$, $(M)''$ and $(M)'''$ ask to repeat application of M until the first argument becomes zero and then to cancel this argument resp. until the first two arguments become equal resp. this followed by cancellation of those two arguments. If C is a class of functions, denote by C°, C' etc. the closure of C against composition, cylindrification and iteration ( ) resp. ( )' etc. . One can now reformulate corollary 2 in these terms obtaining the theorem of Germano and Maggiolo-Schettini [1973] as

<u>Corollary 2'.</u> The class of all partial recursive functions is contained in the classes $\{a,s,0,L\}^{\circ}$, $\{a,s,0\}'$, $\{a,0,L\}''$ and $\{a,0\}'''$.

Indeed the only thing to verify is that a,s,p and ( ) can be obtained. (Remember that $\mathbf{O}$ provides for a new empty register whereas L eliminates one.) In the first case, one has to program $p: xy \Rightarrow yx$: create a new third register, put x into the third one and eliminate the now empty first register, i.e. $p = {}^{cc}\mathbf{O}\ (s_1\ a_3)_1\ L^c$. Since L is used here only to eliminate empty registers, it can be dispensed with in presence of the modified iteration ( )' , and this concludes already the second case because (M) can clearly be simulated with ( )' by providing for a new empty register (untouched by $(M)'$ via cylindrification) in substitution of the one lost at the end of the iteration. In the third class one starts with an obvious program for p : create a new first and fourth register, then add to those until the first two become equal and finally eliminate the first two, i.e. $x\ y \underset{\mathbf{OO}}{\Rightarrow} 0\ x\ y\ 0 \underset{(a_1a_4)''}{\Rightarrow} x\ x\ y\ x \underset{LL}{\Rightarrow} y\ x$; for s, compute $x \underset{\mathbf{OO}}{\Rightarrow} 0\ x\ 0 \underset{(a_1(a_1a_3)'')''}{\Rightarrow\ \ \Rightarrow\ \ \Rightarrow} \begin{cases} 0\ 0\ 0 & \text{if } x=0 \\ x\ x\ x\dot{-}1 & \text{if } 0<x \end{cases} \underset{LL}{\Rightarrow} x\dot{-}1$ and finally (M) can be simulated canonically by the computation $x\ldots \underset{\mathbf{O}}{\Rightarrow} 0x\ldots \underset{(LM\mathbf{O})''}{\Rightarrow\Rightarrow} 00\text{---} \underset{L}{\Rightarrow} (M)(x\ldots)$. As L has been used again only for empty registers, once more one can define p and s in the fourth class as before with ( )''' instead of ( )'' and without L , and for (M) one can write the obvious program

$\mathcal{O}\,(\,\mathcal{O}(a)_1{'''}M\,\mathcal{O}\,){'''}\mathcal{O}$ computing $x\ldots \underset{\mathcal{O}}{\Rightarrow} 0\,x\ldots \underset{\mathcal{O}}{\Rightarrow} 0\,0\,x\ldots \underset{(a_1)'''}{\Rightarrow} x\ldots \underset{M}{\Rightarrow} M(x\ldots)$ etc.

§2. 2-register machines and symbol manipulating systems.

In this paragraph we want to show how the actions of register operators can be viewed in a natural way as symbol manipulations of standard combinatorial systems. To get an intellectually and technically most economical presentation of this it is convenient to make a detour by first simulating arbitrary register operators as in lemma 1 with 2-register machines in the sense of Minsky [1961]. A 2-register machine is a program table of a finite number of instructions $i\ o_i\ p_i\ q_i$ with instruction numbers $0 \leq i, p_i, q_i \leq$ (some) r and elementary operations $o_i \in \{a_1, a_2, s_1, s_2, \text{stop}\}$, interpreted by: in state i, do o_i and then go to instruction p_i, if the register considered was not yet empty, otherwise go to q_i. We assume without loss of generality that the zero test is done only in s-instructions (i.e. $o_i = a_1, a_2 \Rightarrow p_i = q_i$), that stop-instructions have the form i stop $i\ i$ and that the initial instruction is the one with number 0.

By lemma 1 it is sufficient to define for $k = 2,3,5$ a 2-register machine Test[k] (just another exercise!) to have proved the existence of a universal 2-register machine, which we state here as

<u>Corollary 3.</u> Minsky's [1961] theorem on the universality of 2-register machines for the class of all partial recursive functions.

(A program for Test[2] is given for ex. by $0\ s_1\ 1\ 4$, $1\ a_2\ 2\ 2$, $2\ s_1\ 3\ 5$, $3\ a_2\ 0\ 0$, 4: $(s_2\ a_1)_2\ a_2$ stop, 5: $(s_2\ a_1)_2$ stop. Of course we have used the abbreviating operator notation where possible, but as shown in the appendix, not the whole Test[k] could be written as 2-register operator. So the conditioned jump abilities of 2-register machines used in instruction 2 of the above program are crucial for the reduction to 2 registers.)

But now look what 2-register machine programs M are! M is a set of instructions (an algorithm) where each single instruction I_i indicates how one should transform a given M-configuration word $+\,p\,i\,q\,+$ representing "internal state" i and contents p and q of the first resp. second register ($+$ is an endmarker), eventually telling to stop. Taking numbers n as written down in the form $|^n$ of n consecutive occurences of the symbol $|$ one sees that M <u>is</u> (and operates like) a semi-Thue system; to visualize this point just rewrite every I_i with $o_i = a_1$ as "rule" $i \Rightarrow |\,p_i$ and every I_i with $o_i = s_1$

as two rules $|i \Rightarrow p_i$, $+ i \Rightarrow + q_i$, symmetrically for $o_i = a_2, s_2$, cancel all stop-instructions and call M , when interpreted in this way, S_M. S_M is related to M by

(2) $+ p\, i\, q + \underset{M}{\Rightarrow} + r\, j\, s +$ iff $+|^p\, i|^q + \underset{S_M}{\vdash} +|^r\, j|^s +$

and <u>there is nothing to prove</u> because the two sides of the equivalence (2) only formulate from two apparently different points of view the same thing. By the way, S_M is also a Markov algorithm - write M_M instead of S_M when thinking in these terms - related to M by (2) in the same way as S_M , because M is deterministic and so at most one rule of S_M can be applied to a configuration word $+|^p\, i\, |^q +$, and if so then in only one place; by the way M_M has no concluding rules. If you prefer to think in terms of Post normal canonical systems, just look at S_M in this way adding the obvious two frame rules $+ \Rightarrow +$, $| \Rightarrow |$ - call this system P_M - and you will see immedeately that P_M does really the same as M in the sense of (2), it only achieves it in a (microscopically) slightly different way.

We conclude that by a close but natural inspection of what does mean the universality of 2-register machines achieved in corollary 3 one can obtain without effort - using eventually Church's thesis - a lot of the normal form and reduction theorems for semi-Thue, Post canonical systems and Markov algorithms which can be found in the literature (see for ex. Hermes [1965] , Markov [1961] , Minsky [1967] , Post [1943] , Priese [1971]). We limit ourselves to reformulate here corollary 3 in terms of combinatorial systems:

<u>Corollary 3'.</u> Semi-Thue systems, Post normal canonical systems and Markov algorithms are universal.

This contains in particular the recursive unsolvability of the general (and some special) word problems for (some of) the combinatorial systems mentioned (read: of the halting problems of 2-register machines). Aiming at an application to the decision problem of first order predicate logic we extend this by a standard argument to Thue systems. Let S'_M , M'_M , P'_M be S_M, M_M, P_M resp. with the additional cancellation rules $|i \Rightarrow i$, $+ i| \Rightarrow + i$, $+ i + \Rightarrow +$ for every stop-instruction i stop i i of M and define the Thue system T_M as S'_M expanded by all inverse rules $W_k \Rightarrow V_k$ of rules $V_k \Rightarrow W_k$ in S'_M . One now shows for every system Q_M in $\{S'_M, M'_M, P'_M, T_M\}$:

(3) $+ p\, O\, q + \underset{M}{\Rightarrow}$ Stop iff $+|^p\, O\, |^q + \underset{Q_M}{\vdash} +$

<u>Proof.</u> For $Q_M = S'_M, M'_M, P'_M$ there is nothing to show by (2). For the

same reason for $Q_M = T_M$ there is nothing to show in the direction from left to right. For the remaining assume without loss of generality that the initial state cannot be reached from any state in the program of M (i.e. 0 occurs only in the instruction I_o and $p_o, q_o \neq 0$). Then if there exists a deduction from $+|^p 0|^q +$ to $+$ in T_M with at least one application of a rule from $T_M - S'_M$, then there is one such deduction of the form $A_o, \ldots, A_{n-1}, A_n, \ldots, A_m$ with $A_j \neq +$ for $j < m$ and a maximal final part $A_n, \ldots, A_m$ without application of (backward) rules from $T_M - S'_M$. Then one can shorten this deduction eliminating one application of a backward rule without affecting A_o and A_m in the following way: by the maximality of the final part, A_n must have been obtained from A_{n-1} by application of a backward rule $W_k \Rightarrow V_k$ of a (forward) rule $V_k \Rightarrow W_k$ in S'_M. Since every A_j with $j < m$ contains exactly one state symbol of M, A_n must contain such a state symbol i introduced into it by $W_k \Rightarrow V_k$ and A_{n-1} the corresponding successor state symbol p_i or q_i occuring in W_k. Consequently one must have produced A_{n+1} from A_n by an application of the forward rule $V_k \Rightarrow W_k$ because <u>some</u> forward rule has been applied (one is in the final part), at most one is applicable (by the monogenicity of the forward rules system S'_M on configuration words) and $V_k \Rightarrow W_k$ can be applied to A_n as the inverse of the production $W_k \Rightarrow V_k$ applied in going from A_{n-1} to A_n. Therefore $A_{n-1} = A_{n+1}$ and the derivation $A_o, \ldots, A_{n-1}, A_{n+2}, \ldots, A_m$ derives $+$ from $+|^p 0|^q +$ with one application of a backward rule less than in the originally given deduction. This procedure transforms a derivation of $+$ from $+|^p 0|^q +$ in T_M into a derivation in S'_M, so that $+ p 0 q + \vdash_M$ Stop by (2). (3) being proved, one has:

Corollary 4. There are Thue systems, Post normal canonical systems and Markov algorithms with creative word problem.

Corollary 5. Church's [1936] and Turing's [1936] theorem about the creativity of the decision problem of first order predicate logic PL with equality.

It is indeed sufficient (see Rogers [1967: §7.3, Th.V(b)]) to m-reduce the creative set $K = \{x \mid x \in W_x\}$ to these problems: $x \in K$ iff $+ 2^x 0 + \overset{*}{\underset{M}{\Rightarrow}}$ Stop (with a 2-register machine M enumerating K from corollary 3) iff $+ |^{2^x} 0 + \vdash_{T_M} +$ (by (3)) iff the first order formula $\bigwedge_u \bigwedge_v \bigwedge_w ((u \cdot v) \cdot w = u \cdot (v \cdot w)) \wedge \bigwedge_{k \leq l} V_k = W_k \rightarrow$

$\rightarrow + |^{2^x}\ 0 + = +$ is derivable in PL, where $V_k = W_k$ $(k \leq 1)$ are all the defining relations of T_M , $\cdot$ is a binary function symbol and the letters of the alphabet of T_M are individual constants. The idea of this last equivalence is taken from Malcew [1974: §13] : $+ |^{2^x}\ 0 + \vdash_{T_M} +$ means that in every semigroup with a generator for every letter of the alphabet of T_M and all the defining relations of T_M one can derive the identity of $+$ and $+ |^{2^x}\ 0 +$, and this is equivalent to the deducibility in PL of the above formula by the completeness theorem for PL. (But use of the completeness theorem could be avoided by proving by direct inspection (and constructively) the required equivalence between $\vdash_{T_M}$ and $\vdash_{PL}$. Note that this very short and simple argument for corollary 5 needs only minimal knowledge about predicate logic, i.e. the fact that PL can be axiomatized correctly providing for some elementary laws about equality and substitution. Surprisingly enough one needs not even Skolem's theorems on normal form and canonical term models as applied for the first time by Büchi [1962] and later in a still more simplified form by Aanderaa [1971] and Börger [1971] .)

Appendix.

For sake of motivation we want to indicate here why lemma 1 does not hold any more for 2-register operators. Define the depth of register operators by: $d(a_i) = d(s_i) = 0$, $d(MN) = \max(d(M), d(N))$, $d((M)_i) = d(M)+1$. Without loss of generality assume for reason of short formulation that every M has only one stop-instruction, namely I_r .

Lemma2. For every positive natural number n and every 2-register operator M of depth n there exist k,l,m such that either for all x,y,a,b holds

$$x\ 0\ y \vdash_M a\ r\ b \Rightarrow a \leq m\cdot(x+y),\ b = k$$

or for all x,y,a,b holds

$$x\ 0\ y \vdash_M a\ r\ b \Rightarrow a = k,\ b \leq m\cdot(x+y)\ .$$

Proof by induction on n . For $n = 1$ show the assertion first for iteration operators $M \equiv (N)_j$ because the general assertion then follows trivially by induction on the length m of the concatenation $M_1 \ldots M_m$ of elementary or iteration operators M_i of depth 1. Rearrange N in such a way that $N \equiv N_1 N_2$

with N_i consisting only of a_i and / or s_i (suboperators $a_i s_i$ can be eliminated). Then one needs considering only $N \equiv s_1^p s_2^q a_2^s$ for $j = 1$ or $N \equiv s_1^q a_1^s s_2^p$ for $j = 2$ with $0 < p$ (otherwise set $k = 0$, $m = 1$), where one can set $k = 0$, $m = s$. In the inductive step again it is sufficient to prove the assertion for iteration operators $M \equiv (N)_j$ of depth $n+1$. Case 1: $j = 1$. If the first clause of lemma 2 holds for N, then $M(x,y)$ is either undefined or equal $(0,y)$ if $x = 0$ and $(0,k)$ if $0 < x$. If the second clause of lemma 2 holds for N, then $M(x,y)$ is either undefined or with first register $= 0$ and second register $= y$ if $x = 0$ or $\leq m\cdot(x+y)$ else. In both cases therefore lemma 2 holds for M. Case 2: $j = 2$ is symmetric to case 1.

Remember that Барздинь [1963] has established the impossibility of calculating all partial recursive functions f by 2-register machines M without some codification in the sense of $x,0 \underset{M}{\Rightarrow} f(x),0$ if $f(x)$ is defined and $M(x,0)$ undefined else. 2-register machines are universal only modulo some codification of input and output of the type obtained in corollary 3 $(2^x \Rightarrow 2^{f(x)})$. The limitations expressed for 2-register operators by lemma 2 should therefore give no argument against the normalization imposed on register machine programs in restricting attention to the special ones defining operators, because in so far as one is concerned with computability without non linear coding, the two concepts of register machines and of register operators coincide extensionally by lemma 1, especially corollary 1.

We want to conclude by demonstrating the flexibility of the register operator concept still from another point of view which may interest if one does not wish to abandon the concept of Turing machine computability as excellent motivation for the Thesis of Church. It is indeed possible to show that all Turing computable functions are partial recursive without being condemned to the introduction of complicated gödelizations of Turing tables and tape transformation functions (see for ex. Yasuhara [1971: 5.3.] or Rödding [1972.a: 45-49]) by gödelizing only the Turing tapes f in a natural way by two numbers l_f, r_f representing the left and right part of the tape (determined by the position of the scanned cell) and then performing the operations of a Turing machine M on f by a corresponding register operator $\bar{M}$ on l_f, r_f. The proof outlined here is an adaptation of the idea introduced in Rödding [1972.a: Satz 6.3.] for simulation of arbitrary register machines by register operators.

To be explicit, code a tape $f = \ldots a_o a_{i_p} \ldots \underline{a}_{i_o} a_{j_o} \ldots a_{j_q} a_o \ldots$ by $l_f := \langle i_o, \ldots, i_p \rangle$ and $r_f := \langle j_o, \ldots, j_q \rangle$ and code for an arbitrary Turing machine M with alphabet $a_o, \ldots, a_n$ and instructions $i\ k\ o_{i,k}\ j$ $(0 \leq i,j \leq r)$ for some number s (to be specified) an M-configuration (i,f) of internal state i and tape f by the register constellation

$$(i,f)' := 1_f, r_f, 0,\dots,0,1,0,\dots,0,1,0,\dots,0$$

of a s+5+r-register operator with the first indicated copy of 1 in register R_{s+4} and the second in R_{s+5+i}. The register operator computability of Turing computable functions follows from the following

Lemma 3. For every Turing machine M with $r+1$ states one can constuct a number s and a $s+5+r$ -register operator $\bar{M}$ such that for all M-configurations (i,f), (j,g) holds:

$$(i,f) \underset{M}{\Rightarrow} (j,g) \quad \text{iff} \quad (i,f)' \underset{\bar{M}}{\Rightarrow} (j,g)'$$

and

$$(i,f) \underset{M}{\vdash} \text{stopconfiguration with tape } g \text{ iff } \bar{M}((i,f)') = 1_g, r_g, 0\dots$$

Proof. Let M be as above with elementary operations $o_{i,k} \in \{$right, left, print a_1, stop$\}$ and without loss of generality stop state r (i.e. "stop" occurs in the whole instruction sequence $r\ k\ o_{r,k}\ r$ of the internal state r with $k \le n$ and only there). The simulation of M by $\bar{M}$ is done step by step, so we define for $i \le r$ an iteration operator $M_i = (M_i^!)_{s+5+i}$ such that for $i < r$: $(i,f) \underset{M}{\Rightarrow} (j,g)$ by one step iff $M_i((i,f)') = (j,g)'$ and for $i = r$: $M_r((r,f)') = 1_f, r_f, 0,\dots,0$, because then one can define $\bar{M}$ by

$$\bar{M} := a_{s+4}\ a_{s+5}\ (M_o \dots M_r)_{s+4}$$

Define $M_r := (s_{s+4}\ s_{s+5+r})_{s+5+r}$. In $M_i^!$ for $i < r$ one has first to calculate from R_1 which letter a_k M is scanning at this moment by putting 1 into R_{k+1} (with $R_1,\dots,R_{n+1}$ empty). Then the correct one of the suboperators $M_i^o \dots M_i^n$ of $M_i^!$, namely M_i^k, is put into action simulating the tape transformation $o_{i,k}$ and the call of the successive state j by M. One needs 5 auxiliary register operators C, D, m, σ, L having the following effect:

$$x,0,\dots \underset{C}{\Rightarrow} \langle x\rangle,0,\dots \qquad \text{("Coder")}$$

$$x,0,\dots \underset{D}{\Rightarrow} x,(x)_o,0,\dots \qquad \text{("Decoder")}$$

$$x,y,0,\dots \underset{m}{\Rightarrow} x\cdot y,0,\dots \qquad \text{("multiplier")}$$

$$\langle x_1,\dots,x_n\rangle,0,\dots \underset{\sigma}{\Rightarrow} \langle 0,x_1,\dots,x_n\rangle,0,\dots \qquad \text{("set free the component 0")}$$

$$\langle y,x_1,\dots,x_n\rangle,0,\dots \underset{L}{\Rightarrow} \langle x_1,\dots,x_n\rangle,0,\dots \qquad \text{("eliminate the left component")}$$

Such operators do exist because they compute some particular recursive function (see corollary 1). Now define s as maximum of $n+1$ and of the number of registers used by these five operators and define $M_i^!$ by: clean R_{s+5+i}, store away r_f from R_2 into R_{s+2}, calculate k (by D) and store it in R_{s+3}, store l_f from R_1 into R_{s+1}; now using the information k in R_{s+3} add 1 to R_{k+1} cleaning R_{s+3} and then apply $M_i^o \dots M_i^n$, written down in formulae $M_i = (M_i')_{s+5+i}$ with the concatenation M_i' of the following operators:

$$s_{s+5+i}\ (s_2\ a_{s+2})_2\ D\ (s_2\ a_{s+3})_2\ (s_1\ a_{s+1})_1$$

$$a_1\ (s_{s+3}\ s_1\ a_2\ (s_{s+3}\ s_2\ a_3\ (\dots(s_{s+3}\ s_n\ a_{n+1})_{s+3}\dots)_{s+3})_{s+3})_{s+3}$$

$$M_i^o \dots M_i^n$$

There remains the construction of $M_i^k = (N_i^k)_{k+1}$ to describe the operation $o_{i,k}$ of M on the tape l_f, r_f and its call of the successive state j of instruction $i\ k\ o_{i,k}\ j$. <u>Case 1.</u> $o_{i,k} = a_l$: Clean R_{k+1}, bring the left half of the tape $\langle k,\dots\rangle$ from R_{s+1} back to R_1, compute $\langle k,\dots\rangle \underset{L}{\Rightarrow} \langle\dots\rangle \underset{\sigma}{\Rightarrow} \langle 0,\dots\rangle, 0\dots \Rightarrow \langle 0,\dots\rangle, 2^l, 0,,, \underset{m}{\Rightarrow} \langle l,\dots\rangle$ and then bring back the unchanged right half of the tape from R_{s+2} into R_2 and call for the successive state j by a_{s+5+j}, i.e.

$$N_i^k := s_{k+1}\ (s_{s+1}\ a_1)_{s+1}\ L\ \sigma\ a_2^{2^l}\ m\ (s_{s+2}\ a_2)_{s+2}\ a_{s+5+j}\ .$$

<u>Case 2.</u> $o_{i,k}$ = left: clean R_{k+1}, bring the right half $\langle\dots\rangle = r_f$ of the tape from R_{s+2} to R_1, compute $\langle\dots\rangle \underset{\sigma}{\Rightarrow} \langle 0,\dots\rangle \Rightarrow \langle 0,\dots\rangle, 2^k, 0,\dots \underset{m}{\Rightarrow} \langle k,\dots\rangle, 0,\dots$ store the resulting right half of the new tape g from R_1 back into R_{s+2}, call $l_f = \langle k,\dots\rangle$ from R_{s+1} to R_1, compute $\langle k,\dots\rangle \underset{L}{\Rightarrow} \langle\dots\rangle$, bring the new r_g back into R_2 and call for j, i.e.

$$N_i^k = s_{k+1}\ (s_{s+2}\ a_1)_{s+2}\ \sigma\ a_2^{2^k}\ m\ (s_1\ a_{s+2})_1\ (s_{s+1}\ a_1)_{s+1}\ L\ (s_{s+2}\ a_2)_{s+2}\ a_{s+5+j}$$

<u>Case 3.</u> $o_{i,k}$ = right : clean R_{k+1}, take r_f into R_1, calculate the index of the left most letter say a_l in r_f by D and store it away in R_{s+3}. Now transform $r_f = \langle l,\dots\rangle$ into $r_g = \langle\dots\rangle$ by L and store r_g for the moment away into R_{s+2}. Change l in R_{s+3} into 2^l, recall $l_f = \langle\dots\rangle$ from R_{s+1} into R_1 and transform it into $\langle 0,\dots\rangle$ by σ, then into $\langle l,\dots\rangle$ by m with 2^l from R_{s+3}. Finally bring back also R_{s+2} into R_2 and call for j.

Lemma 3 is proved, therefore by corollary 1 all Turing computable functions are shown to be partial recursive. The inverse is very readily established if one simulates a,s,p , concatenation and $(\ )_1$ by Turing "operators" applying corollary 2. So we have given a complete proof for

Corollary 6. The classes of Turing computable and of partial recursive functions coincide.

Acknowledgement. I wish to express here my thanks to the staff of the Institut für mathem. Logik und Grundlagenforschung at the university of Münster for the generous help offered to me in preparing this manuscript.

References

AANDERAA [1971], S.O.: On the decision problem for formulas in which all disjunctions are binary. in: Proc. Second Scand. Logic Symposium, Amsterdam 1971, pp. 1-18.

Барздинь [1963], Я.М.: ОБ ОДНОМ КЛАССЕ МАШИН ТЬЮРИНГА (Машины Минского). in: АЛГЕБРА И ЛОГИКА 1 (1963) 42-51.

BÖRGER [1971], E.: Reduktionstypen in Krom- und Hornformeln. Dissertation, Münster 1971.

BÜCHI [1962], J.R.: Turing-machines and the Entscheidungsproblem. in: Mathematische Annalen 148 (1962) 201-213.

CHURCH [1936], A.: A note on the Entscheidungsproblem. in: The Journal of Symbolic Logic 1 (1936) 40-41. Correction ibid. 101-102.

COHORS-FRESENBORG [1973], E.: Berechenbare Funktionen und Registermaschinen - ein Beitrag zur Behandlung des Funktionsbegriffs auf konstruktiver Grundlage. in: Didaktik der Mathematik 3 (1973) 187-209.

COHORS-FRESENBORG [1974]: Unentscheidbarkeitssätze in einem Leistungskurs "Grundlagen der Mathematik". in: Beiträge zum Mathematikunterricht 1973, Hannover 1974, pp. 74-82.

EILENBERG S. and ELGOT C.C. [1970]: Recursiveness. New York, 1970.

GERMANO G. and MAGGIOLO-SCHETTINI A. [1973]: Quelques caractérisations des fonctions récursives partielles. in: C.R. Acad. Sc. Paris, t. 276 (14 mai 1973), série A, pp. 1325-1327.

HERMES [1965], H.: Enumerability, decidability, computability. Berlin et al. 1965.

MALCEW [1974], A.J.: Algorithmen und rekursive Funktionen. Braunschweig 1974.

MARKOV [1961] , A.A.: Theory of algorithms. Israel Program for Scientific Translations 1961.

MINSKY [1961] M.L.: Recursive unsolvability of Post's problem of "TAG" and othe topics in theory of Turing machines. in: Annals of Math. 74 (1961) 437-455.

MINSKY [1967] , M.L.: Computation: Finite and infinite machines. Englewood Cliffs 1967.

OTTMANN [1974] , Th.: Rekursive Prozeduren und partiell rekursive Funktionen. Karlsruhe 1974, underground.

POST [1943] , E.: Formal reductions of the combinatorial decision problem. in: Amer. J. Math. 65 (1943) 196-215.

PRIESE [1971] , L.: Normalformen von Markov'schen und Post'schen Algorithmen. Report, Institut für math. Logik und Grundlagenforschung of the University of Münster. pp. 124.

RÖDDING [1968] , D.: Klassen rekursiver Funktionen. in: Proc. Summer School in Logic (Leeds 1967). Springer Lecture Notes in Math. Vol. 70, pp. 159-222.

RÖDDING [1972] , D.: Registermaschinen. in: Der Mathematikunterricht 18 (1972) 32-41.

RÖDDING [1972a] , D.: Einführung in die Theorie der berechenbaren Funktionen. Notes of lectures delivered at the Institut für math. Logik und Grundlagenforschung of the University of Münster, Summer 1972.

ROGERS [1967] H.,Jr.: Theory of recursive functions and effective computability. New York et. al. 1967.

SHEPHERDSON J.C. and STURGIS H.E. [1963] : Computability of recursive functions in: J. Assoc. Comp. Mach. 10 (1963) 217-255.

TURING [1936] , A.M.: On computable numbers with an application to the Entscheidungsproblem. in: Proc. London Math. Soc. Ser. 2,42 (1936) 230-265.

YASUHARA [1971] , A.: Recursive function theory and logic. New York, London 1971.

LECTURES ON LARGE CARDINAL AXIOMS

William Boos

Iowa City, Iowa, U. S. A.

The natural office and aspiration of thought is to understand; but a reality that is infinite is necessarily unintelligible. "For want of having contemplated these infinities, men have set forth rashly upon the investigation of nature, as if there were some proportion between it and them." But after they have once truly faced the immensity of even the physical world, they must inevitably be plunged into "an eternal despair of ever knowing either the beginning or the end of things;" they will be certain only that no. . . assurance, no solid knowledge is attainable by them through the use of their natural intellectual powers. . . .

- Pascal, paraphrased by Arthur Lovejoy

L'éternel silence de ces espaces infinis m'effraie.

- Pascal

Contents

0. Introduction

This introductory section provides a brief sketch of the minimal theory of 'small' large cardinals (inaccessible cardinals, Mahlo cardinals, ordinal Π^1_1-indescribable cardinals) needed for sections 1 - 6 to be self-contained. Thorough expositions of this quite extensive theory appear both in the monograph of Devlin in this volume (to which these notes can be read as a sequel), and in the recent book of Drake [Dr] . Unless otherwise noted, we work in ZFC.

0.1. If η is any limit ordinal, $C \subseteq \eta$ is closed iff $\bigcup(\beta \cap C) \in C$ for each limit $\beta < \eta$; C is unbounded (sometimes cofinal) in η iff $\bigcup C = \eta$. We will often abbreviate "closed (and) unbounded" by c.u. $A \subseteq \eta$ is stationary in η iff $A \cap C \neq 0$ for each c.u. $C \subseteq \eta$. The cofinality of η, cf η, is the least ordinal γ such that there is a 1-1 increasing function $f: \gamma \longrightarrow \eta$ whose range is cofinal in η (we can assume the range of f is closed, too, if we wish).
η is regular if cf $\eta = \eta$, singular if cf $\eta < \eta$.
η is a cardinal iff it is an initial ordinal, i. e., there is no (1-1) function f from any $\gamma < \eta$ onto η.

0.2 Proposition. (1) A regular ordinal is a cardinal.
(2) cf η is always regular.
(3) If cf $\eta = \omega$, every unbounded subset of η of order-type ω is closed, and a subset of η is stationary in η iff it includes a terminal segment, i. e., the set of all ordinals $< \eta$ above a given one.
(4) If cf $\eta = \lambda > \omega$, the collection K of closed unbounded subsets of η forms a λ-complete filterbase, i. e., whenever C_α is in K for each $\alpha < \beta < \lambda$, so is $\bigcap_{\alpha<\beta} C_\alpha$.

Proof of (1). Immediate from the definitions.
Proof of (2). If not, the composition of suitably chosen f: cf(cf η) $= \delta \longrightarrow$ cf η and g: cf $\eta \longrightarrow \eta$ would give cf $\eta \leq \delta <$ cf η.
Proof of (3). For each $\beta = \bigcup\beta < \eta$ and unbounded $X \subseteq \eta$ of type ω, $\bigcup(\beta \cap X)$ is the largest element of $X < \beta$. A set S intersects every closed unbounded set iff it intersects every cofinal set (of type ω).
Proof of (4). We do the limit case of the induction over $\beta < \lambda$. Suppose $\bigcap_{\alpha<\delta} C_\alpha$ is c. u. for each sequence $\langle C_\alpha \mid \alpha < \delta \rangle$ of length $< \beta$. Given $\langle C_\alpha \mid \alpha < \beta \rangle$ and arbitrary $\gamma_0 < \eta$, choose $\gamma_\delta > \bigcup_{\alpha<\delta} \gamma_\alpha$ such that $\gamma_\delta \in \bigcap_{\alpha<\delta} C_\alpha$. Then for each $\alpha < \beta$, $\{\gamma_\delta \mid \delta > \alpha\} \subseteq C_\alpha$. so $\gamma = \bigcup_{\delta<\beta} \gamma_\delta$ is an element of C_α, so $\gamma \in \bigcap_{\alpha<\beta} C_\alpha = C$. C is closed, since if $\bigcap_{\alpha<\beta} C_\alpha$ is unbounded in $\zeta = \bigcup\zeta$, so is each C_α.

0.3. Notational convention. A filter $\mathcal{F} \subseteq S(\kappa)$ over infinite cardinal κ (cf., e. g., [Ch-Ke, p. 164 ff.]) will always be assumed to be uniform (every element of $\mathcal{F}$ has power κ), and nonprincipal (no singleton $\{\alpha\}$ is in $\mathcal{F}$), unless stated otherwise, and $\mathcal{J}$ will always denote the ideal dual to $\mathcal{F}$. Thus we might have $\mathcal{F}$, $\mathcal{J}$, $\mathcal{F}^*$, $\mathcal{J}^*$, etc.

0.4. Some of the following definitions are repeated in 1.5 of section 1. If X is a set of ordinals, $f: X \longrightarrow On$ is regressive iff $f(\alpha) < \alpha$ for each $\alpha \in X$. If κ is an uncountable regular cardinal, a filter $\mathcal{F} \subseteq S(\kappa)$ is normal iff whenever $X \notin \mathcal{J}$ (sometimes, in English, "X has positive $\mathcal{F}$-measure") and $f: X \longrightarrow \kappa$ is regressive, there is a $Y \subseteq X$ such that $Y \notin \mathcal{J}$ and $f \restriction Y$ is constant. $\mathcal{F}$ is κ-complete iff $\bigcap_{\alpha<\beta} X_\alpha \in \mathcal{F}$ for each $\beta < \kappa$ and each sequence $\langle X_\alpha \mid \alpha<\beta \rangle$ of elements of $\mathcal{F}$. If $X_\alpha \subseteq \kappa$ for each $\alpha < \kappa$, the diagonal intersection $\Delta_{\alpha<\kappa} X_\alpha$ of the X_α's is $\{\beta \mid \beta \in X_\alpha \text{ for all } \alpha<\beta\}$.

0.5 Theorem (Fodor). Assume κ is an uncountable regular cardinal and $\mathcal{F}$ is a filter on κ. Then
(1) $\mathcal{F}$ is normal iff $\mathcal{F}$ is closed under diagonal intersections, i. e., whenever each $X_\alpha \in \mathcal{F}$ for $\alpha < \kappa$, so is $\Delta_{\alpha<\kappa} X_\alpha$.
(2) The closed unbounded filter (i. e., the κ-complete filter generated by the filterbase of closed unbounded subsets of κ) is normal.

Proof of (1). ($\longrightarrow$). If $\Delta_{\alpha<\kappa} X_\alpha = X \notin \mathcal{F}$, its complement Y in is not in $\mathcal{J}$, and $f: \beta \longmapsto$ (least $\alpha<\beta$ such that $\beta \notin X_\alpha$) is regressive on Y, so there is a $Z \subseteq Y$ with $Z \notin \mathcal{J}$ and an $\alpha < \kappa$ such that $f''Z = \{\alpha\}$, but then $Z \cap X_\alpha = 0$, so $X_\alpha \notin \mathcal{F}$.
($\longleftarrow$). If $f: X \longrightarrow \kappa$ is such that each $f^{-1}(\{\alpha\}) = X_\alpha \in \mathcal{J}$, and $Y_\alpha = \kappa - X_\alpha$, each $Y_\alpha \in \mathcal{F}$, so $\Delta_{\alpha<\kappa} Y_\alpha = Y$ is too, but since $f(\alpha) \geq \alpha$ for each $\alpha \in Y$, either X is disjoint from Y (and so in $\mathcal{J}$), or f is not regressive on X.
Proof of (2). We assume C_α is closed and unbounded for each $\alpha < \kappa$, and $C = \Delta_{\alpha<\kappa} C_\alpha$.
C is unbounded. Let α_0 be arbitrary, and choose α_{n+1} inductively such that $\alpha_{n+1} \in \bigcap_{\beta<\alpha_n} C_\beta$ and $\alpha_{n+1} > \alpha_n$. Then $\alpha = \bigcup_{n<\omega} \alpha_n$ is in C, for if $\beta<\alpha$, $\beta <$ some α_m, so $\{\alpha_n \mid n>m\} \subseteq C_\beta$, so $\alpha = \bigcup\{\alpha_n \mid n>m\} \in C_\beta$ as well.
C is closed. If $C \cap \beta$ is unbounded in $\beta = \bigcup\beta$, and $\alpha<\beta$, $\{\gamma \in C \mid \alpha<\gamma<\beta\} \subseteq C_\alpha$, so $\beta \in C_\alpha$.

0.6. κ is a weak limit cardinal iff $\kappa = \aleph_\alpha$ and $\alpha = \bigcup\alpha$ i. e., whenever $\lambda<\kappa$, so is λ^+.

κ is a strong limit cardinal iff $\kappa = \beth_\alpha$ and $\alpha = \bigcup\alpha$, *i. e.*, whenever $\lambda < \kappa$, so is 2^λ.
κ is inaccessible iff κ is a regular weak limit cardinal.
κ is strongly inaccessible iff κ is a regular strong limit cardinal.
κ is weakly inaccessible iff κ is inaccessible but not strongly inaccessible, *i. e.*, $2^\lambda \geq \kappa$ for some $\lambda < \kappa$.
κ is 1 - Mahlo, or just *Mahlo* iff $\{\lambda < \kappa \mid \lambda \text{ regular}\}$ is stationary in κ.
κ is α-Mahlo for $\alpha > 1$ iff for each $\beta < \alpha$ $\{\lambda < \kappa \mid \lambda \text{ is } \beta\text{ - Mahlo}\}$ is stationary in κ.
κ is ordinal Π^1_0 (Π^1_1) indescribable iff whenever $A \subseteq \kappa$ and a first-order (Π^1_1) φ are given such that $\langle \kappa, \in, A\rangle \models \varphi$, there is a $\lambda < \kappa$ such that $\langle \lambda, \in, A \cap \lambda\rangle \models \varphi$.
κ is weakly compact iff κ is strongly inaccessible and ordinal Π^1_1-indescribable (this is not the original definition, but is known to be equivalent to it; cf. [De3, §3] and [Dr, Theorem 10.2.1]).

0.7 Theorem. *(1) κ is ordinal Π^1_0-indescribable iff κ is regular.* *(2) If κ is Mahlo, $\{\lambda < \kappa \mid \lambda \text{ inaccessible}\}$ is stationary in κ.* *(3) If κ is ordinal Π^1_1-indescribable, κ is α-Mahlo for all $\alpha < \kappa$, $\{\lambda < \kappa \mid \lambda$ is α-Mahlo for all $\alpha < \lambda\}$ is stationary in κ, etc. In fact, whenever φ is Π^1_1, $A \subseteq \kappa$ and $\langle \kappa, \in, A\rangle \models \varphi$, $\{\lambda < \kappa \mid \langle \lambda, \in, A\cap\lambda\rangle \models \varphi\}$ is stationary in κ.*

Proof of (1). We sketch the basic ideas.
($\longrightarrow$). If κ is not regular, let f be a 1-1 increasing function from $\text{cf}\,\kappa = \lambda \longrightarrow \kappa$ which witnesses this. Then $\langle \kappa, \in, \lambda$, the graph of f$\rangle$ satisfies (the domain of f is cofinal in λ), but no $\langle \eta, \in, \lambda, f \cap (\eta \times \eta)\rangle$ can satisfy this. (The use of the binary relation is permissible; see the remarks 1.10 below, and 2.1 and 2.2 of Devlin's monograph in this volume).
($\longleftarrow$). If κ is regular and $\langle \kappa, \in, A\rangle \models \varphi$, let $\langle B_0, \in, A|B\rangle$ be an elementary submodel of $\langle \kappa, \in, A\rangle$ such that $\overline{\overline{(B_0)}} < \kappa$, and let $\langle B_{n+1}, \in, A|B_{n+1}\rangle$ for $n < \omega$ be the smallest elementary submodel of $\langle \kappa, \in, A\rangle$ such that $\bigcup B_n \subseteq B_{n+1}$. If $B = \bigcup_{n<\omega} B_n$, B is an ordinal $\beta < \kappa$ and $\langle \beta, \in, A|\beta\rangle \models \varphi$.
Proof of (2). κ is itself inaccessible, for any singular ordinal σ has a closed cofinal subset of singular ordinals (choose any increasing sequence $\langle \eta_\alpha\ \alpha < \text{cf}\,\sigma\rangle$ of ordinals with range cofinal in σ such that $\eta_0 > \text{cf}\,\sigma$, and η_0 and each $\eta_{\alpha+1}$ are singular), and $\{\eta \mid \eta > \lambda\}$ is such a set if κ is λ^+. Thus $\{\lambda < \kappa \mid \lambda \text{ is a cardinal}\}$ is unbounded in κ, so $\{\lambda < \kappa \mid \lambda = \aleph_\lambda\}$ is a c.u. subset of κ, which must

contain a regular and thus inaccessible λ.

Proof of (3). Note that (I am inaccessible) is Π^1_1, as by induction are (I am α - Mahlo) (in the parameter α), and (I am α - Mahlo for all $\alpha <$ me), so it suffices to prove the last sentence. If C is c.u. u and $\langle \kappa, \in, A, C \rangle \models$ (φ and C is unbounded in me), so does $\langle \lambda, \in, A \cap \lambda, C \cap \lambda \rangle$ for some $\lambda < \kappa$, in C since $\bigcup(\lambda \cap C) = \lambda$.

0.8. Throughout the remainder of the notes, κ, λ, μ and ν will be arbitrary infinite cardinals in the sense of some transitive model or interpretation of ZFC.

1. Subtlety

1.1 Definition (Jensen, Kunen, Baumgartner). If κ is any uncountable cardinal, $X \subseteq \kappa = \bigcup X$ is subtle / almost ineffable / ineffable iff whenever C is a closed unbounded subset of κ and $\langle A_\alpha \mid \alpha \in X \rangle$ is such that $A_\alpha \subsetneq \alpha$ for all $\alpha \in X$, there is a(n) unordered pair $Y \subseteq C \cap X$ / unbounded $Y \subseteq C \cap X$ / stationary $Y \subseteq C \cap X$ such that Y is homogeneous for $\langle A_\alpha \mid \alpha \in X \rangle$, where this means that $A_\alpha = A_\beta \cap \alpha$ for all α, β in Y.

1.2. If in 1.1 one replaces "$A_\alpha \subseteq \alpha$" with "$A_\alpha \subseteq \alpha$ and $(A_\alpha)^= = \bar{\bar{\alpha}}$", "homogeneous" with "weakly homogeneous" and "$A_\alpha = A_\beta \cap \alpha$" with "$(A_\alpha \cap A_\beta)^= = \bar{\bar{\alpha}}$", one has a definition of X is weakly subtle / weakly almost ineffable / weakly ineffable.
κ is a subtle, weakly subtle (etc.) cardinal iff κ is a subtle, weakly subtle (etc.) subset of itself iff there is some unbounded subtle, weakly subtle (etc.) subset of κ.

1.3 Definition (Baumgartner, Ketonen). If κ is any uncountable cardinal, the (weakly) subtle / (weakly) almost ineffable / (weakly) ineffable filter $\mathcal{F} \subseteq S(\kappa)$ is the collection of all complements of non-(weakly) subtle / non- (weakly) almost ineffable / non- (weakly) ineffable subsets X of $\kappa = \bigcup X$.
Notice that if κ is not (say) subtle, the subtle filter on κ is the trivial one $S(\kappa)$.

Jensen and Kunen independently defined ineffable cardinals and proved a number of results about them in [Jen 1]. Ketonen later defined weakly subtle cardinals, which he called "ethereal", and showed in [Ke] tha several results from the theory of subtlety and ineffability have conceptually pleasant 'weak' variants. Without directly naming it he also defined the weakly subtle filter in order to prove 1.7 below. Independently, Baumgartner observed that a wide range of large cardinal definitions generate corresponding normal filters, and used these filters to study what he wryly abstained from calling "subtle properties of cardinals." We give a sample of the structure theory that has accumulated.

1.4 Proposition. (1) A weakly subtle cardinal is inaccessible.
(2) A subtle cardinal is strongly inaccessible.
(3) A strongly inaccessible, weakly subtle (weakly almost ineffable, weakly ineffable) cardinal is subtle (almost ineffable, ineffable).

Proof of (1). If κ is singular, let C be a closed unbounded subset of cardinals $\mu < \kappa$, of order-type $\lambda = \mathrm{cf}\,\kappa$. If $\langle \mu_\alpha \mid \alpha < \lambda \rangle$

enumerates C, let $\tilde{A}_{\mu_\alpha} = \{\alpha\} \times \mu_\alpha$ for $\alpha < \lambda$, and let $A_{\mu_\alpha} = j''\tilde{A}_{\mu_\alpha}$ be the image of $\tilde{A}_{\mu_\alpha}$ under Gödel's pairing function j (the class of ordinals closed under j is closed and unbounded in each uncountable cardinal). A_{μ_α} then violates weak subtlety of κ.
If $\kappa = \lambda^+$, we define a sequence $\langle f_\alpha \mid \alpha < \lambda \rangle$ of functions $f_\alpha : \lambda \longrightarrow \lambda$ whose graphs G_α are almost disjoint, i. e., $\overline{\overline{(G_\alpha \cap G_\beta)}} < \lambda$ for $\alpha \neq \beta < \lambda^+$
Then if $A_\alpha = j''G_\alpha$ for $\alpha \in D = \{$ ordinals $< \lambda^+$ closed under j $\}$, $\langle A_\alpha \mid \alpha \in D \rangle$ contradicts weak subtlety of κ. Let f_α be the constant function α for $\alpha < \lambda$; if $\langle f_\alpha \mid \alpha < \eta \rangle$ is defined for $\eta < \lambda^+$, choose a 1-1 map τ from λ onto η, and define $f_\eta(\beta)$ = the least ordinal not in $\{ f_{\tau(\gamma)}(\delta) \mid \gamma, \delta \leq \beta \}$. For every $\alpha < \eta$, $f_\eta(\beta) \neq f_\alpha(\beta)$ for all $\beta > \tau^{-1}(\alpha)$.

<u>Proof of (2)</u>. If $\kappa \leq 2^\lambda$ for some $\lambda < \kappa$, any enumeration of κ distinct subsets of λ violates subtlety.

<u>Proof of (3)</u>. If κ is strongly inaccessible, there is a bijection $F: R(\kappa) \longrightarrow \kappa$ and a closed unbounded subset D of κ consisting of cardinals λ such that $\overline{\overline{(R(\lambda))}} = \beth_\lambda = \lambda$ and $F \restriction R(\lambda)$ maps $R(\lambda)$ onto λ. If $\langle A_\alpha \mid \alpha \in C \rangle$ is given, let χ_λ for $\lambda \in C \cap D$ be the characteristic function of A_λ, and $A^*_\lambda = \{ F(\chi_\lambda \restriction \alpha) \mid \alpha \in \lambda \}$. If $Y \subseteq C \cap D$ is weakly homogeneous for $\langle A^*_\lambda \mid \lambda \in C \cap D \rangle$, Y is homogeneous for $\langle A_\lambda \mid \lambda \in C \cap D \rangle$.

<u>1.5</u> Recall from 0.4 that a (nonprincipal) <u>filter $\mathcal{F}$</u> on an uncountable regular cardinal κ is <u>normal</u> iff for every X not in the ideal $\mathcal{I}$ dual to $\mathcal{F}$ and regressive f from X into κ, there exists a subset Y of X, also not in $\mathcal{I}$, such that $f \restriction Y$ is constant. Note that a normal filter on κ must be κ-complete.

<u>1.6 Theorem. (Ketonen, Baumgartner)</u>
<u>The (weakly) subtle / (weakly) almost ineffable / (weakly) ineffable filter on a (weakly) subtle / (weakly) almost ineffable / (weakly) ineffable cardinal κ is normal.</u>

<u>Proof.</u> We carry out the argument for weak subtlety. The others are parallel. Suppose $X \subseteq \kappa$ is weakly subtle, $f: X \longrightarrow \kappa$ is regressive, and C a closed unbounded set of ordinals $< \kappa$ closed under j. $X \cap C$ is weakly subtle (immediate from the definition), and we show there is a weakly subtle $S \subseteq X \cap C$ on which f is constant. If not, then for each $\alpha < \kappa$, $X_\alpha = f^{-1}(\{\alpha\}) \cap C$ is not weakly subtle; let c.u. D_α and $\langle A^\alpha_\beta \mid \beta \in D_\alpha \rangle$ witness this. If $D = \triangle_{\alpha<\kappa} D_\alpha$ is the diagonal intersection of the D_α's, D is closed unbounded by 0.5(2), so $W = X \cap C \cap D$ is weakly subtle. If $\langle Y_\beta \mid \beta \in W \rangle$ is now defined by $Y_\beta = j''(\{f(\beta)\} \times A^{f(\beta)}_\beta)$ for $\beta \in W$, weak subtlety of W gives that there

are $\gamma, \delta \in W$ such that $\gamma < \delta$ and $(Y_\gamma \cap Y_\delta)^= = \bar{\bar{\gamma}}$. Then $f(\gamma) = f(\delta) = \alpha$, say, for if $f(\gamma)$ and $f(\delta)$ were different, Y_γ and Y_δ would be disjoint. But then $Y_\gamma = j''(\{\alpha\} \times A^\alpha_\gamma)$ and $Y_\delta = j''(\{\alpha\} \times A^\alpha_\delta)$, so $(A^\alpha_\gamma \cap A^\alpha_\delta)^= = \bar{\bar{\gamma}}$, contrary to the assumption on $\langle A^\alpha_\beta \mid \beta \in D_\alpha \rangle$.

1.7 Corollary (Ketonen, Baumgartner). If κ is weakly subtle, κ is α - Mahlo for all $\alpha < \kappa$.

Proof. We show first that κ is Mahlo, in fact more, that $R = \{\lambda < \kappa \mid \lambda \text{ inaccessible}\}$ is in the weakly subtle filter. If not, the set A of cardinals in $\kappa - R$ is weakly subtle, since we already know κ is inaccessible. The cofinality function is regressive on A, so there is a $\lambda < \kappa$ such that $\{\mu < \kappa \mid \text{cf}\,\mu = \lambda\} = B$ is weakly subtle. Let S_μ be closed unbounded in μ of order-type λ for each $\mu \in B$, and let $\tilde{S}_\mu = j''\{ (\gamma, \beta) \mid \gamma \in S_\mu \text{ and } \beta < \gamma \}$. Since $S_\mu \cap \nu$ is bounded in ν for $\mu, \nu \in B$ with $\nu < \mu$, $\langle \tilde{S}_\mu \mid \mu \in B \rangle$ violates weak subtlety of B.

We show how to proceed through successor stages of the otherwise easy induction over $\alpha < \kappa$ which completes the proof. Let $\mathcal{M}: S(\kappa) \longrightarrow S(\kappa)$ be the Mahlo thinning-operation $\mathcal{M}(A) = \{ \alpha < \kappa \mid A \cap \alpha \text{ is stationary in } \alpha \}$. We show that if A is in the weakly subtle filter, so is $\mathcal{M}(A)$. Suppose not. Then $\kappa - \mathcal{M}(A)$ is weakly subtle. β is in $\kappa - \mathcal{M}(A)$ iff there is a $C_\beta \subseteq \beta$ disjoint from A. $A \cap (\kappa - \mathcal{M}(A)) \cap \{\lambda < \kappa \mid \lambda \text{ inaccessible}\}$ is weakly subtle, since $\kappa - \mathcal{M}(A)$ is, and A and $\{\lambda < \kappa \mid \lambda \text{ inaccessible}\}$ are in the weakly subtle filter. Then there are inaccessible $\lambda, \mu \in A \cap (\kappa - \mathcal{M}(A))$ such that $\lambda < \mu$ and $(C_\lambda \cap C_\mu)^= = \lambda$, so $\lambda \in A \cap C_\mu$, which contradicts the assumption that C_μ is disjoint from A.

1.8 Remark. The weakly subtle filter on a weakly subtle cardinal properly extends the closed unbounded filter on κ; for any regular cardinal $\lambda < \kappa$, the set of cardinals $\mu < \kappa$ such that $\text{cf}\,\mu = \lambda$ is stationary but not weakly subtle.

1.9 (cf. 0.6). If κ is any uncountable cardinal, $X \subseteq \kappa = \bigcup X$ is ordinal Π^n_m- indescribable for integers $n, m \geq 1$ iff whenever φ is Π^n_m, $P \subseteq \kappa$, and $\langle \kappa, \in, P \rangle \models \varphi$, there is a $\lambda \in X$ such that $\langle \lambda, \in, P \cap \lambda \rangle \models \varphi$. Again κ is ordinal Π^n_m- indescribable as a subset of itself iff it contains some unbounded ordinal Π^n_m- indescribable subset. Such a κ must be weakly $L_{\kappa\omega}$ - compact (cf., e. g., [Bo]), and inaccessible and α - Mahlo for all $\alpha < \kappa$ by 0.7(3), but various assumptions allow for κ to be below the continuum (cf. [Ku2] and [Bo]).

If κ is strongly inaccessible and $X \subsetneq \kappa = \bigcup X$ is ordinal Π^n_m-indescribable, it is not hard to prove X is Π^n_m-indescribable, where this means whenever φ is Π^n_m, $P \subseteq R(\kappa)$ (this is the essential difference) and $\langle R(\kappa), \in , P \rangle \models \varphi$, so does $\langle R(\lambda), \in , P \cap R(\lambda) \rangle$ for some $\lambda \in X$ (to verify the equivalence, use the bijection F of the proof of 0.4.3). For readers of [Lé], it may be helpful to remark that "X is an ordinal Π^n_m-indescribable subset of $\kappa = \bigcup X$" expressed in the language of [Lé] and [Dr, Chapter 9.2] is " κ - X is not Π^n_m-enforceable at κ"; "X is Π^n_m-indescribable subset of $\kappa = \bigcup X$" is " κ - X is not weakly Π^n_m-enforceable at κ."

1.10. More generally, if $X \subseteq \kappa = \bigcup X$ and $\alpha > 0$, X is (ordinal) α - indescribable iff whenever φ is first-order, $P \subseteq R(\kappa)$ ($P \subseteq \kappa$), and $\langle R(\kappa + \alpha), \in \rangle \models \varphi[\kappa, P]$, $\langle R(\lambda + \alpha), \in \rangle \models \varphi[\lambda, P \cap R(\lambda)]$ for some $\lambda \in X$. If $\alpha > \kappa$, we will usually write "X is (ordinal) invisible in $R(\alpha)$" instead of "X is (ordinal) α - indescribable". Thus κ is (ordinal) invisible in $R(\alpha)$ iff whenever φ is first-order, $P \subseteq R(\kappa)$ ($P \subseteq \kappa$), and $\langle R(\alpha), \in \rangle \models \varphi[\kappa, P]$, $\langle R(\alpha), \in \rangle \models \varphi[\lambda, P \cap R(\lambda)]$ for some $\lambda < \kappa$, so invisibility of κ in $R(\alpha)$ means $R(\alpha)$ cannot discern what κ is unless it is given enough information so that the recognition is trivial. An obvious totum simul of these notions is the following, which must be phrased in a meta-theory for ZFC such as Morse -- Kelley set theory with choice for sets (which is essentially second-order ZFC), in which a satisfaction predicate is definable for V.
$X \subseteq \kappa = \bigcup X$ is (ordinal) invisible iff whenever φ is first-order, $P \subseteq R(\kappa)$ ($P \subseteq \kappa$) and $\varphi(\kappa, P)$, there is some $\lambda < \kappa$ such that $\varphi(\lambda, P \cap R(\lambda))$.

Perhaps the word "ineffable" should have been held in reserve for this latter notion, which is not unlike the Hanf number in its resemblance to the Cheshire Cat's smile. We will quickly see that there are many invisible (in $R(\beta)$)) cardinals below the least subtle cardinal κ (for each $\beta \geq \kappa$).

Each of the above notions, defined and studied by Hanf, Scott, Lévy, Silver, Jensen and others, can be given a speciously stronger definition in which the 'thin' structures $\langle R(\kappa), \in , P \rangle$ and $\langle \kappa, \in , P \rangle$ are replaced by arbitrary finitary structures with universe $R(\kappa)$ and κ (we have already referred to this in the proof of 0.7). More importantly, each generates a corresponding normal filter, where the elements of the shrdlu filter once again are the complements of non - shrdlu sets (and the elements of the ordinal Π^n_m- indescribable / Π^n_m- indescribable filter are the sets Π^n_m- enforceable / weakly enforceable at κ).

<u>1.11 Theorem (Lévy, Baumgartner).</u> <u>The (ordinal) Π^n_m-indescribable / (ordinal) α - indescribable / (ordinal) invisible (in $R(\beta)$) filter on a(n) (ordinal) Π^n_m- indescribable / (ordinal) α - indescribable / (ordinal) invisible (in $R(\beta)$) cardinal κ is a normal filter for all κ, $l \leq n$, $m < \omega$, $\alpha < \kappa$ and $\beta > \kappa$.</u>

<u>Proof.</u> We sketch Π^n_m-ordinal indescribability; the other arguments are parallel. Suppose $A \subseteq \kappa$, f: $A \longrightarrow \kappa$ is regressive, and no $A_\alpha = f^{-1}(\alpha)$ is Π^n_m indescribable. Let $K_\alpha \subseteq \kappa$ and φ_α witness this for $\alpha < \kappa$, so that $\langle \kappa, \in, K_\alpha \rangle \models \varphi_\alpha$, but $\langle \beta, \in, K_\alpha \cap \beta \rangle \not\models \varphi_\alpha$ for each ordinal $\beta \in A_\alpha$ Using Godel's pairing function j, the binary relation $K = \{ \langle \beta, \alpha \rangle \mid \beta \in K_\alpha \rangle$ and a T which encodes a truth definition over the right sorts of models, we can find a $\Pi^n_m \varphi$ such that for all cardinals $\beta \leq \kappa$ $\langle \beta, \in, j|\beta, K|\beta, T|\beta \rangle \models \varphi$ iff for all $\gamma < \beta$ $\langle \beta, \in, K_\gamma|\beta \rangle \models \varphi_\gamma$. Then $\langle \kappa, \in, j, K, T \rangle \models \varphi$, but $\langle \beta, \in, K_{f(\beta)} \cap \beta \rangle \models \neg\varphi_{f(\beta)}$, and thus $\langle \beta, \in, j|\beta, K|\beta, T|\beta \rangle \models \neg\varphi$ for each $\beta \in A$, so A is not Π^n_m indescribable.

" κ is subtle" is Π^1_1 over $\langle \kappa, \in \rangle$, and " κ is almost ineffable" is Π^1_2, so it follows from the next result that the least subtle cardinal is less than the least almost ineffable, which in turn is less than the least ineffable.

<u>1.12 Theorem (Jensen, Kunen).</u> <u>Let $X \subseteq \kappa = \bigcup X$. Then</u>
<u>(1) If X is almost ineffable, X is Π^1_1 indescribable.</u>
<u>(2) If X is ineffable, X is Π^1_2 indescribable.</u>

<u>Proof of (1)</u>. Suppose we have a $\Pi^1_0 \varphi$, a subset B of κ and $A_\alpha \subseteq \alpha$ such that for all cardinals $\lambda \in X$ $\mathfrak{A}_\lambda = \langle \lambda, \in, A_\lambda, B \cap \lambda \rangle \models \varphi$. Working again with some fixed arithmetization of the first-order language involved, let $Y_\lambda \subseteq \lambda$ code the elementary diagram of $\mathfrak{A}_\lambda$, and let Z_λ be the image under the pairing function j of $\langle A_\lambda, Y_\lambda \rangle$. By almost ineffability let W be such that for all λ, μ in $W \subseteq X$ $Z_\lambda = Z_\mu \cap \lambda$. Then by elementarity $\langle \kappa, \in, A, B \rangle \models \varphi$, where $A = \bigcup_{\lambda \in W} A_\lambda$.

<u>Proof of (2)</u>. Suppose we have $\varphi, B \subseteq \kappa$ and $A_\alpha \subseteq \alpha$ as above, but now such that for all $\lambda \in X$ $\mathfrak{A}_\lambda = \langle \lambda, \in, A_\lambda, B \cap \lambda \rangle \models \forall Y \neg\varphi$. If $S \subseteq X$ is a stationary homogeneous set for $\langle A_\lambda \mid \lambda \in X \rangle$, and

$A = \bigcup_{\lambda\in S} A_\lambda$ $\langle \kappa, \in, A, B \rangle \models \forall Y \neg \varphi$. For if for some $D \subseteq \kappa$ $\langle \kappa, \in, A, B, D \rangle \models \varphi$, there is a closed unbounded $C \subseteq \kappa$ such that for every $\lambda \in C \cap S$, $\langle \mathfrak{A}_\lambda, D \cap \lambda \rangle \models \varphi$, which contradicts for all $\lambda \in X$ $\mathfrak{A}_\lambda \models \forall Y \neg \varphi$.

I verified the following (with "unbounded" rather than "in the subtle filter") in my thesis in order to generalize a method of Solovay for obtaining infinitary compactness below the continuum. The argument is quite simple, and parallels, as it turns out, the proof of a lemma in [Jen 1].

1.13 Theorem. (1) If κ is subtle and β is any ordinal $\geq \kappa$, $\{\lambda < \kappa \mid \lambda$ is invisible in $R(\beta)\}$ is in the subtle filter. (2) (MKC) If κ is subtle, $\{\lambda < \kappa \mid \lambda$ is invisible$\}$ is in the subtle filter.

Proof. Suppose not. Let S be the subtle set of cardinals $\lambda < \kappa$ visible in $R(\beta)$, and let $\langle \varphi_n \mid n < \omega \rangle$ enumerate the formulas in the language of first-order set theory in two free variables. Let $F: R(\kappa) \longrightarrow \kappa$ and D be as in the proof of 1.4(3). If λ is in the subtle set $S \cap D$ and $X \subseteq R(\lambda)$, let $X^* \subseteq \lambda$ be $(F''X \cap (\lambda - \omega)) \cup \{3n \mid n \in F''X \cap \omega\} \cup \{3n+1 \mid \langle R(\beta), \in \rangle \models \varphi_n[\lambda, X]\} \cup \{3n+2 \mid \langle R(\beta), \in \rangle \not\models \varphi_n[\lambda, X]\}$. By assumption we have $A_\lambda \subseteq R(\lambda)$ for each $\lambda \in S \cap D$ such that for all λ, μ in $S \cap D$ with $\lambda < \mu$, $A^*_\lambda \neq A^*_\mu \cap \lambda$, a contradiction. The modification needed for (2) is clear.

Since any element of the subtle filter on κ is stationary, and any λ invisible in $R(\beta)$ is α-Mahlo for all $\alpha < \lambda$, the last result provides another way to see that a subtle κ is α-mahlo for all $\alpha < \kappa$.

Not only are there many invisible λ below a subtle κ, there are, given a closed unbounded $C \subseteq \kappa$ and $\langle A_\alpha \mid \alpha \in C \rangle$, with $A_\alpha \subseteq \alpha$ for $\alpha \in C$, many invisible $X_\lambda \subseteq \lambda = \bigcup X_\lambda$ which are homogeneous for $\langle A_\alpha \mid \alpha \in C \rangle$ as well. The following was proved and applied extensively by Baumgartner in [Ba].

1.14 Theorem (Baumgartner; cf. 4.1 of [Ba]). Suppose κ is subtle, D is a subtle subset of κ, $\langle A_\alpha \mid \alpha \in D \rangle$ is such that $A_\alpha \subseteq \alpha$ for every $\alpha \in D$, and β is any ordinal $\geq \kappa$. Then if $D^* = \{\lambda < \kappa \mid$ there is an $X_\lambda \subseteq \lambda$ which is both invisible in $R(\beta)$ and homogeneous for $\langle A_\alpha \mid \alpha \in D \rangle\}$, $D - D^*$ is not subtle. (Instead of invisible in $R(\beta)$, Baumgartner has Π^1_m-indescribable for each $m < \omega$; the proof is the same).

Proof. If not, let S be a subtle set of cardinals $\lambda \in D$ such that every subset of λ homogeneous for $\langle A_\alpha | \alpha \in D \rangle$ is visible in $R(\beta)$. For $\lambda \in S$ let $B_\lambda = \{\alpha \in S \cap \lambda | A_\alpha = A_\lambda \cap \alpha\}$. Since $B_\lambda \cup \{\lambda\}$ is homogeneous for $\langle A_\alpha | \alpha \in D \rangle$, each B_λ is visible in $R(\beta)$. Let $K_\lambda \subseteq R(\lambda)$ and φ_λ for $\lambda \in S$ be such that $\langle R(\beta), \in \rangle \models \varphi_\lambda [\lambda, K_\lambda]$, but not $\varphi_\lambda [\eta, K_\lambda \cap R(\eta)]$ for any $\eta \in B_\lambda$. Let F, D and $\_^*$ be as in the proof of 1.4(3). Then $\langle H_\lambda | \lambda \in S \cap C \rangle$, where $H_\lambda \subseteq \lambda$ is the image under the pairing function of $\langle A_\lambda, K_\lambda^* \rangle$, contradicts subtlety of $S \cap C$; for if λ, μ are in $S \cap C$ with $\lambda < \mu$ and $A_\lambda = A_\mu \cap \lambda$, $\lambda \in B_\mu$, which contradicts $K_\lambda^* = K_\mu^* \cap \lambda$.

1.14 of course also establishes rather overwhelmingly that if A is subtle, any $\langle A_\alpha | \alpha \in A \rangle$ has homogeneous sets of all order-types $< \kappa$.

<u>1.15. Definition</u>. Two filters <u>$\mathcal{F}$ and $\mathcal{F}'$</u> on an infinite cardinal <u>are coherent</u> iff the least filter on κ including both of them is not trivial, <u>i. e.</u>, all of $S(\kappa)$.

<u>1.16 Theorem (Baumgartner).Let κ be an infinite cardinal. Then (1) $A \subseteq \kappa$ is (almost) ineffable iff $A \cap D$ is Π^1_2 (Π^1_1) indescribable for each D in the subtle filter.</u>
<u>(2) κ is (almost) ineffable iff the subtle and Π^1_2 (Π^1_1) indescribable filters are coherent.</u>

<u>Proof of (1)($\longrightarrow$)</u>: If A is (almost) ineffable, D is in the subtle filter and $D \cap A$ is not Π^1_2 (Π^1_1)-indescribable, A is partitioned into a nonsubtle set (A − D) and a non-Π^1_2 (Π^1_1) -indescribable set (A D) · impossible, since one of the two must be (almost) ineffable, thus both subtle and Π^1_2 (Π^1_1)-indescribable, by 1.12.
($\longleftarrow$): Suppose $\langle S_\alpha | \alpha \in A \rangle$ is given with $S_\alpha \subseteq \alpha$ for each $\alpha \in A$. A is subtle since it intersects every D in the subtle filter, so by 1.14 there is an $A^* \subseteq A$ such that $A - A^*$ is not subtle (so $A^* = A \cap D$ for $D = A^* \cup (\kappa - A)$ in the subtle filter), and each $\lambda \in A^*$ has a stationary subset $K_\lambda \subseteq \lambda$ homogeneous for $\langle S_\alpha | \alpha \in A \rangle$. The datum "there is a stationary (unbounded) set homogeneous for $\langle S_\alpha | \alpha \in A \rangle$" is Σ^1_2 (Σ^1_1) and valid over $\langle \lambda, \in, A \cap \lambda, \{\langle \beta, \alpha \rangle | \beta \in S_\alpha\} \rangle$ for each $\lambda \in A^*$, and by Π^1_2 (Π^1_1) indescribability of A^*, therefore, valid over κ as well.
<u>Proof of (2) from (1)</u>. κ is (almost) ineffable iff every element of subtle filter is Π^1_2 (Π^1_1) indescribable, i.e., no A in the subtle filter is disjoint from any B in the Π^1_2 (Π^1_1) indescribable filter.

Note that there are many λ's which are both subtle and Π^1_2 indescribable below the least almost ineffable κ. For $\{\lambda<\kappa \mid \lambda$ is invisible in $R(\kappa)\}$ is in the subtle filter on κ, and $\{\lambda<\kappa \mid \lambda$ is subtle $\}$ is in the Π^1_1 indescribable filter since " λ is subtle" is Π^1_1.

We close this section with partition-theoretic characterizations of subtlety/almost ineffability/ineffability, and some problems.

1.17 Definition. If A is a set of ordinals, $[A]^n$ is the set of increasing n-sequences of elements of A.
$f\colon [A]^n \longrightarrow \kappa$ is regressive iff $f(\bar{x}) < x_0$ for all $\bar{x}$ in $[A]^n$.
$B \subseteq A$ is homogeneous for f iff $f \restriction [B]^n$ is constant.

The ineffability-case of the following, due to Kunen, was among the earliest results about ineffability.

1.18 (Kunen, Baumgartner). If A is an unbounded subset of an infinite regular cardinal κ, A is (1) subtle / (2) almost ineffable / (3) ineffable iff each regressive $f\colon [A]^2 \longrightarrow \kappa$ has a homogeneous set $X \subseteq A$ which is (1) of order-type ≥ 3 / (2) unbounded in κ / (3) stationary in κ.

Proof. We do the "if" ($\longleftarrow$) direction of all three cases together. Suppose A is not subtle (almost ineffable, ineffable). Let $\langle A_\alpha \mid \alpha \in A \rangle$ witness this, where we assume without loss of generality that $0 \notin A_\alpha$ for $\alpha \in A$, and that A is a stationary set of limit ordinals closed under the pairing function; (if A were nonstationary, there would be a 1-1 regressive $f\colon A \longrightarrow \kappa$ by Fodor's theorem). Define regressive $f\colon [A]^2 \longrightarrow \kappa$ by $f(\alpha,\beta) = j(\gamma,2)$ or $j(\gamma,1)$ if there is a least $\gamma < \alpha$ in $(A_\alpha \cup A_\beta) - (A_\alpha \cap A_\beta)$, $j(\gamma,2)$ if $\gamma \in A_\beta$, $j(\gamma,1)$ if $\gamma \in A_\alpha$; and $f(\alpha,\beta) = 0$ if there is no such γ, i. e., if $A_\alpha = A_\beta \cap \alpha$. There is an $X \subseteq \kappa$ homogeneous for f where X is of order-type ≥ 3 (unbounded, stationary). In each case $f([X]^2) \neq \{0\}$, by the hypothesis on A, so $f([X]^2) = \{j(\gamma,i)\}$ for some $i<2$ and $\gamma > 0$. But then we can derive a contradiction from $\overline{(X)} \geq 3$. For suppose α,β,δ are distinct elements of X, with $\alpha<\beta<\delta$; then $\gamma \in (A_\beta - A_\alpha) \cap (A_\delta - A_\beta)$ or $\gamma \in (A_\alpha - A_\beta) \cap (A_\beta - A_\delta)$, both absurdities.

Proof of (1) and (2), ($\longrightarrow$). Let $f\colon [A]^2 \longrightarrow \kappa$ be regressive, where A is a subtle set of cardinals $<\kappa$, and define S_μ for $\mu \in A$ by $S_\mu = \{(\lambda,\nu) \mid \lambda,\nu \in \mu \cap A$ and $f(\lambda,\mu) = \nu\}$. By 1.14, if $A^* = \{\mu \in A \mid$ there is a stationary $K_\mu \subseteq \mu$ A which is homogeneous for $\langle T_\mu \mid \mu \in A\rangle = \langle j''S_\mu \mid \mu \in A\rangle\}$, $A - A^*$ is not subtle. For μ in A^*, define $g_\mu : K_\mu \longrightarrow \mu$ by $g_\mu(\lambda) = f(\lambda,\mu)$.

A stationary $H_\mu \subseteq \mu$ such that g_μ is constant on H_μ is homogeneous for f.
If A is actually almost ineffable, so is A* (if A - A* is not subtle, neither is it almost ineffable), so there is a $D \subseteq A^*$ such that for all $\tilde{\mu}, \mu \in D$ with $\tilde{\mu} < \mu$, $H_{\tilde{\mu}} = H_\mu \cap \tilde{\mu}$, so that $H = \bigcup_{\mu \in D} H_\mu$ is homogeneous for f.
Proof of (3)($\longrightarrow$). If A is ineffable and $f: [A]^2 \longrightarrow \kappa$ is regressive, let $\langle T_\mu \mid \mu \in A \rangle$ be defined as the proof of (1) and (2)($\longrightarrow$), and let $K \subseteq A$ be a stationary homogeneous set for $\langle T_\mu \mid \mu \in A \rangle$. If $g: K \longrightarrow \kappa$ is defined by $g(\lambda) = f(\lambda, \mu)$ for some (any) $\mu > \lambda$ in K, g is constant on a stationary $H \subseteq K$, which is then homogeneous for f.

1.19 Remarks. (1) We have actually established that $A \subseteq \kappa$ is subtle iff for each regressive $f: [A]^2 \longrightarrow \kappa$, there is a stationary set of regular cardinals $\lambda < \kappa$ which have stationary subsets homogeneous for f.
(2) Kunen actually proved the following refinement of 1.18(3). We include the proof for completeness though it is already available in [Be2].

1.20 Theorem (Kunen). A is ineffable iff every $f: [A]^2 \longrightarrow 2$ has a stationary homogeneous set.

Proof. Only ($\longleftarrow$) remains to be proved. Suppose $\langle A_\alpha \mid \alpha \in A \rangle$ is such that $A_\alpha \subseteq \alpha$ for $\alpha \in A$. Define $f: [A]^2 \longrightarrow 2$ by $f(\alpha,\beta) = 0$ iff $A_\alpha = A_\beta \cap \alpha$, or the least $\gamma < \alpha$ in the symmetric difference of A_α and A_β is in A_β. Let B be stationary such that $f \upharpoonright [B]^2$ is constant. For each η there is a $\tau_\eta \geq \eta$ such that for all $\beta \geq \tau_\eta$ in B, $A_\beta \cap \eta = A_{\tau_\eta} \cap \eta$. (Argue by induction on η. Limit stages are automatic; at a successor stage $\eta + 1$, homogeneity of B for f determines whether η is in, or out of, every A_β for β in a terminal segment of B).
Then $C = \{ \zeta \mid \eta < \zeta \longrightarrow \tau_\eta < \zeta \}$ is closed unbounded in κ, since κ is regular (in fact weakly compact), which makes $B \cap C$ a stationary set homogeneous for $\langle A_\beta \mid \alpha \in A \rangle$.

1.21 Problems. How much of 1.12 1.19, if any, carries over to weak subtlety/almost ineffability/ineffability?

The preceding exposition has been thin on examples of weakly subtle/almost ineffable/ineffable cardinals. Ketonen in [Ke] showed that a κ with a nontrivial, κ-complete, λ-saturated filter for $\lambda < \kappa$ (more about this later) is weakly subtle. Find others.

2. Partitions

2.1. If X is linearly ordered by $\prec$ and α is any ordinal; $[X]^{\alpha}$ ($[X]^{<\alpha}$) is the set of all subsets of X of $\prec$-order-type α ($<\alpha$). In the next three definitions, X is any set of ordinals, well-ordered by the membership relation.

2.2. $X \longrightarrow (\alpha)^{<\omega}_{\lambda}$ ($X \longrightarrow (\alpha)^{m}_{\lambda}$) means that for each $f: [X]^{<\omega} \longrightarrow \lambda$, there is an $H \in [X]^{\alpha}$ such that $f \restriction [H]^{n}$ is constant for each n (for n = m). H is called a homogeneous set or set of indiscernibles, for f.

$X \longrightarrow [\alpha]^{<\omega}_{\lambda}$ ($X \longrightarrow [\alpha]^{m}_{\lambda}$) means that for each f: $[X]^{<\omega} \longrightarrow \lambda$, there is an $H \in [X]^{\alpha}$ such that $f''[H]^{<\omega}$ ($f''[H]^{m}$) is a proper subset of λ.

$X \longrightarrow [\alpha]^{<\omega}_{\lambda,<\mu}$ ($X \longrightarrow [\alpha]^{m}_{\lambda,<\mu}$) means that for each f: $[X]^{<\omega} \longrightarrow \lambda$, there is an $H \in [X]^{\alpha}$ such that $(f''[H]^{<\omega})^{=} < \mu$ ($(f''[H]^{m})^{=} < \mu$).

$X \longrightarrow (\alpha)^{<\omega}$ ($X \longrightarrow (\alpha)^{m}$) abbreviates $X \longrightarrow (\alpha)^{<\omega}_{2}$ ($X \longrightarrow (\alpha)^{m}_{2}$)

If $\mathcal{S} \subseteq S(\kappa)$ is an arbitrary collection of subsets of κ, we can generalize each of the above notions by requiring the homogeneous set H to be in $\mathcal{S}$, obtaining $\kappa \longrightarrow (\mathcal{S})^{<\omega}_{\lambda}$, $\kappa \longrightarrow (\mathcal{S})^{m}_{\lambda}$, $\kappa \longrightarrow [\mathcal{S}]^{<\omega}_{\lambda}$; etc.

2.3. The αth Erdös cardinal, κ_{α}, for limit $\alpha \geqslant \omega$, is the least κ such that $\kappa \longrightarrow (\alpha)^{<\omega}$.

κ is Jonsson iff $\kappa \longrightarrow [\kappa]^{<\omega}_{\kappa}$.

κ is μ-Rowbottom iff $\kappa \longrightarrow [\kappa]^{<\omega}_{\lambda,<\mu}$ for all $\lambda < \kappa$.

κ is Ramsey iff $\kappa \longrightarrow (\kappa)^{<\omega}$.

A μ-Rowbottom cardinal, for any $\mu < \kappa$, is obviously Jonsson.

A Rowbottom cardinal is an $\aleph_1$-Rowbottom cardinal.

Each of these notions is often extended to filters $\mathcal{F} \subseteq S(\kappa)$; thus $\mathcal{F} \subseteq S(\kappa)$ is Jonsson iff $\kappa \longrightarrow [\mathcal{F}]^{<\omega}_{\kappa}$, etc. A few judicious cases of $\kappa \longrightarrow (\alpha)^{n}_{\lambda}$ are theorems of ZFC, such as, Ramsey's theorem ($\omega \longrightarrow (\omega)^{2}_{2}$), and the well-known and useful Erdös-Rado theorem ($(\beth_n(\kappa))^{+} \longrightarrow (\kappa^{+})^{n+1}_{\kappa}$, where $\beth_n(\kappa)$ is the nth 'relative beth cardinal' after κ (cf [Dr, 7.2.4] and [De3, 3.20 and 3.21]). One infinitary generalization of Ramsey's theorem ($\kappa \longrightarrow (\kappa)^{2}_{2}$) is equivalent to weak compactness (cf [Dr, 10.2.1] and [De3, 3.24]).

2.4. There are concomitant model-theoretic definitions of the $<\omega$ versions of these notions, which require some more nomenclature. Suppose $\mathfrak{A}$ is some structure for a countable first-order language including a relation $\prec$ which linearly orders some subset X of the universe A of $\mathfrak{A}$. Then $H \subseteq X$ is a homogeneous set, or set of indiscernibles, for $\mathfrak{A}$ iff for each n and $\bar{x}$, $\bar{y} \in [H]^n$, $\langle \mathfrak{A}, \bar{x}\rangle \equiv \langle \mathfrak{A}, \bar{y}\rangle$. Ehrenfeucht and Mostowski [Eh-Mo] showed essentially that one can graft indiscernibles onto models of any theory by a compactness argument. If one can invariably assume they are already there, large-cardinal properties arise.

2.5. $(\kappa,\lambda) \longrightarrow (\mu, <\nu)$ iff every structure $\langle \mathfrak{A}, U\rangle$ for a countable first-order language with $(A)^= = \kappa$, U unary and $(U)^= = \lambda$ has an elementary substructure $\langle \mathfrak{A}^*, U^*\rangle$ such that $(A^*) = \mu$ and $(U^*)^= < \nu$. Chang's Conjecture is the assertion that $(\aleph_2, \aleph_1) \longrightarrow (\aleph_1, <\aleph_1)$.

2.6. Theorem (Rowbottom). (1) $\kappa \longrightarrow (\alpha)^{<\omega}$ iff whenever $\langle \mathfrak{A}, <\rangle$ is a structure for a countable first-order language in which $<$ well-orders some subset X of A in type $\geq \kappa$, $\langle \mathfrak{A}, <\rangle$ has a homogeneous set $H \subseteq X$ of order-type α.

(2) κ is Jonsson iff every structure $\mathfrak{A}$ for a countable first-order language such that $(A)^= = \kappa$ has a proper elementary substructure $\mathfrak{A}^*$ such that $(A^*)^= = \kappa$.

(3) κ is μ-Rowbottom iff $(\kappa,\lambda) \longrightarrow (\kappa, <\mu)$ for each $\lambda < \kappa$.

(4) If κ is Ramsey, κ is Rowbottom.

Proof of (1). Enumerate the countably many formulae of the language by $\langle \varphi_n | n<\omega\rangle$ such that at most $v_0, \ldots, v_{n-1}$ are free in φ_n, and let $X' \subseteq X$ have $<$-type κ. Define $f: [X']^{<\omega} \longrightarrow 2$ by $f(\langle x_0, \ldots, x_{n-1}\rangle) = 1$ iff $\langle \mathfrak{A}, <\rangle \models \varphi_n[x_0, \ldots, x_{n-1}]$. Any H homogeneous for f is homogeneous in the other sense for $\langle \mathfrak{A}, <\rangle$.

If $f: [\kappa]^{<\omega} \longrightarrow 2$ is given, any H homogeneous for $\mathfrak{A} = \langle \kappa, \in, \langle f \restriction [\kappa]^n \mid n<\omega\rangle\rangle$ is homogeneous for f.

Proof of (2). ($\longrightarrow$) Suppose $\kappa \longrightarrow [\kappa]^{<\omega}_\kappa$ and $\mathfrak{A}$ is given with $A = \kappa$. Let $\mathcal{F} = \{f_n | n<\omega\}$ be a set of Skolem functions for $\mathfrak{A}$ which is closed under composition and enumerated with the aid of dummy variables so that each f_n is n-ary. Let $f(\bar{x}) = f_n(\bar{x})$ for $n<\omega$ and $\bar{x}$ in $[\kappa]^n$. If $H \subseteq \kappa$ is such that $f''[H]^{<\omega}$ is a proper subset of κ, the restriction $\mathfrak{A}^*$ of $\mathfrak{A}$ to $A^* = f''[H]^{<\omega}$ is a proper elementary submodel of $\mathfrak{A}$.

($\longleftarrow$) Apply the assumption to $\mathfrak{A} = \langle \kappa, \in, \langle f \restriction [\kappa]^n | n<\omega\rangle\rangle$ to obtain a proper elementary substructure $\mathfrak{A}^*$ such that $(A^*)^= = \kappa$ and $f''[A^*]^{<\omega} \subseteq A^* \subsetneq \kappa$.

Proof of (3). Essentially the same as (2). For ($\longrightarrow$) suppose $\kappa \longrightarrow [\kappa]^{<\omega}_\lambda$, and $\langle \mathfrak{A}, U\rangle$ is given with $A = \kappa$ and $(U)^= = \lambda$. Form $\mathcal{F}$ as in (2), and set $f(\bar{x}) = f_n(x)$ if $\bar{x} \in [\kappa]^n$ and $f_n(\bar{x}) \in U$; otherwise $f(\bar{x}) = 0$. By hypothesis, there is an H such that $(f''[H]^{<\omega})^= < \mu$. If $A^* = f''[H]^{<\omega}$, and $\langle A^*, U^*\rangle$ is the restriction of $\langle \mathfrak{A}, U\rangle$ to A^*, $(A^*)^= = \kappa$ and $(U^*)^= < \mu$.

For ($\longleftarrow$), let $f: [\kappa]^{<\omega} \longrightarrow \lambda$ be given, and form $\langle \mathfrak{A}, \mathcal{U} \rangle = \langle \kappa, \langle f \restriction [\kappa]^n | n < \omega \rangle, \lambda \rangle$. By hypothesis, there is an $\langle \mathfrak{A}^*, \mathcal{U}^* \rangle \prec \langle \mathfrak{A}, \mathcal{U} \rangle$ such that $(A^*)^= = \kappa$ and $(U^*)^= < \mu$. Since $f"[A^*]^{<\omega} \subseteq \mathcal{U}^*$, $(f"[A^*]^{<\omega})^= \leq (\mathcal{U}^*)^= < \mu$.

Proof of (4). Suppose $f: [\kappa]^{<\omega} \longrightarrow \omega_1$, and H is a set of indiscernibles for $\langle \kappa, \epsilon, \langle f \restriction [\kappa]^n | n < \omega \rangle \rangle$. (x) = f(y) for each $n < \omega$ and x,y in $[H]^n$, since otherwise a sequence $\langle \bar{x}_\alpha | \alpha < \kappa \rangle$ with $\max(\bar{x}_\alpha) < \min(\bar{x}_{\alpha'})$ for all $\alpha < \alpha' < \kappa$ would yield κ elements of ω_1. But then $f"[H]^{<\omega} = \{ \gamma_n | n < \omega \}$, where γ_n is the common value of each $f(\bar{x})$ for $x \in [H]^n$, is a countable subset of ω_1.

A very good journal reference for most of the theory of partition cardinals, including independence results, is [De 1]. In the remainder of this section we survey most of what is known to be provable about Erdös, Jonsson, Rowbottom and Ramsey cardinals in ZFC in the light of section 1, then give in §3 a précis of Silver's theorem on the unintelligibility of κ_{ω_1} in L. Several of the basic results on Erdös cardinals appeared in the unpublished original of Silver's thesis, and are due to Silver and Reinhardt. The exposition given here follows in part unpublished lectures of Jensen [Jen 1].

2.7. Theorem. Suppose $\lambda < \kappa_\alpha$ and $f_\beta: [\kappa_\alpha]^{<\omega} \longrightarrow 2$ for each $\beta < \lambda$. Then there is an $H \in [\kappa_\alpha]^\alpha$ which is homogeneous for each f_β.

Proof. Suppose $g: [\lambda]^{<\omega} \longrightarrow 2$ has no homogeneous set, $\mathfrak{A} = \langle \kappa, \epsilon, \langle g \restriction [\lambda]^n | n < \omega \rangle, \langle \{ (\beta, \bar{\gamma}) | f_\beta(\bar{\gamma}) = 1, n < \omega$ and $\bar{\gamma} \in [\kappa_\alpha]^n \} \rangle \rangle$, and H is a set of indiscernibles for $\mathfrak{A}$ of order-type α. We prove H is homogeneous for f_β by showing for $\beta < \lambda$, $n < \omega$ and $\bar{\gamma}, \bar{\eta}$ in $[H]^n$ with $\bar{\gamma} < \bar{\eta}$ ($\gamma_{n-1} < \eta_0$) that $f_\beta(\bar{\gamma}) = f_\beta(\bar{\eta})$. If this fails for some β and n, fix the least such n, and let $\delta(\bar{\gamma}, \bar{\eta})$ = the least $\beta < \lambda$ such that for some (and therefore all, by indiscernibility of H in $\mathfrak{A}$) $\bar{\gamma}, \bar{\eta}$ in $[H]^n$ with $\bar{\gamma} < \bar{\eta}$, $f_\beta(\bar{\gamma}) \neq f_\beta(\bar{\eta})$. For each $\bar{\gamma}$, $\bar{\gamma}'$, η and $\bar{\eta}'$ from H with $\bar{\gamma} < \bar{\eta} < \bar{\gamma}' < \bar{\eta}'$, $\delta(\bar{\gamma}, \bar{\eta}) \{ \gtreqless \} \delta(\bar{\gamma}', \bar{\eta}')$. $>$ would give an infinite descending ϵ-chain. $<$ would give an increasing sequence $\langle \tau_\beta | \beta < \lambda \rangle$, $\tau_\beta = \delta(\bar{\gamma}_\beta, \bar{\eta}_\beta)$ for each $\beta < \lambda$, whose range would be a subset of λ and homogeneous for g, since $\delta(-,-)$ is definable in $\mathfrak{A}$. Equality doesn't work, for then if β were the common value of all $\delta(\bar{\gamma}, \bar{\eta})$ for $\bar{\gamma}, \bar{\eta}$ in $[H]^n$, we would have $f_\beta(\bar{\gamma}) \neq f_\beta(\bar{\eta})$ for all $\bar{\gamma}, \bar{\eta}$ in $[H]^n$, contradicting $f_\beta(\bar{\gamma}) = 1$ iff $f_\beta(\bar{\eta}) = 1$, required by the model-theoretic homogeneity of H in $\mathfrak{A}$. Since each alternative is impossible $\delta(-,-)$ does not exist, and we are done.

2.8 Corollary. (1) $\kappa_\alpha \longrightarrow (\alpha)_{2^\lambda}^{<\omega}$ for all $\lambda < \kappa_\alpha$.

(2) Whenever $\langle \mathfrak{A}, \prec \rangle$ is a structure for a first-order language of power $<$ κ_α in which $\prec$ orders X in type $\geq \kappa_\alpha$, $\langle \mathfrak{A}, \prec \rangle$ has a homogeneous set H of type
(3) If $\alpha < \beta$, $\kappa_\alpha < \kappa_\beta$.
(4) κ_α is strongly inaccessible for each α.

Proof of (1). If f: $[\kappa_\alpha]^{<\omega} \longrightarrow 2^\lambda$, let $f_\beta(\bar{x}) = 1$ if $(f(\bar{x}))(\beta) = 1$ for $\beta < \lambda$, 0 otherwise. An H homogeneous for each f_β is homogeneous for f.
Proof of (2). Define f: $[\kappa_\alpha]^{<\omega} \longrightarrow$ S (the language of $\mathfrak{A}$) by f(x) = the n-type of $\bar{x} = \{ \varphi \mid \mathfrak{A} \models \varphi[\bar{x}] \}$ for $n \geq 1$ and $\bar{x} \in [X]^n$. An H homogeneous for f is homogeneous for $\mathfrak{A}$.
Proof of (3). If not, for each $\gamma < \kappa_\beta$ let $f_\gamma : [\gamma]^{<\omega} \longrightarrow 2$ have no homogeneous set of type α. Define g: $[\kappa_\beta]^{<\omega} \longrightarrow 2$ by $g(\langle x_0, \ldots, x_n \rangle) = f_{x_n}(\langle x_0, \ldots, x_{n-1} \rangle)$ for $n \geq 1$ and $\bar{x}$ in $[\kappa_\beta]^{n+1}$. If H has type β and is homogeneous for g, $H \cap \gamma$ is homogeneous for f_γ for each $\gamma \in H$. But then $H \cap \gamma$ has type $< \alpha$ for each γ in H, contradicting the assumption that H has type $\beta > \alpha$.
Proof of (4). This is actually subsumed in 2.9(1) below, since we prove that from scratch.
κ_α must be a strong limit cardinal, for $\lambda < \kappa_\alpha$ and $2^\lambda \geq \kappa_\alpha$ would contradict $\kappa_\alpha \not\rightarrow (\alpha)^{<\omega}_{\kappa\alpha}$.
If cf $\kappa_\alpha = \lambda < \kappa_\alpha$ and $\langle \mu_\beta \mid \beta < \lambda \rangle$ is an increasing sequence of cardinals cofinal in κ_α, there are $g_\beta : [\mu_\beta]^{<\omega} \longrightarrow 2$ without homogeneous sets for each $\beta < \lambda$. If H is homogeneous of type α for $\mathfrak{A} = \langle \kappa_\alpha, \in, \langle \mu_\beta \mid \beta < \lambda \rangle, \langle g_\beta \mid \beta < \lambda \rangle \rangle$, $H \subseteq \mu_\beta$ must hold for the least $\beta < \lambda$ such that $H \cap \mu_\beta \neq 0$ by model-theoretic homogeneity of H for $\mathfrak{A}$, contradicting the assumption about g_β.

Though he had not yet isolated the definition of subtlety, Jensen effectively proved the following in [Jen1].

2.9 Theorem (Jensen). (1) Each κ_α is subtle.
(2) A Ramsey cardinal is almost ineffable.

Proof. We do both cases at once. Suppose a closed unbounded $C \subseteq \kappa = \kappa_\alpha$ is given such that $A_\beta \subseteq \beta$ for each $\beta \in C$. If $\alpha < \kappa$ let $g_\beta : [\beta]^{<\omega} \longrightarrow 2$ have no homogeneous set of type α for each $\beta < \kappa_\alpha$, and
$\mathfrak{A} = \langle \kappa, \in, \langle \{ \langle \bar{x}, \beta, i \rangle \mid \bar{x} \in [\beta]^n$ and $g_\beta(\bar{x}) = i \} \mid n < \omega \rangle$, $\{ \langle \eta, \gamma \rangle \mid \gamma \in C$ and $\eta \in A_\gamma \} \rangle$.
If $\alpha = \kappa$, i. e., κ is Ramsey, let $\mathfrak{A} = \langle \kappa, \in, \{ \langle \eta, \gamma \rangle \mid \gamma \in C$ and $\eta \in A_\gamma \} \rangle$. In either case let H be a minimal set of indiscernibles for $\mathfrak{A}$, that is a set of indiscernibles such that $\bigcap H$ is least possible.

(a) $H \subseteq C$. If not let $f(\gamma) = \bigcup(C \cap \gamma) < \gamma$ for γ in H. $f(\gamma) \gtreqqless f(\delta)$ for $\gamma < \delta$ in H. $>$ would yield a descending $\in$-chain. $<$ would contradict the minimality of H, for f is definable in $\mathfrak{A}$, and $\{ f(\gamma) | \gamma \text{ in } H \}$ would be a set of indiscernibles for $\mathfrak{A}$ of order type α. If equality holds, let $\tau = f(\gamma)$ for each γ in H, and $\sigma(\gamma) = \bigcap(C - \gamma)$. Then $\delta < \sigma(\gamma)$ for each $\delta > \gamma$ in H. If β is the common value of the $\sigma(\gamma)$'s for $\gamma \in H$, H remains homogeneous for $\langle \mathfrak{A}, \beta \rangle$ and $H \subseteq \beta$, contradicting the assumption that type H = κ if κ is Ramsey, and the assumption on g_β if not. We finish by showing

(b) $A_\beta = A_\gamma \cap \beta$ for all β, γ in H.

If not, let $\delta(\beta, \gamma)$ be the least element of β in the symmetric difference of A_β and A_γ for β, γ in H with $\beta < \gamma$. Again $\delta(\beta, \gamma) \gtreqqless \delta(\beta', \gamma')$ for all $\beta < \gamma < \beta' < \gamma'$ in H. $>$ would yield a descending $\in$-chain; $<$ would violate minimality of H, for $\delta(\_,\_)$ is definable over $\mathfrak{A}$ and so the range of an ascending α-sequence of $\delta(\beta, \gamma)$'s would be homogenous for $\mathfrak{A}$. But equality is impossible too, since if $\beta < \gamma < \zeta$ in H and $\delta(\beta, \gamma) = \delta(\gamma, \zeta) = \delta$, either $\delta \in (A_\gamma - A_\beta) \cup (A_\zeta - A_\gamma)$ or $\delta \in (A_\beta - A_\gamma) \cup (A_\gamma - A_\zeta)$.

2.10 Remarks. (1) The least Ramsey cardinal is not ineffable, since "I am Ramsey" is Π^1_2.

(2) One can use (b) above to show each member of H is invisible in $R(\kappa_\alpha)$ (this is the lemma of [Jen1] alluded to just before 1.13), and (a) to show $\{ \lambda < \kappa_\alpha | \lambda \text{ invisible in } R(\kappa_\alpha) \}$ is stationary in κ_α This was Jensen's line of argument on pp. 105 - 112 of [Jen1] .

2.11 Theorem (Silver, Reinhardt, Jensen). κ_ω exceeds the least ineffable cardinal.

Proof. We give the traditional proof of this, essentially due to Silver and Reinhardt. A sharper argument constructed along the lines of 2.9 and 2.10 would yield the stronger conclusion that $\{ \lambda < \kappa_\alpha | \lambda \text{ ineffable} \}$ is stationary in each κ_α.

Suppose $\kappa = \kappa_\omega$, $\mathfrak{A} = \langle R(\kappa), \in \rangle$ and $H = \{ \gamma_n | n < \omega \}$ is a set of indiscernibles of $\mathfrak{A}$ of type ω, enumerated in increasing order, and let $\mathfrak{A}^*$ be the Skolem hull of H in $\mathfrak{A}$. (cf [Ch-Ke, pp. 141-2]). The function from H into H which maps each γ_n to γ_{n+1} can be exteded canonically to an elementary embedding $i: \mathfrak{A}^* \longrightarrow \mathfrak{A}^*$. If $\underline{\mathfrak{A}}^*$ is the transitive collapse of $\mathfrak{A}^*$, write $\underline{x}$ for the image of $x \in \mathfrak{A}^*$ under the collapse, $\underline{i}$ for the collapsed counterpart of i, and $\mathcal{B}$ for the image of $\underline{\mathfrak{A}}^*$ under $\underline{i}$. We will show that if λ is the least ordinal in $\underline{\mathfrak{A}}^*$ such that $\underline{i}(\lambda) > \lambda$ (so $\underline{i}(X) \cap \lambda = X$ for each $X \subset \lambda$ in $\mathfrak{A}^*$), λ is ineffable in $\mathfrak{A}^*$, and so if $\lambda = \underline{\mu}$, μ is ineffable in $\mathfrak{A}^*$ and therefore in $\mathfrak{A}$.

Suppose is a λ-sequence in $\mathcal{O}\!\!\!\!\!\!\:\underset{\sim}{}^*$ such that $X(\alpha) \subseteq \alpha$ for each $\alpha < \lambda$. Then $\underset{\sim}{i}(X)(\alpha) \subseteq \alpha$ for each $\alpha < \underset{\sim}{i}(\lambda)$ in B. Let $Y \subseteq \lambda$ be $\underset{\sim}{i}(X)(\lambda)$ and $Z = \{\alpha < \lambda \mid Y \cap \alpha = X(\alpha)\}$ in $\mathcal{O}^*$. Then Z is a homogeneous set for $\langle X(\alpha) \mid \alpha < \lambda \rangle$ which is stationary in λ in $\mathcal{O}^*$. For if $C \in A^*$ is a closed unbounded subset of κ in $\mathcal{O}^*$, $\lambda \in \underset{\sim}{i}(Z) \cap \underset{\sim}{i}(C)$ in B, so Z and C cannot be disjoint in $\mathcal{O}^*$.

By contrast with 2.9 and 2.11, little is yet known about the "size" of Jonsson and Rowbottom cardinals, though we will see in the next section that L thinks they are enormous, to the extent it thinks about them at all. It is not yet known whether a regular Rowbottom cardinal must be Mahlo (there is a singular Rowbottom cardinal in a generic extension of a universe with a measurable cardinal by a theorem of Prikry [Pr]), or whether $\aleph_\omega$ can be Jonsson. What follows is virtually all that is known to be provable in ZFC.

<u>2.12 Theorem.</u> <u>(1) $\aleph_0$ is not Jonsson; if κ is not Jonsson, neither is κ^+.</u>
<u>(2) If κ is λ-Rowbottom, either κ is inaccessible, or cf $\kappa < \lambda$.</u>
<u>(3) (Kleinberg) If $\kappa \longrightarrow [\kappa]^{<\omega}_{\kappa}$, $\kappa \longrightarrow [\kappa]^{<\omega}_{\lambda}$ for some $\lambda < \kappa$.</u>
<u>(4) (Kleinberg) If $\kappa \longrightarrow [\kappa]^{<\omega}_{\lambda}$ and $\lambda \not\longrightarrow [\lambda]^{<\omega}_{\lambda}$, $\kappa \longrightarrow [\kappa]^{<\omega}_{\lambda,<\lambda}$.</u>
<u>(5) (Kleinberg) The least Jonsson cardinal is λ-Rowbottom, where λ is the least cardinal, by (3), such that $\kappa \longrightarrow [\kappa]^{<\omega}_{\lambda}$.</u>

<u>Proof of (1).</u> $\langle \omega, \in, f \rangle$, where $f(n) = n \dot{-} 1$ is a Jonsson model. If $\kappa \not\longrightarrow [\kappa]^{<\omega}_{\kappa}$, let f_α witness $\alpha \not\longrightarrow [\alpha]^{<\omega}_{\alpha}$ for each α with $\kappa \leq \alpha < \kappa^+$. Then $f: \bar{x} \longmapsto f_{x_n}(\langle x_0, \ldots, x_{n-1} \rangle)$ for $n \geq 1$ and $\bar{x} \in [\kappa^+]^{n+1}$ witnesses $\kappa^+ \not\longrightarrow [\kappa^+]^{<\omega}_{\kappa^+}$. For if $\alpha < \kappa^+$ and $\alpha < \beta = \bigcup \beta < \kappa^+$, there is $\bar{y} \in [\beta]^{<\omega}$ with $\alpha = f_\beta(\bar{y}) = f(\bar{y}, \beta)$.
<u>Proof of (2).</u> If $\kappa = \mu^+$ and $f_\alpha : \alpha \xrightarrow[onto]{1-1} \mu$ for $\mu < \alpha < \kappa$, let $R(\alpha, \beta, \gamma)$ iff $f_\alpha(\beta) = \gamma$. If $\langle B, \in, B \cap \mu, R|B \rangle \prec \langle \kappa, \in, \mu, R \rangle$ and $\alpha \in B$ has μ predecessors in B, f_α''(these predecessors) $\subseteq B \cap \mu$ and has cardinality μ. Note that this argument actually gives $\mu^+ \not\longrightarrow [\mu^+]^2_{\mu,<\mu}$. Similarly, if cf $\kappa = \lambda < \kappa$, and $\langle \tau_\beta \mid \beta < \lambda \rangle$ is cofinal in κ, $f: \alpha \longmapsto$ (the least β such that $\alpha < \tau_\beta$) witnesses $\kappa \not\longrightarrow [\kappa]^1_{\lambda,<\lambda}$.
<u>Proof of (3).</u> If $f_\alpha : [\kappa]^{<\omega} \longrightarrow \alpha$ witnesses $\kappa \not\longrightarrow [\kappa]^{<\omega}_{\alpha}$ for $\alpha < \kappa$, $f: \bar{x} \longmapsto f_{x_0}(\langle x_1, \ldots, x_{n-1} \rangle)$ witnesses $\kappa \not\longrightarrow [\kappa]^{<\omega}_{\kappa}$, for if $\beta < \kappa$, $\beta < \alpha_0 \in X \in [\kappa]^\kappa$ and $X' = \{x \in X \mid x > \alpha_0\}$, $\beta = f_{\alpha_0}(\bar{x}) = f(\langle \alpha_0, \bar{x} \rangle)$ for some $\bar{x} \in [X']^{<\omega}$. The least such α is a cardinal, for if $\bar{\bar{\alpha}} = \lambda < \alpha$, let $k : \alpha \xrightarrow[onto]{1-1} \lambda$. Then whenever $f: [\kappa]^{<\omega} \longrightarrow \lambda$, there is an

$X \in [\kappa]^\kappa$ such that $(k \circ f)''\,[X]^{<\omega} \subsetneq \alpha$, so $f''\,[X]^{<\omega} \subsetneq \lambda$.

Proof of (4). Suppose $(\kappa, \lambda) \not\twoheadrightarrow (\kappa, <\lambda)$ and $f: [\lambda]^{<\omega} \longrightarrow \lambda$ witnesses $\lambda \not\twoheadrightarrow [\lambda]^{<\omega}_\lambda$. Then any $\mathcal{B} \prec \mathcal{A} = \langle \kappa, \in, \lambda, \langle f \restriction [\lambda]^n \mid n < \omega \rangle \rangle$ with $(B)^= = \kappa$ has $B \cap \lambda = \lambda$, which violates $\kappa \longrightarrow [\kappa]^{<\omega}_\lambda$, by an argument like that for 2.6(2)($\longrightarrow$). For $(B \cap \lambda)^= = \lambda$, so any $\beta < \lambda$ is $f(\bar{x})$ for some $\bar{x} \in [B \cap \lambda]^{<\omega}$.

Proof of (5). Immediate from (4).

2.13 Conjecture. A regular Rowbottom cardinal is weakly subtle.

The last result of this section will contrast sharply with those of the next.

2.14 Theorem (Jensen, Kunen; Silver, Reinhardt).
(1) Subtlety, almost ineffability and ineffability of κ all relativize to L.
(2) $\kappa \longrightarrow (\alpha)^{<\omega}$ relativizes to L if $\alpha < \omega_1^L$.

Proof of (1). Subtlety is trivial. Suppose κ is (almost) ineffable, $C \in L$, and $\langle A_\alpha \mid \alpha \in C \rangle \in L$ is such that $A_\alpha \subseteq \alpha$ for $\alpha \in C$, and let A be such that for a stationary (unbounded) subset X of κ, $A \cap \alpha = A_\alpha$ iff $\alpha \in X$. The datum " I am a constructible subset of my supremum " is codable by a Σ^1_1 formula (see, e. g. [Be3, Lemma 2.9] , and since it is true of each $A \cap \alpha = A_\alpha$ for $\alpha \in C$, it is true of A as well by the Π^1_1 indescribability of κ. X of course is in L too, since it is definable from $\langle A_\alpha \mid \alpha \in C \rangle$ and A. Finally note that the Π^1_1 definition of "I am stationary" gives (X is stationary) $\longrightarrow$ (X is stationary $)^L$.

Proof of (2). Suppose $f: [\kappa]^{<\omega} \longrightarrow 2$ is constructible and $g \in L$ maps ω 1—1 onto α. Let $\mathcal{P}$ be the set of all finite order-isomorphisms τ from some $\langle g''n, \in \rangle$ into $\langle \kappa, \in \rangle$ such that the range of τ is homogeneous for f; $\mathcal{P}$ is constructible, and it should be clear that $\kappa \longrightarrow (\alpha)^{<\omega}$ just in case $\langle \mathcal{P}, \supseteq \rangle$ is not well-founded. But any constructible partial order is well-founded iff (it is well-founded $)^L$. For suppose $\langle Q, \preccurlyeq \rangle$ is such a constructible partial order, and $\langle Q, \preccurlyeq \rangle$ is well-founded, i. e., there exists no infinite descending $\prec$-chain of elements of Q. Then there is none in L either, so ($\langle Q, \preccurlyeq \rangle$ is well-founded $)^L$.
On the other hand (and this is the direction we need), suppose $\langle Q, \preccurlyeq \rangle$ is not well-founded, so that there is no order-preserving map σ from $\langle Q, \preccurlyeq \rangle$ into the ordinals. Then there is none in L either, so ($\langle Q, \preccurlyeq \rangle$ is not well-founded $)^L$.

2.15. With this result we leave the meadow of moderately large cardinals (defined as those whose definitions relativize to L), and enter the tulgey wood of large cardinals, which make L look rather like an extension by definitions of $\langle On, \in \rangle$. It would be interesting to find more examples of properties in between (for one such, see [$\mathcal{D}r$, pp. 255-6]).

3. a#

In this section, we study the effects of certain partition properties on inner models L(a).

3.1. For the following, let a be a set of ordinals, fixed until further notice, and let b^+, for any set b of ordinals be $\max(\omega_1, (\bigcup b)^+)$. $\mathcal{L}_a$ is the language obtained by adjoining constants $\underset{\sim}{a}$ for a and $\underset{\sim}{\alpha}$ for $\alpha \in \bigcup a$ to the language of ZF, $\mathcal{L}^*_a$ is the associated Skolem expansion (again, cf. [Ch-Ke , pp. 141-2]), and we assume fixed some unnamed coding apparatus for representing each $\mathcal{L}^*_a$ uniformly by a set of ordinals $< a^+$. If $\mathcal{M}$ is a structure for $\mathcal{L}_a$, and $H \subseteq M$, SH(H,$\mathcal{M}$) is the Skolem hull of H in $\mathcal{M}$.

3.2. Then a# exists iff a certain definable closed unbounded class of ordinals, H (whose defining formula φ_H will be derived; this is not an assertion of class theory), has the following properties, for each $\kappa \geq a^+$:
(1) $\kappa \in H$ (2) $H \cap \kappa$ is unbounded in κ and homogeneous for $\mathcal{M}_\kappa(a) = \langle L_\kappa(a), \in, a, \langle \alpha | \alpha \in \bigcup a \rangle\rangle$, and the Skolem hull of $H \cap \kappa$ in $\mathcal{M}_\kappa(a)$ is $L_\kappa(a)$. a# itself is then the set of (codes of) formulas $\mathcal{L}$a satisfied by $\bar{x} \in [H \cap \kappa]^n$ for any $n < \omega$ and $\kappa \geq a^+$.
Note that a# is well-defined and independent of the exact choice of φ_H, so long as H satisfies (1) and (2), since then a# is the set of formulas satisfied in $\mathcal{M}_\kappa(a)$ at some $x \in [K]^{<\omega}$, where $K \subseteq H$ consists of the first ω cardinals $\geq a^+$, and $\kappa = \bigcup K$.
Call a# the Ehrenfeucht - Mostowski (E - M) theory of H in L(a).

3.3. More generally, let $\mathcal{M} = \langle M, E, a, \langle \alpha | \alpha \in \bigcup a \rangle\rangle$ be a (not necessarily well-founded) set-model of ZFC^- (ZFC minus the power - set axiom) + V = L(a) such that $a \cup \{a\} \subseteq M$. Then $\mathcal{M}$ is a $\langle \gamma, T, a \rangle$ - model ($\langle \gamma, T \rangle$ - model when a is clear from context, as here) iff there exists an $H \subseteq On^M$ of type $\gamma = \bigcup \gamma > \bigcup a$ such that 3.2(2) holds with $\mathcal{M}_\kappa(a)$ replaced by $\mathcal{M}$ and $H \cap \kappa$ by H, and T is the set of formulas of $\mathcal{L}_a$ true of increasing finite sequences of elements of H. Call T, similarly, the Ehrenfeucht - Mostowski (E- M) theory of (H in) $\mathcal{M}$. Thus 3.2(2) says $\mathcal{M}_\kappa(a)$ is a $\langle \kappa, a\# \rangle$ - model for each $\kappa \geq a^+$.

3.4. A $\langle \gamma, T, a \rangle$-model $\mathcal{M}$ is called remarkable and T a remarkable E - M theory iff
(1) $\bar{x} < y \longrightarrow t(\bar{x}) < y$ (in the definable "well"-ordering $<$ of M), and

(2) $t(\bar{x},\bar{y}) < y_0 \longrightarrow [t(\bar{x},\bar{y}) = t(\bar{x},\bar{z})$ for all $\bar{z} > \bar{x}]$ for all Skolem terms t, k, $n < \omega$, $\bar{x} \in [H]^k$ and $\bar{y} \in [H]^n$; here $\bar{x} < \bar{y}$ means $x_{k-1} < y_0$, and $t(\bar{x},\bar{y})$ is always written with the understanding that $\bar{x} < y$.

3.5 Theorem (Ehrenfeucht, Mostowski)

(1) If $\mathcal{M}$ is a (remarkable) $\langle \beta, T \rangle$ -model and γ is any other ordinal $> \bigcup a$, there is a (remarkable) $\langle \gamma, T \rangle$ -model $\mathcal{N}$.

(2) If $\gamma, \beta, \mathcal{M}$ and $\mathcal{N}$ are as in (1), any order-embedding of β into γ (or γ into β) can be extended canonically to an elementary embedding from $\mathcal{M}$ into $\mathcal{N}$ ($\mathcal{N}$ into $\mathcal{M}$).

(3) A $\langle \gamma, T \rangle$ -model is unique up to isomorphism.

Proof of (1). If H is the set of indiscernibles in $\mathcal{M}$ and $\gamma < \beta$, we let $\mathcal{N}$ be the Skolem hull of the first γ members of H. If $\gamma > \beta$, we use a compactness argument. Let Σ be the following set of sentences in $\mathcal{L}$a augmented by new constant c_η for $\eta < \gamma$: { ZFC - the power-set axiom } $\cup \{V = L(a)\} \cup \{\underline{\alpha} \in \underline{a} \mid \alpha \in a\} \cup \{\underline{\alpha} \notin \underline{a} \mid \alpha \notin a\} \cup \{\underline{\alpha} \neq \underline{\tilde{\alpha}} \mid \alpha \neq \tilde{\alpha} < \bigcup a\}$ $\cup \{\underline{c}_{\bar{\eta}} < \underline{c}_\eta \mid \bar{\eta} < \eta < \gamma\} \cup \{\underline{c}_{\bar{\eta}} > \bigcup \underline{a} \mid \eta < \gamma\} \cup \{ \varphi(c_{\eta_0}, \ldots, c_{\eta_{n-1}}) \mid \varphi \in T \wedge \bar{\eta} \in [\gamma]^{<\omega}\} \cup \{\neg\varphi(c_{\eta_0}, \ldots, c_{n-1}) \mid \varphi \notin T$ and $\bar{\eta} \in [\gamma]^{<\omega}\}$. Every finite subset $\Gamma \subseteq \Sigma$ has a model, namely $\mathcal{M}$, with members of H interpreting the c_η's mentioned by Γ. If $\mathcal{N}^* \models \Sigma$, $H^* = \{c_\eta^{\mathcal{N}^*} \mid \ < \gamma\}$, and $\mathcal{N}$ is $SH(H^*, \mathcal{N}^*)$, $\mathcal{N}$ is a $\langle \gamma, T \rangle$ -model, and if the remarkability conditions are in T, $\mathcal{N}$ is remarkable too.

Proof of (2). Extend the order-embedding to $\mathcal{M}$ by induction over the Skolem terms.

Proof of (3). The canonical elementary embedding from (2) must be onto.

At this point, we identify what remarkability is good for.

3.6 Theorem (Silver). Suppose $\mathcal{M}$ is a remarkable $\langle \beta, T \rangle$ -model for $\beta = \bigcup \beta > \bigcup a$, generated by a set of indiscernibles H of type β which witnesses this, and let $\prec_m$ be the canonical "well"-ordering" of $\mathcal{M}$. Then (1) H is cofinal in $\langle \mathcal{M}, \prec_m \rangle$.

(2) $\langle \mathcal{M}, \preceq_m \rangle$ is an end extension of $\langle SH(H_\alpha, \mathcal{M}), \prec_m \rangle = \langle \mathcal{N}, \prec_n \rangle$ for any $\alpha = \bigcup \alpha < \beta$, where $H_\alpha = \{$first α elements of H$\}$.

Proof of (1). If $x \in M$, $x = t(\bar{y})$ for some Skolem term t and $\bar{y} \in [H]^{<\omega}$. By (1) of remarkability, whenever $z \in H$ and $\bar{y} < z$, $x = t(\bar{y}) \prec_m z$.

Proof of (2). $\mathcal{N}$ is a remarkable $\langle \alpha, T \rangle$ -model, so H_α is cofinal in $\langle \mathcal{N}, \prec_n \rangle$ by (1). If h is the αth element of H, we show every element y of $\mathcal{M} \prec_m h$ is in $\mathcal{N}$. By (2) of remarkability, $y = t(\bar{u},\bar{v}) \prec_m h$, where

$\bar{u} < h$ and $\bar{v} \geq h$, so $t(\bar{u},\bar{w}) = t(\bar{u},\bar{w}')$ for all $\bar{w}$, $\bar{w}'$ with $\bar{w} > w' > \bar{u}$. Since $\alpha = \bigcup\alpha$, there is such a $\bar{w} < \alpha$, so $y = t(\bar{u},\bar{w}) \in \mathcal{N}$.

3.7 Definition (scheme) (Gloede [$\mathcal{gl}$]). Let $\mathcal{M}$ be a transitive model for ZFC (set or class), and $\kappa, \alpha \in M$. Then $\kappa \xrightarrow{\mathcal{M}} (\alpha)^{<\omega}_{\lambda}$ means each function f in M from $[\kappa]^{<\omega}$ into λ has a homogeneous H, not necessarily in M, of order-type α.

We state Silver's theorem on the existence of indiscernibles for $\mathcal{L}(a)$ in stages below. Silver actually proved a general model-theoretic theorem in his thesis [Si 1] about elementary towers of well-ordered models, from the assumption that κ_{ω_1} exists; the conclusions about the existence of indiscernibles for inner models $\mathcal{L}(a)$ then follow as a corollary. Solovay defined "$0^{\#}$" = $(\emptyset)^{\#}$ in [So 1] and showed it is a Δ^1_3 real (subset of ω) to which every other real is many-one reducible (cf. e.g. Devlin's exposition in [De 2, pp. 197-8]). We prove a version of the result whose sufficient condition is also necessary, following Jensen [Jen 1] and Gloede [$\mathcal{gl}$].

3.8 Theorem (Silver, as modified in [Jen 1] and [$\mathcal{gl}$]). $a^{\#}$ exists iff some $\mathcal{M}_\kappa(a)$ for $\kappa \geq \lambda = a^+$ has a set of indiscernibles of type $\geq\lambda$ iff $\kappa \xrightarrow{\mathcal{L}(a)} (\lambda)^{<\omega}_2$ for some $\kappa \geq \lambda$. In detail,

(1) If for some $\kappa \geq \lambda = a^+$, $\kappa \xrightarrow{\mathcal{L}(a)} (\lambda)^{<\omega}_2$, there exists a set of indiscernibles H for $\mathcal{M}_\kappa(a)$ of type λ.

(2) If for some $\kappa \geq \lambda$ the conclusion of (1) holds, there is a set of indiscernibles H of type λ for $\mathcal{M}_\lambda(a)$ such that SH(H, $\mathcal{M}_\lambda(a)$) = $\mathcal{M}_\lambda(a)$.

(3) If H is as in (2) and T is its $E-M$ theory, $\mathcal{M}_\lambda(a)$ is a remarkable well-founded $\langle\lambda, T\rangle$ -model.

(4) If there is a well-founded remarkable $\langle\lambda, T\rangle$ -model, or if there are well-founded remarkable $\langle\beta, T\rangle$ -models for all limit β with $\bigcup a < \beta < \lambda$, then all the remarkable $\langle\eta, T\rangle$ -models for $\eta > \bigcup a$ are well-founded.

(5) If the conclusion of (3) holds, $a^{\#}$ exists and is equal to T.

(6) If $a^{\#}$ exists, $\lambda \xrightarrow{\mathcal{L}(a)} (\lambda)^{<\omega}$.

Proof of (1). Trace through the arguments of 2.6(1), 2.7, and 2.8(1), (2), keeping in mind that the partitions must be in $\mathcal{L}(a)$, though the homogeneous sets need not be. Everything carries over.

Proof of (2). Suppose κ is the least ordinal such that there is a homogeneous set H* of type λ for $\mathcal{M}_\kappa(a)$. We prove (2) in three stages

(a) There is a homogeneous set H of type λ for $\mathcal{M}_\kappa(a)$ whose Skolem hull in $\mathcal{M}_\kappa(a)$ is $\mathcal{M}_\kappa(a)$. For if $\mathfrak{A}$ = SH(H*, $\mathcal{M}_\kappa(a)$), $\mathfrak{A}$ is isomorphic by the transitive collapse π to some $\mathcal{M}_\beta(a)$ (see the remarks on relative constructibility in 5.6), but β must be κ by minimality. Set $H = \pi''H^*$.

(b) $\bigcup H = \kappa$. If $\bigcup H < \gamma < \kappa$, $\gamma = t(\bar{x})$ for some Skolem term t and $\bar{x} \in [H]^{<\omega}$, but then $H' = \{\eta \in H \mid \eta > \bar{x}\}$ is homogeneous for $\mathcal{M}_\gamma(a)$, contradicting the minimality of κ.

(c) $\lambda = \kappa$. If not, there exists a limit β with $\lambda \le \beta < \kappa$, $\bigcup(H \cap \beta) = \beta$, and $(H \cap \beta)^= < \lambda$, so there is an $\eta < \beta$ not in the Skolem hull of $H \cap \beta$ in $\mathcal{M}_\kappa(a)$. If $\eta = t(\bar{x},\bar{y})$ for some Skolem term t, $\bar{x} < \beta$ and $\bar{y} \ge \beta$, let η_0 be such that $\bar{x}, \eta < \eta_0 \in H \cap \beta$. Then since η is not in the Skolem hull of $H \cap \beta$, $t(\bar{u},\bar{v}) \ne t(\bar{u},\bar{w})$ for all $\bar{u},\bar{v}$ with $\bar{u} < \bar{v}, \bar{w} < \beta$, and therefore for all $\bar{v}$ with $\bar{u} < \bar{v}$, so an increasing $\langle \bar{v}_\xi \mid \xi < \lambda \rangle$ with $\bar{v}_\xi < \bar{v}_{\xi'}$ for all $\xi < \xi' < \lambda$ gives a homogeneous $I = \{ t(\bar{u},\bar{v}_\xi) \mid \xi < \lambda \}$ for $\mathcal{M}_{\eta_0}(a)$, again contradicting minimality of κ (I is a subset of $L_{\eta_0}(a)$, since $\eta = t(\bar{x},\bar{y}) \in L_{\eta_0}(a)$).

Proof of (3). We have actually just done the work of this, in (3)(b) (for clause (1) of 3.4) and (3)(c) (for clause (2)).

(1): Suppose $t(\bar{x}) \ge y$ for some (all) $\bar{x} < y$ in H. Then $H \subseteq t(\bar{x})$, which contradicts the assumption that $|H| = \lambda$.

(2): Suppose $t(\bar{x},\bar{y}) < y_0$, and $t(\bar{x},\bar{y}) \ne t(\bar{x},\bar{z})$ for some, and therefore all, $\bar{y}$, $\bar{z}$ with $x < \bar{y}, \bar{z}$. Then let $\langle \bar{w}_\alpha \mid \alpha < \lambda \rangle$ be an increasing sequence of elements of $[H]^n$, where n = lh $\bar{y}$, such that each $\bar{w}_\alpha$ has first element y_0, and $t(\bar{x}, \bar{w}_\alpha) \gtrless t(\bar{x}, \bar{w}_\beta)$ for all $\alpha < \beta < \lambda$. $>$ would yield an infinite descending chain, and $<$ would contradict the assumption that $|H| = \lambda$.

Proof of (4). If $\bigcup a < \beta < \lambda$, the well-foundedness of the $\langle \beta, T \rangle$ model $\mathcal{M}$ follows by 3.5(2), since an elementary submodel of a well-founded model is well-founded. If $\beta > \lambda$ and $\langle s_n \mid n < \omega \rangle$ is an infinite descending chain, each s_n is $t_n(\bar{x}_n)$ for Skolem terms t_n and $\bar{x}_n \in [K]^{<\omega}$, where K is the set of indiscernibles witnessing that the model $\mathcal{M}$ is a remarkable β-model. If $\mathcal{N}$ is the Skolem hull of $K' = \bigcup_{n<\omega} \bar{x}_n$ in $\mathcal{M}$, $|K'| = \tilde{\beta}$ for some $\tilde{\beta} < \lambda$, so $\mathcal{N}$ is well-founded by the first part of the proof, contradicting the fact that the terms of the descending chain are all in $\mathcal{N}$. Similarly, the equivalence mentioned in passing in the statement of (4) follows from the fact that a descending chain in a $\langle \lambda, T \rangle$ -model with indiscernibles H would have to be in the β-model generated by some initial part of H of type $\beta < \lambda$, since $\lambda = a^+$ is regular.

Proof of (5). We now have a hierarchy of models $\mathcal{M}_\mu$, one for each uncountable cardinal $\mu \ge \lambda$, each of which is a well-founded remarkable $\langle \mu, T \rangle$ model with indiscernibles $H_\mu \subseteq On^{\mathcal{M}_\mu}$ of type μ such that $SH(H_\mu, \mathcal{M}_\mu) = M_\mu$ and T the theory of the H obtained in (2). Without loss of generality, we can assume by 3.5(2) that $H_\mu \subseteq H_{\mu'}$, in fact by 3.5(2) and 3.6(2) that $H_{\mu'}$ end extends H_μ, $\langle \mathcal{M}_{\mu'}, <_{\mathcal{M}_{\mu'}} \rangle$ end-extends $\langle \mathcal{M}_\mu, <_{\mathcal{M}_\mu} \rangle$ for each $\mu < \mu'$ (a recursive definition of a class-function which does this provides the formula φ_H promised in 3.2), and by 3.6(1) and cardinality considerations (i.e., that $(L_\gamma(a))^= = \bar{\bar{\gamma}}$ for each $\gamma \geqslant \lambda$), that each $\mathcal{M}_\mu$ is actually $\mathcal{M}_\mu(a)$, and $\mathcal{M} = \bigcup_{\mu \geqslant \lambda} \mathcal{M}_\mu$ is $\langle L(a), \in, a, \langle \alpha | \alpha \in \bigcup a \rangle \rangle$ Then the union H of the H_μ's for $\mu \geqslant \lambda$ is a class of indiscernibles satisfying the conditions of 3.2. (1) and (2) are immediate. H is closed by (2) of remarkability and 3.6(2).

Proof of (6). If $f: [\lambda]^{<\omega} \longrightarrow 2$ is in L(a) and H is as in the definition of $a^\#$, $f = t(\bar{x})$ for some $\bar{x} \in [H \cap \lambda^+]^n$ where $x_0 < \dots \ x_k < \lambda \leqslant x_{k+1} < \dots < x_{n-1}$. Then $\{ \gamma \in H | x_k < \gamma < \lambda \}$ has type λ and is homogeneous for f.

It is customary at this point to enumerate some of the sad illusions imposed on L(a) by the existence of $a^\#$.

3.9 Theorem. Suppose H is the closed unbounded class of indiscernibles which witnesses that $a^\#$ exists, enumerated as $\{ \gamma_\alpha | \alpha > 0 \}$. Then
(1) $\mathcal{M}_{\gamma_\alpha}(a) \prec \mathcal{M}_{\gamma_\beta}(a)$ for each α, β with $\omega \leqslant \alpha < \beta$. Therefore any element x of L(a) definable from $\{a\} \cup \{ \alpha | \alpha \in \bigcup a \rangle\}$ and ordinals $< \gamma_\alpha$ is in $L_{\gamma_\alpha}(a)$.
(2) There is a satisfaction relation for $\langle L(a), \in, a, \langle \alpha | \alpha \in \bigcup a \rangle \rangle$.
(3) Any $\gamma_\alpha \in H$ is ineffable and invisible in L(a), and there are many $\beta < \bigcap H$ which are ineffable and invisible in L(a).
(4) There is an L-generic set $G \subsetneq P$ for each notion of forcing $\mathbb{P} = \langle P, \preccurlyeq \rangle$ which is (pointwise) definable in $\langle L, \in, \langle \alpha | \alpha < \omega_1 \rangle \rangle$.
The situation can be paraphrased for $a = \emptyset$ by saying that L is a 'countable model of set theory' (cf. (1), (2), (4)) which includes all the ordinals.

Proof of'(1). Since $L_\alpha(a)$ is the Skolem hull of $H \cap \alpha$, and $L_\beta(a)$ of $H \cap \beta$.

Proof of (2). From the fact that $L(a) = \bigcup_{\alpha \in H} L_\alpha(a)$, and $\mathcal{M}_{\gamma_\alpha}(a) \prec \mathcal{M}_{\gamma_\beta}(a)$ for each α, β with $\omega \leqslant \alpha < \beta$.

Proof of (3). If $\lambda = a^+$, say, define an elementary embedding i from L(a) into L(a) by setting $i(\gamma_\alpha) = \gamma_\alpha$ for $\alpha < \lambda$, $i(\gamma_{\lambda+n}) = \gamma_{\lambda+n+1}$ for $n < \omega$, and $i(\gamma_{\lambda+\alpha}) = \gamma_{\lambda+\alpha}$ for $\alpha \geqslant \omega$. Then $(\lambda \text{ is ineffable})^{L(a)}$ follows as in the last part of the proof of 2.11. The same elementary embedding yields invisibility. If $(\varphi(\lambda, A))^{L(a)}$, $(\varphi(i(\lambda), i(A))^{L(a)}$, since $A = t(\bar{x})$ for some $\bar{x} > i(\lambda)$, but $i(A) \cap \lambda = A$, so (there is a $\mu < i(\lambda)$ such that $\varphi(\mu, i(A) \cap \mu))^{L(a)}$, so there is a $\mu < \lambda$ such that $\varphi(\mu, A \cap \mu)^{L(a)}$. The existence of β's $< \bigcap H$ which are ineffable and invisible in L(a) follows by similar arguments.

<u>Proof of (4)</u>. $(S(P) \cap L(a))^{=} = \overline{\overline{P}} = \aleph_0$.

4. Iterated Ultrapowers

4.1 Definition (Kunen). If $\mathcal{M} = \langle M, \in \rangle$ is a transitive model of ZFC (set or class), and $I \in M$ is uncountable in $\mathcal{M}$, $\mathcal{U}$ is an $\mathcal{M}$-ultrafilter (sometimes: $\mathcal{M}$-measure) on I iff

(1) $\mathcal{U}$ is a (proper) subset of $S(I) \cap M$ which contains no singletons.

(2) If x, y are in $S(I) \cap M$, $x \in \mathcal{U}$ and $x \subseteq y$, $y \in \mathcal{U}$.

(3) For each $x \in S(I) \cap M$, either x or $I - x$ is in $\mathcal{U}$.

(4) If $\eta < \overline{\overline{(I)}}^M$, $\langle x_\alpha | \alpha < \eta \rangle \in M$, and each $x_\alpha \in \mathcal{U}$, so is $\bigcap_{\alpha<\eta} x_\alpha$.

(5) If $\langle x_\alpha | \alpha < \overline{\overline{(I)}}^M \rangle$, so is $\{\alpha \mid x_\alpha \in \mathcal{U}\}$.

4.2 Proposition. If there exists an $\mathcal{M}$-ultrafilter on $I \in M$ and $\kappa = \overline{\overline{(I)}}^M$, (κ is weakly compact $)^{\mathcal{M}}$.

Proof. We modify slightly the usual proof that a κ which satisfies the ultrafilter property is weakly compact. See, for example, [$\mathcal{A}$e , Lemmas 8 and 9 and Theorem 17]. (κ is regular $)^M$ follows from 4.1(4). Suppose ($\alpha < \kappa$ and there is an embedding $i: I \longrightarrow {}^{\alpha}2$ $)^M$; define $\mathcal{V}$ on ${}^{\alpha}2$ by $Y \in \mathcal{V}$ iff $i^{-1}(Y) \in \mathcal{U}$. Let i_γ = that element of 2 such that $\{ f \in {}^{\alpha}2 \mid f(\gamma) = i_\gamma \} = A_\gamma \in \mathcal{V}$. Checking that $\langle i_\gamma \mid \gamma < \alpha \rangle$ and thus $\langle A_\gamma \mid \gamma < \alpha \rangle$ are in M by 4.1(5), we apply 4.1(4) to get $\bigcap_{\gamma<\alpha} A_\gamma = \{ \langle i_\gamma \mid \gamma < \alpha \rangle \} \in \mathcal{V}$, contradicting (1). We verify the tree property (cf. [$\mathcal{A}$e3, 3.5]). Suppose we have a tree $\langle \kappa, \prec \rangle$ of height and width κ in M, and we transfer $\mathcal{U}$ to κ. Let $b = \{ \gamma \mid$ the set of $\prec$-successors of γ is in $\mathcal{U} \}$. $b \in M$ by 4.1(5), and b has elements at every level, which are all compatible with each other, so b is a branch.

In general this is the most that can be said, since whenever (κ is weakly compact $)^L$ and $\overline{\overline{(S(\kappa) \cap L)}} = \aleph_0$ (think of $0^\#$), $\mathcal{U} = \bigcup_{n<\omega} \mathcal{W}_n$ is an L- ultrafilter, where $\langle f_n \mid n < \omega \rangle$ enumerates $({}^{\kappa}S(\kappa))^L$ and for all $n < \omega$ ($\mathcal{W}_n$ is a nonprincipal κ-complete ultrafilter in the κ- field generated by $\bigcup \{ \operatorname{ran} f_{\tilde{n}} \mid \tilde{n} \leq n \})^L$.

4.3 Definition. κ is measurable iff there is a V-ultrafilter on κ.

4.4 Notation. (1) If $\mathcal{U}$ is an M-ultrafilter on $I \in M$, $\forall^* i\, \varphi(i)$ means $\{ i \mid \varphi(i) \} \in \mathcal{U}$. Likewise $\varphi(i)$ holds a. e. (almost everywhere).

(2) We use s, t to denote ordinal-indexed sequences, sometimes σ, τ if they are sequences of ordinals. We more often write $\bar{x}$, $\bar{\alpha}$ as before for finite sequences $\langle x_0, \ldots, x_k \rangle$, $\langle \alpha_0, \ldots, \alpha_k \rangle$. The length k will be appropriate to the context. $\langle s, t \rangle$ is the concatenation of s and t.

(3) If X is a set of sequences of length $\alpha + \beta$, f a function with domain $\alpha + \beta$, and s a sequence of length α, $X_{(s)} = \{ \beta$-sequences $t \mid \langle s, t \rangle \in X \}$; $f_{(s)}$ is the function with domain β such that

$f_{(s)}(t) = f(s, t)$.

4.5. If $\mathcal{U}$ is an M-ultrafilter on $\kappa \in M$, we define the Rowbottom M-ultrafilter on κ^n, $\mathcal{U}_n$, as follows, for $1 \leq n < \omega$. $\mathcal{U}_1 = \mathcal{U}$; $\mathcal{U}_{n+1}$ for $n \geq 1$ is the set of all $X \subseteq \kappa^n$ such that $\forall^* \alpha\, X_{(\alpha)} \in \mathcal{U}n$.

If the inductive definition is unravelled, one has $X \in \mathcal{U}_{n+1}$ iff $\forall^* \alpha_0\, \forall^* \alpha_1 \ldots \forall^* \alpha_n \langle \alpha_0, \ldots, \alpha_n \rangle \in X$. Similarly one has the following.

4.6 Lemma. (1) For each $n < \omega$, $\mathcal{U}_n$ is an M-ultrafilter on $\kappa^n \in M$. If $X \in \mathcal{U}$, $X^n \in \mathcal{U}$

(2) For any $m, n < \omega$ with $m < n$, and order-embedding $j: m \longrightarrow n$, $X \in \mathcal{U}_m$ iff $j_{*n}(X) \in \mathcal{U}_n$, where $j_{*n}(X) = \{ \langle \alpha_0, \ldots, \alpha_{n-1} \rangle \mid \langle a_{j(0)}, \ldots, a_{j(m-1)} \rangle \in X \}$.

(3) For all $X \subseteq \kappa^{m+n}$ $X \in \mathcal{U}_{m+n}$ iff $\forall^* s\ X_{(s)} \in \mathcal{U}_n$.

Proof of (1). By induction over n. 3.1(1) - (4) are easy. We check (5). Suppose $\langle X_\alpha \mid \alpha < \kappa \rangle \in M$, where each $X_\alpha \subseteq \kappa^{n+1}$. Then $\langle X_{\alpha(\beta)} \mid \alpha, \beta < \kappa \rangle$ is in M as well (use the pairing function), and so therefore by the inductive hypothesis is $\langle A_\alpha \mid \alpha < \kappa \rangle$, where $A_\alpha = \{ \beta \mid X_{\alpha(\beta)} \in \mathcal{U}_n \}$. Then $\{ \alpha \mid X_\alpha \in \mathcal{U}_{n+1} \} = \{ \alpha \mid A_\alpha \in \mathcal{U} \} \in M$.
For the second sentence, show by induction on n that $\{ \alpha \mid X^{n+1}_{(\alpha)} \in \mathcal{U}_n \} = X \in \mathcal{U}$.

Proof of (2). If $n > 1$, $j: m \longrightarrow n$, $X \in \mathcal{U}_m$, $\tilde{X} = j_{*n}(X)$ and the conclusion holds for all $\hat{j}: \hat{m} \longrightarrow \hat{n} < n$ and $\hat{X} \subseteq \kappa^{\hat{m}}$, $\{ \alpha \mid \tilde{X}_{(\alpha)} \in \mathcal{U}_{n-1} \}$ is either κ or $\{ \alpha \mid X_{(\alpha)} \in \mathcal{U}_{m-1} \} \in \mathcal{U}$, by the inductive hypothesis on $X_{(\alpha)}$ and $\hat{j}: m\text{-}1 \longrightarrow n\text{-}1$ defined for $t < m-1$ by $\hat{j}(t) = j(t+1)$.

Proof of (3). This is (2) for j = the inclusion map: $m \longrightarrow m + n$.

The next series of definitions, slightly condensed from [Ku1], gives Kunen's formulation of the notion of the αth iterated ultrapower $Ult_\alpha(M, \mathcal{U})$ of M by the $\mathcal{M}$-ultrafilter $\mathcal{U}$ on some $\kappa \in M$. This construction generalizes an earlier one due to Gaifman [Ga], and is unusually efficacious in that it unifies arguments about such apparantly diverse phenomena as the existence of indiscernibles of L, and the theory of a very stable inner model for the existence of a measurable cardinal, $L(\mathcal{U})$. (cf. §5)

To define an 'ultrapower' of some kind, one needs a class of function and some sort of reasonable approximation to an ultrafilter on their common domain. For the αth iterated ultrapower $Ult_\alpha(M, \mathcal{U})$, the class of functions below is $Fn(\alpha, \kappa, M)$, the common domain is $(\kappa^\alpha)^{\mathcal{M}}$, and the 'ultrafilter' is $\mathcal{U}_\alpha \subseteq S(\alpha, \kappa, M) \subseteq (S(\kappa^\alpha))^{\mathcal{M}}$.

4.7. If $\mathcal{U}$ is an M-ultrafilter on $\kappa \in M$ and $\alpha \geq 1$, $Fn(\alpha, \kappa, M)$ is the

set of all f: $\kappa^\alpha \longrightarrow M$ such that for some $1 \leq n < \omega$, order embedding $j: n \longrightarrow \alpha$ and $\tilde{f}: \kappa^n \longrightarrow M$.
$f(\langle \tau_\beta \mid \beta < \alpha \rangle) = \tilde{f}(\langle \tau_{j(0)}, \ldots, \tau_{j(n-1)} \rangle)$.
Call j a support of f. Note that $Fn(n, \kappa, M) = M^{\kappa^n}$ for $1 \leq n < \omega$, and that the correspondence $\tilde{f} \longmapsto f$ is an embedding $j_{*\alpha}: Fn(n, \kappa, M) \longrightarrow Fn(\alpha, \kappa, M)$.
Similarly, for $\mathcal{U}$, M and κ as above and α any ordinal ≥ 1, let $S(\alpha, \kappa, M)$ be the set of all $X \subseteq \kappa^\alpha$ such that for some $1 \leq n < \omega$, order-embedding $j: n \longrightarrow \alpha$ and $\tilde{X} \subseteq \kappa^n$, $\langle \tau_\beta \mid \beta < \alpha \rangle \in X$ iff $\langle \tau_{j(0)}, \ldots, \tau_{j(n-1)} \rangle \in \tilde{X}$. Once again j is called a support of X; for $1 \leq n < \omega$, $S(n, \kappa, M) = S(\kappa^n)$, and the correspondence $\tilde{X} \longmapsto X$ determines an embedding, also called $j_{*\alpha}$, from $S(n, \kappa, M)$ into $S(\alpha, \kappa, M)$.

We identify $\langle \alpha \rangle$ and α, and thus $Fn(1, \kappa, M)$ with M and $S(1, \kappa, M)$ with $S(\kappa)$.

4.8. Now define $\mathcal{U}_\alpha \subseteq S(\alpha, \kappa, M)$ for $1 \leq \alpha$ as follows. For $1 \leq n < \omega$, $\mathcal{U}_n$ is the $\mathcal{U}_n$ defined above. For $\alpha \geq \omega$, set $X \in \mathcal{U}_\alpha$ iff for some $1 \leq n < \omega$, support $j: n \longrightarrow \alpha$ and $\tilde{X} \in \mathcal{U}_n$, $X = j_{*\alpha}(\tilde{X})$, i. e., $\langle \tau_\beta \mid \beta < \alpha \rangle \in X$ iff $\langle \tau_{j(0)}, \ldots, \tau_{j(n-1)} \rangle \in \tilde{X}$.

4.9. Divide out $Fn(\alpha, \kappa, M)$ by $\mathcal{U}$ to obtain $Ult_\alpha(M, \mathcal{U}) = \mathcal{N}_\alpha = \langle N_\alpha, E_\alpha \rangle$, for $\alpha \geq 1$, as follows.
For f, g in $Fn(\alpha, \kappa, M)$ set $f \approx_\alpha g$ iff $\{s \mid f(s) = g(s)\}$, which is already in $S(\alpha, \kappa, M)$, is in $\mathcal{U}_\alpha$.
From the Scott equivalence classes $[f]_\alpha = \{g$ of minimal rank $\approx_\alpha f\}$, and set $N_\alpha = \{[f]_\alpha \mid f \in Fn(\alpha, \kappa, M)\}$. For f, g in N_α $[f] E_\alpha [g]$ iff $\{s \mid f(s) \in g(s)\}$, also in $S(\alpha, \kappa, M)$, is in $\mathcal{U}_\alpha$.

If $\mathcal{N}_\alpha$ is well-founded, let M_α be the transitive collapse of N_α, and set $\mathcal{M}_\alpha = \langle M_\alpha, \in \rangle$.

For any α, β with $\alpha < \beta$ and order-embedding $j: \alpha \longrightarrow \beta$ we can define $j_{*\beta}: \mathcal{N}_\alpha \longrightarrow \mathcal{N}_\beta$ much as the finite $j_{*\beta}$ was defined above. First define $j_{*\beta}: Fn(\alpha, \kappa, M) \longrightarrow Fn(\beta, \kappa, M)$ by setting $j_{*\beta}(f) = g$, where $g(\langle \tau_\gamma \mid \gamma < \beta \rangle) = f(\langle \tau_{j(\delta)} \mid \delta < \alpha \rangle)$.
Then pass $j_{*\beta}$ over the equivalence relations by setting $j_{*\beta}([f]_\alpha) = [j_{*\beta}(f)]_\beta$.

If $j: \alpha \longrightarrow \beta$ is the inclusion map, call $j_{*\beta}$ $i_{\alpha\beta}$.

The following lemma is lifted from Lemma 4.6 by means of arguments with finite supports.

4.10 Lemma. (1) If $j: \alpha \longrightarrow \beta$ is an order-embedding and $X \in S(\alpha, \kappa, M)$, $X \in \mathcal{U}_\alpha$ iff $j_{*\beta}(X) \in \mathcal{U}_\beta$.

(2) If $X \in S(\alpha+\beta, \kappa, M)$, $X \in \mathcal{U}_{\alpha+\beta}$ iff $\{s \in \kappa^{\alpha} \mid X_{(s)} \in \mathcal{U}_{\beta}\} \in \mathcal{U}_{\alpha}$.

Proof of (1). Let $k: n \longrightarrow \alpha$ be a finite support of X, so that $X = k_{*\alpha}(Y)$ for some $Y \subseteq [\kappa]^n$. Then $j \cdot k$ is a finite support of $j^*\beta(X)$, and $Y \in \mathcal{U}_n$ iff $k_{*\alpha}(Y) = X \in \mathcal{U}_\alpha$ iff $(j \cdot k)_{*\beta}(Y) = j_{*\beta}(X) \in \mathcal{U}_\beta$.
Proof of (2). As in 4.6, this is the special case $j_{*\beta} = i_{\alpha\beta}$, j = the inclusion map: $\alpha \longrightarrow \alpha+\beta$.

4.11 Lemma. (1) For each φ, $(\varphi([f]_\alpha, \ldots, [f_{n-1}]_\alpha))^{\mathcal{N}_\alpha}$ iff $\{s \in \kappa^\alpha \mid (\varphi(f_0(s), \ldots, f_{n-1}(s)))^{\mathcal{M}}\} \in \mathcal{U}_\alpha$.
(2) For each order-embedding $j: \alpha \longrightarrow \beta$ $j_{*\beta}$ is an elementary embedding: $\mathcal{N}_\alpha \longrightarrow \mathcal{N}_\beta$.
(3) If α is limit ordinal, $\mathcal{N}_\alpha$ is isomorphic to the direct limit of the elementary direct system $\langle\langle \mathcal{N}_\beta \mid \beta < \alpha \rangle, \langle i_{\beta\gamma} \mid \beta < \gamma < \alpha \rangle\rangle$.
(4) $(ZFC)^{\mathcal{N}_\alpha}$ for each α.

Proofs (1), (2), and (4). One uses finite supports to apply the usual ar ments for Łoś's theorem. Notice that (4) for limit α also follows from (4) for $\beta < \alpha$ and (3), since the direct limit of an elementary direct system is an elementary extension of each of the factors (cf. [Sa, theorem 10.1]).
Proof of (3). By verifying the usual arrow-theoretic characterization of the direct limit (again, see [Sa , theorem 10.2]). The $i_{\beta\alpha}$'s for $\beta < \alpha$ give the required embeddings; if $\mathcal{N}^*$ is such that there are elementary embeddings $k_\beta: \mathcal{N}_\beta \longrightarrow \mathcal{N}^*$ commuting with the $i_{\beta\gamma}$'s for $\beta < \gamma < \alpha$, the mapping $k: \mathcal{N}_\alpha \longrightarrow \mathcal{N}^*$ defined by $k([f]_\alpha) = k_n([\tilde{f}]_n)$, where $\tilde{f} \in Fn(n, \kappa, M)$ and finite support $j: n \longrightarrow \alpha$ are such that $j_{*\alpha}([\tilde{f}]_n) = [f]_\alpha$, does what is required.

4.12. Define $\mathcal{U}^{(\alpha)} = \{[f] \mid \{s \mid f(s) \in \mathcal{U}\} \in \mathcal{U}_\alpha\}$. Note that if $\mathcal{U}$ were an element of M (making κ measurable in M), then $\mathcal{U}^{(\alpha)}$ would be $i_{0\alpha}(\mathcal{U})$; also that $[f]$ is in $\mathcal{U}^{(\alpha)}$ iff the graph of f is in $\mathcal{U}_{\alpha+1}$.

The following lemma is quite useful for handwaving, since it often reduces arguments about $Ult_\alpha(M, \mathcal{U})$ and $\mathcal{U}^{(\alpha)}$ to parallel arguments about M and $\mathcal{U}$. In its simplest special case it says there is an isomorphism $e_{11}: Ult_2(M, \mathcal{U}) \longrightarrow Ult_1(N_1, \mathcal{U}^{(1)})$.

4.13 Lemma. If $\mathcal{N}_\alpha$ is well-founded, then:
(1) $\mathcal{U}^{(\alpha)}$ is an N_α-ultrafilter on $i_{0\alpha}(\kappa)$
(2) For any β, there is an isomorphism $e_{\alpha\beta}: \mathcal{N}_{\alpha+\beta} \longrightarrow Ult_\beta(\mathcal{N}_\alpha, \mathcal{U}^{(\alpha)})$;

$e_{\alpha\beta}$ commutes nicely in that if $i_{0\beta}^{(\alpha)}$ is the embedding: $\mathcal{N}_\alpha \longrightarrow \mathrm{Ult}_\beta(\mathcal{N}_\alpha, \mathcal{U}^{(\alpha)})$, $i_{0\beta}^{(\alpha)} = e_{\alpha\beta} \cdot i_{\alpha,\alpha+\beta}$.

Proof of (1). For finite $\alpha = n < \omega$, this follows from the corresponding result for $\mathcal{U}_{n+1}$ in 4.6(1).

For infinite α, argue with finite supports.

E. g., for 4.1 (5), suppose $[f] : i_{0\alpha}(\kappa) \longrightarrow S(i_{0\alpha}(\kappa))$ is in $\mathrm{Ult}_\alpha(M, \mathcal{U})$, and $f = j_{*\alpha}(f')$ for some $f' \in \mathrm{Fn}(n, \kappa, M)$, let $g \in \mathrm{Fn}(n, \kappa, M)$ be such that $[g]_n = \{\beta \in i_{0n}(\kappa) \mid [f'](\beta) \in \mathcal{U}^{(n)}\}$. Then $[j_{*\alpha}(g)]_\alpha = \{\beta \in i_{0\alpha}(\kappa) \mid [f](\beta) \in \mathcal{U}^{(\alpha)}\}$.

Proof of (2). Suppose $f \in \mathrm{Fn}(\alpha+\beta, \kappa, M)$ is supported by $j: n+m \longrightarrow \alpha+\beta$, with $j(n+k) = \alpha + \delta_k$, $j' : \alpha+m \longrightarrow \alpha+\beta$ is defined by $j' \restriction \alpha = \mathrm{id}$ and $j'(\alpha+k) = \alpha + \delta_k$, and $f' \in \mathrm{Fn}(\alpha+m, \kappa, M)$ is such that $j'_{*\alpha}(f') = f$. If $h: s \longrightarrow f'_{(s)}$ for $s \in \kappa^\alpha$, $[h]_\alpha = g' \in \mathrm{Fn}(m, i_{0\alpha}(\kappa), \mathcal{M}_\alpha)$, and $i: m \longrightarrow \beta$ defined by $i(k) = \delta_k$ gives $g = i_{*\beta}(g') \in \mathrm{Fn}(\beta, i_{0\alpha}(\kappa), \mathcal{M}_\alpha)$. Set $e_{\alpha\beta}([f]_{\alpha+\beta}) = [g]_\beta$, where $[\ldots]_\beta$ is the equivalence class of $\mathrm{Fn}(\beta, i_{0\alpha}(\kappa), \mathcal{M}_\alpha)$ determined by $\mathcal{U}_\beta^{(\alpha)}$. Then $e_{\alpha\beta}$ is onto, and $[f]_{\alpha+\beta} \in [f]_{\alpha+\beta}$ iff $\{\langle s, t\rangle \mid s \in \kappa^\alpha$ and $t \in \kappa^\beta$ and $f(\langle s, t\rangle) \in f(\langle s, t\rangle)\} \in \mathcal{U}_{\alpha+\beta}$ iff $\forall^* s\, \forall^* t\; f_{(s)}(t) \in f_{(s)}(t)$ iff $\{r \in (i_{0\alpha}(\kappa))^m \mid g'(r) \in g'(r)\} \in \mathcal{U}_m^{(\alpha)}$ iff $\{r \in (i_{0\alpha}(\kappa))^\beta \mid g(r) \in g(r)\} \in \mathcal{U}_\beta^{(\alpha)}$ iff $[g]_\beta \in [g]_\beta$.

4.14. Theorem. Suppose $\mathcal{N}_\beta$ is well-founded and $\alpha < \beta$. Then

(1) For all $\gamma < i_{0\alpha}(\kappa)$ $i_{\alpha\beta}(\gamma) = \gamma$.

(2) $i_{0\alpha}(\kappa) < i_{0\beta}(\kappa) = i_{\alpha\beta}(i_{0\alpha}(\kappa))$.

(3) If $\beta = \bigcup\beta$, $i_{0\beta}(\kappa) = \bigcup\{i_{0\alpha}(\kappa) \mid \alpha < \beta\}$. (Thus $\{i_{0\beta}(\kappa) \mid \beta > 0\}$ is a closed unbounded class of ordinals if all $\mathcal{N}_\beta$ are well-founded).

(4) (a) ${}^{i_{0\alpha}(\kappa)}M_{\alpha+1} \subseteq M_{\alpha+1}$ and (b) $S(i_{0\alpha}(\kappa)) \cap M_\alpha = S(i_{0\alpha}(\kappa)) \cap M_\beta$, for $\alpha \geq 0$ and $\beta > \alpha$.

(5) $i_{0\beta}(\eta) < [\bar{\bar{\beta}} \cdot ((\bar{\bar{\eta}}^{\kappa})^M)^=]^+$.

(6) $i_{0\mu}((\kappa)) = \mu$ for any cardinal $\mu > (2^\kappa)^M$.

(7) $i_{0\alpha}(\lambda) = \lambda$ if $\alpha < \lambda$, $\mathrm{cf}\,\lambda \neq \kappa$, and $\bar{\bar{\eta}}^\kappa < \lambda$ whenever $\eta < \lambda$ (e. g.: $\lambda = (2^\kappa)^+$).

Proof of (1). Assume by 4.13(2) that $\alpha = 0$. Then for $\gamma < \kappa$ and $i_{0\beta+1}(\gamma)$, if we assume the result for $i_{0\beta}(\gamma)$, the problem reduces to showing that $i_{\beta\beta+1}(\gamma) = \gamma$ for all $\gamma < \kappa$. But again by 4.13(2), this is true by induction on γ for all $\gamma < i_{0\beta}(\kappa)$, for if $[f] < \gamma$, where $[f]$ is with respect to $\mathcal{U}^{(\beta)}$ in $\mathrm{Fn}(1, i_{0\beta}(\kappa), \mathcal{U}^{(\beta)})$, there is an $\eta < \gamma$ such that $\{\delta < i_{0\beta}(\kappa) \mid f(\delta) = \eta\} \in \mathcal{U}^{(\beta)}$. If $\beta = \bigcup\beta$, assume the result holds good for all $\eta \leq \beta$ and $\gamma' < \gamma < \kappa$. Then if $x < i_{0\beta}(\gamma)$, $x = i_{\eta\beta}(\gamma') = i_{\eta\beta}(i_{0\eta}(\gamma')) = i_{0\beta}(\gamma') = \gamma'$ for some $\eta < \beta$ and $\gamma' < \gamma$.

Proof of (2). Again by 4.13, it suffices to assume $\alpha = 0$ and $\beta = 1$, since $i_{0\beta}(\kappa) \geq i_{01}(\kappa)$. The result follows from the fact that $\eta < [\mathrm{id}] < i_{01}(\kappa)$ for each $\eta > \kappa$, where id: $\kappa \longrightarrow \kappa$ is the identity function.

Proof of (3). If $x < i_{0\beta}(\kappa)$, $x = i_{\eta\beta}(\gamma)$ for some $\eta < \beta$ and $\gamma < i_{0\eta}(\kappa)$, so that $x = i_{\eta\beta}(\gamma) = \gamma < i_{0\eta}(\kappa)$ by (1).

Proof of (4). We can again assume by 4.13(2) that $\alpha = 0$.

For (a), let $a_\alpha = [f_\alpha]$; and define $b: \kappa \longrightarrow {}^\kappa M$ by $b(\eta): \alpha \longmapsto f_\alpha(\eta)$. $[b] \in {}^{i_{01}(\kappa)}M_1$; $a = b \upharpoonright \kappa$ in M, for if $\alpha < \kappa$, $\{\eta \mid b(\eta)(\alpha) = f_\alpha(\eta)\} \in \mathcal{U}$.

For (b), argue by induction on β. If $\beta = \gamma + 1$, we can argue as in (a) that $S(\kappa) \cap M_{\gamma+1} = S(\kappa) \cap M_\gamma = S(\kappa) \cap M$, using 4.13.2 for the first equality. A somewhat more straightforward argument for $S(\kappa) \cap M_{\gamma+1} = S(\kappa) \cap M$, or equivalently, by 4.13(2), $S(\kappa) \cap M_1 = S(\kappa) \cap M$, is as follows. Suppose $X \in S(\kappa) \cap M$ and $[h]_1 = \kappa$ in M_1 and $f(\alpha) = X \cap h(\alpha)$ for $\alpha < \kappa$. Then $\xi = i(\xi) \in [f]$ iff $\{\alpha \mid \xi \in X \cap h(\alpha)\} \in \mathcal{U}$ iff $\xi \in X$, for each $\xi < \kappa$. If $\beta = \bigcup \beta$ and $X \in S(\kappa) \cap M$, $X \in M_1$ by $\beta = 1$, and then by $i_{1\beta}(X) = X$ (by (1) for $\alpha = 1$) is in M_β.

Proof of (5). Induction on β, in each case for all η.

$\beta = 1$: ordinals $< i_{01}(\eta)$ are equivalence classes of functions in M from κ into η, so $(i_{01}(\eta))^= < [((\bar{\bar{\eta}})^\kappa)^M]^+$.

$\beta = \alpha + 1$: If $\eta_\xi = i_{0\xi}(\eta)$, $(\eta_{\alpha+1})^= \leq (({}^{\kappa_\alpha}\eta_\alpha)^{\mathcal{M}_\alpha})^= \leq ({}^{\kappa_\alpha}\eta_\alpha)^{=\mathcal{M}_\alpha} = i_{0\alpha}((\bar{\bar{\eta}}^\kappa)^{\mathcal{M}}) < [\bar{\bar{\alpha}} \cdot ((\bar{\bar{\eta}}^\kappa)^{\mathcal{M}})^=]^+$ by inductive hypothesis.

$\bigcup \beta = \beta$. $(\eta_\beta)^= \leq \sum_{\gamma<\beta} (i_{0\gamma}(\eta))^= \leq \sum_{\gamma<\beta} (\bar{\bar{\gamma}} \cdot ((\bar{\bar{\eta}}^\kappa)^{\mathcal{M}})^= = \bar{\bar{\beta}} \cdot ((\bar{\bar{\eta}}^\kappa)^{\mathcal{M}})^=$.

Proof of (6). By (3) and (5), $i_{0\mu}(\kappa) = \bigcup \{i_{0\gamma}(\kappa) \mid \gamma < \mu\}$, and the cardinality of each $i_{0\gamma}(\kappa)$ is $\bar{\bar{\gamma}} \cdot (\kappa^\kappa)^{\mathcal{M}} < \mu$.

Proof of (7). First note that $i_{0\alpha}(\eta) < \lambda$ for all $\alpha, \eta < \lambda$. We show: $i_{0\alpha}(\lambda) = \bigcup_{\eta<\lambda} i_{0\alpha}(\eta)$. $\alpha = 1$: if $[f] < i_{01}(\lambda)$, ran $f \subseteq \eta < \lambda$ since cf $\lambda \neq \kappa$, and so $[f] \leq i_{01}(\eta)$. $\alpha = \beta + 1$: argue as with (1) for $\mathrm{Ult}_1(M_\beta, \mathcal{U}^{(\beta)})$, and transfer by $e_{\beta 1}$ of 4.13(2) to $M_{\beta+1}$.

$\alpha = \bigcup \alpha$: if $\zeta < i_{0\alpha}(\lambda)$, $\zeta = i_{\gamma\alpha}(\zeta')$, some $\zeta' < i_{0\gamma}(\lambda)$, so by the inductive assumption there is an $\eta < \lambda$ such that $\zeta' < i_{0\gamma}(\eta)$. Then $\zeta < i_{\gamma\alpha}(i_{0\gamma}(\eta)) = i_{0\alpha}(\eta)$.

4.15 Theorem. (1) If $\mathrm{Ult}_1(M, \mathcal{U})$ is well-founded, κ is ineffable and Π^1_n indescribable for each $n < \omega$.

(2) If $\mathrm{Ult}_\gamma(\mathcal{M}, \mathcal{U})$ is well-founded, $\alpha < \beta$, $j: \beta \longrightarrow \gamma$ is an order-embedding and $[f]_\beta \neq i_{0\alpha}(\kappa)$, $[j_{*\gamma}(f)]_\gamma \neq i_{0j(\alpha)}(\kappa)$.

(3) If $\alpha \geqslant \omega$ and $\mathrm{Ult}_\alpha(M, \mathcal{U})$ is well-founded, $\{i_{0\gamma}(\kappa) \mid \gamma < \alpha\}$ is a set of indiscernibles for $\langle \mathcal{M}_\alpha, \in, \langle i_{0\alpha}(a) \mid a \in M \rangle\rangle$.
(4) If $\mathrm{Ult}_\alpha(M, \mathcal{U})$ is well-founded, and $\omega \leqslant \alpha < \omega_1^{\mathcal{M}}$, $(\kappa \longrightarrow (\alpha)^{<\omega})^{\mathcal{M}}$.

Proof of (1). Both conclusions follow from 4.14(2) and (4). Ineffability is shown as in the latter part of the proof of 2.11. If φ is Π^1_n, $A \subseteq \kappa$ and $(\langle \kappa, \in, A\rangle \models \varphi)$, $(\langle \kappa, \in, A\rangle \models \varphi)^{\mathcal{M}_1}$ by 4.14(4), so (there is a $\lambda < i_{01}(\kappa)$ such that $\langle \lambda, \in, i_{01}(A) \cap \lambda\rangle \models \varphi)^{\mathcal{M}_1}$, so (there is a $\lambda < \kappa$ such that $\langle \lambda, \in, A \cap \lambda\rangle \models \varphi)^{\mathcal{M}}$.
Proof of (2). If $D(\alpha,\beta)$ is (for all $j: \beta \longrightarrow \gamma$, $[f]_\beta = i_{0\alpha}(\kappa) \longrightarrow [j_*(f)]_\gamma = i_{0j(\alpha)}(\kappa)$) and $C(\alpha)$ is $\forall \beta > \alpha\ D(\alpha,\beta)$,
(*) $C(\alpha)$ for all $\alpha > 0$ follows from $D(0,1)$.
Proof of (*). If $j: \beta \longrightarrow \gamma$ and $\alpha < \beta$ are given, $D(0,1)$ gives that $i_{01}(\kappa)$ in $\mathcal{M}_\beta$ is $k_{*\beta}([f]_1)$, where $[f]_1 = \kappa$, $k: 1 \longrightarrow \beta$ and $k(0) = \alpha$, so $j_{*\gamma}(i_{0\alpha}(\kappa)) = j_*(k_{*\beta}([f]_1)) = i_{0(j\cdot k)(0)}(\kappa)$, again by $D(0,1)$, $= i_{0j(\alpha)}(\kappa)$.
Proof of $D(0,1)$. We have $j: 1 \longrightarrow \gamma$ such that $j(0) = \alpha$, say; we may assume by 4.14(1) that $\gamma = \alpha+1$. If $[j_{*\gamma}(f)]_\gamma = [h]_\alpha < i_{0\alpha}(\kappa)$, $(\forall^* s \in \kappa^\alpha)(\forall^* \eta)$ $f(\eta) = h(s)$, an impossibility since $[f]_1 = \kappa$. If $[j_{*\gamma}(f)]_\gamma > i_{0\alpha}(\kappa) = [g]_\gamma$, $(\forall^* s \in \kappa^\alpha)(\forall^* \eta)$ $f(\eta) > g(\langle s, \eta\rangle)$, so there is an $h: \kappa^\alpha \longrightarrow \kappa$ such that $(\forall^* s \in \kappa^\alpha)(\forall^* \eta)$ $h(s) = g(\langle s, \eta\rangle)$, so $i_{0\alpha}(\kappa) > [h]_\alpha = [i_{\alpha\gamma}(h)]_\gamma = [g]_\gamma = i_{0\alpha}(\kappa)$.
Proof of (3). Suppose $\bar{x} = \langle i_{0\gamma_0}(\kappa), \ldots, i_{0\gamma_{n-1}}(\kappa)\rangle$, $\bar{y} = \langle i_{0\delta_0}(\kappa), \ldots, i_{0\delta_{n-1}}(\kappa)\rangle$. For simplicity, assume there is only one parameter $i_{0\alpha}(a)$, $a \in M$. If $j: n \longrightarrow \alpha$ is such that $j(m) = \gamma_m$ for each $m < n$, $j_{*\alpha}(i_{0n}(a)) = i_{0\alpha}(a)$, and $j_{*\alpha}(i_{0m}(\kappa)) = i_{0\gamma_m}(\kappa)$ by (2). Then $(\varphi(\bar{x}, i_{0\alpha}(a)))^{\mathcal{M}_\alpha}$ iff $(\varphi(i_{01}(\kappa), \ldots, i_{0n-1}(\kappa), i_{0n}(a)))^{\mathcal{M}_n}$. Doing the same for $\bar{y}$ establishes the result.
Proof of (4). Suppose $(f: [\kappa]^{<\omega} \longrightarrow 2)^M$. By (3), $\{i_{0\gamma}(\kappa) \mid \gamma < \alpha\}$ is homogeneous for $i_{0\alpha}(f)$. Since $\alpha < \omega_1^{\mathcal{M}} = \omega_1^{\mathcal{M}_\alpha}$, the argument of 2.14(2) shows $i_{0\alpha}(f)$ has a homogeneous set of type α in $\mathcal{M}_\alpha$. Since $i_{0\alpha}(\alpha) = \alpha$, by elementarity f has a homogeneous set of type α in M.

4.16 Theorem. (1) $\mathcal{N}_\alpha$ is well-founded for all α iff it is well-founded for all $\alpha < \omega_1$.
(2) If countable intersections of elements of $\mathcal{U}$ are in $\mathcal{U}$, $\mathcal{N}_\alpha = \mathrm{Ult}_\alpha(M, \mathcal{U})$ is well-founded for all α.

Proof of (1). Suppose $[f_{n+1}]_\alpha \, E_\alpha [f_n]_\alpha$ for all $n < \omega$, where f_n has support $j_n: m_n \longrightarrow \alpha$. Let $R = \bigcup \mathrm{ran}\, j_n$, and j order-embed the type β of R onto R. Then there are $g_n \in \mathrm{Fn}(\beta, \kappa, M)$ such that $j_*(g_n) = f_n$ and $[g_{n+1}]_\beta \, E_\beta \, [g_n]_\beta$ for all $n < \omega$.
Proof of (2). Suppose again $[f_{n+1}]_\alpha \, E_\alpha [f_n]_\alpha$ for all $n < \omega$, and $X_n = \{ s \mid f_{n+1}(s) \in f_n(s)\}$. We recursively define $s \restriction \gamma: \gamma \longrightarrow \alpha$ for

$\gamma < \alpha$ such that each $X_{n(s \upharpoonright \gamma)} \in \mathcal{U}_{\alpha-\gamma}$. If $s \upharpoonright \gamma$ is already so defined, let $s(\gamma) \in \bigcap_{n<\omega} \{ \eta \mid X_{n(\langle s \upharpoonright \gamma, \eta \rangle)} \in \mathcal{U}_{\alpha-(\gamma+1)} \}$. Then s is in every X_n, since for each finite support $j_n : m_n \longrightarrow \alpha$ of X_n, $s \upharpoonright \operatorname{ran} j_n \in \tilde{X}_n$ in $S(m_n, \kappa, M)$, where $j_{n*}(\tilde{X}_n) = X_n$.

4.17. An $\mathcal{M}$-ultrafilter $\mathcal{U}$ on $\kappa \in M$ is normal iff whenever $X \in \mathcal{U}$ and $f : X \longrightarrow \kappa$ is a regressive function in M, there is a $Y \subseteq X$ in $\mathcal{U}$ such that $f \upharpoonright Y$ is constant.

4.18 Theorem. (1) $\mathcal{U}$ is normal iff (a) $[\mathrm{id}]$ is the κth ordinal of $\mathcal{N}_1$ iff
(b) whenever $X_\gamma \in \mathcal{U}$ for each $\gamma < \kappa$ and $\langle X_\gamma \mid \gamma < \kappa \rangle \in M$, the diagonal intersection $\Delta_{\gamma<\kappa} X_\gamma = \{ \gamma \mid \gamma \in X_\alpha \text{ for all } \alpha < \gamma \} \in \mathcal{U}$.
(2) If $\mathcal{N}$ is a transitive model of ZFC such that $S(\kappa) \cap N = S(\kappa) \cap M$ and $i : \mathcal{M} \longrightarrow \mathcal{N}$ is an elementary embedding such that $i \upharpoonright \kappa = \mathrm{id}$ and $i(\kappa) > \kappa$, $\{ X \in S(\kappa) \cap M \mid \kappa \in i(X) \}$ is a normal M-ultrafilter on κ. In particular, if $\mathcal{N}_1$ is well-founded, a normal $\mathcal{M}$-ultrafilter exists.
(3) More generally, if $\mathcal{N}_1$ has a least ordinal $[f]$ such that $i_{01}(\alpha)\ E_1 [f]$ for all $\alpha < \kappa$, and $X \in \mathcal{V}$ iff $[f]\ E_1\ i_{01}(X)$ iff $f^{-1}(X) \in \mathcal{U}$, $\mathcal{V}$ is a normal M-ultrafilter on κ. Furthermore countable intersections of members of $\mathcal{V}$ are nonempty iff the same is true of $\mathcal{U}$, and each $\mathrm{Ult}_\alpha(M, \mathcal{U})$ is well-founded iff $\mathrm{Ult}_\alpha(M, \mathcal{V})$ is.

Proof of (1). $\mathcal{U}$ is normal iff whenever $[f]\ E_1\ [\mathrm{id}]$, $[f] = i_{01}(\alpha)$ for some $\alpha < \kappa$, iff $[\mathrm{id}]$ is the κth ordinal of $\mathcal{N}_1$. Note that $X \in \mathcal{U}$ iff $\{ \alpha \mid \mathrm{id}(\alpha) \in X \} \in \mathcal{U}$ iff $[\mathrm{id}] \in i(X)$, for any $\mathcal{U}$, normal or not. For the remaining equivalence, argue as in 0.5(1).
Proof of (2). Suppose $X_\gamma \subseteq \kappa$ for $\gamma < \kappa$, and $X = \Delta_{\gamma<\kappa} X_\gamma$. Then $X \in \mathcal{U}$ iff $\kappa \in i(X)$ iff $\kappa \in i(X_\gamma)$ for all $\gamma < \kappa$ iff $X_\gamma \in \mathcal{U}$ for all $\gamma < \kappa$.
Proof of (3). Suppose $Y \in \mathcal{V}$ and $g : Y \longrightarrow \kappa$ is regressive. Then $(g \circ f)(\alpha) < f(\alpha)$ for all $\alpha \in f^{-1}{}''Y \in \mathcal{U}$, i. e., $[g \circ f]\ E_1\ [f]$, so by the assumption on f and 4.1(4) there is an $X^* \subseteq f^{-1}(Y)$ such that $X^* \in \mathcal{U}$ and $(g \circ f)$ is constant on X^*. Then g is constant on $f(X^*) = Y^* \subseteq Y$, and $Y^* \in \mathcal{V}$.

We close this section with a characterization of $a^\#$ in terms of iterated ultrapowers. The basic idea goes back to Gaifman's unpublished work in which he derived the existence of indiscernibles for L from the existence of a measurable cardinal.

4.19 Theorem (Kunen). If a is a set of ordinals, the following are equivalent:
(1) For some cardinal $\kappa \geq \lambda = a^+$, $\mathcal{M}_\kappa(a) = \langle L_\kappa(a), \in, a, \langle \alpha \mid \alpha \in \bigcup a \rangle \rangle$

is not a Jonsson model
(2) For some cardinal $\kappa \geq \lambda$ and $\eta \geq \kappa$, there is an elementary embedding $i: \mathcal{M}_\kappa(a) \prec \mathcal{M}_\eta(a)$ which is not inclusion.
(3) For some $\gamma > \bigcup a$ there is an L(a)-ultrafilter $\mathcal{U}$ on γ such that $\mathrm{Ult}_1(L(a), \mathcal{U})$ is well-founded.
(4) For some $\gamma > \bigcup a$ there is an L(a)-ultrafilter $\mathcal{U}$ on γ such that $\mathrm{Ult}_\alpha(L(a), \mathcal{U})$ is well-founded for every $\alpha > 0$.
(5) $a^\#$ exists.
(6) For any regular cardinal $\kappa \geq \lambda = a^+$, the closed unbounded filter $\mathcal{K}$ is an L(a)-ultrafilter on κ (equivalently, every stationary $X \in S(\kappa) \cap L(a)$ has a closed unbounded subset).

Proof ($1 \longrightarrow 2$). If $\langle M, \in \rangle \not\preceq \mathcal{M}_\kappa(a)$ is such that $(M)^= = \kappa$, and π is the collapsing map, $i = \pi^{-1}$ elementarily embeds $\mathcal{M}_\kappa(a)$ into $\mathcal{M}_\kappa(a)$.
($2 \longrightarrow 3$) Suppose $i: \mathcal{M}_\kappa(a) \prec \mathcal{M}_\eta(a)$ and $\gamma > \bigcup a$ is the least ordinal such that $i(\gamma) > \gamma$. Then $\mathcal{U} = \{X \subseteq \gamma \mid \gamma \in i(X)\}$ is a (normal) L(a)-ultrafiler by 4.18(2), since $S(\gamma) \cap L_\kappa(a) = S(\gamma) \cap L(a)$.
(a) Ult (L(a), $\mathcal{U}$) is well-founded. For if $[g_{n+1}]\ E_1\ [g_n]$ for all $n < \omega$, let $\mu > \bigcup a$ be such that all the g_n's are in $L_\mu(a)$, and let $\langle \mathcal{M}_\delta(a), \in \langle f_n \mid n < \omega \rangle\rangle$ be the transitive collapse of $SH((\gamma+1) \cup \{g_n \mid n < \omega\}, \langle \mathcal{M}_\mu(a), \in, \langle g_n \mid n < \omega\rangle\rangle)$.
Then $\gamma \in i(\{\alpha\ \ g_{n+1}(\alpha) \in g_n(\alpha)\})$ iff $\gamma \in i(\{\alpha \mid f_{n+1}(\alpha) \in f_n(\alpha)\})$ iff $\gamma \in \{\alpha \mid i(f_{n+1}(\alpha)) \in i(f_n(\alpha))\}$ iff $i(f_{n+1}(\gamma)) \in i(f_n(\gamma))$ for all n, an impossibility.
($3 \longrightarrow 4$) γ is the γ of (3), and $\mathcal{U}$ the $\mathcal{U}$ of its proof. We need the following definability result.
(a) If $\gamma > \bigcup a$, $R = \{$the first $\bar{\bar{\gamma}}^+$ cardinals $\lambda = \rho^+$ such that $\rho = \beth_\rho$ and $\mathrm{cf}\rho > \bar{\bar{\gamma}}$ (so $\eta < \lambda \longrightarrow \bar{\bar{\eta}}^{\bar{\bar{\gamma}}} < \lambda$), $\kappa = (\bigcup R)^+$, and X is any element of $S(\gamma) \cap L(a)$, there are $\bar{x} \in [\gamma]^{<\omega}$, $\bar{y} \in [R]^{<\omega}$ and φ in the language of set theory such that $\alpha \in X$ iff $\mathcal{M}_\kappa(a) \models \varphi[\alpha, \bar{x}, \bar{y}, a]$.
Proof of (a). If N is the set of t in $L_\kappa(a)$ (pointwise) definable in $\mathcal{M}_\kappa(a)$ from some $\bar{x}$, $\bar{y}$ or above and a as the unique t in $L_\kappa(a)$ such that $\mathcal{M}_\kappa(a) \models \psi_t[t, \bar{x}, \bar{y}, a]$, $\langle N, \in\rangle \prec \mathcal{M}_\kappa(a)$, so the collapse π takes N to $L_\delta(a)$ for $\delta \geq (\bar{\bar{\gamma}})^+$, and $S(\gamma) \cap L(a) = S(\gamma) \cap L_\delta(a)$ since $\bigcup a < \gamma$. Then any $X \in S(\gamma) \cap L(a)$ is $\pi(\tilde{X})$ for some $\tilde{X} \in N$, so $\alpha \in X$ for $\alpha < \gamma$ iff $\alpha \in \tilde{X}$ iff $\mathcal{M}_\kappa(a) \models (\exists! z\ \psi(z, \bar{x}, \bar{y}, a) \wedge \alpha \in z)$.
(b) If $\mathcal{U}$ and $\mathcal{V}$ are both normal L(a)-ultrafilters on $\gamma > \bigcup a$, $\mathcal{U} \cap L(a) = \mathcal{V} \cap L(a)$. For if $X \subseteq \gamma$, $\gamma \in i_{01}^{\mathcal{U}}(X)$ iff $\gamma \in i_{01}^{\mathcal{V}}(X)$, by (a) and 4.14(7).
(c) We now prove $\mathrm{Ult}_\alpha(L(a), \mathcal{U})$ is well-founded by induction on $\alpha > 0$. If $\alpha = \alpha + 1$, argue as in ($2 \longrightarrow 3$), using the fact that $\mathrm{Ult}_{\alpha+1}(L(a), \mathcal{U}) = \mathrm{Ult}_1(L(a), \mathcal{U}^{(\alpha)})$. If $\alpha = \bigcup \alpha$ and U_α is the class of all x in L(a)

fixed by each $i_{0\beta}$ for $\beta<\alpha$, $\langle U_\alpha,\epsilon\rangle \prec \langle L(a),\epsilon\rangle$, for if $(\exists y\, \varphi(\bar{x}, y))^{L(a)}$ for $\bar{x}\in[U_\alpha]^{<\omega}$, the least such y is definable in L(a) from $\bar{x}$, so is fixed by any elementary embedding which fixes $\bar{x}$. If j: $\langle L(a),\epsilon\rangle \prec \langle L(a),\epsilon\rangle$ is the inverse of the collapsing map $\pi: U_\alpha \longrightarrow L(a)$, $j\restriction\gamma$ is the identity and $j(\gamma)>i_{0\beta}(\gamma)$ for each $\beta<\alpha$. For each $\beta<\alpha$, we now define

<u>(i)</u> k_β: $Ult_\beta(L(a),\mathcal{U}) \prec \langle L(a),\epsilon\rangle$ such that

<u>(ii)</u> $k_\beta \restriction i_{0\beta}(\gamma)$ is the identity, and

<u>(iii)</u> $k_\beta \circ i_{\tilde\beta\beta} = k_{\tilde\beta}$ for each $\tilde\beta<\beta<\alpha$ as follows.

<u>k_0</u> = j. <u>k_β for $\beta = \bigcup\beta$</u> is the "direct limit of $\langle k_{\tilde\beta} \mid \tilde\beta<\beta\rangle$", i. e. the embedding guaranteed by the fact that $Ult_\beta(L(a),\mathcal{U})$ is (isomorphic to) the direct limit of the $Ult_{\tilde\beta}(L(a),\mathcal{U})$'s for $\tilde\beta<\beta$ (cf. [Sa , theorem 10.1]). (iii) is immediate, and (ii) follows from the fact that for each $\eta<i_{0\beta}(\gamma)$, $\eta = i_{\tilde\beta\beta}(\tilde\eta)$ for some $\tilde\eta<i_{0\tilde\beta}(\gamma)$. <u>$k_{\beta+1}$</u> is defined as follows. Identify $\mathcal{N}_{\beta+1}$ with Ult_1 $(L(a), \mathcal{U}^{(\beta)})$, and write <u>i</u> for $i_{01}^{\mathcal{U}(\beta)}$, <u>[f]</u> for $[f]_1^{\beta\,\beta+1}$, <u>E</u> for E_1^β, and <u>γ_β</u> for $i_{0\beta}(\gamma)$. Then <u>$k_{\beta+1}([f]) = k_\beta(f)(\gamma_\beta)$.</u>

<u>(i)</u> $k_{\beta+1}$: $\mathcal{N}_\beta \prec \langle L(a),\epsilon\rangle$. <u>E. g.</u>, $\varphi([f],[g])$ in $\mathcal{N}_{\beta+1}$ iff $\gamma_\beta\in i(\{\alpha<\gamma_\beta \mid \varphi(f(\alpha), g(\alpha))$ in $L(a)\})$ iff $\gamma_\beta\in k_\beta(\{\alpha<\gamma_\beta \mid \varphi(f(\alpha), g(\alpha))$ in $L(a)\})$ (by (a) and 4.14(7)), iff $k_\beta(f)(\gamma_\beta)\in k_\beta(g)(\gamma_\beta)$.

<u>(ii)</u> $k_{\beta+1}\restriction i_{0\beta+1}(\gamma)$ is the identity. First note that $i(f)(\gamma_\beta) = [f]$ for each $f\in Fn(1, \gamma_\beta, L(a))$, by induction over the well-founded relation E. For $[f]\,E[g]$ iff $\{\alpha<\gamma_\beta \mid f(\alpha)\in g(\alpha)\}\in \mathcal{U}^{(\beta)}$ iff $i(f)(\gamma_\beta)\in i(g)(\gamma_\beta)$. But for f: $\gamma_\beta \longrightarrow \gamma_\beta$, $i(f)(\gamma_\beta) = k_\beta(f)(\gamma_\beta)$ by the definability lemma (a) above and 4.14(7).

<u>(iii)</u> $k_{\beta+1}(i(x)) = k_\beta(x)$ by definition. Part (ii) of the inductive hypothesis for k_β gives the rest.

If k_α is now the direct limit of the k_β's for $\beta<\alpha$, k_α: $\mathcal{N}_\alpha \prec \langle L(a),\epsilon\rangle$, so $\mathcal{N}_\alpha$ is well-founded and we are done.

<u>$(4\longrightarrow 5)$</u> $\{i_{0\alpha}(\gamma) \mid \alpha>0\}$ is a closed unbounded class of indiscernibles for L(a) by 4.14(3) and 4.15(3); it follows from 4.14(3) and (6) that $i_{0\alpha}(\gamma)<\kappa$ for each cardinal $\kappa>\lambda = \bar{\bar{\gamma}}^+$ and $\alpha<\kappa = i_{0\kappa}(\gamma)$ so that $\{i_{0\alpha}(\gamma) \mid \alpha<\kappa\}$ is a set of indiscernibles for $\mathcal{M}_\kappa(a)$.

<u>$(5\longrightarrow 6)$</u> If H is the c. u. set of indiscernibles for L(a), any $A\in S(\kappa) \cap L(a)$ is $t(\bar{x},\bar{y})$ for some $t\in\mathcal{L}_a^*$, and $\bar{x}<\kappa\leq\bar{y}$. If C is $\{\gamma \mid \bar{x}<\gamma<\kappa\}$, C is either $\subseteq$ A or disjoint from A, for $\eta\in A$ iff $\eta\in t(\bar{x},\bar{y})$ iff $\zeta\in t(\bar{x},\bar{y})$ iff $\zeta\in A$ for all η,ζ in C. The equivalence and implication both follow if we verify 4.1(5) for $\mathcal{K}$; but $\{\alpha \mid X_\alpha\in\mathcal{K}\}\in M$ if $\langle X_\alpha \mid \alpha<\kappa\rangle$ is, since "I am c.u." is absolute for L(a).

<u>$(6\longrightarrow 5)$</u> As in $(4\longrightarrow 5)$, $\{i_{0\alpha}(\kappa) \mid \alpha>0\}$ is a c.u. set ρf indiscernibles for L(a). The well-foundedness of $Ult_\alpha(L(a),\mathcal{K})$ for each $\alpha>0$

follows more simply this time from the fact that K is closed under arbitrary countable intersections.

(5 ⟶ 1) If $H = \{\gamma_\eta \mid \eta < \lambda\}$ is a c.u. set of indiscernibles for $\mathcal{M}_\lambda(a)$, and $\eta = \bigcup \eta < \lambda$, extend the order-embedding $j: \gamma_\alpha \longmapsto \gamma_\alpha$ for $\alpha < \eta$, $j: \gamma_{\eta+n} \longmapsto \gamma_{\eta+n+1}$ for $n < \omega$, $j: \gamma_\beta \longmapsto \gamma_\beta$ for $\beta \geq \eta + \omega$, to an elementary embedding of $\mathcal{M}_\lambda(a)$ into a proper subset of itself.

4.20 Remark. A proof of 3.5 due to Silver which does not require knowledge of iterated ultrapowers is given in [De 2, pp. 199-204]. Like the proof of (4 ⟶ 5) it uses classes of sets fixed by certain embeddings to define more embeddings, which in turn generate the indiscernibles.

5. Measurability

5.1. Recall that *κ is measurable* iff there is a "V-ultrafilter" on κ, i. e., a κ-complete nonprincipal ultrafilter $\mathcal{U} \subseteq S(\kappa)$; call such an ultrafilter simply a measure on κ, and say that $X \subseteq \kappa$ has *measure 1* (*measure 0*) if $X \in \mathcal{U}$ ($X \notin \mathcal{U}$). Clearly if $\mathcal{M}$ is a transitive model of ZFC and $\mathcal{U}$ is an $\mathcal{M}$-ultrafilter on $\kappa \in M$ with $\mathcal{U} \in M$, (κ is measurable $)^{\mathcal{M}}$. If ($\mathcal{U}$ is normal and $V = L(\mathcal{U}))^{\mathcal{M}}$ for such an $\mathcal{M}$ and $On \subseteq M$, call $\mathcal{M}$ a *κ-model*, with *constructing (normal) ultrafilter/measure* $\mathcal{U}$. Kunen's main application of the techniques of section 4 was to show that the constructing ultrafilter is definable in $\mathcal{M}$ as the unique normal ultrafilter on the only measurable cardinal, and thus that there is only one κ-model. Since if $\mathcal{V}$ is any (normal) measure on κ and $\mathcal{U} = \mathcal{V} \cap L(\mathcal{V})$, ($\mathcal{U}$ is a (normal) measure on κ, and $V = L(\mathcal{U})$ $)^{L(\mathcal{V})}$, a by-product is Con(ZFC + there is a measurable cardinal) $\longrightarrow$ Con(ZFC + there is a measurable cardinal + V = HOD) (if the reader finds the first part of the sentence puzzling, the definitional material and lemmas on $L(\Lambda)$ on pp. 206-209 of [Ae 2] is a good and inexpensive reference).

First some earlier results, several of them subsumed in what we have already done in section 4.

5.2 Theorem. *(1) If κ is the least cardinal such that there exists an ω_1-complete ultrafilter $\mathcal{U} \subseteq S(\kappa)$, κ is measurable.*
(2) If κ is measurable, a normal measure on κ exists.
(3) κ is measurable iff there is an elementary embedding i from V in a transitive $M \subseteq V$ such that $i \upharpoonright \kappa$ is the identity, $i(\kappa) > \kappa$ and ${}^{\kappa}M \subseteq M$.
(4) If $\mathcal{U}$ is a normal measure on κ, $\mathcal{U}$ properly extends the closed unbounded filter on κ.
(5) (Rowbottom) Let $\mathcal{U}$ be a normal measure on κ. Then for each $n < \omega$ (a) $X \in \mathcal{U}_n$ iff $[Y]^n \subseteq X$ for some $Y \in \mathcal{U}$. (b) If $f(\bar{x}) < x_0$ for each $\bar{x} \in X \in \mathcal{U}_n$, there is a $Y \subseteq X$ such that $Y \in \mathcal{U}_n$ and $f \upharpoonright Y$ is constant.
(6) A measurable cardinal is Π^2_1-indescribable and ineffable.
(7) Say that κ is ineffably Ramsey iff each regressive $f: [\kappa]^{<\omega} \longrightarrow \kappa$ has a stationary homogeneous set. If κ is ineffably Ramsey, $\{\mu < \kappa \mid \mu$ is Ramsey$\}$ is stationary in κ.
If $\mathcal{U}$ is a normal measure on κ, κ is ineffably Ramsey, by (5), and $\{\lambda < \kappa \mid \lambda$ is ineffably Ramsey$\} \in \mathcal{U}$ by (6).
(8) If $\mathcal{M}$ is a transitive model of ZFC, $\mathcal{U} \in M$ is an M-ultrafilter on $\kappa \in M$, and $\alpha < \beta \in M$, then, in the notation of section 4,
(a) $\mathcal{U}^{(\alpha)} = i_{0\alpha}(\mathcal{U})$; (b) (Scott) $\mathcal{U}^{(\alpha)} \notin \mathcal{M}_{\alpha+1}$; (c) $\mathcal{M}_\beta \subsetneq \mathcal{M}_\alpha$.
(9) (Scott) If κ is measurable, $V \neq L$.

(10) If $\mathcal{M}$ is a κ-model, (κ is the only measurable cardinal $)^{\mathcal{M}}$
(11) If $\mathcal{U}$ is a normal $\mathcal{M}$-ultrafilter on κ, $(\{\lambda < \kappa \mid 2^{\lambda} = \lambda^{+}\})^{\mathcal{M}} \in \mathcal{U}$ $\rightarrow$ $(2^{\kappa} = \kappa^{+})^{\mathcal{M}}$.

Proof of (1). If $\mathcal{U}$ is not κ-complete, then for some λ with $\omega_1 \leq \lambda < \kappa$ and $\langle B_\beta \mid \beta < \lambda \rangle$, each $B_\beta \in \mathcal{U}$, but $\bigcup_{\beta<\lambda} B_\beta = \kappa$. If λ is the least cardinal with this property, $\mathcal{U}' \subseteq S(\lambda)$ defined by $X \in \mathcal{U}'$ iff $\bigcup_{\beta \in X} B_\beta \in \mathcal{U}$, is an ω_1-complete ultrafilter on λ, for if $\bigcup_{\beta \in X_n} B_\beta \notin \mathcal{U}$ for each $n<\omega$ and $X = \bigcup_{n<\omega} X_n$, $\bigcup_{\beta \in X} B_\beta = \bigcup_{n<\omega} \bigcup_{\beta \in X_n} B_\beta \notin \mathcal{U}$.
Proof of (2). Immediate from 4.18(2) or (3), since $\mathcal{U}$ is ω_1-complete.
Proof of (3). By 4.14(2) and (4), and 4.18(3).
Proof of (4). Any κ-complete normal filter $\mathcal{F}$ extends the closed unbounded filter, for any closed unbounded $C \supseteq \{\eta \mid \eta = \bigcup \eta\} \cap \{\eta \mid \eta >$ the αth element of C for all $\alpha < \eta\}$, and it is an exercise to see that each of these sets must be in $\mathcal{F}$. Alternately, for $\mathcal{U}$, argue that $\kappa \in i_{01}(C)$ in $\mathcal{M}_1$, since $i_{01}(C) \cap \kappa = C$ is unbounded in κ, so $C \in \mathcal{U}$.
Proof of (5). We assume (a) and (b) hold for n, and prove them for n + 1. (b) for 1 is normality, and (a) for 1 a tautology.
Suppose $X \in \mathcal{U}_{n+1}$ and $f: X \longrightarrow \kappa$ is regressive, i. e., $f(\bar{x}) < x_0$ for all $\bar{x} \in X$. Then for each α, the induction hypothesis gives a $\gamma_\alpha < \alpha$ and a $Y_\alpha \in \mathcal{U}_n$ such that $f(\alpha, \bar{y}) = \gamma_\alpha$ for all $\bar{y}$ in Y_α.
Let $g(\alpha) = \gamma_\alpha < \alpha$, and $Z \in \mathcal{U}$ be such that $g''Z$ is $\{\gamma\}$. Then for all $\bar{x} \in W = \{\langle \alpha, \bar{y} \rangle \mid \alpha \in Z$ and $\alpha < \bar{y} \in [Y_\alpha]^n\} \in \mathcal{U}_{n+1}$, $f(\bar{x}) = \gamma$. We now find an $H \in \mathcal{U}$ such that $[H]^{n+1} \subseteq X$. $\{\alpha \mid X_{(\alpha)} \in \mathcal{U}_n\} = K \in \mathcal{U}$. Also, the generalized diagonal intersection $\triangle_{\alpha<\kappa} X_{(\alpha)} = \{\bar{x} \mid \bar{x} \in X_{(\alpha)}$ for all $\alpha < x_0\} \in \mathcal{U}_n$, for if not, by (b) for n $\exists \gamma \forall^* \bar{x}\ \bar{x} \notin X_{(\gamma)}$, which is impossible. By the induction bypothesis, let $I \in \mathcal{U}$ be such that $[I]^n \subseteq \triangle_{\alpha<\kappa} X_{(\alpha)}$, and $H = K \cap I$. Then $[H]^{n+1} \subseteq X$, since if $(\alpha, \bar{x}) \in [H]^{n+1}$, $\alpha < x_0$, $\alpha \in K$ and $\bar{x} \in X_{(\alpha)}$.
Proof of (6). Suppose φ is Π^1_m for some $m < \omega$, and ψ is $\exists \mathcal{X} \varphi$, where $\mathcal{X}$ is third-order. If $\mathcal{U}$ is a normal measure on κ and $\langle \lambda, \in, R \cap \lambda \rangle \models \psi$ for all $\lambda \in X \in \mathcal{U}$, $(\langle \kappa, \in, R \rangle \models \psi)^{\mathcal{M}_1}$, but then $\langle \kappa, \in, R \rangle \models \psi$ by 4.14(4a) or (4b), and the fact that the only third-order quantifier is existential.
Notice that this result, rather than the weaker one of Π^1_m-indescribability for all $m < \omega$ in $\mathcal{M}$, would have been obtainable in 4.15(1) from the assurance that $(S(S(\kappa)))^{\mathcal{M}_1} \subseteq \mathcal{M}$. By 4.15(1), κ is ineffable, and since ineffability is Π^1_3, $\{\lambda < \kappa \mid \lambda$ is ineffable $\} \in \mathcal{U}$.
Proof of (7). κ is ineffably Ramsey: if $f: [\kappa]^{<\omega} \longrightarrow \kappa$ is regressive, $\mathcal{U}$ is a normal measure on κ and $X_n \in \mathcal{U}$ is homogeneous for $f \restriction [\kappa]^n$, $X = \bigcap_{n<\omega} X_n$ is homogeneous for each $f \restriction [\kappa]^n$. Since an ineffably Ramsey κ is ineffable, so Π^1_2 indescribable, and " κ is

Ramsey" is Π^1_2, the second sentence follows. The third follows similarly by (6), since " κ is ineffably Ramsey" is Π^1_3.

Proof of (8). As usual, we can assume $\alpha = 0$.

(a) $[f] \in \mathcal{U}^{(\alpha)}$ iff $\forall^* s\ f(s) \in \mathcal{U}$ iff $[f] \in i_{0\alpha}(\mathcal{U})$.

(b) If $\mathcal{U}$ were in $\mathcal{M}_1$, $(i_{01}(\kappa) \leq 2^{\kappa})^{\mathcal{M}_1}$, since there is an $f : {}^{\kappa}\kappa \xrightarrow{onto} i_{01}(\kappa)$ definable from $\mathcal{U}$ and $S(\kappa) \cap M$, and this would contradict $(i_{01}(\kappa)$ is inaccessible $)^{\mathcal{M}_1}$.

(c) follows from (b), since $\mathcal{M}_1$ is defined from elements of $\mathcal{M}$.

Proof of (9). From (8(b)), if $\mathcal{U}$ is a measure on κ and $\mathcal{M}_1$ is the transitive collapse of $\mathrm{Ult}_1(V, \mathcal{U})$, $\mathcal{U} \notin M_1 \supseteq L$, since $\mathcal{M}_1$ is a transitive inner model of ZFC with $\mathrm{On} \subseteq M_1$.

Proof of (10). (There exists a Ramsey cardinal $> \kappa)^{\mathcal{M}}$ would imply $(\mathcal{U}^{\#}$ exists $)^{\mathcal{M}}$, incompatible with $(V = L(\mathcal{U}))^{\mathcal{M}}$. Or we can argue directly from the premises given, as follows.

Suppose κ, μ are both measurable, $\mathcal{U}$ is a normal measure on κ, $\mathcal{V}$ is any measure on μ, $i = i_{01}^{(\mathcal{V})} : V \longrightarrow \mathcal{M}_1 \cong \mathrm{Ult}_1(V, \mathcal{V})$, and $V = L(\mathcal{U})$.

If $\kappa < \mu$, $\mathcal{U} \in R(\mu)$, so $i(\mathcal{U}) = \mathcal{U}$, and $M = L(\mathcal{U}) = M_1$, which contradicts (8c).

If $\mu < \kappa$, we show $i(\mathcal{U})$ is $\mathcal{U} \cap M$, which again leads to the contradiction $M_1 = L(i(\mathcal{U})) = L(\mathcal{U}) = M$ (note that $L(A) = L(B)$ whenever $A \cap L(B) = B$, or conversely; cf. [De2, pp. 206-7]).

First notice that if $I = \{\lambda \mid \kappa < \lambda \leq \mu$ and λ is strongly inaccessible$\}^{\mathcal{M}}$, $i \upharpoonright I$ is the identity by 4.14(7), and $i(\mathcal{U})$ is on the right cardinal. $i(\mathcal{U}) \supseteq \mathcal{U} \cap M$: if $X \in \mathcal{U} \cap M$, $X \cap I \in i(\mathcal{U})$.

$i(\mathcal{U}) \subseteq \mathcal{U}$: if $X = [f] \in i(\mathcal{U})$, $(\bigcap \mathrm{ran}\, f) \cap I \subseteq X$, and $\bigcap \mathrm{ran}\, f \in \mathcal{U}$.

Proof of (11). If $\mathcal{M}_1 = \mathrm{Ult}_1(M, \mathcal{U})$, $\{\lambda < \kappa \mid (2^{\lambda} = \lambda^+)^{\mathcal{M}} \} \in \mathcal{U}$ iff $(2^{\kappa} = \kappa^+)^{\mathcal{M}_1} \rightarrow (2^{\kappa} = \kappa^+)^{\mathcal{M}}$. The last implication follows from the fact that $({}^{\kappa}\mathrm{On})^{\mathcal{M}_1} = ({}^{\kappa}\mathrm{On})^{\mathcal{M}}$, which implies $S(\kappa) \cap M_1 = S(\kappa) \cap M$ and $\kappa^{+\mathcal{M}_1} = \kappa^{+\mathcal{M}}$.

We shift temporarily to a "weak" variant of measurability.

5.3. Let κ be any uncountable cardinal, $\mathcal{F} \subseteq S(\kappa)$ a filter on κ, and $\omega_1 \leq \lambda$.

(1) If $\lambda \leq 2^{\kappa}$, $\mathcal{F}$ is λ-saturated iff $\mathcal{F}$ is κ-complete and $\mathbb{B} = S(\kappa)/\mathcal{F}$ has the λ-chain condition. Equivalently, for $\lambda \leq \kappa$, there is no pairwise disjoint λ-sequence of sets $A_\alpha \subseteq \kappa$ such that $A_\alpha \notin \mathcal{I}$ for each $\alpha < \lambda$ (the chain condition is immediately equivalent to the same statement with "pairwise disjoint" replaced by "pairwise almost disjoint", i. e., $A_\alpha \cap A_\beta \in \mathcal{I}$ for each $\alpha, \beta < \lambda$. The stronger statement follows by κ-completeness when $\lambda \leq \kappa$).

Write $s(\kappa, \lambda, \mathcal{F})$ as an abbreviation for " $\mathcal{F}$ is a μ-saturated filter on κ for some $\mu \leq \lambda$", s*(κ, λ, $\mathcal{F}$) for [s(κ, λ, $\mathcal{F}$) and not s(κ, λ', $\mathcal{F}'$) for any $\lambda' < \lambda$ and $\mathcal{F}' \subseteq S(\kappa)$. s($\kappa$, λ)(s*(κ, λ)) means $\exists \mathcal{F}(s(\kappa, \lambda, \mathcal{F}))$ ($\exists \mathcal{F}(s^*(\kappa, \lambda, \mathcal{F}))$).

Note that κ is measurable iff s(κ, ω). Little is known about whether s*(κ, λ) can hold for $\kappa^+ < \lambda \leq 2^{\kappa}$.

(3) κ is real-valued measurable iff there is a nonnegative κ-additive real-valued measure μ: $S(\kappa) \longrightarrow [0, 1]$ such that $\mu(\kappa) = 1$, $\mu(\bigcap_{\alpha<\lambda} A_{\alpha}) = \inf \{ \mu(A_{\alpha}) \mid \alpha < \lambda \}$ in $\mathbb{R}$, and $\mu(\{\gamma\}) = 0$ for each $\gamma, \lambda < \kappa$.

Note that κ is measurable iff κ is real-valued measurable for some μ, and there is an $A \subseteq \kappa$ such that $\mu(A) > 0$ and $\mu \restriction (S(A))$ has range $\{0, \mu(A)\}$. Call such an A an atom of μ.

Note further that if κ is real-valued measurable, s*(κ, ω_1, $\mathcal{F}$), where $\mathcal{F}$ is the filter of $A \subseteq \kappa$ such that $\mu(A) = 1$. For if not, there would be ω_1 sets A_{α} and an $n < \omega$ such that $\mu(A_{\alpha} \cap A_{\beta}) = 0$ for all $\alpha \neq \beta < \omega_1$ but each $\mu(A_{\alpha}) \geq 1/n$, violating $\mu(\kappa) = 1$ (In other words, every measure algebra has the ω_1-chain condition).

(4) An ultrafilter $\mathcal{U} \subseteq S(\kappa)$ is (λ, κ)-regular iff there is a sequence $\langle X_{\alpha} \mid \alpha < \kappa \rangle$ of elements of $\mathcal{U}$ such that whenever $\tau : \lambda \longrightarrow \kappa$, $\bigcap_{\beta<\lambda} X_{\tau(\beta)} = 0$, ($\lambda$, κ)-nonregular iff there is no such sequence. (λ, κ)-nonregularity for $\lambda < \kappa$ is a weak form of λ^+-completeness.

5.4 Theorem. (1) (Solovay, Jensen, Fremlin) If $\lambda \leq \kappa^+$ and there is a λ-saturated filter $\mathcal{F}$ on κ, there is one which is normal.

(2) (Ulam) If there is a λ-saturated filter on κ and $2^{\lambda} < \kappa$, there is an $A \subseteq \kappa$ such that $\mathcal{F}_A = \{ B \mid B \subseteq A \text{ and } B \in \mathcal{F} \}$ is a measure on A (call such an A by analogy with the note of (3) an atom of $\mathcal{F}$) and therefore κ is measurable.

(3) (Solovay) A λ-saturated filter on κ for $\lambda < \kappa$ is λ-Rowbottom (recall 2.3).

(4) (Solovay, Jensen, Fremlin) s(κ, κ) implies κ is α-mahlo for all $\alpha < \kappa$ (and more).

(5) (Silver) If $\mathcal{F}$ is a κ-saturated filter on κ and $\mathcal{U}$ any ultrafilter extending $\mathcal{F}$, $\mathcal{U}$ is (λ, κ)-nonregular for each $\lambda < \kappa$.

Proof of (1). Let $\mathcal{I}$ be the ideal dual to $\mathcal{F}$, $\mathcal{B}$ the (complete, cf. [So2]) boolean algebra $S(\kappa)/\mathcal{F}$, and $[X] = \{ Y \subseteq \kappa \mid$ both X - Y and Y - X are in $\mathcal{I} \}$ the element of $\mathcal{B}$ determined by $X \subseteq \kappa$. Extend $(\_)^{\vee}$ to classes A by writing $A^{\vee}(u) = [\![u \in A]\!]^{\mathcal{B}} = \bigvee_{a \in A} [\![u = \check{a}]\!]^{\mathcal{B}}$. Define $\mathcal{U} : \mathrm{dom}(S(\kappa)^{\vee}) \longrightarrow \mathcal{B}$ by $\mathcal{U}(\check{X}) = [\![\check{X} \in \mathcal{U}]\!]^{\mathcal{B}} = [X]$. Let "$\mathcal{U}$ is a weak M-ultrafilter" mean $\mathcal{U}$ satisfies (1) - (4) of 4.1; then $\mathrm{Ult}_1(M, \mathcal{U})$ makes sense, but not necessarily $\mathrm{Ult}_2(M, \mathcal{U})$. Then $[\![\mathcal{U}$ is a weak $\check{V}$-ultrafilter$]\!]^{\mathcal{B}} = \mathbb{1}$: for example, if $\langle X_{\alpha} \mid \alpha < \lambda \rangle$ is given,

$[\![\bigcap_{\alpha<\lambda} X_\alpha \in \mathcal{U}]\!] = [\![\bigcap X_\alpha]\!] = \bigwedge_\alpha [\![X_\alpha]\!] = [\![\text{for all } \alpha<\lambda^\vee\ X_\alpha \in \mathcal{U}]\!]$.
Define further $\underline{E, \approx}$: $(V^\kappa \times V^\kappa)^\vee \longrightarrow \mathcal{B}$ by $[\![f^\vee \{{E \atop \approx}\} g^\vee]\!] =$
$[\![\{ \alpha \mid f(a) \{{\in \atop =}\} g(\alpha) \}]\!]$, and $\underline{\hat{x}}$ for $x \in V$ = (the constant function: $\kappa \longrightarrow x)^\vee$. We show

(a) $\underline{\text{If } [\![x \in (V^\kappa)^\vee]\!] = 1\!\!1, 1\!\!1 = [\![x \approx \check{g}]\!] \text{ for some } g \in V^\kappa}$. By λ-cc there are $\tilde{\lambda} < \lambda$, $\{ f_\alpha \mid \alpha < \tilde{\lambda} \}$ and disjoint $[\![X_\alpha]\!] = [\![x = f^\vee_\alpha]\!]$ such that $\bigvee_{\alpha<\tilde{\lambda}} [\![X_\alpha]\!] = 1\!\!1$, and we can assume $\bigcup_\alpha X_\alpha = \kappa$. If g is such that $g \restriction X_\alpha = f_\alpha \restriction X_\alpha$, $[\![x \approx \check{g}]\!] \geqslant [\![X_\alpha]\!]$ for each α, so $[\![x \approx \check{g}]\!] = 1\!\!1$.

(b) $\underline{[\![\langle (V^\kappa)^\vee, E, \approx \rangle = \mathfrak{A} \text{ is well-founded}]\!] = 1\!\!1}$. For if there are x_n in $V^\mathcal{B}$ such that $[\![x_{n+1} E x_n]\!] \geqslant [\![X]\!] > 0$ for all n, there are f_n in V^κ by (a) such that $\{ \alpha \mid f_{n+1}(\alpha) \in f_n(\alpha) \} \supseteq X$, which is impossible. Therefore there are $\underline{\mathcal{B}\text{-valued classes } k, \mathcal{M}}$ such that $[\![k$ collapses $\mathfrak{A}$ onto $\mathcal{M}]\!] = 1\!\!1$. If $x \in V$, let $\underline{i(x)}$ = that element of $V^\mathcal{B}$ such that $[\![k(\hat{x}) = y]\!] = 1\!\!1$. Then $[\![(\varphi(i(x_0), \dots, i(x_n))]\!]^\mathcal{B} = 1\!\!1$ if $\varphi(x_0, \dots, x_n)$, 0 if not. We let the reader verify $[\![i(\alpha) = \check{\alpha}]\!] = 1\!\!1$ for $\alpha < \kappa$; note $[\![i(\kappa) > \check{\kappa}]\!] = 1\!\!1$, for $[\![\check{\alpha} < k(\check{id}) < i(\kappa)]\!] = 1\!\!1$ for all $\alpha < \kappa$.

By (a), there is an $\underline{f: \kappa \longrightarrow V}$ such that $\underline{[\![k(f) = \check{\kappa}]\!] = 1\!\!1}$ (f is the $\underline{\text{"incompressible function"}}$ of [So 2]). Define $\mathcal{F}^*$ by $X \in \mathcal{F}^*$ iff $[\![k(f^\vee) \in i(X)]\!] = 1\!\!1$ iff $\{ \alpha \mid f(\alpha) \in X \} \in \mathcal{F}$ iff $f^{-1}(X) \in \mathcal{F}$.
Normality: if $\{ \alpha \mid g(\alpha) < \alpha \} \notin \mathcal{I}^*$, $\{ \beta \mid (g \cdot f)(\beta) < g(\beta) \} \notin \mathcal{I}$, so $[\![k(\widecheck{g \cdot f}) < \check{\kappa}]\!] > 0$, so there is a $\gamma < \kappa$ such that $[\![k(\widecheck{g \circ f}) = \check{\gamma}]\!] > 0$, so $\{ \beta \mid (g \circ f)(\beta) = \gamma \} \notin \mathcal{I}$, so $\{ \alpha \mid g(\alpha) = \gamma \} \notin \mathcal{I}^*$.
λ-saturation is immediate. A counterexample with respect to $\mathcal{F}^*$ would translate back to one for f, since if $X \notin \mathcal{I}^*$, $f^{-1}(X) \notin \mathcal{I}$.

$\underline{\text{Proof of (2)}}$. We assume there is no atom for $\mathcal{F}$, and show $2^\lambda \geqslant \kappa$. Construct a tree $T = \langle \{ A_s \mid s \in 2^\alpha \text{ for some } \alpha < \lambda \}, \supseteq \rangle$ of subsets A of κ by induction on $\alpha = \mathit{lh}\, s$ by $A_0 = \kappa$, and $A_{\langle s,0\rangle}, A_{\langle s,1\rangle}$ for A_s known, are $\notin \mathcal{I}$, where $\mathcal{I}$ is again the ideal dual to $\mathcal{F}$, $A_{\langle s,0\rangle} \cap A_{\langle s,1\rangle} = 0$, and $A_{\langle s,0\rangle} \cup A_{\langle s,1\rangle} = A_s$ (possible since there are no atoms), and A_s for $\mathit{lh}\, s = \alpha = \bigcup \alpha = \bigcap_{\beta<\alpha} A_{s \restriction \beta}$ if this is $\notin \mathcal{I}$. T has no branch of length λ, by λ-saturation. Let $S = \{ s \mid \mathit{lh}\, s = \alpha \text{ and } A_s \in \mathcal{I}, \text{ but } A_{s \restriction \beta} \notin \mathcal{I} \text{ for } \beta < \alpha \}$.
Assertion: $\kappa = \bigcup_{s \in S} A_s$, a contradiction unless $\kappa \leqslant \overline{\overline{(S)}} \leqslant 2^{\underline{\lambda}}$.
The assertion follows from the fact that, given $\alpha \in \kappa$, we can choose s inductively so that $\alpha \in A_s$.

$\underline{\text{Proof of (3)}}$. We show $\kappa \longrightarrow [\kappa]^{<\omega}_{\mu, <\lambda}$ for each μ. Let $f: [\kappa]^{<\omega} \longrightarrow \mu$, and $f_n = f \restriction [\kappa]^n$. Defining $\mathcal{F}_n$ from $\mathcal{F}$ as $\mathcal{U}_n$ was from $\mathcal{U}$ in section 4, it is not hard to prove by induction on n that each $\mathcal{F}_n$ is a λ-saturated filter on κ^n (this will not necessarily work if $\lambda = \kappa$), and also the following variant of 5.2(5) for a normal λ-saturated filter $\mathcal{F}$ (retrace the proof given there):

(a) $X \in \mathcal{F}_n$ iff for some $Y \in \mathcal{F}$ $[Y]^n \subseteq X$; (b) if $f(\bar{x}) < x_0$ for all $\bar{x} \in X$ not in the ideal $\mathcal{I}_n$ dual to $\mathcal{F}_n$, then there is a $Y \subseteq X$ with $Y \notin \mathcal{I}_n$ such that $f \restriction Y$ is constant.
Then argue as follows. Let $K_n = \{\beta < \mu \mid X_{n\beta} = f^{-1}(\{\beta\}) \notin \mathcal{I}_n\}$. $(K_n)^= < \lambda$ and $X_n = \bigcup\{X_{n\beta} \mid \beta \in K_n\}$ is in $\mathcal{F}_n$, by λ-saturation of $\mathcal{F}_n$. If $H_n \in \mathcal{F}$ is such that $[H_n]^n \subseteq X_n$ and $H = \bigcap_{n<\omega} H_n$, $f''[H] \subseteq \bigcup_{n<\omega} K_n$.

Proof of (4). Return to the looking-glass $\mathcal{B}$-valued "ultrafilter" $\mathcal{U}$ of (1), where we now assume that $\mathcal{F}$ is normal, so that the $f: \kappa \longrightarrow \kappa$ such that $[\![k(f^\vee) = \check{\kappa}]\!] = \mathbb{1}$ is the identity function id: $\kappa \longrightarrow \kappa$. Then for $X \subseteq \kappa$ $[\![\check{X} \in \mathcal{U}]\!] = [X] = [\![k(\check{id}) \in i(X)]\!] = [\![\check{\kappa} \in i(X)]\!]$, so $X \in \mathcal{F}$ iff $[\![\check{\kappa} \in i(X)]\!] = \mathbb{1}$. We continue (a) and (b) of (1) with (c) $[\![$ M contains all subsets of $\check{\kappa}]\!]^{\mathcal{B}} = \mathbb{1}$ (cf. 4.14(4)).
If $[\![Y \subseteq \check{\kappa}]\!] = \mathbb{1}$ and $[\![\check{\alpha} \in Y]\!] = [X_\alpha]$, define $f: \kappa \longrightarrow S(\kappa)$ by $f(\eta) = \{\alpha \mid \eta \in X_\alpha\}$. Then for each $\alpha < \kappa$ $[\![\check{\alpha} \in k(\check{f})]\!] = [\![i(\alpha) \in k(\check{f})]\!] = [\{\eta \mid \alpha \in f(\eta)\}] = [X_\alpha] = [\![\check{\alpha} \in Y]\!]$, so $\mathbb{1} = [\![Y = k(\check{f}) \cap \check{\kappa}]\!] \leq [\![Y \in M]\!]$.
(d) κ is inaccessible. Since κ is regular and $\mathcal{B}$ has κ-cc, $\mathbb{1} = [\![\check{\kappa}$ is regular $]\!] \leq [\![(\check{\kappa}$ is regular $)^{\mathcal{M}}]\!]$. If κ were λ^+, $[\![(i(\kappa)$ is the successor of $\lambda^\vee)^{\mathcal{M}}]\!]$ would be $\mathbb{1}$, but this is impossible, since $[\![(i(\lambda) = \lambda^\vee$, $\check{\kappa}$ is a regular cardinal and $\check{\kappa} < i(\check{\kappa}))^{\mathcal{M}}]\!] = \mathbb{1}$.
(e) κ is α-Mahlo, in fact $\{\lambda < \kappa \mid \lambda$ is α-Mahlo$\} \in \mathcal{F}$ (equivalently, $[\![(\check{\kappa}$ is $\check{\alpha}$-Mahlo $)^{\mathcal{M}}]\!] = \mathbb{1}$) for each $\alpha < \kappa$. For $\alpha = 0$, $[\![(\check{\kappa}$ is regular$)^{\mathcal{M}}]\!] \geq [\![\check{\kappa}$ is regular $]\!] = \mathbb{1}$ by κ-cc. Now assume the result for α, and suppose $\{\lambda < \kappa \mid$ there is a closed unbounded $K_\lambda \subseteq \lambda$ containing no α-Mahlo $\mu\} = B \notin \mathcal{I}$. For $\delta \leq \alpha < \kappa$ let $D_{\alpha\delta} = \{\lambda > \alpha \mid$ the αth element (in the increasing enumeration) of K_λ is $\delta\}$, $E_\alpha = \{\delta \mid D_{\alpha\delta} \notin \mathcal{I}\}$, $H_\alpha = \bigcup_{\delta \in E_\alpha} D_{\alpha\delta}$, and $Y_\alpha = \bigcup E_\alpha$. By κ-saturation each $(E_\alpha)^= < \kappa$, $Y_\alpha < \kappa$ and $H_\alpha \in \mathcal{F}$, and $\tilde{B} = B \cap \{\lambda \mid \alpha < \lambda \longrightarrow Y_\alpha < \lambda\} \cap \Delta_{\alpha<\kappa} H_\alpha \notin \mathcal{I}$. Then for each λ, μ in B the αth element of K_μ for $\alpha < \lambda$ is in E_α, so $\leq Y_\alpha < \lambda$, so $K_\mu \cap \lambda$ has type λ and $\lambda \in K_\mu$, a contradiction. Therefore B is really in $\mathcal{I}$, and $[\![(\check{\kappa}$ is $(\alpha+1)^\vee$-Mahlo $)^{\mathcal{M}}]\!] = 1$. The induction at limit ordinals follows from κ-completeness of $\mathcal{F}$.

Proof of (5). Let $\langle X_\alpha \mid \alpha < \kappa \rangle$ be any sequence of elements of $\mathcal{U}$, and define a κ-complete nonprincipal filter $\mathcal{G}$ on κ by setting $Y \in \mathcal{G}$ iff $\{\alpha < \kappa \mid X_\alpha - Y \in \mathcal{I}\}$ has complement of cardinality $< \kappa$.
$\mathcal{G}$ is κ-complete, for if $Y_\beta \in \mathcal{G}$ for $\beta < \lambda < \kappa$, and $Z_\beta = \{\alpha < \kappa \mid X_\alpha - Y_\beta \notin \mathcal{I}\}$, $Z = \bigcup_{\beta<\lambda} Z_\beta$ had power $< \kappa$, and whenever $\alpha \in W = \kappa - Z$ and $\beta < \lambda$ $X_\alpha - Y_\beta \in \mathcal{I}$, so $X_\alpha - (\bigcap_{\beta<\lambda} Y_\beta) = \bigcup_{\beta<\lambda}(X_\alpha - Y_\beta) \in \mathcal{I}$.
Note that if $S = \{\alpha_\iota \mid \iota < \sigma\} \subseteq \kappa$ is maximal such that

$(\bigcup_{\iota\leq\tau} X_{\alpha_\tau})-(\bigcup_{\iota<\tau} X_{\alpha_\iota}) \notin \mathcal{I}$ for all $\tau < \sigma$, $\sigma < \kappa$ by κ-saturation and $Y_S = \bigcup_{\alpha_\iota \in S} X_{\alpha_\iota}$ is in $\mathcal{G}$. Use this to define a partition $\langle S_\beta \mid \beta < \kappa \rangle$ of κ into disjoint subsets S_β of power $<\kappa$ such that each $Y_\beta = \bigcup_{\alpha \in S_\beta} X_\alpha$ is in $\mathcal{G}$, and each S_β is a maximal subset of $\kappa - \bigcup_{\tilde\beta<\beta} S_{\tilde\beta}$ in the same sense that $S = S_0$ was a maximal subset of κ above. If $\lambda < \kappa$ is given, choose $\gamma \in \bigcap_{\beta<\lambda} Y_\beta$ by κ-completeness of $\mathcal{G}$, and $\alpha(\beta) \in S_\beta$ such that $\gamma \in X_{\alpha(\beta)}$. Then $\gamma \in \bigcap_{\beta<\lambda} X_{\alpha(\beta)}$. and we are done.

5.5 Remark. If $\mathcal{F}$ is κ-complete, it is not hard to see that $\mathcal{F}$ is Jonsson (cf. 2.3) iff $\kappa \longrightarrow [\mathcal{F}]^{<\omega}_\lambda$ for some (regular) $\lambda < \kappa$ (use the argument of 2.12(3)) , and that $\kappa \longrightarrow [\mathcal{F}]^1_\lambda$ implies $\mathcal{F}$ is λ-saturated. Thus a Jonsson (ultra)filter which is not a λ-saturated filter for some $\lambda < \kappa$ (or a measure on κ , if κ is strongly inaccessible, by 5.4(2)) may not be easy to find. Does one exist? (See the discussion of Jonsson and Rowbottom filters in generic extensions in [De 1]).

5.6. If κ is any uncountable cardinal, a filter $\mathcal{F} \subseteq S(\kappa)$ is bounded iff whenever $X \in \mathcal{F}$, and $f: X \longrightarrow \kappa$ is regressive, there exist $\lambda < \kappa$ and $Y \in S(X) \cap \mathcal{F}$ such that $f''Y \subseteq \lambda$. $\mathcal{F}$ is weakly normal iff whenever $X \notin \mathcal{I}$ and $f: X \longrightarrow \kappa$ is regressive, there exist $\lambda < \kappa$ and $Y \in S(X) - \mathcal{I}$ such that $f''Y \subseteq \lambda$. A bounded filter is weakly normal, but not conversely. There is a close connection between bounded filters and nonregular ultrafilters (cf., e.g., [Pr, 1.44]).

5.7. Let σ_0 be a conjunction of finitely many axioms of ZF which implies that the recursive definition of $L_x(y)$, as a function of x and y, is equivalent to a Σ_1-definition. Then (cf. [De 2])

(1) If $\mathcal{M}$ is any transitive model of σ_0, α an ordinal in M, and $A \in M$, $(L_\alpha(A))^{\mathcal{M}} = L_\alpha(A)$.

(2) If $\mathcal{M}$ is any transitive model of σ_0, α an ordinal in M, and $B \cap M \in M$, $(L_\alpha(B \cap M))^{\mathcal{M}} = L_\alpha(B \cap M) = L_\alpha(B)$

(Note that (2) is a consequence of (1)).

Any $L_\lambda(A)$, where λ is uncountable, and $A \in L_\lambda(A)$, is a model of σ_0.

5.8 Theorem. (1) If (a) (Hajnal) $A \subseteq \kappa^+$, or (b) $A \subseteq S(\kappa)$, $(2^\kappa = \kappa^+)^{L(A)}$.

(2) (Silver) If $\lambda < \kappa = \mathrm{cf}\,\kappa$, and $\mathcal{F} \subseteq S(\kappa)$ is a weakly normal λ^+- Rowbottom filter, $(2^\mu = \mu^+$ for all $\mu \geq \lambda)^{L(\mathcal{F})}$. In particular, the GCH holds in every κ- model $\mathcal{M}$.

(3) (Solovay) If $\mathcal{F}$ is a κ-complete, λ-saturated normal filter on $\kappa > \lambda$, (κ is measurable $)^{L(\mathcal{F})}$.

Later we will see that $\mathcal{U} = \mathcal{F} \cap L(\mathcal{F})$ is in fact the unique normal measure on κ in $L(\mathcal{U})$.

Remark. Silver stated the second sentence of (2); his proof establishes the first. Solovay combined this in [So 2] with his observation 5.4(3) to prove (3).

Proof of (1). (a) It suffices to show each $X \subseteq \kappa$ is in $L_{\kappa^+}(A)$, since this has cardinality κ^+. Suppose $X \in L_\lambda(A)$, and $\mathcal{N} \prec \mathcal{M} = \langle L_\lambda(A), \in \rangle$ is such that $X \in N$ and $(\kappa + 1) \subseteq \kappa^+ \cap N \in \kappa^+$. Such an $\mathcal{N}$ is obtained by letting $\mathcal{N}_0 = SH(\{X\} \cup (\kappa+1), \mathcal{M}), \in)$ $\mathcal{N}_{n+1} =$ $SH(N_n \cup \bigcup(N_n \cap \kappa^+), \mathcal{M})$, for $n < \omega$, and $\mathcal{N} = \bigcup_{n<\omega} \mathcal{N}_n$. Collapse $\mathcal{N}$ to $\langle L_\gamma(A \cap \delta), \in \rangle$, where $\delta = \kappa^+ \cap N =$ the collapse $\pi(\kappa^+)$ of κ^+. $\pi \restriction \delta$ is the identity, so $\pi(X) = X$, and therefore $X \in L_\gamma(A \cap \delta) \subseteq L_{\kappa^+}(A)$. $\mathcal{N}$ collapses to $\langle L_\gamma(A \cap \delta), \in \rangle$ because $\mathcal{N} \models (\sigma_0 \wedge V = L(y))[A]$), and so the transitive collapse $\mathcal{N}^*$ not only thinks it is some $L_\alpha(A \cap N^*)$, it really is since it satisfies σ_0. We will handwave these steps in subsequent arguments.

(b) If $X \subseteq \kappa$, we show $X \in L_{\kappa^+}(A)$ as in (a). If $X \in L_\lambda(A)$, let $\mathcal{N} = SH(\{X, A\} \cup (\kappa+1), \langle L_\lambda(A), \in \rangle)$, and collapse $\mathcal{N}$ to $\langle L_\gamma(\pi''A), \in \rangle = \langle L_\gamma(A \cap L_\gamma(A)), \in \rangle = \langle L_\gamma(A), \in \rangle$, since $\pi \restriction (\kappa+1) =$ the identity, and so $\pi(X) = X$ and $\pi(Y) = Y$ for each $Y \in A \cap N$. Since $\overline{\overline{(N)}} = \kappa$, $\overline{\overline{\gamma}} = \kappa$, and once again we are done.

Proof of (2). By (1b), it suffices to prove the result for $\mu < \kappa$. κ is not a successor by 2.12(1), so $\mu^+ < \kappa$. If the result fails, let $\mathcal{F}^*$ be $\mathcal{F} \cap L(\mathcal{F})$, $\prec$ the canonical well-ordering of $L(\mathcal{F}^*) = L(\mathcal{F})$, and $X \in L_\alpha(\mathcal{F}^*) - \bigcup_{\beta<\alpha} L_\beta(\mathcal{F}^*)$ the $\prec$-least subset of μ such that $\mathcal{S} = S(\mu) \cap \{$the $\prec$-predecessors of $X\}$ has cardinality μ^+. Let $\mathcal{O} = SH(\{\mathcal{F}^*\} \cup \{\alpha\} \cup \{X, \mathcal{S}\} \cup \kappa, \langle L_{\kappa^{++}}(\mathcal{F}^*), \in \rangle)$. Since $\overline{\overline{(A)}} = \kappa$, there is an $H \in \mathcal{F}$ such that $\mathcal{N} = \langle N, \in, N \cap \mathcal{S} \rangle \prec \langle A, \in, \mathcal{S} \rangle$, $\overline{\overline{(N)}} = \kappa$, $\overline{\overline{(N \cap \mathcal{S})}} \leq \mu$, and $N = SH(H \cup \{X, \mathcal{S}\} \cup \{\alpha\} \cup \{\mathcal{F}^*\} \cup \mu, \mathcal{O})$. Let π collapse $\mathcal{N}$ to some transitive $\mathcal{M}$ of power κ. $\pi''N \cap \mathcal{S} = N \cap \mathcal{S}$, since $\mu \subseteq N$, but also $\{\alpha < \kappa \mid \pi(\alpha) < \alpha\} \in \mathcal{J}$, the ideal dual to $\mathcal{F}$, for if not, weak normality of $\mathcal{F}$ gives an $H \notin \mathcal{J}$ such that $\pi'' H \subseteq \nu$ for some $\nu < \kappa$, an impossibility since π is injective. So $\pi(\mathcal{F}^*) = \mathcal{F}^* \cap M$; also $M = L_\gamma(\mathcal{F}^*)$ for some γ, since $L_{\kappa^{++}}(\mathcal{F}^*)$ was a model of σ_0, and X is in $L_{\pi(\alpha)}(\mathcal{F}^*) - [\bigcup_{\beta<\alpha} L_\beta(\mathcal{F}^*)]$. Since $\pi(\alpha) \leq \alpha$, $\pi(\alpha) = \alpha$. But then $\mathcal{S} \cap M = \mathcal{S}$, which contradicts $\overline{\overline{(\mathcal{S} \cap N)}} \leq \mu$, and we are done. The last sentence of (1) follows, since $\mathcal{U}$ is 2-saturated.

Proof of (3). This follows from (2) and from 5.4(2): $(2^\lambda = \lambda^+ < \kappa)^{L(\mathcal{F})}$, so $(\kappa$ is measurable$)^{L(\mathcal{F})}$.

The following collects some of the principal results of [Ku 1].

5.9 Theorem (Kunen)

(1) If $\mathcal{M}, \mathcal{N}$ are both κ-models with constructing ultrafilters $\mathcal{U} \in M$ and $\mathcal{V} \in N$, $\mathcal{U} = \mathcal{V}$, so $\mathcal{M} = \mathcal{N}$.

(2) If $\mathcal{U}$ is a normal measure on κ and $V = L(\mathcal{U})$, $\mathcal{U}$ is unique.

(3) If ρ is the least ordinal such that there exists a ρ-model $\mathcal{M}$ with constructing ultrafilter $\mathcal{U} \in M$, every σ-model $\mathcal{N}$ is the transitive collapse $\mathcal{M}_\alpha$ of some $\mathrm{Ult}_\alpha(M, \mathcal{U})$ for $\alpha \geq 0$.

(4) If $\mathcal{U}$ is a normal measure on κ and $\mathcal{V}$ is any measure on κ, $L(\mathcal{U}) = L(\mathcal{V})$.

In (5) through (7), let $\mathcal{M}$ be the unique κ-model, with constructing ultrafilter $\mathcal{U} \in M$.

(5) (Every measure $\mathcal{V}$ is of the form $\{x \subseteq \kappa \mid \alpha \in i_{0\omega}(\kappa)$ for some α $)^{\mathcal{M}}$.

(6) (Every measure $\mathcal{V}$ is equivalent to some $\mathcal{U}_n$ on κ^n, i. e., $i_{01}^{\mathcal{V}}(\kappa) = i_{01}^{\mathcal{U}_n}(\kappa)$ and $i_{01}^{\mathcal{V}} \restriction S(\kappa) = i_{01}^{\mathcal{U}_n} \restriction S(\kappa)$ $)^{\mathcal{M}}$.

(7) (Every Jonsson cardinal is Ramsey $)^{\mathcal{M}}$.

(8) If $\mathcal{F}$ is a normal λ-saturated filter, where $\lambda < \kappa$, and $\mathcal{U} = \mathcal{F} \cap L(\mathcal{F})$, ($\mathcal{U}$ is the normal measure on κ $)^{L(\mathcal{U})}$.

Proof of (1). Let $i_{\alpha\beta} = i^{\mathcal{U}}_{\alpha\beta}$: $\mathrm{Ult}_\alpha(\mathcal{M}, \mathcal{U}) \longrightarrow \mathrm{Ult}_\beta(\mathcal{M}, \mathcal{U})$.

(a) $x \in \mathcal{U}^{(\alpha)}$ iff $\exists \beta < \alpha$ $\{i_{0\gamma}(\kappa) \mid \beta < \gamma < \alpha\} \subseteq x$.

($\longleftarrow$) if $x = i_{\gamma\alpha}(y)$ and $i_{0\gamma}(\kappa) \in x$, $i_{0\gamma}(\kappa) \in i_{0\gamma+1}(y)$, so $y \in \mathcal{U}^{(\gamma)}$, so $x \in i_{\gamma\alpha}(\mathcal{U}^{(\gamma)}) = \mathcal{U}^{(\alpha)}$.

($\longrightarrow$) If $x \in \mathcal{U}^{(\alpha)}$, $x = i_{\beta\alpha}(y)$ for $y \in \mathcal{U}^{(\beta)}$, and $i_{\beta\gamma}(y) \in \mathcal{U}^{(\gamma)}$ for all $\gamma > \beta$, $\gamma < \alpha$, so $i_{0\gamma}(\kappa) \in i_{\beta\gamma+1}(y) \subseteq x$.

(b) If $\lambda = \mathrm{cf}\,\lambda > \kappa$, and $\mathcal{F}$ is the c.u. filter on κ, $i_{0\lambda}(\kappa) = \lambda$ by 4.14(6), and $\mathcal{F} \cap L(\mathcal{F}) = i_{0\lambda}(\mathcal{U})$, since if $x \in i_{0\lambda}(\mathcal{U})$, x has a c.u. subset by (a) and 4.14(3), and if $x \notin i_{0\lambda}(\mathcal{U})$, $\kappa - x$ does, so x is disjoint from a c.u. subset of $\kappa - x$.

(c) If M is a κ-model and N a $\tilde{\kappa}$-model with $\tilde{\kappa} \leqslant \kappa$, $S(\tilde{\kappa}) \cap N = S(\tilde{\kappa}) \cap M$, since each is $S(\tilde{\kappa}) \cap L(\mathcal{F})$ for the c.u. filter $\mathcal{F}$ on some sufficiently large regular λ.

(d) Whenever $\delta \geqslant \kappa^{+\mathcal{M}}$, $\gamma_0 > \kappa$, $\langle \gamma_\alpha \mid \alpha < \delta \rangle$ is an increasing sequence and μ is a cardinal $> \bigcup_{\alpha<\delta} \gamma_\alpha$, every element of $S(\kappa) \cap M = S(\kappa) \cap N$ is in the Skolem hull of $X = \{\gamma_\alpha \mid \alpha < \delta\} \cup \{\alpha \mid \alpha \leqslant \kappa\}$ in $\langle L_\mu(\mathcal{U}), \in \rangle$. For the transitive collapse of the Skolem hull is $\langle L_\beta(\mathcal{U}), \in \rangle$ for some $\beta \geqslant \kappa^{+(\mathcal{M})}$, so it contains all of $S(\kappa) \cap M$ by 5.8(2).

(e) $\mathcal{U} = \mathcal{V}$. If λ is as in (b), and δ, $\langle \gamma_\alpha \mid \alpha < \delta \rangle$ and μ are as in (d), we can assume each γ_α and δ are fixed by $i_{0\lambda}$ by 4.14(7). We show $\mathcal{U} \subseteq \mathcal{V}$. The converse follws symmetrically.

If $x \in \mathcal{U}$, $x = \{\beta < \kappa \mid \langle L_\mu(\mathcal{U}), \in \rangle \models \varphi[\beta, \bar{\eta}, \kappa, \bar{\gamma}_\alpha]\}$, where φ is in the appropriate Skolem language. If y is the set defined in

$L_\mu(\mathcal{V})$ by the same formula and the same parameters, $y \in \mathcal{V}$, since $i_{0\lambda}(y) = i_{0\lambda}(x) = \{\beta < \lambda \mid \langle L_\mu(\mathcal{F}), \in\rangle \models \varphi[\beta, \bar{\eta}, \lambda, \bar{\gamma}_\alpha]$, which is in $\mathcal{F}$ since x was in $\mathcal{U}$. The conclusion now follows from $x = i^{\mathcal{U}}_{0\lambda}(x) \cap \kappa = i^{\mathcal{V}}_{0\lambda}(y) \cap \kappa = y$.

Proof of (2). Immediate from (1). If $\mathcal{V}$ is a normal measure on κ, $\mathcal{V} \cap L(\mathcal{V}) = \mathcal{U}$, so $\mathcal{V} = \mathcal{U}$ since $V = L(\mathcal{U})$.

Proof of (3). If $\sigma = i_{0\alpha}(\rho)$ for some α, $\mathcal{N}$ = (the transitive collapse of) $\mathrm{Ult}_\alpha(\mathcal{M}, \mathcal{U})$ follows from (1). We show there is no α such that a σ-model exists with $i_{0\alpha}(\rho) < \sigma < i_{0\alpha+1}(\rho)$.

By 4.13, it suffices to show this for $\alpha = 0$. If $\mathcal{N}$ is a σ-model with $\rho < \sigma < i_{01}(\rho)$ and constructing ultrafilter $\mathcal{V}$, let $\mathcal{F}, \langle \gamma_\alpha \mid \alpha < \delta\rangle$, and μ be as in the proof of (1), $\mu > \lambda = \mathrm{cf}\,\lambda > \sigma^+$, and $[f]_1 = \sigma$ in $\mathrm{Ult}_1(M, \mathcal{U})$. By (d) in the proof of (1), $i_{0\lambda}(f)$ is in the Skolem hull of $\{\gamma_\alpha \mid \alpha < \delta\} \cup \{\alpha \mid \alpha \leqslant \kappa\} \cup \{\lambda\}$ in $\langle L_\mu(\mathcal{F}), \in\rangle$. If $j(\alpha) = \alpha + 1$ for $\alpha < \lambda$, $j^{\mathcal{V}}_*$ maps $L(\mathcal{F})$ into $L(\mathcal{F})$ and fixes ρ, μ and all the parameters defining $i_{0\lambda}(f)$, so fixes $i_{0\lambda}(f)(\rho) = \sigma$ (cf. page 4.10, line 22). which contradicts the fact that by 4.14(2), $j^{\mathcal{V}}_*(\sigma) = i^{\mathcal{V}}_{01}(\sigma) > \sigma$.

Proof of (4). By (1), $L(\mathcal{U}) \subseteq L(\mathcal{V})$, since $L(\mathcal{V})$ must contain a κ-model, and it suffices to show that $\mathcal{V} \cap L(\mathcal{U}) \in L(\mathcal{U})$.

Let $\sigma = i^{\mathcal{V}}_{01}(\kappa)$, where $i^{\mathcal{V}}_{01}: V \longrightarrow \mathrm{Ult}(V, \mathcal{V})$, and let $j = i^{\mathcal{V}}_{01} \restriction L(\mathcal{U})$: $L(\mathcal{U}) \longrightarrow$ the σ-model $\mathcal{N}$. Since $\mathcal{N} = \mathrm{Ult}_\alpha(M, \mathcal{U})$ for some α, $k = i^{\mathcal{U}^*}_{0\alpha}$, where $\mathcal{U}^* = \mathcal{U} \cap L(\mathcal{U})$, is also an elementary embedding: $L(\mathcal{U}) \longrightarrow \mathcal{N}$. Since $\mathcal{V} \cap L(\mathcal{U}) = \{x \in S(\kappa) \cap L(\mathcal{U}) \mid \eta \in j(x)\}$, where $\eta = [\mathrm{id}]$, in $\mathrm{Ult}_1(V, \mathcal{V})$, and $\{x \in S(\kappa) \cap L(\mathcal{U}) \mid \eta \in k(x)\} \in L(\mathcal{U})$, it suffices to show $j \restriction S(\kappa) \cap L(\mathcal{U}) = k \restriction S(\kappa) \cap L(\mathcal{U})$. If μ, δ and $\langle \gamma_\alpha \mid \alpha < \delta\rangle$ are again as in the proof of (1), and are fixed by both j and k, any x in $S(\kappa) \cap L(\mathcal{U})$ is $\{\beta < \kappa \mid \langle L_\mu(\mathcal{U}), \in\rangle \models \varphi[\beta, \bar{\eta}, \kappa, \bar{\gamma}_\alpha]$, some $\varphi, \bar{\eta}, \bar{\gamma}_\alpha$, and then $j(x)$ and $k(x)$ are both $\{\beta < \sigma \mid \langle L_\mu(\mathcal{W}), \in\rangle \models \varphi[\beta, \bar{\eta}, \sigma, \bar{\gamma}_\alpha]$, where $\mathcal{W}$ is the constructing ultrafilter on σ in $\mathcal{N}$.

Proof of (5). From the proof of (4), it is clear that for every such $\mathcal{V}$, $i^{\mathcal{V}}_{01}(\kappa) = i^{\mathcal{U}}_{0\beta}(\kappa)$, $i^{\mathcal{V}}_{01} \restriction S(\kappa) = i_{0\beta} \restriction S(\kappa)$, and so $\mathcal{V} = \{x \subseteq \kappa \mid \alpha \in i^{\mathcal{U}}_{0\beta}(x)\}$ for some β and $\alpha < i^{\mathcal{U}}_{0\beta}(\kappa)$.

If $\beta \geqslant \omega$, $i^{\mathcal{U}}_{0\omega}(\kappa)$ is inaccessible in $\mathrm{Ult}_\omega(V, \mathcal{U})$ and therefore in $\mathrm{Ult}_\beta(V, \mathcal{U}) = \mathrm{Ult}_1(V, \mathcal{V})$, an impossibility since $\mathrm{Ult}_1(V, \mathcal{V})$ contains all countable sequences of ordinals.

Proof of (6). We have just done this, in proving (5).

Proof of (7). Arguing in M, we let λ be any Jonsson cardinal, which we know must be between $\aleph_\omega$ and κ.

Suppose $f: [\lambda]^{<\omega} \longrightarrow 2$ is given. If

(*) there is an $x \subseteq \delta < \lambda$ with $x \in L_\beta(\mathcal{U}) - (\bigcup_{\alpha<\beta} L_\alpha(\mathcal{U}))$ and $f \in L_\beta(\mathcal{U})$, then let $\mathcal{M}$ be a transitive set model of "ρ is a measurable cardinal with normal measure $\mathcal{V}$ and V = L($\mathcal{V}$)", for some $\rho \in M$ with $\delta < \rho$, $x \in M$ and $(M)^{=} = \delta$. Then $f \in Ult_\lambda(M, \mathcal{V})$ by 4.14(b), since $Ult_\lambda(M, \mathcal{V})$ is some $\langle L_\gamma(\mathcal{U}), \in\rangle$, $x \in L_\beta(\mathcal{U}) \in L_\gamma(\mathcal{U})$ and $S(\lambda) \cap Ult_\lambda(M, \mathcal{V}) = S(\lambda) \cap L_\gamma(\mathcal{U})$, $i_{0\lambda}(\rho) = \lambda$ is measurable in $Ult_\lambda(M, \mathcal{V})$, so f has a homogeneous set there. (*) of course may fail, but we can reduce the problem for f to one of finding a homogeneous set for an f for which (*) does hold, as follows. Let $\mathcal{M}$ be a transitive set model for "σ is a measurable cardinal, $\mathcal{W}$ is a normal measure on σ, and V = L($\mathcal{W}$)", where $\sigma > \lambda$, $f \in M$ and $(M)^{=} = \lambda$. If F maps λ 1 - 1 onto M, λ Jonsson gives an $A \subsetneq M$ of power λ such that $\langle A, \in, F \restriction A\rangle \prec \langle M, \in, F\rangle$. Since $A \neq M$, the inverse j of the transitive collapse of A onto T must move some least ordinal $\delta < \lambda$, and $j(\lambda) = \lambda$. Since δ is not measurable by 5.2(10), $\{x \in S(\delta) \mid \delta \in j(x)\}$ is not a normal ultrafilter on δ, so $S(\delta) \cap T \subsetneq S(\delta)$. If $j(\tilde{f}) = f$ and $x \in (S(\delta) - T)$, we are in the situation (*), so there is a homogeneous set H for $\tilde{f}$; j"H is then homogeneous for f.

Proof of (8). We know by the theorems of Ulam and Solovay, 5.4(2) and 5.7(3) above, that κ is a union of disjoint atoms a_β for $\mathcal{F}$ in L($\mathcal{U}$), where $\beta < \gamma < \lambda$. If $\mathcal{U}_\beta = \{X \subseteq \kappa \mid X \cap a_\beta \in \mathcal{F}\}$, and $\tilde{\mathcal{U}}$ is the unique normal measure on κ in L($\mathcal{U}$), each $\mathcal{U}_\beta \cap L(\mathcal{U})$ is a normal measure on κ in L($\mathcal{U}$), so must be $\tilde{\mathcal{U}}$, so $\mathcal{F} \cap L(\mathcal{U}) = \tilde{\mathcal{U}}$, and we are done.

We close this section on measurability with a version of Kunen's result on the difficulty of preserving measurability of κ when $2^\kappa > \kappa^+$.

5.10 Theorem. Suppose κ is measurable, $\mathcal{U}$ is a normal measure on κ, $\mathcal{M} = \langle L(\mathcal{U}), \in\rangle$ is the unique κ-model with constructing ultrafilter $\mathcal{U}^* = \mathcal{U} \cap L(\mathcal{U})$, $\mathcal{M}_\alpha$ is (the transitive collapse of) $Ult_\alpha(M, \mathcal{U}^*)$, and $i_{\alpha\beta}$ are the canonical elementary embeddings from $\mathcal{M}_\alpha$ into $\mathcal{M}_\beta$. Then each of the following statements implies the next.

(1) $2^\kappa > \kappa^{+(\mathcal{M})}$

(2) For any measure $\mathcal{V}$ on κ, $i^{\mathcal{V}}_{01}(\kappa) > i_{0\omega}(\kappa)$.

(3) A ρ-model exists for $\rho < \kappa$.

(4) $(\mathcal{U}^*)^\#$ exists, where we identify $\mathcal{U}^*$ with its image in On under the canonical (in fact, definable) well-ordering of L($\mathcal{U}$).

Proofs (1 ⟶ 2) $i_{0\omega}(\kappa) < \kappa^{++(\mathcal{M})} \leq 2^\kappa < i^{\mathcal{V}}_{01}(\kappa)$, since $(i_{0\omega}(\kappa))^{=(\mathcal{M})} = (\omega \cdot (\kappa^\kappa)^{\mathcal{M}}) = \kappa^{+(\mathcal{M})}$, and $i^{\mathcal{V}}_{01}(\kappa)$ for $i^{\mathcal{V}}_{01}: V \longrightarrow Ult(V, \mathcal{V}) \cong \mathcal{N}$, is always $> (2^\kappa)^{\mathcal{N}}$ (since it is inaccessible in $\mathcal{N}$), and $(2^\kappa)^{\mathcal{N}} = 2^\kappa$, since ${}^\kappa N \subseteq N$.

(2 ⟶ 3). Using the notation of (1 ⟶ 2), $a = \{i_{0n}(\kappa) \mid n < \omega\}$ is in N, and so therefore is $\mathcal{F} \cap N$, where $\mathcal{F} = \{x \mid i_{0\omega}(\kappa) \mid$ some terminal segment of a is in x$\}$. By 5.8.1(a) $\mathcal{F} \cap M_\omega$ is $i_{0\omega}(\mathcal{U}^*)$, so

$(\exists \rho < i^{\mathcal{V}}_{01}(\kappa)\, \exists \mathcal{W}\, (\mathcal{W} \cap L(\mathcal{W})$ is a measure on $\rho)^{L(\mathcal{W})})^{\mathcal{M}}$.

(3⟶4). Let $\mathcal{W}$ be the object the ρ-model $P = L(\mathcal{W})$ thinks is a constructing ultrafilter on ρ, and let $\mathcal{V}$ be any measure on κ. Since $i^{\mathcal{V}}_{0\alpha}(\rho) = \rho$ and $i^{\mathcal{V}}_{0\alpha}(\mathcal{W}) = \mathcal{W}$ for all $\alpha > 0$, $i^{\mathcal{V}}_{0\alpha}$ maps P into P, and $\{ i_{0\alpha}(\kappa) \mid \alpha > 0 \}$ is a class of indiscernibles for $\langle P, \in, \langle \alpha \mid \alpha < \rho^{+(P)} \rangle \rangle$, so $\{\lambda \mid \mathrm{cf}\,\lambda > \kappa\}$ is a class of indiscernibles for $\langle M, \in, \langle \alpha \mid \alpha < \kappa^{+(\mathcal{M})} \rangle \rangle$ since $i^{\mathcal{W}}_{0\lambda}(\kappa) = \lambda$ for such λ and $\mathrm{Ult}_{\kappa}(P, \mathcal{W})$ is the κ-model $\mathcal{M}$; this is good enough to show $(\mathcal{U}^*)^{\#}$ exists, by 4.19.

6. The Great Beyond

6.1. We begin with a discussion of filters over $[\lambda]^{<\kappa}$ for $\lambda \geq \mathrm{cf}\,\kappa = \kappa$ (cf. [Jec]). For $p \in [\lambda]^{<\kappa}$, let $\hat{p} = \{q \in [\lambda]^{<\kappa} | q \supseteq p\}$. $D = \{p_\alpha | \alpha < \gamma\} \subseteq [\lambda]^{<\kappa}$ is a chain (is directed) iff for all α, $\beta < \gamma$ with $\alpha < \beta$, $p_\alpha \subseteq p_\beta$ (for all $\alpha \neq \beta < \gamma$, there is a $\delta < \gamma$ such that $p_\alpha \cup p_\beta \subseteq p_\delta$).
$C \subseteq [\lambda]^{<\kappa}$ is closed iff the union of every nonempty chain $D \subseteq C$ of power $< \kappa$ is in C (putting "directed set" for "chain" gives the same definition).
$X \subseteq [\lambda]^{<\kappa}$ is unbounded iff every $p \in [\lambda]^{<\kappa}$ has a superset q in X.
$S \subseteq [\lambda]^{<\kappa}$ is stationary iff S intersects every closed unbounded $C \subseteq [\lambda]^{<\kappa}$. Let $\mathcal{K}$ be the filter generated by the closed unbounded subsets of $[\lambda]^{<\kappa}$. A filter $\mathcal{F}$ on $[\lambda]^{<\kappa}$ is normal iff whenever X is not in the ideal $\mathcal{I}$ dual to $\mathcal{F}$, and $f: [\lambda]^{<\kappa} \longrightarrow \lambda$ is regressive, i. e., $f(p) \in p$ for each $p \in [\lambda]^{<\kappa}$ there is a $Y \subseteq X$ with $Y \notin \mathcal{I}$ such that $f \restriction Y$ is constant. $\mathcal{F}$ is regular iff $\hat{p} \in \mathcal{F}$ for each $p \in [\lambda]^{<\kappa}$.

6.2. κ is λ-compact for $\lambda \geq \kappa$ iff there is a κ-complete, regular ultrafilter $\mathcal{U} \subseteq S([\lambda]^{<\kappa})$.
κ is strongly compact iff κ is λ-compact for all $\lambda \geq \kappa$.
κ is λ-supercompact iff there is a κ-complete, regular and normal ultrafilter $\mathcal{U} \subseteq S([\lambda]^{<\kappa})$.
κ is supercompact iff κ is λ-supercompact for all $\lambda \geq \kappa$.

Notice that an $\alpha < \kappa$, considered as a subset of κ, is in $[\kappa]^{<\kappa}$ also that $S \subseteq \kappa$ is closed unbounded (stationary) in $[\kappa]^{<\kappa}$ iff it is closed unbounded (stationary) as a subset of κ and $f: S \longrightarrow \kappa$ is regressive in the 'new' sense iff it is regressive in the old, so that the case $\lambda = \kappa$ in the following is Fodor's theorem, 0.5.

6.3 Theorem. (1) A filter $\mathcal{F}$ on $[\lambda]^{<\kappa}$ is normal iff it is closed under diagonal intersections, i. e., whenever each $X_\alpha \in \mathcal{F}$, for $\alpha < \lambda$, so is $\Delta_{\alpha<\lambda} X_\alpha = \{p | p \in X_\alpha \text{ for all } \alpha \in p\}$. (2) The closed unbounded filter is a κ-complete regular normal filter on $[\lambda]^{<\kappa}$.

Proof of (1)($\longrightarrow$). If $X = \Delta_{\alpha<\lambda} X_\alpha \notin \mathcal{F}$, $Y = [\lambda]^{<\kappa} - X \notin \mathcal{I}$, and $f: p \longrightarrow$(least $\alpha \in p$ such that $p \notin X_\alpha$) is regressive, so there is a $Z \subseteq Y$ with $Z \notin \mathcal{I}$ and an α such that $f''Z = \{\alpha\}$, i. e., $Z \cap X_\alpha = 0$, so $X_\alpha \notin \mathcal{F}$.
($\longleftarrow$) If $f: [\lambda]^{<\kappa} \longrightarrow \lambda$ is such that each $f^{-1}(\{\alpha\}) \cap X = X_\alpha \in \mathcal{I}$ but $X \notin \mathcal{I}$, $Y_\alpha = [\lambda]^{<\kappa} - X_\alpha \in \mathcal{F}$ for each α so $Y = \Delta_{\alpha<\lambda} Y_\alpha$ is too; but then $f(p) \notin p$ for each p in $X \cap Y \notin \mathcal{I}$.

Proof of (2). Regularity is clear. The limit case of the induction for κ- completeness is done as follows. Assume the result is known for $\alpha < \beta = \bigcup \beta$, and let $\langle C_\alpha | \alpha < \beta \rangle$ be given with $C = \bigcap_{\alpha<\beta} C_\alpha$. C is unbounded, for one can define a sequence $\langle p_\alpha | \alpha < \beta \rangle$ above an arbitrary $p_0 = p$ by the inductive hypothesis so that each $p_\delta \in \bigcap_{\alpha<\delta} C_\alpha$ for $0 < \delta < \beta$, and $p_\alpha \subseteq p_{\alpha'}$ for $\alpha < \alpha' < \beta$; then $p^* = \bigcup_{\alpha<\beta} p_\alpha$ is in C. C is closed, since if $D \subseteq C$ is a chain, $\bigcup D \in C_\alpha$ for each $\alpha < \beta$, so $\bigcup D \in C$.
Normality: If $C_\alpha \in K$ for each $\alpha < \lambda$, and $C = \Delta_{\alpha<\lambda} C_\alpha$, then C is unbounded. If $p = p_0$ is arbitrary, let $p_{n+1} \supseteq p_n$ be such that $p_{n+1} \in \bigcap_{\alpha \in p_n} C_\alpha$ for $n \geq 1$. The $p = \bigcup_{n<\omega} p_n$ is in C, for if $\alpha \in p$, $\alpha \in p_n$ for sufficiently large n, so $p_{n+1} \in C_\alpha$ for sufficiently large n, so $p \in C_\alpha$.
C is closed, for if $D \subseteq C$ is a directed set of power $< \kappa$, and $\alpha \in p = \bigcup D$, $p \in C_\alpha$, since p is also $\bigcup D'$, where $D' = \{ q \in D | \alpha \in q \} \subseteq \{ q \in D | q \in C_\alpha \}$.

For the following, say that a set of sentences $\Sigma \subseteq L_{\kappa\omega}$ is κ-consistent iff every subset of Σ of power $< \kappa$ has a model.

6.4 Theorem. The following are equivalent.
(1) κ is strongly compact.
(2) Every κ-consistent set of sentences $\Sigma \subseteq L_{\kappa\omega}$ has a model.
(3) Every κ-complete filter $\mathcal{F} \subseteq S(\lambda)$ for $\lambda \geq \kappa$ can be extended to a κ-complete ultrafilter.
Remark. The conclusion of (2) follows from $(\Sigma)^=$-compactness of κ, and that of (3) for $\mathcal{F}$ from 2^λ-compactness of κ.

Proof of (1)$\longrightarrow$(2). Let $\mathcal{U}$ be a κ-complete regular ultrafilter over $[\Sigma]^{<\kappa}$, where Σ is assumed well-ordered in type $(\Sigma)^= = \lambda$. For each $p \in [\Sigma]^{<\kappa}$, let $\mathcal{M}_p \models \wedge p$. Then the ultraproduct $Ult(\mathcal{M}_p, \mathcal{U})$ of the $\mathcal{M}_p$'s is a (well-founded) model of Σ, since for each p $\{ p | \mathcal{M}_p \models \wedge p \} \supseteq \hat{p} \in \mathcal{U}$.
Proof of (2)$\longrightarrow$(3). Let Σ be the $L_{\kappa\omega}$ theory of $\langle \lambda, \langle X | X \subseteq \lambda \rangle$ (each $\underset{\sim}{X}$ is a relation symbol), along with sentences $\underset{\sim}{X}(\underset{\sim}{c})$ for each X in $\mathcal{F}$, where $\underset{\sim}{c}$ is a new constant. By κ-completeness of $\mathcal{F}$, Σ is κ-consistent. If $\mathcal{M} = \langle M, \langle X^M | X \subseteq \lambda \rangle, k \rangle$ is a model of Σ, $\{ X \subseteq \lambda | \mathcal{M} \models X^M[k] \} = \mathcal{U}$ extends $\mathcal{F}$, and is a κ-complete ultrafilter, essentially because $\mathcal{M}$ thinks k witnesses $\mathcal{U}$ is principal.
Proof of (3)$\longrightarrow$(1). Extend the closed unbounded filter K to a κ-complete ultrafilter.

6.5 Theorem. The following are equivalent.
(1) κ is λ-supercompact.
(2) There is an elementary embedding i from V into a transitive $M \subseteq V$

such that $i \restriction \kappa$ is the identity, $i(\kappa) > \kappa$ and ${}^{\lambda}M \subsetneq M$.

Proof of (1) $\longrightarrow$ (2). If $\mathcal{U}$ is a κ-complete regular normal ultrafilter on $[\lambda]^{<\kappa}$ let i be the elementary embedding of V into the transitive collapse M of $Ult_1(V, \mathcal{U})$.
$i \restriction \kappa$ is the identity, by κ-completeness: if $[f] < i(\alpha)$, $\forall^* p\ f(p) < \alpha$, so there is a $\gamma < \alpha$ such that $\forall^* p\ f(p) = \gamma$.
$[id] = \{i(\alpha) \mid \alpha < \lambda\} = i''\lambda = H$ in M. $H \subseteq [id]$ for $i(\alpha) \in [id]$ iff $\{p \mid \alpha \in P\} = \{\alpha\}^\wedge \in \mathcal{U}$, and this holds by regularity. $[id] \subseteq H$, for if $[f] \in [id]$, $\forall^* p\ f(p) \in p$, and then by normality $\forall^* p\ f(p) = \gamma$ for some $\gamma < \lambda$, so $[f] = i(\gamma)$.
$i(\kappa) > \lambda$. If $f(p) = |p|$ (the order-type of p) for all $p \in [\lambda]^{<\kappa}$, $[f] = |H| = \lambda$ in M. Since $\forall p\ f(p) < \kappa$, $\lambda < i(\kappa)$.
For ${}^{[\lambda]^{<\kappa}}M \subseteq M$, note first that $[\lambda]^{<\kappa} = [\lambda]^{<\kappa} \cap M \in M$ by 5.2(3), $i \restriction \lambda \in M$ (as the increasing enumeration of H), and normality of $\mathcal{U}$ gives $i(p) = i''p$ for $p \in [\lambda]^{<\kappa}$, so $i \restriction [\lambda]^{<\kappa} \in M$. Then if $x_p = [f_p] \in M$ and $h(q)$: $p \longrightarrow f_p(q)$ for p, $q \in [\lambda]^{<\kappa}$, $dom([h]) = [i(\lambda)]^{<i(\kappa)}$ and $([h]) \circ (i \restriction [\lambda]^{<\kappa})$ maps p to x_p in M for each $p \in [\lambda]^{<\kappa}$.

Proof of (2) (1). Let $H = \{i(\alpha) \mid \alpha < \lambda\} = i''\lambda$, and define $\mathcal{U} \subseteq [\lambda]^{<\kappa}$ by $X \in \mathcal{U}$ iff $H \in i(X)$ in M.
$\mathcal{U}$ is κ-complete, for if $\beta < \kappa$ and $X_\alpha \in \mathcal{U}$ for $\alpha < \beta$, $H \in \bigcap_{\alpha<\beta} i(X_\alpha) = i(\bigcap_{\alpha<\beta} X_\alpha)$, since $i \restriction \kappa = id$, so $\bigcap_{\alpha<\beta} X_\alpha \in \mathcal{U}$.
$\mathcal{U}$ is regular, for if $p \in [\lambda]^{<\kappa}$, $p = \bigcap_{\alpha \in p} \{\alpha\}^\wedge$, and the conclusion follows by κ-completeness if we can show each $\{\alpha\}^\wedge \in \mathcal{U}$. But $i(\alpha) \in H$, so $H \in \{(i(\alpha))\}^\wedge = i(\{\alpha\}^\wedge)$ in M, so $\{\alpha\}^\wedge \in \mathcal{U}$.
$\mathcal{U}$ is normal, for if f: $[\lambda]^{<\kappa} \longrightarrow \kappa$ is regressive, i(f) is regressive on $[i(\lambda)]^{<i(\kappa)}$ in M, so $i(f)(H) \in H$, so $i(f)(H) = i(\gamma)$ for some $\gamma < \lambda$, so $H \in \{p \in i([\lambda]^{<\kappa}) \mid i(f)(p) = (\gamma)\} = i(\{p \mid f(p) = \gamma\})$, so $\{p \mid f(p) = \gamma\} \in \mathcal{U}$.

6.6 Corollary κ is κ-supercompact iff κ is κ-compact iff κ is measurable.
Proof. Immediate from 6.5, 5.2(3) and the definitions.

6.7 Remark By 6.5 if κ is κ-compact every κ-consistent $\Sigma \subseteq L_{\kappa\omega}$ of power κ has a model, but the converse is false. The conclusion is one definition for strongly inaccessible κ of weak compactness (cf [Dr, p. 292 ff.] and [De 3]), and can hold as well below the continuum (cf [Ku 2] and [Bo]).

6.8 Theorem (Vopénka-Hrbáček). If κ is λ^+-compact, $a^\#$ exists for every set of ordinals a such that $\bigcup a < \lambda^+$.

Proof. If $\mathcal{U}$ is a κ-complete ultrafilter $\subseteq$ $S(\lambda^+)$ extending $\mathcal{F} = \{X \subseteq \lambda^+ \mid (\kappa - X)^= \leq \lambda\}$, let N be the transitive collapse of Ult (V , $\mathcal{U}$). (We can still define Fn(1, λ^+, V) as before, and divide this out by $\mathcal{U}$, obtaining an ultrapower which is well-founded by κ-completeness of $\mathcal{U}$). Similarly, we can form $Fn^-(1, \lambda^+, V) = \{ f \mid Fn(1, \lambda^+, V) \mid (\text{ran } f)^= \leq \lambda \}$, and divide this out by $\mathcal{U}$ to obtain $Ult^-(V, \mathcal{U})$, which is still well-founded, and collapses to M . Let i be the resulting embedding from V into M, and j the one from V into N. Let $k: M \longrightarrow N$ be defined by $k([f]^-) = [f]$, where $[f]^-$ is the equivalence class determined by $\mathcal{U}$ in $F_n^-(1, \lambda^+, V)$. We continue to blur the notational distinction between $[f]^-$ in $Ult^-(V, \mathcal{U})$, say, and its image under the transitive collapse in M.
(a) $k \restriction i(\lambda^+)$ is the identity, since whenever $\alpha = [f]^- < i(\lambda^+)$, f can be chosen so that ran f is bounded in λ^+ and $[f]^- = \alpha$ in M iff $[f] = \alpha$ in N, by induction on α.
(b) $i(\lambda^+) = \bigcup \{ i(\alpha) \mid \alpha < \lambda^+\}$ for if $[f]^- < i(\lambda^+)$, with $f \in Fn^-(1, \lambda^+, V)$ $[f]^- \leq \alpha$ in M, where $\alpha = \bigcup \text{ran } f \cap \lambda^+ < \lambda^+$.
(c) $i(\lambda^+) < k(i(\lambda^+) = j(\lambda^+)$, since $j(\alpha) < [id: \lambda^+ \longrightarrow \lambda^+]$ in N for all $\alpha < \lambda^+$, so $i(\lambda^+) = \bigcup \{ i(\alpha) \mid \alpha < \lambda^+ \} = \bigcup \{ j(\alpha) \mid \alpha < \lambda^+\} \leq [id]$. But $[id] < j(\lambda^+)$ in N, since $\{\alpha \mid \lambda^+ > \alpha\} \in \mathcal{U}$.
Also $i(a) = k(i(a)) = j(a)$ by essentially the same argument as for (a). Call the common value $\tilde{a}$. Extend the elementary-embedding notation to a definable class K in the usual way, by writing $k(K) = \bigcup k(K \cap R(\alpha))$. Then $k \restriction L(\tilde{a})$ elementarily embeds $L(\tilde{a})$ in $k(L(\tilde{a})) = L(\tilde{a})$, with $i(\lambda^+)$ the least ordinal moved. Then $(\tilde{a}^\# \text{ exists})^N$ y 4.19 so $a^\#$ exists in V.

Vopénka-Hrbáček stated the weaker conclusion that $V \neq L(a)$, but the proof given above is a trivial extension of their original argument, as interpreted in [Ku 1] .

6.9 Corollary. If κ is strongly compact and $\mathcal{U}$ is a measure on κ, κ^{++} is ineffable; invisible, etc., in $L(\mathcal{U})$.

6.10 Theorem (Solovay).
(1) If $\mu > \kappa$ is regular and κ is μ- compact, $\mu^{\kappa} = \mu$.
(2) If κ, λ, and μ are regular cardinals such that κ is μ - compact and $2^\lambda \leq \mu$, $\mu^\lambda = \mu$.
(3) If ν is a singular strong limit cardinal and κ is ν^+- compact, $2^\nu = \nu^+$.

Proof of (1). By 6.4 let $\mathcal{U}\subseteq S([\mu]^{<\kappa})$ be a κ-complete ultrafilter extending the closed unbounded filter on $[\mu]^{<\kappa}$. If i is the elementary embedding from V into the transitive collapse M of the ultrapower $\mathrm{Ult}(V,\mathcal{U})$, let $[s] = \sigma = \bigcup\{i(\alpha) \mid \alpha<\mu\}$ in M.

(a). $\{p \mid \mathrm{cf}(s(p))<\kappa\} = S \supseteq \{p \mid s(p) = \bigcup s(p) = \bigcup(p\cap s(p)) =_{\mathrm{df}} t(p)\} = T\in\mathcal{U}$. The first containment relation is clear, and we need only show $T\in\mathcal{U}$. But if $\forall^* p\ t(p)<s(p)$, $[t]\leq i(\alpha)$ for some $\alpha<\mu$ in M, an impossibility, since $\hat{\alpha}\in\mathcal{U}$ and $\{p \mid s(p)>\alpha\}\in\mathcal{U}$ for each $\alpha<\mu$.

(b). There exists a sequence $d_\alpha\subseteq\alpha$ for $\alpha\in\tilde{T} = \{\bigcup p \mid p\in T\}$ such that $(d_\alpha)^= \leq \mathrm{cf}\,\alpha$ for each $\alpha\in\tilde{T}$ and $[\mu]^{<\kappa}\subseteq\bigcup_{\alpha<\mu} S(d_\alpha)$, so $([\mu]^{<\kappa})^= = \mu\cdot 2^{\underline{\kappa}} = \mu$. d_α is defined as $h''c_\alpha$, where c_α for $\alpha\in\tilde{T}$ is some closed unbounded subset of α of order-type cf α, and h from μ into μ is defined as follows. Set $c(x) = c_{\bigcup x}$ for $x\in T$, and define $\tau_0 = 0$, $\tau_{\beta+1}$ = the least ordinal $>\tau_\beta$ such that there is a γ in $[c]$ with $i(\tau_\beta)\leq\gamma<i(\tau_{\beta+1})$ in M. Such a γ always exists since $[c]$ and $\{i(\alpha) \mid \alpha<\mu\}$ are both cofinal subsets of σ in M. If $\beta = \bigcup\beta$, set $\tau_\beta = \bigcup_{\bar{\beta}<\beta}\tau_{\bar{\beta}}$. Set $h(\eta) = \beta$ iff $\tau_\beta\leq\eta<\tau_{\beta+1}$. Then $d_\alpha = h''c_\alpha$ satisfies $(d_\alpha)^= \leq \mathrm{cf}\,\alpha$ and $d_\alpha\subseteq\alpha$ since $h''\alpha\subseteq\alpha$, and if $\beta<\mu$, $\{p \mid \beta\in d_{\bigcup p}\} = \{p \mid \beta\in h''c_{\bigcup p}\} = \{p \mid$ for some η in $c_{\bigcup p}$, $h(\eta) = \beta\} = \{p \mid$ for some η in $c_{\bigcup p}$ $\tau_\beta\leq\eta<\tau_{\beta+1}\}\in\mathcal{U}$, by definition of $\tau_{\beta+1}$.

Once we know $\{p \mid \beta\in d_{\bigcup p}\}\in\mathcal{U}$ for each β, we know $\{p \mid y\subseteq d_{\bigcup p}\}\in\mathcal{U}$ for each $y\in[\mu]^{<\kappa}$, by κ-completeness of $\mathcal{U}$, and we are done with the proof that $\mu^{\underline{\kappa}} = \mu$.

Proof of (2). By (1), we can assume $\lambda\geq\kappa$. Let $\mathcal{U}\subseteq S([\lambda]^{<\kappa})$ be a κ-complete ultrafilter extending the closed unbounded filter on $[\lambda]^{<\kappa}$, and let i embed V into the transitive collapse M of $\mathrm{Ult}(V,\mathcal{U})$.

(a). $2^\lambda < i(\kappa) < (2^{(\lambda^{\underline{\kappa}})})^+$ $(= (2^\lambda)^+$ by (1)). Let $g(p) = S(p)$ for $p\in[\lambda]^{<\kappa}$, and $h(p)$ and $r(p)<\kappa$ be such that $h(p)$ maps $S(p)$ 1-1 into $r(p)$. Then $[h]$ maps $[g]$ 1-1 into $[r] = \gamma < i(\kappa)$. If $f_X(p) = X\cap p$ for $X\subseteq\lambda$, $[f_X]\neq[f_Y]$ for $X\neq Y$ by regularity of $\mathcal{U}$, and so $k: X\longmapsto f_X$ embeds $S(\lambda)$ into $[g]$, and $[h]\bullet k$ embeds $S(\lambda)$ in γ. The second inequality follows from $i(\kappa)<i(\lambda)$ and the fact that each $[f]<i(\lambda)$ may be taken to have range $\subseteq\lambda$.

(b). Any $y\subseteq\mu$ with $(y)^= \leq\lambda$ is a subset of $z\in S(\mu)\cap M$ with $(z)^= \leq 2^{(\lambda^{\underline{\kappa}})} = 2^\lambda$ (a weak version of 6.5). For if $y = \{\gamma_\alpha \mid \alpha<\lambda\}$ and $\gamma_\alpha = [f_\alpha]$, let $h(p) = \{f_\alpha(p) \mid \alpha\in p\}$ and $z = [h]$. Each $[f_\alpha]\in z$ in M by regularity of $\mathcal{U}$; $((z)^= < i(\kappa))^M$, since each $z(p)$ has cardinality $<\kappa$, and so $(z)^= \leq (i(\kappa))^= = 2^\lambda$ by (a).

(c). $(\mu^{\underline{i(\kappa)}} = \max(\mu, i(\kappa))^M$, and so $((\mu^{\underline{i(\kappa)}})^M)^= = \mu$. For if $i(\kappa)<\mu$ (which must be the case if $2^\lambda<\mu$, since then $i(\kappa) < (2^\lambda)^+ \leq \mu$), then (1) gives $\mu^{\underline{i(\kappa)}} = \mu$ in M; and if $\mu \leq i(\kappa)$, $\mu^{\underline{i(\kappa)}} \leq i(\kappa)^{\underline{i(\kappa)}} = i(\kappa)$ in M, since $i(\kappa)$ is strongly inaccessible in M.

(d) We now compute $\mu^\lambda = 2^\lambda \cdot ([\mu]^\lambda)^= \leq \mu \cdot ([\mu]^\lambda)^=$.
Each $x \in [\mu]^\lambda$ is a subset of some z such that $z \subseteq \mu$, $(z)^= = 2^\lambda$ and $((z)^= < i(\kappa))^M$. There are $((\mu^{\underline{i(\kappa)}})^M)^= = \mu$ such z's, each containing at most $(2^\lambda)^\lambda = 2^\lambda$ such x's, so $([\mu]^\lambda)^= = 2^\lambda \cdot \mu = \mu$.

Proof of (3). Let cf $\nu = \lambda$, let $(\tau_\sigma \mid \sigma < \lambda)$ be a increasing function with range cofinal in ν, and set $\mu = \nu^+$. Then $2^\nu = 2^{(\sum_{\sigma<\lambda}\tau_\sigma)} = \prod_{\sigma<\lambda} 2^{\tau_\sigma} \leq \prod_{<\lambda} \nu = \nu^\lambda \leq \mu^\lambda = \mu$, the last equality by (2).

After a final series of definitions, we offer Kunen's proof that an elementary-embedding axiom that was taken seriously for a time contradicts the axiom of choice. When large-cardinal axioms begin to approach the rim, they seem to clash with ZFC more than ZF. Even the existence of $0^\#$ can be interpreted as evidence of incompatibility of κ_{ω_1} with a certain Σ_1-definable well-ordering of the universe. Kunen has further shown in [Ku 4] that for uncountable regular μ (there are μ measurable cardinals between μ and ν) implies that no well-ordering of $[\nu]^{<\mu}$ exists inside in Chang's $L_{\mu\mu}$ version C^μ of L. The proof of 6.12 will itself use a simple case of an older refutation in ZFC of a "large-cardinal" axiom (that there are "ω-Jonsson cardinals"), due to Erdös and Hajnal. All this is perhaps wryly amusing when one considers that moderately large cardinals are often "motivated" as infinitary choice/compactness properties. Choice apparently does not take kindly to infinitary "generalizations" of itself.

And who knows, without choice, what happens in the Jabberwocky void of full determinacy, where $\aleph_2$ is measurable and $\aleph_3$ is singular?

The Shadow knows.

6.11. (Kunen) κ is (almost) α-huge for $\alpha \geq 1$ iff there is an elementary embedding i from V into a transitive $M \subseteq V$ such that $i \upharpoonright \kappa$ is the identity, $i(\kappa) > \kappa$ and ${}^{i_\alpha(\kappa)}M \subseteq M$ (${}^\beta M \subseteq M$ for all $\beta < i_\alpha(\kappa)$), where $i_1(\kappa) = i(\kappa)$, $i_{\alpha+1}(\kappa) = i(i_\alpha(\kappa))$ and $i_\alpha(\kappa) = \bigcup_{\beta<\alpha} i_\beta(\kappa)$ if $\alpha = \bigcup\alpha$.

Call a(n) (almost) 1-huge cardinal simply (almost) huge.

(Reinhardt) If $\alpha \geq 1$ κ is α-extendible iff there exists an elementary embedding i from $R(\kappa + \alpha)$ into some $R(\beta)$ such that $i \upharpoonright \kappa$ is the identity and $i(\kappa) > \kappa$ (Recall $R(\kappa + \alpha) = R(\alpha)$ if $\alpha \geq \kappa \cdot \kappa$).

κ is extendible iff κ is α-extendible for all α.

Vopenka's principle, stated in MKC (or over ZFC), is the axiom (scheme): given any (definable) class $\langle \mathfrak{A}_\alpha, \alpha \geq 0 \rangle$ of structures of the same similarity type, a pair α, β exists such that $\alpha \neq \beta$ and $\mathfrak{A}_\alpha \prec \mathfrak{A}_\beta$.

Clearly if $i: V \longrightarrow M$ witnesses that κ is almost huge, κ is λ-supercompact for all $\lambda <$ the strongly inaccessible cardinal $i(\kappa)$, so $R(i(\kappa)) \models$ (ZFC + κ is supercompact); thus, by Gödel's theorem we cannot expect to prove the existence of almost huge cardinals in ZFC + (there is a supercompact). Kunen quite drastically capped the hierarchy of (almost) α-huge cardinals in ZFC by the following.

6.12 Theorem. (Kunen [Ku 3])

(1) (ZFC) There is no ω-huge cardinal.

(2) (MKC) There is no "nontrivial elementary embedding from V into V" — that is, no class i embedding V into V such that $i(x) \neq x$ for some x, yet $\varphi(x)$ iff $\varphi(i(x))$ for all formulas φ in the language of set theory.

Proof of (1). Suppose κ is ω-huge. Then each $i_n(\kappa)$ is measurable by induction on $n < \omega$, and $\lambda = i_\omega(\kappa)$ is a strong limit cardinal of cofinality ω, so $2^\lambda = \lambda^\omega$ (see the proof of 6.10(3)).
For such a λ, it is easy to see that there is a function $F: [\lambda]^\omega \longrightarrow \lambda$ such that whenever $A \in [\lambda]^\lambda$, $F''[A]^\omega = \lambda$:
let $\langle\langle A_\alpha, \gamma_\alpha\rangle \mid \alpha < 2^\lambda\rangle$ enumerate $[\lambda]^\lambda \times \lambda$ (with lots of repetitions in the second coordinate), choose inductively $s_\alpha \in [A_\alpha]^\omega - \{s_\beta \mid \beta < \alpha\}$, and set $F(s_\alpha) = \gamma_\alpha$.
Clearly $i(\lambda) = \bigcup_{n<\omega} i(i_n(\kappa)) = \bigcup_{n<\omega} i_{n+1}(\kappa) = \lambda$, so $i(F): [\lambda]^\omega \longrightarrow \lambda$. Take $A = \{i(\alpha) \mid \alpha < \lambda\} \in [\lambda]^\lambda\}$, and let $t \in [A]^\omega$ be such that $i(F)(t) = \kappa$. $t = \{i(\alpha_n) \mid n < \omega\} = i(s)$, where $s = \{\alpha_n \mid n < \omega\}$, so $i(F)(i(s)) = i(F(s)) = \kappa$, an impossibility since $i(\alpha) = \alpha$ for $\alpha < \kappa$, and $i(\kappa) > \kappa$. We are done.
Proof of (2). Immediate from (1). What M thinks is $S(i_\omega(\kappa))$ cannot be what V thinks it is.

6.13 Theorem. (1) (Solovay) If $\alpha < \kappa < \beth_{\kappa+\alpha} = \mu$ and κ is μ-supercompact, κ is $(\alpha + 1)$-indescribable. In fact there is a normal measure $\mathcal{U}_\alpha \subseteq S(\kappa)$ which extends the $(\alpha + 1)$-indescribable filter, so tha whenever $A \subseteq R(\kappa)$ and $\langle R(\kappa + \alpha + 1), \in\rangle \models \varphi[\kappa, A]$, $\{\lambda < \kappa \mid \langle R(\lambda + \alpha + 1), \in\rangle \models \varphi[\lambda, A \cap R(\lambda)]\} \in \mathcal{U}_\alpha$. In particular, if $\alpha > 0$, $\{\lambda < \kappa \mid \lambda$ is $(\beth_{\lambda+\beta}$-supercompact for all $\beta < \alpha\} \in \mathcal{U}_\alpha$.

(2) (Magidor) If κ is λ-supercompact for all $\lambda < \mu$, and μ is ν-supercompact, κ is ν-supercompact.

(3) (Magidor) If $\alpha = \bigcup\alpha$, $\beta = \bigcup\beta$, i elementarily embeds $R(\alpha)$ into $R(\beta)$, $i \restriction \kappa$ is the identity and $i(\kappa) > \kappa$, κ is λ-supercompact for

all $\lambda < \alpha$.

(4) (Magidor) The following are equivalent:

(a) κ is the least supercompact cardinal.

(b) κ is the least cardinal such that whenever $\beta \geq \kappa$, there is an $\alpha < \beta$ and an elementary embedding i from $R(\alpha)$ into $R(\beta)$.

(c) κ is the least cardinal such that whenever Θ is second-order, $A \subseteq R(\kappa)$ and $\mathcal{A} = \langle S, \in, A \rangle \models \Theta$, there is an elementary submodel $\mathcal{T} = \langle T, \in, A \upharpoonright T \rangle$ of $\mathcal{A}$ such that $\mathcal{T} \models \Theta$.

Proof of (1). Suppose $j: V \longrightarrow M$ witnesses μ-supercompactness of κ, $A \subseteq R(\kappa)$, and $\{\lambda < \kappa \mid \langle R(\lambda+\alpha+1), \in \rangle \models \neg\varphi[\lambda, A \cap R(\lambda)]\} \in \mathcal{U}_\alpha = \{X \subseteq \kappa \mid \kappa \in j(x)\}$, a normal ultrafilter on κ by 4.18(2). Let N be the transitive collapse of $Ult(V, \mathcal{U}_\alpha)$. If $k: [f]_1 \longrightarrow (j(f))(\kappa)$, $k: Ult(V, \mathcal{U}) \longrightarrow M$ is elementary, since (e. g.) $[f]_1 \; E_1 \; [g]_1$ iff $\kappa \in j(\{\alpha \mid f(\alpha) \in g(\alpha)\})$ iff $j(f)(\kappa) \in j(g)(\kappa)$. $k(\kappa) = \kappa$, since $k(\kappa) = (j(id))(\kappa) = \kappa$. Then $(\langle R(\kappa+\alpha+1), \in \rangle \models \neg\varphi[\kappa, A])^N$, so $(\langle R(\kappa+\alpha+1), \in \rangle \models \neg\varphi[\kappa, A])^M$, since $k: N \prec M$ and $k(\kappa) = \kappa$. But then $\langle R(\kappa+\alpha+1), \in \rangle \models \neg\varphi[\kappa, A]$, since $(R(\kappa+\alpha+1))^M = R(\kappa+\alpha+1)$.

Proof of (2). Let $\mathcal{U}$ be normal over $[\nu]^{<\mu}$, and let i embed V into the transitive collapse M of the ultrapower $Ult_1(V,U)$. Since $i(\kappa) = \kappa$, $\nu < i(\mu)$ and κ is λ-supercompact for all $\lambda < \mu$, κ is ν-supercompact in M. If $\tilde{\mathcal{U}} \subseteq S([\nu]^{<\kappa})$ witnesses this in M, $\tilde{\mathcal{U}}$ is a normal ultrafilter in V by 6.5, since ${}^{[\nu]^{<\mu}}M \subseteq M$.

Proof of (3). If $\lambda < \min(\alpha, i(\kappa))$ and $\mathcal{U} \subseteq [\lambda]^{<\kappa}$ is defined by $X \in \mathcal{U}$ iff $\{i(\gamma) \mid \gamma < \lambda\} \in i(X)$, (defined since $\alpha = \bigcup\alpha$, and therefore $S([\lambda]^{<\kappa}) \in R(\alpha)$), $\mathcal{U}$ is a regular normal ultrafilter over $[\lambda]^{<\kappa}$ by the proof of 6.5. If $i_0(\kappa) = \kappa$, $i_1(\kappa) = i(\kappa)$ and $i_n(\kappa) < \alpha$, let $i_{n+1}(\kappa) = i(i_n(\kappa))$. We show first: (*) $\alpha \leq \sigma = \bigcup\{i_{n+1}(\kappa) \mid i_n(\kappa) < \alpha\}$ Suppose not. Then $\sigma < \alpha$, $cf\,\sigma = \omega$, and $i(\sigma) = \sigma$. Let $F: [\sigma]^\omega \rightarrow \sigma$ be as in the proof of 6.11, and $A = \{i(\gamma) \mid \gamma < \sigma\}$. Then $R(\beta) \models (i(F): [\sigma]^\omega \rightarrow \sigma$ and $i(F)''[A]^\omega = \sigma)$, so there is a $t \in [A]^\omega$ such that $i(F)(t) = \kappa$; but $t = i(t')$ where $t' = \{\beta \mid i(\beta) \in t\}$, and so $i(F(t) = \kappa$, an impossibility, and (*) is established. Now we show by induction on n that whenever $i_n(\kappa) < \alpha$ and $\lambda < i_{n+1}(\kappa)$, κ is λ-supercompact; completing the proof of (3), since then κ is λ-supercompact for all $\lambda < \alpha$ by (*). n = 0 was done at the beginning of the proof. If $n \geq 1$ and $\langle R(\alpha), \in \rangle \models \kappa$ is λ- supercompact for all $\lambda < i_n(\kappa)$, $\langle R(\beta), \in \rangle \models i(\kappa)$ is λ- supercompact for all $\lambda < i_{n+1}(\kappa)$, so κ is λ- supercompact for all $\lambda < i_{n+1}(\kappa)$ by (2), and we are done.

Proof of (4). (a) $\longrightarrow$ (b). Suppose κ is supercompact, $\beta \geq \kappa$, $\mu = (R(\beta))^= = \beth_\beta$, $\mathcal{U}$ is normal over $[\mu]^{<\kappa}$, and i embeds V into (the transitive collapse M of) $\mathrm{Ult}_1(V,\mathcal{U})$. By ${}^{\mu}M \subseteq M$, $R(\beta)$ and $i \upharpoonright R(\beta)$ are in M (as is every set hereditarily of cardinality $\le (\mu^{\kappa})^{=} = \mu$, and every subset of M of power $\le \mu$), so there is an ordinal $\beta < i(\kappa)$ and an elementary embedding from $R(\beta)$ into $R(i(\beta))$ in M, so there is an ordinal $\alpha < \kappa$ and an elementary embedding from $R(\alpha)$ into $R(\beta)$.

(a)$\longrightarrow$(c) Assume the universe S of $\mathcal{S}$ is σ and $(R(\sigma+1))^{=} = \mu$. Let $\mathcal{U}$ be normal over $[\mu]^{<\kappa}$, and i embed V into (the transitive collapse M of) $\mathrm{Ult}_1(V,\mathcal{U})$ as before. $\mathcal{S} \in M$ by 6.5, and $i \upharpoonright \sigma$ maps $\mathcal{S}$ onto a substructure of $i(\mathcal{S})$, which satisfies Θ in V and therefore in M, since $S(\sigma) \in M$. Then $i(\mathcal{S})$ has a substructure of cardinality $< i(\kappa)$ satisfying Θ in M, so $\mathcal{S}$ has a substructure of cardinality $< \kappa$ satisfying Θ in V.

(c)$\longrightarrow$(b) If $\beta \ge \kappa$ is given, and $T \subseteq R(\beta)$ codes a truth definition for $R(\beta)$, the following is a second-order (in fact Π^1_1) sentence Θ satisfied by $\langle R(\beta), \in, T \rangle$.

"I am a (well-founded) model of the axiom of extensionality isomorphic to an $R(\alpha)$ for $\alpha > \omega$ and T codes a first-order truth definition for me."

For a transitive M is $R(\alpha)$ for such α iff $\langle M, \in \rangle$ satisfies each of
$\forall x(\, On(x) \longrightarrow \exists y(y = R(x))\,)$, (the axiom of infinity), and
$\forall X \forall x(\, X \subseteq x \longrightarrow \exists y(X = y)\,)$.

Then if $\langle N, \in, T \cap N \rangle$ is a substructure of $\langle R(\beta), \in, T \rangle$ of power $< \kappa$ satisfying Θ, $\langle N, \in \rangle \prec \langle R(\beta), \in \rangle$, and $\langle N, \in \rangle$ is isomorphic to $\langle R(\alpha), \in \rangle$ for some $\alpha < \kappa$.

(b)$\longrightarrow$(a) We first let $\tilde{\kappa}$ be any cardinal satisfying the condition $\varphi(x)$ of (b), and show a supercompact cardinal exists. Let $\beta > \tilde{\kappa}$ with $\mathrm{cf}\,\beta = \omega_1$ reflect supercompactness, that is, $\lambda < \beta$ is supercompact iff λ is α- supercompact for all $\alpha < \beta$.

Let $\eta + 2$ and i be such that $i\colon R(\eta+2) \prec R(\beta+2)$; $i(\omega_1) = \omega_1$ so $\mathrm{cf}\,\eta = \omega_1$, $i(\eta) = \beta$, and the least $\kappa \le \eta$ such that $i(\kappa) > \kappa$ is not η (otherwise $\mathcal{U} \subseteq S(\eta) = \{X \mid \eta \in i(X)\}$ would be a normal ultrafilter on η).

By (2), κ is λ-supercompact for all $\lambda < \eta$, so $R(\beta) \models i(\kappa)$ is λ-supercompact for all $\lambda < \beta$, so $i(\kappa)$ is supercompact by definition of β.

We now assume $\tilde{\kappa}$ is the least cardinal satisfying the condition of (b), and μ is the least supercompact cardinal. By (a)$\longrightarrow$(b) $\tilde{\kappa} \le \mu$. If $\tilde{\kappa}$ were less than μ, we could carry out the above argument in $R(\mu)$, obtaining a cardinal $\tilde{\mu} < \mu$ such that $\tilde{\mu}$ is λ-supercompact for all $\lambda < \mu$, and therefore (by 2) supercompact, a contradiction.

.14 Theorem. (1) (Reinhardt, Solovay) If κ is 1-extendible [extendible], κ is measurable [supercompact], and has a normal easure $\mathcal{U}$ such that $\{\lambda<\kappa \mid \lambda$ measurable [supercompact] $\} \in \mathcal{U}$.

2) (Magidor) κ is extendible iff κ is second-order strongly ompact, i. e., whenever Σ is a set of sentences in the second-order nalogue $L^2_{\kappa\omega}$ of $L_{\kappa\omega}$ such that each subset of power $<\kappa$ has a model, o does Σ.

3) κ is extendible iff for each $\alpha>\kappa$ there exist $\beta>\alpha$ and : $\langle R(\alpha),\in\rangle \prec \langle R(\beta),\in\rangle$ such that $j\restriction\kappa$ is the identity and $j(\kappa)>\alpha$.

4) If an extendible cardinal exists, $\{\lambda \mid \lambda$ is α-extendible$\}$ is a roper class for each $\alpha>0$.

5) (MKC) Vopenka's principle implies that every closed unbounded ubclass of On contains an extendible cardinal.

6) (Powell) If κ is almost huge, $\langle R(\kappa+1),\in\rangle \models$ Vopenka's principle.

Proof of (1). Measurability. Writing $R(\beta)$, etc., for $\langle R(\beta),\in\rangle$, let : $R(\kappa+1) \prec R(\mu+1)$ witness 1-extendibility of κ. Then $\mathcal{U} = \{A \mid \kappa\in i(A)\}$ is a normal measure on κ, so $\{\lambda<\kappa \mid \lambda$ measurable$\}\in\mathcal{U}$. upercompactness. κ is supercompact by 6 13(3). If i: $R(\alpha)\prec R(\beta)$ or some $\alpha = \bigcup\alpha > \kappa$ is such that κ is the least ordinal moved by i, and $\mathcal{U}$ is defined as above, $A = \{\lambda<\kappa \mid \lambda$ is β-supercompact for all $\beta<\kappa\}\in\mathcal{U}$, o each $\lambda\in A$ is supercompact by 6.13(2).

Proof of (2). $(\longrightarrow)$. Let Σ be a κ-consistent set of sentences of $L^2_{\kappa\omega}$, nd let $\mu>\kappa$ be such that $(R(\mu))^{=} = \mu$, $\mathrm{cf}\,\mu = \omega_1$, $\Sigma\in R(\mu)$, and ny $\Gamma\in L^2_{\kappa\omega}$ in $R(\mu)$ has a model in $R(\mu)$ if it has one at all. If : $R(\mu)\prec R(\nu)$ is such that κ is the first ordinal moved by i, there s a least $\tilde{n}<\omega$ by the proof of 6.13(3) such that $i_{\tilde{n}+1}(\kappa)>\mu$, so to how Σ has a model it will suffice to verify P(k) below for each $k\leq\tilde{n}+1$. P(k): Γ has a model for each subset Γ of Σ of power $< i_k(\kappa)$. P(0) is κ-consistency of Σ. If P(k) holds for $k<\tilde{n}+1$, (Δ has a model or each $\Delta\subseteq i(\Sigma)$ of power $< i_{k+1}(\kappa)$ $)^{R(\nu)}$ by elementarity; but if $\Gamma\subseteq\Sigma$ has power $< i_{k+1}(\kappa)$, $\Gamma' = \{i(\varphi) \mid \varphi\in\Gamma\}$ is such a Δ, and as a model iff Γ does. Since (Γ has a model $)^{R(\nu)} \longrightarrow \Gamma$ has model, P(k+1) follows.

$\longleftarrow$). Suppose $\alpha>\kappa$ is given, and let Σ be the following set of entences of $L^2_{\kappa\omega}$, in appropriate constants and relation symbols. a) $\mathrm{Th}(\langle R(\alpha),\in,\langle \underline{t} \mid t\in R(\alpha)\rangle\rangle)$ $\cup$ (b) {the universe is isomorphic o some $R(\beta)$} $\cup$ (c) $\{\forall x\,(x\in\underline{\gamma} \leftrightarrow \bigvee_{\delta<\gamma} x=\underline{\delta}) \mid \gamma<\kappa\}$ $\cup$ (d) $\{c_\eta$ is n ordinal $\mid \eta<\alpha\}\cup\{c_\eta< c_{\tilde\eta} \mid \eta<\tilde\eta<\alpha\}\cup\{\underline{\gamma}<c_\eta \mid \gamma<\kappa,\ \eta<\alpha\}$ $\cup$ (e) $\{c_\eta<\underline{\kappa} \mid \eta<\alpha\}$. The sentence (b) above, which is Π^1_1, appears n the proof of 6.13(4) (c)$\longrightarrow$(b). Σ is κ-consistent, since any $\Gamma\subseteq\Sigma$ f power $<\kappa$ is modelled by $\langle R(\alpha),\in,\langle c_\eta \mid \eta<\tilde\alpha\rangle$ for sufficiently large $\tilde{\alpha}<\kappa$ and increasing $c: \tilde{\alpha} \longrightarrow \kappa$. If $\langle R(\beta), \in, \langle i_t \mid t \in R(\alpha) \rangle, \langle c_\eta \mid \eta < \alpha \rangle \rangle$ models Σ, the correspondence $t \mapsto i_t = i(t)$ elementarily embeds $R(\alpha)$ in $R(\beta)$, and $i(\kappa) > \alpha$ by (e) above.

Proof of (3). Immediate from the proof of (2).

Proof of (4). If $\delta > \kappa$ is arbitrary, let λ be a limit cardinal $> \kappa$ which reflects extendibility, in the sense that ($\tilde{\kappa}$ is η-extendible $)^{R(\lambda)}$ iff $\tilde{\kappa}$ is η-extendible for $\tilde{\kappa}, \eta < \lambda$, and ($\tilde{\kappa}$ is extendible $)^{R(\lambda)}$ iff $\tilde{\kappa}$ is extendible, and let $i: R(\lambda) \prec R(\mu)$ be such that $i \upharpoonright \kappa$ is the identity, and $i(\kappa) > \lambda$; then $i(\kappa)$ is α-extendible for all $\alpha < \mu$.

Proof of (5). (MKC) Suppose C is a closed unbounded class of ordinals, and $S \subseteq C$ is the stationary class of all limit cardinals $\lambda \in C$ such that $\mathrm{cf}\,\lambda = \omega_1$, $\bigcup(\lambda \cap C) = \lambda$ and λ reflects extendibility in the sense of the proof of (4). If $\mathfrak{A}_\nu = \langle R(\nu+2), \in, C \cap \nu \rangle$ for $\nu \in S$, there are $\lambda, \mu \in S$ and $i: \mathfrak{A}_\lambda \prec \mathfrak{A}_\mu$ such that if κ is the least ordinal moved by i (which exists, since $i(\lambda) = \mu$), $\kappa < \lambda$ and κ is α-extendible for all $\alpha < \lambda$ (restrict i to $R(\alpha)$), so extendible in the sense of $R(\lambda)$ since $\lambda \in S$, so extendible, for the same reason.
Now suppose $\sigma = \bigcup(\kappa \cap C) < \kappa$, and $i_{\tilde{n}}(\kappa) < \lambda \leq i_{\tilde{n}+1}(\kappa)$ as in the proof of 6.13(3). By induction on $k \leq n$, $\mathfrak{A}_\mu \models (i_{k+1}(\sigma) = \sigma = \bigcup(i_{k+1}(\kappa) \cap C))$, which contradicts $i_{\tilde{n}+1}(\kappa) \geq \lambda$ and the assumption that $C \cap \lambda$ is unbounded in λ.

Proof of (6). If $i: V \longrightarrow M$ witnesses that κ is almost huge, $\mathcal{A} = \{ \mathfrak{A}_\alpha \mid \alpha < \kappa \}$ is a set of distinct structures of the same type, each in $R(\kappa)$, and $\mathfrak{A} \in i(\mathcal{A}) - \mathcal{A}$, $i \upharpoonright \mathfrak{A} : \mathfrak{A} \prec i(\mathfrak{A})$ is in M, so $\exists \mathfrak{A}_0 \in \mathcal{A}$ $\exists j: \mathfrak{A}_0 \prec \mathfrak{A}$ in V. Since $(R(i(\kappa)))^M = R(i(\kappa))$ and $i(\mathfrak{A}_0) = \mathfrak{A}_0 \in i(\mathcal{A})$, $(R(i(\kappa+1)) \models (\exists \mathfrak{A}_0, \mathfrak{A}_1 \in i(\mathcal{A})$ such that $\mathfrak{A}_0 \neq \mathfrak{A}_1$ and $\exists j: \mathfrak{A}_0 \prec \mathfrak{A}_1))$ so $R(\kappa+1) \models (\exists \mathfrak{A}_0, \mathfrak{A}_1 \in \mathcal{A}$ such that $\mathfrak{A}_0 \neq \mathfrak{A}_1$ and $\exists j: \mathfrak{A}_0 \prec \mathfrak{A}_1$) in V.

6.15 Surmise. ZFC + (There is an extendible cardinal) $\vdash$ $V \neq$ HOD.
Question. Does ZFC refute the existence of an extendible cardinal? For several reasons having to do with the abstrusely impredicative nature of the definition, I think the answer may be yes.

?

6.16 Envoi

" By and by ... by and by ...
There'll be pie in the sky
When you die .... "

- Joe Hill

References

Ba Baumgartner, J., Ineffability properties of cardinals I, Proceedings of the International Colloquium on Infinite and Finite Sets, to appear.

Bo Boos, W., Infinitary compactness without strong inaccessibility, Journal of Symbolic Logic, to appear.

Ch-Ke Chang, C. C. and Keisler, H. J., Model Theory, North Holland (1973).

De 1 Devlin, K., Some weak versions of large cardinal axioms, Annals of Mathematical Logic, 5, 291 - 325 (1973).

De 2 Devlin, K., Aspects of Constructibility, Lecture Notes in Mathematics, Springer (1973).

De 3 Devlin, K., Indescribability Properties and small large cardinals, this volume.

Dr Drake, F., Set Theory: An Introduction to Large Cardinals, North Holland (1973).

Eh-Mo Ehrenfeucht, A. and Mostowski, A., Models of axiomatic theories admitting automorphisms, Fundamenta Mathematicae, 43, 50-68 (1956).

Ga Gaifman, H., Pushing up the measurable cardinal, typescript; a revised version appeared as part of Elementary Embeddings of the Models of Set Theory and Certain Subtheories, Proceedings of Symposia in Pure Mathematics, 13(2), American Mathematical Society, 33-103 (1974).

Gl Gloede, K., Ordinals with partition properties and the constructible hierarchy, Zeitschrift fur Mathematische Logik und Grundlagen der Mathematik, 18, 135-164 (1972).

Jec Jech, T. Some combinatorial problems concerning large cardinals, Annals of Mathematical Logic, 5, 165-198 (1973).

Jen 1 Jensen, R., Grosse Kardinalzahlen, manuscript of lectures at Oberwolfach (1967).

Jen 2 Jensen, R., Some combinatorial properties of L and V, manuscript.

Ke Ketonen, J., Some combinatorial properties, Transactions of the American Mathematical Society, 188, 387-394 (1974).

Ku 1 Kunen, K., Some applications of iterated ultrapowers in set theory, Annals of Mathematical Logic, 1, 179-227.

Ku 2 Kunen, K., Indescribability and the continuum, in Proceedings of Symposia in Pure Mathematics, 13(1), American Mathematical Society, 199-204 (1971).

Ku 3 Kunen, K., Elementary embeddings and infinitary combinatorics, Journal of Symbolic Logic, 36, 407-413 (1971).

Ku 4 Kunen, K., A model for the negation of the axiom of choice, in Lecture Notes in Mathematics, 337, 489-493 (1973).

Lé Lévy, A., The sizes of the indescribable cardinals, in Proceedings of Symposia in Pure Mathematics, 13(1), American Mathematical Society, 205-218 (1971).

Ma Magidor, M., On the role of supercompact and extendible cardinals in logic, Israel Journal of Mathematics, 10, 147-157 (1971).

Pr Prikry, K., Changing measurable into accessible cardinals, Dissertationes Mathematicae, 68 (1970).

Sa Sacks, G., Saturated Model Theory, Benjamin (1972).

Si 1 Silver, J., Some applications of model theory in set theory, Dissertation, University of California (Berkeley) (1966); appeared in revised form in Annals of Mathematical Logic, 3, 45-110 (1971).

Si 2 Silver, J., The consistency of the G. C. H. with the existence of a measurable cardinal, in Proceedings of Symposia in Pure Mathematics, 13(1), 397-428 (1971).

So 1 Solovay, R., A non-constructible Δ^1_3 set of integers, Transactions of the American Mathematical Society, 127, 58-75 (1967).

So 2 Solovay, R., Real-valued measurable cardinals, in Proceedings of Symposia in Pure Mathematics, 13(1), 397-428 (1971).

INDESCRIBABILITY PROPERTIES AND SMALL LARGE CARDINALS

by

Keith J. Devlin (Bonn)

Introduction

These notes were originally written to accompany a course given by the author at the University of Heidelberg during the Summer Term, 1974. The author wishes to thank Professor G.H.Müller for arranging his stay in Heidelberg, and the organisers of the Kiel Conference for extending the invitation to provide these notes with their present resting place.

The material covered is quite standard, though hitherto unpublished in a handy form, and falls naturally into three main sections, §1. Inaccessibility Properties, §2. Indescribability Properties, and §3. Weak Compactness.

We work in ZFC set theory, and use the standard notation and conventions. In particular, V_α denotes the α'th level in the cumulative hierarchy and L_α the α'th level in the constructible hierarchy. We refer the reader to Devlin (1973) for details concerning the constructible hierarchy, and use, as there, $<_L$ to denote the canonical, L-definable well-ordering of L.

With the exception of Theorem 2.18, which is fairly recent, we have avoided the always dangerous practice of crediting the various results. Historians of the field will have no difficulty in obtaining the relevant information elsewhere.

1. Inaccessible Cardinals

The concept of an inaccessible cardinal arises quite naturally in ZF set theory. In fact it is inherent in the intuition motivating the ZF axioms. We think of the universe, V, as being constructed by successive applications of the operations of taking the collection of all sets available at each stage, and forming the collection of all subsets of that collection. The Axiom of Infinity enables us to escape from the finite sets. The Axiom of Replacement facilitates our passing certain limit stages. Suppose then, we introduce a new axiom which says that this process has a closure point; that is, we reach a stage α where V_α is closed under the universe-formation procedure. More precisely, we notice that the crucial axioms in the process are the axioms of power set and replacement, and call a cardinal κ inaccessible iff κ is regular and $(\forall\lambda<\kappa)(2^\lambda<\kappa)$, whence V_κ clearly then satisfies these two axioms. Hence, if κ is inaccessible, V_κ is a model of ZF. But notice now that by closing $\{\emptyset\}$ under a set of skolem functions for V_κ, we can find $\lambda < \kappa$ such that $cf(\lambda) = \omega$ and $V_\lambda \prec V_\kappa$; and since $cf(\lambda) = \omega$, λ is not inaccessible. Thus our notion of κ being inaccessible (i.e. of being a closure point in the cumulative hierarchy) is not quite the same as saying that V_κ is a model of ZF. However, if we consider von Neumann - Bernays - Gödel set theory instead of ZF, we can obtain an exact characterisation in these terms.

Inaccessible Cardinals and NBG Set Theory.

Roughly speaking, NBG set theory is ZF with the notion of a proper class formalised within the theory itself. As the underlying language we take $\mathcal{L}$, our "language of set theory", together with the binary function symbol $\{-,-\}$, only now the variables $v_0, v_1, \ldots$ will range over classes, the primitive objects of NBG. In general we use $A, B, C, \ldots, X, Y, Z$ to denote arbitrary variables or classes. As usual, $A \subseteq B$ abbreviates $\forall X[X \in A \rightarrow X \in B]$. Recalling (or learning) that the German word for "set" is "Menge", we keep in accordance with historical usage and define the predicate M by $M(X) \leftrightarrow \exists Y[X \in Y]$. Note that M is a defined predicate and not a symbol of the theory. Clearly, if $M(X)$, then X is a set. A proper class is a class X such that $\neg M(X)$. We write $a, b, c, \ldots, x, y, z$ to denote sets or "variables" ranging over sets. More precisely, we write "...x..." for "$M(X) \wedge$...X...", "$\forall x$...x..." for "$\forall X[M(X) \rightarrow$ ...X...]", and "$\exists x$...x..." for "$\exists X[M(X) \wedge$...X...]". The axioms of NBG are as follows. (For clarity, we omit non-essential universal quantifiers.)

1. (Extensionality) $\forall x[x \in A \leftrightarrow x \in B] \rightarrow A = B$.
2. (Pairing) $\forall x[x \in \{A,B\} \leftrightarrow M(A) \wedge M(B) \wedge (x = A \vee x = B)]$.
3. (Union) $\exists y \forall x[x \in y \leftrightarrow \exists z(z \in a \wedge x \in z)]$.
4. (Power set) $\exists y \forall x[x \in y \leftrightarrow x \subseteq a]$.
5. (Separation) $\exists y \forall x[x \in y \leftrightarrow x \in a \wedge x \in B]$.
6. (Infinity) $\exists x[\exists y(y \in x) \wedge (\forall y \in x)(\exists z \in x)(y \in z)]$.
7. (Replacement) $\forall x \exists! y[\langle y,x \rangle \in A] \rightarrow \forall a \exists b \forall y[(\exists x \in a)(\langle y,x \rangle \in A) \rightarrow y \in b]$, where $\langle -,- \rangle$ is the function defined by $\langle A,B \rangle = \{\{A,A\},\{A,B\}\}$.
8. (Foundation) $\exists x(x \in A) \rightarrow (\exists x \in A)(\forall y) \neg(y \in x \wedge y \in A)$.
9. (Comprehension) $\exists A \forall x(x \in A \leftrightarrow \varphi(x))$, where $\varphi(x)$ is any formula of the language of NBG which does not contain A, and all of whose bound variables are set variables.

Note that 9. is an infinite axiom schema. It is possible to replace the comprehension schema by a finite set of axioms. Thus NBG is finitely axiomatisable. For details of this, we refer the reader to Gödel (1940).

It should be fairly clear that $Con(NBG) \rightarrow Con(ZF)$. If N is a model of NBG, then $\{x \in N \mid N \models M(x)\}$ is a model of ZF. Conversely, if φ is a sentence in the language of ZF and $NBG \vdash \varphi$, then in fact $ZF \vdash \varphi$. Thus NBG and ZF are "equivalent" with regards to set theory. (For details of this, consult Doets (1969) and Shoenfield (1954).)

Now, for ZF, the "natural models" are those V_κ where κ is inaccessible, or where V_κ is a model of ZF for another reason. A few moments reflection (bearing in mind the definition of "set" in NBG, and the fact that the sets in a model of NBG form a model of ZF) will reveal that a "natural model" of NBG might be a $V_{\kappa+1}$ for some κ (where V_κ is a model of ZF). If we interpret NBG in $V_{\kappa+1}$, then clearly, $V_{\kappa+1} \models M(x)$ iff $x \in V_\kappa$; thus the "proper classes" of this model are those members of $V_{\kappa+1}$ (i.e. those subsets of V_κ) which do not lie in V_κ. Our next theorem shows that V_κ being a model of ZF does not in itself imply that $V_{\kappa+1}$ is a model of NBG, although the converse is of course always true. First a lemma.

Lemma 1

Suppose κ is inaccessible. Let $x \in V_{\kappa+1}$. Then $x \in V_\kappa \leftrightarrow |x| < \kappa$.

Proof: Suppose first that $x \in V_\kappa$. Since $\lim(\kappa)$, $x \subseteq V_\alpha$ for some $\alpha < \kappa$, so it suffices to show that $\alpha < \kappa \rightarrow |V_\alpha| < \kappa$. We do this by induction on α. Firstly, $|V_0| = 0 < \kappa$. Secondly, suppose $|V_\alpha| = \lambda < \kappa$. Then $|V_{\alpha+1}| = |\mathcal{P}(V_\alpha)| = 2^\lambda < \kappa$, since $\mu < \kappa \rightarrow 2^\mu < \kappa$. Finally, suppose $\lim(\alpha)$ and $\beta < \alpha \rightarrow |V_\beta| = \lambda_\beta < \kappa$. Then $|V_\alpha| = |\bigcup_{\beta<\alpha} V_\beta| = \Sigma_{\beta<\alpha}\lambda_\beta < \kappa$, since $cf(\kappa) = \kappa$.

Now suppose that $|x| < \kappa$. Define $f: x \rightarrow \kappa$ by $f(y) = rank(y)$. (Since $x \in V_{\kappa+1}$, $x \subseteq V_\kappa$, so $ran(f) \subseteq \kappa$!) Since $|x| < \kappa$ and $cf(\kappa) = \kappa$, $f''x \subseteq \alpha$ for some $\alpha < \kappa$. But then of course $x \subseteq V_\alpha$, so $x \in V_\kappa$. □

Theorem 2

Let κ be any ordinal. Then κ is an inaccessible cardinal iff $V_{\kappa+1}$ is a model of NBG.

Proof: Suppose $V_{\kappa+1}$ is a model of NBG. Clearly $\lim(\kappa)$. We show that $cf(\kappa) = \kappa$ (and hence, in particular, that κ is a cardinal). Suppose $\lambda < \kappa$ and $f: \lambda \rightarrow \kappa$. Then $f \subseteq V_\kappa$ so $f \in V_{\kappa+1}$. Since $V_{\kappa+1} \models$ "Replacement" and $V_{\kappa+1} \models M(\lambda)$, we have $V_{\kappa+1} \models M(ran(f))$, so $ran(f) \in V_\kappa$. Thus $\sup(f''\lambda) < \kappa$, whence $cf(\kappa) = \kappa$. It remains only to show that $\lambda < \kappa \rightarrow 2^\lambda < \kappa$. Well suppose not. Pick $\lambda < \kappa$ with $\kappa \leq 2^\lambda$. Let $f: \kappa \rightarrow \mathcal{P}(\lambda)$ be one-one, $X = ran(f)$, $g = f^{-1}$. Thus $g: X \leftrightarrow \kappa$. Since $\lambda < \kappa$, $\mathcal{P}(\lambda) \in V_\kappa$, whence $X \in V_\kappa$. Thus $V_{\kappa+1} \models M(X)$. But $g \in V_{\kappa+1}$, so again the replacement axiom in $V_{\kappa+1}$ gives $\kappa = ran(g) \in V_\kappa$, a contradiction.

Conversely, suppose κ is inaccessible. Since $\lim(\kappa)$, it is easily seen that $V_{\kappa+1}$ satisfies all the NBG axioms with the possible exception of the replacement axiom. To verify this axiom, let $x, f \in V_{\kappa+1}$ be such that $V_{\kappa+1} \models M(x) \wedge$ "f is a function with domain x". Since $V_{\kappa+1} \models M(x)$, $x \in V_\kappa$. So by lemma 1, $|x| < \kappa$. Thus $|ran(f)| < \kappa$. So, by lemma 1 again, $ran(f) \in V_\kappa$, whence $V_{\kappa+1} \models M(ran(f))$. □

Weak and Strong Inaccessibility.

We have defined a cardinal κ to be inaccessible, henceforth written $I(\kappa)$, just in case ($\kappa > \omega$ and) $cf(\kappa) = \kappa$ and $\lambda < \kappa \rightarrow 2^\lambda < \kappa$. Such a cardinal must clearly be a limit cardinal. We call κ a weakly inaccessible cardinal, and write $WI(\kappa)$, just in case ($\kappa > \omega$ and) $cf(\kappa) = \kappa$ and $\lambda < \kappa \rightarrow \lambda^+ < \kappa$; i.e. $WI(\kappa)$ means that κ is an uncountable regular limit cardinal. Such a cardinal is clearly a fixed point in the "aleph" sequence (i.e. $WI(\kappa) \rightarrow \omega_\kappa = \kappa$.) In the context of weak inaccessibility, "inaccessibility" is often referred to as "strong inaccessibility", but we shall continue to use our previous termonology here. Clearly, if we assume GCH, then I and WI coincide. However, there are models of ZFC in which they do not; for example, 2^ω can be WI in a model of ZFC (See any introductory treatise on forcing, say Jech (1971).).

Inaccessibility and Indescribability.

Above, we arrived at the notion of an inaccessible cardinal by considering the intuition behind the construction of the cumulative hierarchy. In order to be a closure point in the hierarchy, an "inaccessible" cardinal has to be "much larger" than all smaller cardinals. We can also obtain a notion of "large cardinal" by approaching it from below, using metamathematical notions only.

We say an ordinal α is first-order describable if there are predicates $U_0 \subseteq V_\alpha^{n_0}, \ldots, U_r \subseteq V_\alpha^{n_r}$ such that for some first-order sentence σ in the language $\mathcal{L}(U_0, \ldots, U_r)$, the language obtained from $\mathcal{L}$ by the addition of an n_i-ary predicate letter U_i for each $i \leq r$, $\langle V_\alpha, \in, U_0, \ldots, U_r \rangle \models \sigma$ but for all $\beta < \alpha$, $\langle V_\beta, \in, U_0 \cap V_\beta^{n_0}, \ldots, U_r \cap V_\beta^{n_r} \rangle \models \neg\sigma$. (Thus, α is describable, in terms of some predicates on V_α, as the first ordinal which satisfies a certain first-order sentence.) We say α is first-order indescribable just in case it is not first-order describable.

Theorem 3

Let κ be an ordinal. Then $I(\kappa)$ iff κ is first-order indescribable.

Proof: Suppose first that $I(\kappa)$. Let $U_0 \subseteq V_\kappa^{n_0}, \ldots, U_r \subseteq V_\kappa^{n_r}$ be given, and let σ be a sentence of $\mathcal{L}(U_0, \ldots, U_r)$ such that (with an obvious abuse of notation) $V_\kappa \models \sigma(U_0, \ldots, U_r)$. We seek an $\alpha < \kappa$ such that (with a further abuse of notation) $V_\alpha \models \sigma(U_0, \ldots, U_r)$. Let $\{f_n | n < \omega\}$ be a complete set of skolem functions for the structure $\langle V_\kappa, \in, U_0, \ldots, U_r \rangle$. Let α be the least ordinal such that $(\forall n < \omega)(f_n "V_\alpha^{k(n)} \subseteq V_\alpha)$, where f_n is $k(n)$-ary. Since $I(\kappa)$, such an α is clearly less than κ. But then $\langle V_\alpha, \in, U_0, \ldots, U_r \rangle \prec \langle V_\kappa, \in, U_0, \ldots, U_r \rangle$ (with a similar abuse of notation as before), so α is as sought.

Conversely, suppose κ is first-order indescribable. Then $\lim(\kappa)$, for if $\kappa = \alpha + 1$, then κ is first-order describable by means of the unary predicate $U = \{\alpha\}$ and the sentence $\sigma = \exists x U(x)$. Hence $V_{\kappa+1}$ satisfies all of the axioms of NBG set theory with the possible exception of the replacement axiom. By Theorem 2, therefore, it suffices to show that the NBG replacement axiom does in fact hold in $V_{\kappa+1}$. So let $u, f \in V_{\kappa+1}$ be such that $V_{\kappa+1} \models$ "f is a function with domain the set u". Since u is a set here, $u \in V_\alpha$ for some $\alpha < \kappa$. Let $U_0 = \{\alpha\}$, $U_1 = u$, $U_2 = f$, and let σ be the sentence $\exists x U_0(x) \wedge \forall x \exists y(U_1(x) \rightarrow U_2(y,x))$. Since $V_\kappa \vdash \sigma(U_0,U_1,U_2)$, there is $\gamma<\kappa$ such that $V_\gamma \models \sigma(U_0,U_1,U_2)$. Since $V_\gamma \models \exists x U_0(x)$, $\gamma > \alpha$, whence $U \subseteq V_\gamma$. Hence, as f is a function, $ran(f) = f''u \subseteq V_\gamma$. Thus $ran(f) \in V_\kappa$, which means that $V_{\kappa+1} \models$ "ran(f) is a set". □

<u>Inaccessible Cardinals and L.</u>

Clearly, if $V=L$, then, by GCH, the notions I and WI coincide. However, for this to be at all a meaningful statement, we must check that the existence of such cardinals is compatible with $V=L$. Of course, as with any axiom, an initial act of faith is required concerning the consistency; we <u>assume</u> that the existence of, say, an inaccessible cardinal does not lead to a contradiction with ZFC. With regards to L, however, once the initial act of faith concerning the consistency of $ZFC + \exists\kappa I(\kappa)$ has been made, we can <u>prove</u> the consistency with $ZFC + V=L$. All that we need to know is the trivial fact that Π_1^{ZF} sentences relativise to inner models of ZFC.

<u>Aside</u>: It is the author's viewpoint that consistency is the only point at issue here, and that the question as to the "existence" of inaccessible cardinals is totally meaningless. To us, large cardinal theory is a (worthwhile) structure theory, nothing more.

<u>Lemma 4</u>

The predicates $I(\kappa)$ and $WI(\kappa)$ are Π_1^{ZF} predicates of κ.

<u>Proof</u>: κ is regular $\leftrightarrow \neg(\exists f)(\exists\lambda)(\lambda<\kappa \wedge f:\lambda\rightarrow\kappa \wedge f''\lambda$ is cofinal in $\kappa)$.
Hence the predicate $Reg(\kappa) \leftrightarrow$ " κ is a regular cardinal" is Π_1^{ZF}. But,

$I(\kappa) \leftrightarrow Reg(\kappa) \wedge \neg(\exists f)(\exists\lambda)[\lambda<\kappa \wedge f$ is a one-one function $\wedge\ dom(f)=\kappa\ \wedge$
$(\forall\alpha\in\kappa)(f(\alpha)\subseteq\lambda)]$, and

$WI(\kappa) \leftrightarrow Reg(\kappa) \wedge \neg(\exists f)(\exists\lambda)[\lambda<\kappa \wedge f:\lambda\times\kappa\rightarrow\kappa \wedge (\forall\alpha\in\kappa)(f(-,\alpha)$ maps λ onto $\alpha)]$.

Clearly, both of these predicates are Π_1^{ZF}. □

<u>Theorem 5</u>

(i) If $I(\kappa)$, then $I^L(\kappa)$; if $WI(\kappa)$, then $WI^L(\kappa)$.

(ii) If $WI(\kappa)$, then $I^L(\kappa)$; hence $Con(ZFC + \exists\kappa I(\kappa)) \leftrightarrow Con(ZFC + \exists\kappa WI(\kappa))$. □

<u>Mahlo Cardinals.</u>

Once we consider the existence of inaccessible cardinals, we are led fairly naturally to an entire hierarchy of large cardinals. A cardinal κ is <u>hyperinaccessible</u> if it is an inaccessible fixed-point in the sequence of all inaccessibles (i.e. if $I(\kappa)$ and $|\{\lambda<\kappa \mid I(\lambda)\}| = \kappa$.). Similarly <u>hyper-hyperinaccessible</u>, and so on. However, it is intuitively clear that this procedure does not lead to any cardinals significantly "larger" than the inaccessible cardinals themselves. (In other words, the notion of "largeness" is the same.) A much larger sort of cardinal is obtained by replacing the notion of "unbounded" by that of "stationary" in the definition of hyperinaccessibility.

Let κ be any cardinal. A set $A \subseteq \kappa$ is <u>stationary</u> in κ if $A \cap C \neq \emptyset$ for every closed unbounded subset C of κ. Clearly, every closed unbounded subset of an uncountable regular cardinal κ is stationary in κ. The notion of a stationary subset of a cardinal κ thus falls between the notions of "unbounded" and "closed and unbounded", at least as far as uncountable regular cardinals are concerned.

A cardinal κ is <u>Mahlo</u> iff $I(\kappa)$ and $\{\lambda < \kappa | I(\lambda)\}$ is stationary in κ.

<u>Theorem 6</u>

Let κ be inaccessible. The following are equivalent:

(i) κ is Mahlo.

(ii) $\{\lambda \in \kappa | \lambda$ is regular$\}$ is stationary in κ.

(iii) Every normal function $f: \kappa \to \kappa$ has a regular/inaccessible/hyperinaccessible/ etc. fixed point.

<u>Proof:</u> Trivial. □

Corresponding to Theorem 3 we have:

<u>Theorem 7</u>

Let κ be an ordinal. Then κ is Mahlo iff whenever $U_0 \subseteq V_\kappa^{n_0}, \ldots, U_r \subseteq V_\kappa^{n_r}$ and σ is a first-order sentence in the language $\mathcal{L}(U_0, \ldots, U_r)$ such that $V_\kappa \models \sigma(U_0, \ldots, U_r)$, there is a regular cardinal $\lambda < \kappa$ such that $V_\lambda \models \sigma(U_0, \ldots, U_r)$.

<u>Proof:</u> Assume κ is Mahlo. Let $A = \{\alpha \in \kappa | \langle V_\alpha, \in, U_0, \ldots, U_r\rangle \prec \langle V_\kappa, \in, U_0, \ldots, U_r\rangle\}$. Clearly, A is closed and unbounded in κ, so let $\lambda \in A$ be regular. Then λ is as in the stated property.

Conversely, assume the stated property. Let $A \subseteq \kappa$ be closed and unbounded. Let $\sigma = \forall\alpha \exists\beta(\beta > \alpha \wedge \beta \in A)$. Clearly, $V_\kappa \models \sigma(A)$. Let $\lambda < \kappa$ be regular with $V_\lambda \models \sigma(A \cap \lambda)$. Then $\sup(A \cap \lambda) = \lambda$, so as A is closed, $\lambda \in A$, whence κ is Mahlo. □

And corresponding to Theorem 5 we have:

<u>Theorem 8</u>

If κ is Mahlo, then κ is Mahlo in L.

<u>Proof:</u> Let $A \in L$ be closed and unbounded in κ in the sense of L. Then A really is closed and unbounded in κ, so the result follows from Theorem 5. □

2. <u>Indescribability Properties.</u>

We have already met the notion of first-order indescribability in §1. In this section we shall obtain large cardinal axioms by considering natural generalisations of this concept, based on the Lévy hierarchy of formulas of higher type.

As before, we consider languages of the form $\mathcal{L}(U_0, \ldots, U_r)$, consisting of $\mathcal{L}$ with the additional predicates $U_0, \ldots, U_r$, where U_i is $k(i)$-ary. We also use the same notational conveniences as before. Thus, for instance, we write $\langle V_\kappa, \in,$

$U_0,\ldots,U_r\rangle$ instead of $\langle V_\kappa,\in,U_0\cap V_\kappa^{k(0)},\ldots,U_r\cap V_\kappa^{k(r)}\rangle$. For each language $\mathcal{L}(U_0,\ldots,U_r)$, we let $\mathcal{L}^*(U_0,\ldots,U_r)$ denote the finite-type extension of $\mathcal{L}(U_0,\ldots,U_r)$, obtained by introducing new variables of type 1 to range over properties, new variables of type 2 to range over properties of properties, etc. (Thus $U_0,\ldots,U_r$ are "constants" of type 1.) We assume the reader understands the (obvious) notions of interpretation and satisfaction for $\mathcal{L}^*(U_0,\ldots,U_r)$ sentences in structures of the form $\langle V_\kappa,\in,U_0,\ldots,U_r\rangle$. We use superscripts to indicate the type level of variables, with variables of type 0 being the usual set variables of $\mathcal{L}$. A formula of $\mathcal{L}^*(U_0,\ldots,U_r)$ is Π_n^m if it is of the form $\forall\vec{x}_1^m\exists\vec{x}_2^m\forall\vec{x}_3^m\ldots\vec{x}_n^m\varphi$, where φ contains no bound variables of type greater than m-1, and where the entire formula contains no variables of type greater than m. Dually Σ_n^m.

An ordinal κ is Π_n^m - describable if there are $U_0\subseteq V_\kappa^{k(0)},\ldots,U_r\subseteq V_\kappa^{k(r)}$ and a Π_n^m sentence σ of $\mathcal{L}^*(U_0,\ldots,U_r)$ such that $V_\kappa\models\sigma(U_0,\ldots,U_r)$ but for all $\alpha<\kappa$, $V_\alpha\models\neg\sigma(U_0,\ldots,U_r)$. Otherwise κ is Π_n^m-indescribable. Similarly with Σ_n^m in place of Π_n^m here.

We shall consider Π_n^m - indescribable and Σ_n^m - indescribable ordinals, where $m>0$. By Theorem 1.3, we see that we shall always be dealing with inaccessible cardinals. So, since the formula $\lim(\alpha)$ is first-order, we can always assume that the structures V_α which we come accross have $\lim(\alpha)$. The advantage of this is that V_α is closed under the formation of ordered-pairs if $\lim(\alpha)$, so we can always contract quantifiers without affecting satisfaction within V_α.(e.g. a given Π_n^m formula may always be assumed to have the form $\forall x_1^m\exists x_2^m\forall x_3^m\ldots x_n^m\varphi$, where φ has no bound variables of type greater than m-1, etc.) The exact procedure for achieving this contraction of quantifiers is developed by induction on the type, and we leave it to the reader to supply a suitable definition. By a similar procedure we may in fact always assume that the fixed predicates $U_0,\ldots,U_r$ are all unary, and furthermore that there is in fact only one such predicate. The following lemma sums up all of these remarks, and is "easily" proved.

Lemma 1

Assume $\lim(\alpha)$ and let $U_0\subseteq V_\alpha^{k(0)},\ldots,U_r\subseteq V_\alpha^{k(r)}$. Let σ be a Π_n^m (resp. Σ_n^m) formula of $\mathcal{L}^*(U_0,\ldots,U_r)$. There is a single set $U\subseteq V$ and a Π_n^m (resp. Σ_n^m) formula φ of $\mathcal{L}^*(U)$ of the form $\forall x_1^m\exists x_2^m\forall x_3^m\ldots x_n^m\psi$ (resp. $\exists x_1^m\forall x_2^m\exists x_3^m\ldots x_n^m\psi$) such that for any limit ordinal $\beta\leq\alpha$, $V_\beta\models\sigma(U_0,\ldots,U_r)$ iff $V_\beta\models\sigma(U)$. □

Corollary 2

Let κ be a cardinal. Then κ is Π_m^m (resp. Σ_n^m) - indescribable iff for every $U\subseteq V_\kappa$ and every Π_n^m (resp. Σ_n^m) sentence σ of $\mathcal{L}^*(U)$ (of the above "canonical" form) such that $V_\kappa\models\sigma(U)$, there is a limit ordinal $\alpha<\kappa$ such that $V_\alpha\models\sigma(U)$. □

We shall first concentrate on the case $m=1$, and then see which of our results generalise to the more general case. It turns out that in this case we need only consider the Π^m_n - indescribable cardinals, by virtue of:

Lemma 3

Let κ be a cardinal, $n\in\omega$. Then κ is Σ^1_{n+1} - indescribable iff it is Π^1_n - indescribable.

Proof: One way is trivial. For the other, suppose κ is Π^1_n - indescribable. Let $U \subseteq V_\kappa$ and let σ be a Σ^1_{n+1} sentence of $\mathcal{L}^*(U)$ such that $V_\kappa \models \sigma(U)$. Let φ be the Π^1_n formula such that $\sigma = \exists x^1 \varphi(x^1, U)$. Pick $W \subseteq V_\kappa$ so that $V_\kappa \models \varphi(W,U)$. Then for some $\alpha < \kappa$, $V_\alpha \models \varphi(W,U)$, so $V_\alpha \models \sigma(U)$. Hence κ is Σ^1_{n+1}- indescribable. □

Corollary 4

Let κ be a cardinal. Then κ is Σ^1_1 - indescribable iff $I(\kappa)$.

Proof: By Theorem 1.3. □

For each m,n, we let π^m_n, σ^m_n denote the least Π^m_n, Σ^m_n - indescribable cardinal, respectively. Thus, by lemma 3, $\sigma^1_{n+1} = \pi^1_n$ for all n, and by corollary 4, σ^1_1 is just the first inaccessible cardinal. In order to investigate the relationship between π^1_n and π^1_{n+1}, however, we require some extra metamathematical machinery.

For convenience, we use x,y,z, etc. to denote variables of type 0 and X,Y,Z, etc. for variables of type 1. Fix $n > 0$ from now on. Attach Gödel numbers to the formulas of $\mathcal{L}(U,X_1,\ldots,X_n)$ in some canonical, effective manner, so that ω and the formulas of this language are in one-one correspondence (again in an effective manner).

Let Sat be the (canonically defined) predicate such that Sat(R) iff R is a satisfaction class for $\mathcal{L}(U,X_1,\ldots,X_n)$. More precisely, Sat(R) holds just in case $(\forall r)(\forall \langle\vec{x}\rangle)(R(r,\langle\vec{x}\rangle) \leftrightarrow \ulcorner r \urcorner [\vec{x}])$, where $\ulcorner r \urcorner$ denotes the formula with Gödel number r. There is a first-order formula φ of $\mathcal{L}(R,U,X_1,\ldots,X_n)$, with free variables r,x, both of type 0, such that whenever $\lim(\alpha)$ and $R,U,X_1,\ldots,X_n \subseteq V_\alpha$, $\langle V_\alpha, \in, U, X_1,\ldots,X_n\rangle \models \mathrm{Sat}(R)$ iff $\langle V_\alpha, \in, R, U, X_1,\ldots,X_n\rangle \models \forall r \forall x\, \varphi(r,x)$. Thus Sat(R) is a first-order property in $\mathcal{L}(R,U,X_1,\ldots,X_n)$. One defines the formula φ thus:

$\varphi(r,x) \equiv [\, r\in\omega \wedge x$ is a finite sequence of sets $\wedge \ulcorner r \urcorner$ has at most all its free variables in the set $\{v_0,\ldots,v_k\}$, where $k-1 = lh(x)\,] . \longrightarrow .$

$\ulcorner r \urcorner = U(v_i) \wedge [R(r,x) \leftrightarrow U((x)_i)]$

$\vee \ulcorner r \urcorner = X_1(v_i) \wedge [R(r,x) \leftrightarrow X_1((x)_i)]$

$\vee \ldots$

$\vee \ulcorner r \urcorner = \ulcorner r_1 \urcorner \vee \ulcorner r_2 \urcorner \wedge [R(r,x) \leftrightarrow R(r_1,x) \vee R(r_2,x)]$

$\vee \ulcorner r \urcorner = \neg \ulcorner s \urcorner \wedge [R(r,x) \leftrightarrow \neg R(s,x)]$

$\vee \ulcorner r \urcorner = \exists v_i \ulcorner s \urcorner \wedge [R(r,x) \leftrightarrow \exists u R(s, x(u/i)))$

$\vee \ulcorner r \urcorner = v_i \in v_j \wedge [R(r,x) \leftrightarrow (x)_i \in (x)_j]$

$\vee \ulcorner r \urcorner = v_i = v_j \wedge [R(r,x) \leftrightarrow (x)_i = (x)_j]$,
where $\{v_i | i \in \omega\}$ is the canonical enumeration of the variables of $\mathcal{L}(U,X_1,\ldots,X_n)$, and where we use $x(u/i)$ to denote the sequence obtained from x by substituting u at the i'th place.

We have, of course, not really defined the formula φ, since we have used many clauses which are not 'prima face' first-order in $\langle V_\alpha, \in, R, U, X_1, \ldots, X_n \rangle$, for instance the initial demands on r,x, and the clauses $\ulcorner r \urcorner = \ulcorner r_1 \urcorner \vee \ulcorner r_2 \urcorner$, etc., but these are essentially trivial matters, depending on the actual definition of the language as a set-theoretical structure (we assume, as usual, that our language is a recursive subset of V_ω). In essence, however, we have shown that Sat is a first-order property on a predicate R.

Now, for a given structure $\langle V_\alpha, \in, U, X_1, \ldots, X_n \rangle$, where $\lim(\alpha)$, there is clearly just one set $R \subseteq V_\alpha$ such that $\mathrm{Sat}(R)$, namely the satisfaction relation for $\langle V_\alpha, \ldots \rangle$. Hence the relation (on r) $\exists R(\mathrm{Sat}(R) \wedge R(r,\emptyset))$ and $\forall R(\mathrm{Sat}(R) \rightarrow R(r,\emptyset))$ are equivalent in $\langle V_\alpha, \in, U, X_1, \ldots, X_n \rangle$. And clearly, $\langle V_\alpha, \in, U, X_1, \ldots, X_n \rangle \models \exists R(\mathrm{Sat}(R) \wedge R(r,\emptyset))$ just in case $V_\alpha \models \ulcorner r \urcorner (U, X_1, \ldots, X_n)$. So if we let $\Phi(r)$ denote the $\mathcal{L}^*(U, X_1, \ldots, X_n)$ formula $\exists R(\mathrm{Sat}(R) \wedge R(r,\emptyset))$ and $\Psi(r)$ the $\mathcal{L}^*(U,X_1,\ldots,X_n)$ formula $\forall R(\mathrm{Sat}(R) \rightarrow R(r,\emptyset))$, we obtain:

<u>Lemma 5</u>

There is a Σ^1_1-formula $\Phi(r)$ of $\mathcal{L}^*(U,X_1,\ldots,X_n)$ and a Π^1_1-formula $\Psi(r)$ of $\mathcal{L}^*(U,X_1,\ldots,X_n)$ such that for every sentence φ of $\mathcal{L}(U,X_1,\ldots,X_n)$ there is an integer r such that, whenever $\lim(\alpha)$ and $U,X_1,\ldots,X_n \subseteq V_\alpha$,

$V_\alpha \models \varphi(U,X_1,\ldots,X_n)$ iff $V_\alpha \models \Phi(r,U,X_1,\ldots,X_n)$ iff $V_\alpha \models \Psi(r,U,X_1,\ldots,X_n)$. □

Using lemma 5, we at once obtain:

<u>Lemma 6</u> (Uniform Enumeration)

There is a Π^1_n (resp. Σ^1_m) formula $\Gamma_n(r)$ of $\mathcal{L}^*(U)$ such that whenever φ is a Π^1_n (resp. Σ^1_n) sentence of $\mathcal{L}^*(U)$, there is $r \in \omega$ such that whenever $\lim(\alpha)$ and $U \subseteq V_\alpha$, $V_\alpha \models \varphi(U)$ iff $V_\alpha \models \Gamma_n(r,U)$.

<u>Proof</u>: If n is even, set $\Gamma_n(r) = \forall X_1 \exists X_2 \forall X_3 \ldots \exists X_n \Phi(r)$, with Φ as in lemma 5, and if n is odd, $\Gamma_n(r) = \forall X_1 \exists X_2 \forall X_3 \ldots \forall X_n \Psi(r)$, with Ψ as in lemma 5. □

For our given, fixed $n > 0$, we set $\Gamma(r)$ $\Gamma_n(r)$ from now on.

<u>Lemma 7</u>

There is a Π^1_{n+1} sentence Θ of $\mathcal{L}^*$ such that $V_\kappa \models \Theta$ iff κ is Π^1_n-indescribable.

<u>Proof</u>: Set $\Theta = \forall U \forall r [\Gamma(r) \rightarrow \exists \alpha(\lim(\alpha) \wedge \langle V_\alpha, \in, U \cap V_\alpha \rangle \models \Gamma(r))]$. □

<u>Theorem 8</u>

(i) If κ is Π^1_1 - indescribable, then κ is Mahlo.

(ii) If κ is Π^1_{n+1}-indescribable, then $\{\lambda\in\kappa \mid \lambda$ is Π^1_n-indescribable$\}$ is stationary in κ.

(iii) For all n, $\pi^1_n < \pi^1_{n+1}$.

Proof: (i) Let $C \subseteq \kappa$ be closed and unbounded in κ. Let σ be the obvious Π^1_1 sentence of $\mathcal{L}^*$ such that for all α, $V_\alpha \models \sigma$ iff α is regular. Now, $V_\kappa \models \sigma \wedge \forall\alpha\,\exists\beta(\alpha<\beta \wedge \beta\in C)$ (C), so we can find $\alpha<\kappa$ so that $V_\alpha \models \sigma$ and $\sup(C\cap\alpha)=\alpha$. Since C is closed, $\alpha\in C$. Since $V_\alpha\models\sigma$, α is regular.

(ii) Proceed as in (i), but use the sentence Θ of lemma 7 instead of σ.

(iii) From (i) and (ii). □

Finally, we show that the Π^1_n and Σ^1_n-indescribable cardinals all relativise to L.

Lemma 9

There is a Σ^1_1 sentence, Δ, of $\mathcal{L}^*(X)$ such that for any inaccessible cardinal κ, and any set $X\subseteq L_\kappa$, $X\in L \leftrightarrow V_\kappa\models\Delta(X)$. (Notice that $L_\kappa = V^L_\kappa$ here.)

Proof: Let σ be the conjunction of sufficiently many axioms of ZF + V = L (with the axiom of extensionality explicitly included) so that whenever M is a transitive model of σ, $M = L_\alpha$ for some α.

Let $X\subseteq L_\kappa$ be given. Now, if $X\in L$, then $X\in L_\gamma$ for some $\gamma<\kappa^+$. And as $|\gamma|\leq\kappa$, we can code up L_γ within V_κ. Hence,

$X\in L \longleftrightarrow (\exists E\subseteq V_\kappa\times V_\kappa)[\,E$ is well-founded $\wedge \langle V_\kappa,E\rangle\models\sigma \wedge (\exists F\subseteq V_\kappa\times V_\kappa)(\exists x\in V_\kappa)$ $[(F$ is a function$)\wedge(\mathrm{dom}(F)=\{z\in V_\kappa \mid zEx\})\wedge(\forall z\in\mathrm{dom}(F))(F(z)=\{F(y)\mid y\in V_\kappa \wedge yEz\})\wedge(\mathrm{ran}(F)=X)]]$.

Now, if $I(\kappa)$ and $E\subseteq V_\kappa\times V_\kappa$, $V_\kappa\models$"E is well-founded" iff $V_\kappa\models (\exists G\subseteq V_\kappa\times V_\kappa)$ $(\forall f)[(f:\omega\to V_\kappa)\to(G(f):f''\omega\to\kappa\wedge(\forall x,y\in f''\omega)(xEy\to G(f)(x)<G(f)(y)))]$, which is clearly Σ^1_1 over V_κ.

Also, the sentence $\langle V_\kappa,E\rangle\models\sigma$ can be written out as a (first-order) sentence of $\mathcal{L}(E)$, of course. (Just write out σ with E in place of $\in$!)

Hence, the above expression can be rearranged to give the required sentence Δ. □

Lemma 10

Let κ be a cardinal. Then κ is Π^1_n-indescribable iff, given any $U\subseteq V_\kappa$ and any Π^1_n-sentence σ of $\mathcal{L}^*(U)$ such that $V_\kappa\models\sigma(U)$, there is an inaccessible cardinal $\lambda<\kappa$ such that $V_\lambda\models\sigma(U)$.

Proof: Since there is a Π^1 sentence φ of $\mathcal{L}^*$ such that $V_\alpha\models\varphi$ iff α is an inaccessible cardinal. □

Theorem 11

Let κ be Π^1_n-indescribable. Then κ is Π^1_n-indescribable in the sense of L.

Proof: Since $I(\kappa)$, $I^L(\kappa)$, so $V^L_\kappa = L_\kappa$. In L, let $U\subseteq L_\kappa$, and let σ be a Π^1_n sentence of $\mathcal{L}^*(U)$ such that $L_\kappa\models\sigma(U)$, say $\sigma = \forall X_1\exists X_2\forall X_3\ldots X_n\varphi$. Thus, in V, we have

$V_\kappa \models (\forall X_1 \in L)(\exists X_2 \in L)(\forall X_3 \in L)\ldots(-X_n \in L)\varphi^L(U)$. But by lemma 9, the quantifiers $(\forall X_i \in L)$ are Π^1_1 and the quantifiers $(\exists X_i \in L)$ are Σ^1_1, over V_κ. Hence as κ is Π^1_n-indescribable, there is an inaccessible $\lambda < \kappa$ such that $V_\lambda \models (\forall X_1 \in L)(\exists X_2 \in L)(\forall X_3 \in L)\ldots(-X_n \in L)\varphi^L(U)$. Again, $V^L_\lambda = L_\lambda$, so this means $L_\lambda \models \sigma(U)$, and we are done. □

Let us now turn to the general case, $m > 1$. We shall continue to use $x, y, z, \ldots$ to denote type 0 variables and $X, Y, Z, \ldots$ type 1 variables, with $X^m, Y^m, Z^m, \ldots$ for type m variables.

By a straightforward generalisation of the proof of lemmas 5 and 6, we obtain:

Lemma 12 (Uniform Enumeration)

There is a Π^m_n (resp. Σ^m_n) formula $\Gamma = \Gamma(r)$ of $\mathcal{L}^*(U)$ such that whenever φ is a Π^m_n (resp. Σ^m_n) sentence of $\mathcal{L}^*(U)$, there is $r \in \omega$ such that, whenever $\lim(\alpha)$ and $U \subseteq V_\alpha$, $V_\alpha \models \varphi(U)$ iff $V_\alpha \models \Gamma(r, U)$. □

Corresponding to lemma 7, we have:

Lemma 13

(i) There is a Σ^m_n sentence Θ of $\mathcal{L}^*$ such that $V_\kappa \models \Theta$ iff κ is Π^m_n-indescribable.
(ii) There is a Π^m_n sentence Θ' of $\mathcal{L}^*$ such that $V_\kappa \models \Theta'$ iff κ is Σ^m_n-indescribable.

Proof: (i) As in lemma 7. (Since $m > 1$, the initial "$\forall U$" quantifier can be ignored here.)
(ii) By the same proof as for (i), using the Σ^m_n Γ rather than the Π^m_n Γ. □

Of course, in lemma 7, we did not formulate the corresponding version of part (ii) of the above, having no need for it. This was because Theorem 8 and lemma 3 together gave all the information regarding the relative sizes of π^1_n, π^1_{n+1}, σ^1_n, σ^1_{n+1}, etc. For the case $m > 1$ considered here, however, we shall require all of lemma 13. Firstly, as in Theorem 8, we have:

Theorem 14

(i) If κ is Π^m_{n+1}-indescribable, then $\{\lambda \in \kappa \mid \lambda$ is Π^m_n-indescribable$\}$ and $\{\lambda \in \kappa \mid \lambda$ is Σ^m_n-indescribable$\}$ are stationary in κ.
(ii) If κ is Σ^m_{n+1}-indescribable, then $\{\lambda \in \kappa \mid \lambda$ is Π^m_n-indescribable$\}$ and $\{\lambda \in \kappa \mid \lambda$ is Σ^m_n-indescribable$\}$ are stationary in κ.
(iii) For all n, $\pi^m_n < \pi^m_{n+1}, \sigma^m_{n+1}$ and $\sigma^m_n < \pi^m_{n+1}, \sigma^m_{n+1}$. □

There thus remains only the comparison of the size of π^m_n and σ^m_n. Well, we have, by a simple application of lemma 13:

Lemma 15

For all $n > 0$, $\sigma^m_n \neq \pi^m_n$. □

Unfortunately, it is not at present known whether $\sigma^m_n < \pi^m_n$ or $\pi^m_n < \sigma^m_n$ holds. We thus have the following picture:

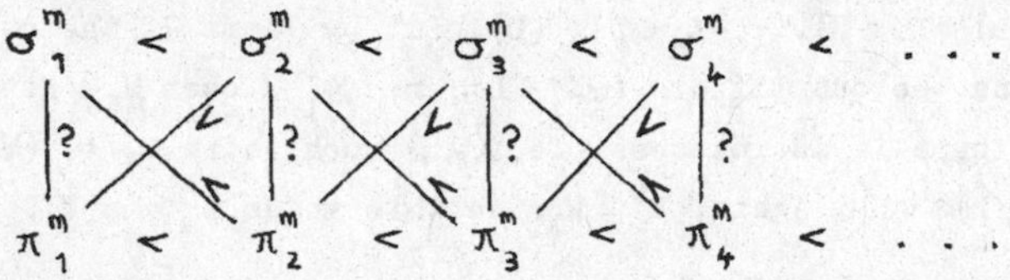

However, following some work of Aanderaa, Moschovakis has answered the question in L. First, of course, we must show that Σ_n^m and Π_n^m-indescribable cardinals relativise to L. Well, corresponding to lemma 9, and proved in an analogous manner, we have:

Lemma 16

There is a Σ_1^m formula $\Delta(X^m)$ of $\mathcal{L}^*$ such that for any inaccessible cardinal κ and any type m object X^m over L_κ, $X^m \in L \leftrightarrow V_\kappa \models \Delta(X^m)$. □

Note that, strictly speaking, the above is proved by induction on m, since we must say that all of the elements of X^m are constructible, etc.

Arguing as in Theorem 11, we then have:

Theorem 17

If κ is Π_n^m-indescribable (resp. Σ_n^m-indescribable), then κ is Π_n^m-indescribable (resp. Σ_n^m-indescribable) in the sense of L. □

Theorem 18 (Moschovakis)

Assume $V=L$. Let $m>1$, $n>0$. Then $\sigma_n^m < \pi_n^m$.

Proof: For the sake of argument we shall take the case $m=2$, $n=1$. The general case is entirely similar.

For each Σ_1^2 formula $\psi(n,X)$ of $\mathcal{L}^*$ and each cardinal κ, let $\mathcal{S}_{\psi,\kappa} = \{(n,X) \mid V_\kappa \models \psi(n,X)\}$.

CLAIM: For each Σ_1^2 formula $\psi(n,X)$ of $\mathcal{L}^*$ there is a Π_1^2 formula $O_\psi(n,X;n',X')$ of $\mathcal{L}^*$ such that whenever κ is an inaccessible cardinal, there is a map $\rho_{\psi,\kappa}: \mathcal{S}_{\psi,\kappa} \to \mathrm{On}$ such that, given $(n',X') \in \mathcal{S}_{\psi,\kappa}$:

(1) $(\forall n,X)[(n,X) \in \mathcal{S}_{\psi,\kappa} \wedge \rho_{\psi,\kappa}(n,X) \leq \rho_{\psi,\kappa}(n',X') . \leftrightarrow . V_\kappa \models O_\psi(n,X;n',X')]$.

We leave the proof of the claim for the time being, and prove the theorem from it. Set $\psi(n,X) = T(n,X) \wedge \forall u \forall \lambda [I(\lambda) \wedge u = V_\lambda \rightarrow u \not\models \neg T(n,X)]$, where T is the universal formula for Σ_1^2 predicates given by lemma 12 (regarded as a formula of $\mathcal{L}^*$ rather than a formula of $\mathcal{L}^*(X)$.).

Suppose the theorem is false. Let $\kappa = \pi_1^2$. Since κ must be Σ_1^2-describable, we can find n,X such that $V_\kappa \models \psi(n,X)$. Pick n,X here with $\rho_{\psi,\kappa}(n,X)$ minimal. Thus, if n',X' are given, and $V_\kappa \models \psi(n',X')$, then $\rho_{\psi,\kappa}(n,X) \leq \rho_{\psi,\kappa}(n',X')$, so by (1), $V_\kappa \models (\forall n',X')[\neg\psi(n',X') \vee O_\psi(n,X;n',X')]$. But this sentence is Π_1^2, so there is an inaccessible $\lambda < \kappa$ with:

(2) $V_\lambda \models (\forall n',X')[\neg\psi(n',X') \vee O_\psi(n,X;n',X')]$.

Again, since $\kappa < \sigma^2_1$, λ is Σ^2_1-describable, so we can find n',X' with:

(3) $V_\lambda \models \psi(n',X')$.

By (2) and (3):

(4) $V_\lambda \models O_\psi(n,X;n',X')$.

By (1), (3), and (4), therefore, $V_\lambda \models \psi(n,X)$. But $\lambda<\kappa$ is inaccessible, so as $V_\kappa \models \psi(n,X)$, this is absurd (Recall the definition of ψ.). This proves the theorem.

We now turn to the proof of the claim. Suppose $\psi(n,X) \equiv \exists \mathcal{R}\,\theta(\mathcal{R},n,X)$, where θ is a type 1 formula, and where we use script letters to denote type 2 variables.

It is convenient not to state O_ψ explicitly, but rather to show that the relation A_κ such that $A_\kappa(n,X;n',X') \leftrightarrow V_\kappa \models O_\psi(n,X;n',X')$ is expressible, <u>in a uniform way with respect to inaccessible κ</u>, as a Π^2_1 statement over V_κ. (This is clearly sufficient to prove the claim.)

Fix κ from now on with $I(\kappa)$. Notice that, by $I(\kappa)$, V_κ is just L_κ, a fact we make frequent use of. Thus from now on, type 0 variables will range over elements of L_κ, type 1 variables over subsets of L_κ, and type 2 variables over collections of subsets of L_κ. In particular, there are pecisely κ type 0 objects, precisely κ^+ type 1 objects, and precisely κ^{++} type 2 objects.

Set $A_\kappa(n,X;n',X') \leftrightarrow$ the $<_L$-least $\mathcal{R}$ such that $V_\kappa \models \theta(\mathcal{R},n,X)$ has constructible rank not greater than that of the $<_L$-least $\mathcal{R}'$ such that $V_\kappa \models \theta(\mathcal{R}',n',X')$. [Hence $\rho_{\psi,\kappa}$ is really just the constructible rank of the first witness which puts a gived (n,X) into $\mathcal{S}_{\psi,\kappa}$.]

Now, all type 1 objects, being subsets of L_κ, are elements of L_{κ^+}. Hence all type 2 objects are subsets of L_{κ^+}, and hence elements of $L_{\kappa^{++}}$. Thus, writing $\theta(\mathcal{R},n,X)$ in place of $V_\kappa \models \theta(\mathcal{R},n,X)$,

$A_\kappa(n,X;n',X') \leftrightarrow (\forall \mathcal{R}')(\forall \gamma<\kappa^{++})[\ \theta(\mathcal{R}',n',X') \wedge \mathcal{R}' \in L_\gamma \longrightarrow (\exists \mathcal{R} \in L_\gamma)\theta(\mathcal{R},n,X)\ ]$.

Now, for $\gamma<\kappa^{++}$, $|L_\gamma| \leq \kappa^+$, so L_γ can be coded as a set of type 1 objects, i.e. as a type 2 object. Thus, letting σ be the sentence of lemma 9, we have:

$A_\kappa(n,X;n',X') \longleftrightarrow \forall \mathcal{R}'\mathcal{D}\mathcal{E}[\langle \mathcal{D},\mathcal{E}\rangle \models \sigma\ \wedge \mathcal{E}$ is well-founded $\wedge \exists \mathfrak{F} \exists X(\ X \in \mathcal{D} \wedge \mathfrak{F}$ collapses X to $\mathcal{R}') \wedge \theta(\mathcal{R}',n',X')\ .\!\!\rightarrow.\ \forall \mathcal{R} \forall \mathfrak{F} \exists X(\ X \in \mathcal{D} \wedge \mathfrak{F}$ collapses X to $\mathcal{R} \rightarrow \theta(\mathcal{R},n,X))\ ]$.

Using the same kind of tricks as in lemma 9, this is easily seen to be Π^2_1 over L_κ (and it is clearly uniform thus for all inaccessible κ). In particular, the requirement "$\langle \mathcal{D},\mathcal{E}\rangle \models \sigma$" can be replaced by the sentence σ itself, only with $\mathcal{E}$ in place of $\in$ and with all variables and quantifiers written as type 1 variables and quantifiers,restricted to $\mathcal{D}$. And the requirement " $\mathcal{E}$ is well-founded" can be replaced by a Σ^2_1 statement as in lemma 9. The statement " $\mathfrak{F}$ collapses X to $\mathcal{R}$" can be written out thus: $\mathfrak{F}$ is a function $\wedge$ dom($\mathfrak{F}$) = $\{Z | Z \mathcal{E} X\} \wedge (\forall Z \in \mathrm{dom}(\mathfrak{F}))(\mathfrak{F}(Z) = \{\mathfrak{F}(Y) \mid Y \mathcal{E} Z\} \wedge \mathrm{ran}(\mathfrak{F}) = \mathcal{R}'$. And this may be written out using type 1 quantifiers only. The proof is complete. □

3. Weakly Compact Cardinals [1]

So far we have encountered two different approaches to obtaining large cardinals. In the first place we can obtain them by postulating the existence of a cardinal which is inaccessible in terms of certain, specified arithmetical operations. In the second place we can obtain them by postulating the existence of a cardinal which is indescribable in a specific, metamathematical sense. Now, large cardinal axioms are also often referred to as "axioms of infinity", for obvious reasons. This terminology suggests a further, not unreasonable method for obtaining large cardinal axioms: namely, to take some property of ω and postulate the existence of an uncountable cardinal with this property. For example, since ω carries a non-principal (ω-complete) ultrafilter (Stone's Theorem), we obtain the notion of a measurable cardinal as an uncountable cardinal κ which carries a κ-complete, non-principal ultrafilter. Now, measurable cardinals are without the scope of these notes, but there are other properties of ω which give rise to interesting large cardinal properties in a similar manner. Interestingly, many of these properties give rise to the same cardinal (at least if we assume inaccessibility as well). The study of weakly compact cardinals thus turns out to be very rich in terms of combinatorial mathematics.

Compactness of Certain Infinitary Languages

A basic property of ω is the compactness theorem for first-order, countable languages. (Perhaps a pause is required here, whilst the reader convinces himself that the compactness theorem is a property of ω. OK? Then we shall continue.) This can be generalised as follows. We first of all generalise the notion of a countable, first-order language.

Let κ be an infinite cardinal. Define the infinitary language $\underline{L}_\kappa$ as follows. As basic symbols we have variables $v_\alpha, \alpha<\kappa$, predicate letters $P_\alpha, \alpha<\kappa$, function letters $F_\alpha, \alpha<\kappa$, and constants $c_\alpha, \alpha<\kappa$. For each $\gamma<\kappa$, κ many of the letters P_α and κ many of the letters F_α are γ-ary. As connectives we have $\wedge, \vee, \neg, \rightarrow, \leftrightarrow, =, \exists, \forall$. Terms are defined in the obvious way. The atomic formulas are defined in the obvious way. In forming the formulas, we start with the atomic formulas and build them up in the usual way, except that we allow conjunctions and disjunctions of all lengths less than κ and quantification over sequences of variables of lengths less than κ. (For example, in $\underline{L}_{\omega_1}$ we can express the notion "$<$ is a well-founded relation" by the sentence $(\forall v_0 v_1 v_2 \ldots)\neg(v_1 < v_0 \wedge v_2 < v_1 \wedge \ldots)$.) The notions of an $\underline{L}_\kappa$-structure and of $\mathfrak{A} \models \varphi[\vec{a}]$, where $\mathfrak{A}$ is an $\underline{L}_\kappa$-structure, φ is an $\underline{L}_\kappa$-formula, and $\vec{a}$ is a κ-sequence from $\mathfrak{A}$, are defined in the obvious way. (As we have already indicated, the idea is that we "lift" everything naively from

1. Much of this section is based on the notes of Rowbottom (1967), although several of the proofs have been changed or rearranged.

ω.) A set, Σ, of $\underline{L}_\kappa$-sentences is <u>κ-satisfiable</u> if every subset Σ' of Σ of cardinality less than κ has a model.

We say κ is <u>weakly compact</u>, and write $WC(\kappa)$, iff $\kappa > \omega$ and every set of $\underline{L}_\kappa$ sentences which is κ-satisfiable is simultaneously satisfiable (i.e. has a model). We remark that the word "weak" here is connected with the fact that we only allow κ many symbols in the language; when we remove this restriction we obtain a notion of "strongly compact", which is not within the scope of these notes.

<u>Lemma 1</u>

$WC(\kappa) \longrightarrow I(\kappa)$.

<u>Proof</u>: (a) Suppose $\lambda = cf(\kappa) < \kappa$. Let $d_\alpha, \alpha < \kappa$, and b be distinct constants of $\underline{L}_\kappa$. Let Σ consist of the $\underline{L}_\kappa$-sentences:

(i) $\bigvee_{\alpha \in I} \bigvee_{\beta < \alpha} [d_\beta = b]$, where I is some cofinal subset of κ of order-type λ.

(ii) $d_\alpha \neq b$, $\alpha < \kappa$.

Clearly, Σ is κ-satisfiable but not simultaneously satisfiable, a contradiction. Hence κ is regular.

(b) Suppose $\lambda < \kappa \leq 2^\lambda$. Let $b_\alpha, \alpha < \lambda$, and $d_{\alpha,i}$, $\alpha < \lambda$ & $i < 2$ be distinct constants of $\underline{L}_\kappa$. Let Σ consist of the $\underline{L}_\kappa$ sentences:

(i) $\bigwedge_{\alpha<\lambda} [b_\alpha = d_{\alpha,0} \vee b_\alpha = d_{\alpha,1}]$.

(ii) $\neg \bigwedge_{\alpha<\lambda} [b_\alpha = d_{\alpha,f(\alpha)}]$, where $f: \lambda \to 2$.

(iii) $d_{\alpha,0} \neq d_{\alpha,1}$, where $\alpha < \lambda$.

Clearly, Σ is κ-satisfiable but not simultaneously satisfiable, a contradiction. Hence $\lambda < \kappa \to 2^\lambda < \kappa$. □

Notice that as an immediate consequence of lemma 1, $\underline{L}_\kappa$ has exactly κ many formulas in case $WC(\kappa)$.

<u>Keisler's V_κ Extension Property. Π^1_1-Indescribability.</u>

In connection with weakly compact cardinals, Keisler has formulated a property of a cardinal κ (which, incidentally, does not hold for ω). We say κ has the <u>V_κ extension property</u>, $EP(\kappa)$, iff $\kappa > \omega$ and, given any structure of the form $\langle V_\kappa, \in, U\rangle$, where $U \subseteq V_\kappa$, there is a transitive set X and a $U' \subseteq X$ such that $X \neq V_\kappa$ and $\langle V_\kappa, \in, U\rangle \prec \langle X, \in, U'\rangle$. We shall show that $WC(\kappa) \leftrightarrow EP(\kappa)$. First let us note the following trivial result:

<u>Lemma 2</u>

Let $\kappa > \omega$, $\langle V_\kappa, \in\rangle \prec \langle X, \in\rangle$, where X is transitive and $X \neq V_\kappa$. Then $\kappa, V_\kappa \in X$. □

<u>Theorem 3</u>

If $WC(\kappa)$, then $EP(\kappa)$.

<u>Proof</u>: Let $U \subseteq V_\kappa$ be given. By lemma 1, $I(\kappa)$, so $|V_\kappa| = \kappa$. Hence we may write as

an $\underline{L}_\kappa$ theory the set of all $\underline{L}_\kappa$ sentences valid in $\langle V_\kappa, \in, U, (x)_{x\in V_\kappa}\rangle$, together with the sentences $c \neq c_x$, $x \in V_\kappa$, where c is some $\underline{L}_\kappa$ constant, and for x in V, c_x is the $\underline{L}_\kappa$ constant denoting x. (Here, and in similar cases, we shall not explicitly describe how the set of sentences concerned is actually written out in $\underline{L}_\kappa$.) This set of sentences is clearly κ-satisfiable (in $\langle V_\kappa, \in, U, (x)_{x\in V_\kappa}\rangle$), and hence has a model, say $\langle M, E, W, (m_x)_{x\in V_\kappa}, m\rangle$. Now, as $\in$ is well-founded, $\langle V_\kappa, \in\rangle \models (\forall_{n<\omega} v_n) \neg \bigwedge_{n<\omega}(v_{n+1} \in v_n)$, so this sentence is true in $\langle M, E\rangle$, whence E is well-founded. Again, as $\in$ is extensional, so is E. Thus we can collapse $\langle M, E\rangle$ to a transitive structure, say $f: \langle M, E\rangle \cong \langle X, \in\rangle$. Let $\langle X, \in, U', (fm_x)_{x\in V_\kappa}, fm\rangle$ be the image of $\langle M, E, W, (m_x)_{x\in V_\kappa}, m\rangle$ under f. By a simple induction on $\in$, we see that $fm_x = x$ for all $x \in V_\kappa$. [For, let $x \in V_\kappa$, and suppose that $fm_y = y$ for all $y \in x$. Clearly, $\langle V_\kappa, \in, \ldots\rangle \models \forall v_0(v_0 \in c_x \leftrightarrow \bigvee_{y\in x} v_0 = c_y)$ (as $I(\kappa)$ and $x \in V_\kappa$, $|x| < \kappa$, so this is an $\underline{L}_\kappa$-sentence). Thus $(\forall z \in M)(zEm_x \leftrightarrow (\exists y \in x)(z = m_y))$, which gives $fm_x = \{f(z) \mid z \in M \wedge zEm_x\} = \{f(z) \mid z \in M \wedge (\exists y \in x)(z = m_y)\} = \{fm_y \mid y \in x\} = \{y \mid y \in x\} = x$.] Hence $V_\kappa \subseteq X$ and $\langle V_\kappa, \in, U\rangle \prec \langle X, \in, U'\rangle$. Moreover, since $fm \notin V_\kappa$, $X \neq V_\kappa$. □

Instead of proving $EP(\kappa) \rightarrow WC(\kappa)$ directly, we shall proceed by means of a long chain of implications, starting with $EP(\kappa)$, and eventually leading back to $WC(\kappa)$. The first property we shall consider needs no introduction, being our old friend Π^1_1-indescribability. The equivalence of WC with Π^1_1-indescribability will show that the "alternative" method of obtaining large cardinals which we mentioned above does not necessarily lead to anything new.

Theorem 4

Suppose $EP(\kappa)$. Then κ is Π^1_1-indescribable.

Proof: Let $U \subseteq V_\kappa$, and suppose that $\langle V_\kappa, \in, U\rangle \models \forall X\varphi(X,U)$, where φ is a first-order sentence of $\mathcal{L}(X,U)$. We must show that there is an $\alpha < \kappa$ with $\langle V_\alpha, \in, U\rangle \models \forall X\varphi(X,U)$. Well suppose not. Then for all $\alpha < \kappa$, $\langle V_\alpha, \in, U\rangle \models \neg\forall X\varphi(X,U)$. Hence, $\langle V_\kappa, \in, U\rangle \models \forall\alpha[\langle V_\alpha, \in, U\cap V_\alpha\rangle \models \neg\forall X\varphi(X,U)]$. By $EP(\kappa)$, there is a transitive set M and a $U' \subseteq M$ such that $\langle V_\kappa, \in, U\rangle \prec \langle M, \in, U'\rangle$. By lemma 2, $\kappa, V_\kappa \in M$, so $\langle M, \in, U'\rangle \models [\langle V_\kappa, \in, U'\cap V_\kappa\rangle \models \neg\forall X\varphi(X,U)]$. But $U' \cap V_\kappa = U$, so $M \models [\langle V_\kappa, \in, U\rangle \models \neg\forall X\varphi(X,U)]$, which is to say $M \models [\langle V_\kappa, \in, U\rangle \models \exists X\neg\varphi(X,U)]$. Pick X with $M \models [\langle V_\kappa, \in, U\rangle \models \neg\varphi(X,U)]$. Since φ is first-order and M is transitive, this means that $\langle V_\kappa, \in, U\rangle \models \neg\varphi(X,U)$, a contradiction. □

The First Ultrafilter Property

Let $\mathcal{F}$ be a field of subsets of some non-empty set X. Recall that $\mathcal{F}$ is κ-complete (or simply a κ-field) if, whenever $A \subseteq \mathcal{F}$ and $0 < |A| < \kappa$, then $\bigcap A$, $\bigcup A \in \mathcal{F}$.

Let X be a non-empty set, $G \subseteq \mathcal{P}(X)$. There is a unique smallest κ-complete field of subsets of X containing G. This field is denoted by $\mathcal{F}_\kappa(G,X)$, and is called the field of subsets of X κ-generated by G; we say G is a set of κ-generators for

$\mathfrak{F}_\kappa(G,X)$. $\mathfrak{F}$ is a <u>(κ,λ)-field</u> if $\mathfrak{F}$ is a κ-complete field which is κ-generated by a set of cardinality at most λ .

We say κ has the <u>first ultrafilter property</u>, $UP_1(\kappa)$, if, whenever $\mathfrak{F}$ is a (κ,κ)-field and U is a κ-complete filter in $\mathfrak{F}$, there is a κ-complete ultrafilter U' in $\mathfrak{F}$ such that $U \subseteq U'$.

Note that $UP(\omega)$ is just Stone's Theorem. Hence UP falls into our category of large cardinal properties which are "lifted" from ω.

<u>Theorem 5</u>

Suppose κ is Π^1_1-indescribable. Then $UP(\kappa)$.

<u>Proof</u>: Let $\mathfrak{F}$ be a (κ,κ)-field of subsets of X. Now, $I(\kappa)$, so it is easily seen that $|\mathfrak{F}| \leq \kappa$. We can assume that $|\mathfrak{F}| = \kappa$, of course, since otherwise the result would be trivial. Let $\wedge, \vee, -, \bigwedge, \bigvee$ be operations defined on V_κ such that there is an isomorphism $\varphi : \mathfrak{F} \cong \mathfrak{F}^* = \langle V_\kappa, \wedge, \vee, -, \bigwedge, \bigvee \rangle$. (Since $I(\kappa)$, $|V_\kappa| = \kappa$, of course.) Let U be a κ-filter on $\mathfrak{F}$, and set $U^* = \varphi'' U$, a κ-filter in $\mathfrak{F}^*$.

Suppose that there were no κ-ultrafilter extending U in $\mathfrak{F}$. Then there would be no κ-ultrafilter extending U^* in $\mathfrak{F}^*$. So:

$\langle V_\kappa, \in, \wedge, \vee, -, \bigwedge, \bigvee \rangle \models$ " $\mathfrak{F}^*$ is a κ-complete field of sets & U^* is a κ-filter in $\mathfrak{F}^*$ & $\forall W [U^* \subseteq W$ & W is an ultrafilter in $\mathfrak{F}^* \rightarrow$ W is not κ-complete "

Now, $I(\kappa)$, so the κ-completeness clauses in the above are all first-order over V_κ (If $A \subseteq V_\kappa$ and $|A| < \kappa$, then $A \in V_\kappa$.). Hence the sentence concerned is Π^1_1 over V_κ. So we can find an inaccessible $\lambda < \kappa$ such that $\langle V_\lambda, \ldots \rangle$ satisfies the same sentence. Let $\bar{\mathfrak{F}}^* = \mathfrak{F}^* \cap V_\lambda$, $\bar{U}^* = U^* \cap V_\lambda$, $\bar{\mathfrak{F}} = \varphi^{-1}{}'' \bar{\mathfrak{F}}^*$, $\bar{U} = \varphi^{-1}{}'' \bar{U}^*$. Then, clearly, $\bar{\mathfrak{F}}$ is a λ-complete subfield of $\mathfrak{F}$ and $\bar{U} \subseteq U$ is a λ-complete filter on $\bar{\mathfrak{F}}$, having no λ-complete prime extension in $\bar{\mathfrak{F}}$. But look, U is κ-complete, so $\bigcap \bar{U} \in U$. Let $x \in \bigcap \bar{U}$, and set $\bar{W} = \{A \in \bar{\mathfrak{F}} \mid x \in A\}$. Clearly, $\bar{W}$ is a λ-ultrafilter in $\bar{\mathfrak{F}}$ which extends $\bar{U}$, a contradiction. This proves the theorem. □

We shall show that $UP_1(\kappa) \rightarrow I(\kappa)$. First some preliminary results.

<u>Lemma 6</u>

Suppose κ is singular. If $\mathfrak{F}$ is a κ-field, then $\mathfrak{F}$ is in fact κ^+-complete. And if U is a κ-filter in $\mathfrak{F}$, U is in fact κ^+-complete.

<u>Proof</u>: Let $f : \lambda \rightarrow \kappa$ be cofinal, with $\lambda < \kappa$. Then, for any family $\{A_\alpha \mid \alpha < \kappa\} \subseteq \mathfrak{F}$, $\bigcap_{\alpha<\kappa} A_\alpha = \bigcap_{\alpha<\lambda} \bigcap_{\beta<f(\alpha)} A_\beta$, etc. □

<u>Lemma 7</u>

Suppose κ is singular. Then there is a (κ,κ)-field $\mathfrak{F}$, $\mathcal{P}_{<\kappa^+}(\kappa^+) \subseteq \mathfrak{F} \subseteq \mathcal{P}(\kappa^+)$, such that every κ-ultrafilter in $\mathfrak{F}$ is principal.

<u>Proof</u>: Let $X \subseteq {}^\kappa\kappa$, $|X| = \kappa^+$. We construct $\mathfrak{F}$ on X rather than on κ^+ itself. Let $A_{\alpha\beta} = \{f \in X \mid f(\alpha) = \beta\}$, $\alpha,\beta < \kappa$. Set $A = \{A_{\alpha\beta} \mid \alpha,\beta < \kappa\}$. Thus $A \subseteq \mathcal{P}(X)$, $|A| \leq \kappa$. Let $\mathfrak{F} = \mathfrak{F}_\kappa(A,X)$. By lemma 6, $\mathfrak{F}$ is κ^+-complete, so $f \in X \rightarrow \{f\} =$

$\bigcap_{\alpha<\kappa} A_{\alpha, f(\alpha)} \in \mathfrak{F}$. Hence by κ^+-completeness again, $\mathcal{P}_{<\kappa^+}(X) \subseteq \mathfrak{F}$.

Now let U be any κ-ultrafilter in $\mathfrak{F}$. By lemma 6, U is in fact κ^+-complete. Hence $(\forall\beta<\kappa)(X - A_{\alpha\beta} \in U) \longrightarrow \emptyset = \bigcap_{\beta<\kappa}(X - A_{\alpha\beta}) \in U$, which means that $(\forall\alpha<\kappa)(\exists\beta<\kappa)(A_{\alpha\beta} \in U)$. Define $f \in {}^{\kappa}\kappa$ by setting $f(\alpha)$ = the least β such that $A_{\alpha\beta} \in U$. Thus, $\{f\} = \bigcap_{\alpha<\kappa} A_{\alpha, f(\alpha)} \in U$, whence U is principal. □

Lemma 8

Suppose $\omega \leq \lambda < \kappa \leq 2^{\lambda}$. Then there is a (κ,λ)-field $\mathfrak{F}$, $\mathcal{P}_{<\kappa}(\kappa) \subseteq \mathfrak{F} \subseteq \mathcal{P}(\kappa)$, such that every κ-ultrafilter in $\mathfrak{F}$ is principal.

Proof: Let $X \subseteq {}^{\lambda}2$, $|X| = \kappa$. Set $A_{\alpha i} = \{f \in X \mid f(\alpha) = i\}$, $\alpha < \lambda$, $i < 2$. Let $A = \{A_{\alpha i} \mid \alpha < \lambda \ \& \ i < 2\}$. Thus $A \subseteq \mathcal{P}(X)$, $|A| \leq \lambda$. Let $\mathfrak{F} = \mathfrak{F}_{\kappa}(A,X)$. As $\lambda < \kappa$, $f \in X \longrightarrow \{f\} = \bigcap_{\alpha<\lambda} A_{\alpha, f(\alpha)} \in \mathfrak{F}$, so $\mathcal{P}_{<\kappa}(X) \subseteq \mathfrak{F}$.

Let U be any κ-ultrafilter in $\mathfrak{F}$. As above, we can define $f \in {}^{\lambda}2$ by $f(\alpha)$ = the least i such that $A_{\alpha i} \in U$, whence $\{f\} \in U$, and U is principal. □

Lemma 9

$UP_1(\kappa) \longrightarrow I(\kappa)$.

Proof: (a) Suppose κ is singular. Let $\mathfrak{F}$ be as given by lemma 7. Let U be the κ-filter $U = \{X \subseteq \kappa^+ \mid |\kappa^+ - X| < \kappa^+\}$. By $UP_1(\kappa)$, let U' be any κ-ultrafilter extending U. Then U must be principal. But as $U \subseteq U'$, this is absurd.

(b) Suppose there is a λ such that $\omega \leq \lambda < \kappa < 2^{\lambda}$. Let $\mathfrak{F}$ be as given by lemma 8. Let U be the κ-filter $U = \{X \subseteq \kappa \mid |\kappa - X| < \kappa\}$. A contradiction follows as in (a). □

We say κ has the weak first ultrafilter property, $WUP_1(\kappa)$, if, whenever $\mathfrak{F}$ is a κ-field of cardinality at most κ and U is a κ-filter in $\mathfrak{F}$, there is a κ-ultrafilter U' in $\mathfrak{F}$ such that $U \subseteq U'$.

Theorem 10

$UP_1(\kappa) \longleftrightarrow I(\kappa) \wedge WUP_1(\kappa)$.

Proof: By lemma 9, $UP_1(\kappa) \longrightarrow I(\kappa) \wedge WUP_1(\kappa)$. For the converse, note that if $I(\kappa)$, then any (κ,κ)-field has cardinality at most κ. □

At this point it is convenient to complete a loop and show that $UP_1(\kappa)$ implies $WC(\kappa)$.

Theorem 11

$UP_1(\kappa) \longrightarrow WC(\kappa)$.

Proof: The idea is to generalise the usual proof of the $\underline{L}_{\omega}$ compactness theorem using ultraproducts. To commence, let Σ be a κ-satisfiable set of $\underline{L}_{\kappa}$ sentences. Since $I(\kappa)$, $|\Sigma| \leq \kappa$. We can assume $|\Sigma| = \kappa$ here, since otherwise there is nothing to prove. Let $X = \{\sigma \subseteq \Sigma \mid |\sigma| < \kappa\}$. Thus $|X| = \kappa$. For each $\sigma \in X$, let $\mathfrak{M}_{\sigma}$ be a model of σ. By an easily proved "downward Löwenheim - Skolem theorem" for $\underline{L}_{\kappa}$, we can assume that $|\mathfrak{M}_{\sigma}| < \kappa$ here. Let F be the set of all $\underline{L}_{\kappa}$ formulas with fewer than κ

free variables. Thus $|F| = \kappa$. Let $J = \{s \mid s$ is a sequence & $\mathrm{dom}(s) = X$ & $(\forall\sigma \in X)(s(\sigma) \in \mathfrak{A}_\sigma)\}$. (Thus $s \in J$ iff $s \in \prod_{\sigma\in X} \mathfrak{A}_\sigma$.) Since $\sigma \in X \rightarrow |\sigma| < \kappa$, $|J| = \kappa$. For each $\varphi \in F$ and each assignment $\vec{s}$ of values from J to the free variables of φ, let $a(\varphi, \vec{s}) = \{\sigma \in X \mid \mathfrak{A}_\sigma \models \varphi[\overrightarrow{s(\sigma)}]\}$, and let G_0 be the set of all such $a(\varphi, \vec{s})$. For each $\sigma \in X$, let $b(\sigma) = \{\sigma' \in X \mid \sigma \subseteq \sigma'\}$, and let G_1 be the set of all such $b(\sigma)$. Let $G = G_0 \cup G_1$. Thus $|G| = \kappa$. Let $\mathcal{F} = \mathcal{F}_\kappa(G, X)$.

Now, for any collection $\{\sigma_\alpha \mid \alpha < \lambda < \kappa\} \subseteq X$, $\bigcap_{\alpha<\lambda} b(\sigma_\alpha) \supseteq b(\bigcup_{\alpha<\lambda} \sigma_\alpha)$, so the set $U = \{c \in \mathcal{F} \mid (\exists\sigma \in X)(b(\sigma) \subseteq c)\}$ is a κ-filter in $\mathcal{F}$. By $UP_1(\kappa)$, let U' be a κ-ultrafilter extending U. We define the ultraproduct $\mathfrak{A} = (\prod_{\sigma\in X} \mathfrak{A}_\sigma)/U'$ in the obvious way.

<u>CLAIM</u>: For each $\varphi \in F$ and each assignment $\overrightarrow{s/U'}$ of values from $\mathfrak{A}$ to the free variables of φ, $\mathfrak{A} \models \varphi[\overrightarrow{s/U'}]$ iff $\{\sigma \in X \mid \mathfrak{A}_\sigma \models \varphi[\overrightarrow{s(\sigma)}]\} \in U'$.

The proof of this claim (and the understood prerequisite that the choice of each representative s from the U'-equivalence class s/U' is irrelevant) is a straightforward generalisation of the usual Łos proof for $\underline{L}_\omega$ ultraproducts. (But note that we need the κ-completeness of U', since $\underline{L}_\kappa$ conjunctions and disjunctions can be of any length less than κ.) We leave the details to the reader. But notice that, although we do not have an ultrafilter on the whole of $\mathcal{P}(X)$, our choice of $\mathcal{F}$ (to contain G_0) ensures that U' contains all the sets it needs to.

We now show that $\mathfrak{A} \models \Sigma$. Let $\varphi \in \Sigma$. Then $\{\varphi\} \in X$, so $b(\{\varphi\}) \in U \subseteq U'$. But $\sigma \in b(\{\varphi\}) \rightarrow \varphi \in \sigma \rightarrow \mathfrak{A}_\sigma \models \varphi$. Hence, by the claim, $\mathfrak{A} \models \varphi$. □

To sum up the situation so far, we have proved:

$$\begin{array}{ccccccc} WC(\kappa) & \xrightarrow{\text{Thm } 3} & EP(\kappa) & \xrightarrow{\text{Thm } 4} & \kappa \text{ is } \Pi^1_1\text{-indescribable} & \xrightarrow{\text{Thm } 5} & UP_1(\kappa) \\ & & & & \text{Thm } 10 \swarrow & & \downarrow \text{Thm } 11 \\ & & & & WUP_1(\kappa) \wedge I(\kappa) & & WC(\kappa) \end{array}$$

Notice that by virtue of Theorem 2.11, we thus have:

<u>Theorem 12</u>

$WC(\kappa) \rightarrow WC^L(\kappa)$. □

<u>The Second Ultrafilter Property.</u>

We say κ has the <u>second ultrafilter property</u>, $UP_2(\kappa)$, if, whenever $\mathcal{F}$ is a (κ, κ)-field such that $\mathcal{P}_{<\omega}(\kappa) \subseteq \mathcal{F} \subseteq \mathcal{P}(\kappa)$, there is a non-principal κ-ultrafilter in $\mathcal{F}$.

<u>Lemma 13</u>

$UP_2(\kappa) \rightarrow I(\kappa)$.

<u>Proof</u>: (a) Suppose κ is singular. Then by lemma 6, $\mathcal{P}(\kappa)$ is a κ^+-complete (κ, κ)-field. Let U be a non-principal κ-ultrafilter in $\mathcal{P}(\kappa)$. By lemma 6 again,

$\emptyset = \bigcap_{\alpha<\kappa}(\kappa-\{\alpha\}) \in U$, which is absurd. Hence κ is regular.
(b) By lemma 8, $UP_2(\kappa)$ implies $(\forall\lambda<\kappa)(2^\lambda < \kappa)$. □

We say κ has the weak second ultrafilter property, $WUP_2(\kappa)$, if, whenever $\mathfrak{F}$ is a κ-field of cardinality κ such that $\mathcal{P}_{<\omega}(\kappa) \subseteq \mathfrak{F} \subseteq \mathcal{P}(\kappa)$, there is a non-principal κ-ultrafilter in $\mathfrak{F}$.

Theorem 14

$UP_2(\kappa) \leftrightarrow I(\kappa) \wedge WUP_2(\kappa)$.

Proof: By lemma 13, as in Theorem 10. □

Theorem 15

$UP_1(\kappa) \rightarrow UP_2(\kappa)$.

Proof: Let $\mathfrak{F}$ be a (κ,κ)-field such that $\mathcal{P}_{<\omega}(\kappa) \subseteq \mathfrak{F} \subseteq \mathcal{P}(\kappa)$. Let $U = \{X \in \mathfrak{F} \mid |\kappa - X| < \kappa\}$, a κ-filter in $\mathfrak{F}$. By $UP_1(\kappa)$, let U' be any κ-ultrafilter which extends U. Clearly, U' is non-principal. □

The Tree Property.

A tree is a partially-ordered set $\underline{T} = \langle T, \leq \rangle$ such that for every $x \in T$, the set $\hat{x} = \{y \in T \mid y < x\}$ is well-ordered by $<$. The order-type of $\hat{x}$ under $<$ is the height of x in $\underline{T}$, ht(x). The α'th level of $\underline{T}$ is the set $T_\alpha = \{x \in T \mid ht(x) = \alpha\}$. $\underline{T}$ is a κ-tree iff (i) $(\forall\alpha<\kappa)(0 < |T_\alpha| < \kappa)$ and (ii) $T_\kappa = \emptyset$. A branch of a tree $\underline{T}$ is a totally-ordered initial segment of $\underline{T}$; it is an α-branch if α is its order-type under $<$.

We say κ has the tree property, $TP(\kappa)$, if every κ-tree has a κ-branch.

By the famous (and trivial) König's Theorem, $TP(\omega)$. By a classical result of Aronszajn, $\neg TP(\omega_1)$. Specker generalised Aronszajn's proof to give: if GCH and κ is regular, then $\neg TP(\kappa^+)$. For details of these results, we refer the reader to Jech (1971a). Specker's assumption of GCH here is necessary, for in Mitchell (1972), Silver and Mitchell prove that $Con(ZFC + \exists\kappa WC(\kappa)) \rightarrow Con(ZFC + 2^\omega = 2^{\omega_1} = \omega_2 + TP(\omega_2))$. [The assumption of a weakly compact cardinal here is also necessary, for in the same paper, Silver shows that if κ is regular and $TP(\kappa^+)$, then $WC^L(\kappa^+)$.]

Jensen (1972) showed that if $V = L$, then TP and WC are entirely equivalent. At the time of writing, we believe it is still not known if we can weaken $V = L$ to GCH here. Now, if GCH, then $WI(\kappa) \leftrightarrow I(\kappa)$. And, as we shall prove below, $TP(\kappa) \rightarrow \kappa$ is regular. Thus, recalling Specker's result above, we see that if GCH is assumed, then $TP(\kappa)$ implies that either $I(\kappa)$ or else $\kappa = \lambda^+$ for some singular λ. But we shall also prove that $TP(\kappa) \wedge I(\kappa) \leftrightarrow WC(\kappa)$ (in ZFC), so the problem stated reduces to the question as to whether one can construct, in ZFC + GCH, for each singular cardinal λ, a λ^+-tree with no λ^+-branches.

Lemma 16

$TP(\kappa) \rightarrow \kappa$ is regular.

Proof: Suppose $\lambda = cf(\kappa) < \kappa$. Let $\langle \kappa_\xi \mid \xi < \lambda \rangle$ be a strictly increasing, continuous, cofinal sequence in κ, with $\kappa_0 = 0$. Define a tree $\underline{T} = \langle \kappa, \leq_T \rangle$ by $\alpha <_T \beta \leftrightarrow \exists \xi (\kappa_\xi \leq \alpha < \beta < \kappa_{\xi+1})$. Clearly, $\underline{T}$ is a κ-tree with no κ-branch. □

Theorem 17

$UP_2(\kappa) \longrightarrow TP(\kappa) \wedge I(\kappa)$.

Proof: By lemma 13, we already know that $UP_2(\kappa) \longrightarrow I(\kappa)$. So, we assume $UP_2(\kappa)$ and verify $TP(\kappa)$. Let $\underline{T}$ be a κ-tree. We may assume $\underline{T} = \langle \kappa, \leq_T \rangle$. For each α, let $[\alpha] = \{\beta \mid \alpha \leq_T \beta\}$. Let $G = \{[\alpha] \mid \alpha \in \kappa\}$, and define $\mathfrak{F} = \mathfrak{F}_\kappa(G \cup \mathcal{P}_{<\omega}(\kappa), \mathcal{P}(\kappa))$. Let U be a non-principal, κ-ultrafilter in $\mathfrak{F}$. We define a κ-branch $\langle \gamma(\xi) \mid \xi < \kappa \rangle$ of $\underline{T}$ by induction on ξ. Now, $|T_0| < \kappa$ and U is κ-complete, so there is a unique $\gamma(0) \in T_0$ such that $[\gamma(0)] \in U$. (Because $\bigcup\{[\gamma] \mid \gamma \in T_0\} = \kappa$.) If $\gamma(\xi) \in T_\xi$ is defined with $[\gamma(\xi)] \in U$, then since $|T_{\xi+1}| < \kappa$ there is a unique $\gamma(\xi+1) \in T_{\xi+1}$ such that $\gamma(\xi) <_T \gamma(\xi+1)$ and $[\gamma(\xi+1)] \in U$. (Because $\{\gamma(\xi)\} \cup \bigcup\{[\gamma] \mid \gamma \in T_{\xi+1} \wedge \gamma(\xi) <_T \gamma\} = [\gamma(\xi)]$.) Finally, suppose $\lim(\xi)$ and $\gamma(\eta) \in T_\eta$, $\eta < \xi$, are defined with $[\gamma(\eta)] \in U$ for all $\eta < \xi$. As U is κ-complete, $X_\xi = \bigcap_{\eta<\xi} [\gamma(\eta)] \in U$. But $|T_\xi| < \kappa$, and again $\bigcup\{[\gamma] \mid \gamma \in T_\xi \wedge (\forall \eta < \xi)(\gamma(\eta) <_T \gamma)\} = X_\xi$, so as before there is a unique $\gamma(\xi) \in T_\xi$ with $[\gamma(\xi)] \in U$. And of course, $\eta < \xi \rightarrow \gamma(\eta) <_T \gamma(\xi)$. This defines a κ-branch, as required. □

Theorem 18

$TP(\kappa) \wedge I(\kappa) \rightarrow UP_1(\kappa)$.

Proof: By Theorem 10, it suffices to prove that $TP(\kappa) \wedge I(\kappa) \longrightarrow WUP_1(\kappa)$. So, assume $TP(\kappa) \wedge I(\kappa)$, and let $\mathfrak{F}$ be a κ-field of subsets of H of cardinality κ. We may assume $|\mathfrak{F}| = \kappa$, of course, since otherwise the result is trivial. Let U be a κ-filter in $\mathfrak{F}$. We seek a κ-ultrafilter U' extending U. Define a tree $\underline{T}$ as follows. The elements of $\underline{T}$ will be α-sequences of 0's and 1's as α ranges over κ. The ordering of $\underline{T}$ will be sequence extension. If $s \in T$ and $t \subseteq s$, we shall have $t \in T$, whence if $\alpha < \kappa$ and $s \in T$, then $s \in T_\alpha$ iff $s \in {}^\alpha 2$. We define $\underline{T}$ by induction on the levels.

Let $U = \{U_\alpha \mid \alpha < \kappa\}$, and let $\langle X_\alpha \mid \alpha < \kappa \rangle$ be a one-one enumeration of $\mathfrak{F} - U$. Let $T_0 = \{\emptyset\}$; and if $\lim(\alpha)$ and T_β, $\beta < \alpha$, are defined, let T_α consist of $\bigcup b$ for each α-branch b of $\bigcup_{\beta<\alpha} T_\beta$. Finally, suppose T_α is defined. Let $s \in T_\alpha$. If

$$X_\alpha \cap [\bigcap_{\beta<\alpha} U_\beta] \cap [\bigcap_{\gamma \in s^{-1}(0)} X_\gamma] \cap [\bigcap_{\gamma \in s^{-1}(1)} (H - X_\gamma)] \neq \emptyset,$$

put $s^\frown\langle 0 \rangle$ into $T_{\alpha+1}$. If,

$$(H - X_\alpha) \cap \quad \dots \quad \dots \quad \dots \quad \dots \quad \neq \emptyset,$$

put $s^\frown\langle 1 \rangle$ into $T_{\alpha+1}$. (So, if both intersections are non-void, s has two extensions in $T_{\alpha+1}$.) Otherwise, s has no extensions in $T_{\alpha+1}$.

This defines $\underline{T}$. Clearly, as $T \subseteq \bigcup_{\alpha<\kappa} {}^\alpha 2$ and $I(\kappa)$, $(\forall \alpha < \kappa)(|T_\alpha| < \kappa)$ and $T_\kappa = \emptyset$. Thus $\underline{T}$ will be a κ-tree providing $(\forall \alpha < \kappa)(T_\alpha \neq \emptyset)$. Well suppose not, and let δ be least such that $T_\delta = \emptyset$. Clearly, $\delta = \lambda + 1$ for some λ, and $T = \bigcup_{\alpha \leq \lambda} T_\alpha$. Since $T_\delta = \emptyset$, we must have $[\bigcap_{\beta<\delta} U_\beta] \cap [\bigcap_{\gamma \in s^{-1}(0)} X_\gamma] \cap [\bigcap_{\gamma \in s^{-1}(1)} (H - X_\gamma)] = \emptyset$ for each

$s \in T_\lambda$. Now, U is κ-complete, so we can pick $x \in \bigcap_{\beta<\delta} U_\beta$. Define $s_0 \in {}^\lambda 2$ by $s_0(\gamma) = 0 \leftrightarrow x \in X_\gamma$. Clearly, $s_0 \in T_\lambda$. But look, $x \in [\bigcap_{\beta<\delta} U_\beta] \cap [\bigcap_{\gamma \in s_0^{-1}(0)} X_\gamma] \cap [\bigcap_{\gamma \in s_0^{-1}(1)} (H - X_\gamma)]$, so we have a contradiction. Hence $\underline{T}$ is a κ-tree. By $TP(\kappa)$, let $s: \kappa \to 2$ be the union of a κ-branch of $\underline{T}$. Let $U' = \{X_\alpha \mid \alpha \in s^{-1}(0)\} \cup U$. Clearly, U' is a κ-ultrafilter in $\mathfrak{F}$ which extends U. □

Since our last diagram, we have proved:

$$WC(\kappa) \xleftrightarrow{\text{Earlier}} UP_1(\kappa) \xrightarrow{\text{Thm 15}} UP_2(\kappa) \xrightarrow{\text{Thm 17}} TP(\kappa) \wedge I(\kappa) \xrightarrow{\text{Thm 18}} UP_1(\kappa)$$

In particular, $UP_1(\kappa) \leftrightarrow UP_2(\kappa)$.

Partition Properties

The study of so-called partition properties was begun by Ramsey, and developed extensively by Erdös, Rado, and others.

Let κ be a cardinal, n a positive integer. We write $[\kappa]^n = \{\sigma \subseteq \kappa \mid |\sigma| = n\}$. Since κ has a natural well-ordering, we can identify $[\kappa]^n$ with $\{\langle\alpha_1,\ldots,\alpha_n\rangle \mid \alpha_1 < \ldots < \alpha_n < \kappa\}$, a frequently used identification. A partition of $[\kappa]^n$ is a map $f: [\kappa]^n \to \mu$ for some ordinal μ. If $f: [\kappa]^n \to \mu$ is a partition, we say a set $X \subseteq \kappa$ is homogeneous for f if there is a single ordinal $\xi \in \mu$ such that for all $\sigma \subseteq X$ with $|\sigma| = n$, $f(\sigma) = \xi$. (Extending our above notation in an obvious way, this can be expressed concisely by: $|f''[X]^n| = 1$.)

For cardinals κ, λ, μ, and positive integers n, we write $\kappa \longrightarrow (\lambda)^n_\mu$ if for every partition $f: [\kappa]^n \to \mu$ there is a homogeneous set X of cardinality λ. (The idea behind this notation is that a valid partition relation $\kappa \longrightarrow (\lambda)^n_\mu$ remains valid if we increase the size of any parameter on the left of the arrow, and decrease the size of any on the right.)

As we shall show below, partition relations are closely connected with properties of trees.

Lemma 19

Let $n \in \omega$ and let $[\kappa]^{n+1} = \bigcup_{\alpha<\mu} C_\alpha$ be a (disjoint) partition of $[\kappa]^{n+1}$ into μ sets. (This is clearly just another way of expressing the notion of a "partition" as defined above!) Then there is a tree $\underline{T} = \langle T, \dashv\rangle$ and a surjection $f: T \to \kappa$ such that:

(i) $x \dashv y \to f(x) < f(y)$;

(ii) $x_1 \dashv \ldots \dashv x_n \dashv y \dashv z$ implies that $\langle f(x_1),\ldots,f(x_n),f(y)\rangle$ and $\langle f(x_1),\ldots,f(x_n),f(z)\rangle$ are in the same C_α;

(iii) if y,z are $\dashv$-incomparable, there are $x_1 \dashv \ldots \dashv x_n \dashv y,z$ such that $\langle f(x_1),\ldots,f(x_n),f(y)\rangle$ and $\langle f(x_1),\ldots,f(x_n),f(z)\rangle$ are in different C_α's;

(iv) for each α, $|T_\alpha| \le \mu^{|\alpha|} + \omega$.

Proof: We define $\underline{T}$ by induction on the elements of κ. The exact nature of $\underline{T}$ is irrelevant, so we shall take $T = \kappa$ for definiteness. To commence, let $0 \to 1 \to \ldots$ $\ldots \to n-1$ and set $f(0) = 0, \ldots, f(n-1) = n-1$. Suppose now that we are at stage α, looking at α for the first time. Let $\underline{T}$ denote that part of the tree so far constructed. For each point x in T such that for all $x_1 \to \ldots \to x_{n+1} \geqq x$, $\langle f(x_1), \ldots, f(x_{n+1})\rangle$ and $\langle f(x_1), \ldots, f(x_n), \alpha\rangle$ are in the same C_ξ and x is maximal in $\underline{T}$ with this property, introduce a new immediate successor, y, of x and set $f(y) = \alpha$. For each maximal branch b of $\underline{T}$ of limit length such that whenever $x_1, \ldots, x_{n+1} \in b$ and $x_1 \to \ldots \to x_{n+1}$, then $\langle f(x_1), \ldots, f(x_{n+1})\rangle$ and $\langle f(x_1), \ldots, f(x_n), \alpha\rangle$ are in the same C_ξ, introduce an immediate successor, y, of b and set $f(y) = \alpha$. Clearly, $\underline{T}$ and f so defined will satisfy (i) - (iii) of the lemma. We prove (iv) by induction on α. For $\alpha = 0, \ldots, n-1$ there is nothing to prove. At limit levels α in $\underline{T}$, at most one point extends each α-branch of $\underline{T}$ on T_α, so $|T_\alpha| \leq$ number of α-branches of $\underline{T} = |\bigcup_{\beta<\alpha} T_\beta|^{|\alpha|} \leq [\Sigma_{\beta<\alpha}(\mu^{|\beta|}+\omega)]^{|\alpha|} \leq \mu^{|\alpha|} + \omega$, by induction hypothesis. Finally, if $x \in T_\alpha$, $y, z \in T_{\alpha+1}$, and $x \to y, z$, then if $y \neq z$ there must be $x_1 \to \ldots \to x_n \to x$ such that (by (iii)) $\langle f(x_1), \ldots, f(x_n), f(y)\rangle$ and $\langle f(x_1), \ldots, f(x_n), f(z)\rangle$ are in different C_ξ's. But there are at most $|\alpha|^n$ choices of $\langle x_1, \ldots, x_n\rangle$ here, and only μ sets C_ξ, so there are at most $\mu^{(|\alpha|^n)}$ distinct immediate successors of x. So, by induction, $|T_{\alpha+1}| \leq |T_\alpha| \cdot \mu^{(|\alpha|^n)} \leq \mu^{|\alpha|} + \omega$.

As an illustration of the way in which lemma 19 gives rise to theorems of ZFC, we derive some classical results of partition calculus. First, the famous "Ramsey Theorem".

Theorem 20

Let $m, n \in \omega$. Then $\omega \longrightarrow (\omega)^{n+1}_m$.

Proof: By induction on n, we prove $(\forall m)[\omega \longrightarrow (\omega)^{n+1}_m]$. For $n = 0$ there is nothing to prove. Let $n = 1$. Let $[\omega]^2 = C_1 \cup \ldots \cup C_m$ be a disjoint partition. Construct $\underline{T}$, $f : T \to \omega$ as in lemma 19. Now, although we did not say so there, it is easily seen that $\underline{T}$ is an ω-tree. By "König's Theorem", let b be an ω-branch of $\underline{T}$. Then, if $x \in b$ and we pick $y, z \in b$ with $x \to y, z$, then $\langle f(x), f(y)\rangle$ and $\langle f(x), f(z)\rangle$ lie in the same C_i. So, for each $x \in b$, let $i(x) \in \{1, \ldots, m\}$ be such that whenever $x \to y$ $\wedge\ y \in b$, $\langle f(x), f(y)\rangle \in C_{i(x)}$. Now, we can find an infinite set $b' \subseteq b$ such that $i''b' = \{k\}$ for some fixed $k \in \{1, \ldots, m\}$. Let $X = f''b'$. By (i) of the lemma, $|X| = \omega$. And clearly, X is homogeneous for the given partition. This proves the result for $n = 1$. Finally, for the general case, let $n \geq 2$ and assume that we have already proved $(\forall m)[\omega \longrightarrow (\omega)^n_m]$. Let a disjoint partition $[\omega]^{n+1} = C_1 \cup \ldots \cup C_m$ be given. Using the lemma much as above, we can find an infinite set $X \subseteq \omega$ such that whenever $\langle x_0, \ldots, x_n\rangle \in [X]^{n+1}$, the C_i to which $\langle x_0, \ldots, x_n\rangle$ belongs depends only upon $\langle x_0, \ldots, x_{n-1}\rangle$. We can then define a partition of $[X]^n$ by putting $\langle y_0, \ldots, y_{n-1}\rangle$ and $\langle y'_0, \ldots, y'_{n-1}\rangle$ into the same partition class iff there is a $y \in X$ with $y_{n-1}, y'_{n-1} < y$ such that $\langle y_0, \ldots, y_{n-1}, y\rangle$ and $\langle y'_0, \ldots, y'_{n-1}, y\rangle$ are in the same C_i (whence this will be true for all such y of course). This partitions $[X]^m$ into at most m sets. So, as $|X| = \omega$, the induction hypothesis gives us an infinite set $Y \subseteq X$ which is homogeneous for this partition. Clearly, Y is homogeneous for our original partition as well. Hence $(\forall m)[\omega \longrightarrow (\omega)^{n+1}_m]$.

Theorem 21 (The Erdős - Rado Theorem)

(i) For any cardinals κ, μ with $\kappa \geq \omega$ and $\kappa \geq \mu$, $(2^\kappa)^+ \longrightarrow (\kappa^+)^2_\mu$.

(ii) For any cardinals κ, μ with $\kappa \geq \omega$ and $\kappa \geq \mu$ and any $n \in \omega$, $\beth_n(\kappa)^+ \longrightarrow (\kappa^+)^{n+1}_\mu$.

Proof: (i) Let $\lambda = (2^\kappa)^+$, and let $[\lambda]^2 = \bigcup_{\xi < \mu} C_\xi$ be a given partition. Let $\underline{T}$, f be given by lemma 19. Since f is onto λ, $|T| \geq \lambda$. It follows that $T_{\kappa^+} \neq \emptyset$ (If not, $T = \bigcup_{\alpha < \kappa^+} T_\alpha$, so $|T| \leq \Sigma_{\alpha < \kappa^+} \mu^{|\alpha|} \leq \kappa^+ \cdot \mu^\kappa = 2^\kappa < \lambda$.) So let $x \in T_{\kappa^+}$ and set $b = \{y \in T \mid y \dashv x\}$, a κ^+-branch of $\underline{T}$. Since $cf(\kappa^+) = \kappa^+ > \mu$, b gives rise to a set $b' \subseteq b$, $|b'| = \kappa^+$, such that f"b' is homogeneous for the partition, exactly as in Theorem 20 (for the case $n = 1$).

(ii) This is proved by induction, much as in Theorem 20. The only difference is that, as indicated by part (i) above, we must increase the size of the given set in order to guarantee a homogeneous set of the required cardinality. And as we saw above, the increase in size required is "successor of the power set": we need a tree of cardinality $\beth_{n+1}(\kappa)^+$ in order to ensure that we have a $\beth_n(\kappa)^+$-branch to which we can apply the induction hypothesis. □

So much for theorems of ZFC. For our present purposes, lemma 19 is useful by virtue of:

Theorem 22

Assume $TP(\kappa) \wedge I(\kappa)$. Then for all $n \in \omega$ and all $\mu < \kappa$, $\kappa \longrightarrow (\kappa)^{n+1}_\mu$.

Proof: Almost identical to the proof of Theorem 20. □

Corollary 23

If $TP(\kappa) \wedge I(\kappa)$, then $\kappa \longrightarrow (\kappa)^2_2$. □

Sierpinski's Order Property.

We say κ has the order property, $OP(\kappa)$, if, whenever $\langle X, \dashv \rangle$ is an ordered set of cardinality κ, there is a $Y \subseteq X$, $|Y| = \kappa$, such that either $\langle Y, \dashv \rangle$ or $\langle Y, \in \rangle$ is a well-ordered set.

Theorem 24

Assume $\kappa \longrightarrow (\kappa)^2_2$. Then $OP(\kappa)$.

Proof: Let $\langle \kappa, \dashv \rangle$ be any ordered set of cardinality κ. Define $f: [\kappa]^2 \longrightarrow 2$ by $f(\langle \alpha, \beta \rangle) = 0 \leftrightarrow \alpha \dashv \beta$. Let $X \subseteq \kappa$, $|X| = \kappa$, be homogeneous for f. If $f"[X]^2 = \{0\}$, then $\langle X, \dashv \rangle$ is well-ordered. If $f"[X]^2 = \{1\}$, then $\langle X, \in \rangle$ is well-ordered. □

Lemma 25

$OP(\kappa) \rightarrow I(\kappa)$.

Proof: (a) Suppose $\lambda = cf(\kappa) < \kappa$. Let $\langle \kappa_\xi \mid \xi < \lambda \rangle$ be strictly increasing, continuous, and cofinal in κ, with $\kappa_0 = 0$. Order κ as follows. If $\exists \xi (\kappa_\xi \leq \alpha < \beta < \kappa_{\xi+1})$, set $\beta \rightarrowtail \alpha$. If $\exists \xi (\alpha < \kappa_\xi \leq \beta)$, set $\alpha \rightarrowtail \beta$. Let $X \subseteq \kappa$, $|X| = \kappa$. Since X must meet some interval $[\kappa_\xi, \kappa_{\xi+1})$ on an infinite set, X has a decreasing ω-chain (under $\rightarrowtail$). And since X must meet infinitely many different intervals $[\kappa_\xi, \kappa_{\xi+1})$, X has an increasing ω-chain. Hence κ is regular.

(b) Suppose $\lambda < \kappa \leq 2^\lambda$, and let λ be the least such. Let $X \subseteq {}^\lambda 2$, $|X| = \kappa$, and let $\rightarrowtail$ be the lexicographic ordering on X. Let $R = \{ f \in {}^\lambda 2 \mid (\exists \alpha < \lambda)(\forall \beta > \alpha)(f(\beta) = 0)\}$. Then $|R| = 2^{\underset{\smile}{\lambda}} < \kappa$, since κ is regular and λ was chosen minimally with $2^\lambda \geq \kappa$. Also, R is dense in ${}^\lambda 2$, so whenever $f, g \in X$ and $f \rightarrowtail g$, there is $h \in R$ with $f \underline{\rightarrowtail} h \underline{\rightarrowtail} g$. Hence X can admit no well-ordered or inversely well-ordered subset of cardinality κ. Hence $\lambda < \kappa \longrightarrow 2^\lambda < \kappa$. □

Theorem 26

$OP(\kappa) \longrightarrow TP(\kappa) \wedge I(\kappa)$.

Proof: By lemma 25, we need only assume $OP(\kappa)$ and verify $TP(\kappa)$. So let $\underline{T} = \langle T, \leq \rangle$ be a κ-tree. For each $\alpha < \kappa$, linearly order T_α by $<_\alpha$. Define an ordering $\rightarrowtail$ of T by setting $x \rightarrowtail y$ iff $x < y$ or $x <_\alpha y$ or $(\exists \beta)(\exists x', y')[x', y' \in T_\beta \wedge x' < x \wedge y' < y \wedge \hat{x}' = \hat{y}' \wedge x' <_\beta y']$. By $OP(\kappa)$, let $\langle x_\xi \mid \xi < \kappa \rangle$ be a well-ordered sequence under $\rightarrowtail$. (The case where we get an inversely well-ordered sequence is entirely similar.) Given $x \in T$, let $[x]$ denote the set $\{ y \in T \mid x \leq y \}$. Set $A = \{ x_\xi \mid \xi < \kappa \}$. Clearly, $A \cap T = [\bigcup_{\beta < \alpha} A \cap T_\beta] \cup [\bigcup_{x \in T_\alpha} A \cap [x]]$, each $\alpha < \kappa$. Fixing $\alpha < \kappa$, we observe that as κ is regular and $|T_\xi| < \kappa$ for all $\xi < \kappa$, $|A \cap [x]| = \kappa$ for some $x \in T_\alpha$. Suppose we could find distinct $x, y \in T_\alpha$ with $|A \cap [x]| = |A \cap [y]| = \kappa$, say $x \rightarrowtail y$. Pick $\xi < \kappa$ with $y \rightarrowtail x_\xi$. (Since there will be κ many such ξ's, this causes no problem.) Suppose now that $x \rightarrowtail x_\eta$. Then by definition of $\rightarrowtail$, $x_\eta \rightarrowtail x_\xi$, so $\eta < \xi$. Hence $|\{ \eta \mid x \rightarrowtail x_\eta \}| \leq |\xi| < \kappa$, which is absurd. Thus, given $\alpha < \kappa$, we can find a unique $x(\alpha) \in T_\alpha$ such that $|A \cap [x(\alpha)]| = \kappa$. It is easily seen that $\langle x(\alpha) \mid \alpha < \kappa \rangle$ is a κ-branch of $\underline{T}$. □

Thus, our final diagram is as follows:

$$
\begin{array}{ccccc}
WC(\kappa) & \overset{\text{Earlier}}{\longleftrightarrow} & TP(\kappa) \wedge I(\kappa) & \overset{\text{Thm 22}}{\longrightarrow} & (\forall n)(\forall \mu < \kappa)[\kappa \longrightarrow (\kappa)^{n+1}_{\mu}] \\
 & & \text{Thm 26} \uparrow & & \downarrow \text{Trivial} \\
 & & OP(\kappa) & \overset{\text{Thm 24}}{\longleftarrow} & [\kappa \longrightarrow (\kappa)^2_2]
\end{array}
$$

REFERENCES

K.J.Devlin (1973) Aspects of Constructibility. Springer:Lecture Notes in Mathematics 354.

H.C.Doets (1969) Novak's Result by Henkin's Method. Fund. Math. 64, 329-333.

K.Gödel (1940) The Consistency of the Axiom of Choice and of the Generalised Continuum Hypothesis. Annals of Math. Studies 3.

T.J.Jech (1971) Lectures in Set Theory. Springer:Lecture Notes in Mathematics 217.

T.J.Jech (1971a) Trees. Journal of Symbolic Logic 36, 1-14.

R.B.Jensen (1972) The Fine Structure of the Constructible Hierarchy. Annals of Math. Logic 4, 229-308.

W.J.Mitchell (1972) Aronszajn Trees and the Independence of the Transfer Property. Annals of Math. Logic 5, 21-46.

F.Rowbottom (1967) Classical Theory of Large Cardinals. Lecture notes, UCLA.

J.R.Shoenfield (1954) A Relative Consistency Proof. Journal of Symbolic Logic 19, 21-28.

MARGINALIA TO A THEOREM OF SILVER

Keith I. Devlin and R. B. Jensen (Bonn)

§ 0 Introduction

The singular cardinals problem, in its simplest form, asks whether the continuum hypothesis can hold below a singular cardinal β and fail at β. A variant of the question is whether we can have $2^{\underset{\smile}{\beta}} = \beta$ and $2^{\beta} > \beta^{+}$. Since forcing is the natural method for producing independence results, set theorists have concentrated on a more specific form of the problem: Given a transitive model M of ZF +   GCH, can a positive solution be obtained by forcing over M with a set of conditions $\mathbb{P}$? This approach suggests a number of related problems: Is there a $\mathbb{P}$ which collapses β^{+} to β? Is there a $\mathbb{P}$ which makes an inaccessible cardinal singular?

Until very recently, there was a widespread assumption among set theorists that such sets of conditions do exist and merely awaited discovery. Then Silver challenged this assumption by proving it false. Specifically, Silver proved - in ZFC - that if the continuum hypothesis holds below a singular cardinal β of uncountable cofinality, then it holds at β. Thus, in many important cases, not only the narrower forcing problem but the general problem itself has a negative solution.

Much of the effort to produce a positive forcing solution centered on the attempt to exploit the properties of special ground models - either L or models containing large cardinals. The latter approach met with some success: Prikry, fr. ins., showed that a measurable cardinal can be turned into an ω cofinal cardinal. Magidor, starting with an elephantine cardinal, produced a model in which $2^{\omega_n} < \omega_\omega$ for $n < \omega$ and $2^{\omega_\omega} > \omega_{\omega+1}$. Jensen's efforts to produce a positive solution over L led to total failure. Silver's work then led him to consider the problem from a new perspective. He discovered that the statement "$0^{\#}$ does not exist" (henceforth abbreviated as $\neg 0^{\#}$) implies a negative solution to all cases of the singular cardinals problem. But then there cannot be a positive forcing solution over L, since every generic extension of L by a set of conditions satisfies $\neg 0^{\#}$.

Throughout this paper we assume ZFC. Our main theorem says, in effect, that if $\neg 0^{\#}$, then the "essential structure" of cardinalities and confinalities in L is retained in V.

Theorem 1. Assume $\neg 0^{\#}$. Let X be an uncountable set of ordinals. Then there is a constructible set Y s.t. $X \subset Y$ and $\overline{\overline{X}} = \overline{\overline{Y}}$.

Remark. By a theorem of Prikry, we cannot replace "uncountable" by "infinite" in Theorem 1.

Corollary 2. Assume $\neg 0^{\#}$. If $\tau \geq \omega_2$ is regular in L, then $cf(\tau) = \overline{\overline{\tau}}$.

Remark. By a theorem of Bukovski, we cannot replace ω_2 by ω_1 in Corollary 2.

The following corollary establishes a totally negative solution of the singular cardinals problem over L.

Corollary 3. Assume $\neg 0^{\#}$. Let β be a singular cardinal. Then

(a) β is singular in L

(b) $\beta^+ = \beta^{+L}$

(c) If $A \subset \beta$ s.t. $H_\beta = L_\beta[A]$, then $\mathcal{P}(\beta) \subset L[A]$.

(d) $cf(\beta) \leq \gamma < \beta \longrightarrow \beta^\gamma = 2^\gamma \cdot \beta^+$

(e) Let $\theta = 2^{\underset{\smile}{\beta}}$. Then

$$2^\beta = \begin{cases} \theta \text{ if } \bigvee \gamma < \beta \ 2^\gamma = \theta \\ \theta^+ \text{ if not.} \end{cases}$$

The proofs of these corollaries are quite straightforward and will be left to the reader.

The above results are due to Jensen and were originally presented in three handwritten notes bearing the title of this paper. The proof given there was developed "piecewise" and contained many redundancies. The present streamlined proof is due chiefly to Devlin.

§ 1 The approach

From now on assume $\neg 0^{\#}$.

Def $\tau > \omega$ is suitable iff either $J_\tau \models$ (There is a largest cardinal) or else there are arbitrarily large $\gamma < \tau$ s.t. $cf(\gamma) > \omega$ and $J_\tau \models$ (γ is regular).

§ 0 Theorem 1 reduces to the statement:

Lemma 1. Let $\tau \geq \omega_2$ be a suitable cardinal in L. Let $X \subset \tau$ be cofinal in τ s.t. $\overline{\overline{X}} < \overline{\overline{\tau}}$. Then there is $Y \supset X$ s.t. $Y \in L$ and $\overline{\overline{Y}}^L < \tau$.

We first show that Lemma 1 implies § 0 Theorem 1. Suppose not. Let X be an uncountable set of ordinals for which the conclusion of §0 Theorem 1 fails. Choose $\tau = \text{lub}(X)$ minimal for such X. Then $\overline{\overline{X}} < \overline{\overline{\tau}}$, since otherwise the conclusion of § 0 Theorem 1 would hold with $Y = \tau$. Hence $\tau > \omega_2$. Now suppose that the conclusion of Lemma 1 held. There would then be $Z \in L$ s.t. $X \subset Z$ and $\overline{\overline{Z}}^L < \tau$. Let $\rho = \overline{\overline{Z}}^L$ and let $f : \rho \longleftrightarrow Z$ be constructible. Set $X' = f^{-1}"X$. Then $X' \subset \rho < \tau$. By the minimal choice of τ there is $Y' \in L$ s.t. $X' \subset Y'$ and $\overline{\overline{X'}} = \overline{\overline{Y'}}$. Hence $Y = f"Y'$ satisfies the conclusion of § 0 Theorem 1. Contradiction! Now suppose the conclusion of Lemma 1 to fail. Then τ is a cardinal in L, since otherwise the conclusion of Lemma 1 would hold with $Y = \tau$. But then τ is not suitable and, in particular, not a successor cardinal in L. Hence there are arbitrarily large $\gamma < \tau$ s.t. $\gamma > \omega_2$ and γ is a successor cardinal in L. But then γ

is suitable and hence $cf(\gamma) = \overline{\overline{\gamma}} > \omega$, since otherwise Lemma 1 would give $Y \subset \gamma$ s.t. $Y \in L$ and $\overline{\overline{Y}}^L < \gamma$, making γ singular in L. Hence τ is suitable. Contradiction! Q E D.

We now outline, very roughly, the method to be used in proving Lemma 1. Let $\tau \geq \omega_2$ be a suitable cardinal in L and let $X \subset \tau$ cofinally s.t. $\overline{\overline{X}} < \overline{\overline{\tau}}$. We can easily construct a map $\pi : J_{\overline{\tau}} \longrightarrow_{\Sigma_1} J_\tau$ s.t.

(*) $\overline{\tau} < \tau$ is suitable

(**) $X \subset rng(\pi)$ (hence $rng(\pi) \cap \tau$ is cofinal in τ).

Suppose that $\overline{\tau}$ is not a cardinal in L. Then there is a least $\overline{\beta} \geq \overline{\tau}$ s.t. $\overline{\tau}$ is not a Σ_ω cardinal in $J_{\overline{\beta}}$ (i.e. there is a $J_{\overline{\beta}}$ definable map of some $\overline{\gamma} < \overline{\tau}$ onto $\overline{\tau}$ (allowing parameters)). But then there is a least $n \geq 1$ s.t. $\overline{\tau}$ is not a Σ_n cardinal in $J_{\overline{\beta}}$ (i.e. there is a $\Sigma_n(J_{\overline{\beta}})$ map from a subset of some $\overline{\gamma} < \overline{\tau}$ onto $\overline{\tau}$). We show that the map π "extends to $\overline{\beta}$" - i.e. there is $\tilde{\pi} \supset \pi$ s.t.

$\tilde{\pi} : J_{\overline{\beta}} \longrightarrow_{\Sigma_n} J_\beta$ for some $\beta \geq \tau$.

By the choice of $\overline{\beta}$, n, there exist $\overline{\gamma} < \overline{\tau}$, $\overline{p} \in J_{\overline{\beta}}$ s.t. each $x \in J_{\overline{\beta}}$ is $\Sigma_n(J_{\overline{\beta}})$ in parameters from $J_{\overline{\gamma}} \cup \{\overline{p}\}$. Let $\gamma = \pi(\overline{\gamma})$, $p = \tilde{\pi}(\overline{p})$. Then $\gamma < \tau$, $p \in J_\beta$. Since τ is a cardinal in L, there must be $\pi' : J_{\beta'} \longrightarrow_{\Sigma_n} J_\beta$ s.t. $\pi' \in L$, $\beta' < \beta$ and $J_\gamma \cup \{p\} \subset rng(\pi')$. But then $rng(\tilde{\pi}) \subset rng(\pi')$ since $\tilde{\pi}''(J_{\overline{\gamma}} \cup \{\overline{p}\}) \subset rng(\pi')$. Hence Lemma 1 holds with $Y = rng(\pi')$.

Now let $\overline{\tau}$ be a cardinal in L. The same proof which showed that π "extends to β" will, in this case, show that π "extends to ∞" - i.e. there is $\tilde{\pi} \supset \pi$ s.t. $\tilde{\pi} : L \longrightarrow_{\Sigma_1} L$. But that is a contradiction by the following well known lemma of Kunen:

<u>Lemma 2.</u> Let $\pi : L \longrightarrow_{\Sigma_1} L$ s.t. $\pi \neq id \restriction L$. Then $0^{\#}$ exists.

The cases: $cf(\tau) > \omega$, $cf(\tau) = \omega$ will be treated separately. The non ω cofinal case is the "natural" one, for we can then show that <u>every</u> $\pi : J_{\overline{\tau}} \longrightarrow_{\Sigma_1} J_\tau$ satisfying (*), (**) has the above extendability properties. In the ω cofinal case we shall have to resort to more or less unsavory legerdemain in order to show that π, $\bar{\tau}$ with the extendability properties exist.

In proving the first extendability property, we shall not work directly with $J_{\bar{\beta}}$ but rather with $\langle J_{\bar{\rho}}, \bar{A}\rangle$, where $\bar{\rho} = \rho^{n-1}_{\bar{\beta}}$, $\bar{A} = A^{n-1}_{\bar{\beta}}$. We show that π extends to $\tilde{\pi} \supset \pi$ s.t. $\tilde{\pi} : \langle J_{\bar{\rho}}, \bar{A}\rangle \longrightarrow_{\Sigma_1} \langle J_\rho, A\rangle$ cofinally for some amenable $\langle J_\rho, A\rangle$. (Where "cofinally" means that $\tilde{\pi}$ "$\omega\bar{\rho}$ is cofinal in $\omega\rho$.) We then prove the existence of β s.t. $\rho = \rho^{n-1}_\beta$, $A = A^{n-1}_\beta$ (the same proof will show that $\tilde{\pi}$ extends to $\pi^* : J_{\bar{\beta}} \longrightarrow_{\Sigma_n} J_\beta$). This latter step is the main concern of § 2.

§ 2 Fine structure lemmas

For the basic theory of the fine structure, the reader is referred to [FS] trough § 4 or [Dev] Ch 7. ρ^n_α, A^n_α, p^n_α denote, as usual, the Σ_n projectum, the Σ_n standard code and the Σ_n standard parameter of α. We recall the following facts:

(1) ρ^n_α = the largest ρ s.t. $\langle J_\rho, A\rangle$ is amenable for all $A \in \Sigma_n(J_\alpha) \cap \mathcal{P}(J_\rho)$.

(2) $R \subset J_{\rho^n_\alpha}$ is $\Sigma_n(J_\alpha)$ iff R is $\Sigma_1(J_{\rho^{n-1}_\alpha}, A^{n-1}_\alpha)$.

(3) $A^0_\alpha = p^0_\alpha = \varnothing$.

(4) Let $n \geq 1$ and let h be the canonical Σ_1 Skolem function for $\langle J_{\rho^{n-1}_\alpha}, A^{n-1}_\alpha\rangle$. Then ρ^n is the least ρ s.t. $J_{\rho^{n-1}_\alpha} = h''(\omega \times J_\rho \times \{p\})$ for some $p \in J_{\rho^n_\alpha}$ and p^n_α is the $<_J$ - least such p.

(5) $R \subset J_{\rho^n_\alpha}$ is $\Sigma_1(J_{\rho^{n-1}_\alpha}, A^{n-1}_\alpha)$ in the parameter p^n_α iff R is rud in $\langle J_{\rho^n_\alpha}, A^n_\alpha\rangle$ (i.e. R is the intersection of $J_{\rho^n_\alpha}$ with a class rudimentary in A^n_α).

(6) Let $\pi : \langle J_{\bar\rho}, \bar A\rangle \longrightarrow_{\Sigma_i} \langle J_{\rho_\alpha^n}, A_\alpha^n\rangle$ ($i \geq 0$). Then there is a unique $\bar\alpha \geq \bar\rho$ s.t. $\bar\rho = \rho_{\bar\alpha}^n$, $\bar A = A_{\bar\alpha}^n$. Moreover, there is a unique $\tilde\pi \supset \pi$ s.t. $\tilde\pi : J_{\bar\alpha} \longrightarrow_{\Sigma_{n+i}} J_\alpha$ and $\tilde\pi(p_{\bar\alpha}^j) = p_\alpha^j (j \leq n)$.

All of these facts are established in [Dev] and [FS]. The next result, though not explicit in our reference articles, does indead follow easily from the above facts.

Def Let $\alpha \leq \beta$, $0 \leq n \leq \omega$: $\omega\alpha$ is a Σ_n cardinal (Σ_n regular) in J_β iff there is no $\Sigma_n(J_\beta)$ function mapping a subset of some $\gamma < \omega\alpha$ onto (cofinally into) $\omega\alpha$.

α is a cardinal (regular) in J_β iff there is no $f \in J_\beta$ mapping a $\gamma < \omega\alpha$ onto (cofinally into) $\omega\alpha$. If α is a cardinal in J_β and $a \in J_\beta$ s.t. $a \subset J_\alpha$, then $\langle J_\alpha, a\rangle$ is amenable.

Clearly, being a cardinal (regular) in J_β is the same as being a Σ_0 cardinal (regular) in J_β.

Lemma 1. Let $n \geq 1$, $\alpha \leq \beta$.

(i) If $\omega\alpha$ is a Σ_{n-1} cardinal but not a Σ_n cardinal in J_β, then $\rho_\beta^n < \alpha \leq \rho_\beta^{n-1}$. Moreover $\omega\rho_\beta^n$ is the least $\gamma < \omega\alpha$ s.t. there is $\Sigma_n(J_\beta)$ map of a subset of γ onto $\omega\alpha$.

(ii) If $\rho_\beta^n < \alpha \leq \rho_\beta^{n-1}$ and α is regular in $J_{\rho_\beta^{n-1}}$, then $cf(\omega\alpha) = cf(\omega\rho_\beta^{n-1})$.

Proof.

(i) $\rho_\beta^0 = \beta \geq \alpha$. Using (4), (2) and the fact that for any ρ there is a $\Delta_1(J_\rho)$ map of $\omega\rho$ onto J_ρ, we get: $\rho_\beta^i \geq \alpha$ for $i < n$ (by induction on i). Hence $\rho_\beta^{n-1} \geq \alpha$. Now let ρ = the least ρ s.t. there is a $\Sigma_n(J_\beta)$ map of a subset of $\omega\rho$ onto $\omega\alpha$. Then $\rho < \alpha$. $\rho_\beta^n \geq \rho$ by (4), (2).

We claim: $\rho^n_\beta = \rho$. Let $f \in \Sigma_1(J_\beta)$ map a subset of $\omega\rho$ onto J_α. Then $f \notin J_\beta$ and hence $\langle J_\gamma, f\rangle$ is not amenable for $\alpha < \gamma \le \beta$. Hence $\rho^n_\beta \le \alpha$. Now set: $a = \{\nu \in \mathrm{dom}(f) \mid \nu \notin f(\nu)\}$. By a diagonal argument, $a \notin J_\alpha$. Hence $\langle J_\gamma, a\rangle$ is not amenable for $\rho < \gamma \le \alpha$. Hence $\rho^n_\beta \le \rho$. QED(i)

(ii) Set: $\rho = \rho^{n-1}_\beta$, $A = A^{n-1}_\beta$, $p = p^n_\beta$, $\gamma = \rho^n_\beta$.

Let h be the canonical Skolem function for $\langle J_\rho, A\rangle$. Define a map f from a subset of J_γ onto J_α by:

$f(\langle i,x\rangle) = h(i,x,p)$ if $x \in J_\gamma$ and $h(i,x,p) \in J_\alpha$

$f(u)$ undefined in all other cases.

Then f is $\Sigma_1(J_\rho, A)$ in a parameter q. Let

$$y = f(x) \longleftrightarrow \bigvee z\, F(z,y,x,q)$$

where F is Σ_0. Let $\lambda = \mathrm{cf}(\omega\rho)$ and let $\langle \xi_\nu \mid \nu < \lambda\rangle$ be a monotone sequence converging to $\omega\rho$. Define $f_\nu (\nu < \lambda)$ by:

$$y = f_\nu(x) \longleftrightarrow y \in S_{\xi_\nu} \wedge \bigvee z \in S_{\xi_\nu}\, F(z,y,x,p).$$

Then $f_\nu \in J_\rho$ and f_ν maps a subset of J_γ into J_α.

Set: $\alpha_\nu = \sup(\mathrm{On} \cap \mathrm{rng}(f_\nu))$. Then $\alpha_\nu < \omega\alpha$ since $\omega\alpha$ is regular in J_ρ. But $\nu \le \eta \longrightarrow \alpha_\nu \le \alpha_\eta$, since $f_\nu \subset f_\eta$. Finally, $\sup_\nu \alpha_\nu = \omega\alpha$ since $\bigcup_\nu f_\nu = f$. QED

Carrying the proof of Lemma 1 (ii) a step further, we get the following rather technical lemma which will be of service to us in § 5.

<u>Lemma 2.</u> Let $\rho^{n-1}_\beta \ge \alpha > \rho^n_\beta$ where $\omega\alpha$ is regular in J_β. Let $\lambda = \mathrm{cf}(\omega\alpha)$. Then there is a sequence $\langle f_\nu \mid \nu < \lambda\rangle$ s.t. $\{f_\nu \mid \nu < \lambda\} \subset J_\alpha$ and if $\pi : J_{\bar\alpha} \longrightarrow_{\Sigma_0} J_\alpha$ s.t. $\{f_\nu \mid \nu < \lambda\} \subset \mathrm{rng}(\pi)$, then:

(a) There are unique $\tilde\pi \supset \pi$, $\bar\rho \ge \bar\alpha$, $\bar A \subset J_{\bar\rho}$ s.t. $p^n_\beta \in \mathrm{rng}(\tilde\pi)$ and $\tilde\pi : \langle J_{\bar\rho}, \bar A\rangle \longrightarrow_{\Sigma_1} \langle J_{\rho^{n-1}_\beta}, A^{n-1}_\beta\rangle$.

(b) There is a unique $\overline{\beta}$ s.t. $\overline{\rho} = \rho^{n-1}_{\overline{\beta}}$, $\overline{A} = A^{n-1}_{\overline{\beta}}$.

(c) $\rho^n_{\overline{\beta}} < \overline{\alpha}$.

(d) If $\pi(\rho^n_{\overline{\beta}}) = \rho^n_\beta$, then $\tilde{\pi}(p^n_{\overline{\beta}}) = p^n_\beta$.

Proof.

We first prove the existence part of (a). Let ρ, A, p, γ, $\langle \xi_\nu | \nu < \lambda \rangle$ h, f, q, $\langle f_\nu | \nu < \lambda \rangle$ be as in the proof of Lemma 1 (ii). We note that $f_\nu \in J_\alpha$, since $f_\nu \in J_\rho$ is bounded in J_α and α is a cardinal in J_ρ. Set $Y = \mathrm{rng}(\pi) \cap J_\gamma$. $X = h''(\omega \times Y \times \{p\})$. Then $X \prec_{\Sigma_1} \langle J_\rho, A \rangle$. It is clear by the definition of f that $X \cap J_\alpha = f''Y$. Using this we get:

<u>Claim</u> $X \cap J_\alpha = \mathrm{rng}(\pi)$.

Proof.

($\subset$) Let $x \in X \cap J_\alpha$. Then $x = f(z)$ for a $z \in Y$. Hence $x = f_\nu(z)$ for some ν. Hence $x \in f''_\nu\, Y \subset \mathrm{rng}(\pi)$.

($\supset$) Let $x \in \mathrm{rng}(\pi)$. Let z = the $<_J$-least z s.t. $x = f(z)$. Then $x = f_\nu(z)$ for some ν. But then $z \in Y$ since z = the $<_J$-least z s.t. $x = f_\nu(z)$.

QED (Claim)

Now let $\tilde{\pi} : \langle J_{\overline{\rho}}, \overline{A} \rangle \xleftrightarrow{\sim} \langle X, A \cap X \rangle$. Then $\overline{\rho} \geq \overline{\alpha}$ and $\tilde{\pi} \upharpoonright J_{\overline{\alpha}} = \pi$ by the claim. This proves the existence part of (a). The uniqueness part of (a) follows by the fact that if $\tilde{\pi} \supset \pi$ s.t. $\tilde{\pi} : \langle J_{\overline{\rho}}, \overline{A} \rangle \longrightarrow_{\Sigma_1} \langle J_\rho, A \rangle$ and $p \in \mathrm{rng}(\tilde{\pi})$, then $\mathrm{rng}(\tilde{\pi}) = h''(\omega \times Y \times \{p\})$. (b) is immediate by fact (6) above. To prove (c), set $J_{\overline{\gamma}} = \pi^{-1}{}''J_\gamma$. Then $\overline{\gamma} < \alpha$. But $\rho^n_{\overline{\beta}} \leq \overline{\gamma}$, since, letting $\overline{h} = h_{\overline{\rho}, \overline{A}}$ be the canonical Σ_1 Skolem function for $\langle J_{\overline{\rho}}, \overline{A} \rangle$ and $\tilde{\pi}(\overline{p}) = p$, we have: $J_{\overline{\rho}} = \overline{h}''(\omega \times J_{\overline{\gamma}} \times \{\overline{p}\})$, since $\mathrm{rng}(\tilde{\pi}) = h''(\omega \times Y \times \{p\}) = \tilde{\pi}''\overline{h}''(\omega \times J_{\overline{\gamma}} \times \{\overline{p}\})$. We now prove (d). We have: $\overline{\gamma} = \rho^n_{\overline{\beta}}$ and $\pi(\overline{\gamma}) = \gamma$. Set $\overline{p}' = p^n_{\overline{\beta}}$, $p' = \tilde{\pi}(\overline{p}')$. Then $\overline{p}' \leq_J \overline{p}$, since $J_{\overline{\rho}} = \overline{h}''(\omega \times J_{\overline{\gamma}} \times \{\overline{p}\})$. But $p \leq_J p'$, since there is $\overline{x} \in J_{\overline{\gamma}}$ s.t. $\overline{p} = \overline{h}(i,\overline{x},\overline{p}')$; hence $p = h(i,x,p')$, where $x = \pi(\overline{x}) \in J_\gamma$. Hence $h''(\omega \times J_\gamma \times \{p'\}) = J_\rho$. QED

The main object of this section is to prove a sort of converse of fact (6) above. First a definition and a preliminary lemma.

<u>Def</u> An imbedding $\sigma : \langle J_\rho, A\rangle \longrightarrow_{\Sigma_1} \langle J_{\rho'}, A'\rangle$ of one amenable structure into another is called <u>strong</u> iff whenever R is a well founded relation on J_ρ which is rud in $\langle J_\rho, A\rangle$ and R' is rud in $\langle J_{\rho'}, A'\rangle$ by the same rud definition, then R' is well founded.

<u>Lemma 3.</u> Let i, n > 0 and suppose $\sigma : \langle J_{\rho^n_{\overline{\beta}}}, A^n_{\overline{\beta}}\rangle \longrightarrow_{\Sigma_i} \langle J_\rho, A\rangle$ is strong.

Then there are η, B, $\tilde{\sigma}$ s.t. $\tilde{\sigma} \supset \sigma$ and:

(i) $\rho = \rho^1_{\eta,B}$, $A = A^1_{\eta,B}$, $\tilde{\sigma}(p^{n-1}_{\overline{\beta}}) = p^1_{\eta B}$.

(ii) $\tilde{\sigma} : \langle J_{\rho^{n-1}_{\overline{\beta}}}, A^{n-1}_{\overline{\beta}}\rangle \longrightarrow_{\Sigma_{i+1}} \langle J_\eta, B\rangle$ is strong.

Proof of Lemma 3.

Set $\overline{\rho} = \rho^n_{\overline{\beta}}$, $\overline{A} = A^n_{\overline{\beta}}$, $\overline{\eta} = \rho^{n-1}_{\overline{\beta}}$, $\overline{B} = A^{n-1}_{\overline{\beta}}$, $\overline{p} = p^{n-1}_{\overline{\beta}}$.

Then $J_{\overline{\eta}} = h_{\overline{\eta},\overline{B}}$ "$(\omega \times J_{\overline{\rho}} \times \{\overline{p}\})$. (We shall generally use $h_{\eta B}$ to denote the canonical Σ_1 Skolem function of an amenable structure $\langle J_\eta, B\rangle$). Define $\overline{h}$ by

$\overline{h}(\langle i,x\rangle) \simeq h_{\overline{\eta},\overline{B}}(i,x,p)$ if $x \in J_{\overline{\rho}}$

$\overline{h}(u)$ undefined otherwise.

Define relations $\overline{D}$, $\overline{E}$, $\overline{I}$, $\overline{B}'$ on $J_{\overline{\rho}}$ by:

$\overline{D} = \mathrm{dom}(\overline{h})$

$\overline{E} = \{\langle x,y\rangle \in \overline{D}^2 \mid \overline{h}(x) \in \overline{h}(y)\}$

$\overline{I} = \{\langle x,y\rangle \in \overline{D}^2 \mid \overline{h}(x) = \overline{h}(y)\}$

$\overline{B}' = \{x \in \overline{D} \mid \overline{h}(x) \in \overline{B}\}$.

Since $\overline{D}$, $\overline{E}$, $\overline{I}$, $\overline{B}'$ are $\Sigma_1(J_{\overline{\eta}},\overline{B})$ in $\overline{p}$, they are rud in $\langle J_{\overline{\rho}}, \overline{A}\rangle$.
Let D, E, I, B' have the same rud definitions in $\langle J_\rho, A\rangle$. Then E is well founded, since $\overline{E}$ is well founded and σ is strong. Set:

$\overline{M} = \langle \overline{D}, \overline{I}, \overline{E}, \overline{B}' \rangle$, $M = \langle D, I, E, B' \rangle$. Let $\overline{T}$ be the Σ_1 satisfaction relation for the model $\overline{M}$. Then $\overline{T}(\phi, \langle \vec{x} \rangle) \longleftrightarrow \models^{\Sigma_1}_{\langle J_{\overline{\eta}}, \overline{B} \rangle} \phi[\overline{h}(\vec{x})]$, so $\overline{T}$ is $\Sigma_1(J_{\overline{\eta}}, \overline{B})$ in $\overline{p}$ and hence rud in $\langle J_{\overline{\rho}}, \overline{A} \rangle$. Let T have the same rud definition in $\langle J_\rho, A \rangle$.

<u>Fact 1.</u> T is the Σ_1 satisfaction relation for M.

Proof of Fact 1. We must show that:

$$T([v \in w], \langle x,y \rangle) \longleftrightarrow x\ E\ y$$

$$T([v = w], \langle x,y \rangle) \longleftrightarrow x\ I\ y$$

$$T(\overset{o}{A}(v), \langle x \rangle) \longleftrightarrow B'\ x$$

$$T(\phi \wedge \psi, \langle \vec{x} \rangle) \longleftrightarrow T(\phi, \langle \vec{x} \rangle) \wedge T(\psi, \langle \vec{x} \rangle)$$

$$T(\neg\phi, \langle \vec{x} \rangle) \longleftrightarrow \neg\, T(\phi, \langle \vec{x} \rangle)$$

$$T(\bigvee v\ \phi, \langle \vec{x} \rangle \longleftrightarrow \bigvee y \in D\ T(\phi, \langle y, \vec{x} \rangle).$$

All but the last equivalence are expressible as Π_1 statements in $\langle J_\rho, A \rangle$ and therefore hold since the corresponding Π_1 statements in $\langle J_{\overline{\rho}}, \overline{A} \rangle$ hold. To see the last equivalence, note that the relation $\models^{\Sigma_1}_{\langle J_{\overline{\eta}}, \overline{B} \rangle} \phi[y, \overline{h}(\vec{x})]$ is $\Sigma_1(J_{\overline{\eta}}, \overline{B})$ in $\overline{p}$; hence

$$\models_{\langle J_{\overline{\eta}}, \overline{B} \rangle} \bigvee v\ \phi[\overline{h}(\vec{x})] \longleftrightarrow \bigvee i < \omega \models_{\langle J_{\overline{\eta}}, \overline{B} \rangle} \phi[\overline{h}(\langle i, \vec{x} \rangle), \langle \vec{x} \rangle]$$

hence:

$$\overline{T}(\bigvee y\ \phi, \langle \vec{x} \rangle) \longleftrightarrow \bigvee i < \omega\ \overline{T}(\phi, \langle \langle i, \vec{x} \rangle, \langle \vec{x} \rangle \rangle).$$

But the last equivalence is expressible as a Π_1 statement in $\langle J_{\overline{\rho}}, \overline{A} \rangle$ (since $\bigvee i < \omega\ \overline{T}(\phi, \langle \langle i, \vec{x} \rangle, \langle \vec{x} \rangle \rangle)$ is rud in $\langle J_{\overline{\rho}}, \overline{A} \rangle$ in the parameter ω) and therefore carries up to $\langle J_\rho, A \rangle$. QED (Fact 1)

Since the satisfaction relations $\overline{T}$, T are rud in $\langle J_{\overline{\rho}}, \overline{A} \rangle$, $\langle J_\rho, A \rangle$ resp. by the same rud definitions and σ is Σ_i preserving, we have: $(\sigma \upharpoonright \overline{D}) : \overline{M} \longrightarrow_{\Sigma_{i+1}} M$. So, in particular, M satisfies the identity axioms and the extensionality axiom, since $\overline{M}$ does. We may thus define the factor

model $\overline{M}^* = \overline{M}/\overline{I} = \langle \overline{D}^*, \overline{E}^*, \overline{B}^* \rangle$, $M^* = M/I = \langle D^*, E^*, B^* \rangle$.
Let $\overline{k} : \overline{M} \longrightarrow \overline{M}^*$, $k : M \longrightarrow M^*$ be the natural projections. $\overline{E}^*$, E^* are both well founded and extensional. Hence we may transitivise the models $\overline{M}^*$, M^* by Mostowski isomorphisms $\overline{l}$, l. Clearly, $\overline{l} : \overline{M}^* \stackrel{\sim}{\longleftrightarrow} \langle J_{\overline{\eta}}, \overline{B} \rangle$ and $\overline{h} = \overline{l}\,\overline{k}$. Let $l : M^* \stackrel{\sim}{\longleftrightarrow} \langle J_\eta, B \rangle$. Set $h = l\,k$. Define $\sigma^* : \overline{M}^* \longrightarrow_{\Sigma_{i+1}} M^*$ by $\sigma^* \overline{k} = k\,\sigma$. Define $\tilde{\sigma} : \langle J_{\overline{\eta}}, \overline{B} \rangle \longrightarrow_{\Sigma_{i+1}} \langle J_\eta, B \rangle$ by $\tilde{\sigma}\,\overline{h} = h\,\sigma$.

Thus:

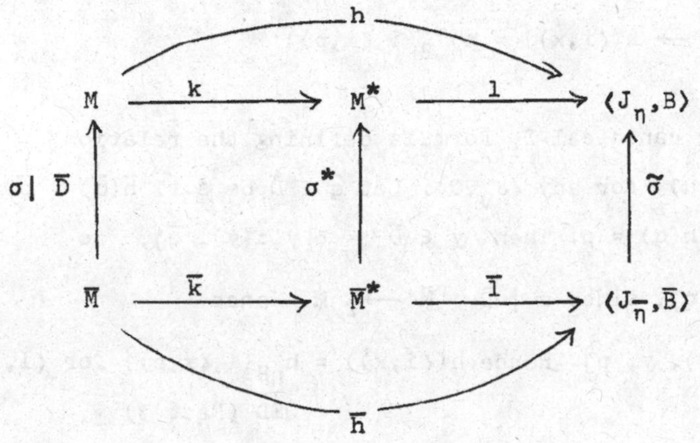

Fact 2. $\tilde{\sigma} \upharpoonright J_{\overline{\rho}} = \sigma$

Proof of Fact 2.

By definition, $\overline{h}$ uniformises the relation
$\{\langle y, \langle i, x \rangle\rangle \mid x \in J_{\overline{\rho}} \wedge \models_{\langle J_{\overline{\eta}}, \overline{B}\rangle} \phi_i[y, \langle x, \overline{p} \rangle]\}$, where $\langle \phi_i \mid i < \omega \rangle$ is some fixed recursive enumeration of the Σ_1 formulae. Let $\phi_{j_0}(y,z)$ be the formula $\bigvee q(\langle y, q \rangle = x)$. Clearly, $\overline{h}(\langle j_0, x \rangle) = x$ for $x \in J_{\overline{\rho}}$. Set: $s(x) = \langle j_0, x \rangle$. Then s is rudimentary. $s \upharpoonright J_\rho$ maps J_ρ into D, since $s \upharpoonright J_{\overline{\rho}}$ maps $J_{\overline{\rho}}$ into $\overline{D}$. For $x, y \in J_\rho$ we have:

$$s(x)\ E\ s(y) \longleftrightarrow x \in y$$
$$s(x)\ I\ s(y) \longleftrightarrow x = y$$

$$z\ E\ s(y) \longleftrightarrow \bigvee x \in y\ \ z\ I\ s(x),$$

since the corresponding formulae hold in $\langle J_{\bar\rho}, \bar A\rangle$.

Thus $k\ s \restriction J_\rho$ maps $\in \restriction J_\rho$ isomorphically onto an initial segment of E^*. But then $h\ s \restriction J_\rho = lk\ s \restriction J_\rho$ maps $\in \restriction J_\rho$ isomorphically onto an initial segment of $\in \restriction J_\eta$. Hence $h\ s \restriction J_\rho = \mathrm{id} \restriction J_\rho$. Clearly, $\bar h s \restriction J_{\bar\rho} = \mathrm{id} \restriction J_{\bar\rho}$. Hence, for $x \in J_{\bar\rho}$, we have $\tilde\sigma(x) = \tilde\sigma\ \bar h\ s(x) = h\ \sigma\ s(x) = h\ s\ \sigma(x) = \sigma(x)$.

QED (Fact 2)

Set $p = \tilde\sigma(\bar p)$.

<u>Fact 3.</u> $\langle i,x\rangle \in D \longrightarrow h(\langle i,x\rangle) = h_{\eta,B}(i,\langle x,p\rangle)$

Proof of Fact 3.

Let $\phi(y,w,u)$ be the canonical Σ_1 formula defining the relation $y = h_{\nu C}((w)_o,\langle (w)_1,u\rangle)$ for any $\langle J_\nu, C\rangle$. Let $\bar q \in \bar D$ be s.t. $\bar h(\bar q) = \bar p$ and set $q = \sigma(\bar q)$. Then $h(q) = p$. Then $\bigwedge y \in \bar D \models_{\bar M} \phi[y,s(y),\bar q]$, so $\bigwedge y \in D \models_M \phi[y,s(y),q]$, since $\sigma \restriction \bar D : \bar M \longrightarrow_{\Sigma_2} M$. Hence $\bigwedge y \in D \models_{\langle J_\eta,B\rangle} \phi[h(y), y, p]$. Hence $h(\langle i,x\rangle) = h_{\eta B}(i,\langle x,p\rangle)$ for $\langle i,x\rangle \in D$

QED (Fact 3)

We recall that by definition, if $\langle J_\nu,C\rangle$ is amenable and $\rho = \rho^1_{\nu,C}$, $p = p^1_{\nu,C}$ then $A^1_{\nu,C} = \{\langle i,x\rangle \mid x \in J_\rho \wedge \models_{\langle J_\nu,C\rangle} \phi_i[x,p]\}$, where $\langle \phi_i\rangle$ is a fixed recursive enumeration of the Σ_1 formulae.

<u>Fact 4.</u> $A = \{\langle i,x\rangle \mid x \in J_\rho \wedge \models_{\langle J_\eta,B\rangle} \phi_i[x,p]\}$.

Proof of Fact 4.

$\bar A = A^1_{\bar\eta,\bar B}$; hence: $\bigwedge i\ x(\bar A(i,x) \longleftrightarrow \models_{\bar M} \phi_i[s(x), \bar q])$.

Since $\sigma : \langle J_{\bar\rho}, \bar A\rangle \longrightarrow_{\Sigma_1} \langle J_\rho, A\rangle$, we conclude:

$\bigwedge i\ x(A(i,x) \longleftrightarrow \models_M \phi_i[s(x),q])$. Hence $A(i,x) \longleftrightarrow \models_{\langle J_\eta,B\rangle} \phi_i[x,p]$.

QED (Fact 4)

<u>Fact 5.</u> $\rho = \rho^1_{\eta,B}$.

Proof of Fact 5.

h is a $\Sigma_1(J_\eta, B)$ map of D onto J_η by Fact 3; but $D \subset J_\rho$, hence $\rho^1_{\eta,B} \leq \rho$. On the other hand, $\langle J_\rho, A\rangle$ is amenable and every $\Sigma_1(J_\eta, B)$ subset P of J_ρ is Σ_1 in parameters from $J_\rho \cup \{p\}$ (by Fact 3), hence rud in $\langle J_\rho, A\rangle$ in parameters from J_ρ (by Fact 4). But then $\langle J_\rho, P\rangle$ is amenable for all such P. Hence $\rho \leq \rho^1_{\eta,B}$. QED (Fact 5)

<u>Fact 6.</u> $p = p^1_{\eta,B}$

Proof.

$p^1_{\eta,B} \leq_J p$ by Facts 3 and 5. Now let $p^1_{\eta,B} <_J p$. Then
$\bigvee i \bigvee q <_J p \bigvee x \in J_\rho\ h_{\eta,B}(i,\langle x,q\rangle) = p$; hence

$\bigvee i \bigvee \bar{q} <_J \bar{p} \bigvee x \in J_{\bar\rho}\ h_{\bar\eta,\bar B}(i,\langle x,\bar q\rangle) = \bar p$ by the fact that $\tilde\sigma$ is Σ_1 preserving and $\mathrm{rng}(\tilde\sigma) \cap J_\rho = \mathrm{rng}(\sigma)$. Hence $p^1_{\bar\eta,\bar B} < \bar p$. Contradiction!

QED (Fact 6)

Facts 4, 5, 6 immediately give:

<u>Fact 7.</u> $A = A^1_{\eta,B}$

All that remains to be proved is

<u>Fact 8.</u> $\tilde\sigma$ is strong.

Proof of Fact 8.

Let $\bar R$ be well founded and rud in $\langle J_{\bar\eta}, \bar B\rangle$. Let R have the same rud definition over $\langle J_\eta, B\rangle$. Set:

$\bar R' = \{\langle x,y\rangle \in \bar D^2 \mid \bar h(x)\ R\ \bar h(y)\}$

$R' = \{\langle x,y\rangle \in D^2 \mid h(x)\ R\ h(y)\}$.

Then $\bar R'$ is $\Sigma(J_{\bar\eta}, \bar B)$ in $\bar p$ and R' is $\Sigma_1(J_\eta, B)$ in p by the same definition. Hence $\bar R'$ is rud in $\langle J_{\bar\rho}, \bar A\rangle$ and R' is rud in $\langle J_\rho, A\rangle$ by the same definition. But σ is strong; hence R' is well founded. But then R is well founded. QED

By finitely many iterations of Lemma 3 we get:

Lemma 4. Let $n > 0$ and suppose $\sigma : \langle J_{\rho^n_{\overline{\beta}}}, A^n_{\overline{\beta}}\rangle \longrightarrow_{\Sigma_1} \langle J_\rho, A\rangle$ is strong, where $\langle J_\rho, A\rangle$ is amenable. Then there is an ordinal β s.t. $\rho = \rho^n_\beta$, $A = A^n_\beta$.

Lemma 4 is the "converse" of Fact (6) announced earlier.

Remark. Though we shall not make use of the fact, notice that β above must be unique and that σ extends to a unique $\tilde{\sigma} : J_{\overline{\beta}} \longrightarrow_{\Sigma_{n+1}} J_\beta$ which preserves the first n standard parameters.

§ 3 The non ω cofinal case

Set $J_\infty = \bigcup_{\nu<\infty} J_\nu = L$.

Lemma 1. Let $\overline{\tau}$ be suitable s.t. $cf(\overline{\tau}) > \omega$. Let $\pi : J_{\overline{\tau}} \longrightarrow_{\Sigma_1} J_\tau$ cofinally (i.e. $\omega\tau = \sup(On \cap rng(\pi))$). Let $\overline{\tau} \leq \overline{\beta} \leq \infty$ where $\overline{\beta}$ is a limit ordinal and $\overline{\tau}$ is a cardinal in $J_{\overline{\beta}}$. Then there are $\beta \geq \tau$, $\tilde{\pi} \supset \pi$ s.t. $\tilde{\pi} : J_{\overline{\beta}} \longrightarrow_{\Sigma_1} J_\beta$ cofinally.

The proof stretches over several sublemmas. Assume for the moment that $1 < \tau \leq \beta \leq \infty$, where β is a limit ordinal.

Def $T = T^{\tau,\beta}$ = the collection of triples $t = \langle \delta_t, \mu_t, u_t\rangle$ s.t. $\delta_t < \tau$, $\mu_t < \beta$, $u_t \subset J_{\mu_t}$, $\overline{\overline{u}}_t < \omega$.

Define a partial ordering on T by $t \leq t' \longleftrightarrow \delta_t \leq \delta_{t'} \wedge \mu_t \leq \mu_{t'} \wedge u_t \subset u_{t'}$. For $t \in T$ set:

X_t = the smallest $x \prec_\Sigma J_{\mu_t}$ s.t. $J_{\delta_t} \cup u_t \subset X$;

hence $X_t = h_{\mu_t}{}''(\omega \times J_{\delta_t} \times \{u_t\})$, where h_μ is the canonical Skolem function for J_μ. Clearly, $t \leq t' \longrightarrow X_t \prec_{\Sigma_0} X_{t'}$. Set:

$\sigma_t : J_{\gamma_t} \stackrel{\sim}{\longleftrightarrow} X_t$; $\sigma_{tt'} = \sigma^{-1}_{t'}\sigma_t$ $(t \leq t')$.

Then $J_{\gamma_t} \xrightarrow[\Sigma_0]{\sigma_{tt'}} J_{\gamma_{t'}} \xrightarrow[\Sigma_0]{\sigma_{t'}} J_\beta$ $(t \leq t')$ and $\langle J_{\gamma_t}\rangle, \langle\sigma_{tt'}\rangle$ is a directed system whose limit is $J_\beta, \langle\sigma_t\rangle$. We note that $\sigma_t \in J_\beta$, since σ_t is the set of pairs $\langle h_{\mu_t}(i,z,u_t), h_{\gamma_t}(i,z, \sigma_t^{-1}(u_t))\rangle$ s.t. $z \in J_{\delta_t}$ and $\langle i,z,u_t\rangle \in \mathrm{dom}(h_{\mu_t})$. If $\mu_t < \tau$, the same argument shows: $\sigma_t \in J_\tau$. But then $\sigma_{tt'} \in J_\tau$ if $\gamma_{t'} < \tau$, since $\sigma_{tt'} = \sigma_s$, where $s = \langle\delta_t,\mu,\sigma_{t'}^{-1}(u_t)\rangle$ and $\mu = \sigma_{t'}^{-1}{}''(X_{t'} \cap \mu_t)$.

We also note that σ_t is describable as the unique $\sigma : J_{\gamma_t} \xrightarrow[\Sigma_0]{} L$ s.t. $\sigma \restriction J_{\delta_t} = \mathrm{id} \restriction J_{\delta_t}$ and $\sigma(\sigma_t^{-1}(u_t)) = u_t$. To see this, note that

(*) $J_{\gamma_t} \models \phi(x) \longrightarrow L \models \phi(\sigma(\vec{x}))$ for all $\vec{x} \in J_{\gamma_t}$ and Σ_1 formulae ϕ.

Now let $h = h_\infty$ be the canonical Σ_1 Skolem function for L. Let $x \in J_{\gamma_t}$. Then $x = h_{\gamma_t}(i,z, \sigma_t^{-1}(u_t))$ for some $i < \omega$, $z \in J_{\delta_t}$. By (*) we have: $\sigma(h_{\gamma_t}(i,z, \sigma_t^{-1}(u_t))) = h(i, z, u_t)$. Hence σ is unique.

Lemma 1.1. $\omega\tau$ is a cardinal in J_β iff $\bigwedge t \in T\ \gamma_t < \tau$.

Proof.

($\longleftarrow$) Suppose $\omega\tau$ is not a cardinal in J_β. Then there are $\mu < \beta$, $f \in J_\mu$ s.t. f maps a $\delta < \tau$ onto J_τ. Hence $J_\tau \subset X_{\langle\delta,\mu,\{f\}\rangle}$ and $\gamma_{\langle\delta,\mu,\{f\}\rangle} \geq \tau$.

QED ($\longleftarrow$)

($\longrightarrow$) We may assume $X_t = J_{\gamma_t}$, since otherwise this holds with t replaced by $t' = \langle\delta_t, \gamma_t, \sigma_t^{-1}(u_t)\rangle$. But then $h_{\mu_t} \in J_\beta$ and $J_{\gamma_t} = h_{\mu_t}{}''(\omega \times J_{\delta_t} \times \{u_t\})$. It follows that an $f \in J_\beta$ maps δ_t onto $\omega\gamma_t$. Hence $\gamma_t < \tau$, since $\omega\tau$ is a cardinal in J_β. QED

Now let $\omega\bar{\tau}$ be a cardinal in $J_{\bar{\beta}}$, $\pi : J_{\bar{\tau}} \longrightarrow_{\Sigma_1} J_\tau$ cofinally, $T = T^{\bar{\tau},\bar{\beta}}$.

For $t, t' \in T$, $t \leq t'$ set:

$$\delta_t^* = \delta_t^{*(\pi)} = \pi(\delta_t)$$

$$\gamma_t^* = \gamma_t^{*(\pi)} = \pi(\gamma_t)$$

$$\sigma_{tt'}^* = \sigma_{tt'}^{*(\pi)} = \pi(\sigma_{tt'}).$$

Then $\sigma^*_{tt'} : J_{\gamma^*_t} \longrightarrow_{\Sigma_0} J_{\gamma^*_{t'}}$, $\sigma^*_{tt'} \restriction J_{\delta^*_t} = \mathrm{id} \restriction J_{\delta^*_t}$ and $\langle J_{\gamma^*_t}\rangle, \langle \sigma^*_{tt'}\rangle$ is a directed system. Define $M = M^{\bar\beta,\pi}$, $\sigma^*_t = \sigma^{*(\bar\beta,\pi)}_t$ $(t \in T)$ by: $M, \langle\sigma^*_t\rangle$ = the direct limit of $\langle J_{\gamma^*_t}\rangle, \langle\sigma^*_{tt'}\rangle$. We assume w. l. o. g. that $\sigma^*_t \restriction J_{\delta^*_t} = \mathrm{id} \restriction J_{\delta^*_t}$ (hence $J_\tau \subset M$ since $\tau = \sup_t \delta^*_t$).

Define $\tilde\pi = \tilde\pi^{(\bar\beta)} : J_{\bar\beta} \longrightarrow_{\Sigma_1} M$ by:

$$\begin{array}{ccc} J_{\gamma^*_t} & \xrightarrow{\sigma^*_t} & M \\ \uparrow {\scriptstyle \pi \restriction J_\gamma} & & \uparrow {\scriptstyle \tilde\pi} \\ J_{\gamma_t} & \xrightarrow[\sigma_t]{} & J_{\bar\beta}\,. \end{array}$$

Then $\tilde\pi \supset \pi$, since for $x \in J_{\bar\tau}$, there is $t \in T$ s.t. $x \in J_{\delta_t}$ and hence $\tilde\pi(x) = \sigma^*_t\,\pi\,\sigma^{-1}_t(x) = \pi(x)$. We note that $\tilde\pi''\omega\bar\beta$ lies cofinally in $\{x \mid M \models x \varepsilon \mathrm{On}\}$, since if $x \in \mathrm{On}$ in M, $x = \sigma^*_t(\eta)$, then $\sigma^*_s(\pi\,\sigma^{-1}_s(\mu_t)) > x$ in M, where $s = \langle \delta_t, \mu_t + 1, u_t \cup \{\mu_t\}\rangle$. We also note that M satisfies the Π_2 statement "I am a J_α", since $M, \langle\sigma^*_t\rangle$ is the limit of $\langle J_{\gamma^*_t}\rangle, \langle\sigma^*_{tt'}\rangle$ and each $J_{\gamma^*_t}$ satisfies it. Hence if M were transitive we could conclude: $\bigvee\beta \le \infty\ M = J_\beta$.

<u>Lemma 1.2.</u> $\{y \mid M \models y \varepsilon x\}$ is a set for $x \in M$.

Proof.

We assume $\bar\beta = \infty$, since otherwise M is a set and there is nothing to prove. We first note:

(1) If $t \in T$, then $\sigma^*_t = \tilde\sigma_t$, where $\tilde\sigma_t = \{\langle y,x\rangle \mid M \models y = \tilde\pi(\sigma_t)(x)\}$.

Proof of (1).

Since M satisfies "I am a J_α", we can define its canonical Σ_1 Skolem function h. Then h, $h_{\gamma^*_t}$ have the same Σ_1 definition. But $J_{\gamma^*_t} =$

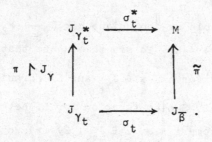

$h_{\gamma_t^*}$ "$(\omega \times J_{\delta_t^*} \times \{\sigma_t^{*-1}(u_t^*)\})$, where $u_t^* = \tilde{\pi}(u_t)$, since $\pi(h_{\gamma_t}) = h_{\gamma_t^*}$ and $\pi(\sigma_t^{-1}(u_t)) = \sigma_t^{*-1}(u_t^*)$.

By our previous argument, we conclude that $\sigma_t^* = \tilde{\sigma}_t$ = the unique $\sigma : J_{\gamma_t^*} \longrightarrow_{\Sigma_0} M$ s.t. $\sigma(\sigma_t^{*-1}(u_t^*)) = u_t^*$. QED (1)

Now set: $\tilde{J}_\kappa = \{y \mid M \models y \in J_{\tilde{\pi}(\kappa)}\}$. It suffices to show that $\tilde{J}_\kappa$ is a set for arbitrarily large κ. We show:

(2) If $\kappa > \tau$ is regular, then $\tilde{J}_\kappa \subset \bigcup_{t \in T \cap J_\kappa} \operatorname{rng}(\sigma_t^*)$

Proof of (2).

Let $t \in T$. We shall construct $t' \in T \cap J_\kappa$ s.t. $\operatorname{rng}(\sigma_t^*) \cap \tilde{J}_\kappa \subset \operatorname{rng}(\sigma_{t'}^*)$. Since κ is regular, there is $\eta < \kappa$ s.t. $\operatorname{rng}(\sigma_t) \cap J_\kappa \subset J_\eta$. Set: $Y = h_{\mu_t}$"$(\omega \times J_\eta \times \{u_t\})$; $\sigma : J_{\mu'} \overset{\sim}{\longleftrightarrow} Y$; $\sigma(u') = u_t$; $t' = \langle \delta_t, \mu', u\rangle$. Then $t' \in T \cap J_\kappa$ and $\operatorname{rng}(\sigma_t) \cap J_\kappa \subset \operatorname{rng}(\sigma_{t'})$. Hence $\operatorname{rng}(\tilde{\sigma}_t) \cap \tilde{J}_\kappa \subset \operatorname{rng}(\tilde{\sigma}_{t'})$ and the conclusion follows by (1). QED

By Lemma 1.2. we may assume w. l. o. g. that the well founded core of M is transitive. Thus M is a transitive class if M is well founded and, in fact, $\bigvee\beta \leq \infty$ $M = J_\beta$, since M satisfies "I am a J_α". We complete the proof of Lemma 1 by showing:

<u>Lemma 1.3.</u> If $\bar{\tau}$ is suitable and $cf(\bar{\tau}) > \omega$, then $\bigvee\beta \leq \infty$ $M = J_\beta$.

Proof of Lemma 1.3.

As remarked, we need only show that M is well founded. Suppose not. Then there are $x_i \in M$ s.t. $x_0 \ni x_1 \ni \ldots$. We may suppose that $x_i \in \operatorname{rng}(\sigma_{t_i}^*)$, where $t_i \leq t_{i+1}$, $\gamma_{t_i} < \delta_{t_{i+1}}$ and $t_i \in u_{t_{i+1}}$. Then the system $\langle J_{\gamma_{t_i}^*}\rangle$, $\langle \sigma_{t_i t_j}^*\rangle$ has a limit which is not well founded. On the other hand, $\langle J_{\gamma_t}\rangle$, $\langle \sigma_{t_i t_j}\rangle$ has a well founded limit, since $\sigma_{t_j} \sigma_{t_i t_j} = \sigma_{t_i}$ and $\sigma_{t_i} : J_{\gamma_{t_i}} \longrightarrow_{\Sigma_0} J_{\bar{\beta}}$, where J_β is well founded. Let N, $\langle \sigma_i \rangle$ = the limit of $\langle J_{\gamma_{t_i}}\rangle$, $\langle \sigma_{t_i t_j}\rangle$. Since N is well founded, we may assume it to be transitive. Hence $N = J_\gamma$ for some γ.

Claim $J_\gamma, \sigma_i \in J_{\bar\tau}$.

Proof.

We first note:

(1) $\sigma_i \in J_\gamma$, since $\sigma_i = \sigma_{t_i'}$ where $t_i' = \sigma_{i+1}\, \sigma^{-1}_{t_{i+1}}(t_i)$.

Since $cf(\tau) > \omega$, we have:

(2) $\sup_i \delta_{t_i} < \bar\tau$.

Let $\delta = \sup_i \delta_{t_i}$. Pick $\rho > \delta$ s.t. $\rho \le \bar\tau$, ρ is regular in $J_{\bar\tau}$ (hence in $J_{\bar\beta}$) and $cf(\rho) > \omega$ (such ρ exists by our assumptions on $\bar\tau$). It is clear that $\gamma \le \sup_i \mu_{t_i} \le \beta$; hence $\sigma_i \in J_{\bar\beta}$ by (1). But $dom(\sigma_i) = J_{\gamma_{t_i}}$ and $\bar{\bar{J}}_{\gamma_{t_i}} = \bar{\bar{\delta}}_{t_i}$ in $J_{\bar\tau}$; hence

(3) $rng(\sigma_i) \cap \rho$ is bounded in ρ, since $\delta_i < \rho$ and ρ is regular in $J_{\bar\tau}$.

Set: $\eta = \bigcup_i rng(\sigma_i) \cap \rho = J_\gamma \cap \rho$. Then $\eta < \rho$ since $cf(\rho) > \omega$.

Hence $\gamma = \eta < \rho$ and $J_\gamma \in J_{\bar\tau}$. Hence $\sigma_i \in J_{\bar\tau}$ by (1).

QED (Claim)

Now set: $\sigma_i^* = \pi(\sigma_i)$, $\gamma^* = \pi(\gamma)$. Then $\sigma_i^* : J_{\gamma^*_{t_i}} \longrightarrow_{\Sigma_0} J_{\gamma^*}$ and $\sigma_j^*\, \sigma^*_{t_i t_j} = \sigma_i^*$.

Hence $\langle J_{\gamma_t^*}\rangle$, $\langle \sigma^*_{t_i t_j}\rangle$ has a well founded limit. Contradiction! QED

This proves Lemma 1. As an immediate corollary we have:

Corollary 2. Let $\bar\tau$ be suitable s.t. $cf(\bar\tau) > \omega$. Let $\pi : J_{\bar\tau} \longrightarrow_{\Sigma_1} J_\tau$ s.t. $\pi \ne id \restriction J_{\bar\tau}$. Then $\bar\tau$ is not a cardinal in L.

Proof.

Suppose not. Then π extends to $\tilde\pi : L \longrightarrow_{\Sigma_1} L$. Hence $\tilde\pi \ne id \restriction L$ and $0^\#$ exists by Kunen's lemma. Contradiction! QED

Note. Corollary 2 could also have been proven by an ultrapower construction. (In Ch. 17 of [Dev] the existence of $0^\#$ is derived from a slightly stronger assumption. That proof can be adapted virtually without change; only the proof that the ultrapower is well founded (p.200) needs amendment.)

Lemma 3. Let $\tau \geq \omega_2$ be a suitable cardinal in L s.t. $cf(\tau) > \omega$. Then the conclusion of § 1 Lemma 1 holds.

Proof.

Let $X \subset \tau$ cofinally s.t. $\overline{\overline{X}} < \overline{\overline{\tau}}$. We wish to construct $Y \in L$ s.t. $X \subset Y$ and $\overline{\overline{Y}}^L < \tau$. Since τ is suitable, we may assume w. l. o. g. That either τ is a successor cardinal in L or there are arbitrarily large $\gamma \in X$ s.t. γ is regular in L and $cf(\gamma) > \omega$. Define sets $Z_i \prec J_\tau (i \leq \omega_1)$ by:

Z_0 = the smallest $Z \prec J_\tau$ s.t. $X \subset Z$

Z_{i+1} = the smallest $Z \prec J_\tau$ s.t. $Z_i \cup Z_i^* \subset Z$

where Z_i^* = the set of limit points $< \tau$ of $\tau \cap Z_i$.

$Z_\lambda = \bigcup_{i<\lambda} Z_i$ for limit λ.

Set $Z = Z_{\omega_1}$. Then

(a) $Z \prec J_\tau$

(b) $\overline{\overline{Z}} = \omega_1 \cdot \overline{\overline{X}} < \overline{\overline{\tau}}$

(c) If $\gamma \in Z$ is a regular cardinal in L, then either $Z \cap \gamma$ is cofinal in γ or else $cf(| Z \cap \gamma |) = \omega_1$.

Let $\pi : J_{\overline{\tau}} \stackrel{\sim}{\longleftrightarrow} Z$. By (b) we have $\overline{\tau} < \overline{\overline{\tau}}$. By (c) and the above assumptions on X we have: $\overline{\tau}$ is suitable. Since π is cofinal in τ and $cf(\tau) > \omega$, we have: $cf(\overline{\tau}) > \omega$. By Lemma 2 it follows that $\overline{\tau}$ is not a cardinal in L. Let $\overline{\beta}$ be the least $\overline{\beta} \geq \overline{\tau}$ s.t. $\overline{\tau}$ is not a Σ_ω cardinal in $J_{\overline{\beta}}$. Let n be the least $n \geq 1$ s.t. $\overline{\tau}$ is not a Σ_n cardinal in $J_{\overline{\beta}}$. Then $\rho^n_{\overline{\beta}} < \overline{\tau} \leq \rho^{n-1}_{\overline{\beta}}$. Set: $\overline{\rho} = \rho^{n-1}_{\overline{\beta}}$, $\overline{A} = A^{n-1}_{\overline{\beta}}$ $\overline{\gamma} = \rho^n_{\overline{\beta}}$, $\overline{p} = p^n_{\overline{\beta}}$. By § 2 Lemma 1 , we have $cf(\overline{\rho}) > \omega$, since $\overline{\gamma} < \eta \leq \overline{\rho}$ for some $\eta \leq \overline{\tau}$ s.t. η is regular in J_β and $cf(\eta) > \omega$. Hence $\overline{\rho}$ is a limit ordinal and Lemma 1 gives us $\rho \geq \tau$, $\tilde{\pi} \supset \pi$ s.t. $\tilde{\pi} : J_{\overline{\rho}} \longrightarrow_{\Sigma_1} J_\rho$ cofinally. Set: $A = \bigcup_{\nu<\overline{\rho}} \tilde{\pi}(\overline{A} \cap \nu)$. Then $\langle J_\rho, A\rangle$ is amenable and $\tilde{\pi} : \langle J_{\overline{\rho}}, \overline{A}\rangle \longrightarrow_{\Sigma_1} \langle J_\rho, A\rangle$ cofinally.

Claim $\quad \bigvee\beta\, (\rho = \rho^{n-1}_\beta \wedge A = A^{n-1}_\beta)$.

If $n = 1$, then $A = \emptyset = A^o_\rho$ and $\rho = \rho^o_\rho$. Now let $n > 1$. By § 2 Lemma 4 it suffices to show that $\tilde{\pi}$ is strong. Suppose not. Then there are $\overline{R}$, R s.t.

(a) $\overline{R} \subset J^2_{\overline{\rho}}$ is rud in $\langle J_{\overline{\rho}}, \overline{A}\rangle$ and $R \subset J^2_\rho$ is rud in $\langle J_\rho, A\rangle$ by the same rud definition.

(b) $\overline{R}$ is well founded but R is not.

Then there are $x_i \in J_\rho$ s.t. $x_{i+1}\ R\ x_i (i<\omega)$. Since $\mathrm{cf}(\overline{\rho}) > \omega$ and $\tilde{\pi}$ is cofinal in ρ, there is $\eta = \tilde{\pi}(\overline{\eta})$ s.t. $\{x_i \mid i < \omega\} \subset J_\eta$. Then $R \cap J_\eta$ is not well founded. But $\overline{R} \cap J_{\overline{\eta}}$ is well founded and $\overline{\rho}$ is admissible. Hence there is $\overline{f} \in J_{\overline{\rho}}$ s.t. $\overline{f} : J_{\overline{\eta}} \longrightarrow \overline{\rho}$ and $x\overline{R}y \longrightarrow \overline{f}(x) < \overline{f}(y)$. Let $f = \tilde{\pi}(\overline{f})$. Then $f : J_\eta \longrightarrow \rho$ and $xRy \longrightarrow f(x) < f(y)$. Hence $R \cap J_\eta$ is well founded. Contradiction ! QED (Claim)

Set: $\gamma = \pi(\overline{\gamma})$, $p = \tilde{\pi}(\overline{p})$. Let h, $\overline{h}$ be the canonical Σ_1 Skolem functions for $\langle J_\rho, A\rangle$, $\langle J_{\overline{\rho}}, \overline{A}\rangle$ resp. Set $Y = h''(\omega \times J_\gamma \times \{p\})$. Then $Y \in L$. Since $\gamma < \tau$ and τ is a cardinal in L, we have: $\overline{\overline{Y}}^L < \tau$. By the definition of $\overline{\gamma}$, $\overline{p}$ we have: $J_{\overline{\rho}} = \overline{h}''(\omega \times J_{\overline{\gamma}} \times \{\overline{p}\})$. But $\pi''J_{\overline{\gamma}} \subset J_\gamma$; hence $X \subset \tilde{\pi}''J_\rho = h''(\omega \times (\pi''J_{\overline{\gamma}}) \times \{p\}) \subset Y$. QED

§ 4 Vicious sequences

Troughout this section we assume that $\nu < \tau$ and $\pi : J_\nu \longrightarrow_{\Sigma_1} J_\tau$ cofinally. We wish to examine more closely the circumstances under which $M^{\beta\pi}$ can fail to be well founded. To this end we define:

Def $\Theta = \Theta(\pi) \simeq$ the least limit ordinal $\Theta \geq \nu$ s.t. ν is a cardinal in J_Θ and $M^{\Theta\pi}$ is not well founded.

§ 3 Lemma 1 says that Θ does not exist if ν is suitable and $\mathrm{cf}(\nu) > \omega$. It is clear that, if Θ does exist, then $\Theta > \nu$. Moreover $\Theta < \infty$, since otherwise $M^{\kappa\pi}$ would be well founded, where κ is the first regular cardinal $> \nu$. But then $M^{\kappa\pi} = J_{\kappa^*}$ for some κ^* and there is $\tilde{\pi} \supset \tilde{\pi}^{(\kappa)} \supset \pi$ s.t. $\tilde{\pi} : J_\kappa \longrightarrow J_{\kappa^*}$ cofinally, contradicting § 3 Corollary 2.

Now suppose that Θ exists. Let $T = T^{\nu,\Theta}$, $M = M^{\Theta,\pi}$, $\sigma^*_t = \sigma^{*(\Theta,\pi)}$, $\sigma^*_{tt'} = \sigma^{*}_{tt'}{}^{(\pi)}$, $\tilde{\pi} = \tilde{\pi}^{(\Theta)}$. There must be a sequence $\langle t_i, x_i\rangle (i < \omega)$ s.t.

(a) $t_i \in T$, $t_i \leq t_{i+1}$, $t_i \in u_{t_{i+1}}$, $\gamma_{t_i} < \delta_{t_{i+1}}$

(b) If $J_\nu \models$ (α is the largest cardinal then $\delta_{t_o} > \alpha$

(c) $x_i \in J_{\gamma^*_{t_i}}$ s.t. $x_{i+1} \in \sigma^*_{t_i t_{i+1}}(x_i)$, (hence $\sigma^*_{t_{i+1}}(x_{i+1}) \in \sigma^*_{t_i}(x_i)$

$(i < \omega)$).

We refer to any sequence satisfying (a) - (c) as a <u>vicious sequence</u> for π. Note that $\sup_i \mu_{t_i}$ is a limit ordinal if $\langle t_i, x_i\rangle$ is vicious, since $\mu_{t_{i+1}} > \mu_{t_i}$. If $\beta \geq \nu$ is any limit ordinal s.t. $\beta \geq \sup_i \mu_{t_i}$ and ν is a cardinal in β, then $M^{\beta\pi}$ is not well founded. But then $\sup_i \mu_{t_i} > \nu$, since otherwise $J_\tau = M^{\nu\pi}$ would not be well founded. Hence $\Theta = \sup_i \mu_{t_i}$ for any vicious $\langle t_i, x_i\rangle$.

We define a <u>canonical vicious sequence</u> $\langle t_i, x_i\rangle = \langle t_i^\pi, x_i^\pi\rangle (i < \omega)$ as follows:

t_i = the $<_J$-least $t \in T$ s.t.

there is a vicious sequence $\langle t'_k, x'_k\rangle$ $(k < \omega)$ with $\langle t'_j, x'_j\rangle = \langle t_j, x_j\rangle$ for $j < i$ and $t'_i = t$.

x_i = the $<_J$-least $x_i \in J_{\gamma^*_{t_i}}$ s.t.

there is a vicious sequence $\langle t'_k, x'_k\rangle$ $(k < \omega)$ with $\langle t'_j, x'_j\rangle = \langle t_j, x_j\rangle$ for $j \leq i$.

<u>Lemma 1.</u> $\bigcup_{i<\omega} \mathrm{rng}(\sigma_{t_i}) = J_\Theta$

Proof.

Set $X = \bigcup_{i<\omega} X_{t_i}$. Since $X_{t_i} \prec_{\Sigma_1} J_{\mu_{t_i}}$ and $\sup_i \mu_{t_i} = \Theta$, we have $X \prec_{\Sigma_1} J_\Theta$. Let $\sigma : J_\lambda \overset{\sim}{\longleftrightarrow} X$. Then $\lambda \leq \Theta$. We know that $t_i \in X$, since $t_i \in u_{t_{i+1}} \subset X_{t_{i+1}}$.

Claim $\sigma^{-1}(t_i) = t_i$.

Proof.

Let $t_i' = \sigma^{-1}(t_i)$. Then $X_{t_i'} = h_{\mu_{t_i'}}"(\omega \times J_{\delta_{t_i}} \times \{\sigma^{-1}(u_{t_i})\}) = \sigma^{-1}"X_t$.

Hence $\gamma_{t_i'} = \gamma_{t_i}$ and $\sigma_{t_i't_j'} = \sigma_{t_it_j}$ $(i \leq j)$. But then $\gamma^*_{t_i'} = \gamma^*_{t_i}$, $\sigma^*_{t_i't_j'} = \sigma^*_{t_it_j}$ and it follows easily that $\langle t_i', x_i\rangle(i < \omega)$ is vicious. But $t_i' \leq_J t_i$. Hence $t_i' = t_i$ by the minimal choice of t_i.

QED (Claim)

But then $\lambda = \sup_i \mu_{t_i} = \Theta$ and $X_{t_i} = h_{t_i}"(\omega \times J_{\delta_{t_i}} \times \{u_{t_i}\}) = \sigma^{-1}"X_{t_i}$.

Hence $J_\Theta = J_\lambda = \sigma^{-1}"X = \bigcup_i \sigma^{-1}"X_{t_i} = \bigcup_i X_{t_i} = X$. QED

Corollary 2. If ν is suitable, then $\sup_i \delta_{t_i} = \nu$.

Proof.

ν is either a successor cardinal or a limit cardinal in J_Θ. In the first case, $\delta_{t_i} > \alpha$ by definition, where ν succeeds α in J_Θ. But then $\eta_i = X_{t_i} \cap \nu$ is transitive. $\eta_i < \nu$, since ν is regular in J_Θ. Clearly $\sup_i \eta_i = \nu$, since $\bigcup_i X_{t_i} = J_\Theta$. But $\delta_i \leq \eta_i \leq \gamma_i < \delta_{i+1}$; hence $\sup_i \delta_i = \nu$.

Now let ν be a limit cardinal in J_Θ. Let $\delta = \sup_i \delta_{t_i} < \nu$. By suitability, there is $\gamma > \delta$ s.t. γ is regular in J_Θ, $\gamma < \nu$ and $\mathrm{cf}(\gamma) > \omega$. Set $\eta_i = \sup(\gamma \cap X_{t_i})$. Then $\eta_i < \gamma$ by the regularity of γ. But then $\eta < \gamma$, where $\eta = \sup_i \eta_i$, since $\mathrm{cf}(\gamma) > \omega$. Hence $\eta \notin \bigcup_i X_{t_i} = J_\Theta$. Contradiction! QED

Remark Using Corollary 2, it would be easy to show that $M^{\Theta,\pi} = \bigcup_i \mathrm{rng}(\sigma^*_{t_i})$ if ν is suitable, but we shall not need this.

Def $v_i = v_i^\pi = \langle \delta_{t_i}, \gamma_{t_i}, \sigma^{-1}_{t_i}(u_{t_i})\rangle$ $(i < \omega)$.

Then $v_i \in J_\nu$ $(i < \omega)$. The sequence $\langle v_i\rangle$ gives "complete information" about J_Θ, since J_Θ, $\langle \sigma_{t_i}\rangle$ = the limit of $\langle J_{\gamma_{t_i}}\rangle$, $\langle \sigma_{t_it_j}\rangle$ and the maps

$\sigma_{t_i t_j}$ are recoverable from the v_i by: $\sigma_{t_i t_j} = \sigma_s$ where $s = \sigma_{t_j}^{-1}(t_i) \in \sigma_{t_j}^{-1}(u_j)$ $(i < j)$.

We use this to prove:

<u>Lemma 3.</u> Let ν be suitable. Let $\overline{\pi} : J_{\overline{\nu}} \longrightarrow_{\Sigma_1} J_\nu$ s.t. $\{v_i \mid i < \omega\} \subset \mathrm{rng}(\overline{\pi})$. Then $\Theta(\pi\overline{\pi})$ exists (hence $\langle t_i^{\pi\overline{\pi}}, x_i^{\pi\overline{\pi}} \rangle$ exists) and $x_i^{\pi\overline{\pi}} = x_i$ $(i < \omega)$.

Proof.

$\sigma_{t_i t_j} \in \mathrm{rng}(\overline{\pi})$ since $\sigma_{t_i t_j}$ is canonically recoverable from the v_i. Set: $\overline{\sigma}_{ij} = \overline{\pi}^{-1}(\sigma_{t_i t_j})$, $\overline{\gamma}_i = \overline{\pi}^{-1}(\gamma_{t_i})$, $\overline{\delta}_i = \overline{\pi}^{-1}(\delta_{t_i})$.

Then $\overline{\sigma}_{ij} : J_{\overline{\gamma}_i} \longrightarrow_{\Sigma_0} J_{\overline{\gamma}_j}$ $(i \leq j < \omega)$ is a directed system s.t. $\overline{\sigma}_{ij} \restriction J_{\overline{\delta}_i} = \mathrm{id} \restriction J_{\overline{\delta}_i}$. Let U, $\langle \overline{\sigma}_i \rangle$ be the limit of $\langle J_{\overline{\gamma}_i} \rangle$, $\langle \overline{\sigma}_{ij} \rangle$. We may assume w. l. o. g. that $\sigma_i \restriction J_{\overline{\delta}_i} = \mathrm{id} \restriction J_{\overline{\delta}_i}$. But $\sup_i \overline{\delta}_i = \overline{\nu}$, since $\sup_i \delta_{t_i} = \nu$ and hence $J_{\overline{\nu}} \subset U$. Define $\hat{\pi} : U \longrightarrow_{\Sigma_0} J_\Theta$ by:

$$\begin{array}{ccc} U & \xrightarrow{\hat{\pi}} & J_\Theta \\ {\scriptstyle \overline{\sigma}_i} \uparrow & & \uparrow {\scriptstyle \sigma_{t_i}} \\ J_{\overline{\gamma}_i} & \xrightarrow[\overline{\pi} \restriction J_{\overline{\gamma}_i}]{} & J_{\gamma_{t_i}} \end{array}$$

It is easily seen that $\hat{\pi} \supset \overline{\pi}$. U is well founded, since $\hat{\pi}$ imbeds it into J_Θ, and satisfies "I am a J_α". Hence we may assume $U = J_{\overline{\Theta}}$ for some $\overline{\Theta}$. Set $\overline{t}_i = \hat{\pi}^{-1}(t_i) = \overline{\sigma}_{i+1}\, \overline{\pi}^{-1}\, \sigma_{t_{i+1}}^{-1}(t_i)$. Since $\sup_i \mu_{t_i} = \Theta$ and $\mu_{t_i} = \hat{\pi}(\mu_{\overline{t}_i})$, it follows that $\hat{\pi}''\overline{\Theta}$ is cofinal in Θ. Hence $\hat{\pi} : J_{\overline{\Theta}} \longrightarrow_{\Sigma_1} J_\Theta$ cofinally. Clearly, $\overline{\Theta} = \sup_i \mu_{\overline{t}_i}$.

(1) If $t \in \overline{T} = T^{\overline{\Theta}\overline{\nu}}$, then $\hat{\pi}(\sigma_t) = \sigma_{\hat{\pi}(t)}$ (hence $\pi(\gamma_t) = \gamma_{\hat{\pi}(t)}$ and $\pi(\sigma_{tt'}) = \sigma_{\hat{\pi}(t)\hat{\pi}(t')}$ for $t, t' \in \overline{T}$, $t \leq t'$).

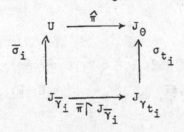

Proof of (1).

$\sigma = \sigma_t$, $u = \sigma^{-1}(u_t)$, $\gamma = \gamma_t$ are uniquely characterized by:

(a) $J_\gamma = h_\gamma''(\omega \times J_{\delta_t} \times \{u\})$

(b) $\sigma : J_\gamma \longrightarrow_{\Sigma_1} J_{\mu_t}$

(c) $\sigma \restriction J_{\delta_t} = \mathrm{id} \restriction J_{\delta_t}$, $\sigma(u) = u_t$. QED (1)

(2) $\overline{\gamma}_i = \gamma_{\overline{t}_i}$, $\overline{\sigma}_i = \sigma_{\overline{t}_i}$ (hence $\overline{\sigma}_{ij} = \sigma_{t_i t_j}$).

Proof of (2)

$\pi(\gamma_{\overline{t}_i}) = \gamma_{t_i} = \pi(\overline{\gamma}_i)$ by (1); hence $\gamma_{\overline{t}_i} = \overline{\gamma}_i$. Set: $\overline{u} = \pi^{-1}\ \sigma_{t_i}^{-1}(u_{t_i})$.

Then $\hat{\pi}\ \overline{\sigma}_i(\overline{u}) = \sigma_{t_i}\pi(\overline{u}) = u_{t_i}$ and $\hat{\pi}\ \sigma_{\overline{t}_i}(\overline{u}) = \hat{\pi}(\sigma_{\overline{t}_i})\pi(\overline{u}) = \sigma_{t_i}\pi(\overline{u}) = u_{t_i}$ by (1). Hence $\overline{\sigma}_i(\overline{u}) = \sigma_{\overline{t}_i}(\overline{u}) = u_{\overline{t}_i}$. But then $\overline{\sigma}_i = \sigma_{\overline{t}_i}$ = the unique $\sigma : J_{\overline{\gamma}_i} \longrightarrow_{\Sigma_0} L$ s.t. $\sigma \restriction J_{\overline{\delta}_i} = \mathrm{id} \restriction J_{\overline{\delta}_i}$ and $\sigma(\overline{u}) = u_{\overline{t}_i}$. QED (2)

(3) $\overline{\Theta} = \Theta(\pi\overline{\pi})$.

Proof.

Let $\overline{\rho} = \Theta(\pi\overline{\pi})$.

($\overline{\Theta} \geq \overline{\rho}$) $\sigma^{*(\pi\overline{\pi})}_{\overline{t}_i\overline{t}_j} = \pi\overline{\pi}(\sigma_{\overline{t}_i\overline{t}_j}) = \pi(\sigma_{t_i t_j}) = \sigma^*_{t_i t_j}$ by (2). But then $\sigma^*_{\overline{t}_{i+1}}(x_{i+1}) \in \sigma^*_{\overline{t}_i}(x_i)$ in $M^{\overline{\Theta},\pi\overline{\pi}}$, where $\sigma^*_{\overline{t}_i} = \sigma^{*(\overline{\Theta},\pi\overline{\pi})}_{\overline{t}_i}$ and $M^{\overline{\Theta},\pi\overline{\pi}}$ is not well founded.

($\overline{\rho} \geq \overline{\Theta}$) suppose not. Let $\langle \overline{s}_i, y_i \rangle$ be vicious for $\pi\overline{\pi}$ and set $\rho = \hat{\pi}(\overline{\rho})$, $s_i = \hat{\pi}(\overline{s}_i)$. By the above argument, $\langle s_i, y_i \rangle$ is vicious for π.
Hence $\Theta(\pi) = \sup_i \mu_{s_i} \leq \rho < \Theta$. Contradiction ! QED (3)

(4) $\langle \overline{t}_i, x_i \rangle = \langle t_i^{\pi\overline{\pi}}, x_i^{\pi\overline{\pi}} \rangle$.

Proof.

$\langle \overline{t}_i, x_i \rangle$ is vicious for $\pi\overline{\pi}$ by the above argument. But (1), (2) and the minimal choice of $\langle t_i, x_i\rangle$, $\langle \overline{t}_i, x_i\rangle$ must be chosen minimally.

QED

<u>Note</u> We could have carried the proof of Lemma 2 a bit further to show:

(a) $M^{\Theta,\overline{\pi}} = J_\Theta$ and $\tilde{\overline{\pi}}(\Theta) = \hat{\pi}$

(b) There is $i : M^{\overline{\Theta},\pi\overline{\pi}} \overset{\sim}{\longleftrightarrow} M^{\Theta,\pi}$ s.t. $\widetilde{\pi\overline{\pi}}(\overline{\Theta}) = i\,\tilde{\pi}\hat{\pi}$.

Thus:

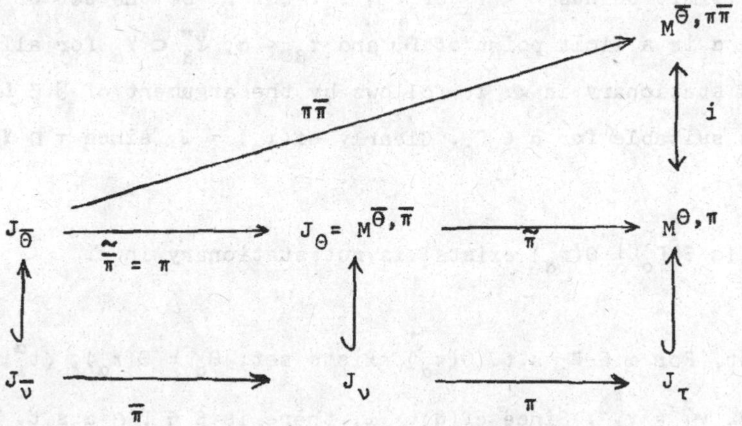

§ 5 <u>The ω cofinal case</u>

Let $\tau \geq \omega_2$ be an ω cofinal suitable cardinal in L. Let $X \subset \tau$ be cofinal in τ s.t. $\overline{\overline{X}} < \overline{\overline{\tau}}$. As before, we suppose w. l. o. g. that, if τ is not a successor cardinal in L, there are arbitrarily large $\gamma \in X$ s.t. γ is regular in L and $\mathrm{cf}(\gamma) > \omega$. We wish to show that there is $Y \in L$ s.t. $X \subset Y$ and $\overline{\overline{Y}}^L < \tau$. Obviously, it suffices to prove this in a generic extension of the universe. Since $\overline{\overline{X}} < \overline{\overline{\tau}}$, $\tau \geq \omega_2$ and τ is singular (in V), there is a regular $\kappa \geq \omega_2$ s.t. $\overline{\overline{X}} < \kappa < \tau$. But we may then assume that $\overline{\overline{\tau}} = \kappa$, since if this is not true already, we can make it true by generically collapsing τ to κ. Now let k map κ onto J_τ. For $\gamma < \kappa$ set: Y_γ = the smallest $Y \prec_{\Sigma_\omega} J_\tau$ s.t. $X \cup \gamma \cup k''\gamma \subset Y$.

Set: $\Gamma = \{\alpha < \kappa \mid \alpha = Y_\alpha \cap \kappa\}$. Then Γ is cub in κ. For $\alpha, \beta \in \Gamma$, $\alpha \leq \beta$ set: $\pi_\alpha : J_{\tau_\alpha} \overset{\sim}{\longleftrightarrow} Y_\alpha$; $\pi_{\alpha\beta} = \pi_\beta^{-1}\,\pi_\alpha$. Then $J_{\tau_\alpha} \xrightarrow[\Sigma_\omega]{\pi_{\alpha\beta}} J_{\tau_\beta} \xrightarrow[\Sigma_\omega]{\pi_\beta} J_\tau$, and J_τ, $\langle\pi_\alpha\rangle$ is the limit of the directed system $\langle J_{\tau_\alpha}\rangle$, $\langle\pi_{\alpha\beta}\rangle$. Clearly, $\pi_\alpha \upharpoonright J_\alpha = \mathrm{id} \upharpoonright J_\alpha$ and $\pi_\alpha(\alpha) = \kappa$. Also, $\overline{\overline{Y}}_\alpha < \kappa$ for $\alpha \in \Gamma$ and κ is regular Hence for each $\alpha \in \Gamma$ there is $\beta \in \Gamma$ s.t. $\tau_\alpha < \beta$ and $Y_\alpha^* \subset Y_\beta$, where Y_α^* is the set of limit points $< \tau$ of $\tau \cap Y_\alpha$. Let Γ_0 be the set of $\alpha \in \Gamma$ s.t. $\mathrm{cf}(\alpha) > \omega$, α is a limit point of Γ, and $\tau_\beta < \alpha$, $Y_\beta^* \subset Y_\alpha$ for all $\beta \in \Gamma \cap \alpha$. Then Γ_0 is stationary in κ. It follows by the argument of § 3 Lemma 3 that τ_α is suitable for $\alpha \in \Gamma_0$. Clearly $\mathrm{cf}(\tau_\alpha) = \omega$, since $\tau \cap Y_\alpha$ is cofinal in τ.

Lemma 1. $\{\alpha \in \Gamma_0 \mid \Theta(\pi_\alpha) \text{ exists}\}$ is not stationary in κ.

Proof.

Suppose not. For $\alpha \in \Gamma_0$ s.t. $\Theta(\pi_\alpha)$ exists set: $\Theta_\alpha = \Theta(\pi_\alpha)$, $\langle t_i^\alpha, x_i^\alpha\rangle = \langle t_i^{\pi_\alpha}, x_i^{\pi_\alpha}\rangle$, $v_i^\alpha = v_i^{\pi_\alpha}$. Since $\mathrm{cf}(\alpha) > \omega$, there is $\beta \in \Gamma \cap \alpha$ s.t. $\{v_i^\alpha \mid i < \omega\} \subset \mathrm{rng}(\pi_{\beta\alpha})$. Let $f(\alpha)$ be the least such β. Then f is regressive and hence there is β_0 s.t. $\Delta = \{\alpha \mid f(\alpha) = \beta_0\}$ is stationary. But then $\{u_i^\beta \mid i < \omega\} \subset \mathrm{rng}(\pi_\alpha)$ for $\alpha, \beta \in \Delta$, $\alpha \leq \beta$ and hence $x_i^\alpha = x_i^\beta$ by § 4 Lemma 3. Set $x_i = x_i^\alpha$ $(\alpha \in \Delta)$. Since $\mathrm{cf}(\kappa) > \omega$ and $J_\tau = \bigcup_{\alpha\in\Delta} \mathrm{rng}(\pi_\alpha)$, there is $\alpha \in \Delta$ s.t. $\{x_i \mid i < \omega\} \subset \mathrm{rng}(\pi_\alpha)$. Then $\sigma_{t_{i+1}^\alpha}\,\pi_\alpha^{-1}(x_{i+1}) \in \sigma_{t_i^\alpha}\,\pi_\alpha^{-1}(x_i)$ $(i < \omega)$ and J_{Θ_α} is not well founded. Contradiction ! QED

Lemma 2. If τ is a limit cardinal in L, then the conclusion of § 1 Lemma 1 holds.

Proof.

Pick $\alpha \in \Gamma_0$ s.t. $\Theta(\pi_\alpha)$ does not exist. Set $\overline{\tau} = \tau_\alpha$, $\pi = \pi_\alpha$.
Then $\overline{\tau}$ is not a cardinal in L, since otherwise $M^{\kappa,\pi}$ would be well founded. But then $M^{\kappa,\overline{\tau}} = J_{\kappa^*}$ for some κ^* and $\tilde{\pi} : J_\kappa \longrightarrow_{\Sigma_1} J_{\kappa^*}$ cofinally,

where $\tilde{\pi} = \tilde{\pi}^{(\kappa)} \supset \pi$, violating § 3 Lemma 2. As in § 3 Lemma 3 set: $\bar{\beta}$ = the least $\bar{\beta} \geq \bar{\tau}$ s.t. $\bar{\tau}$ is not a Σ_ω cardinal in $J_{\bar{\beta}}$; n = the least $n \geq 1$ s.t. $\bar{\tau}$ is not a Σ_n cardinal in $J_{\bar{\beta}}$, $\bar{\rho} = \rho^{n-1}_{\bar{\beta}}$, $\bar{A} = A^{n-1}_{\bar{\beta}}$, $\bar{\gamma} = \rho^n_{\bar{\beta}}$. Then $\bar{\rho} \geq \bar{\tau} > \bar{\gamma}$. By suitability, there must then be η s.t. $\bar{\rho} > \eta > \bar{\gamma}$, η is regular in $J_{\bar{\rho}}$ and $cf(\eta) > \omega$. Hence $cf(\bar{\rho}) > \omega$. We can then finish the proof exactly as in § 3 Lemma 3. QED

Lemma 3. If τ is a successor cardinal in L, then the conclusion of § 1 Lemma 1 holds.

Proof.

Set: $\Gamma_1 = \{\alpha \in \Gamma_o \mid \Theta(\pi_\alpha)$ does not exist$\}$. Then Γ_1 is stationary. As above, τ_α is not a cardinal in L for $\alpha \in \Gamma_1$. Set: β_α = the least $\beta \geq \tau_\alpha$ s.t. τ_α is not a Σ_ω cardinal in J_β; $n = n_\alpha$ = the least $n \geq 1$ s.t. τ_α is not a Σ_n cardinal in J_β; $\rho_\alpha = \rho^{n-1}_{\beta_\alpha}$, $A_\alpha = A^{n-1}_{\beta_\alpha}$, $p_\alpha = p^n_{\beta_\alpha}$ $\gamma_\alpha = \rho^n_{\beta_\alpha}$. Then $\rho_\alpha \geq \tau_\alpha > \gamma_\alpha$ and τ_α is a successor cardinal, hence regular, in J_{ρ_α}. Hence $cf(\omega\rho_\alpha) = cf(\tau_\alpha) = \omega$.

Let f^α_i $(i < \omega)$ be as in § 2 Lemma 2. Since $cf(\alpha) > \omega$ for $\alpha \in \Gamma_1$, there is $\beta \in \Gamma \cap \alpha$ s.t. $\{\gamma_\alpha\} \cup \{f^\alpha_i \mid i < \omega\} \subset rng(\pi_{\beta\alpha})$. Set $g = \langle \beta, \pi^{-1}_{\beta\alpha}(\gamma_\alpha)\rangle$ where β is the least such ordinal. Then g is regressive. Hence there are β', γ' s.t. $\Delta = \{\alpha \in \Gamma_1 \mid g(\alpha) = \langle \beta', \gamma'\rangle\}$ is stationary. But for $\alpha, \beta \in \Delta$, $\alpha \leq \beta$ we then have $n_\alpha = n_\beta$, $\pi(\gamma_\alpha) = \gamma_\beta$ and there is a unique $\tilde{\pi}_{\alpha\beta} \supset \pi_{\alpha\beta}$ s.t. $\tilde{\pi}_{\alpha\beta} : \langle J_{\rho_\alpha}, A_\alpha\rangle \longrightarrow_{\Sigma_1} \langle J_{\rho_\beta}, A_\beta\rangle$ and $\tilde{\pi}_{\alpha\beta}(p_\alpha) = p_\beta$.

By the uniqueness of the $\tilde{\pi}_{\alpha\beta}$ it follows that if $\alpha, \beta, \gamma \in \Delta$, $\alpha \leq \beta \leq \gamma$, then $\tilde{\pi}_{\beta\gamma}\tilde{\pi}_{\alpha\beta} = \tilde{\pi}_{\alpha\gamma}$. Let M, $\langle \tilde{\pi}_\alpha \mid \alpha \in \Delta\rangle$ be the direct limit of $\langle\langle J_{\rho_\alpha}, A_\alpha\rangle \mid \alpha \in \Delta\rangle$, $\langle \tilde{\pi}_{\alpha\beta} \mid \alpha, \beta \in \Delta$ and $\alpha \leq \beta\rangle$.

M is well founded, since if $x_{i+1} \in x_i$ in M $(i < \omega)$, there must be $\alpha \in \Delta$ s.t. $\{x_i \mid i < \omega\} \subset rng(\tilde{\pi}_\alpha)$. But then $\tilde{\pi}_\alpha(x_{i+1}) \in \tilde{\pi}_\alpha(x_i)$ $(i < \omega)$. Contradiction ! M satisfies "I am a J_α" and hence we may assume: $M = \langle J_\rho, A\rangle$ for some ρ. Then $\rho \geq \tau$ and $\langle J_\rho, A\rangle$ is amenable. Fix $\alpha \in \Delta$

and set: $\bar{\tau} = \tau_\alpha$, $\bar{\beta} = \beta_\alpha$, $\bar{\rho} = \rho_\alpha$, $\bar{A} = A_\alpha$, $\bar{\gamma} = \gamma_\alpha$, $\pi = \pi_\alpha$, $\tilde{\pi} = \tilde{\pi}_\alpha$. It is enough to show that $\rho = \rho_\beta^{n-1}$, $A = A_\beta^{n-1}$ for some β, for we can then finish the proof exactly as in § 3 Lemma 3. But for this, it suffices to show that the map $\tilde{\pi}$ is strong. $\tilde{\pi}$ will be strong, however, if α is a chosen sufficiently large. To see this, let $R_n (n < \omega)$ enumerate the relations rud in $\langle J_\rho, A\rangle$ which are not well founded. Let $\bar{R}_n$ have the same rud definition in $\langle J_{\bar{\rho}}, \bar{A}\rangle$. For $n < \omega$ choose $\langle x_n^i \mid i < \omega\rangle$ s.t. $x_n^{i+1}\ R_n\ x_n^i$ $(i < \omega)$. Set $Y = \{x_n^i \mid i, n < \omega\}$. Then $Y \subset \mathrm{rng}(\tilde{\pi})$ for sufficiently large α. But then $\tilde{\pi}^{-1}(x_n^{i+1})\ \bar{R}_n\ \tilde{\pi}^{-1}(x_n^i)$ and $\bar{R}_n$ is not well founded. Now let $\bar{R}$ be well founded and rud in $\langle J_{\bar{\rho}}, \bar{A}\rangle$. Let R be rud in $\langle J_\rho, A\rangle$ by the same rud definition. Then $\bar{R} \neq \bar{R}_n$ and hence $R \neq R_n$ $(n < \omega)$. Hence R is well founded. QED

Bibliography

[Dev] Devlin, Keith. Aspects of Constructibility, Lecture Notes in Mathematics vol. 354 (1973)

[FS] Jensen, R. B. The Fine Structure of the Constructible Hierarchy, Annals of Math. Logic vol. 4, no. 3 pp. 229 - 308 (1972).

COMPUTATION THEORIES : AN AXIOMATIC APPROACH TO RECURSION ON GENERAL STRUCTURES.

Jens Erik FENSTAD
University of Oslo.

This is a brief survey of an axiomatic approach to generalized recursion theory. It is based on a set of lectures to the Kiel Summer Institute in Logic 1974. A first report on the project was given in [6].

An axiomatic study is to a large extent an analysis of existing methods and results. Our debts to those who have worked on general recursion theory is therefore considerable, as the many references to follow will bear witness to. A particular debt is due to Johan Moldestad and Dag Normann who have with great enthusiasm participated in the various investigations reported on in this survey.

Being a survey paper proofs will only occasionally be hinted at. A more comprehensive presentation of the theory will appear as a book in the Spring Verlag series Perspectives in Mathematical Logic.

1. BASIC THEORY.

1.1 Combinatorial part. Our starting point is an analysis of the relation

$$\{a\}(\sigma) \simeq z$$

which is intended to assert that the "computing device" named or coded by a and acting on the input sequence $\sigma = (x_1, \ldots, x_n)$ gives z as output.

Definition 1. A computation domain is a structure

$$\mathfrak{A} = \langle A, C; 0, 1 \rangle$$

where A is a non-empty set, C is a subset of A, and $0,1$ are two designated elements of C. C is called the set of codes.

Definition 2. A computation set Θ over $\mathfrak{A}$ is a set of tuples (a,σ,z) where $a \in C$, $\sigma = (x_1,\dots,x_n)$, where $x_i \in A$, $z \in A$ and $lh(a,\sigma,z) \geq 2$.

At this stage we need not make any requirement of singlevaluedness, hence given a and σ there may be more than one z such that $(a,\sigma,z) \in \Theta$. However, in most cases we will require that Θ is singlevalued.

Let Θ be a computation set over $\mathfrak{A}$. To every $a \in C$ and every natural number $n \geq 0$ we can associate a partial function $\{a\}_\Theta^n$ as follows

$$\{a\}_\Theta^n(\sigma) \simeq z \text{ iff } lh(\sigma) = n \text{ and } (a,\sigma,z) \in \Theta .$$

Definition 3. Let Θ be a computation set over $\mathfrak{A}$. A function f is Θ-computable if for some $\hat{f} \in C$ we have

$$f(\sigma) \simeq z \text{ iff } (\hat{f},\sigma,z) \in \Theta .$$

We call $\hat{f}$ a Θ-code for f and write $f = \{\hat{f}\}_\Theta^n$.

Definition 4. Let Θ be a computation set over $\mathfrak{A}$. A consistent functional φ is called Θ-effective if there exists a code $\hat{\varphi} \in C$ such that for all $e_1,\dots,e_l \in C$ and all $\sigma = (x_1,\dots,x_n)$ from A we have

$$\varphi(\{e_1\}_\Theta^{n_1},\dots,\{e_l\}^{n_l},\sigma) \simeq z \text{ iff } \{\hat{\varphi}\}_\Theta^{l+n}(e_1,\dots,e_l,\sigma) \simeq z$$

We see that φ is Θ-effective if we can calculate φ on Θ-computalbe functions by calculating on the codes of the functions.

We will consider some specific functions and functionals.

(1). <u>Definition by cases</u> (on the code set C) :

$$DC(x,b,c,a) = \begin{cases} 1, & \text{if not all } a,b,c \in C \\ b, & \text{if } x = a \text{ and all } a,b,c \in C \\ c, & \text{if } x \neq a \text{ and all } a,b,c \in C \end{cases}$$

Outright definition by cases makes equality on A Θ-computable. This we may not always want.

(2). <u>Composition</u>:

$$\underset{\sim}{C}^n(f,g,\sigma) = f(g(\sigma),\sigma) \ , \quad \text{where} \quad n = lh(\sigma) \ .$$

(3). <u>Permutation</u>:

$$\underset{\sim}{P}^{m}_{n,j}(f,\sigma,\tau) = f(\sigma^j) \ .$$

Here $n,m \geq 0$, $0 \leq j < n$, and $(x_1,\ldots,x_n)^j = (x_{j+1},x_1,\ldots)$.

Next we consider a property which a computation set Θ on $\mathfrak{A}$ may or may not have.

(4). <u>Iteration property</u>:

For each $m,n \geq 0$ there exists a map (i.e. total and single-valued function) S^n_m such that for all $a,\sigma \in C$ and all $\tau \in A$:

$$\{a\}^{n+m}_{\Theta}(\sigma,\tau) = \{S^n_m(a,\sigma)\}^m_{\Theta}(\tau) \ .$$

<u>Definition 5</u>. Let Θ be a computation set over the domain $\mathfrak{A}$. Θ is called a <u>precomputation</u> <u>theory</u> on $\mathfrak{A}$ if

(i) for each n,j $(0 \leq j < n)$ and m DC , $\underset{\sim}{C}^n$, and $\underset{\sim}{P}^{m}_{n,j}$ are Θ-computable with Θ-codes d,c_n , and $p_{n,j,m}$, respectively;

(ii) Θ satisfies the iteration property, i.e. for each n,m there is a Θ-code $s_{n,m}$ for a mapping S^n_m with property (4) above.

Note that if $\langle N,s\rangle$, i.e. the set of natural numbers with the successo function, is in the structure $\mathfrak{A}$ we may require that the codes c_n, $p_{n,j,m}$ and $s_{n,m}$ are Θ-computable mappings of n,j,m . We also note that we have the following enumeration property: We have a fixed code $p_{n,o,o}$ such that

$$\{p_{n,o,o}\}(a,\sigma) = \{a\}(\sigma)$$

for all $a \in C$.

One may now show that certain functions are Θ-computable. We omit this, only remark that ordered pair exists, and that we have the usual proof of the fixed point theorem.

If we consider precomputation theories Θ over the integers and assume that the successor function $\lambda x.(x+1)$ is Θ-computable, we can be more explicit: Any such theory is closed under the μ-operator, the predecessor function, and primitive recursion. It follows that <u>the (Kleene) partial recursive functions is the minimal precomputation theory over ω</u> .

This result can be extended. In order to have the required uniformity in the various constructions we now assume that the codes c_n, $p_{n,j,m}$, and $s_{n,m}$ are computable functions of the parameters n, j and m . We also assume that all theories are singlevalued.

Given any sequence $\underset{\sim}{f} = f_1,\dots,f_l$ of partial functions over ω one may construct a "least" precomputation theory, which we will denote by $PR[\underset{\sim}{f}]$ (= the prime recursion theory generated by $\underset{\sim}{f}$), in which the functions $\underset{\sim}{f}$ are computable.

<u>Definition 6</u>. Let Θ and H be two precomputation theories over ω. We say that H extends Θ ,

$$\Theta \leq H ,$$

if there is an H-computable mapping $p(a,n)$ such that

$$(a,\sigma,z) \in \Theta \quad \text{iff} \quad (p(a,n),\sigma,z) \in H ,$$

where $n = lh(\sigma)$. If $\Theta \leq H$ and $H \leq \Theta$, we say that Θ and H are equivalent, in symbols $\Theta \sim H$.

<u>Theorem</u>. Let Θ be a precomputation theory over ω . There exists a Θ-computable function f such that

$$\Theta \sim PR[f] .$$

Since Θ has an enumerating function (see the remark following definition 5) it is immediate that there exists a Θ-computable partial f such that $\Theta \leq PR[f]$. It is the converse which requires a closer analysis. The idea is to analyse how we compute a function in $PR[f]$ and to see that this procedure can be carried out inside Θ . One way of doing this is to develop a normal form for functions in $PR[f]$.

Once we have the representation theorem we may lift to arbitrary precomputation theories many of the results valid for ordinary recursion theory over ω . We take the first recursion theorem as an example.

<u>Definition 7</u>. Let Θ be a precomputation theory over ω . A functional $\varphi(f,x)$ is called <u>uniformly Θ-computable</u> if there is a code $\hat{\varphi}$ such that the function $\lambda x.\varphi(f,x)$ is $\Theta[f]$-computable with code $\hat{\varphi}$ for all functions f .

<u>Theorem</u>. Let φ be uniformly Θ-computable: Then φ has a least fixed point f^* , and f^* is Θ-computable.

The proof makes essential use of the fact that Θ is of the form $PR[f]$ for some partial function f .

<u>Remark</u>. The theory developed so far is closely related to the previous

work of Strong [36], Wagner [37], and Friedman [8]. There is also an unpublished study by Aczel [1] on "enumeration systems", a notion essentially equivalent to our precomputation theories. The representation theorem is already stated in Aczel's abstract. The thesis of Sasso [33] also contains a great deal of material relevant for this part of the theory. In particular he has the normal form theorem for a function recursive in a partial function.

1.2. Computation theories. Many arguments of recursion theory seem to require an analysis not only of the computation tuple $(a,\sigma,z) \in \Theta$, but of the whole structure of "subcomputations" of a given computation.

Moschovakis [24] in his analysis emphasized the fact that whatever computations may be, they have a well-defined length which always is an ordinal, finite or infinite. Thus he proposed to add as a further primitive a map from the set Θ of computation tuples to the ordinals, denoting by $|a,\sigma,z|_{\Theta}$ the ordinal associated to the tuple $(a,\sigma,z) \in \Theta$. We suggested in [6] to abstract another aspect of the notion of computation, viz. to add as a further primitive a relation between computation tuples

$$(a',\sigma',z') < (a,\sigma,z)$$

which is intended to express that (a',σ',z') is a "subcomputation" of (a,σ,z), i.e. that the computation (a,σ,z) depends upon (a',σ',z').

Definition 8. Let $\mathfrak{A} = \langle A,C,N;s\rangle$ be a computation domain, i.e. A is a non-empty set, $N \subseteq C \subseteq A$, and $\langle N, s\upharpoonright N\rangle$ is isomorphic to the natural numbers with the successor function. A computation structure $\langle\Theta,<\rangle$ over $\mathfrak{A}$ is a pair where Θ is a computation set and $<$ is a transitive and well-founded relation on Θ.
Thus if $(a,\sigma,z) \in \Theta$ then the set

$$\underset{\sim}{S}_{(a,\sigma,z)} = \{(a',\sigma',z') \mid (a',\sigma',z') < (a,\sigma,z)\}$$

is a well-founded transitive set (the set of "subcomputations" of (a,σ,z)) . $\underset{\sim}{S}_{(a,\sigma,z)}$ has an associated ordinal $|a,\sigma,z|_{\Theta}$, which may be called the "length" of the computation (a,σ,z) .

The notion of computable function carries over unchanged to the present setting. In the definition of Θ-computable functional we make an addition.

Definition 9. Let $\langle\Theta,<\rangle$ be a computation structure on a domain $\mathfrak{A}$. A consistent functional φ is called Θ-computable if there exists a $\hat{\varphi} \in C$ such that for all $e_1,\ldots,e_l \in C$ and all $\sigma = (x_1,\ldots,x_n) \in A$ we have

(a) $\varphi(\{e_1\}_{\Theta}^{n_1},\ldots,\{e_l\}_{\Theta}^{n_l},\sigma) \simeq z$ iff $\{\hat{\varphi}\}_{\Theta}^{l+n}(e_1,\ldots,e_l,\sigma) \simeq z$.

(b) If $\varphi(\{e_1\}_{\Theta}^{n_1},\ldots,\{e_l\}_{\Theta}^{n_l},\sigma) \simeq z$, then there exists functions $g_1,\ldots,g_l$ such that

(i) $g_1 \subseteq \{e_1\}_{\Theta}^{n_1},\ldots,g_l \subseteq \{e_l\}_{\Theta}^{n_l}$ and $\varphi(\underset{\sim}{g},\sigma) \simeq z$;

(ii) for all $i = 1,\ldots,l$, if $g_i(t_1,\ldots,t_{n_i}) \simeq u$, then

$$(e_1,t_1,\ldots,t_{n_i},u) < (\hat{\varphi},e_1,\ldots,e_l,\sigma,z) .$$

We may now state the definition of a computation theory.

Definition 10. A computation structure $\langle\Theta,<\rangle$ on $\mathfrak{A}$ is called a computation theory on $\mathfrak{A}$ if there exist Θ-computable mappings p_1,p_2,p_3,p_4,p_5 such that the following functions and functionals are Θ-computable with Θ-codes as indicated and such that the iteration property holds:

1. $f(x) = s(x)$ $\quad f = \{p_1\}^1_\Theta$.

2. $f(x,b,c,a) = DC(x,b,c,a)$ $\quad f = \{p_2\}^4_\Theta$.

3. $\varphi(f,g,\sigma) = \underset{\sim}{C}^n(f,g,\sigma)$ $\quad \hat{\varphi} = p_3(n)$.

4. $\varphi(f,\sigma,\tau) = \underset{\sim}{P}^m_{n,j}(f,\sigma,\tau)$ $\quad \hat{\varphi} = p_4(n,j,m)$. $0 \leq j < n$.

5. Iteration property: For all n,m $p_3(n,m)$ is a Θ-code for a mapping $S^n_m(a,x_1,\dots,x_n)$ such that for all $a,\sigma \in C$ and all $\tau \in A$, $lh(\tau) = m$:

(i) $\{a\}^{n+m}_\Theta(\sigma,\tau) = \{S^n_m(a,\sigma)\}^m_\Theta(\tau)$;

(ii) if $\{a\}^{n+m}_\Theta(\sigma,\tau) \simeq z$, then

$$(a,\sigma,\tau,z) < (S^n_m(a,\sigma),\tau,z) .$$

We shall state only one main result about computation theories, viz. that if Θ is a computation theory over a domain A and if the set of subcomputations $\underset{\sim}{S}_{(a,\sigma,z)}$ of a computation (a,σ,z) is "finite" in the sense of the theory, then Θ is equivalent to the prime recursion theory associated with a partial type-2 object over A , and this in a way which preserves the structure of subcomputations.

Note that in the present context we add the following clause to definition 6 of the relation $\Theta \leq H$:

$$\text{if } (a',\sigma',z') <_\Theta (a,\sigma,z) \text{, then } (p(a',n'),\sigma',z') <_H (p(a,n),\sigma,z) ,$$

i.e. the function p preserves the subcomputation relation.

Definition 11. Let $\langle\Theta,<\rangle$ be a computation theory on a domain $\mathfrak{A}$. A set $B \subseteq A$ is called _Θ-finite_ if the functional $\underset{\sim}{E}_B$ is Θ-computable:

$$\underset{\sim}{E}_B(f) \simeq \begin{cases} 0 & \text{if } \exists x \in B\ [f(x) \simeq 0] \\ 1 & \text{if } \forall x \in B\ [f(x) \simeq 1] \end{cases}$$

Thus a set is finite if we can computably quantify over it.

Definition 12. A computation theory $\langle\Theta,<\rangle$ is called s-normal if the sets $\underset{\sim}{S}_{(a,\sigma,z)}$ are uniformly Θ-finite for $(a,\sigma,z) \in \Theta$.

Theorem. Let $\langle\Theta,<\rangle$ be an s-normal computation theory on the domain $\mathfrak{A}$. Then there exists a Θ-computable functional Ψ such that $\Theta \sim PR[\Psi]$.

Note that this is an improvement over a result stated in [6] where it was required that Θ has a computable search operator.

A computation theory Θ is also a precomputation theory on the domain. If we are only interested in the Θ-semicomputable relations on $\mathfrak{A}$, this is adequately represented by the semicomputable relations of some theory $PR[f]$, where f is a partial function on $\mathfrak{A}$. What is added in the representation theorem above is that we can preserve the structure of subcomputations, in particular, the length function associated with subcomputations.

2. FINITE THEORIES.

2.1 Finite theories on one type. In this part we will study theories Θ in which the domain A is Θ-finite, i.e. we can Θ-computably quantify over the domain. The resulting theory is a generalization of hyperarithmetic theory or recursion in ^2E over ω .

Let $\langle\Theta,<\rangle$ be a computation theory on a domain $\mathfrak{A} = \langle A,\ldots\rangle$. We will not distinguish between the set of computations Θ and the "coded" set $\{\langle a,\sigma,z\rangle \mid (a,\sigma,z) \in \Theta\}$. For $x \in \Theta$ let $|x|_\Theta$ be the ordinal of the computation x , i.e. the ordinal of the set $\underset{\sim}{S}_x = \{y \in \Theta \mid y < x\}$.

The following prewellordering property is essential in the study of finite theories:

<u>Definition 13</u>. Θ is called <u>p-normal</u> if there is a Θ-computable function p such that if $x \in \Theta$ or $y \in \Theta$, then $p(x,y)\downarrow$, and

$$x \in \Theta \wedge |x|_\Theta \leq |y|_\Theta \Rightarrow p(x,y) = 0$$

$$|x|_\Theta > |y|_\Theta \Rightarrow p(x,y) = 1 .$$

Recursion in ^2E, or more generally, recursion in any normal higher type object over ω, give p-normal theories.

Note that p-normality is under suitable conditions implied by s-normality, e.g. if there is a Θ-semicomputable extension of the relation $<$ to all tuples (a,σ,z). In this case we have the following recursion equations for the function p :

(a) $p(x,y) = 0$ if $\forall x' \in \underset{\sim}{S}_x \exists y'[y' < y \wedge p(x',y') = 0]$

(b) $p(x,y) = 1$ if $\exists x'[x' < x \wedge \forall y' \in \underset{\sim}{S}_y \; p(x',y') = 1]$

The assumption that $<$ is Θ-semicomputable means that whether or not (a,σ,z) represents a "true" computation, i.e. $(a,\sigma,z) \in \Theta$, we should be able to start generating the "subcomputations" of (a,σ,z).

One of the most important consequences of p-normality is the existance of a single-valued selection operator over the integers. This was first proved by Gandy [9] for recursion in a normal type-2 object, later by Moschovakis [22] for recursion in ^3E, and finally extended to all types by Grilliot [11]. Platek has also a similar result in his thesis [30].

We mentioned above that finite theories over one type is a generalization of hyperarithmetic theories. The first systematic study of such generalizations is to be found in Moschovakis work on hyperprojective theories [23]. Recently he has studied the notion of Spector class (see chapter 9 of [26]). This is related to the following class of finite theories:

Definition 14. A Spector theory $\langle\Theta,<\rangle$ on a domain $\mathfrak{A}$ is a computation theory satisfying

(i) Θ is p-normal;

(ii) $A = C$;

(iii) A is Θ-finite.

Note that (ii) implies that equality is Θ-computable. It is also rather straight forward to show that the class of Θ-semicomputable relations in a Spector theory Θ is closed under existential quantifier and disjunction.

Let $en(\Theta)$ denote the class of Θ-semicomputable relations and $sc(\Theta)$ the class of Θ-computable relations. The following result was observed independently of Moschovakis and Moldestad:

Theorem. If Θ is a Spector theory on $\mathfrak{A}$, then $en(\Theta)$ is a Spector class. Conversely, if Γ is a Spector class on $\mathfrak{A}$, then there is a Spector theory Θ on $\mathfrak{A}$ such that $\Gamma = en(\Theta)$.

Spector theories are also related to the theory of inductive definability. We quote one simple, but basic result to show the connection.

Theorem. Let $\langle\Theta,<\rangle$ be a Spector theory on ω and let $\underset{\sim}{R}$ be a sequence of Θ-computable relations:

(a) $Ind(\Sigma_2(\underset{\sim}{R})) \subseteq en(\Theta)$.

(b) $\Theta \sim PR[{}^2E,\underset{\sim}{R}]$ iff $Ind(\Sigma_2(\underset{\sim}{R})) = en(\Theta)$.

This result is basically due to Grilliot [12] and was adapted to the present frame-work by Moldestad. For further information on inductive definitions and the relationship between inductive definability and Spector classes/theories we refer the reader to the papers in [7], to Moschovakis book [26], and to his recent paper [27].

2.2 Finite theories on two types. Moving up in types over ω the notion of finiteness bifurcates. In higher types one must carefully distinguish between a "weak" and a "strong" notions. This is tied up with a phenomenon first observed by Moschovakis [22], viz. that the semi-computable in ${}^3E$ subsets of ${}^{\omega}\omega$ are not closed under $\exists\alpha^1$. A further analysis reveals that higher type theories really can be captured as theories on "two types", i.e. on domains of the form $A = S \cup Tp(S)$ (here $Tp(S) = {}^S\omega$), where S is strongly finite but $Tp(S)$ is only weakly finite. Strong finiteness is finiteness in the sense of definition 11. The weak notion is defined as follows:

Definition 15. Let Θ be a computation theory on a domain $\mathfrak{A}$. A set $B \subseteq A$ is called weakly Θ-finite if the functional $\underset{\sim}{E}'_B$ is Θ-computable:

$$\underset{\sim}{E}'_B(f) \simeq \begin{cases} 0 & \text{if } \forall x \in B\ f(x)\downarrow \wedge \exists x \in B\ f(x) \simeq 0 \\ 1 & \text{if } \forall x \in B\ \exists y \neq 0\ f(x) \simeq y . \end{cases}$$

The distinction between two types can already be found in Moschovakis [25]. It was systematically adopted by Harrington and MacQueen [16] in their proof of the Grilliot selection theorem. It has also been developed by Moldestad [21] as a natural setting for the general plus-2 and plus-1 theorem; we follow his exposition here.

The computation domain will have the form $\mathfrak{A} = \langle A, S, \underset{\sim}{S} \rangle$, where $A = S \cup Tp(S)$ and $\underset{\sim}{S}$ is a coding scheme for S. Let $\underset{\sim}{L}$ be a list $R_1,\dots,R_k$, $\varphi_1,\dots,\varphi_l$, $F_1,\dots,F_m$ where $R_1,\dots,R_k$ are relations on A, $\varphi_1,\dots,\varphi_l$ are partial functions, and $F_1,\dots F_m$ are functionals, i.e. functions from ${}^A A$ to ω. Prime recursion in $\underset{\sim}{L}$, $PR(\underset{\sim}{L}) = PR(\underset{\sim}{R},\underset{\sim}{\varphi},\underset{\sim}{F})$, is introduced via a number of schemata in analogy to the way Kleene introduced recursion in higher types [18].

The schemata are rather standard: Introduction of the characteristic functions for N and S, the successor function, the functions associated with the coding scheme, the functions in the list $\underset{\sim}{L}$. There are schemes for substitution, evaluation $(f(x,y) = x(y)$, if $x \in {}^S\omega$ and $y \in S)$, and the enumerating function for the partial recursive functions.

A scheme is also needed for the following extended case of substitution: If $\lambda x.\{e\}(x,a)$ is total, where $x \in S$ and a is a list of arguments from A , then $f(a) \simeq \{e'\}(\lambda x.\{e\}(x,a),a)$ is partial recursive in L with an index computable from e and e' .

A list $\underset{\sim}{L}$ is called <u>normal</u> if the equality relation on S is recursive in $\underset{\sim}{L}$ and the functional E , where for $f \in {}^A A$

$$E(f) = \begin{cases} 0 & \text{if } \exists x\, f(x) = 0 \\ 1 & \text{if } \forall x\, f(x) \neq 0 , \end{cases}$$

is (weakly) recursive in $\underset{\sim}{L}$. (A functional F is <u>weakly</u> <u>recursive</u> in $\underset{\sim}{L}$ if there is a primitive recursive function $s(e)$ such that

$$\{s(e)\}_{\underset{\sim}{L}}(a) \simeq F(\lambda x.\{e\}_{\underset{\sim}{L}}(x,a))$$

for all e,a.)

Adapting the proofs from MacQueen's thesis [20] Moldestad has verified the following:

<u>Theorem</u>. Let $PR(\underset{\sim}{L})$ be normal on the domain $A = S \cup Tp(S)$. Then the following are true:

(a) $PR(\underset{\sim}{L})$ is p-normal.

(b) A is weakly but not strongly finite, i.e. the $\underset{\sim}{L}$-semicomputable relations are not closed under the existential quantifier over $Tp(S)$.

(c) S is strongly finite, i.e. the $\underset{\sim}{L}$-semicomputable relations are closed under the existential quantifier over S .

Property (c) is a corollary of the Grilliot selection theorem: Let $\underset{\sim}{L}$ be a normal list and $B \subseteq S$ a non-empty set recursively enumerable in $\underset{\sim}{L}, a$. Then there is a non-empty subset $B' \subseteq B$ which is recursive in $\underset{\sim}{L}, a$. This result was first proved in [20], an "abstract" version appears in Harrington and MacQueen [16]. As mentioned above Moldestad's proof is an adaption of the proof in MacQueen [20].

Moldestad verifies in [21] that if $S = Tp(0) \cup \ldots \cup Tp(n-1)$, $n > 0$, then ${}^{S}\omega$ can be identified with $Tp(1) x \ldots x\ Tp(n)$. If F is an object of type-n+2 , there is a list $\underset{\sim}{L}$ such that (Kleene) recursion in F on $Tp(0), \ldots, Tp(n)$ is essentially the same as recursion in $\underset{\sim}{L}$ on $\mathfrak{A} = \langle A, S, \underset{\sim}{S} \rangle$. A converse is also true. Hence results about recursion in higher types can be deduced from the corresponding results for $PR(\underset{\sim}{L})$ on $\mathfrak{A}$.

This is in a more precise way what we meant when we said above that higher type theories really can be captured as theories on two types.

3. INFINITE THEORIES.

The starting point for our analysis of infinite theories is the following fact: If a transitive set $\langle A, \epsilon \rangle$ is resolvable, then A is admissible if and only if every Σ_1 inductive operator over A has a Σ_1 fixed point. From this we shall abstract a notion of admissible prewellordering and characterize this notion in computation theoretic form. We note that the notion of admissible prewellordering was introduced by Moschovakis [24] and that the characterization theorem is essentially an adaption of one of his results to the present setting.

Let $\langle \Theta, < \rangle$ be a computation theory on a domain $\mathfrak{A}$. We assume that Θ is p-normal and that $A = C$. Θ is also taken to be single-valued.

We add the following axioms:

A. There is a Θ-computable prewellordering $\preceq$ of A such that the initial segments of $\preceq$ are (uniformly) Θ-finite.

The finiteness assumption means that the follwing functional is Θ-computable

$$E^{\preceq}(f,x) \simeq \begin{cases} 0 & \text{if } \exists y \prec x.\ f(y) \simeq 0 \\ 1 & \text{if } \forall y \prec x.\ f(y) \simeq 1\ . \end{cases}$$

B. $|\preceq| = \sup\{|a,\sigma,z|_{\Theta} : (a,\sigma,z) \in \Theta\}$.

Here $|a,\sigma,z|_{\Theta}$ is the ordinal of the well-founded set $\underset{\sim}{S}(a,\sigma,z)$ and $|\preceq|$ is the length of the pwo $\preceq$. B is a way of saying that the complexity of the domain matches the complexity of the computations. This is a feature of infinite theories which is missing in the case of finite theories.

C. There is a Θ-computable mapping $p(n)$ such that $\{p(n)\}$ is total and

$$\{p(n)\}(a,\sigma,z,w) = 0 \quad \text{iff} \quad (a,\sigma,z) \in \Theta \wedge |a,\sigma,z|_{\Theta} = |w|\ .$$

Here $n = lh(\sigma)$ and $|w|$ is the ordinal of w in the pwo $\preceq$. Note that it now follows that A is not Θ-finite: if so, we would have $(a,\sigma,z) \in \Theta$ iff $\exists w\ p(a,\sigma,z,w) = 0$, i.e. Θ would be Θ-computable.

Usually the assumption is made that Θ has a (multiple-valued) selection operator. This is needed to have a decent theory for the Θ-semicomputable relations. A closer analysis shows that we get by just by requiring the Θ-semicomputable relations to be closed under existential quantification over A .

D. There is a Θ-computable mapping $q(n)$ such that for all a,σ

$$\exists x.\{a\}(x,\sigma) \simeq 0 \quad \text{iff} \quad \{q(n)\}(a,\sigma) \simeq 0 .$$

From D and p-normality we may now show that the Θ-semicomputable relations are closed under $\vee$ and that a relation R is Θ-computable if and only if R and $\neg R$ are Θ-semicomputable. And our theories are single-valued.

A number of elementary facts can now be established, e.g. Θ-finiteness can be characterized as being Θ-computable and $\precsim$ -bounded. And there exist Θ-computable relations $\underset{\sim}{R}_\Theta$ such that the Θ-semicomputable relations on A are exactly the $\Sigma_1(\precsim, \underset{\sim}{R}_\Theta)$ relations on A.

Definition 16. Let $\precsim$ be a prewellordering on a set A and $\underset{\sim}{R}$ a sequence of relations on A. $(A,\precsim)$ is called $\underset{\sim}{R}$-admissible if for every $\Sigma_1(\precsim, X, \underset{\sim}{R})$ formula θ in which X occurs positively and which has parameters from A, the fixed point X^* of the associated operator $\Gamma_\theta(X) = \{\sigma \mid \theta(\sigma,X)\}$ is a $\Sigma_1(\precsim, \underset{\sim}{R})$ relation.

It is now an immediate consequence of the first recursion theorem for Θ that if $\underset{\sim}{R}$ is any sequence of Θ-computable relations extending the sequence $\underset{\sim}{R}_\Theta$ referred to above, then the structure $(\mathfrak{A}, \precsim)$ is $\underset{\sim}{R}$-admissible. We have the following converse

Theorem. Let $(\mathfrak{A}, \precsim)$ be an $\underset{\sim}{R}$-admissible prewellordering. There exists a p-normal computation theory $\langle \Theta, < \rangle$ on $\mathfrak{A}$ satisfying A to D and such that the Θ-semicomputable relations are exactly the $\Sigma_1(\precsim, \underset{\sim}{R})$ relations.

The theorem has the following refinement. If we start with an infinite theory Θ, there exist suitable Θ-computable relations $\underset{\sim}{R}$ such that the theory Θ^* constructed according to the theorem above from the $\underset{\sim}{R}$-admissible pwo $(\mathfrak{A}, \precsim)$ is equivalent to Θ, i.e. $\Theta \sim \Theta^*$.

Remark. No study of infinite theories can be complete unless degree theoretic arguments are accounted for. This is yet in a preliminary stage; we refer the reader to the recent work of Simpson [34], [35].

4. CLASSIFICATION OF FINITE THEORIES ON ONE TYPE.

4.1 The imbedding theorem. A topic of central importance is the relationship between theories over "finite" and "infinite" domains. The basic example is here the relationship between hyperarithmetic theory over ω and meta-recursion or L_{ω_1}-recursion theory.

We shall in this section describe a result on how finite theories in general can be imbedded into infinite theories. Infinite theories behave very much like ordinary recursion theory, hence one possible application of the imbedding theorem would be to obtain fine structure results for the semi-computable relations of the given finite theory by "pull-back" from the infinite theory in which we imbed.

Another use would be to obtain various classification results for finite theories.

A general imbedding theorem was stated in our paper [6]. For the purpose of these lectures we exhibit the main ideas in the simpler but important case of finite computation theories over ω.

Let $\langle \Theta, < \rangle$ be a Spector theory over ω, i.e. Θ is p-normal and ω is (strongly) Θ-finite. Associated with any such theory we have an ordinal

$$\|\Theta\| = \sup\{|a,\sigma,z|_{\Theta} \; ; \; (a,\sigma,z) \in \Theta\}$$

and a "universal" relation

$$R_{\Theta} x \beta \quad \text{iff} \quad x \in \Theta \wedge |x|_{\Theta} \leq \beta \, .$$

Note that R_{Θ} can be coded as a subset of $\|\Theta\|$.

Theorem. Let $\langle \Theta, < \rangle$ be a Spector theory over ω :

1. $L_{\|\Theta\|}[R_\Theta]$ is R_Θ-admissible and $\|\Theta\|$ is projectible to ω .

2. For any subset $A \subseteq \omega$:

$A \in sc(\Theta)$ iff $A \in L_{\|\Theta\|}[R_\Theta]$.

$A \in en(\Theta)$ iff A is $\Sigma_1(\langle L_{\|\Theta\|}[R_\Theta], \in, R_\Theta \rangle)$.

The ideas implicit in this result can be traced back to the Kreisel-Sacks construction of meta-recursion theory [19] and to the Barwise-Gandy-Moschovakis paper on the next admissible set [5]. The Kreisel-Sacks approach has been analyzed in Aczel [2]. Moschovakis has in chapter 9 of [26] generalized the "next-admissible" construction to a companion theory for Spector classes on transitive sets. We also note that a proof of the theorem is implicit in Sacks [31] when he verifies that the 1-section of a normal higher type object is an abstract 1-section.

Adding the characterization theorem of section 3 to the above result gives the general imbedding theorem as stated in [6].

Theorem. Let $\langle \Theta, < \rangle$ be a Spector theory on a domain $\mathfrak{A}$. It is possible to construct :

(i) a computation domain $\langle \mathfrak{A}^*, \precsim \rangle$ where $\mathfrak{A}^*$ extends $\mathfrak{A}$ and $\precsim$ is a prewellordering on $\mathfrak{A}^*$;

(ii) a relation R on $\langle \mathfrak{A}^*, \precsim \rangle$; and

(iii) a p-normal computation theory $\langle \Theta^*, <^* \rangle$ on $\langle \mathfrak{A}^*, \precsim \rangle$ such that

  (a) $\precsim$ is Θ^*-computable and initial segments of $\precsim$ are uniformly Θ^*-finite,

  (b) Θ^* satisfies A-D of section 3,

  (c) R is Θ^*-computable and Θ^*-semicomputability equals $\Sigma_1(\precsim, R)$ definability over $\mathfrak{A}^*$, and

  (d) a subset $B \subseteq A$ is Θ-semicomputable if and only if it is Θ^*-semicomputable.

The general theorem requires the use of admissible sets with urelements. (For this theory see Barwise [3], [4].)

4.2 <u>Representation in terms of higher type objects over ω</u>. Let Θ be a Spector theory over ω. On general grounds (see section 1.2) we know that Θ is equivalent to the prime recursion theory associated with a <u>partial</u> type-2 functional over the domain. From the above result we also know that Θ can be considered as the ω-part of some $\Sigma_1(R)$-theory on some countable ordinal $\|\Theta\|$.

The relation R_Θ is a relation on $\|\Theta\|$, hence can in a natural way be coded as a <u>total</u> type-2 functional F. What is then the relationship between Θ and $PR[F]$?

There are several difficulties. A basic one is that recursion in F closes off at ω_1^F = the least F-admissible ordinal. But the ordinal $\|\Theta\|$ need not be the least R_Θ-admissible ordinal. There are two ways of correcting this situation.

By forcing it is possible to replace a given $L_\alpha[R]$, where α is R-admissible and countable and $R \subseteq \alpha$, by some $L_\alpha[R']$ such that α is the least R'-admissible ordinal. Since $L_\alpha[R] = L_\alpha[R']$, sections will be preserved; but since R' is obtained by a forcing argument, envelopes will not be preserved.

<u>Theorem</u>. Let Θ be a Spector theory on ω. There exists a normal type-2 object F such that $sc(\Theta) = {}_1sc(F)$.

This result is due to Sacks [31]. Our way of looking at it via the imbedding result is due to Normann [28].

There is another case in which we can modify R_Θ, viz. when $\|\Theta\|$ is not R_Θ-recursively Mahlo then one may modify R_Θ in a

Δ_1-definable way to a set R^* such that $\|\Theta\|$ is the least R^*-admissible ordinal. Then Σ_1-definability over $L_{\|\Theta\|}[R_\Theta]$ will equel Σ_1-definability over $L_{\|\Theta\|}[R^*]$, and with a bit of care it is possible to construct a functional F such that the ${}_1\mathrm{en}(F)$ corresponds to $\Sigma_1(\langle L_{\|\Theta\|}[R^*],\in,R^*\rangle)\cap 2^\omega$.

<u>Theorem</u>. Let Θ be a Spector theory on ω. Then $\Theta \sim PR[F]$ for some normal type-2 object F, if and only if $\|\Theta\|$ is not R_Θ-recursively Mahlo.

This result is due to Simpson and independently to Harrington and Kechris [15].

5. CLASSIFICATION OF FINITE THEORIES ON TWO TYPES.

5.1 <u>Imbedding in higher types</u>. We shall in this section describe an imbedding theory for recursion in higher types due to D. Normann [29]. Let $I = Tp(0)\cup\ldots\cup Tp(n)$ and let Θ be a computation theory on I. To capture the fact that Θ is a computation theory in higher types we assume in addition that Θ is p-normal, that I is weakly Θ-finite (i.e. Θ extends recursion in ${}^{n+2}E$), and that the evaluation maps and the characteristic functions of $Tp(i)$, $i=0,\ldots,n$, are Θ-computable. We also assume that if $(a,\sigma,z)\in\Theta$, then both a and z belongs to ω. Finally we must include the extended case of substitution as described in section 2.2. Note that we could have worked in the framework of two types as outlined in section 2.2. For the moment we need not require that $S = Tp(0)\cup\ldots\cup Tp(n-1)$ be strongly Θ-finite.

Let V_I be the universe of sets with I as urelements. In analogy with the coding of sets in HC as elements of ω^ω it is possible to code certain sets in V_I as subsets of I. Let $a\in I$, by $\Theta[a]$ we understand computations in Θ and the parameter a, i.e. $(e,\sigma,z) \in \Theta[a]$ if and only if $(e,a,\sigma,z) \in \Theta$.

Definition 17. $M_a[\Theta]$ is defined as the collection of elements in V_I which has a code in $n+1\text{-sc}(\Theta[a])$. Let $R_\Theta = \{\langle a,\sigma\rangle : a \in \Theta \wedge |a|_\Theta = \sigma\}$. We call

$$\langle\langle M_a[\Theta]\rangle_{a\in I}, R_\Theta\rangle$$

the spectrum of Θ , and denote it by $\text{Spec}(\Theta)$.

Note that $\text{Spec}(\Theta)$ generalizes the construction of section 4.1. Given Θ we often write M_a instead of $M_a[\Theta]$.

Theorem.

i. Each M_a is countable.

ii. Each M_a is rudimentary closed in R_Θ .

iii. Each M_a satisfies $\Delta_0(R_\Theta)$-separation and $\Delta_0(R_\Theta)$-dependent choice.

iv. Let $x \in M_a$, then $x \subseteq M_a$ iff x is countable in M_a .

We remark that M_a is not necessarily a transitive set. This is connected with "gap phenomena" in computations in normal objects of type ≥ 3 .

Definition 18. Let P be a subset of $M = \bigcup_{a\in I} M_a$. P is called Σ^*-definable if there is a Δ_0 formula φ (without parameters) such that for all $x \in M_a$ and all $a \in I$

$$x \in P \text{ iff } \exists y\, \varphi(x,y) \text{ iff } \exists y \in M_a\, \varphi(x,y) .$$

If P and $M-P$ both are Σ^* P is called Δ^*-definable. This notion may be relativized to R_Θ and to an $a \in I$.

Theorem. (Σ^*-collection) Let $a \in I$. Assume that φ is a $\Delta_0(R_\Theta,a)$-formula and

$$\forall b \in I \; \exists x \in M_{\langle a,b \rangle} \; \varphi(x,b) \; .$$

Then there is a set $u \in M_a$ such that

$$\forall b \in I \; \exists x \in u \cap M_{\langle a,b \rangle} \; \varphi(x,b) \; .$$

With this machinery available Normann is now able to characterize the $k+1$-envelope of Θ in terms of the spectrum of Θ. Harrington has a similar result in his thesis [13].

Theorem. Let Θ be a computation theory on I (Θ is assumed to satisfy the properties listed in the introductory paragraph above.) Let $A \subseteq I$. Then

$$A \in k+1\text{-en}(\Theta) \text{ iff } A \text{ is } \Sigma^*\text{-definable over } \mathrm{Spec}(\Theta) \; .$$

This theorem has many consequences. In particular, it is possible to generalize the characterization theorem of recursion in a normal type-2 object stated in section 4.2.

Definition 19. Let $\langle N_a \rangle_{a \in I}$ be a collection of structures in V_I satisfying: (i) $I \in N_a$, for all $a \in I$, and (ii) $a \in N_b$ iff $N_a \subseteq N_b$. Let $R \subseteq N = \bigcup_{a \in I} N_a$. We call $\langle N_a \rangle_{a \in I}$ nice relative to R if:

i. $x \in N_a$ iff x has a code in N_a.

ii. Each N_a is rudimentary closed in R.

iii. $\langle N_a \rangle_{a \in I}$ satisfies $\Sigma^*(R)$-collection.

It is a basic fact that if $\langle N_a \rangle_{a \in I}$ is a nice collection relative to a type-$n+2$ functional F, then $k+1\text{-sc}(F,a) \subseteq N_a$, for all $a \in I$. A special case of this was already proved by MacQueen [20]. Strengthening the notion of niceness a bit, one obtains a notion of abstract spectrum which characterizes the spectrum of a theory.

Definition 20. Let $\mathrm{Spec}(\Theta) = \langle\langle M_a\rangle_{a\in I}, R_\Theta\rangle$ be given. We say that $\langle A_a\rangle_{a\in I}$ is R_Θ-impenetrable if for all Δ^*-definable functions g mapping each M_a into itself, there is a nice family $\langle N_a\rangle_{a\in I}$ relative to R_Θ such that: (i) $N_a \subseteq M_a$, for all $a \in I$, (ii) $N_a \neq M_a$, for some $a \in I$, and (iii) g is closed in $\langle N_a\rangle_{a\in I}$, i.e. g maps each N_a into itself.

Note that this generalizes the following way of characterizing recursively Mahlo: M is Mahlo if for all Δ_1 functions g there is an admissible $N \subsetneqq M$ such that $g(N) \subseteq N$.

Theorem. The following statements are equivalent:

i. $\mathrm{Spec}(\Theta)$ is not R_Θ-impenetrable.

ii. There is a normal type-k+2 functional F such that Θ is equivalent to (Kleene) recursion in F.

A result essentially equivalent to this theorem has independently been proved by Kechris [17]. Normann also uses his imbedding theory to characterize the recursion theory described by Harrington [14] for the superjump S^{n+3}.

G. Sacks has recently in [32] developed a notion of "abstract k+1-section" corresponding to the notion of an abstract 1-section. And he has proved by a forcing argument a theorem generalizing the main result of his paper [31] (i.e. the first theorem of section 4.2). Normann has verified that the imbedding theory of this section gives a very natural framework for Sacks' forcing argument. But it would lead too far to describe these results here.

5.2 The abstract plus-1 and plus-2 theorem. We shall now return to the setting of two types as described in section 2.2. Let Θ be a computation theory on a domain of the type $A = S \cup {}^S\omega$. We call Θ

normal if the equality relation on S is Θ-computable, if A is weakly Θ-finite, if S is strongly Θ-finite, and if Θ is p-normal. We know from section 2.2 that if F is a normal functional over A, then $PR(F)$ is a normal computation theory.

Theorem. Let Θ be normal. Then there is a normal functional F such that $S\text{-en}(\Theta) = S\text{-en}(F)$ and $sc(\Theta,a) = sc(F,a)$ for all $a \in S$.

Here $S\text{-en}(\Theta) = \{X \subseteq S : X$ is Θ-semicomputable$\}$ and $sc(\Theta,a) = \{X \subseteq A : X$ is $\Theta[a]$-computable$\}$. This result is an abstract version of the plus-2 theorem of Harrington [13] and also of the plus-1 theorem of Sacks [32]. The original plus-2 theorem of Harrington was a reduction result: Starting out with a normal functional of type $> n+2$ he constructed a functional F of type $n+2$ such that $n\text{-en}(G) = n\text{-en}(F)$. The proof in [13] uses the fact that $Tp(n)$ must be strongly finite in G. The theorem above is an improvement in the sense that we start out with a normal computation theory Θ. Hence in the concrete setting of higher types we only assume that $Tp(n)$ is weakly Θ-finite (but it is quite essential that $Tp(n-1)$ should be strongly Θ-finite). Thus the above theorem gives a kind of characterization result. The proof in the setting of two types is due to Moldestad [21]. It is quite similar to Harrington's proof in [13]. There are some necessary modifications, partly suggested by Harrington.

Remark. We cannot enter into the details of the proof in this survey. But we should mention that an essential role is played by a reflection phenomenon which was first studied by Harrington [13] (but see also the discussion in Kechris [17]). The reflection property follows from the strong finiteness of S, one particular important case being the following: Let B be a set of subsets of S, and assume that B is $\Theta[a]$-semicomputable and that B contains an element which is $\Theta[a]$-semicomputable. Then B contains an element (i.e. a subset of S) which is $\Theta[a]$-computable.

REFERENCES.

[1] P. Aczel, An axiomatic approach to recursive function theory on the integers, unpublished abstract, 1969.

[2] P. Aczel, An axiomatic approach to recursion on admissible ordinals and the Kreisel-Sacks construction of meta-recursion theory, Recursive function theory Newsletter, 1974.

[3] K.J. Barwise, Admissible sets over models of set theory, in: Fenstad, Hinman [7], 97-122.

[4] K.J. Barwise, Admissible set theory, Springer Verlag, to appear.

[5] K.J. Barwise, R. Gandy and Y.N. Moschovakis, The next admissible set, J. Symbolic Logic 36 (1971), 108-120.

[6] J.E. Fenstad, On axiomatizing recursion theory, in: Fenstad, Hinman [7], 385-404.

[7] J.E. Fenstad and P. Hinman (eds.), Generalized recursion theory, North-Holland, Amsterdam 1974.

[8] H. Friedman, Axiomatic recursive function theory, in: R. Gandy and C.E.M. Yates (eds.), Logic Colloquium '69, North-Holland, Amsterdam 1971, 113-137.

[9] R. Gandy, General recursive functionals of finite type and hierarchies of functionals, Ann. Fac. Sci. Univ. Clermont-Ferrand 35 (1967), 5-24.

[10] T. Grilliot, Hierarchies, based on objects of finite type, J. Symbolic Logic 34 (1969), 177-182.

[11] T. Grilliot, Selection functions for recursive functionals, Notre Dame Jour. Formal Logic 10 (1969), 225-234.

[12] T. Grilliot, Inductive definitions and computability, Trans. Amer. Math. Soc. 158 (1971), 309-317.

[13] L. Harrington, Contributions to recursion theory in higher types, MIT thesis, 1973.

[14] L. Harrington, The superjump and the first recursively Mahlo ordinal, in: Fenstad, Hinman [7], 43-52.

[15] L. Harrington and A. Kechris, Classifying and characterizing abstract classes of relations, to appear.

[16] L. Harrington and D.B. MacQueen, Selection in abstract recursion theory, to appear.

[17] A. Kechris, The structure of envelopes: a survey of recursion theory in higher types, MIT Logic Seminar Notes, 1973.

[18] S.C. Kleene, Recursive functionals and quantifiers of finite type I, Trans. Amer. Math. Soc. 91 (1959), 1-52.

[19] G. Kreisel and G.E. Sacks, Metarecursive sets, J. Symbolic Logic 30 (1965), 318-338.

[20] D.B. MacQueen, Post's problem for recursion in higher types, MIT thesis, 1972.

[21] J. Moldestad, Recursion theory on two types, Oslo 1974.

[22] Y.N. Moschovakis, Hyperanalytic predicates, Trans. Amer. Math. Soc. 129 (1967), 249-282.

[23] Y.N. Moschovakis, Abstract first order computability, Trans. Amer. Math. Soc. 138 (1969), 427-504.

[24] Y.N. Moschovakis, Axioms for computation theories - first draft, in: R. Gandy and C.E.M. Yates (eds.), Logic Colloquium '69, North-Holland, Amsterdam 1971, 199-255.

[25] Y.N. Moschovakis, Structural characterizations of classes of relations, in: Fenstad, Hinman [7], 53-79.

[26] Y.N. Moschovakis, Elementary induction on abstract structures, North-Holland, Amsterdam.

[27] Y.N. Moschovakis, On non monotone inductive definability, Fund. Math. 1974.

[28] D. Normann, On abstract 1-sections, Synthese 27 (1974), 259-263.

[29] D. Normann, Imbedding of higher type theories, Oslo 1974.

[30] R.A. Platek, Foundations of recursion theory, Stanford thesis, 1966.

[31] G.E. Sacks, The 1-section of a type n object, in: Fenstad, Hinman [7], 81-93.

[32] G.E. Sacks, The k-section of a type n object, to appear.

[33] L.P. Sasso, Degrees of unsolvability of partial functions, Berkeley thesis, 1971.

[34] S.G. Simpson, Degree theory on admissible ordinals, in: Fenstad, Hinman [7], 165-193.

[35] S.G. Simpson, Post's problem for admissible sets, in: Fenstad, Hinman [7], 437-441.

[36] H.R. Strong, Algebraically generalized recursive function theory, IBM J. Res. Devel. 12 (1968), 465-475.

[37] E.G. Wagner, Uniform reflexive structures: on the nature of Gödelizations and relative computability, Trans. Amer. Math. Soc. 144 (1969), 1-41.

CLOSED MODELS AND HULLS OF THEORIES

Robert Fittler
II. Mathematisches Institut
Freie Universität Berlin
1974

Introduction

The theories T considered here are first order theories. Two such theories S and T, having the same languages L(S) = L(T) are called equivalent if their deductive closures S* and T* coincide. The classes $\mathfrak{M}$(T) consisting of all L(T) - structures M which are models of some theory T are called elementary classes (cf. [C.K] p.173). In this paper we investigate for certain classes $\mathfrak{M}$ of L(T) - structures wether they are elementary classes and how the theory Th ($\mathfrak{M}$), consisting of all sentences holding in all structures of $\mathfrak{M}$, can be axiomatized. For the classes we assume that if M $\in \mathfrak{M}$ and N $\cong$ M, then N $\in \mathfrak{M}$. In part I we introduce the classes $\mathcal{F}$(T) of so called F - closed structures with respect to T (cf. I. 4), i.e. structures whose embeddings into certain other L(T) - structures preserve all the formulas of some given set F of formulas (cf. I. 1). Such embeddings are called F-embeddings or embeddings as F-substructures. $\mathcal{F}$(T) depends upon the choice of F. F = ∀1 for example gives rise to the class $\mathcal{E}$(T) of existentially closed structures (cf. I. 7). Other examples are the class $\mathfrak{A}$(T) of algebraically closed structures (cf. I. 8) and the class $\mathcal{J}_\omega$ of ω-injective Λ-modules in the sense of [E.S.] (cf. I. 9). $\mathcal{F}$(T) is closed with respect to taking union of chains (cf. I. 12) and elementary substructures (even ∀1-substructures, cf. I. 13, 14). If $T_\forall$ has the amalgamation property, then

$\mathcal{F}$(T) is closed with respect to taking F-substructures (cf. I. 15). For $F \subseteq \forall 1$ the class $\mathcal{F}$(T) is always mutually model-consistent with $\mathcal{M}$(T) (cf. I. 18). In part II we introduce the syntactical notion of the G-hull T^G of some theory T, where G is some set of (closed) formulas (cf. II. 1). This is a generalization of Kaisers "inductive hull" (cf. [K], see also II. 4). New examples are the algebraic hull (cf. II. 5), and the ω-injective hull of the theory T_Λ of Λ-modules (cf. II. 6). Another application is P. Henrard's syntactical construction of the finite forcing companion (cf. [H]), which turns out to be an iterated hull construction (cf. II. 7). The close connection between G-hulls of theories and F-closed models is given by theorem II. 8, stating that Th ($\mathcal{F}$(T)) $\cap \forall \neg F$ is equivalent to $T^{\forall\neg F}$, provided that $\mathcal{F}$(T) is model-consistent with $\mathcal{M}$(T). Notice that $\forall \neg F$ is the set of closed fomulas of the form $\forall \vec{x} \neg \varphi(\vec{x})$, where $\varphi(\vec{x}) \in F$. For $F = \forall 1$ this yields corollary 7.16 of [E.S.] (cf. II 9.a). Other applications arise in the case of algebraic hulls, and algebraically closed structures (cf. II 9.b) as well as for ω-injective hulls and ω-injective Λ-modules (cf. II 9.c).

Part III concerns elementary classes of F-closed structures. Theorem III. 3 roughly states that if F is a subset of F' and $\mathcal{F}'$(T) is an elementary class then so is $\mathcal{F}$(T). This generalizes corollary 7.14 of [E.S.] (cf. III. 4,5 d) and yields a new characterization of coherent rings (cf. III. 5 c)

For elementary classes $\mathcal{F}$(T), the theory Th ($\cdot\mathcal{F}$ (T)) coincides with the $\forall \neg$ F-hull $T^{\forall\neg F}$ (provided that $F \subseteq \forall 1$ and that $T_\forall$ have the amalgamation property, cf. theorem III. 6). For $F = \forall 1$ this reflects the well known fact that for the inductive hull $T^{\forall\exists} \sim$ Th ($\mathcal{E}$ (T)) $\sim$ model-companion of T (provided that the class $\mathcal{E}$(T) of existentially closed structures is an elementary class, cf. III. 7). For a coherent ring Λ we get an alternative to Eklof-Sabbaghs description of the theory Th ($\mathcal{O}$(T_Λ)) of all absolutely pure Λ-modules (cf. III.8. and [E.S.]).

In part IV we consider theories T, where the inductive hull is complete. For any $F \subseteq \forall 1$, Th ($\mathcal{F}$(T)) then coincides with $T^{\forall\neg F}$ (cf.theorem IV. 4); e.g. the theory Th ($\mathcal{E}$(T)) of the class of all existentially closed structures is complete (cf. IV.3). It turns out that the inductive hull of the theory T_Λ of Λ-modules is complete, without further restrictions on Λ (cf. proposition IV.9). This yields an explicit syntactical description of the theory Th($\mathcal{J}_\omega$) of ω- injective modules, for arbitrary rings Λ (cf. IV.12.). Notice that Th ($\mathcal{J}_\omega$) has already been described in [E.S.] for coherent rings Λ. I am indebted to Martin Ziegler for considerably simplifying my original proof of theorem IV.8.

I. Closed Structures

1. Sets F of formulas

By F we denote any set of formulas of the language L (T) of some theory T, which contains all quantifier free formulas and which is closed with respect to conjunction and substitution.

Let $\forall\neg F$ be the set of all formulas of the form $\forall\vec{x}\neg\varphi(\vec{x})$, where $\varphi(\vec{x}) \in F$.

2. Examples:

a) For F consisting of all quantifier free formulas the set $\forall\neg F$ consists of all universal formulas (i.e. formulas whose only quantifiers are universal ones).

Thus $\forall\neg F = \forall 1$ (up to logical equivalence of formulas). In general we say that some formula ψ belongs to $\forall n$ if ψ is some formula which has at most n blocks of quantifiers, the first one consisting of universal ones. The set $\forall 1$ will also be denoted by $\forall$, and $\forall 2$ by $\forall\exists$. $\forall 0$ is the set of quantifier free formulas.

b) Let F be the set L (T) of all formulas.
Then $\forall\neg F = L(T)$.

c) Let F be the set $\forall 1$.
Then $\forall\neg F = \forall 2$.
In general : If $F = \forall n$ then $\forall\neg F = \forall(n+1)$, $n \in \mathbb{N}$

d) Let $F = A$ consist of conjunctions of quantifier free formulas and formulas of the form
$\forall\vec{y}\neg\psi(\vec{x},\vec{y})$, where the $\psi(\vec{x},\vec{y})$ are quantifier free positive formulas.
Then the set $\forall\neg A$ consists of formulas of the form
$\forall\vec{x}\exists\vec{y}\ (\chi(\vec{x},\vec{u}) \Rightarrow \psi(\vec{x},\vec{y},\vec{v}))$
(where χ is quantifier free) and of all universal formulas.

e) Let $F = P$ be the smallest set of formulas having the properties required in 1 which contains all formulas of the form $\neg\varphi(\vec{x})$, where φ

is positive, i.e. there is no negation involved in φ. Then $\forall\neg P$ is the set of all formulas having no negation in the scope of existential quantifiers.

3. F - embeddings

Let M, N be L-structures.
An embedding $M \subseteq N$ of M into N as a substructure is called an F-embedding if it preserves the validity of the formulas of F, i.e. if for $\varphi(\vec{x}) \in F$, $M \models \varphi(\vec{m})$ implies $N \models \varphi(\vec{m})$.
We will denote this by $M \underset{F}{\subseteq} N$ and say that M is embedded as an F-substructure of N.
In the special case that $F = \forall n$ the F-embeddings $M \underset{F}{\subseteq} N$ are denoted by $M \prec_n N$.
Any embedding $M \subseteq N$ is a $\forall 0$ -embedding since $\forall 0$ consists of all quantifier free formulas, i.e. $M \prec_o N$.

4. F-closed structures

An L-structure M is called F-closed with respect to the theory T if it is a model of $T^* \cap \forall 1$ and if all embeddings of M into models of $T^* \cap \forall\neg F$ are F-embeddings.
By $T^* \cap \forall 1$ we denote the set consisting of all universal sentences which are provable in T. $T^* \cap \forall 1$ will also be denoted by $T_\forall$. Let $\mathcal{F}(T)$ denote the class of all F-closed structures.

5. Remarks

a) An L-structure is a model of $T^* \cap \forall\neg F$ if and only if it is an F-substructure of some model of T (cf.[F] , Theorem 8).
b) The theory $T^* \cap \forall\neg F$ for the "test-structures" N, which serve to test whether some M is F-closed, can be slightly modified under certain circumstances:
Let $F \subseteq \forall 1$ and $M \models T_\forall$, then the following statements are equivalent

(i) M is F-closed.
(ii) $M \subseteq N \models T$ implies $M \underset{F}{\subseteq} N$.

(iii) $M \subseteq N \models T_\forall$ implies $M \underset{F}{\subseteq} N$.

Proof : (i) ⇒ (ii) and (iii) ⇒ (i) are trivial, the latter since $T_\forall \sim T_\forall \cap \forall \neg F$.

We prove (ii) ⇒ (iii) (see also [Si_1] , Theorem 2.4.)

Let $M \subseteq N \models T_\forall$ and $M \models \varphi(\vec{m})$, $\varphi(\vec{x}) \in F$.

It is to be shown that $N \models \varphi(\vec{m})$. Since $N \models T_\forall$, there exists $N' \models T$ such that $N \subseteq N'$ (cf. 5 a). Thus $N' \models \varphi(\vec{m})$, because of (ii). But then obviously $N \models \varphi(\vec{m})$, since $\varphi \in \forall 1$.

Q.E.D.

6. Completing Models

Let F be the set L(T) (cf. 2b). The L (T)-closed structures (with respect to T) coincide with the so called completing models of T (cf. [Ba.R.]).

7. Existentially closed structures

Let F be the set ∀1 (cf. 2 c). In this case the F-closed structures are the so called existentially closed structures (cf. [E.S.] ,[M_1] [Si_1] [R_2]). The class of existentially closed structures will be denoted by $\mathcal{E}(T)$.

8. Algebraically closed structures

If F = A is the set of example 2 d, then the A-closed structures are the so called algebraically closed structures (cf. [M_1] , [E.S.]) E.g. let T be the theory of Λ-left modules over some unitary, associative ring Λ with $0 \neq 1$ (cf. [E.S.]), then the algebraically closed structures are the absolutely pure modules. The class of algebraically closed structures will be denoted by $\mathfrak{A}(T)$.

9. ω -injective Λ-modules

Let T be the theory of Λ-modules (cf. 8). Let F = I be the smallest set having the properties required in 1 which contains all formulas of the form $\forall x \neg(\bigwedge_{i=1}^{n} \lambda_i x = y_i)$ where $\lambda_i \in \Lambda$, $n \in \mathbb{N}$. Then the I-closed structures coincide with the ω-injective Λ-modules of [E.S.]. The class of ω-injective Λ-modules will be denoted by $\mathcal{J}_\omega$. $\mathcal{J}_\omega$ contains the class of injective models as a subclass (cf. [E.S.]).

10. P-closed structures

Let P be as in example 2 e. The class of P-closed structures will be called $\mathcal{P}(T)$. E.g. if T is the theory of Λ-modules (cf.8) the injective Λ-modules are elements of $\mathcal{P}(T)$. This holds since for an injective Λ-module M and any model N of $(T^* \cap \forall\neg P)^* = T^*$, if $M \subseteq N$, then there is a retraction $g : N \to M$. But this implies immediatly that if for a positive formula $\varphi(\vec{x})$, $N \models \varphi(\vec{n})$ holds, then $M \models \varphi(g(\vec{n}))$. Hence $M \models \psi(\vec{m})$ implies $N \models \psi(\vec{m})$, for any positive formula $\psi(\vec{x})$, as well as for any quantifier free formula. Thus $M \models \chi(\vec{m})$ implies $N \models \chi(\vec{m})$ for any formula $\chi(\vec{x}) \in P$.
The following characterization of F-closed models is a straight forward generalization of theorem 2.1 of $[Si_1]$.

11. Lemma

For any model M of $T_\forall$ the following statements are equivalent:
(i) M is F-closed with respect to T.
(ii) If $M \models \varphi(\vec{m})$, $\varphi \in F$, then
$(T^* \cap \forall\neg F) \cup \Delta_M \vdash \varphi(\vec{m})$, where Δ_M is the diagram of M.
(iii) If $M \models \varphi(\vec{m})$, $\varphi(\vec{x}) \in F$, then there is some existential formula $\psi(\vec{x})$ such that $M \models \psi(\vec{m})$ and $T^* \cap \forall\neg F \vdash \forall \vec{x}\ (\psi(\vec{x}) \Rightarrow \varphi(\vec{x}))$.
Proof : By modification of Simmons proof.

12. Corollary

$\mathcal{F}(T)$ is closed with respect to unions of chains.
Proof : Use Lemma 11 (ii).

13. Corollary (cf. also [E.S.] Corollary 7.7)

Let $F \subseteq \forall 1$. Any $\forall 1$-substructure of an F-closed structure is F-closed.
Proof: Statement (iii) of Lemma 11 carries over to $\forall 1$-substructures

since $F \subseteq \forall 1$.

Q.E.D.

14. Corollary

Any elementary substructure of an F-closed structure (with respect to T) is F-closed (with respect to T).

For a similar statement about F-substructures of F-closed structures we need the amalgamation property of $T_\forall$ (which means that for any models $M' \overset{i}{\supseteq} M \overset{j}{\subseteq} M''$ of T there is a model Q of T and embeddings $M \overset{r}{\subseteq} Q \overset{s}{\supseteq} M''$ such that $r\,i = s\,j$, cf. [B.S.] p.203).

15. Lemma

Let $T_\forall$ have the amalgamation property and let $F \subseteq \forall 1$. Then F-substructures of F-closed structures are F-closed.
(cf. [E.S.] remark 2 p. 286).
Proof : Let M be F-closed and $N' \supseteq N \underset{F}{\subseteq} M$. It is to be shown that the embedding $N' \supseteq N$ is an F-embedding, for $N' \models T_\forall$ (cf. remark 5b). The amalgamation property of $T_\forall$ guarantees the existence of $Q \models T_\forall$ and embeddings such that

$$\begin{array}{ccc} N & \underset{F}{\subseteq} & M \\ \cap & & \cap \\ N' & \subseteq & Q \end{array}$$

commutes.
Let $N \models \varphi(\vec{n})$ for $\varphi \in F$, then $M \models \varphi(\vec{n})$, hence $Q \models \varphi(\vec{n})$ (since M is F-closed). Thus $N' \models \varphi(\vec{n})$ because $F \subseteq \forall 1$.

Q.E.D.

16. Model-consistency

According to $[R_2]$ a theory T is called model-consistent with the theory S if every model of S can be embedded into a model of T.

This definition which refers directly to the model classes $\mathfrak{M}(T)$ and $\mathfrak{M}(S)$ of all models of T and S has been extended to arbitrary classes $\mathfrak{M}, \mathfrak{N}$ of L-structures in the following way (cf. [C]) :

A class $\mathfrak{M}$ of L-structures is called model-consistent with the class $\mathfrak{N}$ of L-structures if any structure in $\mathfrak{N}$ can be embedded into a structure of $\mathfrak{M}$. It is obvious that T is model-consistent with S if and only if $\mathfrak{M}(T)$ is model-consistent with $\mathfrak{M}(S)$.
From remark 5a we conclude that T is model-consistent with S if and only if $T_\forall \subseteq S_\forall$.
T and S are called mutually model-consistent if T is model-consistent with S and S is model-consistent with T. The analog definition holds for classes, .
It follows that T and S are mutually model-consistent if and only if $T_\forall = S_\forall$. The following properties are trivial

(a) For any theory T, $T_\forall$ and T are mutually model-consistent.
(b) If $\mathfrak{M} \subseteq \mathfrak{N}$ and $\mathfrak{M}$ is model -consistent with $\mathfrak{N}$ then $\mathfrak{M}$ and $\mathfrak{N}$ are mutually model-consistent.

17. Theorem

The class $\mathcal{E}(T)$ of existentially closed models is model-consistent with $\mathfrak{M}(T_\forall)$.
Hence $\mathcal{E}(T)$ and $\mathfrak{M}(T_\forall)$ (and $\mathfrak{M}(T)$!) are mutually model-constistent.

Proof: cf. [Si_1] Theorem 2.3

18. Corollary

Let $F \subseteq \forall 1$.
The class $\mathcal{F}(T)$ of F-closed structures (with respect to T) is mutually model-consistent with $\mathfrak{M}(T_\forall)$ (and with $\mathfrak{M}(T)$), for $F \subseteq \forall 1$.
Proof : $\mathcal{E}(T) \subseteq \mathcal{F}(T) \subseteq \mathfrak{M}(T_\forall)$ and $\mathcal{E}(T)$ and $\mathfrak{M}(T_\forall)$ are mutually model-consistent. Hence $\mathcal{F}(T)$ is mutually model-consistent with both $\mathcal{E}(T)$ and $\mathfrak{M}(T_\forall)$.

Q.E.D.

The following compactness property will be used later.

19. Lemma

Any finite subset of T is model-consistent with S if and only if T is model-consistent with S.

Proof:
If T is model-consistent with S then any finite subset is obviously model-consistent with S. Now, assume that T is not model-consistent with S. We want to show that some finite subset $T' \subseteq T$ is not model-consistent with S. Since T is not model-consistent with S there is some model M of S which cannot be embedded into any model N of T. I. e. the theory $T \cup \Delta M$ where ΔM is the diagram of M is inconsistent. Hence some finite subset $T' \cup \Delta' M$ of $T \cup \Delta M$ is inconsistent, $T' \subseteq T$, $\Delta' M \subseteq \Delta M$. Then T' is not model-consistent with S.

Q.E.D.

II. Hulls of Theories

1. G-hulls

Let T be any theory and $G \subseteq L(T)$ be some set of formulas, which at least contains all closed $\forall 1$-formulas.
The G-hull T^G of T is defined as the theory S which fulfils

(a) $T^* \cap G \subseteq S^*$
(b) $S \subseteq G$
(c) S and T are mutually model consistent
(d) S contains any theory fulfilling (a), (b) and (c).

Notice that if T^G exists, it is uniquely determined by T and G (up to equivalence of theories). T^G does not always exist. For example let T be the theory of dense orderings (without specifications about extreme elements) and let $G = L(T)$. Let S_1 be the theory of dense orderings having no extreme elements and S_2 be the theory of dense orderings, each having, say, a smallest but no greatest element. Then S_1 and S_2 fulfil (a), (b) and (c) but there cannot possibly exist any consistent theory T^G containing S_1 and S_2, since S_1 and S_2 contradict each other.

2. Lemma

If T is the $\forall n$-hull of S and if for $\sigma, \tau \in \forall(n+1)$, $T \cup \{\sigma\}$ and $T \cup \{\tau\}$ are both model consistent with T, then $T \cup \{\sigma\} \cup \{\tau\}$ is model consistent with T.

Proof:
T is equivalent to $T^* \cap \forall n$ because T is $\forall n$-hull of S. Any extension of T which is model-consistent with T contains the same $\forall n$-sentences as T^*. Hence
$T^* \cap \forall n = (T \cup \{\sigma\})^* \cap \forall n = (T \cup \{\tau\})^* \cap \forall n$.
By application of remark (5 a) to the case $F = \forall(n-1)$, $\forall \neg F = \forall n$ it follows that any model of T is $\forall(n-1)$ - substructure of some model of $T \cup \{\sigma\}$, and vice versa. Thus we can find for any model M of T a chain of $\forall(n-1)$-substructures
$M \prec_{n-1} M_1 \prec_{n-1} N_1 \prec_{n-1} M_2 \prec_{n-1} N_2 \prec_{n-1} \ldots$
where $M_i \models (T \cup \{\sigma\})^* \cap \forall n$ and $N_i \models (T \cup \{\tau\})^* \cap \forall n$, $i > 0$.
Since $T \cup \{\sigma\}$ is $\forall(n+1)$-axiomatizable it follows that $\bigcup_{i=1}^{\infty} M_i$ is a model of $T \cup \{\sigma\}$ (cf. [C.K], lemma 3.1.15).

Similarly $\bigcup_{i=1}^{\infty} N_i$ is a model of $T \cup \{\tau\}$. Since $\bigcup_{i=1}^{\infty} M_i = \bigcup_{i=1}^{\infty} N_i$ it follows that any $M \models T$ is contained in some model $\bigcup_{i=1}^{\infty} M_i$ of $T \cup \{\sigma\} \cup \{\tau\}$, i.e. $T \cup \{\sigma\} \cup \{\tau\}$ is model-consistent with T.

Q.E.D.

3. Corollary

If T is the $\forall n$-hull of S, then the $\forall(n+1)$-hull $T^{\forall(n+1)}$ of T exists and $T^{\forall(n+1)} = \{\sigma \in \forall(n+1) \mid T \cup \{\sigma\}$ is model-consistent with T$\}$.

Proof: Any finite subset of $T' = \{\sigma \in \forall(n+1) \mid T \cup \{\sigma\}$ is model-consistent with T$\}$ is model-consistent with T, according to lemma 2. Then T' is model-consistent with T (cf. Lemma I.19). Thus conidtion II 1(c) is obviously fulfilled. Conditons II 1(a) and (b) hold, too. It is also obvious that any theory fulfilling II 1(a), (b), (c) is a subtheory of T' (cf. II 1(d)). Thus $T' = T^{\forall(n+1)}$.

Q.E.D.

4. Inductive hull $T^{\forall\exists}$ of T.

Let G be the set of all $\forall\exists$-sentences of L (T).
Then $T^{\forall\exists}$ coincides with the inductive hull of Kaiser (cf. [K])

$T^{\forall\exists}$ exists always and can be described by

$T^{\forall\exists} = \{\sigma \in \forall\exists \mid T_\forall \cup \{\sigma\}$ is model-consistent with $T\}$.

This follows from corollary 3, considering that $T^{\forall\exists} = (T_\forall)^{\forall\exists}$

and that $T_\forall$ is the $\forall 1$-hull of T.

5. Algebraic hull $T^{\forall\neg A}$ of T

For $F = A$ (cf. I. 2d and I.8) the $\forall\neg A$-hull of T will be called the algebraic hull of T. Since $\forall\neg A \subseteq \forall\exists$ it follows from II.4 that

$T^{\forall A} = \{\sigma \in \forall A \mid T_\forall \cup \{\sigma\}$ is model-consistent with $T_\forall\}$

i.e. $T^{\forall\neg A} = T^{\forall\exists} \cap \forall\neg A$

6. ω-injective hull $T^{\forall\neg I}$ of T_Λ

For $F = I$ (cf. I.9) the $\forall\neg I$-hull of the theory of Λ-left-modules will be called the ω-injective hull of T_Λ.

Similarly to II.5 we get

$T^{\forall\neg I} = T^{\forall\exists} \cap \forall\neg I = \{\sigma \in \forall\neg I \mid T_\Lambda \cap \{\sigma\}$ is model-consistent with $T_\Lambda\}$.

7. The finite forcing companion

The finite forcing companion T^f (cf. [Ba.-R]) of any theory T can be constructed by an iteration process on theories (cf. [H]), at each step taking the appropriate hull of the theory already constructed:

Let T be any theory

a) Recursive definition of T_n:

Set $T_1 = T_\forall$ i.e. T_1 is the $\forall 1$-hull of T and

$T_{n+1} = \forall (n+1)$-hull of T_n.

The existence of T_n, $n > o$, follows from corollary 3.

b) The theory $\bigcup_{n=1}^{\infty} T_n$ is model-consistent with T. This follows from lemma 1.19.

c) It has been shown by P. Henrard that $T^f = \bigcup_{n=1}^{\infty} T_n$. (cf. [H])

A semantical descriptien of the theories T^f and T_n, in the special case that there are "enough" generic structures, has been given in $[Si_1]$. It is closely connected to the following theorem 8.

8. Theorem

If the class $\mathcal{F}(T)$ of F-closed structures with respect to T is model-consistent with the model class $\mathfrak{M}(T)$ of T then the $\forall\neg$ F-hull $T^{\forall\neg F}$ of T exists and fulfils:

$T^{\forall\neg F} \sim Th(\mathcal{F}(T)) \cap \forall\neg F$ where $Th(\mathcal{F}(T))$ denotes the set of all sentences holding in all F-closed structures.

Proof : Set $S = Th(\mathcal{F}(T)) \cap \forall\neg F$.

Verification of condition 1 (a) : $T^* \cap \forall\neg F \subseteq S^*$.

Assume that $\varphi(\vec{x}) \in F$ and that $\forall \vec{x} \neg \varphi(\vec{x}) \in T^* \cap \forall\neg F$. If $\forall \vec{x} \neg \varphi(\vec{x})$ were not in S, then there would exist some F-closed structure M with $M \models \neg\forall \vec{x} \neg \varphi(\vec{x})$ i.e. $M \models \varphi(\vec{m})$ for some $\vec{m}$ in M. Since there is some model $N \models T$ with $M \subseteq N$, and M is F-closed we would have $N \models T$ and $N \models \varphi(\vec{m})$, in contradiction to $\forall \vec{x} \neg \varphi(\vec{x}) \in T^* \cap \forall\neg F$.

Condition 1(b): $S \subseteq \forall\neg F$ is obviously fulfilled.

Verification of condition 1(c) :

According to I. 16 it suffices to show that $T_\forall = S_\forall$.

$T_\forall \subseteq S_\forall$ holds, since any F-closed structure is a model of $T_\forall$.

$S_\forall \subseteq T_\forall$ follows from the fact that $\mathcal{F}(T)$ is model-consistent with $\mathfrak{M}(T)$, i.e. with $\mathfrak{M}(T_\forall)$.

It remains to be shown that any theory T' meeting the conditions 1 (a), (b), (c) is contained in S*.

Since T' is $\forall\neg$ F-axiomatizable, it suffices to verify that any sentence $\forall \vec{x} \neg \varphi(\vec{x}) \in (T)^*$, $\varphi \in F$, holds in every F-closed structure M. If not, we would have $M \models \varphi(\vec{m})$ for some F-closed structure M, $\vec{m}$ in M. Since $M \subseteq N$ for some model N of T' (condition 1 (c)) and $T^* \cap \forall\neg F \subseteq T'$ (condition 1 (a)) we would have $N \models \varphi(\vec{m})$, in contradiction to $\forall \vec{x} \neg \varphi(\vec{x}) \in (T)^*$.

Q.E.D.

9. Examples

a) For the inductive hull and existentially closed structures (cf. I.7., II. 4) we get

$$T^{\forall\exists} = \mathrm{Th}\ (\mathcal{E}(T)) \cap \forall\exists$$

according to I.17 and II.8. This has already been shown in [E.S.] (cf. corollary 7.16).

b) For the algebraic hull and the algebraically closed structures we get

$$T^{\forall\neg A} = \mathrm{Th}\ (\mathcal{O}\!\mathcal{l}(T)) \cap \forall\neg A = \mathrm{Th}\ (\mathcal{E}(T)) \cap \forall\neg A = T^{\forall\exists} \cap \forall\neg A,$$

according to I.17, II.5 and II.8.

c) For the ω-injective hull $T_\Lambda^{\forall\neg I}$ of T_Λ we get $T_\Lambda^{\forall\neg I} = \mathrm{Th}\ (\mathcal{J}_\omega) \cap \forall\neg I = T^{\forall\exists} \cap \forall\neg I$. $\mathcal{J}_\omega$ is model-consistent with $\mathfrak{M}(T_\Lambda)$ since the subclass of $\mathcal{J}_\omega$ consisting of injective Λ-modules is model-consistent with $\mathfrak{M}(T_\Lambda)$ (cf. II.6).

d) The P-hull $T_\Lambda^{\forall\neg P}$ of T_Λ fullfills

$$T_\Lambda^{\forall\neg P} = \mathrm{Th}\ (\mathcal{P}(T_\Lambda)) \cap \forall\neg P.$$

$\mathcal{P}(T_\Lambda)$ is model-consistent with $\mathfrak{M}(T_\Lambda)$, since the subclass of $\mathcal{P}(T_\Lambda)$ consisting of injective modules is model-consistent with $\mathfrak{M}(T_\Lambda)$ (cf. I. 2e, I.1o).

III. Elementary class of closed structures

1. Elementary classes

A class $\mathfrak{M}$ of L-structures is called an elementary class if it consists of all models of some theory T (cf. [C.K.] p. 173). It is well known that $\mathfrak{M}$ is an elementary class if and only if it is closed with respect to ultraproducts and elementary substructures.

2. Lemma

The class $\mathcal{F}(T)$ of F-closed structures with respect to T is an elementary class if and only if ultraproducts of F-closed structures are F-closed, too.

Proof : This follows from III.1. considering that elementary substructures of F-closed structures are F-closed (cf. I.14).

Q.E.D.

3. Theorem

Let F and F' be given such that $\forall 1 \supseteq F \subseteq F'$, and assume that $T_\forall$ has the amalgamation property. Then the class $\mathcal{F}(T)$ of F-closed structures is an elementary class, provided that the class $\mathcal{F}'(T)$ of F'-closed structures is an elementary class which is model-consistent with $\mathcal{F}(T)$ (i.e. with $\mathfrak{M}(T)$).

Proof : According lemma III.1. it is to be shown that $\mathcal{F}(T)$ is closed with respect to ultraproducts. For this purpose let $\{M_i \mid i \in I\}$ be a family of F-closed structures, D an ultrafilter on I, $\varphi(\vec{x}) \in F$ such that $\prod_I M_{i/D} \models \varphi(\vec{m})$ and $N \models T$ such that $\prod_I M_{i/D} \subseteq N$. We are going to veryfy that $N \models \varphi(\vec{m})$.

Since $\vec{\mathcal{F}}(T)$ is model-consistent with $\mathcal{F}(T)$ there is an F-closed structure $M_i' \models T_\forall$ such that $M_i \underset{F}{\subseteq} M_i'$. According to Los' theorem we have $\prod_I M_{i/D} \underset{F}{\subseteq} \prod_I M'_{i/D}$. Since $T_\forall$ has the amalgamation property, there exists $Q \models T_\forall$ such that

$$\begin{array}{ccc} \prod_I M_{i/D} & \underset{F}{\subseteq} & \prod_I M'_{i/D} \\ \text{\rotatebox{90}{$\supseteq$}} & & \text{\rotatebox{90}{$\supseteq$}}\, f \\ N & \subseteq & Q \end{array}$$

is commutative.

where the embedding f is an F'-embedding (as $\vec{\mathcal{F}}(T)$ is closed with respect to ultraproducts). We have $Q \models \varphi(\vec{m})$ because $\varphi \in F \subseteq F'$. Thus $N \models \varphi(\vec{m})$, since $F \subseteq \forall 1$.

Q.E.D.

4. Remark

Theorem 3, as well as its proof, has been condensed from corollary 7.14 of [E.S.], (cf. next example 5 a).

5. Examples

a) If $\mathcal{E}(T)$ is an elementary class then $\mathfrak{A}(T)$ is so, too, provided that $T_\forall$ has the amalgamation property.
E.g. If T has a model-completion then T has the amalgamation property (cf. [E.S.] lemma 2.1) and $\mathcal{E}(T)$ is an elementary class since it is the model class of the model-completion. Thus $\mathfrak{A}(T)$ is an elementary class.

b) If $\mathcal{E}(T_\Lambda)$ is an elementary class (T_Λ has the amalgamation property) then the class $\mathcal{J}_\omega$ of ω-injective structures is an elementary class.

c) In [E.S.] it is shown that $\mathcal{E}(T_\Lambda)$ is an elementary class if and only if $\mathfrak{A}(T_\Lambda)$ is an elementary class, if and only if $\mathcal{J}_\omega$ is an elementary class, if and only if Λ is coherent. Furtheremore, $\mathfrak{A}(T_\Lambda)$ coincides with $\mathcal{J}_\omega$ provided that Λ is coherent.

d) $\mathcal{P}(T_\Lambda)$ is an elementary class if and only if Λ is coherent.
Proof : Let Λ be coherent. Then M is ω-injective if and ony if $M \prec M'$ for some injective M' (cf. [E.S.] lemma 3.17.2). But $M \prec M'$, M' injective, implies M is P-closed (cf. I.1o, I.13).
Conversely if M is P-closed it is obviously ω-injective, since $I \subseteq P$.
Hence M is ω-injective if and only if it is P-closed. Then $\mathcal{P}(T_\Lambda) = \mathcal{J}_\omega$ which is an elementary class (cf. c).
Conversely let $\mathcal{P}(T_\Lambda)$ be an elementary class, then $\mathcal{J}_\omega$ is an elementary class, according to theorem 3. Hence Λ is coherent (cf. c).

Q.E.D.

6. Theorem : Let $F \subseteq \forall 1$ and $T_\forall$ have the amalgamation property. Then $Th(\mathcal{F}(T)) \sim T^{\forall\neg F}$, provided that $\mathcal{F}(T)$ is an elementary class.
Proof : We have $T^{\forall\neg F} \sim Th(\mathcal{F}(T)) \cap \forall\neg F$ according to theorem II.8 and theorem I.17. It remains to be shown that $Th(\mathcal{F}(T))$ is $\forall\neg F$ -axiomatizable. This holds, according to remark I.5, if $\mathcal{F}(T) = \mathfrak{M}(Th(\mathcal{F}(T)))$ is closed with respect to taking F-substructures.

But this follows from lemma I.15.

Q.E.D.

7. Remark

The conclusion of Theorem 6 holds for $F = \forall\exists$, without assuming the amalgamation property for $T_\forall$, since Th ($\mathcal{E}$ (T)) is the model-companion of T (provided that $\mathcal{E}$(T) is an elementary class, cf. [E.S.] Corollary 7.13).

8. Examples

If the class $\mathfrak{A}$(T) of all algebraically closed structures is an elementary class then Th ($\mathfrak{A}$(T)) can be axiomatized by all those formulas of the form $\forall \vec{x} \exists \vec{y} (\chi (\vec{x}) \Rightarrow \psi (\vec{x},\vec{y}))$ (where χ, ψ are quantifierfree and ψ positive) which belong to the inductive hull $T^{\forall\exists}$, provided that $T_\forall$ has the amalgamation property (cf. also II.7.b).

For $T = T_\Lambda$, where Λ is a coherent ring, this gives us a set of axioms for the absolutely pure Λ-modules (cf. I.8 and III. 5 c), which then actually coincide with the ω-injective modules (cf. [E.S.], proposition 3. 23).

IV. Non elementary classes of closed structures

1. Lemma

Let $\mathfrak{M}$ be any class of L-structures. Any model N of Th ($\mathfrak{M}$) is an elementary substructure of some ultraproduct of structures in $\mathfrak{M}$ (cf. [S_2], proposition 3).

2. Lemma

Let $F \subseteq \forall 1$. Any model $M \models$ Th ($\mathcal{F}$(T)) is an F-substructure of some $N \models$ Th ($\mathcal{E}$ (T)).

Proof : According to lemma 1 there are F-closed structures M_i, $i \in I$ and an ultrafilter D on I, such that $M \prec \prod_I M_i/_D$. Furthermore for each $i \in I$ there is some existentially closed structure N_i such that $M_i \subsetneq N_i$, hence $M_i \underset{F}{\subseteq} N_i$. According to Los' theorem

$\prod_I M_i/_D \underset{F}{\subseteq} \prod_I N_i/_D$.

Since $N = \prod_I N_i/_D \models Th(\mathcal{E}(T))$ we have $M \underset{F}{\subseteq} \models Th(\mathcal{E}(T))$

Q.E.D.

3. Lemma

If $T^{\forall\exists}$ is complete then $T^{\forall\exists} \sim Th(\mathcal{E}(T))$.

Proof : We have

$T^{\forall\exists} \sim Th(\mathcal{E}(T)) \cap \forall\exists \sim Th(\mathcal{E}(T))$.

The first equivalence holds because of theorem II.7. ,the second one because $Th(\mathcal{E}(T)) \cap \forall\exists$ is a complete subtheory of $Th(\mathcal{E}(T))$.

Q.E.D.

4. Theorem

Let $F \subseteq \forall 1$ and $T^{\forall\exists}$ be a complete theory. Then $Th(\mathcal{F}(T)) \sim T^{\forall\neg F}$

Proof : $M \models Th(\mathcal{F}(T))$ if and only if $M \underset{F}{\subseteq} N \models Th(\mathcal{E}(T))$

(according to lemma IV.2 and I.14). The latter holds if and only if $M \models (Th(\mathcal{E}(T)) \cap \forall\neg F$ (cf. I.5) i.e.

$M \models T^{\forall\exists} \cap \forall\neg F$ (cf. IV.3).

But $T^{\forall\exists} \cap \forall\neg F \sim T^{\forall\neg F}$ according to II.5.

Q.E.D.

5. Lemma

(cf. [S_1] , corollaire 1, p. 911). Two Λ -modules are elementary equivalent if and only if both fulfill precisely the same set of $\forall\exists$ -sentences.

6. Lemma

(cf. [S_3] , corollaire 1, p. 1291). Any Λ -module is existentially closed if and only if it is finitely generic.

7. Corollary

Th ($\mathcal{E}(T_\Lambda)$) is complete.

Proof : Th ($\mathcal{E}(T_\Lambda)$) is equivalent to the finite forcing companion T^f_Λ of T_Λ , according to lemma 6 (cf. [Si_2] theorem 1). The latter is complete since $\mathfrak{M}(T_\Lambda)$ has the joint embedding property (cf. [Ba.R] theorem 4.6).

Q.E.D.

8. Theorem

The theory Th ($\mathcal{E}(T_\Lambda)$) of the class $\mathcal{E}(T_\Lambda)$ of all existentially closed Λ -modules is inductive, i.e. Th ($\mathcal{E}(T)$) ~ Th ($\mathcal{E}(T)) \cap \forall\exists$.

Proof : Let $M\forall\exists N$ ($M\exists N$) abreviate the statements that N fulfills the $\forall\exists$ sentences ($\exists$-sentences) which hold in M. Let M be existentially closed and let N be such that $M\forall\exists N$. If we can show that $N\forall\exists M$, then it follows that $M \equiv N$, according to lemma 5, and N is a model of Th ($\mathcal{E}(T_\Lambda)$). Hence Th ($\mathcal{E}(T_\Lambda)$) is $\forall\exists$-axiomatizable.
Thus it is left to be shown that $N\forall\exists M$. Since $M\forall\exists N$ we habe $M\exists N$. Hence there exists an N' such that $M \subseteq N' \succ N$ (cf. [B.S] lemma 9.38). Since M is existentially closed we have $N'\forall\exists M$. Hence $N\forall\exists M$.

Q.E.D.

9. Proposition

The inductive hull $T_\Lambda^{\forall\exists}$ of the theory T_Λ of Λ-modules is a complete theory.

Proof :

We have $\mathrm{Th}\ (\mathcal{E}(T_\Lambda)) \sim \mathrm{Th}\ (\mathcal{E}(T_\Lambda)) \cap \forall\exists$ (cf. theorem 8) and

$\mathrm{Th}\ (\mathcal{E}(T_\Lambda)) \cap \forall\exists \sim T_\Lambda^{\forall\exists}$ (cf. II.7) Q.E.D.

Combining theorem 4 and proposition 9 we get

1o. Theorem

For the theory T_Λ of Λ-modules we have $\mathrm{Th}\ (\mathcal{F}(T_\Lambda)) \sim T_\Lambda^{\forall\neg F}$, for $F \subseteq \forall 1$.

11. Absolutely pure modules

The set of axioms for the theory $\mathrm{Th}\ (\mathfrak{A}(T_\Lambda)$ of all absolutely pure Λ-modules which we got in III. 8 does the job also without the assumption of Λ being coherent, according to theorem 1o.

12. ω -injective modules

According to theorem 1o we have

$\mathrm{Th}\ (\mathcal{J}_\omega) \sim T^{\forall\neg I} \sim T^{\forall\exists} \cap \forall\neg I$ (cf. II. 7c). One gets a set of axioms for the theory $\mathrm{Th}\ (\mathcal{J}_\omega)$ of ω-injective Λ-modules by taking all the formulas

$\forall x_1 \ldots \forall x_n \exists y_1 \ldots \exists y_m (\chi(x_1,\ldots,x_n,y_1,\ldots,y_m) \Rightarrow \psi(x_1,\ldots,x_n,y_1,\ldots,y_m))$

which belong to $T^{\forall\exists}$, where χ is quantifierfree and $\psi\ (x_1, \ldots, x_n, y_i, \ldots, y_m)$ is either a conjunction of some formulas $\lambda_{ij}\, x_i = y_j$, $\lambda_{ij} \in \Lambda$, or of the form $x_1 = x_1$.

Another axiom system, in the special case that Λ is coherent, has been given in [E.S.] p. 263.

Notice that our technique works for α-injectives too, if $\alpha < \omega$.

References

[B.S.] Bell, J.L and Slomson, A.B.
Models and Ultraproducts.
North Holland Publ. Co. 1969

[Ba.R.] Barwise, J. and Robinson, A.
Completing Theories by Forcing.
Ann. Math. Logic 2 (2), 197o, 119-142

[C.] Cherlin, G.L.
The Model Companion of a Class of Structures.
The Journ. of Symb. Logic 27 (3), 1971, 546-556

[C.K.] Chang, C.C. and Keisler, H.J.
Model Theory.
North Holland Publ. Co. 1973

[E.S.] Eklof, P. and Sabbagh, G.
Model Completions and Modules.
Ann. Math. Logic 2 (3), 1971, 251-295

[F.] Fittler, R.
Some Categories of Models.
Arch. math. Logik Grundlagenforsch. 15, 1972, 179-189

[H.] Henrard, P.
Le forcing-compagnon sans forcing.
C.R. Acad. Sc. Paris, 276, Serie A, 1973, 821-822

[K.] Kaiser, K.
Über eine Verallgemeinerung der Robinsonschen Modellvervollständigung I.
Z. Math. Logik Grundl. Math. 15, 1969, 37-48

[M.] Macintyre, A.
On Algebraically Closed Groups.
Ann. of Math. 96, 1972, 53-97

[M_2] Macintyre, A.
Lecture Notes on Forcing in Model Theory.
Freie Universität Berlin, 1972 (mimeographed).

[R_1] Robinson, A.
Introduction to Model Theory and the Metamathematics of Algebra.
North Holland Publ. Co. 1965

[R_2] Robinson, A.
Infinite Forcing in Model Theory.
Proc. Sec. Scand. Logic Symp. 197o
North Holland Publ. Co 1971

[S_1] Sabbagh, G.
Aspects logiques de la pureté dans les mudules.
C.R. Acad. Sc. Paris, 271, Série A, 197o, 9o9-912

[S_2] Sabbagh, G.
A Note on the Embedding Property,
Math. Z. 121, 1971, 239-242

[S_3] Sabbagh, G.
Sous-modules purs, existentiellement clos et élémentaires.
C.R. Acad. Sc. Paris, 272, Série A, 1971, 1289-1292

[Si_1] Simmons, H.
Existentially Closed Structures.
The Journ. of Symb. Logic 37 (2), 1972, 293-31o

[Si_2] Simmons, H.
A Possible Characterization of Generic-Structures.
Math. Scand. 31, 1972, 257-261

AXIOMS OF CHOICE IN MORSE-KELLEY CLASS THEORY

by T. B. Flannagan

SECTION 1
INTRODUCTION

Morse-Kelley class theory M, first formulated in Wang [14] as a natural extension of von Neumann-Bernays-Gödel set-theory NBG, is written in the first-order language (with equality) whose only predicate is the 2-place predicate $\in$. We denote this language by $\mathcal{L}_M$. Variables (called class-variables) are denoted by capital Roman letters X, Y, Z with or without subscripts. Those variables restricted to the predicate $Z(\cdot)$ defined by $Z(X) \leftrightarrow \exists Y(X \in Y)$ are called set-variables and denoted by small Roman letters x, y, z, u, v with or without subscripts. Formulae of $\mathcal{L}_M$ are denoted by capital Greek letters Φ, Ψ etc..

The non-logical axioms of M are: I. the axiom of extensionality; II. the axiom of pairs for sets; III. the sum-set axiom; IV. the power-set axiom; V. the axiom of infinity; VI. the axiom of foundation; VII. the axiom of replacement, and the following impredicative comprehension schema.

VIII. $$\forall X_1,\dots,X_n \exists Y \forall z(z \in Y \leftrightarrow \Phi(z,X_1,\dots,X_n))$$

As usual, we denote by $\{z : \Phi(z)\}$ the unique class Y such that $\forall z(z \in Y \leftrightarrow \Phi(z))$.

The word impredicative is used to describe VIII since VIII is a schema which asserts the existence of a class even when the defining formula Φ itself contains bound class-variables. Thus, M may be regarded as an extension of NBG obtained from NBG by allowing bound class-variables to appear in the specified formula in the comprehension schema.

The superscript ^{o} will always denote the absence of the axiom of foundation (VI), and the superscript $^{-}$ will always denote the absence of the power-set axiom. Thus, for example, $M = M^{o} + VI$ and $NBG = NBG^{o-} + VI + IV$.

We do not regard the following strong axiom of choice (axiom IX of Marek [10]) as an axiom of M:

C^V : $\forall x \exists Y \Phi(x,Y) \rightarrow \exists Z \forall x \Phi(x,Z^{(x)})$,

where $Z^{(x)}$ is defined as $\{y : \langle x,y\rangle \in Z\}$.

The following definitions, like the one above, are fundamental to the sequel.

DEFINITION 1.1 (X is a <u>pair-class</u> or <u>relation</u>) $\mathrm{Rel}(X) \leftrightarrow \forall x \in X \exists y,z(x = \langle y,z\rangle)$.

DEFINITION 1.2 The <u>domain</u> of X , $\mathcal{D}(X) = \{x : \exists y(\langle x,y\rangle \in X)\}$.

DEFINITION 1.3 The <u>range</u> of X , $\mathcal{R}(X) = \{y : \exists x(\langle x,y\rangle \in X)\}$.

DEFINITION 1.4 The <u>field</u> of X , $\mathcal{F}(X) = \mathcal{D}(X) \cup \mathcal{R}(X)$.

DEFINITION 1.5 $X \eta Y \leftrightarrow \exists x(X = Y^{(x)})$

DEFINITION 1.6 $X \mathrm{Inc}\, Y \leftrightarrow \forall Z(Z \eta X \rightarrow Z \eta Y)$

DEFINITION 1.7 $X \simeq Y \leftrightarrow X \mathrm{Inc}\, Y \wedge Y \mathrm{Inc}\, X$

If X is a relation, then since for any $y \in \mathcal{D}(X)$ there is a unique Y such that $Y = X^{(y)}$, we shall refer to X as a <u>class-valued function</u>, $X^{(y)}$ being the <u>class-valued image</u> of y under X. Moreover, if $\mathcal{D}(X)$ is a set, then we call X a <u>set of classes</u> ; but in general we call it a <u>class of classes</u>. Despite the confusion with Definition 1.14 below, when it is clear that X is a class-valued function, we shall often just call it a function. The letters F, G, H, J and K will be used as variables for relations.

Although in M we cannot form the totality of classes which satisfy a given formula, we often use the notation $\{X : \Phi(X)\}$ when talking informally about the collection of classes X such that $\Phi(X)$ holds.

DEFINITION 1.8 If Φ is a formula with one free variable and $\exists X \forall Y(Y \eta X \leftrightarrow \Phi(Y))$ holds, then we say that the collection $\{Y : \Phi(Y)\}$, or simply Φ, is coded by X. Of course X is only unique up to $\simeq$.

DEFINITION 1.9 $F \approx G \leftrightarrow \mathrm{Rel}(F) \wedge \mathrm{Rel}(G) \wedge \mathcal{D}(F) = \mathcal{D}(G) \wedge \forall x \in \mathcal{D}(F)(F^{(x)} \simeq G^{(x)}$

DEFINITION 1.10 (X is a well-ordering) $\mathrm{Bord}(X) \leftrightarrow \mathrm{Rel}(X) \wedge \forall y \in \mathcal{F}(X)(\langle y,y\rangle \in X)$ $\wedge \forall y,z(\langle y,z\rangle \in X \wedge \langle z,y\rangle \in X \rightarrow y = z) \wedge \forall x,y,z(\langle x,y\rangle \in X \wedge \langle y,z\rangle \in X \rightarrow$ $\langle x,z\rangle \in X) \wedge \forall y,z(y \in \mathcal{F}(X) \wedge z \in \mathcal{F}(X) \rightarrow \langle y,z\rangle \in X \vee \langle z,y\rangle \in X) \wedge$ $\forall Y \subseteq \mathcal{F}(X)(Y \neq \emptyset \rightarrow \exists x \in Y \forall y \in Y(\langle x,y\rangle \in X))$.

The letter T will always denote a well-ordering.

DEFINITION 1.11 $x <_T y \leftrightarrow \langle x,y\rangle \in T$

DEFINITION 1.12 (The initial segment of $\mathcal{F}(T)$ determined by x.) $0_T(x) = \{y : y \in \mathcal{F}(T) \wedge y <_T x\}$

DEFINITION 1.13 $T \upharpoonright x = T \cap (0_T(x))^2$

DEFINITION 1.14 (X is a function.) $\mathrm{Func}(X) \leftrightarrow \mathrm{Rel}(X) \wedge \forall y \in \mathcal{D}(X) \exists! z(\langle y,z\rangle \in X)$

DEFINITION 1.15 $(F : X \rightarrow Y) \leftrightarrow \mathrm{Func}(F) \wedge \mathcal{D}(F) = X \wedge \mathcal{R}(F) \subseteq Y$

DEFINITION 1.16 $(F : X \overset{1-1}{\longleftrightarrow} Y) \leftrightarrow (F : X \rightarrow Y) \wedge \mathcal{R}(F) = Y \wedge$ $\mathrm{Func}(\{\langle x,y\rangle : \langle y,x\rangle \in F\})$

DEFINITION 1.17 $T_1 \overset{H}{\sim} T_2 \leftrightarrow (H : \mathcal{F}(T_1) \overset{1-1}{\longleftrightarrow} \mathcal{F}(T_2)) \wedge \forall x,y(\langle x,y\rangle \in T_1 \leftrightarrow$ $\langle H(x),H(y)\rangle \in T_2)$

DEFINITION 1.18 $T_1 \sim T_2 \leftrightarrow \exists H(T_1 \overset{H}{\sim} T_2)$

DEFINITION 1.19 $T_1 \overset{H}{<} T_2 \leftrightarrow \exists x(T_1 \overset{H}{\sim} T_2 \upharpoonright x)$ and $T_1 < T_2 \leftrightarrow \exists H(T_1 \overset{H}{<} T_2)$

DEFINITION 1.20 $T_1 \leqslant T_2 \leftrightarrow T_1 < T_2 \vee T_1 \sim T_2$

DEFINITION 1.21 Let X be a relation and T a well-ordering such that $\forall y \in \mathcal{D}(X)(\mathrm{Bord}(X^{(y)}) \wedge X^{(y)} < T)$ and $\forall T' < T \exists y(T' \leqslant X^{(y)})$. Then

T is said to have the property Sup X, and we write T = Sup X.

DEFINITION 1.22 If $T_1 \overset{H}{\sim} T_2$, Rel(F), Rel(G), $\mathcal{D}(F) = \mathcal{F}(T_1)$ and $\mathcal{D}(G) = \mathcal{F}(T_2)$, then we write (i) $F \cong G$ if and only if $\forall x \in \mathcal{F}(T_1)(F^{(x)} = G^{(H(x))})$ and (ii) $F \approxeq G$ if and only if $\forall x \in \mathcal{F}(T_1)(F^{(x)} \simeq G^{(H(x))})$.

The following notation will also be useful.

DEFINITION 1.23 Let $\Phi(\cdot)$ be a formula with one free variable. Then in analogy with Definition 1.15 above we write $F : X \longrightarrow \Phi$ to denote that F is a relation, $\mathcal{D}(F) = X$ and $\forall y \in X\, \Phi(F^{(y)})$.

DEFINITION 1.24 If $\approxeq$ defines some equivalence relation between classes, then we write $\overset{\approxeq}{\exists}!X\Phi(X)$ to denote that there is an X such that $\Phi(X)$ and that X is unique up to $\approxeq$; that is $\exists X(\Phi(X) \wedge \forall Y(Y \approxeq X \leftrightarrow \Phi(Y)))$.

DEFINITION 1.25 We write $\tilde{\exists}_\mu T\Phi(T)$ to denote that, to within $\sim$, there is a least well-ordering T such that $\Phi(T)$; that is,

$$\exists T[\Phi(T) \wedge \forall T' < T(\neg\Phi(T')) \wedge \forall T'(T' \sim T \rightarrow \Phi(T'))].$$

The plan of the paper is as follows: In Section 2 we show that class-valued functions can be defined by recursion on well-orderings in much the same way that set-theoretic functions can be defined in Zermelo-Fraenkel set-theory ZF by recursion on ordinals. We also show that functional formulae $\Psi(T,X)$ can be defined by recursion on all the well-orderings just as functional relations $F(\alpha,x)$ can be defined in ZF by recursion on the ordinals.

In Section 3 we formulate some axioms of choice and prove various implications between them.

In Section 4 we show that Fraenkel-Mostowski-Specker methods can be used in M^0 to prove independence results for some of the axioms of choice in Section 3.

In Section 5 we consider the theory M^{0}_{ε} which is obtained from M^{0} by adjoining Hilbert's ε-symbol to $\mathcal{L}_{M}$ and admitting ε-terms $\varepsilon X \Phi$ to the impredicative comprehension schema. In particular, we show that in M^{0}_{ε} it is provable that the universe of sets (V) is well-orderable.

In Section 6 we prove a conservative extension result, the proof of which strengthens a result of Prof. Mostowski. Finally, we ask some questions.

We are grateful to U. Felgner and W. Guzicki for many valuable conversations and to the Alexander von Humboldt-Stiftung for its generous finantial support.

"What more felicity can fall to creature,
Than to enjoy delight with liberty?"

SECTION 2

DEFINITIONS BY RECURSION

In this section we prove two theorems in $M^o + C^V$. The first says that class-valued functions can be defined by recursion on any given well-ordering. The second says that certain formulae $\Psi(T,X)$ which, in a natural sense, are functional on the second coordinate, can be defined by recursion on all the well-orderings.

THEOREM 2.1 Let T be a well-ordering and $\Delta(\cdot)$ a formula with one free variable. Let $\Gamma_T(F)$ denote the formula

$$\forall x \in \mathcal{F}(T)[\mathcal{D}(F) = O_T(x) \wedge \forall z <_T x\, \Delta(F^{(z)}) \quad ,$$

and let $\Phi(\cdot,\cdot)$ be a formula with the following properties:

P1. $\forall X,Y(\Phi(X,Y) \rightarrow \Gamma_T(X) \wedge \Delta(Y))$,

P2. $\forall F,G,X(F \approx G \wedge \Phi(F,X) \rightarrow \Phi(G,X))$,

P3. $\forall F,X,Y(\Phi(F,X) \wedge \Phi(F,Y) \rightarrow X \simeq Y)$ (Φ is functional on the second coordinate with respect to $\simeq$) and

P4. $\forall F(\Gamma_T(F) \rightarrow \exists Y \Phi(F,Y))$.

Then there is a class-valued function F, which is unique up to $\approx$, such that

$$\mathcal{D}(F) = \mathcal{F}(T) \wedge \forall x \in \mathcal{F}(T)\Phi(F\restriction O_T(x), F^{(x)}).$$

<u>Proof.</u> 1. (Uniqueness up to $\approx$). Suppose there are functions F and G which both satisfy the theorem, and that $F \not\approx G$. Then $\exists y \in \mathcal{F}(T)(F^{(y)} \not\simeq G^{(y)})$. Let y_o be the T-le st such y. Then $\forall z <_T y_o(F^{(z)} \simeq G^{(z)})$. Hence, $F\restriction O_T(y_o) \approx G\restriction O_T(y_o)$, so by P3, $F^{(y_o)} \simeq G^{(y_o)}$ — a contradiction.

2. (Existence) Let $\mathcal{J}$ denote the class

$$\{x \in \mathcal{F}(T) : \exists F(\mathcal{D}(F) = O_T(x) \wedge \forall y <_T x\ \Phi(F \restriction O_T(y), F^{(y)}))\}.$$

We first show that $\mathcal{J}$ is an initial segment of $\mathcal{F}(T)$. Let $y <_T x$ and $x \in \mathcal{J}$ and suppose

$$\mathcal{D}(F) = O_T(x) \wedge \forall z <_T x\, \Phi(F \restriction O_T(z), F^{(z)}).$$

Let $G = F \restriction O_T(y)$. Then $\mathcal{D}(G) = O_T(y)$ and

$$\forall z <_T y(G^{(z)} = F^{(z)} \wedge G \restriction O_T(z) = F \restriction O_T(z)),$$

so

$$\forall z <_T y\ \Phi(G \restriction O_T(z), G^{(z)}).$$

Hence, $y \in \mathcal{J}$ and $\mathcal{J}$ is an initial segment of $\mathcal{F}(T)$.

Now

$$\forall x \in \mathcal{J} \exists F(\mathcal{D}(F) = O_T(x) \wedge \forall y <_T x\ \Phi(F \restriction O_T(y), F^{(y)})),$$

so by C^V, there is an H such that

2.1.1 $$\forall x \in \mathcal{J}\ (\mathcal{D}(H^{(x)}) = O_T(x) \wedge \forall y <_T x\ \Phi(H^{(x)} \restriction O_T(x), H^{(x)}(y))).$$

REMARK 2.1.2 Notice that the use of C^V is essential here, for although F is unique up to $\approx$, it is not strictly unique.

Now by 2.1.1 and P4 we have $\forall x \in \mathcal{J}\ \exists X\ \Phi(H^{(x)}, X)$, so by C^V,

2.1.3 $$\exists Y \forall x \in \mathcal{J}\ \Phi(H^{(x)}, Y^{(x)});$$

and by 2.1.1 and uniqueness up to $\approx$,

$$\forall z <_T x(H^{(x)} \restriction O_T(z) \approx H^{(z)}).$$

Hence, by 2.1.1, 2.1.2, P2 and P3,

$$\forall z <_T x(Y^{(z)} \simeq H^{(x)}(z));$$

that is, $Y \restriction O_T(x) \approx H^{(x)}$. Hence, by 2.1.3 and P2,

2.1.4 $$\forall x \in \mathfrak{J}\ \Phi(Y \restriction O_T(x), Y^{(x)}).$$

Now $\mathfrak{J}$ is an initial segment of $\mathfrak{F}(T)$, so it is either $O_T(y)$ for some $y \in \mathfrak{F}(T)$ or $\mathfrak{F}(T)$ itself; but if $\mathfrak{J}$ is $O_T(y)$, then by 2.1.4, $y \in \mathfrak{J}$ – a contradiction. Hence, $\mathfrak{J}$ is $\mathfrak{F}(T)$ and Y is the required function. Q.E.D.

REMARK 2.3 It is easy to see that if " $\simeq$ " in P3, and hence " $\approx$ " in P2, is replaced by " = ", then C^V need not be used in the above proof. However, note that even if the initial segments of $\mathfrak{F}(T)$ are sets, then the proof remains impredicative so long as some of the "values" of the function F being defined are proper classes. Thus, for example, the proof is impredicative if F is defined by recursion on ω with "values" $F^{(n)}$ which are proper classes; but if $\mathfrak{F}(T)$ is On (the class of all ordinals) and the values $F^{(\alpha)}$ of F are sets, then the proof reduces to the usual proof in NBG^o.

In order to prove the next theorem, we need the following lemma:

LEMMA 2.4 Let Φ be a formula such that $\exists T\, \Phi(T)$ and $T_1 \sim T_2 \wedge \Phi(T_1) \rightarrow \Phi(T_2)$. Then $\tilde{\exists}_\mu T\, \Phi(T)$.

<u>Proof</u>. Suppose $\Phi(T_o)$. If there is no least T such that $\Phi(T)$, then since Φ is closed under $\sim$, there exists $x \in \mathfrak{F}(T_o)$ such that $\Phi(T_o \restriction x)$, but no least such x – a contradiction since T_o is a well-ordering.

THEOREM 2.5 Let $\Delta(\cdot)$ be a formula with one free variable and $\Gamma(F)$ denote the formula

$$\exists T(\mathfrak{D}(F) = \mathfrak{F}(T) \wedge \forall y \in \mathfrak{F}(T)\, \Delta(F^{(y)})).$$

Let $\Phi(\cdot,\cdot)$ be a formula with the following properties:

P1'. $\forall X,Y(\Phi(X,Y) \rightarrow \Gamma(X) \wedge \Delta(Y))$,

P2'. $\forall F,G,X(F \cong G \wedge \Phi(F,X) \rightarrow \Phi(G,X))$

P3'. $\forall F,X,Y(\Phi(F,X) \wedge \Phi(F,Y) \rightarrow X \simeq Y)$ and

P4'. $\forall F(\Gamma(F) \rightarrow \exists Y\,\Phi(F,Y))$.

Then there is a formula $\Psi(\cdot,\cdot)$ with the following properties:

Q1. $\forall T,X(\Psi(T,X) \leftrightarrow \exists F(\Phi(F,X) \wedge \mathfrak{D}(F) = \mathcal{F}(T) \wedge \forall x \in \mathcal{F}(T)\Psi(T\ x,F^{(x)})))$,

Q2. $\forall T,T',X(T \sim T' \wedge \Psi(T,X) \rightarrow \Psi(T',X))$ and

Q3. $\forall T,X,Y(\Psi(T,X) \wedge \Psi(T,Y) \rightarrow X \simeq Y)$.

<u>Proof.</u> By Theorem 2.1 we have

2.5.1 $$\forall T \overset{\approx}{\exists}! F(\mathfrak{D}(F) = \mathcal{F}(T) \wedge \forall x \in \mathcal{F}(T)\Phi(F \upharpoonright O_T(x),F^{(x)}).$$

Let $\Psi(T,X)$ be the formula

$$\exists F(\Phi(F,X) \wedge \mathfrak{D}(F) = \mathcal{F}(T) \wedge \forall x \in \mathcal{F}(T)\Phi(F \upharpoonright O_T(x),F^{(x)}))$$

Q3 now follows trivially from 2.5.1 and P3'.

<u>The proof of Q2</u>: Assume $T \overset{H}{\sim} T'$ and

$$\Phi(F,X) \wedge \mathfrak{D}(F) = \mathcal{F}(T) \wedge \forall x \in \mathcal{F}(T)\Phi(F \upharpoonright O_T(x),F^{(x)})$$

and put $G = \{\langle H(x),y\rangle : \langle x,y\rangle \in F\}$. Then

2.5.2 $$\mathfrak{D}(G) = \mathcal{F}(T'),$$

2.5.3 $$\forall x \in \mathcal{F}(T)(F^{(x)} = G^{(H(x))})$$

2.5.4 $$F \equiv G \quad \text{and}$$

2.5.5 $$\forall x \in \mathcal{F}(T)(F \upharpoonright O_T(x) \equiv G \upharpoonright O_{T'}(H(x))).$$

By 2.5.3 and P2', $\Phi(G,X)$ holds; and by 2.5.3, 2.5.5 and P2',

$$\forall x \in \mathcal{F}(T)\Phi(G\restriction O_{T'}(H(x)),G^{(H(x))}).$$

Hence, $\Psi(T',X)$ holds; so Q2 holds.

It remains to prove Q1. Let $\Theta(T,X)$ denote the formula

$$\Psi(T,X) \leftrightarrow \Delta^*(T,X),$$

where $\Delta^*(T,X)$ denotes

$$\exists F(\Phi(F,X) \wedge \mathcal{D}(F) = \mathcal{F}(T) \wedge \forall x \in \mathcal{F}(T)\Psi(T\restriction x,F^{(x)}).$$

<u>Claim 2.5.6</u> $T \sim T' \wedge \Delta^*(T,X) \rightarrow \Delta^*(T',X)$

<u>Proof</u>. Assume $T \sim T'$ and

$$\Phi(F,X) \wedge \mathcal{D}(F) = \mathcal{F}(T) \wedge \forall x \in \mathcal{F}(T)\Psi(T\restriction x,F^{(x)}).$$

Then

$$\forall x \in \mathcal{F}(T)(T\restriction x \sim T'\restriction H(x)).$$

Define G as in the proof of Q2. Then, as above, $\Phi(G,X)$ holds. Moreover, by Q2 and 2.5.3,

$$\forall x \in \mathcal{F}(T)\Psi(T'\restriction H(x),G^{(H(x))}),$$

so

$$\forall x \in \mathcal{F}(T')\Psi(T'\restriction x,G^{(x)}).$$

Hence, $\Delta^*(T',X)$ holds; so the claim is proved.

It now follows trivially from Q2 that

2.5.7 $$T \sim T' \wedge \Theta(T,X) \rightarrow \Theta(T',X)$$

By induction on T, we show now that Q1 holds, that is, $\forall T,X\,\Theta(T,X)$.

Case (i). $T = \emptyset$. This is trivial by the definition of Ψ.

Case (ii). $T > \emptyset$. If $\neg\forall T,X\,\Theta(T,X)$, then by 2.5.7 and Lemma 2.4, there is a least $T > \emptyset$, T_o say, such that $\exists X \neg\theta(T,X)$; and T_o is unique to within $\sim$. Then

2.5.8 $$\forall T < T_o \forall X \theta(T,X).$$

Now by 2.5.1, there is a function F_{T_o}, unique up to $\approx$, such that

2.5.9 $$\mathfrak{D}(F_{T_o}) = \mathcal{F}(T_o) \wedge \forall x \in \mathcal{F}(T_o) \Phi(F_{T_o} \upharpoonright O_{T_o}(x), F_{T_o}{}^{(x)}).$$

Moreover,

2.5.10 $$\forall x \in \mathcal{F}(T_o) \overset{\approx}{\exists!} F(\mathfrak{D}(F) = O_{T_o}(x) \wedge \forall y <_{T_o} x\, \Phi(F \upharpoonright O_{T_o}(y), F^{(y)}).$$

Hence, by C^V, there is a J such that

2.5.11 $$\forall x \in \mathcal{F}(T_o)(\mathfrak{D}(J^{(x)}) = O_{T_o}(x) \wedge \forall y <_{T_o} x\, \Phi(J^{(x)} \upharpoonright O_{T_o}(y), J^{(x)(y)})).$$

By uniqueness to within $\approx$, it follows from 2.5.8 and 2.5.10 that for all $x \in \mathcal{F}(T_o)$,

2.5.12 $$J^{(x)} \approx F_{T_o} \upharpoonright O_{T_o}(x).$$

Claim 2.5.13. $\forall x \in \mathcal{F}(T_o) \forall X(\Psi(T_o \upharpoonright x, X) \leftrightarrow \Phi(F_{T_o} \upharpoonright O_{T_o}(x), X))$

Proof. By induction on x. Let x_o be the T_o-least element of $\mathcal{F}(T_o)$.

Case (i)': $x = x_o$. This is trivial by the definition of Ψ.

Case (ii)': $x_o <_{T_o} x$. Assume

$$\forall y <_{T_o} x \forall X(\Psi(T_o \upharpoonright y, X) \leftrightarrow \Phi(F_{T_o} \upharpoonright O_{T_o}(y), X)).$$

Then by 2.5.8, for all $x \in \mathcal{F}(T_o)$ and all X,

$$\Psi(T_o \upharpoonright x, X) \leftrightarrow \exists F(\Phi(F,X) \wedge \mathfrak{D}(F) = O_{T_o}(x) \wedge \forall y <_{T_o} x\, \Phi(F_{T_o} \upharpoonright O_{T_o}(y), F^{(y)}))$$

Now suppose

$$\Phi(F,X) \wedge \mathcal{D}(F) = O_{T_o}(x) \wedge \forall y <_{T_o} x\, \Phi(F_{T_o} \upharpoonright O_{T_o}(y), F^{(y)}).$$

Then by uniqueness up to $\approx$, $F \approx F_{T_o} \upharpoonright O_{T_o}(x)$; so by P2', $\Phi(F_{T_o} \upharpoonright O_{T_o}(x),X)$ holds. Hence,

$$\Psi(T_o \upharpoonright x,X) \rightarrow \Phi(F_{T_o} \upharpoonright O_{T_o}(x),X).$$

The converse is immediate from 2.5.11, 2.5.12 and the definition of Ψ. Case (ii)', and hence Claim 2.5.13, is thus proved.

We now complete the proof of Case (ii) by showing that $\forall x\, \theta(T_o,X)$ holds. Assume $\Psi(T_o,X)$. Then by the definition of Ψ, there is an F such that

2.5.14 $\Phi(F,X) \wedge \mathcal{D}(F) = \mathcal{F}(T_o) \wedge \forall x \in \mathcal{F}(T_o)\Phi(F \upharpoonright O_{T_o}(x),F^{(x)})$

Hence, by 2.5.1 and 2.5.9, $F \approx F_{T_o}$, so

$$\forall x \in \mathcal{F}(T_o)(F \upharpoonright O_{T_o}(x) \approx F_{T_o} \upharpoonright O_{T_o}(x)).$$

By P2', it therefore follows that

$$\forall x \in \mathcal{F}(T_o)\Phi(F_{T_o} \upharpoonright O_{T_o}(x),F^{(x)})$$

Hence,

5.15 $\Psi(T_o,X) \rightarrow \exists F(\Phi(F,X) \wedge \mathcal{D}(F) = \mathcal{F}(T_o) \wedge \forall x \in \mathcal{F}(T_o)\Phi(F_{T_o} \upharpoonright O_{T_o}(x),F^{(x)})$

Now assume

2.5.16 $\Phi(F,X) \wedge \mathcal{D}(F) = \mathcal{F}(T_o) \wedge \forall x \in \mathcal{F}(T_o)\Phi(F_{T_o} \upharpoonright O_{T_o}(x),F^{(x)})$.

Then by P3',

$$\forall x \in \mathcal{F}(T_o)(F^{(x)} \simeq F_{T_o}{}^{(x)}),$$

so $F \cong F_{T_o}$. Hence, by P2' and 2.5.16,

$$\forall x \in \mathcal{F}(T_o)\Phi(F \restriction O_{T_o}(x), F^{(x)}),$$

so the converse of 2.5.15 holds. Hence, by Claim 2.5.13, $\theta(T_o, X)$ holds, so Q1 is proved. Q.E.D.

REMARK 2.6. (i) By Remark 2.3, it follows that if "$\simeq$" is replaced by " = " in P3' and the definition of $F \cong G$ (Defn. 1.22), then C^V need not be used in the proof of Theorem 2.5. As far as we know, this has not been previously noticed.

(ii) If, in addition to having the properties P1' - P4', Φ has the property

$$\forall F, X, Y(X \simeq Y \wedge \Phi(F, X) \rightarrow \Phi(F, Y)),$$

then it follows trivially from the definition of Ψ that

Q4. $\forall T, X, Y(X \simeq Y \wedge \Psi(T, X) \rightarrow \Psi(T, Y))$.

Furthermore, if Φ is functional on the first coordinate with respect to $\cong$, that is,

$$\forall F, G, X(\Phi(F, X) \wedge \Phi(G, X) \rightarrow F \cong G),$$

then it follows trivially from Q1 that

Q5. $\forall T, T', X(\Psi(T, X) \wedge \Psi(T', X) \rightarrow T \sim T')$.

If both Q4 and Q5 hold, then Ψ is what Marek [10] calls a <u>sequence</u>.

SECTION 3

IMPLICATIONS BETWEEN SOME AXIOMS OF CHOICE

We first formulate, as schemata in $\mathcal{L}_M$, two strong principles of dependent choices: $DCColl_1^T$ and $DCColl_2^T$. We then show them to be equivalent in NBG^o, and so refer to them both as $DCColl^T$. The formulation of these schemata is more intimidating than the idea expressed in both, which is roughly that if T is a well-ordering and $\{X : \Phi(X)\}$ is a collection of classes, then T-many dependent choices of classes can be made from the collection. Later, in Remark 3.4 (v), we shall see that $DCColl^T$ is the broadest possible generalization in $\mathcal{L}_M$ of Lévy's well-known principle of dependent choices DC^α, which involves making α-many dependent choices of elements from a set, where α is a cardinal.

Most of the implications in this section are proved in NBG^o, so they also hold in M^o.

$DCColl_1^T$: Let T be a well-ordering, x_o the T-least element of the field of T, $\{X : \Phi(X)\}$ a collection of classes, and $\pi(\cdot,\cdot)$ a formula with two free variables. Then

$$\forall x \in \mathcal{F}(T) \forall F[(F : O_T(x) \rightarrow \Phi) \rightarrow \exists Y(\Phi(Y) \wedge \pi(F,Y))] \rightarrow$$
$$\forall X[\Phi(X) \rightarrow \exists F[(F : \mathcal{F}(T) \rightarrow \Phi) \wedge F^{(x_o)} = X \wedge \forall x \in \mathcal{F}(T)(x_o <_T x \rightarrow$$
$$\pi(F \restriction O_T(x), F^{(x)}))]]$$

$DCColl_2^T$: Let T and x_o be as above, $\{X : \Phi(x,X)\}_{x \in \mathcal{F}(T)}$ be a collection (indexed by $\mathcal{F}(T)$) of collections of classes, and $\prec$ be a definable partial-ordering of the collection $\{X : \Phi^*(X)\}$, where $\Phi^*(X)$ denotes $\exists x \in \mathcal{F}(T)\Phi(x,X)$. Then

$$\forall x \in \mathcal{F}(T)\forall F[(F : O_T(x) \to \Phi^*) \wedge \forall y,z(y <_T x \wedge z <_T x \to (\Phi(y,F^{(y)}) \wedge (y <_T z \to F^{(y)} \prec F^{(z)}))) \to \exists X(\Phi(x,X) \wedge \forall y <_T x(F^{(y)} \prec X))] \to$$

$$\forall X[\Phi(x_o,X) \to \exists F[(F : \mathcal{F}(T) \to \Phi^*) \wedge F^{(x_o)} = X \wedge \forall x,y(x_o <_T x \wedge x_o <_T y \to (\Phi(x,F^{(x)}) \wedge (x <_T y \to F^{(x)} \prec F^{(y)})))]]$$

LEMMA 3.1 $NBG^o \vdash DCColl_2^T \to DCColl_1^T$

<u>Proof</u>. Assume the hypothesis of $DCColl_1^T$ and suppose $\Phi(X_o)$ holds. For $x \in \mathcal{F}(T)$, define

3.1.1 $$\Psi(x,F) \leftrightarrow_{Df} \begin{cases} F = \{\langle x_o,y\rangle : y \in X_o\} & \text{if } x = x_o; \\ (F : O_T(x) \to \Phi) \wedge \forall y <_T x\,\Pi(F \restriction O_T(y),F^{(y)}) & \text{if } x \neq x_o \end{cases}$$

By the hypothesis of $DCColl_1^T$, there is an X, X_1 say, such that $\Phi(X) \wedge \Pi(\emptyset,X)$. Define

$$F \prec G \leftrightarrow_{Df} \exists x_1,x_2(x_1 <_T x_2 \wedge \Psi(x_1,F) \wedge \Psi(x_2,F) \wedge \forall y \in \mathcal{D}(F)(F^{(y)} = G^{(y)}));$$

that is,

$$F \prec G \leftrightarrow \exists x_1,x_2(\Psi(x_1,F) \wedge \Psi(x_2,G) \wedge \text{" } F \text{ is an initial segment of } G \text{ "}).$$

Clearly, $\prec$ is a partial-ordering of the collection $\{F : \Psi^*(F)\}$, where $\Psi^*(F)$ denotes $\exists x \in \mathcal{F}(T)\Psi(x,F)$. Now suppose

3.1.2 $$(F : O_T(x) \to \Psi^*) \wedge \forall y,z(y <_T x \wedge z <_T x \to (\Psi(y,F^{(y)}) \wedge (y <_T z \to F^{(y)} \prec F^{(z)})))$$

In order to show that the hypothesis of $DCColl_2^T$ is satisfied, it suffices to show that

3.1.3 $$\exists G(\Psi(x,G) \wedge \forall y <_T x(F^{(y)} \prec G)).$$

Case (i): $O_T(x)$ has a T-greatest element x_m. By 3.1.2, $\Psi(x_m, F^{(x_m)})$ holds, so by the hypothesis of $DCColl_1^T$, there is an X, say X_2, such that

3.1.4 $$\Phi(X) \wedge \Pi(F^{(x_m)}, X)$$

Define $G : O_T(x) \to \Phi$ as follows:

$$G^{(y)} = \begin{cases} F^{(y+1)}(y) & \text{if } y <_T x_m ; \\ X_2 & \text{if } y = x_m, \end{cases}$$

where $y+1 =_{Df}$ the T-least element $z \in \mathcal{F}(T)$ such that $y <_T z$. Suppose $y <_T z <_T x_m$. Then

3.1.5 $$G^{(y)} = F^{(y+1)}(y) = F^{(z)}(y) = F^{(x_m)}(y)$$

since, by 3.1.2, $F^{(y+1)} \prec F^{(z)} \prec F^{(x_m)}$. Hence,

3.1.6 $$G \restriction O_T(z) = F^{(z)} = F^{(x_m)} \restriction O_T(z)$$

and

3.1.7 $$F^{(z)} \prec G.$$

Now $\Psi(x_m, F^{(x_m)})$ holds by 3.1.2, so by 3.1.1, 3.1.5 and 3.1.6,

3.1.8 $$\forall z <_T x_m \Pi(G \quad O_T(z), G^{(z)});$$

and since $G^{(x_m)} = X_2$, it follows from 3.1.4 and 3.1.6 that

$$\Pi(G \restriction O_T(x_m), G^{(x_m)}).$$

Hence, by 3.1.7 and 3.1.8, 3.1.4 holds in Case (i).

Case (ii): $O_T(x)$ has no greatest element. The argument is similar to the one above, except that G is defined as follows:

$$G^{(y)} = F^{(y+1)}(y).$$

Now applying $DCColl_2^T$, there is an H such that

$$(H : \mathcal{F}(T) \to \Psi^*) \wedge H^{(x_o)} = X_1 \wedge \forall x,y(x_o <_T x \wedge x_o <_T y \to$$

$$(\Psi(x,H^{(x)}) \wedge (x <_T y \to H^{(x)} \prec H^{(y)}))).$$

As above, we consider the cases (i)': T is a successor well-ordering, and (ii)': T is a limit well-ordering.

Case (i)': Let x_m be the T-greatest element of $\mathcal{F}(T)$. Since $\Psi(x_m,H^{(x_m)})$ holds, it follows from the hypothesis of $DCColl_1^T$ that there is an X, X_3 say, such that

$$\Phi(X) \wedge \Pi(H^{(x_m)},X).$$

Define $F : \mathcal{F}(T) \to \Phi$ as follows:

$$F^{(y)} = \begin{cases} X_o & \text{if } y = x_o, \\ H^{(y+1)}(y) & \text{if } x_o <_T y <_T x_m, \\ X_3 & \text{if } y = x_m. \end{cases}$$

As above, it can be shown that

$$\forall x \in \mathcal{F}(T)(x_o <_T x \leq_T x_m \to \Pi(F \restriction O_T(x),F^{(x)}).$$

Case (ii)': Define $F : \mathcal{F}(T) \to \Phi$ as follows:

$$F^{(y)} = \begin{cases} X_o & \text{if } y = x_o, \\ H^{(y+1)}(y) & \text{if } x_o <_T y. \end{cases}$$

Then again it follows that

$$\forall x \in \mathcal{F}(T)(x_o <_T x \to \pi(F \upharpoonright O_T(x), F^{(x)});$$

so in either case, we obtain the conclusion of $DCColl_1^T$. Q.E.D.

LEMMA 3.2 $\quad NBG^o \vdash DCColl_1^T \to DCColl_2^T$

Proof. Assume the hypothesis of $DCColl_2^T$. Let $\Phi^*(X)$ denote $\exists x \in \mathcal{F}(T)\Phi(x,X)$, suppose $\Phi(x_o,X_o)$ holds, and define $\pi(F,X)$ as follows:

$$\pi(F,X) \leftrightarrow (\mathcal{D}(F) = \emptyset \wedge X = X_o) \vee \exists x \in \mathcal{F}(T)[x_o <_T x \wedge (F : O_T(x) \to \Phi^*) \wedge$$
$$[\forall y,z(y <_T x \wedge z <_T x \to \Phi(y,F^{(y)}) \wedge (y <_T z \to F^{(y)} \prec F^{(z)}))$$
$$\to (\Phi(x,X) \wedge \forall y(y <_T x \to F^{(y)} \prec X))]] .$$

Clearly the hypothesis of $DCColl_2^T$ now implies the hypothesis of $DCColl_1^T$. Hence, by $DCColl_1^T$, there is an $F : \mathcal{F}(T) \to \Phi^*$ such that $F^{(x_o)} = X_o$ and

3.2.1 $$\forall x \in \mathcal{F}(T)(x_o <_T x \to \pi(F \upharpoonright O_T(x), F^{(x)}).$$

It remains to show that

3.2.2 $$\forall x \in \mathcal{F}(T)(\Phi(x,F^{(x)}) \wedge \forall y <_T x(F^{(y)} \prec F^{(x)})).$$

The proof is by induction on x. If $x = x_o$, 3.2.2 is trivial. Suppose $x_o <_T x$ and assume

3.2.3 $\forall y <_T x(\Phi(y,F^{(y)}) \wedge \forall z <_T y(F^{(z)} \prec F^{(y)}))$.

From 3.2.1 and 3.2.3 it clearly follows that

$$\Phi(x,F^{(x)}) \wedge \forall y <_T x(F^{(y)} \prec F^{(x)}),$$

so we are done.

By Lemmas 3.1 and 3.2, we have for each well-ordering T,

THEOREM 3.3 $NBG^{\circ} \vdash DCColl_1^T \leftrightarrow DCColl_2^T$

We now consider the following axioms of choice:

DCC^T : This is formulated like $DCColl^T$, except that the collection of classes $\{X : \Phi(X)\}$ is now required to be a class.

INJ^X : If $\{Z : \Phi(Z)\}$ is an uncodable collection of classes and $X \neq \emptyset$, then X can be "injected" into it; that is,

$$\exists F\,(F : X \to \Phi) \wedge \forall x,y(x \in X \wedge y \in X \wedge x \neq y \to F^{(x)} \neq F^{(y)}) .$$

$(INJ^X)'$: If $\{Z : \Phi(Z)\}$ cannot be coded by a set of classes and $X \neq \emptyset$, then X can be injected into it.

Inj^X : If $X \neq \emptyset$, then X can be injected into every proper class.

C^X : $\forall y \in X \exists Y \Phi(y,Y) \to \exists Z \forall y \in X \Phi(y,Z^{(y)})$

SC^X : $\forall y \in X \exists z \Phi(y,z) \to \exists F(Func(F) \wedge \forall y \in X \Phi(y,F(y)))$

$Proj^X$: If $X \neq \emptyset$, then every proper class can be projected onto it.

N (von Neumann's axiom) : There is a bijection between any two proper classes.

WO(V) : V (the universe of sets) is well-orderable.

E (Gödel's axiom of choice) : $\exists F(Func(F) \wedge \forall x \neq \emptyset(F(x) \in x))$

AC (the ZF-form of choice) : $\forall x \exists f(\mathrm{Func}(f) \wedge \forall y \in x(y \neq \emptyset \rightarrow f(y) \in y))$

REMARK 3.4 (i) If T is a well-ordering, we write $\ldots^{T}$ for $\ldots^{\mathcal{F}(T)}$. For example, we write INJ^{T} for $\mathrm{INJ}^{\mathcal{F}(T)}$.

(ii) If α is an ordinal, then DCColl^{α} and DCC^{α} are like DCColl^{T} and DCC^{T} respectively, except that $\mathcal{F}(T)$ is replaced by α. DCColl^{On} and DCC^{On} are obtained similarly.

(iii) $(\mathrm{DCColl}^{T})^{*}$, $(\mathrm{INJ}^{X})^{*}$ and $(\mathrm{INJ}^{X})'^{*}$ are like DCColl^{T}, INJ^{X} and $(\mathrm{INJ}^{X})'$ respectively, except that the specified collection of classes is required to be disjoint.

(iv) $(C^{X})^{*}$ is like C^{X} except that for each $y \in X$, the collection $\{Y : \Phi(y,Y)\}$ is required to be disjoint.

(v) Lévy's formulation of DC^{α} required α to be a cardinal, so the simplest and most natural class-form of DC^{α} is the following: Let κ be a cardinal, C a class, and $R(x,y)$ a relation between subsets x of C and elements y of C. Then, denoting the cardinality of x by $\bar{\bar{x}}$,

$$\forall x \subset C(\bar{\bar{x}} < \kappa \rightarrow \exists y(y \in C \wedge R(x,y)) \rightarrow$$

$$\exists f[(f : \kappa \rightarrow C) \wedge \forall \beta < \kappa \; R(f''\beta, f(\beta))].$$

Let us denote this axiom by $(\mathrm{DCC}^{\kappa})'$. It is easy to see that

$$\mathrm{NBG}^{0} \vdash \mathrm{DCC}^{\kappa} \leftrightarrow (\mathrm{DCC}^{\kappa})'$$

for each cardinal κ, but from Theorem 3.8 below, it also follows that

$$\mathrm{NBG}^{0} \vdash \forall \alpha \, \mathrm{DCC}^{\alpha} \leftrightarrow \forall \kappa \, \mathrm{DCC}^{\kappa},$$

where κ denotes a cardinal. Hence, there is no need to restrict ourselves to cardinals in the formulation of DCC^{α} or DCColl^{α}.

THEOREM 3.5 The following implications hold in NBG^{0} for any X.

(i) $(C^X)^* \longrightarrow SC^X$,

(ii) $(INJ^X)'^* \xrightarrow{1} Inj^X \xrightarrow{2} Proj^X$, and

(iii) $Proj^X + SC^X \rightarrow Inj^X$.

<u>The proof of (i)</u>. Assume $\forall y \in X \exists z \Phi(y,z)$. Put $C_y = \{z : \Phi(y,z)\}$ and $C_y^* = \{\{z\} \times z : z \in C_y\}$. Then C_y^* is disjoint and $\forall y \in X \exists u(u \in C_y^*)$; so by $(C^X)^*$, there is a function F such that $\forall y \in X(F^{(y)} \in C_y^*)$. Now define $G : X \rightarrow \bigcup_{y \in X} C_y$ by putting $G(y)$ = the unique element in the domain of $F^{(y)}$. Clearly, G is the required function.

<u>The proof of (ii)</u>. Implication 2 is trivial and the proof of 1 uses an argument like the one above.

<u>The proof of (iii)</u>. Let C be a proper class, $X \neq \emptyset$, and F a projection of F onto X. For $y \in X$, put $F^{-1}(y) = \{x \in C : F(x) = y\}$. Then $F^{-1}(y) \neq \emptyset$, so by SC^X, there is a function G such that $G : X \rightarrow C$ and $G(y) \in F^{-1}(y)$ for all $y \in X$. Clearly G is an injection. The proof of the theorem is complete.

THEOREM 3.6 For every class X, $NBG^o \vdash (INJ^X)' \longrightarrow INJ^X$

The proof is trivial.

THEOREM 3.7 For every well-ordering T, the following implications hold in NBG^o.

(i)
$$DCColl^T \begin{array}{l} \xrightarrow{1} C^T \\ \xrightarrow{2} INJ^T \end{array}$$

(ii)
$$(DCColl^T)^* \begin{array}{l} \xrightarrow{1} (C^T)^* \\ \xrightarrow{2} DCC^T \xrightarrow{3} SC^T \\ \xrightarrow{4} (INJ^T)^* \end{array}$$

<u>The proof of (i)</u>. 1. Assume $\forall x \in \mathcal{F}(T) \exists Y \Phi(x,Y)$. Define $\Phi^*(Y)$ as $\exists y \in \mathcal{F}(T) \Phi(y,Y)$ and $X \prec Y$ as $\Phi^*(X) \wedge \Phi^*(Y)$. The hypothesis of $DCColl_2^T$ is clearly satisfied, so there is an F such that $F : \mathcal{F}(T) \rightarrow \Phi^*$ and $\forall y \in \mathcal{F}(T) \Phi(y, F^{(y)})$. Hence, C^T holds.

2. Let $\{X : \Phi(X)\}$ be an uncodable collection of classes. Define $\Pi(F,X)$ as $\neg(X \eta F)$. Since Φ is uncodable, the hypothesis of $DCColl_1^T$ is satisfied, so there is an F such that $F : \mathcal{F}(T) \rightarrow \Phi$ and $\forall y \in \mathcal{F}(T) \neg(F^{(y)} \eta\ F \restriction y)$. Clearly, F is an injection.

<u>The proof of (ii)</u>. The proofs of implications 1 and 3 are like the proof of (i)1 and the proof of 4 is like the proof of (i)2. The proof of 2 uses a disjointing argument like the one in the proof of Theorem 3.5 (i).

THEOREM 3.8 Let T_1 and T_2 be well-orderings and $T_1 < T_2$. Then the following hold in NBG^o.

(i) $DCColl^{T_2} \rightarrow DCColl^{T_1}$,

(ii) $(DCColl^{T_2})^* \rightarrow (DCColl^{T_1})^*$,

(iii) $DCC^{T_2} \rightarrow DCC^{T_1}$.

The proofs are all similar, so we just prove (i), as follows:

Suppose $T_1 \overset{H}{\sim} T_2 \restriction x_o$. Let y_{1o} be the T_1-least element of $\mathcal{F}(T_1)$ and y_{2o} the T_2-least element of $\mathcal{F}(T_2)$.

If X is an initial segment of $\mathcal{F}(T_1)$ and $F : X \rightarrow \Phi$, then define $F^* : H''X \rightarrow \Phi$ as follows: $F^{*(H(y))} = F^{(y)}$ for $y \in X$. Similarly, if Y is an initial segment of $\mathcal{F}(T_2)$ and $G : Y \rightarrow \Phi$, then define $G^{**} : H^{-1}''Y \rightarrow \Phi$ by putting $G^{**(H^{-1}(x))} = G^{(x)}$ for $x \in Y$. Clearly, $(F^*)^{**} = F$ and $(G^{**})^* = G$.

Now assume the hypothesis of $DCColl^{T_1}$, that is,

$$3.8.1 \qquad \forall x \in \mathcal{F}(T) \forall F[(F : O_{T_1}(x) \to \Phi) \to \exists X(\Phi(X) \wedge \pi(F,X))].$$

Define $\pi^*(F,X)$ as follows:

$$\pi^*(F,X) \leftrightarrow_{Df} \exists x[x \in \mathcal{F}(T_2) \wedge (F : O_{T_2}(x) \to \Phi) \wedge [(x <_{T_2} x \wedge \pi(F^{**},X)) \vee (x_o \leq_{T_2} x \wedge \Phi(X))]]$$

Clearly, by 3.8.1, the hypothesis of $DCColl^{T_1}$ is satisfied; so for any X such that $\Phi(X)$, there is an F such that $F : \mathcal{F}(T_2) \to \Phi$, $F^{(y_{2o})} = X$ and

$$\forall y \in \mathcal{F}(T_2)(y_{2o} <_{T_2} y \to \pi^*(F \upharpoonright O_{T_2}(y), F^{(y)})).$$

Hence, setting $G = (F \upharpoonright O_{T_2}(x_o))^{**}$, it follows that $G^{(y_{1o})} = X$ and

$$\forall x \in \mathcal{F}(T_1)(y_{1o} <_{T_1} x \to \pi(G \upharpoonright O_{T_1}(x), G^{(x)}));$$

so $DCColl^{T_1}$ holds. Q.E.D.

THEOREM 3.9 For every X and Y such that $Y \subseteq X$, the following implications hold in NBG^o.

(i) $C^X \to C^Y$,

(ii) $(C^X)^* \to (C^Y)^*$,

(iii) $SC^X \to SC^Y$,

(iv) $INJ^X \to INJ^Y$,

(v) $(INJ^X)^* \to (INJ^Y)^*$,

(vi) $(INJ^X)' \to (INJ^Y)'$,

(vii) $(INJ^X)'^* \to (INJ^Y)'^*$,

(viii) $Inj^X \to Inj^Y$ and

(ix) $Proj^X \to Proj^Y$.

The proofs ______ are all left to the reader. The proof of the next theorem is not.

THEOREM 3.10 All the following implications except (vii)2 hold in NBG^o, but (vii)2 holds in M^o.

(i) $\forall\alpha\, DCC^{\alpha} \xrightarrow{1} \forall x\, SC^{x} \xrightarrow{2} AC$; $\quad \forall\alpha\, DCC^{\alpha} \xrightarrow{3} \forall x\, Inj^{x} \xrightarrow{4} \forall x\, Proj^{x} \xrightarrow{5} AC$

(ii) $\forall\alpha\, DCColl^{\alpha} \xrightarrow{1} \forall x\, C^{x}$; $\quad \forall\alpha\, DCColl^{\alpha} \xrightarrow{2} \forall x (INJ^{x})'$

(iii) $\forall\alpha (DCColl^{\alpha})^* \xrightarrow{1} \forall x (C^{x})^*$; $\quad \forall\alpha (DCColl^{\alpha})^* \xrightarrow{2} \forall x (INJ^{x})'^*$

(iv) $\forall T(DCColl^{T}) + WO(V) \xrightarrow{1} C^{V} \xrightarrow{3} DCC^{On}$; $\quad \forall T(DCColl^{T}) + WO(V) \xrightarrow{2} INJ^{V}$

(v) $\forall T(DCColl^{T})^* + WO(V) \xrightarrow{1} (C^{V})^*$; $\quad \forall T(DCColl^{T})^* + WO(V) \xrightarrow{2} (INJ^{V})^*$

(vi) $Inj^{V} \xrightarrow{1} Proj^{V} \xrightarrow{2} N \xrightarrow{3} Inj^{V}$

(vii) $N \xrightarrow{1} WO(V) \xrightarrow{2} \forall T(DCC^{T})$; $\quad WO(V) \xrightarrow{3} SC^{V} \xrightarrow{4} \forall\alpha\, DCC^{\alpha}$; $\quad SC^{V} \xrightarrow{5} E \xrightarrow{6} AC$

The proof of (i). Implication 3 is proved in Felgner [5] and is like the proof of Theorem 3.7 (i)3. 4 and 5 are proved in Rubin and Rubin [12]. The proof of 1 is like the proof of Theorem 3.7 (i)1, and the proof of 2 is trivial.

The proofs of (ii)1 and (ii)2 are like the proofs of (i)1 and (i)3 respectively; so are the proofs of (iii)1 and (iii)2. The proofs of (iv) and (v), except (iv)3, are again similar to those in (ii) and (iii), but, whereas it follows from $\forall\alpha\, DCC^{\alpha}$ (and hence $\forall\alpha\, DCColl^{\alpha}$ and $\forall\alpha(DCColl^{\alpha})^*$) that every set is well-orderable, in (iv)1, (iv)2 and (v) we need to assume that V is well-orderable since it seems, although this has not yet been proven, that in M^{o}, $\forall T(DCColl^{T})$ does not imply WO(V). The reason for this is that in M there is no power-class axiom which says that the collection of subclasses of each class is codable. We shall have more to say about this in section 5.

<u>The proof of (iv)3</u>. Let C be a class and assume that

$$\forall\alpha\forall f((f:\alpha\to C)\to\exists x(x\in C\wedge R(f,x))) .$$

Then by C^{V} (in fact, SC^{V} will do here),

$$\forall\alpha\,\exists G\,\forall f\left[(f:\alpha\to C)\to G^{(f)}\in C\wedge R(f,G^{(f)})\right],$$

so again by C^{V},

$$\exists H\forall\alpha\,\forall f\left[(f:\alpha\to C)\to H^{(\alpha)(f)}\in C\wedge R(f,H^{(\alpha)(f)})\right].$$

Now let $x_o\in C$ and define $K: On\to C$ by recursion on On, as follows:

$$K(\alpha)=\begin{cases}x_o & \text{if } \alpha=0\\ H^{(\alpha)}(K\restriction\alpha) & \text{if } \alpha>0.\end{cases}$$

Clearly, K is the required function.

QUESTION 3.11 Does $(C^{V})^*$ imply DCC^{On} or C^{V} imply DCC^{T}, where $T>On$? The above proof suggests that the answer to both questions is negative.

<u>The proof of (vii)</u>. The proofs of 1, 3 and 5 are trivial, and 6 is well-known. The proof of 4 is like the proof of (iv)3.

The proof of 2 is given in M^o, as follows: Let T' be a well-ordering of V and C be a class. Assume the hypothesis of DCC^T, that is,

$$\forall x \in \mathcal{F}(T) \forall F \left[(F : O_T(x) \rightarrow C) \rightarrow \exists y (y \in C \wedge R(f,y)) \right] .$$

Let x_o be the T-least element of $\mathcal{F}(T)$, x_1 be the T'-least element of C, and $\Phi(X,y)$ denote the formula

$$\exists x \in \mathcal{F}(T)(X : O_T(x) \rightarrow C) \wedge \left[(\mathcal{D}(X) = \emptyset \wedge y = x_1) \vee (\mathcal{D}(X) \neq \emptyset \wedge \right.$$

$$\left. \vee \quad (\mathcal{D}(X) \neq \emptyset \wedge y = \text{the T'-least } z \in C \text{ such that } R(X,z)) \right] .$$

Φ clearly satisfies the conditions P1 - P4 on page 00 above, so by Theorem 2.1 and Remark 2.3, there is a strictly unique F such that $F : \mathcal{F}(T) \rightarrow C$ and

$$F(x_o) = \begin{cases} x_1 & \text{if } x = x_o, \\ \text{the T'-least } z \in C \text{ such that } R(F x,z), & \text{if } x_o <_T x. \end{cases}$$

F is clearly the required function. This completes the proof of Theorem 3.10.

One naturally asks if any of the implications in Theorems 3.5 - 3.110, apart from those in Theorem 3.10 (vi), can be reversed in NBG^o or even M^o. In the next section, we give some examples of those which cannot be reversed.

However, it is still not yet known that they are all irreversible, although intuitively, this appears to be the case.

It was shown in Rubin and Rubin[12] that in ZF (= ZF^o + VI), AC implies $\forall x Inj^x$. In fact, it is not difficult to see that ZF $\vdash$ AC $\rightarrow \forall\alpha DCC^\alpha$. (Note that DCC^α and Inj^x are schemata in ZF.) Thus, in the presence of xiom VI, all the above forms of choice which can be formalized in ZF are equivalent. However, we can see no way of showing, for example, that

(i) $M + AC \vdash Proj^{On}$,

(ii) $M + E \vdash \forall x C^x$ or $\forall x(INJ^x)'$,

(iii) $M + C^V + INJ^V \vdash \forall\alpha(DCColl^\alpha)$ or

(iv) $M + C^V + \forall T(DCColl^T)^* \vdash \forall x(INJ^x)$.

Nevertheless, Theorem 3.12 below extends the known results.

THEOREM 3.12 (i) $NBG + AC \vdash \forall\alpha(DCColl^\alpha)^*$

(ii) $M^o + WO(V) \vdash \forall T(DCColl^T)^*$

and (iii) $NBG^o + WO(V) \vdash (DCColl^{On})^*$

Proof. (i) Let $\{X : \Phi(X)\}$ be a disjoint collection of classes. For each X such that $\Phi(X)$, let X' be the subset of X consisting of those elements of X of least rank. Then $\{X' : \Phi(X)\}$ is a disjoint class, C say, and if $x \in C$, there is a unique X, x^* say, such that $\Phi(X) \wedge X' = x$.

If x is a set and $F : x \to \Phi$, define $F' : x \to C$ by putting $F'(y) = (F^{(y)})'$; and if $f : x \to C$, then define $f^* : x \to \Phi$ by $f^{*(y)} = (f(y))^*$.

Now assume the hypothesis of $(DCColl^\alpha)^*$, viz.,

$$\forall\beta < \alpha\, \forall F[(F : \beta \to \Phi) \to \exists X(\Phi(X) \wedge \pi(F,X))],$$

and define $\pi'(f,x)$ as $\pi(f^*,x^*)$. Then clearly,

$$\forall\beta < \alpha\, \forall f[(f : \beta \to C) \to \exists x(x \in C \wedge \pi'(f,x))];$$

so by DCC^α, which follows from AC in NBG, for all X such that $\Phi(X)$, there is an $f : \alpha \to C$ such that $f(0) = X'$ and $\forall\beta < \alpha\, \pi(f \restriction \beta, f(\beta))$. Hence, $f^{*(0)} = X$ and $\forall\beta < \alpha\, \pi(f^* \restriction \beta, f^{*(\beta)})$, so $(DCColl^\alpha)^*$ holds. Q.E.D.

(ii) Let T be a well-ordering of V and $X : (X)$ be a disjoint collection of classes. Assume the hypothesis of $(\mathrm{DCColl}^T)^*$, viz,

$$\forall x \in \mathcal{F}(T) \forall F\ (F: O_T(x) \rightarrow \Phi) \rightarrow \exists Y(\Phi(Y) \wedge \Pi(F,Y)) .$$

Let $C_{x,F}$ be the class $\{Y_T : \Phi(Y) \wedge \Pi(F,Y)\}$, where Y_T is the T-least element of Y. Let $l_{x,F}$ denote the T-least element of $C_{x,F}$ and $Y_{x,F}$ be the unique Y such that $\Phi(Y) \wedge \Pi(F,Y) \wedge Y_T = Y_{x,F}$. Remember that this Y is unique since the collection $\{Y : \Phi(Y)\}$ is disjoint. Suppose $\Phi(X)$ holds. We may define $F : \mathcal{F}(T) \rightarrow \Phi$ by recursion on T as follows:

$$F^{(x)} = \begin{cases} X & \text{if } x = x_o , \\ Y_{x,F} \upharpoonright O_T(x) & \text{if } x_o <_T x , \end{cases}$$

where x_o is the T-least element in $\mathcal{F}(T)$. Clearly F is the required function and $(\mathrm{DCColl}^T)^*$ holds.

The proof of (iii) is left to the reader.

SECTION 4

FRAENKEL-MOSTOWSKI-SPECKER METHODS IN M^o

In this section we show that the well-known Fraenkel-Mostowski-Specker methods for proofs in NBG^o of the independence of the axiom of foundation and various forms of the axiom of choice carry over easily to M^o. We also prove some elementary results in M^o for certain forms of choice mentioned in the previous section. It would be tedious to give details, especially since they are well-known and can in any case be found in Felgner [3] or Jech [8].

First, let F be a permutation of the universe of sets V, that is, a 1-1 mapping of V onto itself. We write $X \in_F Y$ for $Z(X) \wedge F(X) \in Y$, as usual.

For any formula Φ, let Φ_F be obtained from Φ by replacing every occurrence of $\in$ by $\in_F$. It is shown in Boffa [1] that $\ldots_F$ is a syntactic model of NBG^o in NBG^o, that is, $NBG^o \vdash \Phi_F$ for every axiom Φ of NBG^o. In order to show that it is also a syntactic model of M^o in M^o, it therefore suffices to show that the relativized version of the impredicative comprehension schema holds in M^o; that is, that for any formula Φ with one free set-variable x,

$$M^o \vdash \forall X_1 \ldots X_n \exists Y \forall x (x \in_F Y \leftrightarrow \Phi_F(x, X_1, \ldots, X_n)),$$

and by taking Y as $\{F(x) : \Phi_F(x, X_1, \ldots, X_n)\}$, this is trivial to verify.

It follows that it is possible to violate the axiom of foundation in M^o in all the usual ways (see Felgner [3], p. 50 - 51). For example, the same proof as in Boffa [1] shows that if $M^o + N$ is consistent, then $M^o + N +$ "there is a proper class of reflexive sets" is also consistent. (A reflexive set x, or atom, has the property $x = \{x\}$.) Note that in Tharp [13], it is shown that if M^o is consistent, then so is $M + V = L$, where $V = L$ is Gödel's axiom of constructibility for sets.

Hence, $M^o + N$ is consistent relative to M^o. In fact, Tharp proved the stronger result that if M^o is consistent, then so is $M^o + \mathcal{V} = \mathcal{L}$, where $\mathcal{V} = \mathcal{L}$ is the axiom of constructibility for classes, and he showed that one can define a well-ordering of the universe of constructible classes. By elementary arguments like those at the beginning of the next section, it therefore follows that if M^o is consistent, then so, for example, is $M^o + \forall T(DCColl^T) + N$; and it is straightforward to show that $M^o + \forall T(DCColl^T) \vdash \forall T(DCColl^T)_F$ for any permutation F of V. Thus, for example, by Boffa's proof, the theory $M^o + \forall T(DCColl^T) + N$ + "there is a proper class of reflexive sets", is also consistent.

Now let s be a <u>set</u> of atoms. The relative von Neumann hierarchy for s is defined as follows: $R_o(s) = s$, $R_\alpha(s) = \bigcup_{\beta<\alpha} \mathcal{P}(R_\beta(s))$. We put $W(s) = \bigcup_{\alpha \in On} R_\alpha(s)$. If A is a class of atoms, then we put $W(A) = \bigcup_{s \subseteq A} W(s)$. By interpreting sets as elements of $W(A)$ and classes as subclasses of $W(A)$, one obtains by standard arguments the relative consistency of, for example, $M^o + \forall T(DCColl^T) + N$ + "there is a proper class A of reflexive sets such $WF(A)$", where $WF(A)$ is

$$\forall x(x \neq \emptyset \rightarrow \exists y \in x(y \cap x = \emptyset \vee y \in A))$$

It is well-known that in this theory it is provable that $V = W(A)$. By similar means, one obtains the relative consistency of theories like $M^o + \forall T(DCColl^T) + N$ + "there is countable set s of reflexive sets such that $WF(s)$".

We now show that the so-called permutation models $\mathfrak{M}[G, \mathcal{F}]$, described in Felgner [3] and Jech [6] are models of M^o.

We shall assume for definiteness that A is a proper class of atoms, but the argument is exactly the same if A is a set.

By a <u>permutation</u> of A, we mean a <u>set</u> π which is a permutation of some subset s of A. By $\in$-recursion, π can be extended to a unique

$\in$-automorphism π^* of $W(s)$; so if $\hat{\pi}$ is defined as $\pi^* \cup \mathrm{Id} \restriction (V \setminus W(A))$, then π extends to a unique $\in$-automorphism $\hat{\pi}$ of V, which in this case, is W(A). For notational convenience, however, we write π for $\hat{\pi}$.

Now let $\mathcal{G}$ be the group of all permutations of A and $\mathcal{F}$ be a normal filter (in the sense of Felgner [3], p.53) of subgroups of a subgroup G of $\mathcal{G}$. Note that $\mathcal{G}$ is a class, so $\mathcal{F}$ is not a class, but a collection of classes. Nevertheless, in all permutation models $\mathfrak{M}[G,\mathcal{F}]$, G and are defined collections. Indeed, the right definitions of G and $\mathcal{F}$ are the alpha and omega in the search for such models. Hence, we can safely use the shorthand "$X \in \mathcal{F}$", meaning that X satisfies the formula which defines $\mathcal{F}$.

For any class, let $H[X] =_{Df} \{\pi \in G : \pi''X = X\}$. We say that X is <u>symmetric</u> if $H[X] \in \mathcal{F}$ and <u>hereditarily symmetric</u> if, in addition, $H[y] \in \mathcal{F}$ for every set $y \in TC(X)$, where $TC(X)$, the <u>transitive closure</u> of X, is $X \cup \bigcup X \cup \bigcup\bigcup X$ .... We write $\mathfrak{M}(X)$, or say that X is an $\mathfrak{M}$-class, if X is hereditarily symmetric; and for any formula Φ, we write $\Phi_{\mathfrak{M}}$ for the formula obtained from Φ by replacing every quantifier $\exists X$ by $\exists X(\mathfrak{M}(X) \wedge$ .... Thus, the universe of the model $\mathfrak{M}[G,\mathcal{F}]$ is the collection of hereditarily symmetric classes and the membership relation of the model is simply the real membership relation $\in$. It is easy to see that $Z_{\mathfrak{M}}(X) \leftrightarrow Z(X) \wedge \mathfrak{M}(X)$; that is that the $\mathfrak{M}$-sets are just the hereditarily symmetric sets. (The proof uses the fact that for any set x, $H[\{x\}] = H[x]$.) We shall write $X \subset \mathfrak{M}$ for $\forall y(y \in X \rightarrow \mathfrak{M}(y))$.

As usual, it is easy to see that $\mathfrak{M}$ has the following properties :

M1. $\mathfrak{M}$ is transitive: $x \in Y \wedge \mathfrak{M}(Y) \rightarrow \mathfrak{M}(x)$,

M2. $X \subset \mathfrak{M} \rightarrow (\mathfrak{M}(X) \leftrightarrow H[X] \in \mathcal{F})$.

The proof that the impredicative comprehension schema holds in $\mathfrak{M}[G,\mathcal{F}]$ is a simple consequence of the following lemmata:

LEMMA 4.1 $\mathfrak{M}(X) \wedge \pi \in G \rightarrow \mathfrak{M}(\pi``X)$

<u>Proof</u>. Assume $\mathfrak{M}(X) \wedge \pi \in G$. A proof like that on page 55 of Felgner [3] shows that

4.1.1 $$\pi \in G \rightarrow \pi . H[X] . \pi^{-1} \leq H[\pi``X] .$$

Hence, $H[\pi``X] \in \mathcal{F}$. Furthermore, a proof like that on page 54 <u>loc cit</u>, shows that $\mathfrak{M}(y) \wedge \pi \in G \rightarrow \mathfrak{M}(\pi(y))$. Hence, $\pi``X \subset \mathfrak{M}$. Now by 4.1.1, $H[\pi``X] \in \mathcal{F}$, so by M2, $\mathfrak{M}(\pi``X)$. Q.E.D.

LEMMA 4.2 For any formula $\Phi(X_1,\ldots,X_n)$,

$$\mathfrak{M}(X_1) \wedge \ldots \wedge \mathfrak{M}(X_n) \wedge \pi \in G \rightarrow [\Phi_{\mathfrak{M}}(X_1,\ldots,X_n) \leftrightarrow \Phi_{\mathfrak{M}}(\pi``X_1,\ldots,\pi``X_n)].$$

The proof is by induction on the length of Φ and uses Lemma 4.1.

Now let $\Phi(x,X_1,\ldots,X_n)$ be a formula with one free set-variable x, and suppose that $X_1,\ldots,X_n$ are $\mathfrak{M}$-classes. Let $Y = \{x : \mathfrak{M}(x) \wedge \Phi_{\mathfrak{M}}(x,X_1,\ldots,X_n)\}$. In order to prove that the impredicative comprehension schema holds in $\mathfrak{M}[\in,\mathcal{F}]$, it suffices, by M1, to show that Y is an $\mathfrak{M}$-class. Now $Y \subset \mathfrak{M}$, so by M2, it suffices to show that $H[Y] \in \mathcal{F}$.

Now for $i = 1\ldots n$, $H[X_i] \in \mathcal{F}$, since X_i is an $\mathfrak{M}$-class. Hence, putting $H_0 = \bigcap_{i=1}^{n} H[X_i]$, it follows that $H_0 \in \mathcal{F}$. We show that $H_0 \leq H[Y]$, for then $H[Y] \in \mathcal{F}$.

Let $\pi \in H_0$ and $x \in Y$. Then x, and hence $\pi(x)$, is an $\mathfrak{M}$-set. Moreover, $\Phi_{\mathfrak{M}}(x,X_1,\ldots,X_n)$ holds, so by Lemma 4.2, $\Phi_{\mathfrak{M}}(\pi(x), \pi``X_1,\ldots, \pi``X_n)$ also holds. However, $\pi \in H[X_i]$ for every $i = 1\ldots n$, since $\pi \in H_0$; so $\pi``X_i = X_i$. Hence, $\pi(x) \in Y$ and $\pi``Y \subseteq Y$. Now $\pi \in H_0 \rightarrow \pi^{-1} \in H_0$, since H_0 is a group; so $\pi^{-1}``Y \subset Y$. Hence, $\pi``Y = Y$ and $\pi \in H[Y]$. Q.E.D.

<u>Applications of the FMS methods.</u>

It follows from the above result that all the non-implications proved in Felgner-Jech [6] and Felgner [5] also hold in M^0, in particular,

(i) $AC \not\to Proj^{\omega}$

(ii) $\forall x Proj^x \not\to Inj^{\omega}$

(iii) $\forall x Inj^x \not\to DCC^{\omega}$ and

(iv) $E \not\to WO(V)$

From the theorems in Section 3, there are clearly many more non-implications which we could try to prove in M^0, but in this section we prove only a few since we are chiefly interested in showing that FMS methods can be used to give independence results in M^0.

THEOREM 4.3 $M^0 + \forall\alpha DCColl^{\alpha} + \neg Proj^{On} + \neg SC^{On} + \neg E$ is consistent.

Proof. We find a permutation model $\mathfrak{M}$ of the above theory. Let A be a proper class of atoms and G be the group of all permutations of A. A subgroup H of G consisting of all permutations which leave a certain subset s of A pointwise fixed, is called a set-support subgroup of G. s is called a support for H. Let $\mathcal{F}$ be the normal filter whose filter-base consists of all set-support subgroups of G. A class X is thus symmetric if and only if there is a subset s of A such that every permutation π which leaves s pointwise fixed, leaves X fixed en bloc. s is also called a support for X. The $\mathfrak{M}$-classes are the hereditarily-symmetric classes.

The proof that $\forall\alpha DCColl^{\alpha}$ holds in $\mathfrak{M}$: Suppose that the hypothesis of $DCColl^{\alpha}$ holds in $\mathfrak{M}$. Then in the surrounding theory,

$$\forall\beta<\alpha\,\forall F\left[\mathfrak{M}(F) \wedge (F : \beta \to \Phi_{\mathfrak{M}}) \to \exists X(\mathfrak{M}(X) \wedge \Phi_{\mathfrak{M}}(X) \wedge \Pi_{\mathfrak{M}}(F,X))\right]$$

Now let F be any pair-class such that for all $\gamma<\beta$, $\mathfrak{M}(F^{(\gamma)})$ holds. By $\forall x SC^x$ in the surrounding theory, there is then a function f such that for every $\gamma<\beta$, $f(\gamma)$ is a support for $F^{(\gamma)}$. Put $s = \bigcup_{\gamma<\beta} f(\gamma)$. Then any permutation π which leaves s pointwise fixed leaves $F^{(\gamma)}$ fixed en bloc for every $\gamma < \beta$. Moreover, every permutation acts as the identity on the well-founded part of V, in particular, on the ordinals. Hence, π leaves F fixed en bloc; but $F \subset \mathfrak{M}$, so by M2, F is an $\mathfrak{M}$-class. Hence,

$$\forall\beta < \alpha \forall F[(\mathrm{Rel}(F) \wedge \mathcal{D}(F) = \beta \wedge \forall\gamma < \beta(\mathfrak{M}(F^{(\gamma)}) \wedge \Phi_{\mathfrak{M}}(F^{(\gamma)})) \rightarrow$$

$$\exists X[\mathfrak{M}(X) \wedge \Phi_{\mathfrak{M}}(X) \wedge \Pi_{\mathfrak{M}}(F,X)]] .$$

Now suppose $\mathfrak{M}(X) \wedge \Phi_{\mathfrak{M}}(X)$ holds. Applying DCColl^{α} in the surrounding theory, there is a G such that

$$G^{(0)} = X \wedge \forall\beta(0 < \beta < \alpha \rightarrow \mathfrak{M}(G^{(\beta)}) \wedge \Phi_{\mathfrak{M}}(G^{(\beta)}) \wedge \Pi_{\mathfrak{M}}(G \restriction \beta, G^{(\beta)})) \quad ;$$

and an argument like the one above shows that G is an $\mathfrak{M}$-class. Hence, DCColl^{α} holds in $\mathfrak{M}$.

<u>The proof that $\mathrm{Proj}^{\mathrm{On}}$ fails in $\mathfrak{M}$</u>: Since every atom is clearly an $\mathfrak{M}$-set, it follows that A is an $\mathfrak{M}$-class. Suppose that in $\mathfrak{M}$, A can be projected onto On by a function F. For each $\alpha \in \mathrm{On}$, let $A_\alpha = \{x \in A : F(x) = \alpha\}$ and let s be a support for F. Since A is a proper class, we can choose $\alpha, \beta \in \mathrm{On} \setminus F''s$ and x and y such that $F(x) = \alpha$ and $F(y) = \beta$. Let π be a permutation which leaves s pointwise fixed but interchanges x and y. Then π leaves F fixed en bloc, but $\pi(\langle x,\alpha\rangle) = \langle y,\alpha\rangle$, and $\langle y,\alpha\rangle \notin F$ – a contradiction.

<u>The proof that $\mathrm{SC}^{\mathrm{On}}$ fails in $\mathfrak{M}$</u>: First note that for any class X the kernel of X, Ker(X), is defined as $\mathrm{TC}(X) \cap A$, where TC(X) is the transitive closure of X.

For a cardinal α let $C_\alpha = \{x : \mathfrak{M}(x) \wedge \mathrm{Ker}(x) \text{ has cardinality } \alpha\}$. Clearly every C_α is an $\mathfrak{M}$-class. In fact every permutation leaves every C_α fixed en bloc. Suppose there is an $\mathfrak{M}$-class F such that for every cardinal α,

$F(\alpha) \in C_\alpha$. Let e be a support for F and have cardinality α. Let $x \in Ker(F(\alpha^+)) \setminus e$ and $y \in A \setminus (Ker(F(\alpha^+)) \cup e)$. Let π be a permutation which interchanges x and y and acts as the identity elsewhere. Then $\pi(F(\alpha^+))$ $\pi(F(\alpha^+)) \neq F(\alpha^+)$ and $\pi(F(\alpha^+)) \in C_{\alpha^+}$. Hence, $\pi''F \neq F$. However, π leaves e pointwise fixed; so $\pi''F = F$ – a contradiction. Hence, SC^{On} fails in $\mathfrak{M}$.

The proof that E fails in $\mathfrak{M}$: Clearly every subset of A is an $\mathfrak{M}$-set. Assume that in $\mathfrak{M}$ there is a function F such that for every subset s of A, $F(s) \in s$. Let s_0 be a support for F and choose s_1 with at least two elements such that $s_1 \cap s = \emptyset$. Choose $x \in s_1 \sim \{F(s_1)\}$ and let π be a permutation which interchanges x and $F(s_1)$ and is the identity elsewhere. Then π leaves s_1 and F fixed en bloc; but $\pi(\langle s_1, F(s_1)\rangle) = \langle s_1, x\rangle$ and $\langle s_1, x\rangle \notin F$ – a contradiction. This completes the proof of Theorem 4.3.

THEOREM 4.4 $M^0 + E + \neg Proj^\omega + \neg SC^\omega$ is consistent.

Proof. We use the model $\mathfrak{M}$ of Lemma 8 of Felgner [5]. Felgner shows that E holds in $\mathfrak{M}$. The $\mathfrak{M}$-sets are the sets x whose kernel, $Ker(x)$, is finite, $Ker(x)$ being $TC(\{x\}) \cap A$, where, in this case, A is a countable set of atoms. It follows that $\mathfrak{M}$ does not have the property M2, so it is not a model of the form $\mathfrak{M}[G, \mathcal{F}]$. Nevertheless, a proof similar to the one we gave above shows that it also is a model of M^0.

The proof that $Proj^\omega$ fails in $\mathfrak{M}$: Suppose that in $\mathfrak{M}$ there is a projection F of the atoms A onto ω. (Notice that although A is a set in the sense of the surrounding theory, it is a proper class in the sense of $\mathfrak{M}$.) Let s be a finite support for F and for each $n \in \omega$, let $s_n = \{x \in A: F(x) = n\}$. Choose $n, m \in \omega$ such that $s_n \cap s = s_m \cap s = \emptyset$ and choose $x \in s_n$ and $y \in s_m$. There is clearly an order-preserving permutation π which maps x to y and leaves s pointwise fixed. Thus, π leaves F fixed en bloc; but $\pi(\langle x,n\rangle) = \langle y,n\rangle \notin F$ – a contradiction.

The proof that SC^{ω} fails in $\mathfrak{M}$: For $n \in \omega$, let

$$M_n^* = \{x : \mathfrak{M}(x) \wedge \text{Ker}(x) \text{ has cardinality } n\}.$$

It is clear that M_n^* is a proper class in the sense of $\mathfrak{M}$, but in $\mathfrak{M}$, there is no function f such that $\forall n \in \omega(f(n) \in M_n^*)$, for such a function has an infinite kernel. Hence, SC^{ω} fails in $\mathfrak{M}$.

SECTION 5

THE THEORY M^o_ε

DEFINITION 5.1 Let M^O_ε be the theory obtained from M^o by adding Hilbert's ε-symbol to the language and allowing ε-terms $\varepsilon X\Phi$ to figure in the comprehension schema. M_ε is, of course, M^o_ε + VI.

In writing $M^o_\varepsilon \vdash \Psi$, we mean that there is a deduction of Ψ from M^o_ε in Hilbert's ε-calculus (see, for example, Leisenring [9] or Flannagan [7]).

This section is devoted to the proof of the following theorem :

THEOREM 5.2 $M^o_\varepsilon \vdash$ WO(V)

First we note the following lemma, the proof of which is left to the reader.

LEMMA 5.3 (i) $M^o_\varepsilon \vdash C^V$

and (ii) $M^o_\varepsilon \vdash \forall T(\mathrm{DCColl}^T)$

The proof of Theorem 5.2 is similar to the usual ZF-proof of the well-ordering theorem

WO : " Every set is equipotent to an ordinal."

Remember, however, that there is no "power-class" axiom in M in the sense already mentioned on page 214 , but that every proof of WO in ZF makes essential use of the power-set axiom. WO can, however, be proved without the power-set axiom by invoking a stronger choice principle than AC. For instance, it is easy to see that

(i) $NBG^{O-} + DCC^{On} \vdash WO$,

(ii) $NBG^{O-} + E \vdash WO$,

and that if ZF^{O-}_{τ} is obtained from ZF^{O} by adjoining the 1-place operator τ to the language of ZF, adding the axiom $\forall x(x \neq \emptyset \rightarrow \tau(x) \in x)$, and allowing τ to appear in the replacement schema, then

(iii) $ZF^{O-}_{\tau} \vdash WO$.

Likewise, if ZF^{O-}_{ε} is obtained from ZF^{O-} by admitting ε-terms of the form $\varepsilon x \Phi$ to the replacement schema, then

(iv) $ZF^{O-}_{\varepsilon} \vdash WO$

Hence, even without the power-set axiom, WO can be proved if there is a universal choice operator available which selects elements from sets and which is allowed to figure in the relevant schemata. It is not surprising, therefore, that $M^{O}_{\varepsilon} \vdash WO(V)$. Indeed, one might suspect that $NBG^{O-}_{\sigma} \vdash WO(V)$, where NBG^{O-}_{σ} is obtained from NBG^{O-} by adjoining the 1-place operator σ, adding Bernay's axiom

(*) $$\forall X(X \neq \emptyset \rightarrow \sigma(X) \in X)$$

and allowing σ to figure in the (predicative) comprehension schema. However, this does not seem to be the case since the proof of Theorem 5.2 is highly impredicative. Moreover, the proof makes essential use of C^V, so, since we see no way of showing that $M_{\sigma} \vdash C^V$, where M_{σ} is defined in the obvious way, it even seems that $M^{O}_{\sigma} \nvdash WO(V)$. Nevertheless, the proof of Theorem 5.2 does show that $(M^{O} + C^{V})_{\sigma} \vdash WO(V)$, where $(M^{O} + C^{V})_{\sigma}$ is obtained from $M^{O} + C^{V}$ by adding (*) and allowing σ to appear in both the comprehension schema and C^{V}.

<u>The proof of Theorem 5.2</u> : Let x_o be a fixed set and $\Phi(F,X)$ be the formula

$$(\mathcal{D}(F) \neq \emptyset \wedge X = x_0) \vee \exists T\left[T \neq \emptyset \wedge \mathcal{D}(F) = \mathfrak{F}(T) \wedge \left[(\exists y \forall z \in \mathfrak{F}(T)(y \neq F^{(z)}) \wedge\right.\right.$$

$$\left.\left. x = \varepsilon y \forall z \in \mathfrak{F}(T)(y \neq F^{(z)})) \vee (\forall y \exists z \in \mathfrak{F}(T)(y = F^{(z)}) \wedge X = V)\right]\right]$$

Φ clearly satisfies conditions P1' - P4' of Theorem 2.5 with $X = Y$ in place of $X \simeq Y$ and $F \equiv G$ in place of $F \cong G$ (see Definition 1.22). Hence, by Theorem 2.5, there is a formula $\Psi(T,X)$ with the properties Q1, Q2 and

Q3' : $$\forall T,X,Y(\Psi(T,X) \wedge \Psi(T,Y) \rightarrow X = Y).$$

Furthermore, it is clear that

5.4.1 $$\Psi(T,X) \rightarrow X \in V \vee X = V.$$

<u>Claim 5.4.2:</u> $$T' < T \wedge \Psi(T',x) \wedge \Psi(T,y) \rightarrow y \neq x.$$

<u>Proof</u>: Assume the left-hand side. Since $T > \emptyset$, it follows from Q1 that

.4.3 $$\Psi(T,y) \leftrightarrow \exists F\left[\mathcal{D}(F) = \mathfrak{F}(T) \wedge \exists u \forall v \in \mathfrak{F}(T)(u \neq F^{(v)}) \wedge y = \varepsilon u \forall v \in \mathfrak{F}(T)(u \neq F^{(v)}) \wedge \forall v \in \mathfrak{F}(T)\Psi(T \restriction v, F^{(v)})\right].$$

Now $T' < T \rightarrow T' \sim T \restriction z$ for some $z \in \mathfrak{F}(T)$, so by 5.4.3 and Q2, $\Psi(T', F^{(z)})$ for some F. holds. But $\Psi(T',x)$ holds, so by Q3', $x = F^{(z)}$. Hence, by 5.4.3, $y \neq x$, so the claim is proved.

We now show that

5.4.4 $$\neg \forall T \exists x\, \Psi(T,x).$$

Assume the contrary and let $C = \{x : \exists T\, \Psi(T,x)\}$. By Claim 5.4.2 it follows that

5.4.5 $$\forall x \in C \tilde{\exists}! T\, \Psi(T,x)$$

Now we do not have strict uniqueness here, so we must use C^V to pick a T for each x in C. Thus, there is an H such that

5.4.6 $$\forall x \in C\, \Psi(H^{(x)},x),$$

and since $H^{(x)}$ is a well-ordering which is unique up to $\sim$, it follows from Q2 that

5.4.7 $$x \in C \wedge y \in C \wedge x \neq y \rightarrow (H^{(x)} \not\sim H^{(y)}).$$

We now define a relation $\prec$ on C as follows:

$$x \prec y \leftrightarrow H^{(x)} < H^{(y)}.$$

We claim that $\prec$ well-orders C. The proof is as follows. First, by Lemma 0.21 in Marek [10], $\prec$ is a total ordering. Let D be a non-empty subclass of C. By Q2,

5.4.8 $$T_1 \sim T_2 \wedge \exists x \in D \Psi(T_1,x) \rightarrow \exists x \in D \Psi(T_2,x),$$

so by Lemma 2.4, $\tilde{\exists}_{\mu} T \exists x \in D\Psi(T,x)$. Let T' be such a least T and suppose $\Psi(T',x)$ holds. By Q3', x is unique. Furthermore, $T' \sim H^{(x)}$ by the uniqueness of $H^{(x)}$ to within $\sim$, so x is the $\prec$-least member of D and the claim is proved.

Thus, by Theorem 0.25 in Marek [10], there is a well-ordering, say T_o, with the property Sup H. Now by our assumption, and by 5.4.5 and 5.4.6,

5.4.9 $$\forall T \exists x \in C(T \sim H^{(x)});$$

but this means that $\forall T(T < T_o)$, which is absurd, so 5.4.4 holds. Hence, by 5.4.1 there is a T such that $\Psi(T,V)$; and by Q 2 and Lemma 2.4 there is a least such T, say T_1, which is unique up to $\sim$. Hence, by 5.4.1,

5.4.10 $$T < T_1 \rightarrow \exists x \Psi(T,x)$$

since trivially $\forall T \exists X \Psi(T,X)$.

<u>Claim</u> 5.4.12. For any well-ordering T', $\Psi(T',V) \rightarrow \forall x \exists T < T' \Psi(T,x)$.

<u>Proof</u>. By R1,

$$\Psi(T',V) \rightarrow \exists F[\exists T(T \neq \emptyset \wedge \mathcal{D}(F) = \mathcal{F}(T) \wedge \forall x \exists y \in \mathcal{F}(T)(x = F^{(y)}))$$
$$\wedge\ \mathcal{D}(F) = \mathcal{F}(T') \wedge \forall y \in \mathcal{F}(T')\Psi(T' \upharpoonright y, F^{(y)})]$$

$$\rightarrow \exists F[\mathcal{D}(F) = \mathcal{F}(T') \wedge \forall x \exists y \in \mathcal{F}(T')(x = F^{(y)})$$
$$\wedge \forall y \in \mathcal{F}(T')\Psi(T' \upharpoonright y, F^{(y)})]$$
$$\rightarrow \forall x \exists y \in \mathcal{F}(T')\Psi(T' \upharpoonright y, x)$$
$$\rightarrow \forall x \exists T < T' \Psi(T, x).$$

Thus, if $V \neq \{x : \exists T < T_1 \Psi(T,x)\}$, then $\exists x \Psi(T_1, x)$. But by Q3' this contradicts $\Psi(T_1, V)$. Hence $V = \{x : \exists T < T_1 \Psi(T,x)\}$. Now just as it was proved that C is well-orderable, it can be proved that V is well-orderable, and in fact that T_1 well-orders V. Q.E.D.

Finally in this section we ask the following question:

QUESTION 5.5 Can it be shown that $M^{o}_{\varepsilon} \vdash N$?

At present we can see now way of answering this question, but strongly suspect that the answer is negative.

SECTION 6

THE THEORY M_R^o AND A CONSERVATIVE EXTENSION RESULT

DEFINITION 6.1 Let $R(\cdot,\cdot)$ be a new 2-place predicate and $\mathcal{L}_M(R)$ be the language obtained from $\mathcal{L}_M$ by adjoining R. M_R^o is the theory, written in $\mathcal{L}_M(R)$, which is obtained from M^o by admitting R to the impredicative comprehension schema and adding the following axioms:

R1. $\forall T \exists X R(T,X)$

R2. $\forall X \exists T R(X,T)$

R3. $\forall T,X,Y(R(T,X) \wedge R(T,Y) \rightarrow X = Y)$

R4. $\forall T,T',X(R(T,X) \wedge R(T',X) \rightarrow T \sim T')$

R5. $\forall T,T',X(T \sim T' \wedge R(T,X) \rightarrow R(T',X))$

DEFINITION 6.2 $R^*(X,Y) \leftrightarrow \forall T,T'(R(T,X) \wedge R(T',Y) \rightarrow T < T')$.

The next two lemmas show that R^* well-orders the universe of classes in such a way that every initial segment is coded by a class. In particular, it well-orders V.

LEMMA 6.3 If $\{X : \Phi(X)\}$ is a non-empty collection of classes, then there is a R^*-least member of it.

Proof. Let T_o be the least T (to within $\sim$) such that $\exists X(\Phi(X) \wedge R(T,X))$. Then the unique X such that $R(T_o,X)$ is the R^*-least member of $\{X : \Phi(X)\}$.

LEMMA 6.4 Every R^*-initial segment of the universe of classes is codable; that is, $\forall X \exists Y \forall Z(Z \,\eta\, Y \leftrightarrow R^*(Z,X))$.

Proof. Let X be fixed. Define $\Phi(Z)$ as $R^*(Z,X)$ and suppose $R(T,X)$ holds. Then clearly

6.4.1 $$\Phi(Z) \rightarrow \exists! x \in \mathcal{F}(T) R(T \restriction x, Z)$$

and

6.4.2 $$\forall x \in \mathcal{F}(T)(R(T \restriction x, Z) \rightarrow \Phi(Z))$$

Now define a class-valued function F with domain $\mathcal{F}(T)$ as follows: $F^{(x)}$ = the unique Z such that $R(T \restriction x, Z)$. By 6.4.1 and 6.4.2, F is easily seen to be a code for Φ.

REMARKS 6.5 (i) By property Q5, F is not the only code for Φ.

(ii) By interpreting $\varepsilon X \Phi$ as the R^*-least X such that $\Phi(X)$, it is easy to see that every ε-free theorem of M^o_ε, that is, every theorem of M^o_ε which does not contain the ε-symbol, is an R-free theorem of M^o_R.

(iii) Proofs like those of Lemma 5.3 (i) and (ii) show that C^V and $\forall T(\text{DCColl}^T)$ also hold in M^o_R, where the formulae specified in these schemata may now contain the predicate R.

Most of the rest of this section is devoted to the proof of the following theorem.

THEOREM 6.6 M^o_R is a conservative extension of $M^o + WO(V) + \forall T(\text{DCColl}^T)$, where DCColl^T is R-free.

The following corollary is an immediate consequence of Remarks 6.5 (i) and (ii).

COROLLARY 6.7 M^o_ε is also a conservative of $M^o + WO(V) + \forall T(\text{DCColl}^T)$.

The proof of Theorem 6.6 is a simple forcing proof similar to the proof in Mostowski [11]. The two proofs were discovered independently but this is not surprising since they both stemmed from the argument in Felgner [4]. It was shown in Mostowski [11] that if $(A_2)_R$ is obtained from second-order arithmetic A_2 in exactly the same way as M^o_R is obtained from M^o, then every ω-model $\mathfrak{M}$ of A_2 + DCColl can be expanded by a relation $\bar{R}$ to a model $(\mathfrak{M}, \bar{R})$ of $(A_2)_R$, where DCColl is the following principle of dependent choices:

$$\forall X[\Phi(X) \rightarrow \exists Y(\Phi(Y) \wedge \Pi(X,Y))] \rightarrow \exists F \forall n \in \omega[\Phi(F^{(n)}) \wedge \Pi(F^{(n)},F^{(n+1)})] .$$

By an argument similar to Lévy's proof that $ZF^{\circ} \vdash DC \leftrightarrow DC^{\omega}$ (see Felgner[3],p. 147) but using finite sequences of classes instead of finite sequences, (a finite sequence of classes being a class-valued function whose domain is a natural number) one easily sees that DCColl is equivalent to $DCColl^{\omega}$. Hence, the proof of Theorem 6.6 carries over almost word-for-word to yeild the following strengthening of Mostowski's result:

THEOREM 6.8 $(A_2)_R$ is a conservative extension of A_2 + DCColl.

The idea of the proof of Theorem 6.6 is as follows: Let θ be an R-free theorem of M_R° and let $\mathcal{T}$ denote the theory $M^{\circ} + WO(V) + \forall T(DCColl^{T})$. If $\mathcal{T} \nvdash \theta$ then $\mathcal{T} + \neg\theta$ is consistent and so has a countable model $\mathfrak{M}$, which, of course, need not be a standard model. That is, the membership relation in $\mathfrak{M}$ need not be the real membership relation. Forcing is used not in order to construct what Mostowski calls a C-extension of $\mathfrak{M}$ by adding new classes, but in order to define a relation $\bar{R}$ in $\mathfrak{M}$ so that the expanded structure $(\mathfrak{M},\bar{R})$ is a model of $M_R^{\circ} + \neg\theta$, thus contradicting the assumption that $M_R^{\circ} \vdash \theta$. Since no new classes are added to $\mathfrak{M}$, we do not need a forcing language.

DEFINITION 6.9 If X is a relation, then it will be called a (class-valued) <u>bijection</u> if and only if $\forall Y \eta X \exists! x \in \mathfrak{D}(X)(Y = X^{(x)})$

<u>The proof of Theorem 6.6.</u> Let $\mathcal{C}(X)$ be the formula: X is a relation with domain 2 , $X^{(0)}$ is a well-ordering, and $X^{(1)}$ is a class-valued bijection whose domain is the field of $X^{(0)}$. Classes which satisfy $\mathcal{C}$ will be called <u>conditions</u> and be denoted by the letters P and Q.

DEFINITION 6.10 (i) Let F and G be relations whose domains are $\mathcal{F}(T)$ and $\mathcal{F}(T')$ respectively, where T and T' are well-orderings; and suppose

$T \overset{H}{\leq} T'$. We write $F \preccurlyeq G$ to denote that $\forall x \in \mathcal{F}(T)(F^{(x)} = G^{(H(x))})$.

(ii) For conditions P and Q, $P \sqsubset Q \leftrightarrow P^{(0)} \leq Q^{(0)} \wedge P^{(1)} \preccurlyeq Q^{(1)}$ and $P \approx Q \leftrightarrow P^{(0)} \sim Q^{(0)} \wedge P^{(1)} \equiv Q^{(1)}$.

(iii) If P is a condition and $x \in \mathcal{F}(P^{(0)})$, then $P \upharpoonright x$ will denote the condition Q defined by: $Q^{(0)} = P^{(0)} \upharpoonright x$ and $Q^{(1)} = P^{(1)} \upharpoonright O_{P^{(0)}}(x)$.

LEMMA 6.11 Let T be a well-ordering and suppose $F : \mathcal{F}(T) \to \mathcal{C}$, where $x <_T y \to F^{(x)} \sqsubset F^{(y)}$. Then there is a condition P such that

(i) $\forall x \in \mathcal{F}(T)(F^{(x)} \sqsubset P)$,

(ii) $\forall x \in \mathcal{F}(P^{(0)}) \exists y \in \mathcal{F}(T)(P \upharpoonright x \sqsubset F^{(y)})$

and (iii) P is unique up to $\approx$.

P shall be called the least upper bound of F and we shall write $P \equiv \text{lub } F$.

Proof.

Case (i): T is a successor well-ordering. Let x_m be the T-greatest element in $\mathcal{F}(T)$. It clearly suffices to take P as $F^{(x_m)}$.

Case (ii): T is a limit well-ordering. Define F_0 and F_1 with domain $\mathcal{F}(T)$ as follows:

$$F_0^{(x)} = F^{(x)(0)} \text{ and } F_1^{(x)} = F^{(x)(1)}.$$

Let $T' = \text{Sup } F_0$, which exists by Theorem 0.25 of Marek [10]. Then T' is a limit well-ordering and

$$y \in \mathcal{F}(T') \to \exists x \in \mathcal{F}(T)(T' \upharpoonright y \leq F_0^{(x)})$$

Let 1_y be the T-least such x and suppose $T' \upharpoonright y \overset{H_y}{\leq} F_0^{(1_y)}$. Since $y \in \mathcal{F}(T' \upharpoonright y+1)$, it follows that

$$H_{y+1}(y) \in \mathcal{F}(F_0^{(1_{y+1})}) = \mathcal{D}(F_1^{(1_{y+1})}) .$$

Now define G with domain $\mathcal{F}(T')$ as follows: $G^{(y)} = F_1^{(l_{y+1})}(H_{y+1}(y))$.
We show that G is a class-valued bijection. Suppose it is not; then

6.11.1 $$\exists x,y \in \mathcal{F}(T')(x <_{T'} y \wedge G^{(x)} = G^{(y)}).$$

Clearly $x <_{T'} y \rightarrow l_{x+1} \leq_T l_{y+1}$, so $F_o^{(l_{x+1})} \overset{K}{\leq} F_o^{(l_{y+1})}$ for some K.
But $F_1^{(l_{x+1})} \leq F_1^{(l_{y+1})}$, so

6.11.2 $$\forall z \in \mathcal{F}(F_o^{(l_{x+1})})(F_1^{(l_{x+1})(z)} = F_1^{(l_{y+1})(K(z))}).$$

Hence,

6.11.3 $$F_1^{(l_{x+1})}(H_{x+1}(x)) = F_1^{(l_{y+1})}(K(H_{x+1}(x))).$$

Now $T' \upharpoonright x+1 \overset{H_{x+1}}{\leq} F_o^{(l_{x+1})} \overset{K}{\leq} F_o^{(l_{y+1})}$ and $T' \upharpoonright y+1 \overset{H_{y+1}}{\leq} F_o^{(l_{y+1})}$, so by uniqueness, $H_{y+1} \upharpoonright x+1 = K \cdot H_{x+1}$. Hence, $H_{y+1}(x) = K(H_{x+1}(x))$, so by 6.11.3,

6.11.4 $$F_1^{(l_{x+1})}(H_{x+1}(x)) = F_1^{(l_{y+1})}(H_{y+1}(y)).$$

Hence, by 6.11.1,

6.11.5 $$F_1^{(l_{y+1})}(H_{y+1}(x)) = F_1^{(l_{y+1})}(H_{y+1}(y)).$$

This, however, contradicts the bijectiveness of F_1, so G is a bijection.

Now define P as follows: $P^{(0)} = T'$ and $P^{(1)} = G$.

In order to prove (i) it suffices to show that

6.11.6 $$\forall x \in \mathcal{F}(T)(F_1^{(x)} \leq G).$$

Let $z \in \mathcal{F}(F_o^{(x)}) = \mathcal{D}(F_1^{(x)})$. Since $F_o^{(x)} < T'$, there is a unique similarity K' such that $F_o^{(x)} \overset{K'}{<} T'$. Then

$$F_o^{(x)} \overset{K'}{\simeq} T' \upharpoonright t_x,$$

where t_x is the least strict upper bound of the class $\{K'(u) : u \in \mathcal{F}(F_o^{(x)})\}$.
Now

$$T' \upharpoonright t_x+1 \overset{H_{t_x+1}}{\leqslant} F_o^{(1_{t_x+1})},$$

so

$$F_1^{(x)} \preccurlyeq F_1^{(1_{t_x+1})}$$

and

$$F_1^{(x)}(z) = F_1^{(1_{t_x+1})}(H_{t_x+1}(K(z))) = G^{(K(z))}.$$

Hence, $F_1^{(x)} \preccurlyeq G$; so 6.11.6, and hence (i), holds.

Now $T' \upharpoonright x \overset{K''}{\leqslant} F_o^{(1_{x+1})}$, say, so in order to prove (ii), it suffices to show that

6.11.7 $\quad \forall z <_{T'} x (G^{(z)} = F_1^{(1_{x+1})}(K''(z)))$.

Let $z <_{T'} x$. Since $T' \upharpoonright z+1 \overset{H_{z+1}}{\leqslant} F_o^{(1_{z+1})} \overset{K^*}{\leqslant} F_o^{(1_{x+1})}$ for a unique K^*, it is clear that $K'' \upharpoonright z+1 = K^* \cdot H_{z+1}$ and $K''(z) = K^*(H_{z+1}(z))$. But $F_1^{(1_{z+1})} \preccurlyeq F_1^{(1_{x+1})}$, so

$$G^{(z)} = F_1^{(1_{z+1})}(H_{z+1}(z)) = F_1^{(1_{x+1})}(K^*(H_{z+1}(z))) = F_1^{(1_{x+1})}(K''(z)),$$

so 6.11.7 is proved.

(iii) follows immediately from the fact that T' is unique to within $\sim$; so the proof of Lemma 6.11 is complete.

Now let C be the set $\{Y \in |\mathfrak{M}| : \mathfrak{M} \models \mathcal{C}(Y)\}$. A subset X of C is said to be <u>dense</u>$_{\mathfrak{M}}$ in C if $\forall Y \in C \exists Z \in X(\mathfrak{M} \models Y \sqsubset Z)$.

Since $\mathfrak{M}$ is denumerable, the set of formulae of $\mathcal{L}_M$ with one free variable and parameters in $\mathfrak{M}$ is also denumerable. Hence, the set D of dense$_{\mathfrak{M}}$ subsets of C, which are definable in $\mathfrak{M}$ by a formula with one free variable and parameters in $\mathfrak{M}$, is denumerable. Let $(D_n)_{n \in \omega}$ be an enumeration of D.

We now define, by recursion on ω, a sequence of elements of such that $\forall n \in \omega(Y_n \in D_n)$. Let Y_0 be any member of D_0 and Y_{n+1} be any member of D_{n+1} such that $\mathfrak{M} \models Y_{n+1} \sqsupset Y_n$. (Notice that each Y_n can be chosen since D_n is $\text{dense}_{\mathfrak{M}}$.)

Let G be the set $\{Y_n : n \in \omega\}$. Then $\forall n \in \omega(\mathfrak{M} \models Y_n \sqsubset Y_{n+1})$.

Now let Γ be a new 1-place predicate and let $\mathcal{T}'$ be the theory obtained from $\mathcal{T} + \neg\theta$ by adjoining Γ to the language $\mathcal{L}_{M'}$, <u>not</u> allowing Γ to figure in the impredicative comprehension schema, and adding the following axioms and schema:

$\Gamma 1.$ $\forall X(\Gamma(X) \rightarrow \mathscr{C}(X))$

$\Gamma 2.$ $\forall X,Y(\Gamma(X) \wedge \Gamma(Y) \rightarrow X \sqsubset Y \vee Y \sqsubset X)$

$\Gamma 3.$ $\forall X_1,\ldots,X_n[\forall X(\Phi(X,X_1,\ldots,X_n) \rightarrow \mathscr{C}(X)) \wedge \forall X(\mathscr{C}(X) \rightarrow \exists Y(\mathscr{C}(Y) \wedge X \sqsubset Y \wedge \Phi(Y,X_1,\ldots,X_n)) \rightarrow \exists X(\Gamma(X) \wedge \Phi(X,X_1,\ldots,X_n))]$, where Φ does not contain Γ (i.e., is Γ-free).

REMARK 6.12 The hypothesis of $\Gamma 3$ says that Φ defines a <u>dense</u> subcollection of $\{X : \mathscr{C}(X)\}$; so $\Gamma 3$ says that every dense subcollection of $\{X : \mathscr{C}(X)\}$, which is defined by a Γ-free formula, meets Γ.

LEMMA 6.13 $\mathcal{T}'$ is consistent.

<u>Proof</u>. First, we know that $\mathfrak{M} \models \mathcal{T} + \neg\theta$. By interpreting $\Gamma(X)$ as $X \in G$, it is clear that $\Gamma 1$ and $\Gamma 2$ are satisfied. The interpretion of the hypothesis of $\Gamma 3$ says that $\{X \in |\mathfrak{M}| : \mathfrak{M} \models \Phi(X)\}$ is a dense subset of C (defined by a Γ-free formula), so by the definition of G, it meets Γ. Thus, $\Gamma 3$ is also satisfied, so the lemma is proved.

Henceforth we work in the theory $\mathcal{T}'$. Our aim is to find an interpretation I of $M_R^o + \neg\theta$ in $\mathcal{T}'$ and hence obtain the required contradiction.

In $\mathcal{T}'$ we define a relation $\bar{R}(T,X)$ as follows:

$$\bar{R}(T,X) \leftrightarrow \exists P,x(\Gamma(P) \wedge x \in \mathcal{F}(P^{(0)}) \wedge T \sim P^{(0)} \upharpoonright x \wedge P^{(1)}(x) = X).$$

The interpretation Φ_I of a formula Φ is obtained by replacing every occurrence of R in Φ by $\bar{R}$

LEMMA 6.14 For any well-ordering T, the collection of conditions P such that $T < P^{(0)}$ is dense.

<u>Proof</u>. Let P be any condition and T' be a well-ordering $> \max\{T, P^{(0)}\}$. Suppose $P^{(0)} \overset{H}{\sim} T' \upharpoonright x$ and let $\mathcal{E}$ denote the collection

$$\{ P^{(1)}(u)(v) \ ; \ u \in \mathcal{F}(P^{(0)}) \wedge v \in \mathcal{F}(T') \setminus O_{T'}(x) \}.$$

Since $\mathcal{E}$ is codable by the class

$$\{ \langle\langle u,v\rangle,y\rangle : y \in P^{(1)}(u)(v) \wedge u \in \mathcal{F}(P^{(0)}) \wedge v \in \mathcal{F}(T') \setminus O_{T'}(x) \},$$

there is a class A such that A is not in $\mathcal{E}$. Put

$$B = \{ \langle v, \langle v,y\rangle\rangle : v \in \mathcal{F}(T') \setminus O_{T'}(x) \wedge y \in A \}.$$

Clearly B is a bijection and for all $j \in \mathcal{F}(T') \setminus O_{T'}(x)$, $\neg(B^{(v)} \eta \ P^{(1)})$. Now putting

$$P^{(1)*} = \{ \langle H(y),z\rangle : \langle y,z\rangle \in P^{(1)} \},$$

it follows that $P^{(1)*} \equiv P^{(1)}$ (see Defn. 1.22), so

$$j \in \mathcal{F}(T') \setminus O_{T'}(x) \rightarrow \neg(B^{(v)} \eta \ P^{(1)*}).$$

The required extension Q of P is now obtained by putting

$$Q^{(0)} = T'$$

and $$Q^{(1)} = P^{(1)*} \cup B.$$

LEMMA 6.15 For any X, the collection $\{P : X \eta P^{(1)}\}$ is dense.

Proof. Let P be any condition and suppose $\neg(X \eta P^{(1)})$. Let T be a successor of $P^{(0)}$, x_m be the T-greatest element in $\mathcal{F}(T)$ and suppose $T \upharpoonright x_m \overset{H}{\sim} P^{(0)}$. Define F with domain $\mathcal{F}(T)$ as follows:

$$F^{(x)} = \begin{cases} P^{(1)(H(x))} & \text{if } x <_T x_m; \\ X \quad \text{if } x = x_m. \end{cases}$$

Now define a condition Q as follows: $Q^{(0)} = T$ and $Q^{(1)} = F$. Clearly $P \sqsubset Q$ and $X \eta Q^{(1)}$, so the lemma is proved.

LEMMA 6.16 Let X be any relation whose domain is well-orderable. Then the collection $\{P : X \text{ Inc } P^{(1)}\}$ is dense. (See page 191 for the definition of Y Inc Z .)

Proof. Suppose T is a well-ordering of $\mathfrak{D}(X)$. Let P_o be any condition and x_o the T-least element of (X). For $x \in \mathfrak{D}(X)$, let $\mathcal{P}_x$ denote the collection $\{P : X^{(x)} \eta P^{(1)}\}$. By Lemma 6.15, $\mathcal{P}_x$ is dense for every $x \in \mathfrak{D}(X)$, and clearly if P is in $\mathcal{P}_x$ and $P \sqsubset Q$, then Q is in $\mathcal{P}_x$.

We now use DCColl^T to find an extension Q of P_o such that Q is in $\mathcal{P}_x$ for every $x \in \mathfrak{D}(X)$.

Let $x \in \mathfrak{D}(X)$ and suppose

$$(F : O_T(x) \rightarrow \mathcal{C}) \wedge \forall y,z <_T x(X^{(y)} \eta F^{(y)(1)} \wedge (y <_T z \rightarrow F^{(y)} \sqsubset F^{(z)})).$$

By Lemma 6.11 there is a condition P_x such that $P_x = \text{lub } F$. Since $\mathcal{P}_x$ is dense there is a P in $\mathcal{P}_x$ such that $P_x \sqsubset P$. Then clearly,

$$\forall y <_T x(F^{(y)} \sqsubset P),$$

so the hypothesis of DCColl^T is satisfied. Hence, choosing P' in $\mathcal{P}_{x_o}$

such that $P_o \sqsubset P'$ (which is possible since $\mathcal{P}_{x_o}$ is dense), there is a G such that

$$(G : \mathcal{D}(X) \to \mathcal{C}) \wedge G^{(x_o)} = P' \wedge \forall y,z \in \mathcal{D}(X)(X^{(y)} \eta\, G^{(y)(1)} \wedge (y <_T z \to G^{(y)} \sqsubset G^{(z)})).$$

Let $Q = \text{lub}\, G$. Then $P_o \sqsubset Q$ and for every $x \in \mathcal{D}(X)$, Q is in $\mathcal{P}_x$. Hence, X Inc $Q^{(1)}$ and the lemma is proved.

LEMMA 6.17 The interpretations of R1 - R5 hold in $\mathcal{J}'$.

Proof.

(i) The proof of $(R1)_I$: Let T be any well-ordering. By Lemma 6.14 the collection of conditions P such that $T < P^{(0)}$ is dense, so by $\Gamma 3$ it meets Γ. Hence, suppose $\Gamma(P)$, $T < P^{(0)}$, $T \sim P^{(0)} \restriction x$ and $P^{(0)(x)} = X$. Then $\bar{R}(T,X)$ holds.

(ii) The proof of $(R2)_I$: Let X be any class. By Lemma 6.14 and $\Gamma 3$, there is a P such that $\Gamma(P)$ and $X \,\eta\, P^{(1)}$. Let $X = P^{(1)(x)}$. Then $\bar{R}(P^{(0)} \restriction x, X)$ holds trivially, so $(R2)_I$ holds.

(iii) The proof of $(R3)_I$: Suppose $\bar{R}(T,X) \wedge \bar{R}(T,Y)$ holds. Then there exist P and Q such that $\Gamma(P)$ and $\Gamma(Q)$, and there exist $x \in \mathcal{F}(P^{(0)})$ and $y \in \mathcal{F}(Q^{(0)})$ such that $T \sim P^{(0)} \restriction x \sim Q^{(0)} \restriction y$, $P^{(0)(x)} = X$ and $Q^{(0)(y)} = Y$. By $\Gamma 2$ we may suppose $P \sqsubset Q$. Then for some H, $P^{(0)} \overset{H}{\leqslant} Q^{(0)}$ and $P^{(1)(x)} = Q^{(1)(H(x))}$. But $P^{(0)} \restriction x \overset{H \restriction x}{\sim} Q^{(0)} \restriction H(x)$, so $H(x) = y$ and $X = Y$. Thus, $(R3)_I$ holds.

(iv) The proof of $(R4)_I$: Suppose $\bar{R}(T,X) \wedge \bar{R}(T',X)$ holds. Then there exist P and Q which satisfy Γ and $x \in \mathcal{F}(P^{(0)})$ and $y \in \mathcal{F}(Q^{(0)})$ such that $T \sim P^{(0)} \restriction x$, $P^{(1)(x)} = X$, $T' \sim Q^{(0)} \restriction y$, $Q^{(1)(y)} = X$. By $\Gamma 2$ we may suppose $P \sqsubset Q$. Then for some H, $P^{(0)} \overset{H}{\leqslant} Q^{(0)}$ and $P^{(1)(x)} = Q^{(1)(H(x))} = Q^{(1)(y)}$. Since $Q^{(1)}$ is a bijection, it follows that $H(x) = y$. But $P^{(0)} \restriction x \overset{H \restriction x}{\sim} Q^{(0)} \restriction H(x) = Q^{(0)} \restriction y$, so $T \sim T'$ and $(R4)_I$ holds.

(v) The proof of $(R5)_I$ is trivial.

Now $(\neg\theta)_I$ is $\neg\theta$ since θ is R-free. Hence, in order to show that $\mathcal{J}' \vdash (M_R^o + \neg\theta)_I$, that is, $\mathcal{J}' \vdash \Phi_I$ for every axiom Φ of $M_R^o + \neg\theta$, it only remains, after the above lemma, to show that $\mathcal{J}' \vdash (\mathrm{VIII}_R)_I$, that is, $\mathcal{J}' \vdash \Phi_I$ for every instance Φ of the impredicative comprehension schema which contains the predicate R. For this purpose we now define the forcing relation $P \Vdash \Phi$ (P <u>forces</u> Φ) between conditions P and formulae Φ of $\mathcal{L}_M(R)$. The definition is by induction on the length of Φ.

$$P \Vdash X = Y \leftrightarrow X = Y,$$

$$P \Vdash X \in Y \leftrightarrow X \in Y,$$

$$P \Vdash R(T,X) \leftrightarrow \exists x \in \mathcal{F}(P^{(0)})(T \sim P^{(0)} \restriction x \wedge P^{(1)}(x) = X),$$

$$P \Vdash \neg\Phi \quad \leftrightarrow \neg\exists Q(P \sqsubset Q \wedge Q \Vdash \Phi),$$

$$P \Vdash \Phi \wedge \Psi \leftrightarrow P \Vdash \Phi \wedge P \Vdash \Psi,$$

and $$P \Vdash \exists X\Phi \leftrightarrow \exists X(P \Vdash \Phi(X)).$$

Notice that for any formula Φ of $\mathcal{L}_M(R)$, the formula $P \Vdash \Phi$ is R-free. We write $P \parallel \Phi$ (P <u>decides</u> Φ) for $P \Vdash \Phi \vee P \Vdash \neg\Phi$. The usual forcing lemmas follow.

LEMMA 6.18 For no formula Φ is there a condition P such that $P \Vdash \Phi \wedge \neg\Phi$.

The proof is immediate from the definition of $P \Vdash \neg\Phi$.

LEMMA 6.19 If Φ is R-free, then $(P \Vdash \Phi) \leftrightarrow \Phi$.

The proof is by induction on the length of Φ. So is the proof of the next lemma.

LEMMA 6.20 $P \Vdash \Phi \wedge P \sqsubset Q \rightarrow Q \Vdash \Phi$.

LEMMA 6.21 For any formula Φ, the collection $\{P : P \parallel \Phi\}$ is dense.

The proof is by the definition of $P \Vdash \neg\Phi$.

LEMMA 6.22 For every formula Φ, $\Phi_I \leftrightarrow \exists P(\Gamma(P) \wedge P \Vdash \Phi)$.

Proof. By induction on the length of Φ.

The proof is trivial if Φ is atomic.

Suppose Φ is $\neg\Psi$. If Ψ_I holds, then by the induction hypothesis, there is a P such that $\Gamma(P) \wedge P \Vdash \Psi$. Hence, by $\Gamma 2$ and Lemmas 6.18 and 6.20, there is no Q such that $\Gamma(Q) \wedge Q \Vdash \Phi$; so $\exists P(\Gamma(P) \wedge P \Vdash \Phi) \rightarrow \Phi_I$. Assume $\neg\exists P(\Gamma(P) \wedge P \Vdash \Phi)$. By Lemma 6.21 and $\Gamma 3$, there is a Q such that $\Gamma(Q) \wedge Q \parallel \Psi$. Hence, $Q \Vdash \Psi$, so by the induction hypothesis, Ψ_I holds. Thus, $\Phi_I \leftrightarrow \exists P(\Gamma(P) \wedge P \Vdash \Phi)$.

Suppose Φ is $\Phi_1 \wedge \Phi_2$. Then

$$\begin{aligned}(\Phi_1 \wedge \Phi_2)_I &\leftrightarrow (\Phi_1)_I \wedge (\Phi_2)_I \\ &\leftrightarrow \exists P(\Gamma(P) \wedge P \Vdash \Phi_1) \wedge \exists Q(\Gamma(Q) \wedge Q \Vdash \Phi_2) \\ &\leftrightarrow \exists P(\Gamma(P) \wedge P \Vdash \Phi_1 \wedge \Phi_2)\end{aligned}$$

by $\Gamma 2$ and Lemma 6.20.

Suppose Φ is $\exists X \Psi$. For every X,

$$(\Psi(X))_I \leftrightarrow \exists P(\Gamma(P) \wedge P \Vdash \Psi(X))$$

by the induction hypothesis; so clearly,

$$\begin{aligned}(\exists X \Psi)_I &\leftrightarrow \exists X(\Psi(X))_I \\ &\leftrightarrow \exists P(\Gamma(P) \wedge P \Vdash \exists X \Psi).\end{aligned}$$

The proof of the lemma is complete.

LEMMA 6.23 Let $\Phi(x)$ be any formula with one free set-variable x and possibly parameters. Then the collection $\{P : \forall x(P \parallel \Phi(x))\}$ is dense.

<u>Proof</u>. By WO(V) there is a well-ordering T of V. Let P_0 be any condition. For any x let $\Delta(x,P)$ denote the formula $P_0 \sqsubset P \wedge P \parallel \Phi(x)$. By Lemma 6.21, the collection $\{P : \Delta(x,P)\}$ is dense for every x, and by Lemma 6.20, if $\Delta(x,P)$ holds and $P \sqsubset Q$, then $\Delta(x,Q)$ also holds.

We now use DCColl^T to find a condition Q such that $\Delta(x,Q)$ holds for every x. The lemma follows immediately.

Let x be any set and suppose

$$(F : O_T(x) \rightarrow \mathcal{C}) \wedge \forall y,z(y <_T x \wedge z <_T x \rightarrow \Delta(y,F^{(y)}) \wedge (y <_T z \rightarrow F^{(y)} \sqsubset F^{(z)})).$$

Put $P_x = \mathrm{lub}\, F$, which exists by Lemma 6.11. Since $\{P : \Delta(x,P)\}$ is dense, there is a P such that $\Delta(x,P) \wedge P_x \sqsubset P$. Clearly $\forall y <_T x(F^{(y)} \sqsubset P)$, so the hypothesis of DCColl_2^T is satisfied. Hence, there is a G such that

$$(G : \mathcal{F}(T) \rightarrow \mathcal{C}) \wedge \forall y,z(\Delta(y,G^{(y)}) \wedge (y <_T z \rightarrow G^{(y)} \sqsubset G^{(z)})).$$

Put $Q = \mathrm{lub}\, G$. Then Q is the required condition.

It can now be shown that $\mathcal{J}' \vdash (\mathrm{VIII}_R)_I$.

Let $\Phi(x)$ be any formula of $\mathcal{L}_M(R)$ with one free set-variable and possibly parameters. We need to show that

$$(*) \qquad \exists Y \forall x(x \in Y \leftrightarrow (\Phi(x))_I)$$

By Lemma 6.22. $(\Phi(x))_I \leftrightarrow \exists P(\Gamma(P) \wedge P \Vdash \Phi(x))$ for every x, and by Lemma 6.23 and $\Gamma 3$, there is a Q such that $\Gamma(Q) \wedge \forall x(Q \parallel \Phi(x))$. We show that

$$(**) \qquad \forall x(\exists P(\Gamma(P) \wedge P \Vdash \Phi(x)) \leftrightarrow Q \Vdash \Phi(x))$$

for then (*) follows trivially from Lemma 6.22 and the impredicative comprehension schema.

Since $\Gamma(Q)$ holds, the implication from right to left in (**) is trivial. Suppose $\Gamma(P) \wedge P \Vdash \Phi(x)$. By $\Gamma 2$, $P \sqsubset Q$ or $Q \sqsubset P$. If $P \sqsubset Q$, then by Lemma 6.20, $Q \Vdash \Phi(x)$. If $Q \sqsubset P$ and $\neg(Q \Vdash \Phi(x))$, then $Q \Vdash \neg\Phi(x)$ since $Q \parallel \Phi(x)$. Hence, by Lemma 6.20, $P \Vdash \neg\Phi(x)$, which is impossible since $P \Vdash \Phi(x)$. (**) is thus proved and the proof of Theorem 6.6 is complete.

EPILOG

ANOTHER CONSERVATIVE EXTENSION RESULT

Let ZF^{o-}_R be obtained from ZF^{o-} (= ZF without the axiom of foundation and without the power-set axiom) by adding a new 2-place predicate R to the language of ZF, allowing R to figure in the replacement schema, and adjoining the following axioms, the conjunction of which says that R establishes a bijection between On and V.

1. $\forall\alpha\exists xR(\alpha,x)$

2. $\forall x\exists\alpha R(\alpha,x)$

3. $\forall\alpha,x,y(R(\alpha,x)\wedge R(\alpha,y)\rightarrow x=y)$

4. $\forall\alpha,\beta,x(R(\alpha,x)\wedge R(\beta,x)\rightarrow\alpha=\beta)$

By taking conditions as bijections whose domains are ordinals or as pairs $\langle\alpha,f\rangle$, where f is a bijection whose domain is α, one easily sees that the proof of Theorem 6.6 can be modified to yield the following result:

THEOREM 7.1 ZF^{o-}_R is a conservative extension of $ZF^{o-} + WO + \forall\alpha DCC^{\alpha}$.

COROLLARY 7.2 (Immediate)

(i) $NBG^{o-} + N$ is a conservative extension of $NBG^{o-} + WO + \forall\alpha DCC^{\alpha}$ with respect to ZF-sentences.

(ii) $NBG^{o} + N$ is a conservative extension of $NBG^{o} + \forall\alpha DCC^{\alpha}$ with respect to ZF-sentences.

(iii) NBG + E is a conservative extension of NBG + AC with respect to ZF-sentences.

Corollary 7.2 (iii) is well-known and was proved, for example, in

Felgner [4].

We conclude with some questions.

QUESTION 7.3 By adapting the forcing method in Chaqui [2], U. Felgner has shown that M + E is a conservative extension of M + AC with respect to ZF-sentences, but can it be shown that M^o + N is a conservative extension of $M^o + \forall\alpha DCC^\alpha$ with respect to ZF-sentences ?

QUESTION 7.4 (i) Is M^o + E a conservative extension of M^o + AC with respect to ZF-sentences ?

(ii) Is NBG^o + E a conservative extension of NBG^o + AC with respect to ZF-sentences ?

In Section 3 it was shown that, in the absence of the axiom of foundation, E is at least as weak as any other class-form of choice mentioned in that section. (See, for example, Theorem 3.10(vii).) Moreover, we have sought in vain a formula which is a theorem of M^o + E (for example) but which is not a theorem of M^o + AC. It would not surprise us, therefore, if the answer to Question 7.4 is affirmative.

QUESTION 7.5 Is $M^o + C^V$ a conservative extension of $M^o + \forall x C^x$ with respect to ZF-sentences ?

QUESTION 7.6 Is $M^o + SC^V$ a conservative extension of $M^o + \forall x SC^x$ with respect to ZF-sentences ?

REFERENCES

[1] M. BOFFA: Graphes extensionnels et axiome d'universalite. Zeit. f. math. Logik und Grund. d. Math., vol. 14 (1968), p. 329 - 334.

[2] R.CHUAQUI: Forcing for the impredicative theory of classes. The Jour. of Sym. Logic, vol. 37 (1972), p. 1 - 18.

[3] U. FELGNER: Models of ZF-set theory. Lecture Notes in Math., vol. 223, Springer Verlag, Berlin, Heidelberg, New York, 1971.

[4] U. FELGNER: Comparison of the axioms of local and universal choice. Fund. Math., vol. 71 (1971), p. 43 - 62.

[5] U. FELGNER: Choice functions on sets and classes. To appear in the Bernays Festschrift, North Holland.

[6] U. FELGNER, T. JECH: Variants of the axiom of choice in set theory with atoms. Fund. Math., vol. 79 (1973), p. 79 - 85.

[7] T.B. FLANNAGAN: Set theories incorporating Hilbert's ε-symbol. Dissertationes Math., vol. 114, Warsaw, 1974.

[8] T. JECH: The axiom of choice. Studies in Logic, vol. 75, North Holland, 1973.

[9] A. LEISENRING: Mathematical logic and Hilbert's ε-symbol. University Math. Series, MacDonald, London, 1969.

[10] W. MAREK: On the metamathematics of impredicative set theory. Dissertationes Math., vol. 98, Warsaw, 1973.

[11] A. MOSTOWSKI: Models of second order arithmetic with definable skolem functions. Fund. Math., vol. 75 (1972), p. 223 - 234.

[12] H. RUBIN, J. RUBIN: Equivalents of the axiom of choice. Studies in Logic, North Holland, 1^63.

First-order logic and its extensions

J. Flum

We start with Lindström's theorem which tells us that there is no language more expressive than the language of first-order logic and still satisfying the compactness theorem and the Löwenheim-Skolem theorem. We analyze two aspects of the proof of this theorem, namely the use of the algebraic characterization of elementary eqivalence and the way the compactness theorem is applied.

The algebraic characterization of elementary equivalence, due to Fraissé, leads to an algebraic understanding of the expressive power of first-order logic and is a powerful tool in model theory (see §.1). In particular, Fraissé's notion of n-isomorphism, $\simeq_n$, is a first-order definable approximation of the isomorphism relation $\simeq$. The sequence $\simeq_n$ converges to $\simeq$, if convergence is defined in a way suggested by Lindström's proof. Similarly, we may introduce first-order definable finite approximations for other algebraic relations between structures, such as being an extension or a homomorphic image. We apply the convergence notion to obtain a general interpolation and preservation theorem for first-order logic (§.3).

Let L be a language satisfying the compactness theorem. Let Σ be a set of L-sentences containing a binary predicate symbol $<$. If Σ has a model $\mathfrak{A}$ where $<^A$ is well-ordered of order type ω, then, by the compactness theorem, Σ has a model where $<^A$ is not well-ordered. The application of the compactness property in Lindström's proof is of this form. It turns out that the non-definability of

$(\omega,<)$ in a language L is equivalent to the compactness theorem. Thus, if ξ is any ordinal the statement "$(\xi,<)$ is not L-definable" may be regarded as a weak compactness property. The smallest non L-definable ξ is called the well-ordering number of L. We use the well-ordering number of $L_{\omega_1\omega}$ to derive some preservation theorems for $L_{\omega_1\omega}$ (§.4) and to show that the class of (first-order) minimal structures is not $L_{\omega_1\omega}$-axiomatizable. For a large class of logics (including $L_{\kappa\omega}$, $L(Q_\kappa)$ and admissible logics), we apply the well-ordering number to obtain upward Löwenheim-Skolem theorems (§.6).

§.1 Fraïssé's theorem

Two structures $\mathfrak{A}$ and $\mathfrak{B}$ are elementarily equivalent if they are models of the same first-order formulas. In this section we derive Fraïssé's purely algebraic characterization of elementary equivalence [18].

We denote similarity types by $K, K_1, K_2, \ldots$. They are sets of predicate symbols $(P,Q,R,\ldots)$ and function symbols $(f,g,\ldots)$. Sometimes 0-placed function symbols are denoted by $c,d,\ldots$. Let T^K be the set of terms corresponding to K and L^K the set of formulas of first-order logic. Assume we have a fixed enumeration $v_o, v_1, v_2, \ldots$ of the variables. For $n \in \omega$ let L_n^K be the set of formulas whose free variables are among $v_o, \ldots, v_{n-1}$, i.e.

$$L_n^K = \{\varphi \mid fr(\varphi) \subset \{v_o, \ldots, v_{n-1}\}\},$$

where $fr(\varphi)$ is the set of free variables of φ. We denote structures of type K by $\mathfrak{A}, \mathfrak{B}, \ldots$. $A, B, \ldots$ are the corresponding universes.

A one-to-one function p is a partial isomorphism from $\mathfrak{A}$ to $\mathfrak{B}$ if the domain of $p, do(p)$, is a subset of A, the range of $p, rg(p)$, is a subset of B, and for each $P \in K$ and $a_o, \ldots, a_{n-1} \in do(p)$

$$P^{\mathfrak{A}} a_o \ldots a_{n-1} \quad \text{iff} \quad P^{\mathfrak{B}} p(a_o) \ldots p(a_{n-1}) ,$$

and for each $f \in K$ and $a_o, \ldots, a_n \in do(p)$

$$(1) \qquad f^{\mathfrak{A}}(a_o, \ldots, a_{n-1}) = a_n \quad \text{iff} \quad f^{\mathfrak{B}}(p(a_o), \ldots, p(a_{n-1})) = p(a_n) .$$

Thus in case $a_o,\dots,a_{n-1} \in do(p)$ and $f^{\mathfrak{A}}(a_o,\dots,a_{n-1}) \notin do(p)$, (1) only gives us the information that $f^{\mathfrak{B}}(p(a_o),\dots,p(a_{n-1}))$ does not lie in rg(p). Note that $p = \emptyset$, the empty function, is in $P(\mathfrak{A},\mathfrak{B})$ in case K does not contain any o-ary predicate symbol. Let $P(\mathfrak{A},\mathfrak{B})$ be the set of partial isomorphisms from $\mathfrak{A}$ to $\mathfrak{B}$.

1.1 Definition. We write $I: \mathfrak{A} \simeq_p \mathfrak{B}$ and say that $\mathfrak{A}$ and $\mathfrak{B}$ are partially isomorphic via I, if I is a non-empty set of partial isomorphisms with the back and forth property,

forth property: for every $p \in I$ and $a \in A$ there is a $q \in I$ with $p \subset q$ and $a \in do(q)$,

back property: for every $p \in I$ and $b \in B$ there is a $q \in I$ with $p \subset q$ and $b \in rg(q)$.

We write $\mathfrak{A} \simeq_p \mathfrak{B}$, if there is an I such that $I: \mathfrak{A} \simeq_p \mathfrak{B}$.

The following well-known generalization of a theorem of Cantor describes the relation between $\simeq$ and $\simeq_p$.

1.2 Theorem.

(i) If $\mathfrak{A} \simeq \mathfrak{B}$ then $\mathfrak{A} \simeq_p \mathfrak{B}$.

(ii) If $\mathfrak{A} \simeq_p \mathfrak{B}$ and A and B are denumerable, then $\mathfrak{A} \simeq \mathfrak{B}$.

For future reference we give a proof of (ii).

Assume $I: \mathfrak{A} \simeq_p \mathfrak{B}$ where $A = \{a_o, a_1, \dots\}$ and $B = \{b_o, b_1, \dots\}$. Given $p_o \in I$ let

p_{2n+1} be some extension p of p_{2n} in I with $a_n \in do(p)$,

p_{2n+2} be some extension p of p_{2n+1} in I with $b_n \in rg(p)$.

Then $p = \bigcup p_n$ has domain A and range B. Hence, $\mathfrak{A} \simeq \mathfrak{B}$.

1.3 Exercises.

(a) If $\mathfrak{A}$ and $\mathfrak{B}$ are dense linear orderings (without endpoints), then $\mathfrak{A} \simeq_p \mathfrak{B}$. (Take $I = \{p \mid p \in P(\mathfrak{A},\mathfrak{B}),\ do(p)$ finite$\}$).

(b) If $\mathfrak{A}$ and $\mathfrak{B}$ are ω-saturated, algebraically closed fields of the same characteristic, then $\mathfrak{A} \simeq_p \mathfrak{B}$. (Take $I = \{p \mid p \in P(\mathfrak{A},\mathfrak{B}), do(p)$ finitely generated subfield of $\mathfrak{A}\}$).

(c) If $\mathfrak{A}$ and $\mathfrak{B}$ are ω-saturated discrete linear orderings, then $\mathfrak{A} \simeq_p \mathfrak{B}$. (Take $I = \{p \mid p \in P(\mathfrak{A},\mathfrak{B}), do(p)$ finite$\}$).

By the preceeding theorem $\simeq_p$ may be viewed as a countable approximation of the relation of isomorphism. We need for our purposes still weaker notions of isomorphsm.

We use ξ,η,ζ to denote ordinals.

1.4 Definition. We write $(I_\eta)_{\eta<\xi}: \mathfrak{A} \simeq_\xi \mathfrak{B}$ and say that $\mathfrak{A}$ and $\mathfrak{B}$ are ξ-partially isomorphic via $(I_\eta)_{\eta<\xi}$, if

(i) $I_o \supset I_1 \supset \ldots \supset I_\eta \supset \ldots$, where for each $\eta < \xi$, I_η is a non-empty set of partial isomorphisms.

(ii) (back and forth property) for each $\eta + 1 < \xi$, $p \in I_{\eta+1}$ and $a \in A$ (resp. $b \in B$) there is a $q \in I_\eta$ with $p \subset q$ and $a \in do(q)$ (resp. $b \in rg(q)$).

Write $\mathfrak{A} \simeq_\xi \mathfrak{B}$ if there is $(I_\eta)_{\eta<\xi}$ such that $(I_\eta)_{\eta<\xi}: \mathfrak{A} \simeq_\xi \mathfrak{B}$.

1.5 Theorem.

(i) If $\mathfrak{A} \simeq_p \mathfrak{B}$ then for all ξ, $\mathfrak{A} \simeq_\xi \mathfrak{B}$.

(ii) If $\mathfrak{A} \simeq_{n+2} \mathfrak{B}$ and A has n elements, then $\mathfrak{A} \simeq \mathfrak{B}$.

(ii) tells us that for $n \in \omega$, $\simeq_n$ may be viewed as a <u>finite approximation</u> of the isomorphism relation.

1.6 <u>Exercises</u>.

(a) If $\mathfrak{A} = (A,R)$ and $\mathfrak{B} = (B,S)$ are ordered structures, we denote by $\mathfrak{A} \otimes \mathfrak{B}$ the structure $(A \times B, [)$, where $[$ is the order defined by

$$(a,b) \,[\, (a',b') \quad \text{iff} \quad (a < a') \text{ or } (a = a' \text{ and } b < b').$$

Assume that ξ_o and η_o are ordinals such that $\omega^{\eta_o} < \xi_o$ and whenever $\xi,\eta < \xi_o$ then $\xi + \eta < \xi_o$. Show: If $\mathfrak{A} = (A,R)$ is any ordered structure with first element, then

$$(\xi_o,<) \simeq_{\eta_o} \mathfrak{A} \otimes (\xi_o,<).$$

In particular, $(\lambda,<) \simeq_\lambda \mathfrak{A} \otimes (\lambda,<)$ for any uncountable cardinal (cf.[22]).

(b) Let $(Z,+)$ be the additive group of integers. If $\mathfrak{G}$ is a torsion-free abelian group and $\mathfrak{G}/_{p\mathfrak{G}} \simeq (Z,+)/_{p(Z,+)}$ for each prime p, then $\mathfrak{G} \simeq_\omega (Z,+)$.

The following lemma establishes the first connection between the purely algebraic concept of isomorphism and the model-theoretic notion of elementary equivalence. It tells us that if $\mathfrak{A} \simeq_n \mathfrak{B}$, then $\mathfrak{A}$ and $\mathfrak{B}$ satisfy the same formulas of quantifier rank $< n$. More precisely: we define the quantifier rank of a formula φ, $qr(\varphi)$, by induction on φ as follows:

$$qr(\varphi) = 0, \text{ if } \varphi \text{ is atomic},$$

$$qr(\neg\varphi) = qr(\varphi),$$

$$qr(\varphi \vee \psi) = \max\{qr(\varphi), qr(\psi)\},$$

$$qr(\exists x \varphi) = qr(\varphi) + 1 .$$

1.7 Lemma. Let K be a set of predicate symbols. Assume $(I_m)_{m<n} : \mathfrak{A} \simeq_n \mathfrak{B}$. If $p \in I_m$, $\overset{r}{a} \in do(p)$ and $\varphi \in L_r^K$ has quantifier rank $\leq m$, then

$$\mathfrak{A} \models \varphi[\overset{r}{a}] \quad \text{iff} \quad \mathfrak{B} \models \varphi[p(\overset{r}{a})]$$

($\overset{r}{a}$ denotes $a_o, \ldots, a_{r-1}$, and $p(\overset{r}{a})$ denotes $p(a_o), \ldots, p(a_{r-1})$).
The proof is by induction on m. Use the back and forth property in case $\varphi = \exists x \psi$. Thus, if $p \in I_m$ we may pass through m quantifiers.

In case K contains function symbols the lemma remains true, if we take $rk(\varphi)$ instead of $qr(\varphi)$, where $rk(\varphi)$ is defined by:

if φ is atomic, then $rk(\varphi)$ is the number n of occurences of function symbols in φ, unless φ is of the form $t_1 = t_2$ in which case $rk(\varphi) = n \dot{-} 1$
$rk(\neg \varphi = rk(\varphi)$,
$rk(\varphi \vee \psi) = \max\{rk(\varphi), rk(\psi)\}$,
$rk(\exists x \varphi) = rk(\varphi) + 1$.

1.8 Corollary. If $\mathfrak{A} \simeq_\omega \mathfrak{B}$ (or for each n, $\mathfrak{A} \simeq_n \mathfrak{B}$), then $\mathfrak{A} \equiv \mathfrak{B}$.

Our aim is to prove the converse of 1.8. For this purpose we have to assume that K is finite. Then we can express in first-order logic that $\mathfrak{A}$ contains a finite subset of a given isomorphism type. Hence, if $\mathfrak{A}$ and $\mathfrak{B}$ are elementarily equivalent, $\mathfrak{B}$ also contains such a subset. We put the corresponding partial isomorphism in I_o. Similarly, given a set T of isomorphism types of n+1 elements we may express in first-order logic that $\mathfrak{A}$ contains a subset of n elements that may be extended, adjoining one element, just to the types in T. Hence, if $\mathfrak{A} \equiv \mathfrak{B}$ then $\mathfrak{B}$ also contains such a subset. Put the corresponding partial isomorphism in I_1. Going on in this way we obtain a sequence $(I_n)_{n<\omega}$ such that $(I_n)_{n<\omega}: \mathfrak{A} \simeq_\omega \mathfrak{B}$.

More precisely, for $r \in \omega$ let

$$\Psi_r = \{\psi \mid \psi \in L^K_r, \psi \text{ is of the form } Px_o\ldots x_{s-1}, x_o = x_1 \text{ or } fx_o\ldots x_{s-1} = x_s\}.$$

1.9 <u>Lemma</u>. Given $\mathfrak{A}$ and $\mathfrak{B}$, $\overset{r}{a} \in A$ and $\overset{r}{b} \in B$ the following are equivalent.

(i) For all $\psi \in \Psi_r$: $\mathfrak{A} \models \psi[\overset{r}{a}]$ iff $\mathfrak{B} \models \psi[\overset{r}{b}]$.

(ii) $p(a_i) = b_i$ for $i < r$ defines a partial isomorphism.

Thus the isomorphism type of an r-tupel $\overset{r}{a} \in A$ is determined by the set of formulas of Ψ_r that it satisfies, or, equivalently, by

$$(2) \qquad \varphi^o_{\overset{r}{a}} = \bigwedge_{\substack{\psi \in \Psi_r \\ \mathfrak{A} \models \psi[\overset{r}{a}]}} \psi \wedge \bigwedge_{\substack{\psi \in \Psi_r \\ \mathfrak{A} \models \neg\psi[\overset{r}{a}]}} \neg\psi.$$

Call $\varphi^o_{\overset{r}{a}}$ the <u>0-isomorphism type of $\overset{r}{a}$ (in $\mathfrak{A}$)</u> (cf.[21]) and put

$$IT^o_r = \{\varphi^o_{\overset{r}{a}} \mid \mathfrak{A} \text{ K-structure, } \overset{r}{a} \in A\}.$$

Note that IT^o_r is finite, since K is finite.

We define by induction on n the <u>n-isomorphismtype</u> $\varphi^n_{\overset{r}{a}}$ of $\overset{r}{a}$ and the set IT^n_r of <u>n-isomorphism types of r-tuples</u>. Assume that $n > 0$ and that IT^{n-1}_s is finite for any $s \in \omega$.

Put

(3) $$\varphi^n_{\overset{r}{a}} = \bigwedge_{a \in A} \exists v_r \varphi^{n-1}_{\overset{r}{a}a} \wedge \forall v_r \bigvee_{a \in A} \varphi^{n-1}_{\overset{r}{a}a}$$

and

(4) $$IT^n_r = \{\varphi^n_{\overset{r}{a}} \mid \mathfrak{A} \text{ K-structure, } \overset{r}{a} \in A\} .$$

Thus $\varphi^n_{\overset{r}{a}}$ tells us to which (n-1)-isomorphism types we can extend $\overset{r}{a}$ adjoining one element. The proof of the following lemma is by induction on n.

1.10 Lemma.

(i) IT^n_r is finite.

(ii) If $\varphi \in IT^n_r$ then $\varphi \in L^K_r$ and $qr(\varphi) = rk(\varphi) = n$.

(iii) Given $\mathfrak{A}$ and $\overset{r}{a} \in A$, $\varphi^n_{\overset{r}{a}}$ is the only $\varphi \in IT^n_r$ such that $\mathfrak{A} \models \varphi[\overset{r}{a}]$.

(iv) $\models \varphi^n_{\overset{r}{a}} \rightarrow \varphi^n_{\overset{s}{a}}$ for $s < r$, and $\models \varphi^n_{\overset{r}{a}} \rightarrow \varphi^{n-1}_{\overset{r}{a}}$ for $n > 0$.

In case $r = 0$ denote $\varphi^n_{\overset{r}{a}}$ by $\varphi^n_{\mathfrak{A}}$. $\varphi^n_{\mathfrak{A}}$ is a sentence; it tells us how we may extend the empty subset n times.

If $\mathfrak{A}$ and $\mathfrak{B}$ are K-structures, put

(5) $$I_n = \{p \mid p \in P(\mathfrak{A},\mathfrak{B}), \text{ there is } r \in \omega \text{ and } \overset{r}{a} \in A \text{ such that } do(p) = \{a_o,\dots,a_{r-1}\} \text{ and } \mathfrak{B} \models \varphi^n_{\overset{r}{a}}[p(\overset{r}{a})]\} .$$

Then

1.11 Lemma.

(i) $I_n \supset I_{n+1}$.

(ii) $(I_n)_{n < \omega}$ has the back and forth property.

(iii) $I_n \neq \emptyset$ iff $\mathfrak{B} \models \varphi^n_{\mathfrak{A}}$.

The proof is a simple consequence of the definitions. Use 1.7 to derive (iii).-

From 1.7 and 1.11 we obtain

1.12 Corollary. Let K be finite. Then

$$\mathfrak{A} \simeq_{n+1} \mathfrak{B} \quad \text{iff} \quad \mathfrak{B} \models \varphi^n_{\mathfrak{A}}.$$

1.13 Fraïssé's theorem. Assume that K is finite. Then the following are equivalent.

(i) $\mathfrak{A} \equiv \mathfrak{B}$.

(ii) For all $n \in \omega$, $\mathfrak{B} \models \varphi^n_{\mathfrak{A}}$.

(iii) There is $(I_n)_{n<\omega}$ containing only partial isomorphisms with finite domain such that $(I_n)_{n<\omega} : \mathfrak{A} \simeq_\omega \mathfrak{B}$.

(iv) $\mathfrak{A} \simeq_\omega \mathfrak{B}$.

Though the back and forth technique has been applied extensively in the study of infinitary languages (see, for example, [3], [13]), we believe that it has been neglected in first-order logic. The formulas $\varphi^n_{\bar{a}}$ have a clear algebraic content. Since each first-order formula is equivalent to a disjunction of a finite set of these formulas (see the exercises below), we obtain an algebraic understanding of the expressive power of first-order logic. Thus, if model theory is viewed as satisfying the equation "universal algebra + logic = model theory" (see [9]), the importance of the back and forth technique for first-order logic is not surprising.

We use this technique in §.2 to prove Lindström's theorem, and in §.3 to give a uniform treatment of some interpolation theorems. We list some more applications in the exercises to show its wide applicability.

1.14 Exercises.

(a) Call a set Σ of L_o^K complete, if any two models of Σ are elementarily equivalent. Show: If K is at most countable, then Σ is complete iff any two ω-saturated models of Σ are partially isomorphic. Hence the following theories are complete (see 1.3): the theory of dense linear orderings, the theory of discrete linear orderings, the theory of algebraically closed fields of fixed characteristic.

(b) [33] Assume that Q is an infinite subset of ω, $Q = \{q_o, q_1, \ldots\}$ with $q_o < q_1 < \ldots$. Q has increasing distances, if for $n < m \in \omega$, $q_{n+1} - q_n < q_{m+1} - q_m$.

Show:

(i) If Q has increasing distances, then there is a complete set Σ_Q recursive in Q, such that $(\omega,<,Q) \models \Sigma_Q$. In particular, if Q is recursive, $Th((\omega,<,Q))$ is decidable.

(ii) For any Turing degree $\underline{a}$ there is a $Q \subset \omega$, $Q \in \underline{a}$, such that $Th((\omega,<,Q))$ has degree $\underline{a}$ (Hint: use part(i)).

(c) Let K be finite and let $\varphi \in L_r^K$ be of rank $\leq n$.

Then $\models \varphi \leftrightarrow \bigvee\{\varphi^n_{\overset{r}{a}} \mid \mathfrak{A}$ K-structure, $\overset{r}{a} \in A$, $\mathfrak{A} \models \varphi[\overset{r}{a}]\}$.

(d) Assume that Σ is a complete set in a countable language. Then the following are equivalent:

(i) Each formula φ is equivalent in Σ to a formula of rank $\leq n$.

(ii) If $\mathfrak{A}$ and $\mathfrak{B}$ are ω-saturated models of Σ, then $I: \mathfrak{A} \simeq_p \mathfrak{B}$ for $I = \{p \mid$ there is $\overset{r}{a} \in A$ such that $do(p) = \{a_o, \ldots, a_{r-1}\}$ and $(\mathfrak{A}, \overset{r}{a}) \simeq_{n+1} (\mathfrak{B}, p(\overset{r}{a}))\}$.

(e) Show that Fraïssé's theorem is not true without the assumption that K is finite. In case K is countable, say $K = \bigcup K_n$ with finite K_n, derive a theorem analogous to Fraïssé's theorem by putting into I_n partial isomorphisms with respect to K_n.

(f) By (c), the following may be viewed as an algebraic version of the compactness theorem:

If $\mathfrak{A}_0, \mathfrak{A}_1, \ldots$ is a sequence of K-structures and $\mathfrak{A}_n \simeq_n \mathfrak{A}_m$ for $n < m$, then there is a $\mathfrak{B}$ such that for all $n < \omega$, $\mathfrak{A}_n \simeq_n \mathfrak{B}$.

Give a purely algebraic proof of this fact, i.e. a proof that does not use a first-order language [32,34]. Similarly, find an algebraic version and proof of the omitting types theorem.

§. 2 Lindström's theorem.

In the last 20 years many languages that are more expressive than the language of first-order logic, have been introduced and studied. But none of them happened to satisfy both the compactness theorem and the Löwenheim-Skolem theorem. Later Lindström [24] proved that there is no logic that strengthens first-order logic and has these two model-theoretic properties. Lindström's theorem was the first important result of that branch of mathematical logic which is now called "soft model theory" or "abstract model theory". Soft model theory deals with properties of logics and studies their relations (see [2],[4],[14],[15]).

In this section we state and prove Lindström's theorem. But first we list some examples of extensions of first-order logic:

second-order logic L^2,

weak second-order logic L^{2w},

logics with added quantifiers

e.g. $L(Q_\varkappa)$ with the unary quantifier $Q_\varkappa$ where $\mathfrak{A} \models Q_\varkappa x\varphi(x)$ means that there are at least $\varkappa$ many $a \in A$ such that $\mathfrak{A} \models \varphi[a]$.

$L(Q_\varkappa^2)$ with the binary quantifier $Q_\varkappa^2$ where $\mathfrak{A} \models Q_\varkappa^2 xy\varphi(x,y)$ means that there is $A_o \subset A$, $|A_o| \geq \lambda$ such that for distinct $a,b \in A_o$, $\mathfrak{A} \models \varphi[a,b]$.

$L(Q_H)$ with the added quantifier Q_H where $\mathfrak{A} \models Q_H x\varphi\psi$ means $|\{a|\mathfrak{A} \models \varphi[a]\}| = |\{a|\mathfrak{A} \models \psi[a]\}|$.

infinitary logics

e.g. $L_{\infty\omega}$ (arbitrary conjunctions and disjunctions), $L_{\varkappa\omega}$ (conjunctions and disjunctions of fewer than $\varkappa$ formulas, $\varkappa$ an infinite regular cardinal).

Thus $L_{\omega\omega}$ "is" first-order logic. $L_{\omega\omega}$ will be our notation for first-order logic in this paragraph.

We choose some properties that are shared by all these logics as defining axioms of a logic. Denote by Str(K) the class of K-structures.

2.1 Definition. By a logic (or a model-theoretic language) $\mathfrak{L}$ we mean a pair $(L,\models_L)$ (!) where L is a function which associates to any type K a class L^K, the class of L-sentences of type K, and where $\models_L$ is a binary relation satisfying the following conditions:

(i) If $K_1 \subset K_2$ then $L^{K_1} \subset L^{K_2}$.

(ii) If $\mathfrak{A} \models_L \varphi$ then, for some K, $\mathfrak{A} \in \mathrm{Str}(K)$ and $\varphi \in L^K$.

(iii) (Isomorphism property) If $\mathfrak{A} \simeq \mathfrak{B}$ and $\mathfrak{A} \models_L \varphi$, then $\mathfrak{B} \models_L \varphi$.

(iv) (Reduct property) If $\varphi \in L^K$, $K \subset K'$ and $\mathfrak{A} \in \mathrm{Str}(K')$, then

$$\mathfrak{A} \models_L \varphi \quad \text{iff} \quad \mathfrak{A} \upharpoonright K \models_L \varphi$$

(where $\mathfrak{A} \upharpoonright K$ denotes the K-reduct of $\mathfrak{A}$).

To simplify notation, where possible we shall omit the subscrit L in $\models_L$ and write L for a logic $\mathfrak{L}$. If $\varphi \in L^K$ let

$$\mathrm{Mod}(\varphi) = (\mathrm{Mod}^K_{\mathfrak{L}}(\varphi)) = \{\mathfrak{A} \mid \mathfrak{A} \in \mathrm{Str}(K), \mathfrak{A} \models \varphi\}.$$

Suppose that L and L' are logics. We say that <u>L is contained in L'</u>, $L \leq L'$, if for each type K and $\varphi \in L^K$ there is a $\psi \in L'^K$ such that $\mathrm{Mod}(\varphi) = \mathrm{Mod}(\psi)$. L and L' are <u>equivalent</u>, $L \sim L'$, if $L \leq L'$ and $L' \leq L$.

For example, $L_{\omega\omega} \leq L(Q_\varkappa)$, $L(Q_{\aleph_0}) \leq L_{\omega_1\omega}$ and $L(Q_\varkappa) \leq L(Q^2_\varkappa)$.

We introduce some further properties of logics. We write <u>Boole(L)</u> and say that L is closed under finite Boolean operations if the following hold

(i) for each $\varphi \in L^K$ there is a $\chi \in L^K$ with

$$\mathrm{Mod}(\chi) = \mathrm{Str}(K) - \mathrm{Mod}(\varphi),$$

(ii) for each $\varphi, \psi \in L^K$ there is a $\chi \in L^K$ with

$$\mathrm{Mod}(\chi) = \mathrm{Mod}(\varphi) \cup \mathrm{Mod}(\psi).$$

If Boole(L), a χ satisfying (i) resp. (ii) is denoted by $\neg\varphi$ resp. $\varphi \vee \psi$.

Suppose $\mathfrak{A}$ is a K-structure and B is a subset of A. If B is <u>K-closed</u>, i.e. non-empty and closed under f^A for each $f \in K$, we write $[B]^{\mathfrak{A}}$ for the substructure of $\mathfrak{A}$ whose universe is B. By relativizing the quantifiers it is easy to show that all logics listed at the beginning of this section have the following property which we denote by Relat(L).

Relat(L) iff for each K, $\varphi \in L^K$ and each unary predicate U there is a $\psi \in L^{K \cup \{U\}}$ such that $(\mathfrak{A}, U^A) \models \psi$ iff $[U^A]^{\mathfrak{A}} \models \varphi$ holds for all $\mathfrak{A} \in Str(K)$ and all K-closed $U^A \subset A$ (a sentence ψ satisfying this condition is denoted by φ^U).

We call a logic L <u>regular</u>, if $L_{\omega\omega} \leq L$, Relat(L) and Boole(L).

We list some more properties of logics.

Löw(L) : if $\varphi \in L^K$ has a model, then it has a denumerable model.

Comp(L) : if every finite subset of $\Sigma \subset L^K$ has a model, then Σ has a model.

$Comp_\lambda$(L): if every finite subset of a set Σ of $\leq \lambda$ L-sentences has a model, then Σ has a model.

For example, Löw($L_{\omega\omega}$), Löw($L_{\omega_1\omega}$), Comp($L_{\omega\omega}$), $Comp_{\aleph_o}$ ($L(Q_{\aleph_1})$) (cf.[19]) and, if V = L then $Comp_{\aleph_o}$ ($L(Q^2_{\aleph_1})$) (cf. [28]).

2.2 <u>Lindström's theorem</u>. Suppose that L is a regular logic with Comp(L) and Löw(L). Then

$$L \sim L_{\omega\omega}.$$ [1]

[1] Lindström used in [24] a renaming property (see §.6) instead of the relativization property which is part of the definition of regularity. Though the renaming property seems to be more natural, it is easier to work with the relativization property. The idea of the proof becomes more transparent.

The idea of the proof presented in the next three lemmas is the following. We show using the compactness property that each L-sentence depends only on a finite set of symbols (lemma 1). By the Löwenheim property we need to examine only countable models. By L-compactness a countable model may be regarded as the limit of its finite parts. But all one can say about a finite part of a structure (of finite type) can be expressed in first-order logic.

2.3 <u>Lemma</u>. Let L be a regular logic with Comp(L), and $\varphi \in L^K$. Then there is a finite $K_o \subset K$ such that for any K-structures $\mathfrak{A}$ and $\mathfrak{B}$, $\mathfrak{A} \restriction K_o \simeq \mathfrak{B} \restriction K_o$ implies ($\mathfrak{A} \models \varphi$ iff $\mathfrak{B} \models \varphi$).

<u>Proof</u>. Take new unary predicate symbols U and V and a new unary function symbol h. Let Σ be a set of $L_{\omega\omega}$-sentences, and hence, by regularity, of L-sentences, saying that the U-part and the V-part are K-closed and that the restriction of h to the K-reduct on U is an isomorphism onto the K-reduct on V. Then

$$\Sigma \models \varphi^U \leftrightarrow \varphi^V.$$

By Comp(L) there is a finite Σ_o such that

$$\Sigma_o \models \varphi^U \leftrightarrow \varphi^V.$$

Choose a finite $K_o \subset K$ such that $\Sigma_o \subset L_{\omega\omega}^{K_o \cup \{U,V,h\}}$. - Suppose that $\mathfrak{A} \restriction K_o \simeq \mathfrak{B} \restriction K_o$. We may assume $A \cap B = \emptyset$. There is $(\mathfrak{C}, U^C, V^C, h^C)$ such that

$$[U^C]^{\mathfrak{C}} = \mathfrak{A}, [V^C]^{\mathfrak{C}} = \mathfrak{B} \text{ and } (\mathfrak{C}, U^C, V^C, h^C) \models \Sigma_o.$$

Thus $(\mathfrak{C}, U^C, V^C, h^C) \models \varphi^U \leftrightarrow \varphi^V$, hence by Relat(L)

$$\mathfrak{A} = [U^C]^{\mathfrak{C}} \models \varphi \quad \text{iff} \quad \mathfrak{B} = [V^C]^{\mathfrak{C}} \models \varphi .$$

We have proved

2.3' Lemma. Suppose that L is regular and Comp(L) holds. Then for each $\varphi \in L^K$ there is a finite $K_o \subset K$ such that for any $\mathfrak{C} \in \mathrm{Str}(K)$ and any K_o-closed $U^C, V^C \subset C$,

if $[U^C]^{\mathfrak{C} \restriction K_o} \simeq [V^C]^{\mathfrak{C} \restriction K_o}$, then $(\mathfrak{C}, U^C, V^C) \models \varphi^U \leftrightarrow \varphi^V$.

2.4 Lemma. Assume that L is regular with Comp(L). If for any $\mathfrak{A}$ and $\mathfrak{B}$, $\mathfrak{A} \equiv_{\omega\omega} \mathfrak{B}$ implies $\mathfrak{A} \equiv_L \mathfrak{B}$, then $L \sim L_{\omega\omega}$.

Proof. We show $L \leq L_{\omega\omega}$. Let φ be in L^K. If $\mathfrak{A} \models \varphi$, by hypothesis $\mathrm{Th}_{\omega\omega}(\mathfrak{A}) \models \varphi$ where $\mathrm{Th}_{\omega\omega}(\mathfrak{A}) = \{\psi \mid \psi \in L^K_{\omega\omega}, \mathfrak{A} \models \psi\}$. By Comp(L) there is $\chi_{\mathfrak{A}} \in L^K_{\omega\omega}$ such that

$$\chi_{\mathfrak{A}} \models \varphi \quad \text{and} \quad \mathfrak{A} \models \chi_{\mathfrak{A}} .$$

Hence

$$\mathrm{Mod}(\varphi) = \bigcup_{\mathfrak{A} \models \varphi} \mathrm{Mod}(\chi_{\mathfrak{A}}) .$$

Use Comp(L) to get for some $\mathfrak{A}_1, \ldots, \mathfrak{A}_n$,

$$\mathrm{Mod}(\varphi) = \bigcup_{i=1}^{n} \mathrm{Mod}(\chi_{\mathfrak{A}_i}) = \mathrm{Mod}(\chi_{\mathfrak{A}_1} \vee \ldots \vee \chi_{\mathfrak{A}_n}) .$$

But $\chi_{\mathfrak{A}_1} \vee \ldots \vee \chi_{\mathfrak{A}_n} \in L^K_{\omega\omega}$.

The next lemma finishes the proof of Lindström's theorem.

2.5 Lemma. Assume L is a regular logic with Comp(L) and Löw(L).

If $\mathfrak{A} \equiv_{\omega\omega} \mathfrak{B}$ then $\mathfrak{A} \equiv_L \mathfrak{B}$.

Proof. By contradiction assume that $\mathfrak{A} \models \varphi$ but non $\mathfrak{B} \models \varphi$ holds for some $\varphi \in L^K$. Choose a finite K_o as in 2.3'. By Fraïssé's theorem, $(I_n)_{n \in \omega} : \mathfrak{A} \restriction K_o \simeq_{\omega} \mathfrak{B} \restriction K_o$

for some sequence $(I_n)_{n\in\omega}$. Take new predicate symbols P,U,V (unary), <,I (binary) and G (ternary). Let $\mathfrak{C}$ be a structure such that

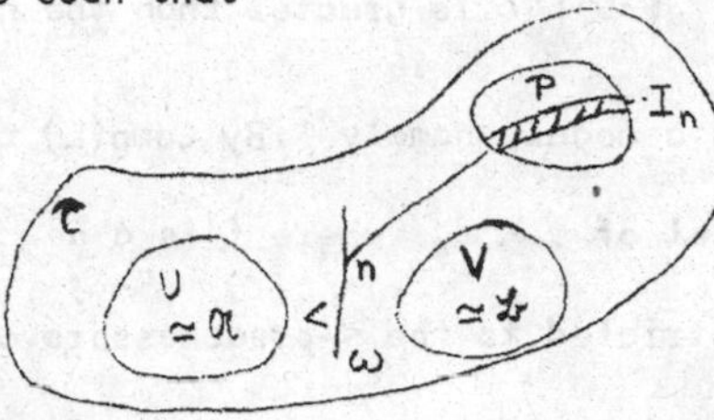

(1) $U^{\mathfrak{C}} = A$ and $[U^{\mathfrak{C}}]^{\mathfrak{C}\restriction K} = \mathfrak{A}$,

(2) $V^{\mathfrak{C}} = B$ and $[V^{\mathfrak{C}}]^{\mathfrak{C}\restriction K} = \mathfrak{B}$

(assume that $A \cap B = \emptyset$),

(3) $\omega \subset C$ and $<^{\mathfrak{C}}$ is the natural ordering on ω,

(4) $P^{\mathfrak{C}}p$ iff $p \in \bigcup_{n\in\omega} I_n$,

(5) $I^{\mathfrak{C}}mp$ iff $m \in \omega$ and $p \in I_m$,

(6) $G^{\mathfrak{C}}pab$ iff there is $m \in \omega$ such that $p \in I_m$ and $p(a) = b$.

Then $\mathfrak{C}$ is a model of the conjunction ψ of the following sentences:

$\forall p(Pp \to \forall x \forall y(Gpxy \to Ux \wedge Vy))$

$\forall p(Pp \to \forall x,x'\forall y,y'(Gpxy \wedge Gpx'y' \to (\neg\, x = x' \leftrightarrow \neg\, y = y')))$

$\forall p(Pp \to$ "p preserves predicate and function symbols $\in K_o$")

$\forall x(\exists y(x < y \vee y < x) \to \exists p(Pp \wedge Ixp))$

$\forall p \forall x \forall y(y < x \wedge Ixp \to Iyp)$

Forth property: $\forall x \forall y \forall p \forall u(y < x \wedge Ixp \wedge Uu \to \exists q \exists v$

$(Iyq \wedge Gquv \wedge \forall x'\forall y'(Gpx'y' \to Gqx'y')))$

"Back property"

"< is a discrete ordering with first element"

"U is K_o-closed"

"V is K_o-closed"

φ^U

$(\neg\,\varphi)^V$

If c is a new constant any finite subset of

$$\psi \cup \{\text{"c is greater than the n-th element of} <\text{"} \mid n \in \omega\}$$

has a model, namely $\mathfrak{C}$. By Comp(L) there is a model of the whole set, thus a model of $\psi \wedge \varphi_f$ where f is a new unary function symbol and φ_f says that f restricted to the <-predecessors of c is an injective but not surjective function into the <-predecessors of c. By Löw(L) there is a countable model $\mathfrak{D}$ of $\psi \wedge \varphi_f$. Let $(d_n)_{n \in \omega}$ be an infinite descending sequence of $<^D$. U^D and V^D are K_o-closed and for

$$I = \{p \mid I^D d_n p \text{ for some } n \in \omega\},$$

we have

$$I: [U^D]^{\mathfrak{D}} \upharpoonright K_o \simeq_p [V^D]^{\mathfrak{D}} \upharpoonright K_o \quad .$$

Thus $[U^D]^{\mathfrak{D}} \upharpoonright K_o \simeq [V^D]^{\mathfrak{D}} \upharpoonright K_o$. Hence by 2.3', $\mathfrak{D} \models \varphi^U \leftrightarrow \varphi^V$, a contradiction as $\mathfrak{D} \models \psi$, in particular $\mathfrak{D} \models \varphi^U$ and $\mathfrak{D} \models (\neg\varphi)^V$.

2.6 Problem. Let φ_o be an $L_{\omega_1\omega}$-sentence. Let L be the least logic that contains φ_o and first-order logic and is closed under the (finite) Boolean operations. Does $\mathrm{Comp}_{\aleph_o}$ (L) imply that φ_o is equivalent to a first-order formula? Note that in general L does not satisfy Relat(L).

§.3 Interpolation theorems.

We take the formula ψ of the proof of the last lemma of the preceeding section. If for $n \in \omega$, we have a model $\mathfrak{C}_n$ of

$$\psi \cup \{\text{"c is greater than the m-th element of <"} \mid m < n\},$$

then the U-part of $\mathfrak{C}_n$, say $\mathfrak{A}_n$, and the V-part of $\mathfrak{C}_n$, say $\mathfrak{B}_n$, are n-partially isomorphic, $\mathfrak{A}_n \simeq_n \mathfrak{B}_n$. Using the compactness theorem we obtain a model of

$$\psi \cup \{\text{"c is greater than the n-th element of <"} \mid n \in \omega\},$$

i.e. a model such that $\mathfrak{A} \simeq_p \mathfrak{B}$, where $\mathfrak{A}$ is the U-part and $\mathfrak{B}$ is the V-part. This leads us to the definition of the convergence of relations given below. In particular, the sequence of relations $\simeq_n$ will converge to $\simeq_p$. We apply this notion of convergence to obtain a general form of the interpolation theorem (cf. [25] and [29] for similar techniques).

All types will be finite in this section and all formulas will be first-order formulas.

3.1 Definition. Suppose that $R \subset Str(K_1) \times Str(K_2)$. R is a PC_δ-relation, if for some type $\overline{K} \supset K_1 \cup K_2$ there is a set $\Sigma \subset L_o^{\overline{K}}$ (i.e. a set of first-order sentences) and there are unary predicate symbols $U, V \in \overline{K}$ such that the following holds for all $\mathfrak{A} \in Str(K_1)$ and $\mathfrak{B} \in Str(K_2)$

$\mathfrak{A} R \mathfrak{B}$ iff there is a $\mathfrak{C} \in Str(\overline{K})$ such that $\mathfrak{C} \models \Sigma$, U^C is K_1-closed, V^C is K_2-closed, $[U^C]^{\mathfrak{C} \upharpoonright K_1} \simeq \mathfrak{A}$ and $[V^C]^{\mathfrak{C} \upharpoonright K_2} \simeq \mathfrak{B}$.[1)]

1) Many-sorted languages are more natural for the study of relations among structures (see [14],[16]). Nevertheless we do not introduce many-sorted languages as we are not going to use them at any further point.

We say that Σ (or more precisely $(\bar{K},\Sigma,U,V)$) defines R.

Let R and R_n, for $n \in \omega$, be PC_δ-relations on $Str(K_1) \times Str(K_2)$. We write $R_n \to R$ and say that R_n first-order converges to R, if there are $\bar{K},U,V$ and Σ_n such that for $n \in \omega$, $(\bar{K},\Sigma_n,U,V)$ defines R_n, $\Sigma_n \subset \Sigma_{n+1}$ and $(\bar{K},\cup\Sigma_n,U,V)$ defines R.

3.2 Examples.

(a) For $K_1 = K_2$ let $\mathfrak{A} R_n \mathfrak{B}$ hold if $\mathfrak{A} \simeq_n \mathfrak{B}$, and $\mathfrak{A} R \mathfrak{B}$ if $\mathfrak{A} \simeq_p \mathfrak{B}$. R_n and R are PC_δ-relations and the proof of 2.5 shows that $R_n \to R$, i.e. $\sim_n \to \simeq_p$.

(b) More generally, let K be $K_1 \cap K_2$ and define R_n, $R \subset Str(K_1) \times Str(K_2)$ by

$$\mathfrak{A} R_n \mathfrak{B} \quad \text{iff} \quad \mathfrak{A} \upharpoonright K \simeq_n \mathfrak{B} \upharpoonright K \quad \text{and}$$

$$\mathfrak{A} R \mathfrak{B} \quad \text{iff} \quad \mathfrak{A} \upharpoonright K \simeq_p \mathfrak{B} \upharpoonright K.$$

Then $R_n \to R$.

The proof of 2.5 is a special case of

3.3 Convergence lemma. Suppose that for $n \in \omega$ $\mathfrak{A}_n R_n \mathfrak{B}_n$, and that $R_n \to R$. Then there are $\mathfrak{A}^*$ and $\mathfrak{B}^*$ such that

$$\mathfrak{A}^* R \mathfrak{B}^*, \ |A^*| \le \aleph_0, \ |\mathfrak{B}^*| \le \aleph_0,$$

for any $\varphi \in L_o^{K_1}$, if for all $n \in \omega$ $\mathfrak{A}_n \models \varphi$, then $\mathfrak{A}^* \models \varphi$

and

for any $\varphi \in L_o^{K_2}$, if for all $n \in \omega$ $\mathfrak{B}_n \models \varphi$, then $\mathfrak{B}^* \models \varphi$.

Proof. Take a denumerable model of

$$\cup\Sigma_n \cup \{\varphi^U \mid \varphi \in L_o^{K_1}, \mathfrak{A}_n \models \varphi \text{ for all } n \in \omega\} \cup \{\varphi^V \mid \varphi \in L_o^{K_2}, \mathfrak{B}_n \models \varphi \text{ for all } n \in \omega\}.$$

The relation of isomorphism coincides with $\simeq_p$ in countable models, $\simeq_n$ may be regarded as finite approximation. The following definitions give the corresponding approximations for some other algebraic relations.

A function p is called a <u>partial homomorphism from $\mathfrak{A}$ to $\mathfrak{B}$</u>, if $do(p) \subset A$, $rg(p) \subset B$ and p preserves predicate and function symbols, i.e.

if $P^{\mathfrak{A}}\overset{n}{a}$ and $\overset{n}{a} \in do(p)$, then $P^{\mathfrak{B}}p(\overset{n}{a})$,

if $f^{\mathfrak{A}}(\overset{n}{a}) = a_n$ and $\overset{n+1}{a} \in do(p)$, then $f^{\mathfrak{B}}(p(\overset{n}{a})) = p(a_n)$.

p is called a <u>homomorphism</u>, if in addition $do(p) = A$.

3.4 <u>Definition</u>. Suppose that $\mathfrak{A}$ and $\mathfrak{B} \in Str(K)$.

(i) We write $\mathfrak{A} \to \mathfrak{B}$ and say that $\mathfrak{B}$ "is" an <u>extension</u> of $\mathfrak{A}$, if $\mathfrak{A}$ is isomorphic to a substructure of $\mathfrak{B}$.

(ii) Suppose that $\varepsilon \in K$ is a binary predicate symbol. Then $\mathfrak{B}$ is called an <u>end-extension</u> of $\mathfrak{A}$, $\mathfrak{A} \underset{\varepsilon}{\to} \mathfrak{B}$, if $\mathfrak{A}$ is isomorphic to a substructure $\mathfrak{B}'$ of $\mathfrak{B}$ with the following property,

if $b' \in B'$, $b \in B$ and $b \, \varepsilon^B \, b'$, then $b \in B'$.

(iii) We write $\mathfrak{A} \geq \mathfrak{B}$ and say that $\mathfrak{B}$ is a <u>homomorphic image</u> of $\mathfrak{A}$, if there is a homomorphism from $\mathfrak{A}$ onto $\mathfrak{B}$.

In the proof that $\mathfrak{A} \simeq_p \mathfrak{B}$ implies $\mathfrak{A} \simeq \mathfrak{B}$ for countable models (see 1.2) we used the <u>forth property</u> of a set I with $I: \mathfrak{A} \simeq_p \mathfrak{B}$ to obtain a mapping <u>defined on all of $\mathfrak{A}$</u> and the <u>back property</u> to ensure that this mapping is <u>onto $\mathfrak{B}$.</u> This leads to the following definitions of <u>partial extension</u> $\to_p$, <u>partial end-extension</u> $\underset{\varepsilon p}{\to}$ and <u>partial homomorphic image</u> $\geq_p$.

3.5 Definition.

(i) $\mathfrak{A} \to_p \mathfrak{B}$ if for some I, I: $\mathfrak{A} \to_p \mathfrak{B}$, where I: $\mathfrak{A} \to_p \mathfrak{B}$ means that I is a non-empty set of partial isomorphisms with the forth property.

(ii) $\mathfrak{A} \underset{\varepsilon p}{\to} \mathfrak{B}$ if for some I, I: $\mathfrak{A} \underset{\varepsilon p}{\to} \mathfrak{B}$, where I: $\mathfrak{A} \underset{\varepsilon p}{\to} \mathfrak{B}$ means that I is a non-empty set of partial isomorphisms with the forth property and the ε-back property.

for all $p \in I$, $b' \in rg(p)$ and $b \in B$, if $b \,\varepsilon^B\, b'$ then $b \in rg(q)$ for some $q \in I$ with $q \supset p$.

(iii) $\mathfrak{A} \geq_p \mathfrak{B}$ if for some I, I: $\mathfrak{A} \geq_p \mathfrak{B}$, where I: $\mathfrak{A} \geq_p \mathfrak{B}$ means that I is a non-empty set of partial homomorphisms with the back and forth property.

3.6 Lemma. Suppose that $\mathfrak{A}$ and $\mathfrak{B}$ are at most countable K-structures. Then

(i) $\mathfrak{A} \to_p \mathfrak{B}$ implies $\mathfrak{A} \to \mathfrak{B}$,

(ii) $\mathfrak{A} \underset{\varepsilon p}{\to} \mathfrak{B}$ implies $\mathfrak{A} \underset{\varepsilon}{\to} \mathfrak{B}$,

(iii) $\mathfrak{A} \geq_p \mathfrak{B}$ implies $\mathfrak{A} \geq \mathfrak{B}$.

Proof: We prove (ii). Let I: $\mathfrak{A} \underset{\varepsilon p}{\to} \mathfrak{B}$ and $A = \{a_o, a_1, \ldots\}$. Let $b_o, b_1, \ldots$ be an enumeration of B such that each element of B appears infinitely many times. Given $p_o \in I$ choose

p_{2n+1} = some extension p of p_{2n} in I with $a_n \in do(p)$,

p_{2n+2} = some extension p of p_{2n+1} in I with $b_n \in rg(p)$, if there is $b \in rg(p_{2n+1})$ such that $b_n \,\varepsilon^B\, b$.

Put $p = \bigcup p_n$. Then there is a $\mathfrak{B}' \subset \mathfrak{B}$ with $p: \mathfrak{A} \simeq \mathfrak{B}'$ and whenever $b' \in B'$, $b \in B$ and $b \,\varepsilon^B\, b'$, then $b \in B'$. Thus $\mathfrak{A} \underset{\varepsilon}{\to} \mathfrak{B}$.

It now should be clear how we define the relations $\mathfrak{A} \to_\xi \mathfrak{B}$, $\mathfrak{A} \underset{\varepsilon\,\xi}{\to} \mathfrak{B}$ and $\mathfrak{A} \geq_\xi \mathfrak{B}$. By an obvious modification of the proof of 2.5 we may show that each of the above algebraic relations is the limit of its finite approximations, i.e.

3.7 Lemma. For fixed K and all $n \in \omega$ the relations $\rightarrow_n$, $\vec{\varepsilon}_n$, $\geq_n$ and $\rightarrow_p$, $\vec{\varepsilon}_p$, $\geq_p$ are PC_δ. Moreover

$$\rightarrow_n \longrightarrow \rightarrow_p, \ \vec{\varepsilon}_n \longrightarrow \vec{\varepsilon}_p \ \text{ and } \geq_n \longrightarrow \geq_p .$$

3.8 Exercise. Extend 3.2(b) to the above relations.

If we look at the (n+1)-isomorphism type $\varphi^{n+1}_{\overset{r}{a}}$ of $\overset{r}{a}$ in $\mathfrak{A}$ (see §.1),

$$\varphi^{n+1}_{\overset{r}{a}} = \underbrace{\bigwedge_{a \in A} \exists v_r \varphi^n_{\overset{r}{a}a}}_{\text{forth property}} \wedge \underbrace{\forall v_r \bigvee_{a \in A} \varphi^n_{\overset{r}{a}a}}_{\text{back property}}$$

the first part expresses the forth property with respect to $\mathfrak{A}$ and the second part the back property. This tells us how to define the corresponding formulas for the other algebraic relations.

Suppose that $\mathfrak{A}$ is a K-structure and $\overset{r}{a} \in A$. Define, by induction on n, the n-extension type $\alpha^n_{\overset{r}{a}}$, the n-$\varepsilon$-extension type $\beta^n_{\overset{r}{a}}$ and the n-homomorphism type $\gamma^n_{\overset{r}{a}}$ as follows:

$\alpha^o_{\overset{r}{a}} = \beta^o_{\overset{r}{a}} = \varphi^o_{\overset{r}{a}}$ (see §. 1.(2) for the definition of $\varphi^o_{\overset{r}{a}}$),

$\gamma^o_{\overset{r}{a}} = \bigwedge\{\varphi \mid \varphi \text{ atomic}, \varphi \in L^K_r, \mathfrak{A} \models \varphi[\overset{r}{a}]\}$,

and

$$\alpha^{n+1}_{\overset{r}{a}} = \bigwedge_{a \in A} \exists v_r \alpha^n_{\overset{r}{a}a}$$

$$\beta^{n+1}_{\overset{r}{a}} = \bigwedge_{a \in A} \exists v_r \beta^n_{\overset{r}{a}a} \wedge \underbrace{\bigwedge_{i < r} \forall v_r \varepsilon v_i \bigvee_{a \varepsilon A} \beta^n_{\overset{r}{a}a}}_{\varepsilon\text{-back property}}$$

$$\gamma^{n+1}_{\underset{a}{r}} = \bigwedge_{a \in A} \exists v_r \gamma^n_{\underset{aa}{r}} \wedge \forall v_r \bigvee_{a \in A} \gamma^n_{\underset{aa}{r}}.$$

Denote $\alpha^n_{\emptyset}$ by $\alpha^n_{\mathfrak{A}}$, $\beta^n_{\emptyset}$ by $\beta^n_{\mathfrak{A}}$ and $\gamma^n_{\emptyset}$ by $\gamma^n_{\mathfrak{A}}$.

3.9 Lemma.

(i) $\mathfrak{A} \rightarrow_{n+1} \mathfrak{B}$ iff $\mathfrak{B} \models \alpha^n_{\mathfrak{A}}$

$\mathfrak{A} \underset{\varepsilon n+1}{\rightarrow} \mathfrak{B}$ iff $\mathfrak{B} \models \beta^n_{\mathfrak{A}}$

$\mathfrak{A} \geq_{n+1} \mathfrak{B}$ iff $\mathfrak{B} \models \gamma^n_{\mathfrak{A}}$

(ii) For each $n \in \omega$ the sets $\{\alpha^n_{\mathfrak{A}} \mid \mathfrak{A} \in Str(K)\}$, $\{\beta^n_{\mathfrak{A}} \mid \mathfrak{A} \in Str(K)\}$ and $\{\gamma^n_{\mathfrak{A}} \mid \mathfrak{A} \in Str(K)\}$ are finite.

(iii) $\alpha^n_{\underset{a}{r}}$ is an existential formula, i.e. built up from atomic formulas and negations of atomic formulas with $\wedge, \vee$ and $\exists$.

$\beta^n_{\underset{a}{r}}$ is a restricted -existential formula, i.e. built up from atomic formulas and negations of atomic formulas with $\wedge, \vee, \exists$ and $\forall x \varepsilon y$.

$\gamma^n_{\underset{a}{r}}$ is a positive formula, i.e. does not contain the negation symbol.

Thus for each of the above algebraic relations R the formulas that describe their finite approximations R_n have a special syntactic form. Using the fact that $R_n \rightharpoonup R$ we show that φ is equivalent to a formula of this form, if the class of models of φ is closed under R.

3.10 Preservation theorem. Suppose that

(i) $R_n, R \subset Str(K) \times Str(K)$ and $R_n \rightharpoonup R$.

(ii) for $\mathfrak{A} \in Str(K)$ and $n \in \omega$ there is $\psi^n_{\mathfrak{A}} \in L^K_o$ such that $\mathfrak{A} \models \psi^n_{\mathfrak{A}}$ and for all $\mathfrak{B} \in Str(K)$

$$\mathfrak{B} \models \psi^n_{\mathfrak{A}} \quad \text{iff} \quad \mathfrak{A} R_n \mathfrak{B}.$$

(iii) $\{\psi^n_{\mathfrak{A}} \mid \mathfrak{A} \in \mathrm{Str}(K)\}$ is finite for $n \in \omega$.

(iv) φ is an L^K-sentence such that $\mathrm{Mod}(\varphi)$ is R-closed, i.e. $\mathfrak{A} R \mathfrak{B}$ and $\mathfrak{A} \models \varphi$. implies $\mathfrak{B} \models \varphi$.

Then φ is equivalent to a disjunction of the $\psi^n_{\mathfrak{A}}$.

Proof. For $n \in \omega$ put $\chi^n = \bigvee\{\psi^n_{\mathfrak{A}} \mid \mathfrak{A} \models \varphi\}$. By (i), $\models \chi^{n+1} \rightarrow \chi^n$. By (ii), $\models \varphi \rightarrow \chi^n$. Thus, it suffices to show that

$$\{\chi^n \mid n \in \omega\} \models \varphi.$$

Take a model $\mathfrak{B}$ of $\{\chi^n \mid n \in \omega\}$. Then for each $n \in \omega$ there is $\mathfrak{A}_n$ such that

$$\mathfrak{A}_n \models \varphi \quad \text{and} \quad \mathfrak{B} \models \psi^n_{\mathfrak{A}_n}.$$

By (ii), $\mathfrak{A}_n R_n \mathfrak{B}$. Hence by the convergence lemma, there are denumerable $\mathfrak{A}^*$ and $\mathfrak{B}^*$ with

$$\mathfrak{A}^* R \mathfrak{B}^*, \mathfrak{A}^* \models \varphi \text{ and } \mathfrak{B}^* \equiv \mathfrak{B}.$$

By (iv), $\mathfrak{B}^* \models \varphi$; hence $\mathfrak{B} \models \varphi$.

Remark. We did not need (iv) in its full generality but only for denumerable models $\mathfrak{A}$ and $\mathfrak{B}$.

3.11 Corollary.

(i) If $\mathrm{Mod}(\varphi)$ is closed under extensions, then φ is equivalent to an existential formula.

(ii) If $\mathrm{Mod}(\varphi)$ is closed under end-extensions, then φ is equivalent to a restricted-existential formula.

(iii) If Mod(φ) is closed under homomorphic images then φ is equivalent to a positive formula.

We leave it to the reader to extend 3.10 to m-ary relations R_n and R and to apply it for other algebraic relations.

The preservation theorem is a special case of the interpolation theorem 3.12 below. As the proof in the general case 3.12 is similar, we assume that $K = K_1 \cap K_2$ and that $R_n, R \subset \mathrm{Str}(K_1) \times \mathrm{Str}(K_2)$ are given by:

$$\mathfrak{A} R_n \mathfrak{B} \quad \text{iff} \quad \mathfrak{A} \restriction K \simeq_n \mathfrak{B} \restriction K,$$

$$\mathfrak{A} R \mathfrak{B} \quad \text{iff} \quad \mathfrak{A} \restriction K \simeq_p \mathfrak{B} \restriction K.$$

For $\varphi \in L_o^{K_1}$ and $n \in \omega$ put

$$\chi^n = \bigvee\{\varphi^n_{\mathfrak{A} \restriction K} \mid \mathfrak{A} \in \mathrm{Str}(K_1),\ \mathfrak{A} \models \varphi\}.$$

Thus $\chi^n \in L_o^K$, $\models \chi^{n+1} \rightarrow \chi^n$ and $\models \varphi \rightarrow \chi^n$.

$\{\chi^n \mid n \in \omega\}$ is a set of K-interpolating formulas for φ, i.e. $\psi \in L_o^{K_2}$ and $\models \varphi \rightarrow \psi$ implies $\models \varphi \rightarrow \chi^n$ and $\models \chi^n \rightarrow \psi$ for some $n \in \omega$.

It suffices to show that $\{\chi^n \mid n \in \omega\} \models \psi$. Thus suppose that $\mathfrak{B}$ is a K_2-structure and $\mathfrak{B} \models \{\chi^n \mid n \in \omega\}$. Then for $n \in \omega$ there is an $\mathfrak{A}_n \in \mathrm{Str}(K_1)$ such that $\mathfrak{A}_n \models \varphi$ and $\mathfrak{B} \models \varphi^n_{\mathfrak{A}_n \restriction K}$. Hence $\mathfrak{A}_n R_n \mathfrak{B}$. As $R_n \longrightarrow R$ there are denumerable structures $\mathfrak{A}^*$ and $\mathfrak{B}^*$ such that

$$\mathfrak{A}^* R \mathfrak{B}^*,\ \mathfrak{A}^* \models \varphi \ \text{ and } \ \mathfrak{B}^* \equiv \mathfrak{B}.$$

Thus $\mathfrak{A}^* \restriction K \simeq \mathfrak{B}^* \restriction K$. Let $\mathfrak{C}$ be a $K_1 \cup K_2$-structure with $\mathfrak{C} \restriction K_1 \simeq \mathfrak{A}^*$ and $\mathfrak{C} \restriction K_2 \simeq \mathfrak{B}^*$. Then $\mathfrak{C} \models \varphi$, $\mathfrak{C} \models \psi$, $\mathfrak{B}^* \models \psi$ and hence, $\mathfrak{B} \models \psi$.

Note that if $\mathfrak{A}$ is a K_1-structure and $\mathfrak{B}$ is K_2-structure and $\mathfrak{A} \restriction K \rightarrow \mathfrak{B} \restriction K$

where $K = K_1 \cap K_2$ then $\mathfrak{A} \to \mathfrak{B}_1$ for some K_1-structure $\mathfrak{B}_1$ with $\mathfrak{B}_1 \upharpoonright K = \mathfrak{B} \upharpoonright K$. But in general there will be no K_2-structure $\mathfrak{A}_1$ such that $\mathfrak{A}_1 \to \mathfrak{B}$ and $\mathfrak{A}_1 \upharpoonright K = \mathfrak{A} \upharpoonright K$, if $K_2 - K$ contains function symbols. This leads to the assumptions on S in 3.12. If we consider the possibility of extending a homomorphism the role of $\mathfrak{A}$ and $\mathfrak{B}$ are reversed. This leads to the assumptions on T in 3.12.

3.12 Interpolation theorem. Assume $K = K_1 \cap K_2$ and that

(i) $R_n, R \subset \mathrm{Str}(K_1) \times \mathrm{Str}(K_2)$ and $R_n \longrightarrow R$,

(ii) $S \subset \mathrm{Str}(K_1) \times \mathrm{Str}(K_1)$ and for denumerable $\mathfrak{A}$ and $\mathfrak{B}$ with $\mathfrak{A}R\mathfrak{B}$ there is $\mathfrak{B}_1$ such that $\mathfrak{A}S\mathfrak{B}_1$ and $\mathfrak{B}_1 \upharpoonright K = \mathfrak{B} \upharpoonright K$

(resp. $T \subset \mathrm{Str}(K_2) \times \mathrm{Str}(K_2)$ and for denumerable $\mathfrak{A}$ and $\mathfrak{B}$ with $\mathfrak{A}R\mathfrak{B}$ there is $\mathfrak{A}_1$ such that $\mathfrak{A}_1 T \mathfrak{B}$ and $\mathfrak{A}_1 \upharpoonright K = \mathfrak{A} \upharpoonright K$),

(iii) for $\mathfrak{A} \in \mathrm{Str}(K_1)$ and $n \in \omega$ there is $\psi^n_{\mathfrak{A}} \in L^K_o$ such that $\mathfrak{A} \models \psi^n_{\mathfrak{A}}$ and for all $\mathfrak{B} \in \mathrm{Str}(K_2)$,

$$\mathfrak{B} \models \psi^n_{\mathfrak{A}} \quad \text{iff} \quad \mathfrak{A} R_n \mathfrak{B},$$

(iv) $\{\psi^n_{\mathfrak{A}} \mid \mathfrak{A} \in \mathrm{Str}(K_1)\}$ is finite for $n \in \omega$,

(v) $\varphi \in L^{K_1}_o, \psi \in L^{K_2}_o$ and $\models \varphi \to \psi$,

(vi) $\mathrm{Mod}(\varphi)$ is S-closed, i.e. $\mathfrak{A}S\mathfrak{B}$ and $\mathfrak{A} \models \varphi$ implies $\mathfrak{B} \models \varphi$ for denumerable $\mathfrak{A}$ and $\mathfrak{B}$

(resp. $\mathrm{Mod}(\psi)$ is T-closed).

Then there is a disjunction χ of the $\psi^n_{\mathfrak{A}}$ ($n \in \omega, \mathfrak{A} \in \mathrm{Str}(K_1)$) such that

$$\models \varphi \to \chi \quad \text{and} \quad \models \chi \to \psi.$$

3.13 Corollary. Let $K = K_1 \cap K_2$, $\varphi \in L^{K_1}_o$, $\psi \in L^{K_2}_o$ and $\models \varphi \to \psi$.

(i) If $\mathrm{Mod}(\varphi)$ is closed under extensions resp. end-extensions, then there is an existential resp. restricted-existential $\chi \in L^K_o$ such that $\models \varphi \to \chi$ and $\models \chi \to \psi$.

(ii) If $\mathrm{Mod}(\varphi)$ is closed under homomorphic images, then there is a positive $\chi \in L_o^K$ such that $\models \varphi \to \chi$ and $\models \chi \to \psi$.

3.14 Exercises:

(a) Derive an interpolation and a preservation theorem relative to a set Σ of sentences.

(b) Assume that $K = K_1 \cap K_2$, that K contains a constant symbol and that for $n \geq 1$, K_1 and K_2 contain the same n-ary function symbols. Let $\varphi \in L_o^{K_1}$ be universal and $\psi \in L_o^{K_2}$ existential. Then there is a quantifier-free $\chi \in L_o^K$ such that $\models \varphi \to \chi$ and $\models \chi \to \psi$ (Hint: Define $R_n \subset \mathrm{Str}(K_1) \times \mathrm{Str}(K_2)$ by: $\mathfrak{A} R_n \mathfrak{B}$ iff $\mathfrak{A} \upharpoonright K$ and $\mathfrak{B} \upharpoonright K$ satisfy the same quantifier-free formulas of rank $\leq n$).

(c) Let $K_1 = K \cup \{P_1,\ldots,P_r\}$ and assume that φ is a second-order K_1-sentence of the form $\exists P_1 \ldots \exists P_r \psi$ where ψ is a first-order formula. Show that φ is equivalent to a set of first-order formulas of type K.

(d) Let L be a <u>many-sorted</u> first-order logic. By considering the back and forth properties for each sort derive Feferman's many-sorted preservation theorems (cf. [16],[15]).

§. 4 $L_{\omega_1\omega}$ and the well-ordering number

We return to the proof of the last lemma of Lindström's theorem. There we used the compactness theorem to obtain from a model where the ordering $<$ was well-ordered, a model that contained an infinite descending $<$-chain, i.e. a non-well-ordered model. The next theorem shows that each application of the compactness theorem (in a countable language) may be rewritten such that it becomes an application of this form.

4.1 Theorem. For a regular logic L the following are equivalent.

(i) If each finite subset of a set Σ of L^K-sentences, where $|\Sigma|, |K| \leq \aleph_o$, has a model, then Σ has a model.

(ii) Let Σ be a countable set of L^K-sentences, K at most countable, and $< \in K$ a binary relation. Suppose that for $n \in \omega$, Σ has a model $\mathfrak{A}$ such that $<^A$ well-orders its field and the order type of $<^A$ is $\geq n$. Then Σ has a model $\mathfrak{B}$ such that $<^{\mathfrak{B}}$ is not a well-ordering.

Proof. (i) $\Rightarrow$ (ii): By compactness, $\Sigma \cup \{c_{n+1} < c_n \mid n \in \omega\}$ has a model.
(ii) $\Rightarrow$ (i): Let $\varphi_o, \varphi_1, \ldots$ be countably many L-sentences and suppose that for $n \in \omega, \{\varphi_o, \ldots, \varphi_n\}$ has a model. Let U be a new unary and $<$ be a new binary predicate symbol. Then apply (ii) to the set

$$\Sigma = \{\text{"U is f closed"} \mid f \in K\} \cup \{\text{"if the field of } < \text{ has more than n elements, then } \varphi_n^{U}\text{"} \mid n \in \omega\}$$

to get a model of $\{\varphi_n \mid n \in \omega\}$.

Assume that L is a logic and ξ is an ordinal such that whenever a set Σ of L-sentences containing a predicate < has, for each $\eta < \xi$, a well-ordered model of length at least η, it has a non-well-ordered model. We call the least ξ with this property the well-ordering number w(L) of L (see §.6 for the precise definition). Hence $w(L_{\omega\omega}) = \omega$. By the preceeding theorem the statement "the well-ordering number of L is ξ" may be viewed as a generalization of the compactness property, and indeed, for some logics it turns out to be, as the compactness theorem of first-order logic, a useful model-theoretic result. In this section we determine the well-ordering number of $L_{\omega_1\omega}$ and we use it to derive preservation theorems, results on non axiomatizability (§.5) and upward Löwenheim-Skolem theorems (§.6).

Let us first state some well-known results for future reference (cf.[23]).

4.2 $L_{\varkappa^+\omega}$-Löwenheim theorem. Assume that $\mathfrak{A}$ is a model of the $L_{\varkappa^+\omega}$- sentence φ. Let $A_0 \subset A$ and $|A_0| + \varkappa \leq \mu \leq |A|$. Then there is a $\mathfrak{B}$ such that

$$\mathfrak{B} \subset \mathfrak{A},\ \mathfrak{B} \models \varphi,\ A_0 \subset B \ \text{ and } \ |B| = \mu.$$

In particular, if $\varphi \in L_{\omega_1\omega}$ has an infinite model, then it has a countable model.

4.3 Scott's isomorphism theorem. Let $\mathfrak{A}$ be a denumerable structure of denumerable type. There is $\varphi_{\mathfrak{A}} \in L^K_{\omega_1\omega}$ such that for all denumerable $\mathfrak{B}$

$$\mathfrak{B} \simeq \mathfrak{A} \quad \text{iff} \quad \mathfrak{B} \models \varphi_{\mathfrak{A}}$$

$\varphi_{\mathfrak{A}}$ is called a Scott sentence of $\mathfrak{A}$.

4.4 Exercises.

(a) Assume $\mathfrak{A}$ is a denumerable structure and $|K| > \aleph_o$. Show: There is a countable $K_o \subset K$ such that the predicates and functions in $K - K_o$ are $L^{K_o}_{\omega_1\omega}$-definable with quantifier-free formulas, i.e. if $P \in K - K_o$ is n-ary then

$$\mathfrak{A} \models \forall \overset{n}{v}(P\overset{n}{v} \leftrightarrow \varphi(\overset{n}{v}))$$

for some quantifier-free $\varphi(\overset{n}{v}) \in L^{K_o}_{\omega_1\omega}$; similarly for function symbols.

(b) Using (a) extend Scott's theorem to uncountable K.

(c) Prove the analogue of (a) for $L_{\kappa^+\omega}$.

4.5 $L_{\omega_1\omega}$ - interpolation theorem. Assume that $\varphi \in L^{K_1}_{\omega_1\omega}$, $\psi \in L^{K_2}_{\omega_1\omega}$ and $K = K_1 \cap K_2$. If $\models \varphi \to \psi$ then $\models \varphi \to \chi$ and $\models \chi \to \psi$ for some $\chi \in L^{K}_{\omega_1\omega}$.

4.6 Lemma. If $\mathfrak{A} \simeq_{\kappa^+} \mathfrak{B}$ then $\mathfrak{A} \equiv_{\kappa^+\omega} \mathfrak{B}$.

The proof is a straightforward generalization of the proof of 1.7 and uses the fact that every $L_{\kappa^+\omega}$ - sentence has quantifier-rank $< \kappa^+$.

Let $<$ be a binary predicate symbol and $K = \{<\}$. It is easy to define for each ordinal ξ, by induction, formulas $\varphi_\xi(x)$ and ψ_ξ such that

(1) $\varphi_\xi(x), \psi_\xi \in L^K_{\omega\omega}$ for $\xi < \omega$,

$\varphi_\xi(x), \psi_\xi \in L_{|\xi|^+\omega}$ for $\xi \geq \omega$,

(2) for all $\mathfrak{A}$ and $a \in A$,

$$\mathfrak{A} \models \varphi_\xi[a] \quad \text{iff} \quad (\{b \mid b \in A, b <^A a\}, <^A) \simeq (\xi, <),$$

(3) for all $\mathfrak{A}$,

$$\mathfrak{A} \models \psi_\xi \quad \text{iff} \quad \mathfrak{A} \simeq (\xi,<).$$

In contrast to (1) - (3) we obtain

4.7 <u>Lemma</u>. Let $\varphi \in L^{\{<\}}_{\varkappa^+\omega}$ and $\varkappa^+ \leq \eta$.

(i) If for all $\xi < \varkappa^+$, $(\xi,<) \models \varphi$, then $(\eta,<) \models \varphi$.

(ii) If for all $\xi < \varkappa^+$, $(\xi,<) \models \varphi$, then φ has a model that is not well-ordered and even contains a copy of the rationals.

<u>Proof</u>. (i) If $(\eta,<) \models \neg\varphi$, then by the $L_{\varkappa^+\omega}$ - Löwenheim theorem there is $(B,<) \subset (\eta,<)$, $|B| = \varkappa$ and $(B,<) \models \neg\varphi$. But $(B,<) \simeq (\xi,<)$ for some $\xi < \varkappa^+$.

(ii) By (i), $(\varkappa^+,<) \models \varphi$. Let Q^+ be the set of rationals ≥ 0. By 1.6.a and 4.6, $(\varkappa^+,<) \equiv_{\varkappa^+\omega} (Q^+,<) \otimes (\varkappa^+,<)$. Thus $(Q^+,<) \otimes (\varkappa^+,<) \models \varphi$.

Note that we used the assumption $(\xi,<) \models \varphi$ only for ξ with $|\xi| = \varkappa$.

4.8 <u>Exercise</u>. Assume that A is an admissible set and $\varphi \in L^{\{<\}}_A$. Show: If for all $\xi \in A$, $(\xi,<) \models \varphi$, then φ has a model that contains a copy of the rationals (Hint: The set of ordinals in A is closed under ordinal exponentiation. Note that $qr(\varphi) \in A$ and apply 1.6.a).

Using 4.7 and the interpolation theorem we show that only countable well-orderings are $L_{\omega_1\omega}$-definable <u>in</u> a model:

4.9 <u>Theorem</u>. Let φ be an $L^K_{\omega_1\omega}$ - sentence and $<$ a binary predicate symbol. Suppose that for all $\xi < \omega_1$, φ has a model $\mathfrak{A}$ such that $<^A$ is a well-ordering of its field of order type $\geq \xi$. Then φ has a model that is not well-ordered, and even contains a copy of the rationals.[1)]

Proof. We may assume that for all $\xi < \omega_1$ there is a model $\mathfrak{A}$ of φ with well-ordered $\overset{A}{<}$ of order type ξ (take as φ the formula: $\varphi \wedge$ "$\prec$ is a subordering of $<$", where $\prec$ is not in φ). Let R, P and f be new symbols, R and P binary and f unary. Let $\bar{\varphi} \in L_{\omega_1\omega}^{K \cup \{R,f\}}$ be

$\varphi \wedge$ "R is an ordering of the universe" $\wedge$ "f is an isomorphism of the universe with the ordering R onto (field of $<$,$\prec$)".

Let $\bar{\psi} \in L^{\{R,P\}}$ be

" if P is a subordering of R, then P is not a dense ordering".

Assume by contradiction that φ has only well-ordered models. Then $\models \bar{\varphi} \rightarrow \bar{\psi}$. Thus by the interpolation theorem 4.5, there is $\chi \in L_{\omega_1\omega}^{\{R\}}$ such that

(a) $\models \bar{\varphi} \rightarrow \chi$ and (b) $\models \chi \rightarrow \bar{\psi}$.

By (a), for all countable $\xi < \omega_1$, $(\xi,<) \models \chi$. By 4.7. (ii) this contradicts (b).

4.10 Exercise. Let A be a countable admissible set and φ an L_A - sentence containing $<$. If for all $\xi \in A$, φ has a well-ordered model of order type $\geq \xi$, then φ has a non-well-ordered model containing a copy of the rationals (Hint: see the proof of 4.9, use the L_A - interpolation theorem and 4.8).

By the preceeding theorem the well-ordering number of $L_{\omega_1\omega}$ is ω_1. We use this fact to derive some preservation theorems for $L_{\omega_1\omega}$ in a way analogous

---

1) Note that 4.9 is an immediate consequence of the Lusin-Sierpinski result about Σ_1^1 classes in descriptive set theory.

to that for first-order predicate logic in §.3. Compare [8] for another proof of these theorems using back and forth techniques.

Let $\varphi_\xi(x)$ be the formula given below 4.6. Arguing as in the proof of 4.1 we may restate the definition of $R_n \rightharpoonup R$:

Let R_n, $R \subset Str(K) \times Str(K)$. Then $R_n \rightharpoonup R$ if for some $\bar{K}, \Sigma, U$ and V, where $K \subset \bar{K}$ and $\Sigma \subset L^{\bar{K}}_{\omega\omega}$, we have

(i) $(\bar{K}, \Sigma \cup \{\exists x \varphi_n(x)\}, U, V)$ defines R_n and

(ii) $\mathfrak{A} R \mathfrak{B}$ iff $\mathfrak{A}$ and $\mathfrak{B}$ are the U-part resp. V-part of a model of φ with non-well-ordered $<$.

Hence the corresponding convergence notion of $L_{\omega_1\omega}$ is:

4.11 Definition. Let R_ξ (for $\xi < \omega_1$) and $R \subset Str(K) \times Str(K)$, and assume that K is at most countable.

$R_\xi \rightharpoonup R$ if for some type $\bar{K} \supset K$ and some $U, V \in \bar{K}$ there is $\varphi \in L^{\bar{K}}_{\omega_1\omega}$ such that

(i) $(\bar{K}, \varphi \wedge \exists x \varphi_\xi(x), U, V)$ defines R_ξ and

(ii) $\mathfrak{A} R \mathfrak{B}$ iff for some model $\mathfrak{C}$ of φ with K-closed $U^{\mathfrak{C}}$ and $V^{\mathfrak{C}}$ and non-well-ordered $<^{\mathfrak{C}}$ we have

$$[U^{\mathfrak{C}}]^{\mathfrak{C}\restriction K} \simeq \mathfrak{A} \quad \text{and} \quad [V^{\mathfrak{C}}]^{\mathfrak{C}\restriction K} \simeq \mathfrak{B} .$$

4.12 Examples. $\simeq_\xi \rightharpoonup \simeq_p$; $\rightarrow_\xi \rightharpoonup \rightarrow_p$;

$$\vec{\epsilon}_\xi \rightharpoonup \vec{\epsilon}_p \ ; \ \geq_\xi \rightharpoonup \geq_p .$$

4.13 $L_{\omega_1\omega}$ - convergence lemma. Suppose $R_\xi \rightharpoonup R$ and for $\xi < \omega_1$, $\mathfrak{A}_\xi R_\xi \mathfrak{B}_\xi$. Let φ_1 and φ_2 be $L^K_{\omega_1\omega}$ - sentences such that $\mathfrak{A}_\xi \models \varphi_1$ and $\mathfrak{B}_\xi \models \varphi_2$ for all $\xi < \omega_1$. Then there are denumerable $\mathfrak{A}^*$ and $\mathfrak{B}^*$ such that

$$\mathfrak{A}^*R\mathfrak{B}^*,\ \mathfrak{A}^* \models \varphi_1 \quad \text{and} \quad \mathfrak{B}^* \models \varphi_2.$$

Proof. We use 4.9 to get a model of $\varphi \wedge \varphi_1^U \wedge \varphi_2^V$ with non-well-ordered < (here φ is the formula of 4.11).

Assume that $\mathfrak{A}$ is a countable K-structure and $\overset{r}{a} \in A$. We define for $\xi < \omega_1$, by induction on ξ, the "ξ-extension type" of $\overset{r}{a}$:

$$\alpha^{\xi}_{\overset{r}{a}} = \varphi^{o}_{\overset{r}{a}},$$

$$\alpha^{\xi+1}_{\overset{r}{a}} = \bigwedge_{a \in A} \exists v\, \alpha^{\xi}_{\overset{r}{a}\,\overset{r}{a}} \quad \text{and}$$

$$\alpha^{\xi}_{\overset{r}{a}} = \bigwedge_{\eta < \xi} \alpha^{\eta}_{\overset{r}{a}}, \quad \text{if } \xi \text{ is a limit ordinal.}$$

Denote $\alpha^{\xi}_{\emptyset}$ by $\alpha^{\xi}_{\mathfrak{A}}$.

4.14 Lemma.

(i) $\alpha^{\xi}_{\overset{r}{a}}$ is an existential $L_{\omega_1\omega}$ - formula.

(ii) $\mathfrak{A} \to_{\xi} \mathfrak{B}$ iff for all $\eta < \xi$ $\mathfrak{B} \models \alpha^{\eta}_{\mathfrak{A}}$

4.15 Theorem. Assume that $\mathfrak{A}$ and $\mathfrak{B}$ are countable structures and that every existential $L_{\omega_1\omega}$ - sentence holding in $\mathfrak{A}$ holds in $\mathfrak{B}$. Then $\mathfrak{A}$ is embeddable in $\mathfrak{B}$.

Proof. By assumption and 4.14, $\mathfrak{A} \to_{\xi} \mathfrak{B}$ for $\xi < \omega_1$. Let $\varphi_{\mathfrak{A}}$ resp. $\varphi_{\mathfrak{B}}$ be a Scott-sentence of $\mathfrak{A}$ resp. $\mathfrak{B}$ (see 4.3). By the $L_{\omega_1\omega}$ - convergence lemma there are countable $\mathfrak{A}^*$ and $\mathfrak{B}^*$ such that $\mathfrak{A}^* \to \mathfrak{B}^*$, $\mathfrak{A}^* \models \varphi_{\mathfrak{A}}$ and $\mathfrak{B}^* \models \varphi_{\mathfrak{B}}$. Hence $\mathfrak{A} \simeq \mathfrak{A}^*$, $\mathfrak{B} \simeq \mathfrak{B}^*$, and $\mathfrak{A} \to \mathfrak{B}$.

Given $\varphi \in L_{\omega_1\omega}$ and $\xi < \omega_1$ the set $\{\alpha^{\xi}_{\mathfrak{A}} \mid \mathfrak{A} \models \varphi\}$, in general, is not countable, thus the disjunction of this set is <u>not</u> an $L_{\omega_1\omega}$ - formula. Therefore we do not get using the convergence notion, the preservation theorem 3.12 for $L_{\omega_1\omega}$ in an analogous way as for first-order logic. Using back and forth techniques that <u>depend on a given φ</u> one obtains for $L_{\omega_1\omega}$ a purely model-theoretic proof of some interpolation and preservation theorems. (See [30], [35]).

4.16 <u>Exercises.</u>

(a) Let $\mathfrak{A}$ be a countable structure and φ an $L_{\omega_1\omega}$ - sentence. Show: If $\mathfrak{A}$ is a model of every universal $\psi \in L_{\omega_1\omega}$ such that $\models \varphi \to \psi$, then $\mathfrak{A}$ can be embedded in some model of φ (use the $L_{\omega_1\omega}$-convergence lemma.)

(b) Prove similar results for $\underset{\varepsilon}{\to}$ an $\geq$.

(c) Let $\mathfrak{A} = (A,<)$ be any ordering. Show: Every existential $L_{\omega_1\omega}$ - sentence that holds in $\mathfrak{A}$ holds in $(\omega_1,<)$. In particular, 4.14 is not true in case $\mathfrak{B}$ is uncountable. (Hint: Let $(C,<)$ be an ordered structure with first element such that $(A,<) \subset (C,<)$. Then $(C,<) \otimes (\omega_1,\omega) \underset{\omega_1\omega}{\equiv} (\omega_1,<)$).

<u>§.5 A non-axiomatizability result</u>

We show in this section that the class of minimal structures is <u>not</u> $L_{\omega_1\omega}$-axiomatizable. We first derive a characterization of minimal structures that enables us to show that arbitrarily large countable well-orderings are impliciti-ly definable in the class of minimal structures. Then we apply the fact that the well-ordering number of $L_{\omega_1\omega}$ is ω_1 to obtain the above non-axiomatizability result. The results of this section are due to Deißler [11].

Assume throughout that K is <u>at most countable</u>. A structure $\mathfrak{A}$ is called <u>minimal</u> if it has no proper elementary substructures, i.e. if $\mathfrak{B} \prec \mathfrak{A}$ then $\mathfrak{B} = \mathfrak{A}$. Of course, every minimal structure is at most countable and every finite structure is minimal.

5.1 <u>Examples</u>.

(i) $(\mathbb{Z},<)$ is minimal and prime.

(ii) $(\mathbb{Q},<)$ is prime but not minimal.

(iii) $(\mathbb{Z},+)$ is minimal but not prime. (To show that $(\mathbb{Z},+)$ is not prime, let $(\mathbb{Z}_{(p)},+)$ be the group of all rationals whose denominators are not divisible by the prime p. Then, by 1.6.b, $(\mathbb{Z},+) \equiv \sum_p(\mathbb{Z}_{(p)},+)$, but $(\mathbb{Z},+)$ cannot be elementarily embedded in $\sum_p(\mathbb{Z}_{(p)},+)$ since each non-zero element of $\sum_p \mathbb{Z}_{(p)}$ is divisible by some prime while $1 \in (\mathbb{Z},+)$ is not (see [1]).

$(\omega,<)$ is minimal as each element is first-order definable, $(\mathbb{Z},<)$ and $(\mathbb{Z},+)$ are minimal structures as each element is first-order definable using any element of $\mathbb{Z}$ as parameter, i.e. each element is definable "after one step". We shall prove that $\mathfrak{A}$ is minimal just in case each element of $\mathfrak{A}$ is definable "after finitely many steps".

Denote first-order logic by L. If $\mathfrak{A}$ is a K-structure and $A_o \subset A$ denote by $K \cup A_o$ the type $K \cup \{c_a \mid a \in A_o\}$, and by $(\mathfrak{A},A_o)$ the $K \cup A_o$-structure where c_a is interpreted by a. If $\varphi \in L_1^K$, i.e. $\varphi = \varphi(v_o)$, put

$$\varphi^{\mathfrak{A}} = \{a \mid a \in A, \mathfrak{A} \models \varphi[a]\} .$$

5.2 Definition. The rank in $\mathfrak{A}$ of an element $a \in A$ over a subset $A_o \subseteq A$, $rk(a,A_o,\mathfrak{A})$, is defined by induction on the ordinals:

$rk(a,A_o,\mathfrak{A}) = 0$, if for some $\varphi \in L_1^{K \cup A_o}$

$\mathfrak{A} \models \exists^{=1} v_o \varphi$ and $\mathfrak{A} \models \varphi[a]$,

and for $\xi > 0$

$rk(a,A_o,\mathfrak{A}) = \xi$, if $rk(a,A_o,\mathfrak{A}) = \eta$ does not hold for any $\eta < \xi$, and for some $\varphi \in L_1^{K \cup A_o}$ with $\varphi^{(\mathfrak{A},A_o)} \neq \emptyset$ and all $b \in \varphi^{(\mathfrak{A},A_o)}$, $rk(a,A_o \cup \{b\},\mathfrak{A}) < \xi$ holds.

In case there is no ξ such that $rk(a,A_o,\mathfrak{A}) = \xi$ we put $rk(a,A_o,\mathfrak{A}) = \infty$. By convention, let $\xi < \infty$ for each ordinal ξ. Let $rk(a,\mathfrak{A}) = rk(a,\emptyset,\mathfrak{A})$ and $rk(\mathfrak{A}) = \sup \{rk(a,\mathfrak{A}) + 1 \mid a \in A\}$. Thus $rk(a,A_o,\mathfrak{A})$ gives us the number of steps we need to define a in $\mathfrak{A}$ using parameters of A_o.

5.3 Example. $rk((\omega,<)) = 1$, $rk((\mathbb{Z},<)) = rk((\mathbb{Z},+)) = 2$ and $rk((\mathbb{Q},<)) = \infty$.

The next lemma is a simple consequence of definition 5.2.

5.4 Lemma. Let $\mathfrak{A}$ be countable and $a \in A$.

(i) If $A_o \subset A_1 \subset A$ then $rk(a,A_1,\mathfrak{A}) \leq rk(a,A_o,\mathfrak{A})$.

(ii) If $A_o \subset A$ and $rk(a,A_o,\mathfrak{A}) < \infty$, then $rk(a,A_o,\mathfrak{A}) < \omega_1$.

(iii) Let $A_o \subset A$ and $rk(a,A_o,\mathfrak{A}) = \infty$. If $\varphi \in L_1^{K \cup A_o}$ and $\varphi^{(\mathfrak{A},A_o)} \neq 0$, then $rk(a,A_o \cup \{b\},\mathfrak{A}) = \infty$ for some $b \in \varphi^{(\mathfrak{A},A_o)}$.

5.5 Theorem. Let $\mathfrak{A}$ be countable and $a \in A$. Then

(i) $rk(a,\mathfrak{A}) < \infty$ iff a is contained in each elementary submodel of $\mathfrak{A}$.

(ii) $rk(\mathfrak{A}) < \infty$ iff $\mathfrak{A}$ is minimal.

<u>Proof</u>. (ii) follows immediately from (i).

(i): Assume $rk(a,\mathfrak{A}) < \infty$ and let $\mathfrak{B}$ be an elementary substructure of $\mathfrak{A}$. Prove by induction on ξ,

$$\text{if } B_o \subset B, b \in A \text{ and } rk(b,B_o,\mathfrak{A}) = \xi, \text{ then } b \in B.$$

In particular $a \in B$.

Now assume $rk(a,\mathfrak{A}) = \infty$. Let $\varphi_o, \varphi_1, \ldots$ be an enumeration of the formulas of L^K such that $\varphi_n \in L^K_{n+1}$. Define a sequence $(b_n)_{n \in \omega}$ of elements of A by induction. Assume that b_m has been defined for $m < n$ such that $rk(a,\{b_o,\ldots,b_{n-1}\},\mathfrak{A}) = \infty$. If $(\mathfrak{A},\{b_o,\ldots,b_{n-1}\}) \models \exists v_o \varphi(v_o, c_{b_o},\ldots,c_{b_{n-1}})$ choose, using 5.4.(iii), b_n such that

(1) $\quad (\mathfrak{A},\{b_o,\ldots,b_{n-1}\} \models \varphi(b_n, c_{b_o},\ldots,c_{b_{n-1}})$ and $rk(a,\{b_o,\ldots,b_n\},\mathfrak{A}) = \infty$.

Otherwise put $b_n = b_{n-1}$. Let $\overline{B} = \{b_n \mid n \in \omega\}$. Then, by (1) and Tarski's lemma there is a $\mathfrak{B}$ such that $\mathfrak{B} < \mathfrak{A}$ and $B = \overline{B}$. As for $n \in \omega$ $rk(a,\{b_o,\ldots,b_n\},\mathfrak{A}) = \infty$, a is not contained in the elementary submodel $\mathfrak{B}$.

<u>Formalization of the rank in $L_{\omega_1\omega}$.</u>

Let $U, <, c_o, R_n (n \in \omega)$ be symbols not in K, U unary, < binary and R_n (n+2)-ary. Let $K' = K \cup \{U,<,c_o\} \cup \{R_n \mid n \in \omega\}$. Let φ_{rk} be the conjunction of the following $L^{K'}_{\omega_1\omega}$ - sentences:

(a) "< is an ordering, U is the field of <, c_o is the first element of <".

(b) $\bigwedge_{n \in \omega} \forall x \forall \overset{n}{x} (R_n x \overset{n}{x} c_o \leftrightarrow \bigvee_{\varphi(x,\overset{n}{x}) \in L^K_{\omega\omega}} \exists^{=1} y \varphi(y,\overset{n}{x}) \wedge \varphi(x,\overset{n}{x})$

("$rk(x,\overset{n}{x},\cdot) = 0$ iff x is first-order definable over $\overset{n}{x}$) .

(c) $\bigwedge_{n \in \omega} \forall x \forall \overset{n}{x} \forall u(Uu \wedge \neg u = c_o \rightarrow (R_n x \overset{n}{x} u \leftrightarrow (\neg \exists v(Uv \wedge v < u \wedge R_n x \overset{n}{x} v) \wedge$

$\bigvee_{\varphi(\overset{n}{x},y) \in L^K_{\omega\omega}} (\exists y \varphi(\overset{n}{x},y) \wedge \forall y(\varphi(\overset{n}{x},y) \rightarrow \exists v(v < u \wedge R_{n+1} x \overset{n}{x} y v)))))$

(" if $u > o$, then $rk(x,\overset{n}{x},\cdot) = u$ iff ... ").

(d) $\bigwedge_{n \in \omega} \forall x \forall \overset{n}{x} \exists u(Uu \wedge R_n x \overset{n}{x} u)$

("each element has a rank over any finite set") .

(e) $\forall u(Uu \rightarrow \exists x \exists v(u \leq v \wedge R_o xv))$

("the rank of elements are cofinal in U").

Clearly

5.6 <u>Lemma</u>. Let $\mathfrak{A}$ be a countable <u>minimal</u> K-structure. Then

(i) $\mathfrak{A}' \models \varphi_{rk}$ holds for some K'-structure $\mathfrak{A}'$ with $\mathfrak{A}' \restriction K = \mathfrak{A}$,

(ii) if the K'-structure $\mathfrak{A}'$ is a model of φ_{rk} and $\mathfrak{A}' \restriction K = \mathfrak{A}$, then $<$ is a well-ordering of type $rk(\mathfrak{A})$.

Below we show (for some K)

(*) for all $\xi < \omega_1$ there is a minimal K-structure $\mathfrak{A}_\xi$ with $rk(\mathfrak{A}_\xi) \geq \xi$.

5.7 <u>Theorem</u>. The class of (countable) minimal models is not the class of (countable) models of an $L_{\omega_1\omega}$-sentence.

<u>Proof</u>. Otherwise for some $\varphi \in L^K_{\omega_1\omega}$,

$$\{\mathfrak{A} \mid \mathfrak{A} \in Str(K), \mathfrak{A} \models \varphi, |A| = \aleph_o\} = \{\mathfrak{A} \mid \mathfrak{A} \in Str(K), \mathfrak{A} \text{ minimal}, |A| = \aleph_o\}.$$

By (*) and (i) of the preceeding lemma $\varphi \wedge \varphi_{rk}$ has, for each $\xi < \omega_1$, a well-ordered model of type $\geq \xi$, hence, by 4.9 a non-well-ordered model $\mathfrak{A}'$.Let

$\mathfrak{A} = \mathfrak{A}' \upharpoonright K$, so $\mathfrak{A} \models \varphi$. But then, by part (ii) of the preceeding lemma, $\mathfrak{A}$ is not a minimal structure.

The construction of a structure $\mathfrak{A}_\xi$ with $rk(\mathfrak{A}_\xi) \geq \xi$ is by induction on ξ. We sketch the idea:

We know that $rk(\mathbb{Z},<) = 2$ and that we need one step to define any $a \in \mathbb{Z}$. By attaching to each element of $(\mathbb{Z},<)$ a copy of $(\mathbb{Z},<)$ we obtain a structure $\mathfrak{B}$ such that we need two steps to define any $b \in B$. More precisely: Let K be $\{<,P\}$ where $<$ and P are binary. Let $\mathfrak{B}$ be the K-structure with

(a) $B = \mathbb{Z} \cup (\mathbb{Z} \times \mathbb{Z})$.

(b) $P^B z_1(z_1,z_2)$ for $z_1, z_2 \in \mathbb{Z}$.

(c) $z_1 <^B z_2$ for $z_1, z_2 \in \mathbb{Z}$, $z_1 < z_2$.

(d) $(z,z_1) <^B (z,z_2)$ for $z, z_1, z_2 \in \mathbb{Z}, z_1 < z_2$.

(e) $P^B a_1 a_2$ or $a_1 <^B a_2$ only holds, if required by (b), (c) or (d).

Using a refinement of this construction we obtain the following lemma (see [11] for a proof).

5.8 Lemma. Assume that for $n \in \omega$, $\mathfrak{A}_n$ is a minimal K_n-structure. Then for some K there is a K-structure $\mathfrak{A}$ such that

$$\infty > rk(\mathfrak{A}) \geq rk(\mathfrak{A}_o) + rk(\mathfrak{A}_1) + \ldots$$

5.9 Corollary. For each $\xi < \omega_1$ there is an $\mathfrak{A}$ such that $\xi \leq rk(\mathfrak{A}) < \infty$.

5.10 Exercise. Show that for $\xi < \omega_1$ there is a $\varphi_\xi \in L^K_{\omega_1\omega}$ such that the class of models of φ_ξ contains exactly the minimal K-structures of rank ξ.

(Hint: Define by induction on ξ a formula $\psi_\xi(v_o, v_1, \ldots, v_n)$ that says the rank of v_o over $\{v_1, \ldots, v_n\}$ is ξ). Thus, if we restrict our attention to countable models, $\bigcup_{\xi < \aleph_1} \mathrm{Mod}(\varphi_\xi)$ is the class of minimal structures.–The straightforwarded formalization of the definition of a minimal structure is

$$(*) \quad \forall P(\bigwedge_{n \in \omega} \forall \overset{n}{v}(Pv_o \wedge \ldots \wedge Pv_{n-1} \rightarrow \bigwedge_{\varphi(\overset{n}{v}) \in L_{\omega\omega}} (\varphi^P \leftrightarrow \varphi)) \rightarrow \forall x Px) .$$

The formula is of the form $\forall P\psi$ where ψ is in $L_{\omega_1\omega}$. Vaught [35] has shown that (over countable models) the class of models of any formula χ of this form is the union $\bigcup_{\xi < \omega_1} \mathfrak{K}_\xi$ of $\aleph_1$ $L_{\omega_1\omega}$ - elementary classes. $\mathfrak{K}_\xi$ contains just those models of χ that have an implicitly definable well-ordering of length ξ. Thus in the preceeding exercise we obtain Vaughts result for the formula (*).

5.11 Exercise. An abelian group G is reduced if it has no non-trivial divisible subgroups. G is a p-group, p a prime, if every element has order p^n for some $n \in \omega$. Show: The class of reduced p-groups is not the class of models of some $L_{\omega_1\omega}$ - sentence. (Hint: Define $p^\xi G$ by induction: $p^o G = G$ and for $\xi > 0$ $p^\xi G = \bigcap_{\eta < \xi} p(p^\eta G)$. If G is a reduced p-group there must be a smallest ξ such that $p^\xi G = 0$. ξ is called the length of G. For any $\xi < \omega_1$ there exists a reduced p-group of length ξ. Apply 4.9 to get the conclusion).

§.6 Hanf number

An easy application of the compactness theorem shows that each set of first-order sentences with an infinite model has arbitrarily large models (i.e. for any cardinal κ there is a model of power $\geq \kappa$). The corresponding upward Löwenheim-Skolem theorem for $L_{\omega_1\omega}$ is not obvious. We will obtain it using the fact that the well-ordering number of $L_{\omega_1\omega}$ is ω_1. Since most of the results are true for logics in the sense of §.2, we start in the general framework.

6.1 Definition. Let L be a logic.
The Hanf number of L, $h(L)$, is the least cardinal κ such that if an L-sentence has a model of power $\geq \lambda$, then it has arbitrarily large models (provided such a κ exists). More general, for a cardinal λ let $h_\lambda(L)$ be the least κ such that if a set of $\leq \lambda$ many L-sentences has a model of power $\geq \kappa$, then it has arbitrarily large models. Thus $h(L) = h_1(L)$.

$$h_\lambda(L_{\omega\omega}) = \aleph_o, \; h(L_{\omega_1\omega}) > \aleph_o, \; h(L(Q_\kappa)) > \kappa$$

(take the sentence that says "< is an ordering of the universe of power at least κ and each proper initial segment has power $< \kappa$"). $h(L_{\infty\omega})$ does not exist (consider the sentences ψ_ξ of §.4(3)).

Assume that L is a logic and L^K is a set: For $\Sigma \subset L^K$, $|\Sigma| \leq \lambda$, define κ_Σ by

$$\varkappa_\Sigma = \begin{cases} 0 & \text{, if } \Sigma \text{ has arbitrarily large models.} \\ \sup\{|A| \mid \mathfrak{A} \models \Sigma\}, & \text{otherwise.} \end{cases}$$

Put $\varkappa_\lambda^K = \sup\{\varkappa_\Sigma^+ \mid \Sigma \subset L^K, |\Sigma| \leq \lambda\}$. By the axiom of replacement, $\varkappa_\lambda^K$ is a cardinal[1]. $\varkappa_\lambda^K$ depends on K and $\sup\{\varkappa_\lambda^K \mid K \text{ a type}\}$ might not exist. But all familiar logics satisfy a kind of renaming property that implies the existence of a K_o such that for all $K, \varkappa_\lambda^K \leq \varkappa_\lambda^{K_o}$.

An injective mapping α of a type K_1 into a type K_2 is a <u>name changer</u>, if it maps n-ary predicate (function) symbols onto n-ary predicate (function) symbols. Such a name changer induces a natural transformation of a K_2-structure $\mathfrak{A}$ into a K_1-structure $\mathfrak{A}^\alpha$. We write Renam(L) and say that L has the <u>renaming property</u>,

if for each name changer $\alpha: K_1 \to K_2$ and each $\varphi \in L^{K_1}$ there is $\psi \in L^{K_2}$ such that for $\mathfrak{A} \in \mathrm{Str}(K_2)$

$$\mathfrak{A} \models \psi \quad \text{iff} \quad \mathfrak{A}^\alpha \models \varphi .$$

6.2 <u>Theorem</u> (cf. [20]). Assume that L is a logic with the following properties.

(i) Renam(L),

(ii) the number of symbols occuring in an L-sentence is bounded, i.e. there is a cardinal μ such that for each L-sentence $\varphi, \varphi \in L^K$ for some K with $|K| \leq \mu$.

---

1) Note that the existence of the Hanf number of second-order logic is not provable in Zermelo set theory $+\Sigma_1$-replacement (cf.[5] for the precise statement).

(iii) L^K is a set for all K.

Then $h_\lambda(L)$ exists for all λ.

Proof. Let K_o contain for each $n \in \omega$, $\mu \cdot \lambda$ n-ary predicate symbols and $\mu \cdot \lambda$ n-ary function symbols. Using (i) and (ii) it is easy to show that $\varkappa_\lambda^K \le \varkappa_\lambda^{K_o}$ holds for any K.

All logics listed at the beginning of §.2 satisfy (i), and all but $L_{\infty\omega}$ satisfy (i) and (ii). Thus the Hanf number exists for all these logics.

We will show that for some logics the Hanf number is related to its well-ordering number. Let us say that a cardinal μ is captured by the L-sentence φ, if φ has a model of power $\ge \mu$ but not arbitrarily large models. Then the Hanf number of L is the smallest cardinal that may not be captured by an L-sentence. Similarly, let us say that an ordinal η is captured by the L-sentence φ containing a binary predicate symbol $<$, if φ has a model with well-ordered $<$ of order type at least η but φ has no non-well-ordered models. Then the well-ordering number of L is the least ordinal that may not be captured by an L-sentence. More precisely:

6.3 Definition. The well-ordering number of the logic L, w(L), is the least ordinal ξ (provided such an ordinal exists), such that if an L-sentence φ containing a binary predicate symbol $<$ has a model $\mathfrak{A}$ with well-ordered $<^A$ of order type $\ge \xi$, then it has a model $\mathfrak{B}$ such that $<^B$ is not well-ordered. Similarly, $w_\lambda(L)$ is defined.

6.4 Exercise. Let L be a regular logic.[1)]

(i) If the compactness theorem holds for sets of sentences of power $\le \lambda$, then $w_\lambda(L) = \omega$.

(ii) $w_\lambda(L)$ is a limit ordinal closed under ordinal addition and multiplication.

By (i), $w_\lambda(L_{\omega\omega}) = \omega$, $w_{\aleph_0}(L(Q_{\aleph_1})) = \omega$, and if V = L then $w_{\aleph_0}(L(Q^2_{\aleph_1})) = \omega$. By 4.9, $w(L_{\omega_1\omega}) = \omega_1$. The sentences ψ_ξ of §.4 show that $w(L_{\infty\omega})$ does not exist and that $w(L_{\varkappa^+\omega}) \ge \varkappa^+$. It is shown in [27] that $w(L_{\varkappa^+\omega})$ exists. Compare [7] for the value of $w(L_{\varkappa^+\omega})$. By [26], $w(L(Q_H))$ does not exist (see §.2 for the definition of $L(Q_H)$).

Define $\beth^\lambda_\xi$ (λ a cardinal, ξ an ordinal) by induction on ξ:

$$\beth^\lambda_0 = \lambda \quad \text{and} \quad \beth^\lambda_\xi = \sup\{2^{\beth^\lambda_\eta} \mid \eta < \xi\} \quad \text{for } \xi > 0.$$

Denote $\beth^0_\xi$ by $\beth_\xi$. Obviously $\beth_{\lambda+\xi} \ge \beth^\lambda_\xi$. Thus

(1) if $\lambda + \xi = \xi$, then $\beth^\lambda_\xi = \beth_\xi$.

By definition, $\beth_\omega = \aleph_0$. Hence we have

---

1) Whenever in this section we speak of a regular logic L we assume that L satisfies the following relativization property, Relat*(L), that is stronger than that introduced in § 2:

Relat*(L) iff for any L-sentence φ there is a K_0 such that $\varphi \in L^{K_0}$ and for all unary $U \notin K_0$ there is a ψ (denoted by φ^U), $\psi \in L^{K_0 \cup \{U\}}$ such that for any $\mathfrak{A} \in \mathrm{Str}(K_0 \cup \{U\})$,

$$\mathfrak{A} \models \varphi^U \quad \text{iff} \quad U^A \text{ is } K_0\text{-closed and } [U^A]^{\mathfrak{A} \upharpoonright K_0} \models \varphi.$$

$$h(L_{\omega\omega}) = \beth_{w(L_{\omega\omega})}.$$

We shall see that a similar relation between the Hanf number and the well-ordering number holds for some logics.

If φ_o is an L-sentence having only models of the infinite power κ, then the L-sentence

$$\varphi_o^U \wedge \forall x \forall y(\forall u(Uu \rightarrow (Eux \leftrightarrow Euy)) \rightarrow x = y),$$

where E and U are new symbols, has a model of power $2^{\kappa} (= \beth_1^{\kappa})$ but not a larger one. Iterating the power set operation ξ times with a well-ordering of order type ξ, we obtain an L-sentence that has a model of power $\beth_{\xi}^{\kappa}$ but none larger. We use this idea to show:

6.5 <u>Lemma.</u> Let $1 \leq \lambda \leq \kappa$ and assume that L has the following properties:

(i) L is regular and $w_{\lambda}(L)$ exists.

(ii) There is an L-sentence φ_o that has only models of power κ.

(iii) If Σ is a set of $\leq \lambda$ L-sentences, then for some K with $\Sigma \subset L^K$, any $\mathfrak{A} \in Str(K)$ and any $A_o \subset A$, if $\mathfrak{A} \models \Sigma$ then $\mathfrak{B} \models \Sigma$ for some $\mathfrak{B}$ with $\mathfrak{B} \subset \mathfrak{A}$, $A_o \subset B$ and $|B| \leq |A_o| + \kappa$.

Then

$$h_{\lambda}(L) \geq \beth^{\kappa}_{w_{\lambda}(L)}.$$

<u>Proof.</u> $w_{\lambda}(L)$ is a limit ordinal (see 6.4(ii)), hence it suffices to show that for $\xi < w_{\lambda}(L)$ there is a set Σ of $\leq \lambda$ many L-sentences that has not arbitrarily large models but has a model of power $\beth_{\xi}^{\kappa}$. Since $\xi < w_{\lambda}(L)$ there is $\bar{\Sigma}, |\bar{\Sigma}| \leq \lambda$, and $<$ such that $\bar{\Sigma}$ only has well-ordered models and one of order type ξ. Let

U,V,E and g be new symbols. Let Σ consist of the sentences:

φ_o^U

φ^V for $\varphi \in \overline{\Sigma}$

$\forall x(Vgx \wedge$ "$gx \in fd(<)$")

$\forall x \forall y(Eyx \rightarrow g(y) < g(x))$

$\forall x(Ux \rightarrow$ "gx is the first element of <")

$\forall x \forall y(\neg Ux \wedge \neg Uy \wedge \forall z(Ezx \leftrightarrow Ezy) \rightarrow x = y)$

6.6 Corollary. $h(L_{\varkappa^+\omega}) \geq \beth^{\varkappa^+}_{w(L_{\varkappa^+\omega})}$ and $h(L(Q_\varkappa)) \geq \beth^{\varkappa}_{w(L(Q_\varkappa))}$.

Proof. For $L = L_{\varkappa^+\omega}$ take as φ_o the sentence $\psi_\varkappa$ of §.4 and for $L = L(Q_\varkappa)$ take as φ_o the sentence

$\psi^\varkappa$ = "< is a total ordering" $\wedge\ Q_\varkappa xx = x \wedge \forall x \neg Q_\varkappa yy < x$" .

We call $\mathfrak{A} = (A,<)$ a $\varkappa$-like ordering, if $\mathfrak{A}$ is a model of $\psi^\varkappa$.

Next we prove inequalities of the kind $h(L) \leq \beth^{\cdots}_{w(L)}$ for a class of logics called $\mathfrak{M}$-logics. The result for $\mathfrak{M}$-logics will imply the corresponding result for many familiar logics. We are lead to $\mathfrak{M}$-logics by the following observations:

If $\bigvee_{n \in \omega} \varphi_n$ is an $L_{\omega_1\omega}$ - formula and a unary predicate symbol U is interpreted in $\mathfrak{A}$ by the set of natural numbers then, in $\mathfrak{A}$, we will see that we may replace the infinite disjunction $\bigvee_{n \in \omega}$ by a first-order existential quantification over U. Similarly, if $Q_\varkappa x\varphi$ is an $L(Q_\varkappa)$-sentence, $\mathfrak{A}$ is a structure and $(U^A, <^A)$ is a $\varkappa$-like ordering, then using a new unary function symbol f we may express $Q_\varkappa x\varphi$ by the first-order sentence "f maps U^A 1-1 into $\varphi^{\mathfrak{A}}$" and $\neg Q_\varkappa x\varphi$ by the <u>first-order</u> sentence "f maps a proper initial segment of U^A onto $\varphi^{\mathfrak{A}}$". - Thus $L_{\omega_1\omega}$ and $L(Q_\varkappa)$ may be translated into first-order logic in structures with an adequate U-part. Such a restriction of admitted models is essential for $\mathfrak{M}$-logics (cf.[17]):

Let $\mathfrak{M}$ be a non-empty class of K_o-structures, and let U be a unary predicate symbol <u>not</u> in K_o. If $\mathfrak{A}$ is a K-structure, then we call $\mathfrak{A}$ an <u>$\mathfrak{M}$-model</u>, if $K_o \cup \{U\} \subset K$ and the U-part of $\mathfrak{A}$ lies in $\mathfrak{M}$, more precisely: U^A is K_o-closed and $[U^A]^{\mathfrak{A} \upharpoonright K_o} \simeq \mathfrak{C}$ for some $\mathfrak{C} \in \mathfrak{M}$. The <u>$\mathfrak{M}$-logic</u> is the pair $(L(\mathfrak{M}), \models_{\mathfrak{M}})$ where

$$L(\mathfrak{M})^K = \begin{cases} \emptyset & \text{, if } K_o \cup \{U\} \not\subset K \\ L^K_{\omega\omega} & \text{, if } K_o \cup \{U\} \subset K \end{cases}$$

and where for $\mathfrak{A} \in Str(K)$ and $\varphi \in L(\mathfrak{M})^K$

$$\mathfrak{A} \models_{\mathfrak{M}} \varphi \quad \text{iff} \quad \mathfrak{A} \text{ is an } \mathfrak{M}\text{-model and } \mathfrak{A} \models \varphi .$$

In particular, for $\mathfrak{M} = \{(\omega, (n)_{n \in \omega})\}$ we obtain ω-logic.

For the rest of this section assume that <u>all structures in $\mathfrak{M}$ have the same infinite cardinality $\varkappa$</u>. Then

(a) $\forall x Ux$ has only $\mathfrak{M}$-models of power $\varkappa$,

(b) if Σ has an $\mathfrak{M}$-model and $|\Sigma| \leq \varkappa$,
then Σ has an $\mathfrak{M}$-model of power $\varkappa$,

(c) $w_\lambda(L(\mathfrak{M}))$ exists ("embed" $L(\mathfrak{M})$ in some $L_{\mu\omega}$ and use the fact that $w(L_{\mu\omega})$ exists).

Though $L(\mathfrak{M})$ is not regular in the sense of §.2 the following lemma is proved as lemma 6.5.

6.8 Lemma. For all $\lambda \leq \varkappa$, $h_\lambda(L(\mathfrak{M})) \geq \beth^{\varkappa}_{w_\lambda(L(\mathfrak{M}))}$.

In order to prove $h_\lambda(L(\mathfrak{M})) \leq \beth^{\varkappa}_{w_\lambda(L(\mathfrak{M}))}$ we need the following partition theorem of Erdös and Rado. For a set X and $n \in \omega$ let $[X]^n$ be the set of all subsets of X of power n. For cardinals λ,μ and ρ, and $n \in \omega$, $1 \leq n$, let $\rho \to (\lambda)^n_\mu$ mean that for all $f:[\rho]^n \to \mu$ there is a subset X of ρ of power λ such that f is constant on $[X]^n$.

6.9 Theorem. (cf. [12])

(i) $\beth^{\varkappa}_{r+n+1} \to (\beth^{\varkappa}_r)^n_{2^\varkappa}$ for $r,n \geq 1$.

(ii) $\beth^{\varkappa}_{\omega(\xi+1)} \to (\beth^{\varkappa}_{\omega\xi})^n_{2^\varkappa}$ for $n \geq 1$ and any ξ.

6.10 Theorem. For all $\lambda \leq \varkappa$

$$h_\lambda(L(\mathfrak{M})) = \beth^{\varkappa}_{w_\lambda(L(\mathfrak{M}))}.$$

Proof. Assume that the set Σ of $\leq \lambda$ sentences has an $\mathfrak{M}$-model $\mathfrak{A}$ of power $\geq \beth^{\varkappa}_{w_\lambda(L(\mathfrak{M}))}$. We must show that Σ has arbitrarily large $\mathfrak{M}$-models. First we sketch the idea of the proof:

If we find an $\mathfrak{M}$-model $\mathfrak{C}$ of Σ containing an infinite subset X of elements indiscernible with respect to U (i.e. any two finite sequences satisfy the same formulas with parameters in U), then we obtain an $\mathfrak{M}$-model of Σ of a given cardinality $\mu > |C|$ throwing into $\mathfrak{C}$ μ new elements that behave as the elements in X. We get such a $\mathfrak{C}$ with an infinite set X of indiscernibles as follows: Since $\mathfrak{A} \models_{\mathfrak{M}} \Sigma$ and $|A| \geq \beth^{\varkappa}_{w_\lambda(L(\mathfrak{M}))}$ we find using the partition theorem 6.9, for each r, a subset X_r of A such that any two r-tuples in X_r satisfy the same formulas. Moreover $X_{r+1} \subset X_r$. We code this fact in $\mathfrak{A}$ with a well-ordering of length $w_\lambda(L(\mathfrak{M}))$ associating to each $<$ - descending sequence of length r the set X_r. By definition of the well-ordering number there is a non-well-ordered $\mathfrak{M}$-model $\mathfrak{B}$ elementarily equivalent to $\mathfrak{A}$. An infinite $<^B$ - descending chain will give rise to a set X where any two finite sequences of X satisfy the same formulas.

We start the proof with an $\mathfrak{M}$-model $\mathfrak{A}$ of Σ, where $|\Sigma| \leq \lambda$, of cardinality $\geq \beth^{\varkappa}_{w_\lambda(L(\mathfrak{M}))}$.

<u>Case 1</u>: λ is infinite and $w_\lambda(L(\mathfrak{M})) > \omega$.

Choose $\bar{K} \subset K$ such that $\Sigma \subset L^{\bar{K}}_{\omega\omega}$, $U \in \bar{K}$ and $|\bar{K}| \leq \lambda$. Assume that Σ is a set of universal sentences (introduce Skolem functions). Let $<^A$ be any ordering of A. We define, by induction on r, for any $\xi_0,\ldots,\xi_{r-1} \in w_\lambda(L(\mathfrak{M}))$ with $\xi_{r-1} < \ldots < \xi_0$ a subset $X_{\overset{r}{\xi}}$ such that

(a) $X_{\overset{r-1}{\xi}} \supset X_{\overset{r}{\xi}}$,

(b) $|X_{\overset{r}{\xi}}| = \beth^{\varkappa}_{\omega\xi_{r-1}}$

(c) if $\overset{r}{a}, \overset{r}{b} \in X_{\overset{r}{\xi}}$, $a_0 <^A \ldots <^A a_{r-1}$, $b_0 <^A \ldots <^A b_{r-1}$ and $\psi(\overset{2r}{v}) \in L^{\bar{K}}_{\omega\omega}$, then $\mathfrak{A} \models \forall \overset{r}{v}(Uv_0 \wedge \ldots \wedge Uv_{r-1} \rightarrow (\psi(\overset{r}{v},\overset{r}{a}) \leftrightarrow \psi(\overset{r}{v},\overset{r}{b})))$.

Suppose $X_{\overset{r}{\xi}}$ is defined and $\xi_r < \xi_{r-1}$. For $a_0 <^A \ldots <^A a_r \in X_{\overset{r}{\xi}}$ define the function

$s_{\underset{a}{r+1}}: (U^A)^{r+1} \times L^{\bar{K}}_{2(r+1)} \rightarrow \{0,1\}$ by

$$s_{\underset{a}{r+1}}(\overset{r+1}{u}, \psi(\overset{2(r+1)}{v})) = \begin{cases} 1, & \text{if } \mathfrak{A} \models \psi[\overset{r+1}{u}, \overset{r+1}{a}] \\ 0, & \text{if } \mathfrak{A} \models \neg\psi[\overset{r+1}{u}, \overset{r+1}{a}]. \end{cases}$$

Hence $s:[X_{\underset{\xi}{r}}]^{r+1} \rightarrow 2^{(U^A)^{r+1} \times L^{\bar{K}}_{2(r+1)}}$,

where $|X_{\underset{\xi}{r}}| = \beth^{\varkappa}_{\omega\xi_{r-1}}$ and $|2^{(U^A)^{r+1} \times L^{\bar{K}}_{2(r+1)}}| = 2^{\varkappa}$.

Therefore, by 6.9 (ii), there is a subset $X_{\underset{\xi}{r+1}}$ of $X_{\underset{\xi}{r}}$ of cardinality $\beth^{\varkappa}_{\omega\xi_r}$ on which s is constant.

Thus $X_{\underset{\xi}{r+1}}$ satisfies (a),(b) and (c).

W.l.o.g assume that $w_\lambda(L(\mathfrak{M})) \subset A$ and let $\prec^A$ be the ϵ-relation on $w_\lambda(L(\mathfrak{M}))$. For $r \in \omega$ we define the (r+1)-ary relation P^A_r on A by

$$P^A_r \overset{r}{\xi} a \quad \text{iff} \quad a \in X_{\underset{\xi}{r}}.$$

Put $\bar{\bar{K}} = \bar{K} \cup \{<,\prec\} \cup \{P_r \mid r \in \omega\}$ and

$\Sigma_1 = \{\varphi \mid \varphi \in L^{\bar{\bar{K}}}_{\omega\omega}, (\mathfrak{A}, \prec^A, <^A, (P^A_r)_{r \in \omega}) \models \varphi\}$.

In particular, $\Sigma \subset \Sigma_1$ and $|\Sigma_1| \leq \lambda$. As $\prec^A$ has order type $w_\lambda(L(\mathfrak{M}))$, Σ_1 has a non-well ordered $\mathfrak{M}$-model $\mathfrak{B}$. Let $(\xi_n)_{n \in \omega}$ be an infinite descending sequence in $\prec^B$. For $r \in \omega$, put $B_r = \{b \mid b \in B, P^B \overset{r}{\xi} b\}$. Since $\mathfrak{B} \models \Sigma_1$, we have:

($\bar{a}$) $B \supset B_0 \supset B_1 \ldots,$

($\bar{b}$) B_r is infinite,

(c) if $\vec{a},\vec{b} \in B_r, a_0 <^B \ldots <^B a_{r-1}, b_0 <^B \ldots <^B b_{r-1}, \vec{u} \in U^B$ and $\psi \in L^{\bar{K}}_{2r}$, then

$$\mathfrak{B} \models \psi[\vec{u},\vec{a}] \quad \text{iff} \quad \mathfrak{B} \models \psi[\vec{u},\vec{b}] .$$

Now, let μ be any cardinal. We show that Σ has an $\mathfrak{M}$-model of power $\geq \mu$. Choose any ordering $(X,<)$ of power μ. By the compactness theorem the following set Σ_2 of first-order sentences has a model $\mathfrak{C}$.

$$\Sigma_2 = \{\neg c_y = c_z \mid y,z \in U^B \cup X, y \neq z\} \cup$$
$$\{\psi(c_{u_0},\ldots,c_{u_{r-1}},c_{x_0},\ldots,c_{x_{r-1}}) \mid r \in \omega, \vec{u} \in U^B,$$
$$x_0 < \ldots <_{r-1} \in X, \psi(\overset{2r}{v}) \in L^{\bar{K}}_{\omega\omega} \text{ and } \mathfrak{B} \models \psi[\vec{u},\vec{a}]$$
$$\text{for some } a_0 <^B \ldots <^B a_{r-1} \in B_r\}.$$

Assume that $c_u^C = u$ for $u \in U^B$, $c_x^C = x$ for $x \in X$. Let $\mathfrak{D}$ be the $\bar{K}$-substructure of $\mathfrak{C} \upharpoonright \bar{K}$ generated by $U^B \cup X$. Then $|D| \geq \mu$ and since Σ is a set of universal sentences, $\mathfrak{D} \models \Sigma$. We prove that $\mathfrak{D}$ and $\mathfrak{B}$ have the same U-part. This will finish our proof: Since $\mathfrak{B}$ is an $\mathfrak{M}$-model, $\mathfrak{D}$ (or an expansion of $\mathfrak{D}$) will be an $\mathfrak{M}$-model too, hence $\mathfrak{D}$ is an $\mathfrak{M}$-model of Σ of power $\geq \mu$.

Clearly $U^B \subset U^D$. If $d \in U^D$ then $d = t^{\mathfrak{D}}[\vec{u},\vec{x}]$ for some $\bar{K}$-term $t(\overset{2r}{v})$, $\vec{u} \in U^B$ and $x_0 < \ldots < x_{r-1} \in X$. Thus $\mathrm{U}t(c_{u_0},\ldots,c_{u_{r-1}},c_{x_0},\ldots,c_{x_{r-1}}) \in \Sigma_2$. Hence, for any $a_0 <^B \ldots <^B a_{r-1}$ in B_{r+1}, $\mathfrak{B} \models \mathrm{U}t[\vec{u},\vec{a}]$. Choose $u \in U^B$ such that $u = t^{\mathfrak{B}}[\vec{u},\vec{a}]$. Then $c_u = t(c_{u_0},\ldots,c_{u_{r-1}},c_{x_0},\ldots,c_{x_{r-1}}) \in \Sigma_2$, thus $u = t^{\mathfrak{D}}[\vec{u},\vec{x}] = d$. Hence $U^D = U^B$. - Similarily, if $P \in K_0 \cap \bar{K}$ then P^D and P^B coincide on $U^D = U^B$. If $P \in K_0 - \bar{K}$ then define P^D such that $P^D = P^B$ on U. Do the same for function symbols, and put $\mathfrak{D}_1 = (\mathfrak{D}, (P^D)_{P \in K_0, P \notin \bar{K}}, (f^D)_{f \in K_0, f \notin \bar{K}})$. Then

$[U^{D_1}]^{\mathfrak{D}_1 \upharpoonright K_o} = [U^B]^{\mathfrak{B} \upharpoonright K_o}$, thus $\mathfrak{D}_1$ is an $\mathfrak{M}$-model.

Case 2: λ is infinite and $w_\lambda(L) = \omega$.

The proof is similar. We now use part (i) of 6.9 to obtain for $\xi_o, \ldots, \xi_{r-1} \in \omega$ with $\xi_{r-1} < \ldots < \xi_o$ a subset $X_{\bar{\xi}}$ of r-indiscernibles.

Case 3: λ is finite, or, equivalently, $\lambda = 1$.

In this case $\bar{K}$ is finite. If we code descending sequences of ordinals we do not need infinitely many predicate symbols P_r to express the properties $(\bar{a})$, $(\bar{b})$ and $(\bar{c})$. We may replace Σ_1 by a recursive set of sentences. But a recursive set of sentences in a finite type is equivalent, by a theorem of Craig and Vaught (see [10]), to a finite set of sentences using additional predicates. .

Some applications.

A) $L_{\varkappa^+\omega}$

Put $K_o = \{c_\xi \mid \xi < \varkappa\}$ and $\mathfrak{M} = \{(\varkappa, (\xi)_{\xi < \varkappa})\}$.

(i) There is $\varphi_o = L^{K_o}_{\varkappa^+\omega}$ such that $\mathrm{Mod}(\varphi_o) = \mathfrak{M}$

(more precisely, $\mathrm{Mod}(\varphi_o) = \{\mathfrak{A} \mid \mathfrak{A} \simeq (\varkappa, (\xi)_{\xi < \varkappa})\}$).

Take $\varphi_o = \bigwedge_{\xi < \eta < \varkappa} \neg\, c_\xi = c_\eta \wedge \forall x \bigvee_{\xi < \varkappa} x = c_\xi$.

(ii) For each $\varphi \in L_{\varkappa^+\omega}$, $\mathrm{Mod}(\varphi)$ is the class of relativized reducts of $\mathfrak{M}$-models of some set of $\leq \varkappa$ many first-order sentences; more precisely, if $\varphi \in L_{\varkappa^+\omega}$ then for some $K' \supset K$, some $\Sigma \subset L^{K'}_{\omega\omega}$, $|\Sigma| \leq \varkappa$ and some unary predicate symbol $V \in K'$,

$$\mathrm{Mod}(\varphi) = \{[V^A]^{\mathfrak{A} \upharpoonright K} \mid \mathfrak{A} \models_{\mathfrak{M}} \Sigma\} .$$

Proof. Define $\psi^* \in L_{\omega\omega}$ for $\psi \in L_{\varkappa^+\omega}$ by induction:

$\psi^* = \psi$, if ψ is atomic,

$(\neg\psi)^* = \neg\psi^*$, $(\psi_1 \vee \psi_2)^* = \psi_1{}^* \vee \psi_2{}^*$,

$(\exists x\psi)^* = \exists x(Vx \wedge \psi^*)$.

If $\psi(\overset{n}{v})$ is $\bigvee_{\xi<\varkappa} \psi_\xi$ choose a new $(n+1)$-ary predicate symbol P_ψ. Put

$$\psi^* = \exists x(Ux \wedge P_\psi x v_0 \ldots v_{n-1}),$$

and

$$D_\psi = \{\forall \overset{n}{v}(P_\psi c_\xi v_0 \ldots v_{n-1} \leftrightarrow \psi_\xi^*(\overset{n}{v})) \mid \xi < \varkappa\}.$$

To prove (ii), take for Σ the set

$$\{\varphi^*\} \cup \bigcup\{D_\psi \mid \psi \text{ a subformula of } \varphi,\ \psi \text{ an infinite disjunction}\}.$$

Now, using (i) and (ii), it is easy to show that

$h(L_{\varkappa^+\omega}) = h_\varkappa(L(\mathfrak{M}))$ and $w(L_{\varkappa^+\omega}) = w_\varkappa(L(\mathfrak{M}))$, therefore $h(L_{\varkappa^+\omega}) = \beth^{\varkappa}_{w(L_{\varkappa^+\omega})}$.

Since $w(L_{\varkappa^+\omega}) \geq \varkappa^+$ we have $\varkappa + w(L_{\varkappa^+\omega}) = w(L_{\varkappa^+\omega})$ and hence,

$\beth^{\varkappa}_{w(L_{\varkappa^+\omega})} = \beth_{w(L_{\varkappa^+\omega})}$.

6.11 Theorem $h(L_{\varkappa^+\omega}) = \beth_{w(L_{\varkappa^+\omega})}$. In particular, $h(L_{\omega_1\omega}) = \beth_{\omega_1}$.

B) $L(Q_\varkappa)$.

Put $K_0 = \{<\}$ and $\mathfrak{M} = \{\mathfrak{A} \mid \mathfrak{A} \in \mathrm{Str}(K_0), \mathfrak{A}$ a $\varkappa$-like ordering$\}$. Then

(i) There is $\varphi_o \in L^{K_o}(Q_\varkappa)$ such that $Mod(\varphi_o) = \mathfrak{M}$.

(ii) For each $\varphi \in L(Q_\varkappa)$, $Mod(\varphi)$ is the class of relativized reducts of $\mathfrak{M}$-models of a first-order sentence (see [19]).

Thus $h_\lambda(L(Q_\varkappa)) = h_\lambda(L(\mathfrak{M}))$ and $w_\lambda(L(Q_\varkappa)) = w_\lambda(L(\mathfrak{M}))$.

6.12 <u>Theorem.</u> For $\lambda \leq \varkappa$, $h_\lambda(L(Q_\varkappa)) = \beth^\varkappa_{w_\lambda(L(Q_\varkappa))}$. Compare [34] for the value of $h_\lambda(Q_\varkappa)$ for some λ and $\varkappa$.

C) <u>Weak second-order logic L^{2w}</u>

Put $K_o = \{<\}$ and $\mathfrak{M} = \{(\omega,<)\}$. Prove the statements corresponding to (i) and (ii) in B. Since $w(L^{2w}) = \omega_1^c$, where ω_1^c is the first non-recursive ordinal (cf.[31]), and $\beth^{\aleph_o}_{\omega_1^c} = \beth_{\omega_1^c}$, we obtain

6.13 <u>Theorem.</u> $h(L^{2w}) = \beth_{\omega_1^c}$.

D) <u>Admissible logics.</u>

Let A be an admissible set and assume, <u>for simplicity</u>, that A is countable and $\omega \in A$. For any $\mathfrak{M}$-logic let the A-Hanf number $h_A(L(\mathfrak{M}))$ be the least cardinal ρ such that if an A-recursive set of <u>first-order</u> sentences has a model of power $\geq \rho$, then it has arbitrarily large models. Similarly define the well-ordering number $w_A(L(\mathfrak{M}))$. Then the proof of theorem 6.10 shows that $h_A(L(\mathfrak{M})) = \beth^\varkappa_{w_A(L(\mathfrak{M}))}$, since the set of sentences that ensure that $(\bar{a})$, $(\bar{b})$ and $(\bar{c})$ hold is recursive.

For any transitive $a \in A$, put $K_a = \{c_b \mid b \in a\}$ and let $\mathfrak{M}_a$ be the class containing only the K_a-structure

$$(a, \epsilon \upharpoonright a \times a, (b)_{b \in a}).$$

Since this structure is L_A-characterizable, we have

$$h(L_A) \geq \sup\{h_A(L(\mathfrak{M}_a)) \mid a \in A \;\;, a \text{ transitive}\} \quad \text{and}$$

$$w(L_A) \geq \sup\{w_A(L(\mathfrak{M}_a)) \mid a \in A \;\;, a \text{ transitive}\} .$$

Using a similar reduction as in A) (ii), we obtain
$h(L_A) \leq \sup\{h_A(L(\mathfrak{M}_a)) \mid a \in A, a \text{ transitive}\}$ and
$w(L_A) \leq \sup\{w_A(L(\mathfrak{M}_a)) \mid a \in A, a \text{ transitive}\}$.

Since for countable $a \in A$, $h_A(L(\mathfrak{M}_a)) = \beth_{w_A(L(\mathfrak{M}_a))}^{\aleph_0}$, we get $h(L_A) = \beth_{w(L_A)}^{\aleph_0}$.

Put $\alpha_A = \sup\{\xi \mid \xi \in A\}$. By 4.10, $w(L_A) = \alpha_A$ holds for countable A.

6.14 <u>Theorem</u> [6]. Assume that A is a countable admissible set. Then

$$h(L_A) = \beth_{\alpha_A} .$$

6.15 <u>Exercise</u> [17]. For a logic L let $h^*(L)$ be the least cardinal ρ such that if an L-sentence has for each $\mu < \rho$ a model of power $\geq \mu$, then it has arbitrarily large models. Let $w^*(L)$ be the least ordinal ξ such that if an L-sentence containing a binary symbol $<$ has for each $\eta < \xi$ a well-ordered model of order type $\geq \eta$, then it has a non-well-ordered model. Show:

(i) $h(L) \leq h^*(L) \leq (h(L))^+$ and $w(L) \leq w^*(L) \leq w(L) + 1$.

(ii) $w(L(Q_\varkappa)) = w^*(L(Q_\varkappa))$ for any $\varkappa$.

(iii) If $\mathfrak{M}$ contains only one structure, then $w(L(\mathfrak{M})) = w^*(L(\mathfrak{M}))$.

(iv) If $w(L(\mathfrak{M})) = w^*(L(\mathfrak{M}))$ (and all structures in $\mathfrak{M}$ have the same cardinality), then $h(L(\mathfrak{M})) = h^*(L(\mathfrak{M}))$.

(v) Both Hanf numbers and well-ordering numbers coincide for $L_{\varkappa^+\omega}$, $L((Q_\varkappa)$, L^{2w} and L_A (Hint: use (ii), (iii) and (iv)).

(vi) Give an example of a class $\mathfrak{M}$ of structures (all of cardinality $\varkappa$) such that $w(L(\mathfrak{M})) \neq w^*(L(\mathfrak{M}))$ and $h^*(L(\mathfrak{M})) \neq \beth^{\varkappa}_{w^*(L(\mathfrak{M}))}$ (assume that for $\lambda = \beth^{\varkappa}_{w(L(\mathfrak{M}))}$, $2^\lambda > \lambda^+$).

The result 6.10 that relates the Hanf number and the well-ordering number for $\mathfrak{M}$-logics is fairly general, since, as we have seen, most of the logics for which $h(L) = \beth^{\cdot^{\cdot^{\cdot}}}_{w(L)}$ is known to hold "are" $\mathfrak{M}$-logics. But, if we look on big models, an $\mathfrak{M}$-logic behaves like first-order logic as it is first-order logic outside of a subset of fixed cardinality. What happens to logics that are more close to second-order logic and for which the well-ordering number exists? In particular, if V = L we know that $w(L(Q^2_{\aleph_1})) = \omega$, but is it true that $h(L(Q^2_{\aleph_1})) = \beth^{\aleph_1}_{\omega}$?

References

[1] Baldwin, J.T., Blass, A.R., Glass, A.M., Kueker, D.W.: A note on the elementary theory of finite abelian groups, Algebra universalis 3 (1973), 156-159.

[2] Barwise, K.J.: Axioms for abstract model theory, Annals of math. logic. (1974)

[3] Barwise, K.J.: Back and forth thru infinitary logics, in: MAA-Studies in Mathematics 8 (1973), 5-34.

[4] Barwise, K.J.: Absolute logic and $L_{\infty\omega}$, Annals of Math. Logic 4 (1972), 309-340.

[5] Barwise, K.J.: The Hanf number of second order logic, Jour. Symb. Logic 37 (1972), 588-594.

[6] Barwise, K.J.: Infinitary logic and admissible sets, Doctoral Dissertation, Stanford (1967).

[7] Barwise, K.J. - Kunen, K.: Hanf numbers for fragments of $L_{\kappa^+\omega}$, Israel J. Math. 10 (1971), 306-320.

[8] Chang, C.C.: Some remarks on the model theory of infinitary languages, in: The Syntax and Semantics of Infinitary languages, Berlin (1968), 36-63.

[9] Chang, C.C. - Keisler, H.J.: Model theory Amsterdam (1974).

[10] Craig, W. - Vaught, R.L.: Finite axiomatizability using additional predicats, Jour. Symb. Logic 23 (1958), 289-308.

[11] Deißler, R.: Untersuchungen über Minimalmodelle, Dissertation, Univ. of Freiburg (1974).

[12] Erdös, P.-Rado,R.: A partition calculus in set theory, Bull.Amer.Math. Soc. 62 (1956), 427-489.

[13] Feferman, S.: Infinitary properties, local functors and systems of ordinal functions, in: Conference in Mathematical Logic-London 1970, Berlin (1972),63-97.

[14] Feferman, S.: Two notes on abstract model theory I. Properties invariant on the range of definable relations between structures, Fund.Math. 82 (1974), 153-164.

[15] Feferman, S.: Applications of many-sorted interpolation theorems, to appear in: Proceedings of the Tarski Symposium.

[16] Feferman, S.: Lectures on proof theory, in: Proceedings of the Summer School in Logic, Leeds 1967, Berlin (1968), 1- 107.

[17] Flum, J.: Hanf numbers and well-ordering numbers, Arch.Math.Logic 15 (1972), 164-178.

[18] Fraïssé, R.: Sur quelques classifications des systèmes de relations, Publ. Sci. Univ. Alger Ser. A 1, 35-182.

[19] Fuhrken,G.: Skolem-type normal forms for first-order languages with a generalized quantifier, Fund.Math.54 (1964), 291-302.

[20] Hanf,W.: Incompleteness in languages with infinitely long expressions, Fund. Math. 53 (1964), 303-323.

[21] Hintikka, J.: Distributive normal forms in first-order logic, in: Formal systems and recursive functions, Amsterdam (1965), 48 - 91.

[22] Karp, C.: Finite quantifier equivalence, in: The Theory of Models, Amsterdam (1965), 407-412.

[23] Keisler, H.J.: Model theory for infinitary logic, Amsterdam (1971).

[24] Lindström, P.: On extensions of elementary logic, Theoria 35 (1969), 1-11.

[25] Lindström, P.: On relations between structures, Theoria 32 (1966), 172-185.

[26] Lindström, P.: First order predicate logic with generalized quantifiers, Theoria 32 (1966), 186-195.

[27] López-Escobar, E.G.K.: On definable well-orderings, Fund. Math. 59 (1965), 13-21.

[28] Magidor,M. - Malitz, J.: Compact extensions of L(Q), to appear.

[29] Makkai,M.: Svenonius sentences and Lindström's theory on preservation theorems, Fund. Math. 73 (1972), 219-233.

[30] Makkai,M.: Vaught sentences and Lindström's regular relations, in Cambridge Summer School in Mathematical Logic, Berlin (1973), 622-660.

[31] Morley, M.: The Hanf number for ω-logic, Jour. Symb.Logic 32 (1967),437.

[32] Schönfeld, W.: Ehrenfeucht-Fraïssé Spiel. Disjunktive Normalform und Endlichkeitssatz. Dissertation, Stuttgart (1974).

[33] Thomas, W.: Decision problems for some extensions of successor arithmetic, to appear.

[34] Vaught, R.L.: The Löwenheim Skolem theorem, in: Proc. of 1964 Int'l. Cong.Logic,Meth. and Phil. Sci., Amsterdam (1965),81-89.

[35] Vaught, R.L.: Descriptive set theory in $L_{\omega_1\omega}$, in Cambridge Summer School in Mathematical Logic, Berlin (1973), 574-598.

[36] Wolf, Th.: Diplomarbeit Freiburg (1975).

SET THEORY IN INFINITARY LANGUAGES

K. Gloede
Mathematisches Institut
der Universität Heidelberg

CONTENTS

SUMMARY (and OUTLOOK)

In this paper we introduce infinitary languages $\mathcal{L}_{A\,B}$ which are determined by given sets (or classes) A and B of a suitably chosen metasystem and which can be regarded as a generalization of both the languages $\mathcal{L}_A$ of BARWISE 1969a and the languages $\mathcal{L}_{\kappa\lambda}$ of KARP 1964. The languages $\mathcal{L}_{A\,B}$ will (informally) defined in Chapter I and will be used in Chapter II to set up set theoretical systems axiomatized by means of these languages. The resulting set theoretical systems include and possibly extend the usual finitary theories KP of KRIPKE-PLATEK and ZF of ZERMELO-FRAENKEL. It will be shown that many results carry over from the finitary to the infinitary case. Additional results (e.g. the existence of standard models of ZF) can be obtained in the infinitary case which usually are proven in a second order set theoretical system which is stronger than the infinitary system we are using. Moreover, we can make use of infinitary logical axioms in order to prove set theoretical results which are obtained in the usual first order ZF set theory only upon the assumption of the (set theoretical) axiom of choice.

We believe that one of the advantages of the infinitary set theoretical systems introduced in this paper is the fact that they naturally incorporate a schema of reflection for the language in which they are formalized whereas in the case of a second (or higher) order set theoretical system like QUINE-MORSE set theory the corresponding principles of reflection lead to very strong axioms of infinity which are not provable in the original systems.

This paper is the first part of the author's Habilitationsschrift (GLOEDE 1974). The second part (to be published elsewhere) deals with the following topics:

1) A formal definition of the infinitary language $\mathcal{L}_{A\,B}$ and various syntactical and semantical notions related to this language. In particular it is shown how the notions of relativization and truth for infinitary formulas can be related to each other in a suitably chosen infinitary system. (Such a relationship is well-known to hold in the finitary case, but it is not at all obvious how it extends to the infinitary case, since the relativization of an infinitary formula is again an infinitary formula whereas the notion of truth for infinitary formulas is usually defined by means of a finitary formula.)

2) Various hierarchies of sets which are "constructible" with respect to a hierarchy of infinitary languages (possibly a fixed infinitary language) and which define an inner model of ZF. The main results are the following: Consider the hierarchy of sets $\langle M_\alpha \mid \alpha \in On\rangle$ defined by recursion as follows:

$$M_0 = 0,\quad M_{\alpha+1} = \mathrm{Def}^{C_\alpha}_{\mathcal{L}_\alpha}(M_\alpha)\quad,\quad M_\alpha = \bigcup_{\xi<\alpha} M_\xi \quad \text{if } \alpha \text{ is a limit ordinal,}$$

$$M = \bigcup_{\alpha\in On} M_\alpha\quad,$$

where (intuitively) $\mathrm{Def}^{C}_{\mathcal{L}}(a)$ is the set of subsets $b \subseteq a$ which are definable in $\langle a, \in\rangle$ by means of a formula of $\mathcal{L}$ and using C-finitely many parameters from a.

Then one can prove in ZF (the language $\mathcal{L}^*_{AB}$ will be defined in I.2 as a sublanguage of $\mathcal{L}_{AB}$ which is closed under forming the universal closure of a formula):

1. $M = L$

   if $C_\alpha = HF$ for all α and

   (a) $\mathcal{L}_\alpha = \mathcal{L}_{\omega\omega} = \mathcal{L}_{ZF}$ for all α (GÖDEL) ; or

   (b) $\mathcal{L}_\alpha = \begin{cases} \mathcal{L}^*_{M_\alpha \omega} & \text{if } M_\alpha \text{ is admissible ,} \\ \mathcal{L}_{\omega\omega} & \text{otherwise} \end{cases}$ for all α ; or

   (c) $\mathcal{L}_\alpha = \mathcal{L}^*_{M^+_\alpha \omega}$ for all α .

2. $M = C^\kappa$ (where κ is a regular cardinal $\geq \omega$)

   if $\mathcal{L}_\alpha = \mathcal{L}_{\kappa\kappa}$, $C_\alpha = H(\kappa)$ for all α (CHANG)

   (here we assume the axiom of choice).

   Moreover, the theory ZF can be generalized to a theory $ZF_{\kappa\kappa}$ in the language $\mathcal{L}_{\kappa\kappa}$ (containing a name $\bar{\kappa}$ for κ) such that one can prove in $ZF_{\kappa\kappa}$ (the schema) that $C^{\bar{\kappa}}$ is the least inner model of $ZF_{\kappa\kappa}$.

3. $M = HOD$

   if $C_\alpha = HF$ for all α and

   (a) $\mathcal{L}_\alpha = {*\mathcal{L}}_{\alpha^+\alpha^+}$ for all α ; or

   (b) $\mathcal{L}_\alpha = \mathcal{L}_{M^+_\alpha M^+_\alpha}$ for all α .

   (${*\mathcal{L}}_{\alpha^+\alpha^+}$ is a suitably defined set of $\mathcal{L}_{\alpha^+\alpha^+}$-formulas, and again we assume the axiom of choice.)

4. If $\mathcal{L}_\alpha = \mathcal{L}_{M_\alpha M_\alpha}$, $C_\alpha = HF$ for all α ,

   then M is a transitive model of ZFC (in the sense of relativization), $L \subseteq M \subseteq HOD$, M has a definable well-ordering and M contains (e.g.) every Σ^1_n-subset of ω .

CHAPTER I INFINITARY LANGUAGES FOR SET THEORY

1 *Introduction*

Among various approaches to set theory we shall adopt here the axiomatic one, and among the axiomatic foundations of set theory our starting point will be the theory ZF of ZERMELO-FRAENKEL set theory. This theory can be used both as a formal system for set theory and also as a metatheory for mathematical theories which have been formalized within the first order predicate calculus (like ZF set theory itself) or within higher order or infinitary languages.

If ZF is taken as a formal system for set theory, then there are various questions of intrinsic set theoretical interest which cannot be solved on the basis of its axioms. Some of these open problems have been decided upon the addition of axioms which either restrict the universe of all sets or else claim the existence of new sets (or "large" cardinals). Although most of these axioms can and will be formalized within the language of ZF set theory, they often involve notions which are related to metamathematical concepts of (possibly extended) set theoretical languages. As an example of the former type of axioms we refer to the axiom $V = L$ (GÖDEL 1940) which decides many set theoretical questions but which usually is regarded as being too restrictive as an axiom for set theory. Since L, GÖDEL's class of constructible sets, is the least inner model of ZF (i.e. the least class M (definable by a formula of ZF set theory) such that M is transitive, contains the class On of all the ordinals and such that the relativization to M of every axiom of ZF is provable in ZF), one might wish to extend ZF set theory in such a way that the least inner model of the resulting theory is a "larger" (or "wider") inner model M which still can be characterized by means of a suitably chosen hierarchy of sets like GÖDEL's class L of constructible sets. It is well-known that M is an inner model of ZF iff there is a hierarchy of sets $\langle M_\alpha \mid \alpha \in On \rangle$ such that $M = \bigcup_{\alpha \in On} M_\alpha$ and for each ordinal α:

$$M_\alpha \text{ is transitive,} \quad M_\alpha \subseteq M_\beta \quad \text{if } \alpha \leq \beta,$$

(3) $$M_\alpha = \bigcup_{\xi<\alpha} M_\xi \quad \text{if } \alpha \text{ is a limit number, and}$$

$$Def(M_\alpha) \subseteq M_{\alpha+1} \subseteq \mathfrak{P}(M_\alpha),$$

where (intuitively) $Def(a)$ is the set of $b \subseteq a$ such that b is first order definable in $\langle a, \in \rangle$ using finitely many elements from a as parameters. In fact, if M is an inner model of ZF we may put

$$M_\alpha = V_\alpha^M \quad \text{for each } \alpha,$$

where $\langle V_\alpha \mid \alpha \in On \rangle$ is the von NEUMANN hierarchy, in order to obtain a hierarchy of sets satisfying the above requirements. However, if e.g. $M = HOD$, the class of hereditarily ordinal-definable sets (cp. SCOTT-MYHILL 1971), this hierarchy does not yield any essentially new information about M. Therefore we propose to take the opposite approach and define a hierarchy $< M_\alpha | \alpha \in On >$ by specifying (besides (3))

(1) $M_0 = 0$ (say) and

(2) $M_{\alpha+1} = \mathrm{Def}_{\mathcal{L}_\alpha}(M_\alpha)$

where for each α , $\mathcal{L}_\alpha$ is a suitably chosen infinitary language which may depend on M_α (for examples see the summary), and (intuitively) $\mathrm{Def}_{\mathcal{L}}(a)$ is the set of subsets $b \subseteq a$ which are definable in $<a, \in>$ by means of a formula of $\mathcal{L}$ and constants for elements of a.

The requirement (2) may be thought of as a description of the powerset in M, and it allows to investigate the fine structure of M by means of the hierarchy $< M_\alpha | \alpha \in On >$ (as in the case of L) in terms of the process which leads from M_α to $\mathcal{L}_\alpha$. With respect to this process it seems to be particularly interesting to investigate the case of a language $\mathcal{L}_\alpha$ which is built up from M_α and the sets in M_α in a "constructive" manner (in a sense to be made more precise), since if e.g. $\mathcal{L}_\alpha$ is the language $\mathcal{L}_{\alpha^+ \alpha^+}$ of KARP 1964 and we assume the axiom of choice, then we will simply obtain $M_\alpha = V_\alpha$. Thus we shall primarily consider languages $\mathcal{L}$ which are built up from a given set A and which do not only contain constants for elements of A but which may also contain infinitary conjunctions and disjunctions of any length a and possibly quantifications over sequences of any length a provided a is an element of A. (See §2 for more details.)

Once having introduced infinitary languages for the purpose of defining inner models of set theory it seems to be natural to consider axiomatic set theories formalized in this language. This idea has been put forward by MOSTOWSKI and some of his students in Warsaw (according to a remark in CHANG 1971), but has also been discussed in private conversations between Prof. G.H. Müller and the author in Heidelberg.

As a first step in this direction we will use a finitary set theoretical system (**M**) (like ZF) as a metatheory and any given set (or even class) A of this metatheory to define a (possibly infinitary) language $\mathcal{L}$ which will be used (in Chapter II) in order to set up a formal set theoretical system (**F**) the axioms of which are formalized by means of the language $\mathcal{L}$.

Although ZF set theory has been used frequently as a metatheory for ZF set theory itself and for the formalization of metamathematical concepts related to infinitary and higher order languages, as far as the author knows it has not yet been fully employed in order to deal with infinitary systems of set theory. Of course, one might argue that any result obtainable in an infinitary set theoretical system will finally be interpreted in its metasystem and hence will appear as a result in a suitably chosen finitary system; however, even in this case the detour via infinitary languages can be instructive (as it gives, e.g., a better understanding of CHANG's C^{κ} as well as a new characterization of HOD) and, moreover, we believe that it might be worthwile to consider infinitary statements as they are - and as it is often done even in usual mathematics (though in general not in an explicit manner).

2 *Infinitary set theoretical languages*

The usual first order language with at most countably many non-logical symbols is usually defined within a system of elementary number theory as its metatheory which has to be replaced by some suitable system of set theory when dealing with various semantical notions. Instead we choose to start at the very beginning with a set theoretical system, the KRIPKE-PLATEK theory KP of admissible sets (cp. BARWISE 1969a) which will be taken as our metatheory in order to deal simultaneously with the case of a finitary and an infinitary language resp. If the system KP is used as a metatheory we will refer to it as Meta-KP in order to distinguish it from the formal theory which may again be KP or an extension of it.

Throughout this paper the metatheory will be used mostly informally. In order to make the distinction between a formal theory and its metatheory more apparent we shall use the following notations:

	Metatheory (M)	Formal theory (F)
variables:	$a, b, c, \ldots, m, n, \ldots, x, y, z$ (possibly with indices)	u, v, w, ... (possibly with indices)
equality:	$=$	$\equiv$
membership:	$\in$	ε
logical symbols:	$\neg, \wedge, \vee, \rightarrow, \leftrightarrow, \forall, \exists$ (when formalized)	$\neg, \bigwedge, \forall$ ($\wedge, \vee, \bigvee, \exists, \rightarrow, \leftrightarrow$ as defined symbols)
class terms:	$A, B, C, \ldots, F, \ldots M, \ldots$	A, B, ..., F,.., R,...

2.1 Let A and B be any class terms in a suitably chosen metatheory (possibly denoting the universe of all sets in the metatheory). We use A and B to built up an infinitary language $\mathcal{L}_{A,B}$ as follows: Its <u>symbols</u> are for every $a \in A$, $b \in B$:

variables v_a, individual constants c_a,
b-place relation symbols P^b_a, b-place function symbols F^b_a,
negation $\neg$, (infinitary) conjunction $\bigwedge$ and (infinitary) universal quantification $\forall$.

The class of <u>terms</u> of $\mathcal{L}_{AB}$ is the least class $\mathcal{J}$ such that

(T1) v_a and c_a are in $\mathcal{J}$ (for every $a \in A$).

(T2) If $\mathfrak{f}$ is a function, the domain of $\mathfrak{f}$ is $\mathfrak{b}$ and the range of $\mathfrak{f}$ is included in $\mathcal{T}$, then $F_{\mathfrak{a}}^{\mathfrak{b}}\mathfrak{f}$ is in $\mathcal{T}$ (for every $\mathfrak{a} \in \mathbf{A}$, $\mathfrak{b} \in \mathbf{B}$, $\mathfrak{f} \in \mathbf{A}$).

The class of formulas of $\mathcal{L}_{AB}$ is the least class $\mathcal{F}$ such that

(F1) if $\mathfrak{f}$ is a function, the domain of $\mathfrak{f}$ is $\mathfrak{b}$ and the range of $\mathfrak{f}$ is icluded in $\mathcal{T}$, then $P_{\mathfrak{a}}^{\mathfrak{b}}\mathfrak{f}$ is in $\mathcal{F}$ (for every $\mathfrak{a} \in \mathbf{A}$, $\mathfrak{b} \in \mathbf{B}$, $\mathfrak{f} \in \mathbf{A}$),

(F2) if φ is in $\mathcal{F}$, then $\neg(\varphi)$ is in $\mathcal{F}$,

(F3) if $\{\varphi_x \mid x \in \mathfrak{a}\}$ is a set in $\mathbf{A}$ and for each $x \in \mathfrak{a}$, φ_x is in $\mathcal{F}$, then $\bigwedge\{\varphi_x \mid x \in \mathfrak{a}\}$ is in $\mathcal{F}$ (for every $\mathfrak{a} \in \mathbf{A}$),

(F4) if φ is in $\mathcal{F}$, $0 \neq \mathfrak{b}$, and if $\mathfrak{w}$ is any function on $\mathfrak{b}$ into the class of variables of $\mathcal{L}_{AB}$, then $\forall \mathfrak{w}\, \varphi$ is in $\mathcal{F}$ (for every $\mathfrak{b} \in \mathbf{B}$, $\mathfrak{w} \in \mathbf{A}$).

Additional symbols $\wedge, \vee, \bigvee, \exists, \rightarrow$ and $\leftrightarrow$ are assumed to be defined as usual by means of the basic logical symbols of $\mathcal{L}_{AB}$. Formulas will be denoted by $\varphi, \psi, \ldots$ We also write $\bigwedge_{x \in \mathfrak{a}} \varphi_x$ for $\bigwedge\{\varphi_x \mid x \in \mathfrak{a}\}$, and similarly for $\bigvee$. If $\mathfrak{w}$ is a function the range of which consists of a single variable v, we write $\forall v$ in place of $\forall \mathfrak{w}$, similarly for $\exists$.

2.2 The above definitions of the class of formulas and the class of terms are given only informally; actually the condition

(+) $\{\varphi_x \mid x \in \mathfrak{a}\}$ is a set in $\mathbf{A}$

occurring in (F3) does not make any sense unless the formulas themselves are sets. (A similar remark applies to the condition " $\mathfrak{w}$ is a set in $\mathbf{A}$ " and the condition " $\mathfrak{f} \in \mathbf{A}$ ".) In a subsequent paper (cf. GLOEDE 1974) we shall show how the above concepts can be formalized within a suitably chosen metatheory Meta-KP($\mathbf{A}$, $\mathbf{B}$). In particular it will be shown that the symbols, terms and formulas of $\mathcal{L}_{AB}$ can be identified with suitably chosen sets in the metatheory in such a way that $\mathcal{T}$ and $\mathcal{F}$ will become definable class terms in the metatheory. For this purpose we will always assume that

$$\mathbf{B} \subseteq \mathbf{A}, \ \mathbf{B} \text{ is admissible and } \mathbf{A} \text{ is } \mathbf{B}\text{-admissible}$$

holds in the metatheory. This has several - though mainly technical - advantages, in particular, every formula of $\mathcal{L}_{AB}$ can be denoted by an element of $\mathbf{A}$. (If at the present stage of development the reader does not want to be referred to a formalization of $\mathcal{L}_{AB}$ as indicated, he may think of $\mathbf{A}$ and $\mathbf{B}$ as being restricted to sets of the form $H(\kappa)$ for some regular cardinal κ (cp. 2.3 below) or as being equal to the class of all sets of the metatheory; in these cases the restriction (+) can be replaced by the simple condition $\mathfrak{a} \in \mathbf{A}$.

2.3 The language $\mathcal{L}_{AB}$ can roughly be characterized as follows: it is an extension of the corresponding first order language (since A and B are assumed to be admissible and hence contain the hereditarily finite sets) which in addition allows to take (infinitary) conjunctions over every set (in A) of $\mathcal{L}_{AB}$-formulas and universal quantifications over functions of variables (in A) provided the domain of the function is in B. Thus if B is HF, the class of hereditarily finite sets (and hence the least admissible set), and if A is any admissible set, then $\mathcal{L}_{AB}$ is the infinitary language of BARWISE 1969a (disregarding some minor modifications), and if $A = H(\kappa)$[1], $B = H(\lambda)$ (where κ, λ are regular cardinals, $\omega \leq \lambda \leq \kappa$) then $\mathcal{L}_{AB}$ is (again up to some modifications) the infinitary language $\mathcal{L}_{\kappa\lambda}$ of KARP 1964 (with respect to the obvious similarity type). Therefore we simply write $\mathcal{L}_{\kappa\lambda}$ for $\mathcal{L}_{H(\kappa)\,H(\lambda)}$ and $\mathcal{L}_A$ or $\mathcal{L}_{A\omega}$ for $\mathcal{L}_{A\,HF}$.

2.4 In order to use the language $\mathcal{L}_{AB}$ as a formal language for set theory we do not need all its symbols but only the binary predicate symbols $\equiv$ and ε (denoting equality and membership resp.). However, at several places we shall make use of the availability of additional symbols which will be added as being required by the context. E.g. we will later need a formalization of the syntax and semantics of the infinitary language in a suitably chosen metatheory. This metatheory may again be an infinitary system (F) in the set theoretical language $\mathcal{L}_{AB}$ (which in turn will be formalizable in a finitary metasystem (M)). In this case we will need two additional unary predicate constants $\dot{A}$, $\dot{B}$ (or respective set constants $\bar{A}$, $\bar{B}$) in the language of (F) in order to be able to describe in (F) the language $\mathcal{L}_{AB}$, now taken as the language of an infinitary system (FF) to be formalized within (F). Finally, at several places we will need constants $\bar{a}$ ($a \in A$) as terms of the language $\mathcal{L}_{AB}$ denoting the set a. These constants correspond to the numerals in the finitary case and can be eliminated in a suitably chosen infinitary system (cp. II.4). Thus we will actuallly need the following <u>set theoretical languages</u>:

(a) the "pure" set theoretical languages

$\mathcal{L}_{AB}$ (no individual constants, function symbols and relation symbols except $\equiv$ and ε),

$\mathcal{L}_{\omega\omega}$ (as above) in the finitary case $A = B = HF$;

(b) $\mathcal{L}_{AB}$ with constants from A

(like $\mathcal{L}_{AB}$ in (a) but with additional set constants $\bar{a}$ for each $a \in A$);

---

1) $|a|$ denotes the cardinality of a (in this context we always assume the axiom of choice). $TC(a)$ is the transitive closure of a, $H(\kappa) := \{x \mid |TC(x)| < \kappa\}$, $HF := H(\omega)$.

(c) $\mathcal{L}_{AB}(\dot{A},\dot{B})$ and $\mathcal{L}_{AB}(\bar{A},\bar{B})$

(like the resp. languages of type (a) with additional predicate symbols $\dot{A}$, $\dot{B}$ (or set constants $\bar{A}$, $\bar{B}$, resp.) and set constants $\bar{a}$ for each $a \in A$; we will write $v \varepsilon \dot{A}$, $v \varepsilon \dot{B}$ in place of $\dot{A}v$, $\dot{B}v$, resp.).

Usually we shall not distinguish between $\mathcal{L}_{AB}$ (as in (a)) and $\mathcal{L}_{AB}$ with constants from A (as in (b)). We shall use the letter $\mathcal{L}$ to denote any of the set theoretical languages of type (a) - (c) (suppressing reference to A and B if no confusion is likely to occur); many results, however, remain valid if $\mathcal{L}_{AB}$ is the full language of 2.1 or any of its sublanguages.

2.5 If B is a proper subclass of A, e.g. if $B = HF \in A$, then there are formulas φ of $\mathcal{L}_{AB}$ such that the universal closure of φ is a formula of $\mathcal{L}_{AA}$ but no longer a formula of $\mathcal{L}_{AB}$. Thus it seems to be advisable to introduce a subclass $\mathcal{L}^*$ of formulas of $\mathcal{L}$ which is closed under universal closure of formulas and still "as big as possible". If $\mathcal{L}$ is any of the set theoretical languages of 2.4, then we define $\mathcal{L}^*$ to be the sublanguage of $\mathcal{L}$ which is the least class (with respect to inclusion) $\mathcal{F}^*$ of $\mathcal{L}$-formulas such that

(F1*) every atomic formula φ of $\mathcal{L}$ such that $\mathcal{F}v(\varphi) \in B$, is in $\mathcal{F}^*$,

(F2*) if φ is in $\mathcal{F}^*$, then $\neg(\varphi)$ is in $\mathcal{F}^*$,

(F3*) if $\psi = \bigwedge\{\varphi_x \mid x \in a\}$ is an $\mathcal{L}$-formula, for each $x \in a$ φ_x is in $\mathcal{F}^*$ and $\mathcal{F}v(\varphi_x) \in B$, and if $\mathcal{F}v(\psi) \in B$, then ψ is in $\mathcal{F}^*$,

(F4*) if φ is in $\mathcal{F}^*$, $\mathcal{F}v(\varphi) \in B$ and if $\mathfrak{w}$ is a function in A, the domain of $\mathfrak{w}$ is a set in B and the range of $\mathfrak{w}$ is contained in the class of variables $\{v_x \mid x \in B\}$, then $\forall \mathfrak{w}\, \varphi$ is in $\mathcal{F}^*$.

Here $\mathcal{F}v(\varphi)$ denotes the set of indices x of variables v_x which are free in φ. Since we always assume that B is admissible, we have:

(a) if φ is in $\mathcal{F}^*$ then $\mathcal{F}v(\varphi) \in B$.

(b) If φ is in $\mathcal{F}^*$ and ψ is a subformula of φ, then ψ is in $\mathcal{F}^*$.

(c) If φ is in $\mathcal{F}^*$, then $a \in B$ for every variable v_a occurring in φ.

(d) If φ is in $\mathcal{F}^*$, $b = \mathcal{F}v(\varphi)$, $\mathfrak{w} = \langle v_x \mid x \in b \rangle$, then the universal closure of φ, $\forall \mathfrak{w}\, \varphi$, is in $\mathcal{F}^*$.

(e) If $\mathcal{F}_0$ is any class of $\mathcal{L}$-formulas satisfying (F1*), (a) and (b) (with $\mathcal{F}_0$ in place of $\mathcal{F}^*$), then $\mathcal{F}_0 \subseteq \mathcal{F}^*$.

Thus $\mathcal{F}^*$ can also be characterized as the largest class (w.r.t. inclusion) $\mathcal{F}^*$ such that $\mathcal{F}^*$ contains every atomic formula φ of $\mathcal{L}$ such that $\mathcal{F}v(\varphi) \in B$ and $\mathcal{F}^*$ satisfies (a) and (b). Note that $\mathcal{L}^*_{AA}$ is again $\mathcal{L}_{AA}$;

$\mathcal{L}^*_{A\omega}$ is the class of $\mathcal{L}_{A\omega}$-formulas which contain at most finitely many free variables from $\{v_x \mid x \in HF\}$ (which may be identified with $\{v_n \mid n<\omega\}$), whereas $\mathcal{L}_{A\omega}$-formulas may contain variables from $\{v_x \mid x \in A\}$, possibly infinitely many:

2.6 Example. Let A be the least admissible set such that $\omega \in A$, and let $\langle \varphi_n(\mathfrak{w}_n) \mid n<\omega \rangle$ be a suitable enumeration of the formulas of the language of ZF set theory with free variables as indicated. Then

$$\bigwedge_{n<\omega} \forall \mathfrak{w}_n \varphi_n \quad \text{is a formula of } \mathcal{L}^*_{A\omega}, \text{ but}$$

$$\bigwedge_{n<\omega} \varphi_n(\mathfrak{w}_n)$$

is a formula of $\mathcal{L}_{A\omega}$ which is not a formula of $\mathcal{L}^*_{A\omega}$, since it contains infinitely many free variables.[1)]

3 *Introducing class terms*

Let $\mathcal{L}$ be any of the set theoretical languages of 2.4. Then $\mathcal{L}$ and $\mathcal{L}^*$ can be extended as usual by introducing class terms $\{v \mid \varphi\}$, e.g. by adding the clause

(F5) If φ is in $\mathcal{F}$, then so is $u \varepsilon \{v \mid \varphi\}$.

to the formation rules for formulas (with the requirement that u and v be variables of $\mathcal{L}$ (from $\{v_x \mid x \in B\}$ in case of $\mathcal{L}^*$)) and by adding the (generalized)

CHURCH-Schema: $u \varepsilon \{v \mid \varphi(v, \ldots)\} \leftrightarrow \varphi(u, \ldots)$

(provided u is not bound in $\varphi(v, \ldots)$)

to the logical axioms for $\mathcal{L}$. Hence class terms can be eliminated, and therefore we will in general not distinguish between $\mathcal{L}$, $\mathcal{L}^*$ and their respective class term extensions.

In a language with class terms we further extend the language by defining

$$\mathfrak{a} = \mathfrak{b} \ :\leftrightarrow \ \forall v \, (v \varepsilon \mathfrak{a} \leftrightarrow v \varepsilon \mathfrak{b}),$$

$$\mathfrak{a} \varepsilon \mathfrak{b} \ :\leftrightarrow \ \exists v \, (v = \mathfrak{a} \wedge v \varepsilon \mathfrak{b})$$

if $\mathfrak{a}$, $\mathfrak{b}$ are class terms or sets, but not both are sets. (If both $\mathfrak{a}$ and $\mathfrak{b}$ are sets, then $\mathfrak{a} = \mathfrak{b}$ and $\mathfrak{a} \varepsilon \mathfrak{b}$ are atomic formulas and the above equivalences will later be derivable by means of the axiom of extensionality and suitably chosen logical axioms.)

---

1) In the following we will often use C to denote A in case of the languages $\mathcal{L}_{AB}$ and B in case of the languages $\mathcal{L}^*_{AB}$.

4 *Sequences of variables and relativization*

4.1 As indicated in 2.2, we will later denote terms and formulas of the infinitary language $\mathcal{L}$ by suitably chosen sets in the corresponding metatheory. Though at the present stage of development it is not strictly necessary, we will adopt this formal approach with respect to the functions of variables.

The ordered pair $\langle 0,\alpha\rangle$ will be assigned to the variable v_α, and hence functions of variables will become sets (in the metatheory) which are functions with range included in $\{\langle 0,a\rangle \mid a\in A\}$. We will denote these functions by $\vec{u}, \vec{v}, \vec{w},\ldots$ (possibly with indices). The domain of $\vec{v}$ will be denoted by $d(\vec{v})$, and as usual, $\vec{v}(x)$ denotes the value of $\vec{v}$ at $x\in d(\vec{v})$. (Of course, even if we identify v_x with $\langle 0,x\rangle$, $\vec{v}(x)$ need not be equal to v_x.) Note that the domain of $\vec{v}$ need not be well-ordered (but e.g. in case of KARP's languages $\mathcal{L}_{\kappa\lambda}$ it may be assumed to be well-ordered by some ordinal $<\kappa$). In spite of this fact we prefer to use the term "sequence of variables" instead of the more precise expression "function of variables".

The concatenation $\vec{v} * \vec{w}$ of two sequences of variables $\vec{v}, \vec{w}$ is defined to be the function $\vec{u}$ determined by

$$d(\vec{u}) = (d(\vec{v})\times\{0\}) \cup (d(\vec{w})\times\{1\}),$$

$$\vec{u}(\langle x,0\rangle) = \vec{v}(x) \text{ if } x\in d(\vec{v}),\quad \vec{u}(\langle x,1\rangle) = \vec{w}(x) \text{ if } x\in d(\vec{w}).$$

The one-term sequence $\langle v_a\rangle = \langle v_x \mid x=a\rangle$ will often be identified with v_a, and u, v, w,... will also be used to denote single variables. If we wish to indicate that the variables in the range of $\vec{u}$ occur as free variables in φ, we often write $\varphi(\vec{u})$ or $\varphi(\vec{v},\vec{w})$ or $\varphi(\vec{v},\ldots)$ if $\vec{u} = \vec{v} * \vec{w}$. (If v and $\vec{v}$ both occur within a formula it does not mean that there is any relationship between them; v simply denotes a variable v_a and $\vec{v}$ a sequence of variables among which v_a may occur or not.) We hope that the use of further notations not explicitly defined will be clear from the context.

4.2 If $\vec{v}, \vec{w}$ are sequences of variables, $d(\vec{v}) = d(\vec{w}) = d$, and if t is any term (possibly a class term) then we use the following abbreviations:

$$\vec{v} \equiv \vec{w} :\leftrightarrow \bigwedge_{x\in d} \vec{v}(x) \equiv \vec{w}(x),$$

$$\vec{v} \varepsilon t :\leftrightarrow \bigwedge_{x\in d} \vec{v}(x)\, \varepsilon\, t .$$

With regard to the latter definition, note that $\vec{v}\varepsilon t$ is a formula of some infinitary language whereas e.g. $\vec{v}\in A$ is a metamathematical statement expressing that $\vec{v}$ (as a sequence of sets denoting variables) is an element of A. The formula $\vec{v}\,\varepsilon\, t$ is a generalization of an abbreviation which is commonly used, viz.

$$v_0,\ldots,v_n \,\varepsilon\, t \leftrightarrow v_0 \,\varepsilon\, t \wedge \ldots \wedge v_n \,\varepsilon\, t .$$

Similarly

$$\forall \vec{v} \varepsilon \mathfrak{t}\ \varphi \qquad \text{and} \qquad \exists \vec{v} \varepsilon \mathfrak{t}\ \varphi$$

are used to abbreviate the respective formulas

$$\forall \vec{v}\ (\ \vec{v}\ \varepsilon\ \mathfrak{t} \rightarrow \varphi\ ) \qquad \text{and} \qquad \exists \vec{v}\ (\ \vec{v}\ \varepsilon\ \mathfrak{t} \wedge \varphi\ ),$$

and similarly for a single variable v in place of $\vec{v}$.

4.3 If φ is a formula and $\mathfrak{t}$ is a term (possibly a class term) then the <u>relativization of φ to $\mathfrak{t}$</u>, denoted by $\varphi^{\mathfrak{t}}$ or $\mathrm{Rel}(\mathfrak{t}, \varphi)$, is defined to be the formula obtained from φ by replacing every quantifier $\forall \vec{v}$ occurring in φ by $\forall \vec{v} \varepsilon \mathfrak{t}$. (If v is a variable occuring as a bound variable in φ or in the range of $\vec{v}$ and if v occurs in $\mathfrak{t}$, we assume that is has been suitably renamed in φ before performing the process of relativization.) Here we assume that φ is a formula containing no class terms and no defined symbol; otherwise we define $\varphi^{\mathfrak{t}}$ or $\mathrm{Rel}(\mathfrak{t}, \varphi)$ to be $\mathrm{Rel}(\mathfrak{t}, \psi)$ where ψ is obtained from φ by eliminating the class terms and the defined symbols occurring in φ in a suitable manner. As regards the existential quantifiers this amounts to replacing $\exists \vec{v}$ by $\exists \vec{v} \varepsilon \mathfrak{t}$, as regards class terms, $\{v \mid \psi_0\}$ will be replaced by $\{v \mid \mathrm{Rel}(\mathfrak{t}, \psi_0)\}$ (whereas predicate symbols will not be changed under the process of relativization).

As in the finitary case there is a close connection between $\mathrm{Rel}(\mathfrak{t}, \varphi)$ and (a formalized notion of) truth of φ in the structure $\langle \mathfrak{t}, \varepsilon \rangle$. In fact, defining $\vec{v}\ \varepsilon\ \mathfrak{t}$ and $\forall \vec{v} \varepsilon \mathfrak{t}$ in the way described above, we were lead by the usual interpretation of the quantifier $\forall \vec{v}$ in a structure of the form $\langle \mathfrak{t}, \varepsilon \rangle$.

5 *Hierarchies of formulas*

5.1 The usual classification of finitary set theoretical formulas (LEVY 1965) can easily be extended to the infinitary case. Let $\mathcal{F}$ be a class of formulas of an infinitary language $\mathcal{L}$ where the formulas are now supposed to contain $\wedge, \exists, \vee, \bigvee$ in addition to $\neg, \bigwedge, \forall$ as primitive logical symbols as well as symbols for restricted quantifications (i.e. quantifiers of the form $\forall \vec{v} \varepsilon u$ and $\exists \vec{v} \varepsilon u$).

$\Delta_0(\mathcal{F})$ is the class of formulas φ in $\mathcal{F}$ such that φ does not contain any unrestricted quantifiers (but possibly restricted quantifiers of any kind).

$\Sigma_1(\mathcal{F})$ is the class of formulas φ in $\mathcal{F}$ which are of the form

$$\exists \vec{v}\ \psi \quad \text{for some } \Delta_0(\mathcal{F})\text{-formula } \psi .$$

$\Sigma(\mathcal{F})$ is the class of formulas φ in $\mathcal{F}$ such that the negation symbol occurs in φ only in front of atomic formulas and φ contains no unrestricted universal quantifiers (but possibly restricted universal and|or (restricted or unrestricted) existential quantifiers).

$\Pi_1(\mathcal{F})$ and $\Pi(\mathcal{F})$ are defined similarly, and one may also define classes of formulas $\Sigma_n(\mathcal{F})$, $\Pi_n(\mathcal{F})$ for $n \geq 2$.

In general we shall not distinguish between formulas which are logically equivalent; in particular φ will be called an $\mathcal{F}$-formula whenever it is logically equivalent to a formula in $\mathcal{F}$. If T is a theory (usually identified with its set or class of axioms) then φ is an $\mathcal{F}$-formula in T iff there is a formula ψ in $\mathcal{F}$ such that $T \vdash \varphi \leftrightarrow \psi$. (If equivalence is taken in the sense of formal derivability, these notions depend on the underlying logical system (e.g. with or without infinitary axioms of choice, distributive laws etc.), thus in general the latter notions should be understood in the semantical sense, cp. also II.1.)

Δ_o and Σ are the respective classes of formulas $\Delta_o(\mathcal{F})$, $\Sigma(\mathcal{F})$ where $\mathcal{F}$ is the set of formulas of the finitary set theoretical language, similarly for $\Delta_o(\dot{A})$, $\Delta_o(\dot{A},\dot{B})$ etc., where $\mathcal{F}$ is the set of formulas of the finitary set theoretical language with additional predicate constant $\dot{A}$ and predicate symbols $\dot{A}$, $\dot{B}$ resp. This classification will be used for the language of the metatheory, too, and also if $\dot{A}$ is being replaced by a class term A. In this case e.g. $\Delta_o(A)$ is the set of formulas φ of the usual finitary language of set theory with additional atomic formulas of the form $x \in A$ such that φ does not contain any unrestricted quantifiers.

5.2 The above notions will also be needed relativized to some class A. For our purposes it is most convenient to use the following definitions:

Let $\mathcal{L}_{\omega\omega}(A,B)$ be the (finitary) language of some set theoretical system (M) in which A and B (which will be treated like unary predicate constants) denote two class terms such that $B \subseteq A$ holds in (M).

An $m+1$-place relation $R \subseteq A^{m+1}$ will be called $\Sigma(B)$ over A iff there is a $\Sigma(B)$-formula $\Phi(x_o,\ldots,x_m)$ of $\mathcal{L}_{\omega\omega}(A,B)$ with free variables as indicated such that

$$R = \{\langle x_o,\ldots,x_m\rangle \mid x_o,\ldots,x_m \in A \wedge \mathrm{Rel}(A,\Phi(x_o,\ldots,x_m))\} .$$

Similarly, $R \subseteq A^{m+1}$ will be called $\underset{\sim}{\Sigma}(B)$ over A iff there is some n, a $\Sigma(B)$-formula $\Phi(x_o,\ldots,x_m,x_{m+1},\ldots,x_{m+n})$ with free variables as indicated and n elements $a_1,\ldots,a_n \in A$ such that

$$R = \{\langle x_o,\ldots,x_m\rangle \mid x_o,\ldots,x_m \in A \wedge \mathrm{Rel}(A,\ \Phi(x_o,\ldots,x_m,a_1,\ldots,a_n))\}.$$

(Note that here B is treated as a unary predicate symbol, and hence when forming $\mathrm{Rel}(A,\Phi)$ from Φ, B will not be relativized but will be left unchanged.)

CHAPTER II AXIOMATICS FOR INFINITARY SYSTEMS

1 *Logical axioms and derivations*

Let $\mathcal{L}_{AB}$ be the infinitary language introduced in I.2.1 (containing possibly A-many predicate and function symbols and A-many individual constants, P_o^2 denoting equality). It is well-known that there are complete logical systems for $\mathcal{L}_{AB}$ only for particular sets A and B (cp. 11.3.3 - 11.3.5 of KARP 1964 for the languages $\mathcal{L}_{\kappa\lambda}$ and Theorem 2.7 of BARWISE 1969a for the languages $\mathcal{L}_A = \mathcal{L}_{A\omega}$, where $A \subseteq HC = H(\omega_1)$). For our purposes, viz. derivations from the set theoretical axioms to be described below, we generally need only suitably chosen axioms and rules of inference which can be obtained by strict analogy to the finitary case (e.g. take the axioms and rules of the basic formal system $P_{\alpha\beta}(L)$ of KARP 1964, §11.1, suitably rewritten for $\mathcal{L}_{AB}$). We shall not bother here to write them down explicitly, since in general we shall not describe formal derivations but argue informally. However, we shall try to point out the cases which require to take into account additional axioms and rules which are peculiar to the infinitary case (e.g. distributive laws, axioms and rules of choice, cp. §9 below). Hence whenever we use the symbol $\vdash$ for formal derivability we leave it to the reader to check which particular axioms and rules are needed for a particular formal derivation. Alternatively, one may either assume that the language under consideration is one of the languages $\mathcal{L}_{\kappa\lambda}$ of KARP or $\mathcal{L}_A$ of BARWISE having a complete axiomatization, or one may else replace $\vdash$ by its semantical counterpart.

A theory will often be identified with its set (or class) of non-logical axioms. If (F) is a theory formalized in the language $\mathcal{L}_{AB}$ we shall frequently make use of the following principle which will be referred to as "<u>A-induction for (F)</u>":

If $\langle \varphi_x \mid x \in A \rangle$ is a sequence of formulas of $\mathcal{L}_{AB}$ such that for each $a \in A$ $\bigwedge_{x \in a} \varphi_x$ is a formula of $\mathcal{L}_{AB}$, then:

$$(F) \vdash (\bigwedge_{x \in a} \varphi_x) \rightarrow \varphi_a \text{ for each } a \in A \text{ implies } (F) \vdash \varphi_a \text{ for each } a \in A.$$

In the finitary case, i.e. $A = B = HF$, this principle (which must not be confused with the ω-rule) usually does not receive any attention since it can easily be justified using (informal) induction on the natural numbers. Similarly in the infinitary case: Once we have specified the metatheory (M) for (F) (as will be done in a subsequent paper) one has to check that the notion of derivability of a formula from the axioms in (F) is expressible in (M) by a predicate for which the schema of foundation holds in (M). This can in fact be done for the theories (F) which we will consider in the following and their respective metatheories.

2 *Axioms in infinitary set theoretical languages*

2.1 We start with a list of the well-known axioms of ZF set theory which will be referred to as follows:

Ext: $\forall u\, (u \in v \leftrightarrow u \in w) \rightarrow v \equiv w$ *(extensionality)*

Null: $\exists u\, \forall v\, (\neg\, v \in u)$ *(empty set)*

Pair: $\exists u\, \forall w\, (w \in u \leftrightarrow w \equiv v_0 \vee w \equiv v_1)$ *(axiom of pairing)*

Sum: $\exists u\, \forall w\, (w \in u \leftrightarrow \exists v \in v_0\, (w \in v))$ *(axiom of sum set)*

Pow: $\exists u\, \forall w\, (w \in u \leftrightarrow \forall v \in w\; v \in v_0)$ *(axiom of power set)*

Inf: $\exists u\, [\, \exists w\, (w \in u\,) \wedge \forall w \in u\; \exists v \in u\; \forall v_0 (v_0 \in v \leftrightarrow v_0 \equiv w \vee v_0 \in w\,)]$ *(axiom of infinity)*

ReplS: $\forall u\, \forall v\, \forall w\, (\varphi\, (u,v,\ldots) \wedge \varphi\, (u,w,\ldots) \rightarrow v \equiv w\,) \rightarrow$
$\rightarrow \exists u_0 \forall w\, (w \in u_0 \leftrightarrow \exists v \in v_0\; \varphi\, (v,w,\ldots))$ *(axiomschema of replacement)*

Fund: $\exists u\, (u \in v) \rightarrow \exists u\, (u \in v \wedge \forall w \in u\, (\neg\, w \in v\,))$ *(axiom of foundation (Fundierung, regularity))*

The empty set will be denoted by 0, the (unordered) pair of u and v by $\{ u,v\}$; $\{u\} := \{u,u\}$; $< u,v> := \{ \{ u\} ,\{ u,v\} \}$ denotes the ordered pair of u and v. $\mathfrak{P}(u)$ denotes the power set of u, $\bigcup u$ the sum of u; $A \subseteq B$, $A \cup B$, $A \cap B$, $A \times B$, $\cup A$, $\cap A$ also have their usual meanings (A, B denoting class terms as in I.3). $V := \{ u \mid u \equiv u\}$ is the universe of all sets.

$Rel(A) :\leftrightarrow \forall u \in A\, \exists v\, \exists w\, (u \equiv\, < v,w >)$ (relation),

$dom(R) :\equiv \{ v \mid \exists u < v,u > \in R \}$ (domain), $rng(R) :\equiv \{ u \mid \exists v < v,u > \in R \}$ (range),

$Ft(F) :\leftrightarrow Rel(F) \wedge \forall u\, \forall v\, \forall w\, (< u,v> \in F \wedge < u,w> \in F \rightarrow v \equiv w\,)$ (function),

$F \restriction A :\equiv F \cap (A \times V)$ (restriction).

If F is a function, $A \subseteq dom(F)$, then $F \restriction A$ will also be denoted by $< F(u) \mid u \in A >$, and $\bigcup_{u \in A} F(u) := \bigcup \{ F(u) \mid u \in A \}$.

${}^u A :\equiv \{ v \mid Ft(v) \wedge dom(v) \equiv u \wedge rng(v) \subseteq A \}$,

$trans(A) :\leftrightarrow \forall u \in A\, \forall v \in u\; v \in A$ "A is transitive".

The von NEUMANN <u>ordinals</u> are defined by

$Ord(u) :\leftrightarrow trans(u) \wedge \forall v\, \forall w\, (v,w \in u \rightarrow v \in w \vee v \equiv w \vee w \in v\,)$,

$On :\equiv \{u \mid Ord(u)\}$, $u+1 :\equiv u \cup \{u\}$,

$Lim(u) :\leftrightarrow Ord(u) \wedge u \not\equiv 0 \wedge \forall v \varepsilon u\ (v+1 \varepsilon u)$ (limit ordinal).

Ordinals will be denoted by lower case greek letters $\alpha, \beta, \gamma, \ldots, \kappa, \lambda, \ldots, \xi, \eta, \ldots$; ω is the least infinite ordinal (and the set of natural numbers). *Cardinals* are identified with the initial ordinals.

The theory obtained from ZF by adding the *axiom of choice*

AC: $\forall u \varepsilon v\ \exists w\ (w \varepsilon u) \rightarrow \exists f\ (Ft(f) \wedge dom(f) \equiv v \wedge \forall w \varepsilon v\ f(w) \varepsilon w)$,

will be denoted by ZFC.

The set theoretical definitions introduced above will also be used in the metatheory (except with $=$, $\in$ in place of $\equiv$, ε and other notations for variables).

2.2 When generalizing the axioms of ZF by allowing infinitary formulas to occur in the respective schemata we restrict ourselves to the set theoretical languages $\mathcal{L}$ of I.2.4 and I.2.5. More generally, let $\mathcal{F}$ be any class of $\mathcal{L}$-formulas. Then we introduce the following axiomschemata in which φ is supposed to be an arbitrary formula of $\mathcal{F}$ with free variables as indicated and such that u does not occur in φ [1], except in the $\mathcal{F}$-ReplS we require that in addition $\forall \vec{v}\ \varphi(\vec{v},w,\vec{w})$ or $\exists \vec{v}\ \varphi(\vec{v},w,\vec{w})$ be a formula of $\mathcal{F}$ (the reason for this restriction will become apparent in §6); moreover the variables and sequences of variables are assumed to be elements of A [1].

$\mathcal{F}$- *AusS:* $\exists u\ \forall w\ (w \varepsilon u \leftrightarrow w \varepsilon v \wedge \varphi(w,\vec{v}))$ *($\mathcal{F}$- Aussonderungsschema)*

$\mathcal{F}$- *ReplS:* $\forall \vec{v}\ \forall v\ \forall w\ (\varphi(\vec{v},v,\vec{w}) \wedge \varphi(\vec{v},w,\vec{w}) \rightarrow v \equiv w) \rightarrow$

$\rightarrow \exists u\ \forall w\ (w \varepsilon u \leftrightarrow \exists \vec{v} \varepsilon v_0\ \varphi(\vec{v},w,\vec{w}))$

(Schema of replacement for formulas in $\mathcal{F}$)

$\mathcal{F}$- *FundS:* $\exists w\ \varphi(w,\vec{w}) \rightarrow \exists w\ [\varphi(w,\vec{w}) \wedge \forall v \varepsilon w\ \neg \varphi(v,\vec{w})]$

(Schema of foundation for formulas in $\mathcal{F}$)

$\mathcal{F}$- *ReflS:* $\exists u[trans(u) \wedge v \varepsilon u \wedge \forall \vec{v} \varepsilon u\ (\varphi(\vec{v}) \leftrightarrow \varphi^u(\vec{v}))]$

(Schema of complete reflection for $\mathcal{F}$)

$\mathcal{F}$- *PReflS:* $\varphi(\vec{v}) \rightarrow \exists u\ [trans(u) \wedge \vec{v} \varepsilon u \wedge \varphi^u(\vec{v})]$

(Schema of partial reflection for $\mathcal{F}$)

---

[1] Restrictions of this kind will not always be mentioned explicitly in the sequel.

Note that if $\mathcal{F}$ is either $\mathcal{L}_{AB}$ or $\mathcal{L}_{AB}$, then $\mathcal{F}$ -ReplS is a schema of formulas in $\mathcal{F}$, but $\mathcal{F}$-ReflS and $\mathcal{F}$- PReflS may consist of formulas which are not necessarily in $\mathcal{F}$.

Finally we introduce the following generalization of the axiom of pairing where C is either A or B (the case $\mathsf{C} = \mathsf{H}(\kappa)$ has been considered in KEISLER-SILVER 1971, p. 181):

$$\text{Pair}^{\mathsf{C}} : \quad \exists u \, \forall w \, (w \varepsilon u \leftrightarrow \bigvee_{x \in c} w \equiv v_x)$$

$$\text{b-Pair}^{\mathsf{C}} : \bigwedge_{x \in c} v_x \varepsilon v \rightarrow \exists u \, \forall w \, (w \varepsilon u \leftrightarrow \bigvee_{x \in c} w \equiv v_x)$$

where in both axiom(-schemata) c ranges over the non-empty elements of C (and for each c , u , w, v are suitably chosen variables). Thus Pair^{C} and b-Pair^{C} are schemata (even in the case $\mathsf{C} = \mathsf{HF}$ in which case, however, Pair^{C} can be reduced to the single axiom of pairing, Pair). (Further remarks on these axiomschemata will be given in §5 below.)

For each of the languages $\mathcal{L}$ of I.2.4 (referring to fixed classes A and B) we define

$\mathcal{L}$ -Pair to be Pair^{A},

$\mathcal{L}^*$ -Pair to be Pair^{B},

and similarly for b-Pair. (Cp. footnote 1) on p. 1o.)

2.3 <u>Remark</u>

In 6.9 we shall show that several of the above axiomschemata can be strengthened by allowing sequences of variables in place of single variables at appropriate places (using a suitable infinitary axiom of pairing). However, it should be noted that this kind of generalization is not applicable in each case, e.g. the Aussonderungsschema cannot be strengthened to yield the schema

$$\exists u \, \forall \vec{w} \, (\vec{w} \varepsilon u \leftrightarrow \vec{w} \varepsilon v \wedge \varphi(\vec{w},\vec{v}))$$

since this schema is inconsistent even in the case $\mathbf{d}(\vec{w}) = \mathbf{2}$:

$$\exists u \, \forall \vec{w} \, (\vec{w} \varepsilon u \leftrightarrow \vec{w} \varepsilon \{0,1\} \wedge \vec{w}(\mathbf{0}) \, \varepsilon \{0\} \wedge \vec{w}(\mathbf{1}) \, \varepsilon \, \{1\})$$

is equivalent to the formula

$$\exists u \, \forall w_0 \, \forall w_1 \, (w_0, w_1 \, \varepsilon \, u \leftrightarrow w_0 \equiv 0 \wedge w_1 \equiv 1)$$

which clearly cannot hold. Thus the axioms of finitary set theory cannot be generalized to the infinitary case in a too simple-minded way (cp. however Theorem 6.12).

3 *Infinitary set theoretical systems*

3.1 Let $\mathcal{L}'$ be any of the languages $\mathcal{L}$ of I.2.4 or its restriction $\mathcal{L}^*$ of I.2.5. We denote by $KP_{\mathcal{L}'}$ the theory consisting of the following axioms:

$$\text{Ext} + \Delta_0(\mathcal{L}')\text{-AusS} + \Sigma(\mathcal{L}')\text{-PReflS} + \mathcal{L}'\text{-FundS},$$

and by $ZF_{\mathcal{L}'}$ the theory

$$\text{Ext} + \Delta_0(\mathcal{L}')\text{-AusS} + \mathcal{L}'\text{-ReflS} + \mathcal{L}'\text{-Pair} + \text{Pow} + \text{Fund}.$$

In both cases, if $\mathcal{L}'$ contains the special predicate symbols $\dot{A}$, $\dot{B}$ (or corresponding set constants $\bar{A}, \bar{B}$) the theories $KP_{\mathcal{L}'}$ and $ZF_{\mathcal{L}'}$ are supposed to contain in addition axioms stating that

(1) $\dot{B} \subseteq \dot{A}$, $\dot{B}$ is admissible , $\dot{A}$ is $\dot{B}$ -admissible

and the sentences

(2) $\bar{a} \varepsilon \dot{A}$, $\bar{b} \varepsilon \dot{B}$ for every $a \in A$, $b \in B$.

(Similarly in case of set constants $\bar{A}$, $\bar{B}$ in place of $\dot{A}$, $\dot{B}$)
If $\mathcal{L}$ is the pure set theoretical language $\mathcal{L}_{AB}$ then we simply write

KP_{AB} for $KP_{\mathcal{L}_{AB}}$, and ZF_{AB} for $ZF_{\mathcal{L}_{AB}}$,

and similarly in the other cases, e.g.

$KP^*_{AB}(\dot{A}, \dot{B})$ is $KP_{\mathcal{L}^*_{AB}(\dot{A}, \dot{B})}$.

In particular, if $A = B = HF$, KP_{AB} and ZF_{AB} are (essentially) the respective finitary theories KP and ZF.

3.2 In a subsequent paper we will use formal set theoretical systems like those introduced above as metasystems for a formal system of axioms formalized in the language $\mathcal{L}_{AB}$ whereas throughout this chapter the metatheory will be used mainly in an informal manner. However, since there are some results referring to a particularly chosen metatheory, we will append some notes on the use of the above systems as metasystems.

According to our intention outlined in I.1, A and B are regarded as class terms definable in the language of ZF set theory (now regarded as a metalanguage) or in a suitably extended language, possibly by using parameters, e.g. $A = H(\kappa)$ for some regular cardinal $\kappa > \omega$, $B = HF$. The metatheory is supposed to contain the axioms of Meta-KP (instead of number theory) and axioms expressing (1) (possibly as a schema). Moreover, we shall need in (M) e.g. the following instances of the Aussonderungsschema:

(3) $\exists x\, \forall y\, (y \in x \leftrightarrow y \in a \wedge \varphi(y, a_0, \ldots, a_n))$

where φ is a formula built up from formulas of the form

$x = y, \ x \in y \ , \ x \in A \ , \ x \in B$

by means of the propositional connectives and restricted quantifiers. I.e., we need the $\Delta_0(A, B)$-Aussonderungsschema where A and B are treated like predicate symbols. (For, the formula φ in (3) need no longer be a Δ_0-formula if we eliminate all subformulas of the form $x \in A$, $y \in B$ by means of the formulas defining A and B resp.) Thus our metatheory will be Meta-KP(A,B) (or an extension of it) where A and B are regarded as predicate symbols, and this is one reason for introducing $KP_{AB}(\dot{A}, \dot{B})$ and $ZF_{AB}(\dot{A}, \dot{B})$ with the additional predicate symbols $\dot{A}$ and $\dot{B}$. For particular applications, i.e. if A and B are specified to be particular sets or classes, Meta-KP(A,B) can be further specified, e.g. to be KP , ZF , ZFC or ZFC + In , where In is the axiom stating the existence of a strongly inaccessible cardinal $> \omega$.

4 *Representation of A-finite sets*

4.1 The way the language $\mathcal{L}$ has been introduced as a generalization of the usual finitary language suggests to use the notion " A-finite set" just for the elements of A (although we do not wish at this place to enter into a discussion whether this is the proper generalization of the notion of finiteness).

It is well-known that the possibility of defining numerals (and possibly representing recursive functions) in a formal system has far reaching consequences for the metamathematical properties of these systems but also for working within these systems. In the infinitary case the (metamathematical) definition of numerals $0^{(n)}$ can easily be extended to a definition of $0^{(\alpha)}$ for each ordinal $\alpha \in A$. In fact, a more general method is known to "describe" every set $c \in C$ by a formula of $\mathcal{L}'$ where $C = A$ if $\mathcal{L}' = \mathcal{L}_{AB}$, $C = B$ if $\mathcal{L}' = \mathcal{L}^*_{AB}$:

We define a formula ψ_c for each set c by $\in$-recursion (in Meta-KP) as follows:

$$\psi_c(v_0) \ = \ \forall v_1 \ (\ \neg \ v_1 \varepsilon \ v_0 \ ) \qquad \text{if} \quad c = 0,$$

$$\psi_c(v_0) \ = \ \forall u \ (\ u \varepsilon \ v_0 \leftrightarrow \bigvee_{x \in c} \psi_x(u)) \qquad \text{if} \quad c \neq 0$$

(where u is a suitably chosen variable, $v_{\{c\}}$ say). Using a suitable coding of infinitary formulas, the function assigning to each set c the formula ψ_c can be regarded as a primitive recursive set function (in the sense of JENSEN-KARP 1971), in particular for every admissible C :

$$\langle \psi_x \mid x \in c \rangle \in C \qquad \text{for every} \quad c \in C.$$

Hence for each $c \in C$, ψ_c is an $\mathcal{L}'$ -formula. Moreover, under the usual interpretation of infinitary formulas, for every transitive set S such that $A \subseteq S$ and for every set $c \in S$:

$\langle S, \in \rangle \models \psi_a[c]$ iff $c = a$.

4.2 Lemma. For every $a \in A$:

(1) $\mathrm{Ext} \vdash \psi_a(u) \wedge \psi_a(v) \leftrightarrow u \equiv v$,

(2) $\mathrm{Ext} \vdash \mathrm{trans}(u) \rightarrow \forall v \varepsilon u\, (\psi_a(v) \leftrightarrow \psi_a^{\,u}(v))$.

Proof: (1) is obvious. (2) is proved by A -induction for the system Ext (considered as a theory in the language $\mathcal{L}_{A\omega}$) in Meta-KP(A) = Meta-KP(A, A) as its metatheory (cp. §1). □

4.3 Lemma.

$KP_{A\omega} \vdash \exists v\ \psi_a(v)$ for every $a \in A$, $KP^*_{AB} \vdash \exists v\ \psi_b(v)$ for every $b \in B$.

Proof (again by A -induction for $KP_{A\omega}$ in the metatheory Meta-KP(A)): Suppose $a \in A$ and

(i) $\bigwedge_{x \in a} \exists v\ \psi_x(v)$

is provable in $KP_{A\omega}$. Since (i) is a $\Sigma(\mathcal{L}_{A\omega})$-formula, by the partial reflection principle of $KP_{A\omega}$ there is a transitive set u such that

$\bigwedge_{x \in a} \exists v \varepsilon u\ \psi_x^{\,u}(v)$, and hence by Lemma 4.2 (2):

(ii) $\bigwedge_{x \in a} \exists v \varepsilon u\ \psi_x(v)$.

By the $\Delta_0(\mathcal{L}_{A\omega})$-AusS there is a set w such that

(iii) $\forall v\, (v \varepsilon w \leftrightarrow v \varepsilon u \wedge \bigvee_{x \in a} \psi_x(v))$.

Because of (ii) and (1) of Lemma 4.2, the conjunctive member " $v \varepsilon u$ " can be deleted in (iii). Hence we obtain $\psi_a(w)$. Thus we have proved in $KP_{A\omega}$:

$\bigwedge_{x \in a} \exists v\ \psi_x(v) \rightarrow \exists v\ \psi_a(v)$.

The conclusion now follows using the principle of A -induction. □

By Lemma 4.2 and 4.3 we can prove in $KP_{A\omega}$ (even without using the axiom-schema of foundation) for every $a \in A$ the existence and uniqueness of a set v satisfying $\psi_a(v)$. Therefore we can introduce a set constant $\bar{a}$ for each $a \in A$ denoting the unique set v satisfying $\psi_a(v)$.

Let T_A be the theory in the language $\mathcal{L}^*_{A\omega}$ consisting of the following class of formulas as axioms:

$\forall v\, (v \varepsilon \bar{a} \leftrightarrow \bigvee_{x \in a} v \equiv \bar{x})$ for each $a \in A$.

As we have seen, T_A can be interpreted in $KP_{A\omega}$ and T_B can be interpreted in KP^*_{AB} . Hence we shall henceforth assume that if $\mathcal{L}$ is any of the languages of I.2.4 , then $\mathcal{L}$ contains the constants $\bar{a}$ ($a \in A$) and $T_A \subseteq KP_{\mathcal{L}}$ and $T_A \subseteq ZF_{\mathcal{L}}$, similarly with B in place of A in case of the languages $\mathcal{L}^*$ of I.2.5.

4.4 Lemma. For every $a, b \in A$:

$KP_{A\omega} \vdash \bar{a} \varepsilon \bar{b}$ if $a \in b$,

$KP_{A\omega} \vdash \bar{a} \equiv \bar{b}$ if $a = b$,

$KP_{A\omega} \vdash \bar{a} \not\varepsilon \bar{b}$ if $a \notin b$,

$KP_{A\omega} \vdash \bar{a} \not\equiv \bar{b}$ if $a \neq b$. Similarly for KP^*_{AB} and $a, b \in B$.

Proof: again by A -induction. □

4.5 Lemma. For every $a \in A$, $a \neq 0$, and every $\mathcal{L}_{AB}$ -formula $\varphi(v, \ldots)$:

$$T_A \vdash \forall v \varepsilon \bar{a} \ \varphi(v, \ldots) \leftrightarrow \bigwedge_{x \in a} \varphi(\bar{x}, \ldots) .$$

Proof:
$$\begin{aligned} \forall v \varepsilon \bar{a} \ \varphi(v, \ldots) &\leftrightarrow \forall v \varepsilon \bar{a} \ (\bigvee_{x \in a} v \equiv \bar{x} \rightarrow \varphi(v, \ldots)) \\ &\leftrightarrow \forall v \varepsilon \bar{a} \bigwedge_{x \in a} (v \equiv \bar{x} \rightarrow \varphi(v, \ldots)) \\ &\leftrightarrow \forall v \varepsilon \bar{a} \bigwedge_{x \in a} (v \equiv \bar{x} \rightarrow \varphi(\bar{x}, \ldots)) \\ &\leftrightarrow \bigwedge_{x \in a} \varphi(\bar{x}, \ldots) \quad \text{since } \forall v \varepsilon \bar{a} \bigvee_{x \in a} v \equiv \bar{x} . \square \end{aligned}$$

The corresponding result for sequences of variables already requires a *distributive law:*

$$\textit{Dist}_{AB} : \bigwedge_{x \in a} \bigvee_{y \in b} \varphi_{xy} \longrightarrow \bigvee_{f \in {}^a b} \bigwedge_{x \in a} \varphi_{x\, f(x)}$$

for every sequence $\langle \varphi_{xy} \mid x \in a \wedge y \in b \rangle$ of $\mathcal{L}_{AB}$-formulas such that the above formula is a formula of $\mathcal{L}_{AB}$.

4.6 Lemma. Assuming a suitable logical system for $\mathcal{L}_{AB}$ containing the distributive laws of the schema Dist_{AB} , if $a \in A$, $b \in B$, ${}^b a \in A$ and if φ is a formula of $\mathcal{L}_{AB}$, then:

$$T_A \vdash \forall \vec{v} \varepsilon \bar{a} \ \varphi(\vec{v}, \ldots) \leftrightarrow \bigwedge_{f \in {}^b a} \varphi(\bar{f}, \ldots) ,$$

where $\varphi(\bar{f}, \ldots)$ results from $\varphi(\vec{v}, \ldots)$ by replacing all the free occurrences of $\vec{v}(x)$ by $\bar{f}(x)$ for each $x \in d(\vec{v}) = b$.

Proof: Similar to the proof of Lemma 4.5 except that the following distributive law is required:

$$\bigwedge_{x\in b}\ \bigvee_{y\in a} \vec{v}(x) \equiv \bar{y} \quad \rightarrow \bigvee_{f\in b_a}\ \bigwedge_{x\in a} \vec{v}(x) \equiv \overline{f(x)} \ ,$$

and also a distributive law of the simple form

$$(\bigwedge_{f\in b_a} \varphi_f) \vee \psi \leftrightarrow \bigwedge_{f\in b_a} (\varphi_f \vee \psi) \ . \ \square$$

4.7 Lemma. For every $a, b \in A$ one can prove in KP + T_A :

(1) $\overline{\{a\}} \equiv \{\bar{a}\}$

(2) $\overline{\{a,b\}} \equiv \{\bar{a},\bar{b}\}$

(3) $\overline{\langle a,b\rangle} \equiv \langle \bar{a},\bar{b}\rangle$

(4) $\overline{\bigcup a} \equiv \bigcup \bar{a}$

(5) $\bar{n} \equiv \Delta_n$ for every natural number $n \in \omega \leq A$, where Δ_n denotes the class term of KP defining the nth numeral, i.e. $\Delta_0 = \{u \mid u \notin u\}$, $\Delta_{n+1} = \Delta_n \cup \{\Delta_n\}$,

(6) $\overline{TC(a)} = TC(\bar{a})$.

Proof of (1):

$$v \in \overline{\{a\}} \leftrightarrow \bigvee_{x\in\{a\}} v \equiv \bar{x} \leftrightarrow v \equiv \bar{a} \leftrightarrow v \in \{\bar{a}\} \ .$$

(2) - (5) are proved similarly. (6) is proved by A -induction for KP + T_A : Suppose $a \in A,\ a \neq 0$, and let φ_x denote the formula

$$\overline{TC(x)} \equiv TC(\bar{x}) \ .$$

Assuming $\bigwedge_{x\in a} \varphi_x$ we have to show: φ_a . This can be done as follows:

$$\begin{aligned} v \in \overline{TC(a)} &\leftrightarrow \bigvee_{x\in TC(a)} v \equiv \bar{x} \leftrightarrow \bigvee_{x\in a} v \equiv \bar{x} \ \vee \bigvee_{y\in \bigcup\{TC(x) \mid x\in a\}} v \equiv \bar{y} \\ &\leftrightarrow v \in \bar{a} \ \vee \bigvee_{x\in a}\ \bigvee_{y\in TC(x)} v \equiv \bar{y} \\ &\leftrightarrow v \in \bar{a} \ \vee \bigvee_{x\in a} v \in TC(\bar{x}) \qquad \text{(by assumption)} \\ &\leftrightarrow v \in \bar{a} \vee \exists u \in \bar{a}\ v \in TC(u) \qquad \text{by Lemma 4.5} \\ &\leftrightarrow v \in TC(\bar{a}). \quad \square \end{aligned}$$

Similarly one can prove:

4.8 Lemma. Suppose $a \in A$, $f \in A$ and $\mathrm{Ft}(f) \wedge \mathrm{dom}(f) = a$. Then

$$\mathrm{KP} + T_A \vdash \forall u\,(u \varepsilon \bar{f} \leftrightarrow \bigvee_{x \in a} u \equiv \langle \bar{x}, \overline{f(x)} \rangle) \wedge \mathrm{Ft}(\bar{f}) \wedge \mathrm{dom}(\bar{f}) \equiv \bar{a} \wedge$$
$$\wedge \bigwedge_{x \in a} \bar{f}(\bar{x}) = \overline{f(x)} \quad . \quad \square$$

4.9 Corollary. Suppose $a \in A$ and $f = \{\langle x, \langle 0, x \rangle\rangle \mid x \in a\}$. Then:

$$\mathrm{KP} + T_A \vdash \mathrm{Ft}(\bar{f}) \wedge \mathrm{dom}(\bar{f}) \equiv \bar{a} \wedge \bar{f} \equiv \{\langle u, \langle 0, u \rangle\rangle \mid u \varepsilon \bar{a}\} \wedge$$
$$\wedge \mathrm{rng}(f) \equiv \{\langle 0, u \rangle \mid u \varepsilon \bar{a}\} \equiv \{\langle 0, v \rangle \mid \bigvee_{x \in a} v \equiv \bar{x}\} \quad . \square$$

4.1o. Lemmata 4.7 - 4.9 can be viewed as particular cases of a more general theorem on the representation of A -recursive functions in a suitably chosen infinitary theory formalizable in the language $\mathcal{L}_{A\omega}$. In fact, many theorems related to this subject carry over from the finitary to the infinitary case. However, since a detailed presentation of these results is beyond the scope of this paper, we will restrict ourselves to some results which will be needed later (cp. Chapter III of GLOEDE 1974).

First of all, note that the axioms of T_A correspond to the axioms Ω_4 of ROSSER's system $\mathbf{R}$ of number theory (cp. TARSKI-MOSTOWSKI-ROBINSON 1953, p. 53), and hence it is not surprising that one can generalize the theorem on the representability of recursive (numbertheoretic) functions in $\mathbf{R}$.

Let $T_A(\dot{A}, \dot{B})$ denote the theory consisting of the following axioms:

$\forall v\,(v \varepsilon \bar{a} \leftrightarrow \bigvee_{x \in a} v \equiv \bar{x})$ for every $a \in A$ (i.e. the axioms of T_A),

$\bar{a} \not\equiv \bar{b}$ for every $a, b \in A$ such that $a \neq b$,

$\bar{a} \not\varepsilon \bar{b}$ for every $a, b \in A$ such that $a \notin b$,

$\bar{a} \varepsilon \dot{A}$ for every $a \in A$,

$\bar{b} \varepsilon \dot{B}$ for every $b \in B$, and

$\forall u\, \forall v\,(u \varepsilon v \wedge v \varepsilon \dot{A} \rightarrow u \varepsilon \dot{A})$ (i.e. trans($\dot{A}$)).

(Note that $T_A(\dot{A}, \dot{B})$ can be regarded as a subtheory of $\mathrm{KP}_{AB}(\dot{A}, \dot{B})$.)

4.11 THEOREM.

Suppose that $R \subseteq A^n$ is an n-place relation on A which is $\Sigma(B)$ over A with B occurring positively only. Then there is a $\Sigma(\dot{B})$-formula φ with $\dot{B}$ occurring positively only such that for all $a_1, \dots, a_n \in A$: if $\langle a_1, \dots, a_n \rangle \in R$, then $T_A(\dot{A}, \dot{B}) \vdash \varphi^{\dot{A}}(\bar{a}_1, \dots, \bar{a}_n)$.

Proof by induction on the logical complexity of the formula Φ defining R: Since all the other cases are trivial we consider only the case

$$\Phi(a_1,\dots,a_n) = \forall x \in a_i \quad \Psi(x, a_1,\dots,a_n) .$$

Thus suppose $\Phi(a_1,\dots,a_n)$ holds for given sets $a_1,\dots,a_n \in A$. Then by induction hypothesis we have for all $x \in a_i$:

$$T_A(\dot{A}, \dot{B}) \vdash \psi^{\dot{A}}(\bar{x}, \bar{a}_1,\dots,\bar{a}_n)$$

where ψ is a formula claimed to exist for the relation defined by Ψ). Thus

$$T_A(\dot{A}, \dot{B}) \vdash \bigwedge_{x \in a_i} \psi^{\dot{A}}(\bar{x}, \bar{a}_1,\dots,\bar{a}_n) ,$$

hence by Lemma 4.5:

$$T_A(\dot{A}, \dot{B}) \vdash \forall u \in \bar{a}_i \ \psi^{\dot{A}}(u, \bar{a}_1,\dots,\bar{a}_n) .$$

Thus we may take $\varphi = \forall u \in v_i \ \psi(u, v_1,\dots,v_n)$ as the formula claimed to exist for the relation defined by Φ. □

5 *Infinitary pairing axioms*

We recall that by the infinitary axiom of pairing, Pair^{C}, we denote the class of axioms

$$\exists u \forall w \, (w \in u \leftrightarrow \bigvee_{x \in c} w \equiv v_x)$$

for every $c \in C$, $c \neq 0$.

Intuitively, Pair^{C} (b-Pair^{C}) means that for every sequence $\langle v_x | x \in c \rangle$ of length $c \in C$ (in the metamathematical sense) there is a set u such that the elements of u are exactly the elements v_x for $x \in c$ (provided the sets v_x are "bounded" by some set v, i.e. $v_x \in v$ for each $x \in c$). This set u (which is uniquely determined if we assume the axiom of extensionality) will be denoted by $[\vec{v}]$:

$$w \in [\vec{v}] \leftrightarrow \bigvee_{x \in c} w \equiv \vec{v}(x) .$$

Clearly, the set $[\vec{v}]$ must be distinguished from the metamathematical set $\{ \vec{v}(x) | x \in c \} = \mathrm{rng}(\vec{v})$ just as in the finitary case one has to distinguish between the unordered pair $\{ v_0, v_1 \}$ (as in object in a formal set theoretical system) and the set of variables $\{ v_i | i = 0,1 \}$ (in the corresponding metamathematical system). This distinction will also become apparent if we write down the conditions expressing that a structure $\langle S, \in \rangle$ satisfies $\mathrm{Pair}^{H(\kappa)}$:

Under the usual definition of satisfaction for infinitary formulas it means that s is κ-closed, i.e.

$$\forall f\, [\mathrm{Ft}(f) \wedge \mathrm{dom}(f) \in H(\kappa) \wedge \mathrm{rng}(f) \subseteq s \rightarrow \mathrm{rng}(f) \in s\;] ,$$

or equivalently (since in this context we always assume the axiom of choice):

$$\forall x\, [x \subseteq s \wedge |x| < \kappa \rightarrow x \in s\;].$$

Here we used the fact that the <u>interpretation</u> of $[\vec{v}]$ in $\langle s, \in \rangle$ is the range of f if f is assigned to the sequence of variables $\vec{v}$.

5.1 <u>THEOREM</u>

(1) Ext + $\Delta_o(\mathcal{L})$-AusS + $\Delta_o(\mathcal{L})$-PReflS $\vdash$ $\mathcal{L}$-Pair ,

(2) Ext + $\Delta_o(\mathcal{L})$-AusS $\vdash$ $\mathcal{L}$-bPair.

The same holds for $\mathcal{L}^*$ in place of $\mathcal{L}$.

Proof of (1): For every sequence $\vec{v} = \langle v_x \mid x \in c \rangle$ where $0 \neq c \in C$ (where C is A in case of $\mathcal{L}$, B in case of $\mathcal{L}^*$) one can apply the $\Delta_o(\mathcal{L})$-PReflS to the formula

$$\vec{v} \equiv \vec{v} \qquad (\text{i.e. } \bigwedge_{x \in c} v_x \equiv v_x \quad)$$

obtaining the existence of a set v such that $\vec{v} \,\varepsilon\, v$, i.e.

$$\bigwedge_{x \in c} v_x \;\varepsilon\, v \;.$$

By $\Delta_o(\mathcal{L})$-AusS there is a set u such that

$$\forall w\, (\, w \,\varepsilon\, u \longleftrightarrow w \,\varepsilon\, v \wedge \bigvee_{x \in c} w \equiv v_x \;) \qquad , \text{hence}$$

$$\forall w\, (\, w \,\varepsilon\, u \longleftrightarrow \bigvee_{x \in c} w \equiv v_x \;) \;.$$

The last part of this proof also proves (2). □

5.2 <u>Corollary</u>

$KP_{\mathcal{L}} \vdash \mathcal{L}$-Pair and $KP_{\mathcal{L}^*} \vdash \mathcal{L}^*$-Pair . □

Throughout this section $\mathcal{L}$ is either $\mathcal{L}_{AB}$ and $C = A$ or $\mathcal{L}$ is $\mathcal{L}^*_{AB}$ and $C = B$ (hence $\mathcal{L}$-Pair is Pair^C).
Both languages are supposed to contain the set constants $\bar{c}$ ($c \in C$), and hence by results of §4 we may assume $T_C \subseteq KP_{\mathcal{L}}$.

Whereas in the finitary case, more exactly if $d(\vec{v}) = 2$, $[\vec{v}]$ corresponds to the unordered pair of $\vec{v}(0)$ and $\vec{v}(1)$, in order to obtain an analogue of the ordered pair in the infinitary case, we use a method which in the finitary case is employed in order to represent ordered pairs of classes: Using the constants $\bar{c}$ we define the analogue of the ordered pair $\langle \vec{v} \rangle$ as follows:

$$w \varepsilon \langle \vec{v} \rangle \leftrightarrow \bigvee_{x \in c} w \equiv \langle \bar{x}, \vec{v}(x) \rangle \qquad \text{where} \quad c = \mathbf{d}(\vec{v}).$$

Thus we also have:

5.3 <u>Corollary</u>

$$KP_{\mathcal{L}} \vdash \forall \vec{v} \; \exists u \, \forall w \, (w \varepsilon u \leftrightarrow \bigvee_{x \in c} w \equiv \langle \bar{x}, v_x \rangle) \quad \text{, i.e.}$$

$$KP_{\mathcal{L}} \vdash \forall \vec{v} \; \exists u \quad u \equiv \langle \vec{v} \rangle$$

for every sequence of variables $\vec{v} = \langle v_x \mid x \in c \rangle$, where $c \in C$, $c \neq 0$. □

Let us define for $\vec{v}$ a sequence in A, $\mathbf{d}(\vec{v}) = c \in C$, $c \neq 0$:

$Rp_c(u,\vec{v}) :\leftrightarrow u \equiv \langle \vec{v} \rangle$ " u represents the sequence $\vec{v}$" .

(Again one has to distinguish between the sequence of variables $\vec{v}$ as a metamathematical object and the object $\langle \vec{v} \rangle$ which represents $\vec{v}$ in the formal theory in the above sense.) By results of §4 and Cor. 5.3 we have:

5.4 <u>Lemma</u>. If $\vec{v}, \vec{w}$ are sequences in A, $\mathbf{d}(\vec{v}) = \mathbf{d}(\vec{w}) = c \in C$, $c \neq 0$, then the following formulas are provable in $KP_{\mathcal{L}}$:

(1) $\forall \vec{v} \; \exists u \; Rp_c(u,\vec{v})$,

(2) $Rp_c(u,\vec{v}) \rightarrow Ft(u) \wedge dom(u) \equiv \bar{c} \wedge \bigwedge_{x \in c} u(\bar{x}) \equiv \vec{v}(x)$,

(3) $Rp_c(u,\vec{v}) \wedge Rp_c(w,\vec{v}) \rightarrow u \equiv w$,

(4) $Ft(u) \wedge dom(u) = \bar{c} \rightarrow \exists \vec{v} \; Rp_c(u,\vec{v})$,

(5) $Rp_c(u,\vec{v}) \wedge Rp_c(u,\vec{w}) \rightarrow \vec{v} \equiv \vec{w}$. □

Thus for every $c \in C$, $c \neq 0$, there is in $KP_{\mathcal{L}}$ a one-to-one correspondence between sequences of variables $\vec{v}$ such that $\mathbf{d}(\vec{v}) = c \in C$ and functions u such that $dom(u) \equiv \bar{c}$. (Note, however, that this correspondence is not a two-place relation between sequences and sets, but a c-place function $F(\vec{v}) \equiv \langle \vec{v} \rangle$.)

We now further assume that the language $\mathcal{L}$ has been extended so as to contain class terms (cp. I.3). Just as ordered pairs are used in the finitary case in order to deal with relations by reduction to classes we now use the infinitary analogue of the ordered pair in order to define class terms (in a wider sense) representing c-place relations for every $c \in C$:

$$\{\vec{v} \mid \varphi(\vec{v},\ldots)\} := \{ w \mid \exists \vec{v} \, (w \equiv \langle \vec{v} \rangle \wedge \varphi(\vec{v},\ldots))\}$$

which may also be denoted by

$$\{\langle\vec{v}\rangle \mid \varphi(\vec{v},\ldots)\} .$$

Similarly we can define binary relations between sequences by

$$\{\langle\vec{v},\vec{w}\rangle \mid \varphi(\vec{v},\vec{w},\ldots)\} := \{u \mid \exists v\ \exists w\ \exists\vec{v}\ \exists\vec{w}\ (\ v \equiv \langle\vec{v}\rangle \wedge\ w \equiv \langle\vec{w}\rangle \wedge u \equiv \langle v,w\rangle \wedge \varphi(\vec{v},\vec{w},\ldots))\}$$

$$= \{\langle\langle\vec{v}\rangle,\langle\vec{w}\rangle\rangle \mid \varphi(\vec{v},\vec{w},\ldots)\} .$$

If $\mathfrak{t}$ is either a class term or a set, $\vec{v}$, $\vec{w}$ are sequences of variables, $d(\vec{v}) = c$, $d(\vec{w}) = d$, then we define

$$\vec{v}\ \varepsilon_c \mathfrak{t} :\leftrightarrow \langle\vec{v}\rangle \,\varepsilon\, \mathfrak{t} , \qquad \langle\vec{v},\vec{w}\rangle\ \varepsilon_{c,d}\mathfrak{t} \leftrightarrow \langle\langle\vec{v}\rangle, \langle\vec{w}\rangle\rangle \,\varepsilon\, \mathfrak{t} .$$

From these definitions we immediately obtain:

5.5 <u>Lemma</u>. If $\vec{u}$, $\vec{v}$, $\vec{w}$, $\vec{w}_0$ are sequences of variables, $d(\vec{u}) = d(\vec{v}) = c$, $d(\vec{w}) = d(\vec{w}_0) = d$, $c, d \in C$, $c, d \neq 0$, then in $KP_{\mathcal{L}}$:

$$\vec{u}\,\varepsilon_c\{\vec{v} \mid \varphi(\vec{v},\ldots)\} \leftrightarrow \varphi(\vec{u},\ldots) ,$$

$$\langle\vec{v},\vec{w}\rangle\ \ \varepsilon_{c,d}\{\langle\vec{u},\vec{w}_0\rangle \mid \varphi(\vec{u},\vec{w}_0,\ldots)\} \leftrightarrow \varphi(\vec{v},\vec{w},\ldots) \quad . \ \Box$$

5.6 <u>Lemma</u>. If $\vec{v}$ is a sequence of variables, $d(\vec{v}) = c \neq 0$, $c \in C$, then

$$KP_{\mathcal{L}} \vdash \vec{v}\ \varepsilon\ [\vec{v}] \wedge \vec{v}\ \varepsilon \bigcup\bigcup \langle\vec{v}\rangle . \quad \Box$$

Note, however, that $\vec{v}\ \ \varepsilon_c[\vec{v}]$ need not hold, and hence ε and ε_c need not coincide. -

The axiom of pairing, Pair^{C} , can also be used in order to deal with the composition of C -finitely many functions (cp. Chapter IV of GLOEDE 1974). We also have an analogue of the cartesian product:

5.7 <u>THEOREM</u>. For every $b \in B$, $b \neq 0$, and every sequence $\vec{v}$ such that $d(\vec{v}) = b$:

(1) $KP^{*}_{AB} \vdash \exists u\ \forall w\ (\ w \varepsilon u \leftrightarrow \exists\vec{v} \varepsilon v \quad w \equiv \langle\vec{v}\rangle)$,

(2) $KP^{*}_{AB} \vdash \exists u \forall w (\ w \varepsilon u \leftrightarrow \exists v_0 (Ft(v_0) \wedge dom(v_0) \equiv \bar{c} \wedge rng(v_0) \subseteq v\ ))$.

Proof: Apply the $\Sigma(\mathcal{L}^{*}_{AB})$-PReflS to the formula

$$\forall\vec{v}\,\varepsilon\, v \quad \exists w (\ w \equiv \langle\vec{v}\rangle\ )$$

which holds in KP^{*}_{AB} by Cor. 5.3. Thus there is a set w_0 such that

$$v\,\varepsilon\, w_0 \wedge trans(w_0) \wedge \forall\vec{v}\,\varepsilon\, v\, \exists w \varepsilon\, w_0 (\ w \equiv \langle\vec{v}\rangle\ ) .$$

(1) then follows from the $\Delta_o(\mathcal{L}^*_{AB})$-AusS. (2) is a consequence of (1) and Lemma 5.4. □

5.8 Remark

1. Lemma 5.6 shows that in $KP_{\mathcal{L}}$ the quantifiers of the form $\forall \vec{v} \varepsilon_c w$ and $\exists \vec{v} \varepsilon_c w$ can be regarded as restricted quantifiers.

2. The definitions and results described above show that the sequences of variables $\vec{v}$ such that $d(\vec{v}) = c \in C$ can be treated in $KP_{\mathcal{L}}$ like objects (a fact which is mainly due to the infinitary axiom of pairing), provided one maintains the necessary distinction between objects in a formal system and the related metamathematical notions. (Examples will be provided in the following section, cp. 6.9.)

6 *Schemata of reflection and replacement*

The theory ZF can be axiomatized either by means of a schema of complete reflection or else by the schema of replacement and the axiom of infinity. There is a similar result for the infinitary case (at least for ZF_{AA}); however, since the proper generalization of the axiom of infinity for infinitary set theoretical systems seems to be difficult to guess at, it seems to be preferable to start with an axiomatization based on a schema of reflection, which has a straightforward generalization to infinitary languages.

First we shall show how to derive various set theoretical axioms from suitably chosen reflection principles. - The following result is well-known (and easily provable):

6.1 Lemma

(1) Ext + Δ_o-AusS + Δ_o-PReflS $\vdash$ Pair + Sum ,

(2) Ext + Δ_o-AusS + Π_2-PReflS $\vdash$ Inf ,

(3) Ext + Δ_o-AusS + Δ_o-ReflS $\vdash$ Null + $\forall v \exists w (w \equiv \{v\})$ + Sum .

(Note that Theorem 5.1 (1) is the infinitary analogue of the first part of Lemma 6.1 (1).) □

In order to prove the partial reflection principle from the corresponding principle of complete reflection one seemingly needs an axiom of pairing (cp. however Cor. 6.7 (1) and Remark 6.8 below). - The following results can be proved just as in the finitary case:

6.2 THEOREM

Ext + $\Delta_o(\mathcal{L})$-AusS + $\mathcal{L}$-ReflS + $\mathcal{L}$-Pair $\vdash$ $\mathcal{L}$-ReplS + $\mathcal{L}$-PReflS + $\mathcal{L}$-AusS;

and the same holds for $\mathcal{L}^*$ in place of $\mathcal{L}$. (We shall see later that $\mathcal{L}$-Pair is redundant if $\mathcal{L} = \mathcal{L}_{AA}$ or $\mathcal{L} = \mathcal{L}^*_{AB}$ (Cor. 6.6).) □

6.3 THEOREM

Ext + Null + $\mathcal{L}$-ReplS $\vdash$ $\mathcal{L}$-AusS ;

the same holds for $\mathcal{L}^*$ in place of $\mathcal{L}$. □

6.4 THEOREM

Ext + $\Delta_0(\mathcal{L})$-AusS + $\mathcal{L}$-ReflS + Pair $\vdash$

$$\exists u [\ \mathrm{trans}(u) \wedge v \varepsilon u \wedge \bigwedge_{x \in a} \forall \vec{v} \varepsilon u\ (\ \varphi_x(\vec{v}) \longleftrightarrow \varphi_x^u(\vec{v}))\ ]$$

for every $a \in A$, $a \neq 0$, and every sequence $\langle \varphi_x(\vec{v}) \mid x \in a \rangle$ of $\mathcal{L}$-formulas with free variables as indicated such that $\bigwedge_{x \in a} \varphi_x(\vec{v})$ is a formula of $\mathcal{L}$. The same holds for $\mathcal{L}^*$ in place of $\mathcal{L}$ if $a \in B$.

Proof: Suppose w is a variable not occurring in the range of $\vec{v}$ and consider the formula

$$\varphi(w,\vec{v}) := \bigvee_{x \in a} (\ w \equiv \overline{x} \ \wedge \ \varphi_x(\vec{v}))\ .$$

Using Pair and applying the $\mathcal{L}$-ReflS to the formula φ , we obtain for every set v the existence of a transitive set u such that

$$\overline{a},\ v \varepsilon u \wedge \mathrm{trans}(u) \wedge \forall w \varepsilon u\ \forall \vec{v} \varepsilon u\ [\quad \varphi(w,\vec{v}) \longleftrightarrow \varphi^u(w,\vec{v})\ ]\ .$$

The desired result is now immediate from Lemma 4.2 (2). □

6.5 THEOREM

Ext + $\Delta_0(\mathcal{L})$-AusS + $\mathcal{L}$-ReflS $\vdash$ $\mathcal{L}$-ReplS ;

the same holds for $\mathcal{L}^*$ in place of $\mathcal{L}$.

Proof: First we modify the proof of Theorem 6.4 in order to dispense with the axiom of pairing; e.g. if we define the numerals by

$$0^{(0)} \equiv 0\ , \qquad 0^{(k+1)} \equiv \{ 0^{(k)} \} \qquad \text{(using Lemma 6.1 (3))}$$

we obtain the conclusion of Theorem 6.4 for finite sets a without the conjunct "$v \varepsilon u$". This weaker form of 6.4 can be applied to the formulas

$\varphi_1 := \ \ \varphi(\vec{v},w,\vec{w})$,

$\varphi_2 := \ \ \exists w \ \ \varphi(\vec{v},w,\vec{w})$,

$\varphi_3 := \ \ \psi(v_0,\vec{w})$,

$\varphi_4 := \ \ \forall \vec{w} \forall v_0 \ \ \psi(v_0,\vec{w})$,

where $\psi(v_0,\vec{w})$ is the instance of the $\mathcal{L}$-ReplS corresponding to the $\mathcal{L}$-formula φ . φ_4 can now be proved just as in the finitary case. □

6.6 Corollary. $\mathrm{Ext} + \Delta_o(\mathcal{L}^*_{AB})\text{-AusS} + \mathcal{L}^{\frown}_{AB}\text{-ReflS} \vdash \mathrm{Pair}^B$.

Proof: By Theorem 6.5 we can apply the $\mathcal{L}^*_{AB}$-ReplS to the formula

$$\varphi(v,w,\vec{v}) := \bigvee_{x\in b} (v \equiv \bar{x} \wedge w \equiv \vec{v}(x))$$

(where $\vec{v}$ is a suitable sequence of variables such that $0 \neq b = d(\vec{v}) \in B$).
Thus we obtain the existence of a set u such that

$$\forall w \, (w \varepsilon u \leftrightarrow \exists v \varepsilon \bar{b} \;\; \varphi(v,w,\vec{v})),$$

$$\forall w \, (w \varepsilon u \leftrightarrow \bigvee_{x\in b} w \equiv \vec{v}(x)) .$$

(Here we use Lemma 4.3 which is easily seen to hold with the theory under discussion in place of KP^*_{AB}.) □

6.7 Corollary. (1) $\mathrm{Ext} + \Delta_o(\mathcal{L}^*)\text{-AusS} + \mathcal{L}^*\text{-ReflS} \vdash \mathcal{L}^*\text{-PReflS}$,

(2) $\mathrm{Ext} + \Delta_o(\mathcal{L})\text{-AusS} + \mathcal{L}\text{-Pair} + \mathcal{L}\text{-ReflS} \vdash \mathcal{L}\text{-PReflS}$. □

6.8 Remark

Assuming the axioms $\mathrm{Ext} + \Delta_o(\mathcal{L})$-AusS, the axiomschemata $\mathcal{L}$-Pair and $\mathcal{L}$-ReflS can be combined into the following strengthened schema of reflection:

$$\mathcal{L}\text{-ReflS}': \quad \exists u \, [\, \mathrm{trans}(u) \wedge \vec{w} \varepsilon u \wedge \forall \vec{v} \varepsilon u \, (\varphi(\vec{v}) \leftrightarrow \varphi^u(\vec{v})) \,]$$

where $\varphi(\vec{v})$ is an arbitrary formula of $\mathcal{L}$ with free variables as indicated and $\vec{w}$ is a sequence of variables such that $d(\vec{w}) = d(\vec{v})$.

$\mathcal{L}$-ReflS' immediately implies the corresponding schema of partial reflection (in contradistinction to $\mathcal{L}$-ReflS). The same remark also applies to the language $\mathcal{L}^*$ in place of $\mathcal{L}$; however, in this case $\mathcal{L}^*$-ReflS and $\mathcal{L}^*$-ReflS' are equivalent by Cor. 6.6 (just as in the finitary case).

6.9 We now consider the theory $KP_{\mathcal{L}}$. Just as in the case of ZF and its infinitary analogue $ZF_{\mathcal{L}}$, the basic properties provable in KP carry over to $KP_{\mathcal{L}}$, e.g. we have:

6.1o THEOREM

The following schema of replacement holds in $KP_{\mathcal{L}}$:

$$\forall \vec{v} \varepsilon u \, \forall v \, \forall w [\, \varphi(\vec{v},v,\vec{w}) \wedge \varphi(\vec{v},w,\vec{w}) \rightarrow v \equiv w \,] \wedge \forall \vec{v} \varepsilon u \, \exists w \;\; \varphi(\vec{v},w,\vec{w})$$
$$\rightarrow \exists v \forall w \, (w \varepsilon v \leftrightarrow \exists \vec{v} \varepsilon u \;\; \varphi(\vec{v},w,\vec{w}))$$

for every $\Sigma(\mathcal{L})$-formula φ with free variables as indicated such that $\exists \vec{v} \;\; \varphi(\vec{v},w,\vec{w})$ is a formula of $\mathcal{L}$.

The same result holds for $\mathcal{L}^*$ in place of $\mathcal{L}$. □

We shall now present some examples which show how the infinitary axiom of pairing and the results of §5 (cp. Remark 5.8) can be used to extend some of the above theorems stated for functions having sets as arguments and values to the case of many-place functions having many-place sequences as values. (We use sequences $\vec{v}_o$, $\vec{u}_o$ at those places where single variables occured in the previous results.)

6.11 <u>THEOREM</u>

The following schema of replacement holds in $KP_{\mathcal{L}}$:

$$\forall \vec{v} \varepsilon u \; \forall \vec{v}_o \; \forall \vec{u}_o [\; \varphi(\vec{v},\vec{v}_o,\vec{w}) \wedge \varphi(\vec{v},\vec{u}_o,\vec{w}) \rightarrow \vec{v} \equiv \vec{u} \;] \wedge \forall \vec{v} \varepsilon u \; \exists \vec{v}_o \; \varphi(\vec{v},\vec{v}_o,\vec{w})$$

$$\rightarrow \exists v \; \forall \vec{v}_o (\; \vec{v}_o \varepsilon_d v \longleftrightarrow \exists \vec{v} \varepsilon u \quad \varphi(\vec{v},\vec{v}_o,\vec{w}))$$

for every $\Sigma(\mathcal{L})$-formula with free variables as indicated such that $\exists \vec{v} \; \exists \vec{v}_o \quad \varphi(\vec{v},\vec{v}_o,\vec{w})$ is a formula of $\mathcal{L}$ and $d(\vec{v}) = c$, $d(\vec{v}_o) = d(\vec{u}_o) = d$ (in particular $c, d \in \mathcal{B}$). The same holds for $\mathcal{L}^*$ in place of $\mathcal{L}$. □

6.12 <u>THEOREM</u>

The following schema of Aussonderung holds in $KP_{\mathcal{L}}$:

$$\forall \vec{v}_o (\quad \varphi(\vec{v}_o,\vec{w}) \longleftrightarrow \neg \psi(\vec{v}_o,\vec{w})) \rightarrow \exists u \; \forall \vec{v}_o (\; \vec{v}_o \varepsilon_b u \leftrightarrow \vec{v}_o \varepsilon_b v \wedge \varphi(\vec{v}_o,\vec{w}))$$

where $\varphi(\vec{v}_o,\vec{w})$ and $\psi(\vec{v}_o,\vec{w})$ are $\Sigma(\mathcal{L})$-formulas such that $d(\vec{v}_o) = b$, $b \in \mathcal{B}$, $b \neq 0$. The same holds for $\mathcal{L}^*$ in place of $\mathcal{L}$. □

Theorems 6.11 and 6.12 can be proved either directly or by reduction to to the single-valued case as follows: In order to prove e.g. Theorem 6.11 by using Theorem 6.1o one may proceed as follows:

Let φ be a formula as referred to in Theorem 6.11 and assume the hypothesis expressing existence and uniqueness. Then consider the formula

$$\varphi_o(v,v_o,\vec{w}) := \quad \exists \vec{v} \; \exists \vec{v}_o (\; v \equiv \langle \vec{v} \rangle \wedge v_o \equiv \langle \vec{v}_o \rangle \wedge \; \varphi(\vec{v},\vec{v}_o,\vec{w})).$$

Applying the axiomschema of replacement of Theorem 6.1o to φ_o and the class

$$u_1 :\equiv \{ \langle \vec{v} \rangle \mid \vec{v} \varepsilon u \} \qquad \text{(which is a set by Theorem 5.7)}$$

in place of u , we obtain the existence of a set v_1 such that

$$\forall v_o (\; v_o \varepsilon v_1 \longleftrightarrow \exists v \varepsilon u_1 \quad \varphi_o(v,v_o,\vec{w})).$$

Hence:

$$\forall \vec{v}_o (\; \vec{v}_o \varepsilon_d v_1 \longleftrightarrow \exists v_o (\; v_o \equiv \langle \vec{v}_o \rangle \quad \wedge \quad v_o \varepsilon v_1)$$

$$\forall \vec{v}_o (\; \vec{v}_o \varepsilon_d v_1 \longleftrightarrow \exists v_o (\; v_o \equiv \langle \vec{v}_o \rangle \quad \wedge \exists v \varepsilon u_1 \quad \varphi_o(v,v_o,\vec{w}))$$

$$\forall \vec{v}_o\ (\ \vec{v}_o\ \varepsilon_d v_1 \leftrightarrow \exists \vec{v}\,\varepsilon\, u\quad \varphi(\vec{v},\vec{v}_o,\vec{w}))$$, q.e.d.

A further variant of Theorem 6.1o can be obtained by replacing in 6.11 $\vec{v}\,\varepsilon\, u$ by $\vec{v}\,\varepsilon_c u$. This form of 6.11 is proved by the same method except using

$$u_2 :\equiv \{ \langle \vec{v} \rangle \mid \vec{v}\,\varepsilon_c u \}$$

in place of u_1 (u_2 is a set, too, by Theorem 5.7 and Remark 5.8 (1)).

6.13 Applications

Suppose that A is an admissible set such that $\omega \in A$, and let T be either KP or ZF . If $\langle \varphi_n \mid n<\omega \rangle$ is a suitable enumeration of the axioms of T such that $\bigwedge_{n<\omega} \varphi_n$ is a sentence of $\mathcal{L}^*_{A\omega}$, then one can prove:

$$T + \mathcal{L}^*_{A\omega}\text{-PReflS} \vdash \exists u\ [\ \text{trans}(u) \wedge v\,\varepsilon\, u \wedge \bigwedge_{n<\omega} \varphi_n^u\ ]\ ,$$

i.e. one can prove in this theory the existence of "arbitrarily large" standard models of T .

Similarly, if $\langle \psi_n \mid n<\omega \rangle$ is a suitable enumeration of the formulas of the language of ZF set theory, then by Theorem 6.4 one has:

$$ZF_{AA} \vdash \exists u\ [\ \text{trans}(u) \wedge v\,\varepsilon\, u \wedge \bigwedge_{n<\omega} \forall \vec{v}\,\varepsilon\, u(\ \psi_n(\vec{v}) \leftrightarrow \psi_n^u(\vec{v}))\ ]$$

where $\vec{v} = \langle v_n \mid n<\omega \rangle$. Hence if e.g. $A = L_{\omega_1^{CK}} = HF^+$ is the least admissible set A' such that $\omega \in A'$, then one can prove in ZF_{AA} the existence of arbitrarily large transitive sets u such that $\langle u, \varepsilon \rangle$ is an elementary subsystem of $\langle V, \varepsilon \rangle$ (with respect to the finitary formulas of ZF set theory), and even in $ZF^*_{A\omega}$ (or $ZF + KP^*_{A\omega}$) one can prove the existence of a (countable) standard transitive model $\mathfrak{M}$ of ZFC and hence the existence of any model $\mathfrak{N}$ obtained from $\mathfrak{M}$ by the usual methods of COHEN-forcing. Similar results hold for standard models of ZF_{AB} which can be proved to exist in ZF_{A^+B} (or even in $ZF_{AB} + KP_{A^+B}$) where for every admissible set A , A^+ is the least admissible set A' such that $A \in A'$ (cp. BARWISE-GANDY-MOSCHOWAKIS 1971 for the existence and properties of the "next" admissible set). (For the existence of natural models of the form $\langle V_\alpha, \varepsilon \rangle$ cp. Theorem 8.4 below.)

7 *Critical points of normal functions*

In this section we will be concerned with some technical results (which will mainly be used in a subsequent paper).

7.1 Definitions

$OSq(F) :\leftrightarrow Ft(F) \wedge (dom(F) \equiv On \vee dom(F) \varepsilon On) \wedge rng(F) \subseteq On$

"F is a sequence of ordinals"

$ASq(F) :\leftrightarrow OSq(F) \wedge \forall \xi, \eta \ \varepsilon \ dom(F) (\xi < \eta \rightarrow F(\xi) \leq F(\eta))$

"F is an ascending sequence of ordinals"

$\nearrow Sq(F) :\leftrightarrow OSq(F) \wedge \forall \xi, \eta \ \varepsilon \ dom(F) (\xi < \eta \rightarrow F(\xi) < F(\eta))$

"F is a strictly increasing sequence of ordinals"

$conf(\alpha, \beta) :\leftrightarrow \exists u (\nearrow Sq(u) \wedge dom(u) \equiv \beta \wedge \bigcup rng(u) \equiv \alpha)$

" α *is cofinal with* β "

$cf(\alpha) := \mu \beta (conf(\alpha, \beta))$ (where μ denotes the usual minimum operator) is the *cofinality* of α.

7.2 Lemma (KP)

$$Lim(\lambda) \wedge cf(\lambda) > \alpha \leftrightarrow \lambda \not\equiv 0 \wedge \forall u (Ft(u) \wedge dom(u) \equiv \alpha \wedge rng(u) \subseteq \lambda \rightarrow \exists \xi < \lambda \ rng(u) \subseteq \xi). \quad \square$$

We shall later need a generalization of the predicate $cf(\lambda) > \alpha$ to the case where α is not necessarily an ordinal but may be any set:

7.3 Definition

$$gcf(\lambda, v) :\leftrightarrow \lambda \not\equiv 0 \wedge \forall u (Ft(u) \wedge dom(u) \equiv v \wedge rng(u) \subseteq \lambda \rightarrow \exists \xi < \lambda \ rng(u) \subsetneq \xi).$$

Thus $gcf(\lambda, \alpha) \leftrightarrow Lim(\lambda) \wedge cf(\lambda) > \alpha$ by Lemma 7.2.

7.4 Lemma (KP)

$$OSq(f) \wedge ASq(g) \wedge dom(f) \equiv \gamma \wedge \forall \eta < \lambda \ f(\eta) < \alpha \wedge \alpha \equiv \bigcup_{\eta < \gamma} f(\eta) \wedge \alpha \subseteq dom(g)$$

$$\rightarrow \bigcup_{\xi < \alpha} g(\xi) = \bigcup_{\eta < \gamma} g(f(\eta)).$$

Proof: We have to show:

$$\exists \xi < \alpha (\beta < g(\xi)) \leftrightarrow \exists \eta < \gamma (\beta < g(f(\eta))) .$$

The part " $\leftarrow$ " follows from the assumption $\forall \eta < \gamma \ f(\eta) < \alpha$, the converse implication follows from

$$\forall \xi (\xi < \alpha \leftrightarrow \exists \eta < \gamma (\xi < f(\eta)) \quad \text{and} \quad ASq(g). \quad \square$$

7.5 Corollary (KP)

$$\uparrow Sq(f) \wedge ASq(g) \wedge dom(f) \equiv \lambda \wedge Lim(\lambda) \wedge \alpha \equiv \bigcup_{\eta<\lambda} f(\eta) \subseteq dom(g)$$

$$\rightarrow \bigcup_{\xi<\alpha} g(\xi) \equiv \bigcup_{\eta<\lambda} g(f(\eta)) \quad . \quad \square$$

7.6 THEOREM (KP)

$$gcf(\lambda, v) \wedge \uparrow Sq(f) \wedge dom(f) \equiv \lambda \wedge \alpha \equiv \bigcup_{\eta<\lambda} f(\eta) \rightarrow gcf(\alpha, v) \quad .$$

Proof: Suppose $Ft(g) \wedge dom(g) \equiv v \wedge rng(g) \subseteq \alpha$.

We have to show: $\exists \xi < \alpha \;\; rng(g) \subseteq \xi$. Suppose not. Then:

(i) $\forall \xi < \alpha \quad rng(g) \nsubseteq \xi$, i.e.

$$\forall \xi < \alpha \quad \exists \eta \quad (\eta \in rng(g) \wedge \xi \leq \eta < \alpha).$$

Let us define a function h such that $dom(h) \equiv v$ and for each $w \in v$

$$h(w) = \mu \eta \quad (g(w) < f(\eta)) .$$

Then $rng(h) \subseteq \lambda$. We shall obtain a contradiction by proving

(ii) $\delta < \lambda \rightarrow \exists w \in v \quad (\delta < h(w))$.

Thus let $\delta < \lambda$. Then $f(\delta) < \alpha$, hence by (i) there is some $w \in v$ such that

$$f(\delta) \leq g(w) .$$

Put $\delta_0 := h(w)$. Then we have by the definition of h:

$$f(\delta) \leq g(w) < f(\delta_0) \text{ , hence } \delta < \delta_0 \text{ since } \uparrow Sq(f) .$$

Thus we have proved (ii) which contradicts our assumption $gcf(\lambda, v)$. $\square$

7.7 Corollary (KP)

$$\forall u \in v \;\; gcf(\lambda, u) \wedge \lambda < \gamma \rightarrow \exists \xi \;\; (\gamma < \xi \wedge \forall u \in v \;\; gcf(\xi, u)). \quad \square$$

The definition of gcf involves quantification over functions which can be reduced to a restricted quantification over sets when using infinitary languages. More precisely, let us define:

7.8 Definitions

$$gcf_c(\lambda) :\leftrightarrow \lambda \neq 0 \wedge \forall \vec{v} \, (\vec{v} \in \lambda \rightarrow \exists \xi < \lambda \;\; \vec{v} \in \xi)$$

where $\vec{v}$ is a sequence of variables such that $d(\vec{v}) = c$, $c \neq 0$.

$$SNft(F) :\leftrightarrow ASq(F) \wedge dom(F) \equiv On \wedge \forall \xi \, (\xi \leq F(\xi)) \wedge \forall \lambda \, (Lim(\lambda) \rightarrow$$
$$\rightarrow F(\lambda) \equiv \bigcup_{\xi<\lambda} F(\xi)) \quad \textit{"F is a semi-normal function"}$$

$$SNft_c(F) :\leftrightarrow ASq(F) \wedge dom(F) \equiv On \wedge \forall \xi \, (\xi \leq F(\xi)) \wedge \forall \lambda \, (gcf_c(\lambda) \rightarrow$$
$$\rightarrow F(\lambda) \equiv \bigcup_{\xi<\lambda} F(\xi))$$

$$Nft(F) :\leftrightarrow \uparrow Sq(F) \wedge dom(F) \equiv On \wedge \forall \lambda \, (Lim(\lambda) \rightarrow F(\lambda) \equiv \bigcup_{\xi<\lambda} F(\xi))$$

"F is a normal function"

$$\mathrm{Nft}_c(F) :\leftrightarrow \nearrow \mathrm{Sq}(F) \wedge \mathrm{dom}(F) \equiv \mathrm{On} \wedge \forall \lambda \; (\mathrm{gcf}_c(\lambda) \rightarrow F(\lambda) \equiv \bigcup_{\xi<\lambda} F(\xi)).$$

By Lemma 5.4 and 7.2 these notions are related to each other as follows:

7.9 Lemma

The following statements hold in $KP_{\mathcal{L}}$ for every $c, d \in A$, $0 \neq c \subseteq d$, and in $KP_{\mathcal{L}^*}$ for every $c, d \in B$, $0 \neq c \subseteq d$:

(1) $\mathrm{gcf}(\lambda, \bar{c}) \leftrightarrow \mathrm{gcf}_c(\lambda)$,

(2) $\mathrm{gcf}_d(\lambda) \rightarrow \mathrm{Lim}(\lambda) \wedge \mathrm{gcf}_c(\lambda)$,

(3) $\mathrm{Lim}(\lambda) \leftrightarrow \mathrm{gcf}_1(\lambda) \leftrightarrow \mathrm{gcf}_n(\lambda)$ for each $n<\omega$, $n \neq 0$,

(4) $\mathrm{SNft}(F) \leftrightarrow \mathrm{SNft}_1(F)$, $\mathrm{SNft}(F) \rightarrow \mathrm{SNft}_c(F)$,

(5) $\mathrm{Nft}(F) \leftrightarrow \mathrm{Nft}_1(F)$, $\mathrm{Nft}(F) \rightarrow \mathrm{Nft}_c(F)$,

(6) $\mathrm{SNft}_c(F) \wedge \mathrm{Lim}(\lambda) \rightarrow \bigcup_{\xi<\lambda} F(\xi) \leq F(\lambda)$,

(7) $\mathrm{Nft}_c(F) \rightarrow \mathrm{SNft}_c(F)$. □

From 7.5 and 7.6 we obtain the following theorem which shows that seminormal and normal functions are continuous in a more general sense:

7.1o THEOREM (In $KP_{\mathcal{L}}$ for every $c \in A$, $c \neq 0$; in $KP_{\mathcal{L}^*}$ for every $c \in B$, $c \neq 0$)

$$\mathrm{SNft}_c(F) \wedge \mathrm{gcf}_c(\lambda) \wedge \nearrow \mathrm{Sq}(g) \wedge \mathrm{dom}(g) \equiv \lambda \wedge \alpha \equiv \bigcup_{\xi<\lambda} g(\xi)$$

$$\rightarrow \mathrm{gcf}_c(\alpha) \wedge F(\alpha) \equiv \bigcup_{\xi<\alpha} F(\xi) \equiv \bigcup_{\eta<\lambda} F(g(\eta)). \quad \square$$

7.11 THEOREM (In $ZF_{\mathcal{L}}$ for every $c \in A$, $c \neq 0$; in $ZF_{\mathcal{L}^*}$ for every $c \in B$, $c \neq 0$)

Suppose that F is a class term and G is defined by recursion as follows:

$$G(\alpha) \equiv F(\beta), \text{ where } \beta \equiv \mu\eta \; (\forall \xi<\alpha \quad G(\xi) < F(\eta))$$

(i.e. G is the function enumerating the range of F in increasing order). Then:

$$\mathrm{SNft}_c(F) \rightarrow \mathrm{Nft}_c(G) \wedge \forall \xi \; (\xi \leq F(\xi) \leq G(\xi)).$$

Proof: Assume $\mathrm{SNft}_c(F)$. Then G is obviously a strictly increasing function. Let λ be an ordinal such that $\mathrm{gcf}_c(\lambda)$ and put $\beta := \bigcup_{\xi<\lambda} G(\xi)$.

We have to show: $G(\lambda) \leq \beta$.

Since $\forall \xi<\lambda \quad G(\xi) = F(g(\xi))$ for some g such that $\nearrow \mathrm{Sq}(g) \wedge \mathrm{dom}(g) \equiv \lambda$

(in fact: $g(\delta) \equiv \mu\eta \; (\forall \xi<\delta \quad G(\xi) < F(\eta))$ for $\delta<\lambda$),

we have by Cor. 7.5:

$$\beta \equiv \bigcup_{\eta<\lambda} G(\eta) \equiv \bigcup_{\eta<\lambda} F(g(\eta)) \equiv \bigcup_{\xi<\alpha} F(\xi), \text{ where } \alpha :\equiv \bigcup_{\eta<\lambda} g(\eta).$$

Since $gcf_c(\alpha)$, by Theorem 7.6 and since $SNft_c(F)$:

$$\beta \equiv F(\alpha).$$

Moreover, since $G(\lambda) \equiv F(\delta)$ for some δ, namely

$$\delta \equiv \mu\eta(\forall\xi < \lambda \quad G(\xi) < F(\eta)),$$

we have $\delta \leq \alpha$, since $\forall\xi < \lambda \quad G(\xi) < F(\alpha)$.

Therefore $G(\lambda) \equiv F(\delta) \leq F(\alpha) \equiv \beta$. □

Similarly, by using again Theorem 7.6 and 7.10 we obtain:

7.12 <u>THEOREM</u> (In $ZF_{\mathcal{L}}$ for every $c \in A$, $c \neq 0$; in $ZF_{\mathcal{L}^*}$ for every $c \in B$, $c \neq 0$)

$$Nft_c(F) \wedge Nft_c(G) \rightarrow Nft_c(F \circ G),$$

where $F \circ G := \{u,v \mid \exists w(\langle u,w\rangle \varepsilon G \wedge \langle w,v\rangle \varepsilon F\}$ denotes the composition of the functions F and G. □

The existence of critical points (or fixed points) of normal functions is usually proved by means of the axiom of infinity and the axiomschema of replacement, but can also be proved by applying the principle of complete reflection. The latter method is simpler and can easily be extended to the infinitary case.

7.13 <u>THEOREM</u>

$$KP_{\mathcal{L}^*} + \mathcal{L}^*\text{-ReflS} \vdash SNft_b(F) \rightarrow \exists\xi \ (\alpha < \xi \wedge gcf_b(\xi) \wedge F(\xi) \equiv \xi)$$

for every $b \in B$, $b \neq 0$, and every $\mathcal{L}^*$-definable class term F.

Proof: Suppose $F \equiv \{u,v \mid \varphi(u,v,\vec{v})\}$ for some $\mathcal{L}^*$-formula φ, and assume $SNft_b(F)$. We apply the principle of complete reflection (generalized to finitely many formulas as in Theorem 6.4) to the formulas

$$\varphi_1 := \varphi(u,v,\vec{v}),$$
$$\varphi_2 := \forall\xi\,\exists\eta\ \varphi(\xi,\eta,\vec{v}),$$
$$\varphi_3 := \forall\vec{w}\,\exists v\,(\vec{w}\,\varepsilon\,v)$$

where $\vec{w}$ is a sequence of variables such that $d(\vec{w}) = b$, $\varphi_3 \in \mathcal{L}^*$, and to the set $\{v,\alpha\}$ where v is a set such that $\vec{v}\,\varepsilon\,v$ (which exists because of $Pair^B$). Thus we obtain the existence of a transitive set u such that

$$\alpha, \vec{v}\,\varepsilon\,u \quad \text{and}$$

(i) $\forall\xi\,\varepsilon u\,\exists\eta\,\varepsilon u \quad \eta = F(\xi)$,

(ii) $\forall\vec{w}\,\exists v\,(\vec{w}\,\varepsilon\,v) \leftrightarrow \forall\vec{w}\,\varepsilon u\,\exists v\,\varepsilon u\,(\vec{w}\,\varepsilon\,v)$.

Since $\forall\vec{w}\,\exists v(\vec{w}\,\varepsilon\,v)$ is provable in the theory under consideration,

(ii) implies

$$\mathrm{gcf}_{\flat}(\gamma), \quad \text{where} \quad \gamma :\equiv u \cap On .$$

By (i) we have:

$$\forall \xi < \gamma \quad F(\xi) < \gamma \quad , \text{hence}$$

$$\gamma \leq F(\gamma) \equiv \bigcup_{\xi < \gamma} F(\xi) \leq \gamma \quad , \text{since} \quad \mathrm{gcf}_{\flat}(\gamma) \wedge \mathrm{SNft}_{\flat}(F).$$

Thus we have

$$\alpha < \gamma \wedge F(\gamma) \equiv \gamma \quad . \square$$

In particular, by Theorem 7.13 every normal function F has arbitrarily large critical points. In the finitary case we obtain:

7.14 THEOREM (ZF)

$$\mathrm{Nft}(F) \wedge \forall u \varepsilon v \ \mathrm{gcf}(\lambda, u) \rightarrow \exists \xi \quad (\alpha < \xi \wedge \forall u \varepsilon v \ \mathrm{gcf}(\xi, u) \wedge F(\xi) \equiv \xi)$$

for every class term F definable in the language of ZF set theory.

Proof: Let G be the normal function enumerating the critical points of F in increasing order. Using Cor. 7.7, we can choose for each α an ordinal $\gamma > \alpha$ such that $\forall u \varepsilon v \ \mathrm{gcf}(\gamma, u)$.
Consider $\delta :\equiv G(\gamma)$. Then by Theorem 7.6:

$$\forall u \varepsilon v \ \mathrm{gcf}(\delta, u) \quad \text{and} \quad F(\delta) \equiv \delta \quad . \square$$

8 *Infinitary analogues of the axiom of infinity*

In the usual finitary theory of KRIPKE-PLATEK (and a fortiori in ZF set theory) the axiom of infinity implies the following statements (each of which, in fact, is equivalent to the axiom of infinity):

Inf^1: $\exists u \forall v (v \varepsilon u \longleftrightarrow \mathrm{Nn}(v))$,

where $\mathrm{Nn}(v) : \longleftrightarrow (v \equiv 0 \vee \exists w (v \equiv w+1) \wedge \forall w \varepsilon v (w \equiv 0 \vee \exists u (w \equiv u+1)))$ is the usual definition of a *natural number;*

Inf^2: $\exists u (0 \varepsilon u \wedge \forall v \ w \varepsilon u \ \{v, w\} \varepsilon u)$,

Inf^3: $\exists \lambda \ \mathrm{Lim}(\lambda)$,

Inf^4: $\exists \lambda \ \forall v (\mathrm{Nn}(v) \rightarrow \mathrm{gcf}(\lambda, v))$.

We shall try to find analogues of the axiom of infinity for the infinitary systems $KP_{\mathcal{L}}$ and $ZF_{\mathcal{L}}$. Intuitively, such an axiom should imply the existence of a set $\bar{A}$ (or $\bar{B}$ resp. in case of the language $\mathcal{L}^*$) such that one can prove for these constants in the formal theory the formal analogues of the basic properties of the corresponding set A (or B resp.) of the metasystem. Instead of giving a precise meaning to the notion "basic property" (by using e.g. notions from recursion theory on admissible sets) we shall try to obtain the analogues of Inf^1 - Inf^4.

One of the difficulties we encounter in the infinitary case is the following problem: In general the infinitary axiom of pairing, Pair^C, constitutes a schema of formulas of the language $\mathcal{L}$ which cannot be replaced by a single axiom of this language (except for the case $A = H(\kappa)$, κ a singular or a successor cardinal). In addition, we shall show later that for every inaccessible cardinal $\kappa > \omega$, $\langle V_\kappa, \in \rangle$ is a model of $ZF_{\kappa\kappa}$, and hence one cannot prove in $ZF_{\kappa\kappa}$ the existence of the set V_κ (this is expressed in a somewhat vague sense since in this context κ is used in two different meanings; the statement can be made more precise by specifying κ to be (e.g.) the least inaccessible cardinal $> \omega$).

Let us first consider

Inf^3_C : $\exists \lambda \ \mathrm{gcf}_c(\lambda)$ for each $c \in C$, $c \neq 0$,

$\mathcal{L}\text{-}Inf^3 := Inf^3_A$, $\mathcal{L}^*\text{-}Inf^3 := Inf^3_B$,

as an infinitary analogue of Inf^3. Clearly, this is a schema of sentences, but this fact is not surprising since $\mathcal{L}$-Pair is in general a schema, too. If $C = HF$, Inf^3_C reduces to the usual axiom of infinity Inf^3 (by (3) of Lemma 7.9).

8.1 <u>THEOREM</u>

$$ZF_{\mathcal{L}} \vdash \mathcal{L}\text{-}\mathrm{Inf}^3 \quad \text{and} \quad ZF_{\mathcal{L}^*} \vdash \mathcal{L}^*\text{-}\mathrm{Inf}^3 .$$

Proof (the result already follows from Theorem 7.13, but in this case the proof can be simplified): Let C be A in the case of $\mathcal{L}$ and B in the case of $\mathcal{L}^*$. Then by Pair^C, for every $c \in C$, $c \neq 0$, we have

$$\forall \vec{v} \ \exists w (\vec{v} \varepsilon w) \quad \text{where} \quad d(\vec{v}) = c .$$

Hence by the complete reflection principle (cp. Lemma 6.1 (2)) there is a transitive set u such that

$$u \neq 0 \wedge \forall \vec{v} \varepsilon u \ \exists w \varepsilon u (\vec{v} \varepsilon w) .$$

Put $\beta :\equiv u \cap On$. Then $\mathrm{gcf}_c(\beta)$. □

8.2 Corollary. If $\mathcal{B} \in \mathcal{A}$, then

$$ZF_{\mathcal{AB}} \vdash \exists \lambda \bigwedge_{0 \neq x \in \mathcal{B}} gcf_x(\lambda) \wedge \exists \lambda \quad \forall u \in \overline{\mathcal{B}} \quad gcf(\lambda, u) .$$

Proof: By $Pair^{\mathcal{B}}$ we have:

$$\bigwedge_{0 \neq x \in \mathcal{B}} \forall \vec{v}_x \ \exists w \ (\vec{v}_x \varepsilon w) \quad \text{where} \quad d(\vec{v}_x) = x \quad \text{for } x \in \mathcal{B} .$$

Now proceed as in the proof of 8.1: by reflection there is some ordinal $\beta > 0$ such that

$$\bigwedge_{0 \neq x \in \mathcal{B}} gcf_x(\beta)$$

and hence also $\forall u \varepsilon \overline{\mathcal{B}} \ gcf(\beta, u)$ by 7.9 and 4.5. □

This proof shows that we also have an infinitary analogue of Inf^2:

$$Inf^2_{\mathcal{B}} \ : \ \exists u (0 \varepsilon u \wedge \bigwedge_{0 \neq x \in \mathcal{B}} \forall \vec{v}_x \varepsilon u \ [\vec{v}_x] \varepsilon u)$$

holds in $ZF^{*}_{\mathcal{AB}}$ if $\mathcal{B} \in \mathcal{A}$.

8.3 Corollary. If $\mathcal{B} \in \mathcal{A}$, then

$$ZF^{*}_{\mathcal{AB}} \vdash Nft(F) \rightarrow \exists \lambda \ (\alpha < \lambda \wedge \bigwedge_{x \in \mathcal{B}} gcf_x(\lambda) \wedge F(\lambda) \equiv \lambda) .$$

Proof: Proceed as in the proof of Theorem 7.13 taking for φ_3 the sentence

$$\bigwedge_{0 \neq x \in \mathcal{B}} \forall \vec{v}_x \ \exists w \ (\vec{v}_x \varepsilon w) . \ \square$$

In a subsequent paper we shall prove (cp. GLOEDE 1974, IV. 1.12 and IV.1.9 (1)):

8.4 THEOREM

$$ZF + Pair^{\mathcal{A}} + T_{\mathcal{A}} + \mathcal{L}\text{-ReplS} + \mathcal{L}\text{-Inf}^3 \vdash \exists \lambda [\alpha < \lambda \wedge \forall \vec{v} \varepsilon V_{\lambda} (\varphi(\vec{v}) \leftrightarrow \varphi^{V_\lambda}(\vec{v}))]$$

for every $\mathcal{L}$-formula $\varphi(\vec{v})$ with free variables as indicated.

In $ZF_{\mathcal{AA}}$ the above schema holds for every $\mathcal{L}_{\mathcal{AA}}$-formula φ with free variables as indicated. □

8.5 Corollary

The theories $ZF_{\mathcal{L}}$ and $ZF + \mathcal{L}\text{-Pair} + \mathcal{L}\text{-ReplS} + \mathcal{L}\text{-Inf}^3$ are equivalent if $\mathcal{A} = \mathcal{B}$. □

Since in Cor. 8.5 ZF may be taken as the theory of ZF axiomatized (as usually done) by means of the axiomschema of replacement rather than the schema of complete reflection, Cor. 8.5 is a generalization of the result (just mentioned) which is well-known in the finitary case, viz. in ZF the axiomschema of replacement and infinity are equivalent to the axiomschema of complete reflection and the (Δ_o-)-Aussonderungsschema. Since Cor. 8.5 is a generalization of this result to the infinitary case, we may take it as a justification for regarding $\mathcal{L}$-Inf^3 as a suitable infinitary analogue of the axiom of infinity.

8.6 Remark

Let us assume the axioms of Meta-ZFC = Meta-(ZF + AC) for our metatheory (M) and let A, B be sets in (M) such that

$B \subseteq A$, B is admissible and A is B-admissible.

Using the notion of satisfaction for formulas of $\mathcal{L}$ in a structure of the form $\langle M, \in \rangle$ (where M is a set), we have the following results:

Suppose $A = H(\kappa)$ for some infinite cardinal κ and for every ordinal δ let δ^+ denote the least infinite cardinal $> \delta$. If M is a transitive set, $M \neq 0$, then:

$\langle M, \in \rangle$ satisfies the axiomschema $\mathcal{L}_{AA}$-Inf^3 iff

$\forall x \in A \quad \exists \xi \in M \, \forall f \in {}^x\xi \, \exists \eta < \xi \;(\mathrm{rng}(f) \subseteq \eta)$ iff

$\forall \delta < \kappa \quad \exists \xi \in M \, \forall f \in {}^\delta\xi \, \exists \eta < \xi \;(\mathrm{rng}(f) \subseteq \eta)$ iff

$\forall \delta < \kappa \quad \exists \xi \in M \,(\mathrm{Lim}(\xi) \wedge \mathrm{cf}(\xi) > \delta)$ (by Lemma 7.2) iff

$\forall \delta < \kappa \quad \delta^+ \in M$.

Thus

$\langle M, \in \rangle$ satisfies $\mathrm{Inf}^3_{H(\lambda^+)}$ iff $\lambda^+ \in M$,

but if κ is a limit cardinal $> \omega$:

$\langle V_\alpha, \in \rangle$ satisfies $\mathrm{Inf}^3_{H(\kappa)}$ iff $\kappa \leq \alpha$.

Thus if e.g. κ is inaccessible ($\mathrm{in}(\kappa)$) and $> \omega$, then $\mathrm{Inf}^3_{H(\kappa)}$ does not guarantee that every transitive model of this axiomschema contains κ, since e.g. $\langle V_\kappa, \in \rangle$ is a standard model of $\mathrm{Inf}^3_{H(\kappa)}$.

Therefore we proceed to replace $\mathcal{L}$-Inf^3 by a (possibly) stronger axiom which can be regarded as a generalization of Inf^4 :

Inf^4_C : $\quad \bar{a} \in \bar{C}$ for each $a \in C$, and

$\exists \lambda \quad \forall u \in \bar{C} \;\; \mathrm{gcf}(\lambda, u)$.

We use the notation $\mathcal{L}(\bar{C})$ to indicate that $\mathcal{L}$ is supposed to be any of the languages of I.2.4 or I.2.5 containing set constants $\bar{a}$ ($a \in C$) and a set constant $\bar{C}$ (possibly in place of the unary predicate symbol $\dot{C}$ if $\mathcal{L}$ contains this symbol). Then we define

$$\mathcal{L}(\bar{A})\text{-}\mathrm{Inf}^4 := \mathrm{Inf}^4_A \;,\quad \mathcal{L}^*(\bar{B})\text{-}\mathrm{Inf}^4 := \mathrm{Inf}^4_B \;,$$

$$\mathrm{ZFI}_{\mathcal{L}(\bar{A})} := \mathrm{ZF}_{\mathcal{L}(\bar{A})} + \mathcal{L}(\bar{A})\text{-}\mathrm{Inf}^4 ,\quad \mathrm{ZFI}_{\mathcal{L}^*(\bar{B})} := \mathrm{ZF}_{\mathcal{L}^*(\bar{B})} + \mathcal{L}^*(\bar{B})\text{-}\mathrm{Inf}^4 .$$

8.7 THEOREM

If $A = B$, then in $\mathrm{ZFI}_{\mathcal{L}(\bar{A})}$ the principle of complete reflection can be strengthened to yield the following schema of complete reflection:

$$\exists \lambda [\alpha < \lambda \wedge \forall u \in \bar{A}\ \ \mathrm{gcf}(\lambda, u) \wedge \forall \vec{v} \in V_\lambda\ (\ \varphi(\vec{v}) \longleftrightarrow \varphi^{V_\lambda}(\vec{v}))]$$

for every $\mathcal{L}(\bar{A})$-formula $\varphi(\vec{v})$ with free variables as indicated.

(This theorem follows from Lemma 7.14 and results of GLOEDE 1974, Ch. IV.) □

8.8 Remark

In the finitary case (where $\mathcal{L}$ is the finitary language of ZF set theory with additional set constants $\bar{x}$ ($x \in HF$) and a set constant $\overline{HF}$) $\mathrm{ZFI}_{\mathcal{L}}$ can be identified with ZF. If $B \in A$, then

$$\mathrm{ZF}_{\mathcal{L}(\bar{B})} \vdash \mathrm{Inf}^4_B \qquad \text{by Cor. 8.2 and Lemma 4.4,}$$

hence $\mathrm{ZF}_{\mathcal{L}}$ and $\mathrm{ZFI}_{\mathcal{L}(\bar{B})}$ are equivalent (note that $\bar{B}$ is definable in the former theory since $B \in A$).

In the general case we can only prove that

$$\mathrm{ZF}_{\mathcal{L}(\bar{A})} + \mathrm{AC} \quad \text{and} \quad \mathrm{ZFI}_{\mathcal{L}(\bar{A})} + \mathrm{AC}$$

as well as

$$\mathrm{ZF}_{\mathcal{L}^*(\bar{B})} + \mathrm{AC} \quad \text{and} \quad \mathrm{ZFI}_{\mathcal{L}^*(\bar{B})} + \mathrm{AC}$$

are equivalent. Here we need the axiom of choice, since in ZFC one can prove the existence of arbitrarily large regular cardinals and hence

(+) $\quad \exists \lambda\ \forall u \in v\ \ \mathrm{gcf}(\lambda, u)$

is provable in ZFC. On the other hand, by Theorem 8.1 (+) is provable in

$\mathrm{ZF}_{\mathcal{L}}$ for every set v of the form $\bar{a}$ for some $a \in A$, and in

$\mathrm{ZF}_{\mathcal{L}^*}$ for every set v of the form $\bar{b}$ for some $b \in B$

<u>without</u> using the axiom of choice. E.g. in $\mathrm{ZF}_{\omega_1\omega_1}$ one can prove the existence of ordinals α such that $\mathrm{cf}(\alpha) > \omega$.

8.9 An important consequence of the axiom of infinity is the result that for every set u there exists a set v consisting just of the finite sequences of elements of u. This set is also the set of assignments for finitary formulas with respect to the definitión of truth in a structure with universe u, and hence it is of particular importance for a formalization of truth. Since here we are interested in the infinitary case let us define

$$S(u,v) := \{ w \mid Ft(w) \wedge dom(w) \varepsilon u \wedge rng(w) \subseteq v \} .$$

Note that in $ZF_{\mathcal{L}}$ one can prove (using the axiom of power set):

$\mathcal{L}$-Inf[5]: $\forall v \exists w (w \equiv S(\bar{A}, v))$

provided $\mathcal{L}$ contains the set constant $\bar{A}$.

8.1o Lemma (ZF)

If $v \not\equiv 0$, then:

(1) $v \varepsilon V_\alpha \wedge \forall u \varepsilon v \; gcf(\alpha, u) \rightarrow S(v, V_\alpha) \subseteq V_\alpha$,

(2) $S(v, V_\alpha) \subseteq V_\alpha \rightarrow \forall u \varepsilon v \; gcf(\alpha, u)$.

Proof of (1): Suppose $v \varepsilon V_\alpha \wedge \forall u \varepsilon v \; gcf(\alpha, u)$, and let $w \varepsilon S(v, V_\alpha)$, i.e. $Ft(w) \wedge dom(w) \varepsilon v \wedge rng(w) \subseteq V_\alpha$.

Let $\rho(u) :\equiv \mu \xi \; (u \varepsilon V_{\xi+1})$ be the (von NEUMANN) rank of u, and define

$w_o :\equiv \{ \langle u, \rho(w(u)) \rangle \mid u \varepsilon dom(w) \}$. Then:

$$Ft(w_o) \wedge dom(w_o) \varepsilon v \wedge rng(w_o) \subseteq \alpha \text{ , hence}$$

$$\exists \xi < \alpha \;\; rng(w_o) \subseteq \xi \qquad \text{since} \qquad gcf(\alpha, dom(w_o)) .$$

Since α must be a limit number, we have $w_o \varepsilon V_\alpha$ and hence $w \varepsilon V_\alpha$, too.

Proof of (2): Suppose $S(v, V_\alpha) \subseteq V_\alpha$.

Since $v \not\equiv 0$, α is a limit number. Now suppose

$Ft(w) \wedge dom(w) \varepsilon v \wedge rng(w) \subseteq \alpha$. Then

$w \varepsilon V_\alpha$ by assumption, hence

$rng(w) \subseteq V_\beta$ for some $\beta < \alpha$, and therefore

$rng(w) \subseteq \alpha$. □

Theorem 8.7 and Lemma 8.1o suggest to consider the theory $ZF_{\mathcal{L}(\bar{A})}(\bar{A})$ obtained from $ZF_{\mathcal{L}(\bar{A})}$ by adding the axioms

$$\bar{a} \varepsilon \bar{A} \quad \text{for} \quad a \in A$$

and replacing the $\mathcal{L}(\bar{A})$-ReflS by the stronger principle

$\mathcal{L}(\bar{A})$-ReflS$(\bar{A})$: $\exists u\,[\,\text{trans}(u) \wedge v \in u \wedge S(\bar{A},u) \subseteq u \wedge$

$$\wedge\ \forall \vec{v} \in u \ (\ \varphi(\vec{v}) \leftrightarrow \varphi^{u}(\vec{v}))\ ]$$

for every formula $\varphi(\vec{v})$ of $\mathcal{L}(\bar{A})$ with free variables as indicated (similarly with B and $\bar{B}$ in place of A and $\bar{A}$ in case of the languages $\mathcal{L}^*$).

8.11 THEOREM

(1) $ZF_{\mathcal{L}(\bar{A})}(\bar{A}) \vdash \mathcal{L}(\bar{A})\text{-Inf}^4$, $ZF_{\mathcal{L}^*(\bar{B})}(\bar{B}) \vdash \mathcal{L}^*(\bar{B})\text{-Inf}^4$.

(2) If $A = B$, then $ZF_{\mathcal{L}(\bar{A})}(\bar{A})$ and $ZFI_{\mathcal{L}(\bar{A})}$ are equivalent.

Proof of (1): An application of the $\mathcal{L}(\bar{A})$-ReflS$(\bar{A})$ yields the existence of a transitive set u such that

$$S(\bar{A},u) \subseteq u \quad \wedge \forall w \in u \quad \bigcup \text{rng}(w) \in u \ .$$

If we put $\beta :\simeq u \cap On$, then (cp. the proof of Lemma 8.1o (2)):

$$\forall w \in \bar{A} \quad \text{gcf}(\beta, w) \ ,$$

and similarly in case of the language $\mathcal{L}^*$.

Proof of (2):

$$ZF_{\mathcal{L}(\bar{A})}(\bar{A}) \vdash ZFI_{\mathcal{L}(\bar{A})} \qquad \text{by (1)},$$

and $ZFI_{\mathcal{L}(\bar{A})} \vdash \mathcal{L}(\bar{A})$-ReflS$(\bar{A})$

follows from Theorem 8.7 and Lemma 8.1o. □

8.12 Remark

1. By Lemma 8.1o, in $ZFI_{\mathcal{L}(\bar{A})}$ we may replace the axiom

$$\exists \lambda \quad \forall u \in \bar{A} \quad \text{gcf}(\lambda, u) \quad \text{by}$$

by

$$\exists \lambda \quad S(\bar{A}, V_\lambda) \subseteq V_\lambda \ ,$$

and similarly for $ZFI_{\mathcal{L}^*(\bar{B})}$.

2. If κ is regular, $\text{Inf}^4_{H(\kappa)}$ is a more satisfactory axiom of infinity as $\text{Inf}^3_{H(\kappa)}$ as far as the existence of $H(\kappa)$ is concerned:

First note that by results of §4 the theory $T_{H(\kappa)}$ is interpretable in $ZF_{\kappa\omega}$, and hence every model of $ZF_{\kappa\omega}$ is isomorphic to an end-extension of $\langle H(\kappa), \in, x\rangle_{x\in H(\kappa)}$. Secondly, under the assumptions of Remark 8.6, for a transitive set M we have:

$\langle M, \in, x, H(\kappa)\rangle_{x\in H(\kappa)}$ is a model of $\text{Inf}^4_{H(\kappa)}$ iff

$\exists \lambda \in M \ \forall \delta < \kappa \ (\ \text{Lim}(\lambda) \wedge \text{cf}(\lambda) > \delta \ )$ iff

$$\exists \lambda \in M \ (\mathrm{Lim}(\lambda) \wedge \mathrm{cf}(\lambda) \geq \kappa) .$$

Thus if κ is regular, $\kappa \leq \alpha$, then

$\langle V_\alpha, \in, x, H(\kappa) \rangle_{x \in H(\kappa)}$ is a model of $\mathrm{Inf}^4_{H(\kappa)}$ iff $\kappa < \alpha$

(just as in the finitary case $\kappa = \omega$).

8.13 Lemma (Meta-ZFC)

Let κ be an infinite cardinal, $a \neq 0$, a transitive. Then:

(1) $\forall f \ \forall \xi < \kappa (\mathrm{Ft}(f) \wedge \mathrm{dom}(f) = \xi \wedge \mathrm{rng}(f) \subseteq a \rightarrow \mathrm{rng}(f) \in a)$
$\rightarrow H(\kappa) \subseteq a$,

(2) if a is admissible:
$\forall f \ (\mathrm{Ft}(f) \wedge \mathrm{dom}(f) \in a \wedge \mathrm{rng}(f) \subseteq a \rightarrow \mathrm{rng}(f) \in a)$
$\leftrightarrow \exists \lambda \ (\mathrm{reg}(\lambda) \wedge a = H(\lambda))$,

where $\mathrm{reg}(\lambda) :\leftrightarrow \mathrm{cf}(\lambda) = \lambda > \omega$ ("λ is regular");

(3) $\forall f \ \forall \xi < \kappa (\mathrm{Ft}(f) \wedge \mathrm{dom}(f) = \xi \wedge \mathrm{rng}(f) \subseteq H(\kappa) \rightarrow \mathrm{rng}(f) \in H(\kappa))$
$\leftrightarrow \mathrm{reg}(\kappa)$.

Proof of (1) by $\in$-induction:

Suppose $0 \neq b \in H(\kappa)$. Then $b \subseteq a$ by induction hypothesis. Since $b \in H(\kappa)$, there is some function f such that $\mathrm{dom}(f) < \kappa \wedge \mathrm{rng}(f) = b$. By assumption, $b = \mathrm{rng}(f) \in a$.

Proof of (2): Let a be admissible and assume the closure condition

(i) $\forall f \ (\mathrm{Ft}(f) \wedge \mathrm{dom}(f) \in a \wedge \mathrm{rng}(f) \subseteq a \rightarrow \mathrm{rng}(f) \in a)$

holds. Then we obviously have

(ii) $\forall x, y \ (x \subseteq y \wedge y \in a \rightarrow x \in a)$ (i.e. $\mathrm{strans}(a)$).

moreover, if $b \in a$, $f : b \longrightarrow a$, then $f \subseteq b \times \mathrm{rng}(f) \in a$ by (i), and hence $f \in a$ by (ii). Thus we have shown:

(iii) $b \in a \wedge f : b \longrightarrow a \rightarrow f \in a$.

Next we claim:

(iv) $a \subseteq H(\lambda)$ where $\lambda := \mathrm{On} \cap a$.

Proof: Suppose $b \in a$ and let $c := \mathrm{TC}(b)$. Then $c \in a$, since a is admissible. We have to prove: $|c| < \lambda$. Suppose not, then there is a function $f : c \rightarrow \lambda$

such that $rng(f) = \lambda$, hence $\lambda \in \alpha$ by (i), a contradiction.

(v) $H(\lambda) \subseteq \alpha$ (cp. the proof of (1).)

Finally: λ is regular:

Suppose $\lambda = \bigcup rng(f)$ for some function f such that $dom(f) = \delta < \lambda$ and $rng(f) \subseteq \lambda$. Then $rng(f) \in \alpha$ by (i), hence

$$\lambda = \bigcup rng(f) \in \alpha \text{, a contradiction.}$$

Conversely, if $\alpha = H(\lambda)$ for some regular cardinal λ, then (i) clearly holds for α.

(3) is a consequence of (2). □

8.14 Remark

Let the metatheory be again Meta-ZFC, M a transitive set in the metatheory, and let κ be an infinite cardinal. Then by Lemma 8.13:

(1) if $< M, \in >$ satisfies $Pair^{H(\kappa)}$, then $H(\kappa) \subseteq M$;

(2) $< H(\kappa), \in >$ satisfies $Pair^{H(\kappa)}$ iff κ is regular;

(3) $< \alpha, \in >$ is a transitive model of $KP + Pair^{\alpha}$ iff $\alpha = H(\lambda)$ for some regular cardinal λ;

(4) $< V_\alpha, \in >$ is a model of $KP + Pair^{V_\alpha}$ iff α is inaccessible

(since V_α is admissible iff $H(\alpha) = V_\alpha$ iff $\alpha = \beth_\alpha$).

On the other hand, it is easy to check that

(5) for every inaccessible $\kappa > \omega$:

$< V_\kappa, \in >$ is a model of $ZF + Pair^{H(\kappa)} + \mathcal{L}_{\kappa\kappa}\text{-ReplS} + Inf^3_{H(\kappa)}$

(cp. (4) above and BOWEN 1973 for the schema $\mathcal{L}_{\kappa\kappa}$-ReplS), and hence

$< V_\kappa, \in >$ is a model of $ZF_{\kappa\kappa}$ by Cor. 8.5,

$< V_\kappa, \in, x, V_\lambda >_{x \in V_\lambda}$ is a model of $ZFI_{\lambda\lambda}$ for every regular $\lambda < \kappa$.

(This also follows from GLOEDE 1974, Chapter IV.) Finally, again by (4):

(6) $< V_\kappa, \in >$ is a model of $ZF_{\kappa\kappa}$ iff κ is inaccessible (if $\kappa > \omega$).

(By "inaccessible cardinal" we always mean "strongly inaccessible cardinal".)

Thus we obtain the following result which has also been proved by BOWEN 1973 (independently and by different methods):

If κ is inaccessible, $\omega \leq \beta \leq \alpha < \kappa$, α, β regular, then there are κ-many $\gamma < \kappa$ such that

$$< V_\gamma, \in > \prec_{\mathcal{L}_{\alpha\beta}} < V_\kappa, \in > .$$

(Cp. Theorem 8.7 and 6.4.) In fact, by results of GLOEDE 1974 (IV.1.1o) we have the following stronger conclusion:

There is a function $f:\kappa \to \kappa$ which is strictly increasing and continuous and there are arbitrarily large ordinals $\gamma < \kappa$ such that

$$\mathrm{cf}(\gamma) \geq \alpha \quad \text{and} \quad f(\gamma) = \gamma \qquad \text{(cp. Theorem 7.13)}$$

and for every such γ :

$$< V_\gamma, \in > \prec_{\mathcal{L}_{\alpha\alpha}} < V_\kappa, \in > \quad .$$

Finally note that for every regular α :

(7) if $< V_\gamma, \in > \prec_{\mathcal{L}_{\alpha\omega}} < V_\kappa, \in >$, $\gamma \leq \kappa$, κ inaccessible, then $\mathrm{cf}(\gamma) \geq \alpha$,

and hence the assumption on the cofinality of γ made above is necessary. For, if $\delta < \alpha$, $f: \delta \to \gamma$, then $\mathrm{rng}(f) \in V_\kappa$, and since

$< V_\kappa, \in >$ satisifes the following instance of $\mathrm{Pair}^{H(\kappa)}$ under the assignment f :

$$\exists u \forall v (v \varepsilon u \leftrightarrow \bigvee_{\xi<\delta} v \equiv \vec{v}(\xi)) \quad ,$$

$< V_\gamma, \in >$ satisfies the same formula under the assigment f , i.e.

$$\mathrm{rng}(f) \in V_\gamma \quad , \text{ and hence } \quad \exists \xi < \gamma \ \mathrm{rng}(f) \subseteq \xi \ .$$

Therefore $\mathrm{cf}(\gamma) \geq \delta$ for every $\delta < \alpha$, and thus $\mathrm{cf}(\gamma) \geq \alpha$.

8.15 <u>THEOREM</u> (Meta-ZFC)

Suppose κ is Π^1_1-indescribable (i.e. inaccessible and weakly compact). Then the set of ordinals $\gamma < \kappa$ such that

$$\mathrm{in}(\gamma) \wedge < V_\gamma, \in > \prec_{\mathcal{L}_{\gamma\gamma}} < V_\kappa, \in >$$

is stationary in κ .

Proof: A set $X \subseteq \kappa$ is called *stationary in* κ (or *dense in* κ) iff it intersects each closed unbounded subset of κ . By results of GLOEDE 1974 , Chapter IV (referred to in the preceeding remark 8.14 following statement (6)) for each $\alpha < \kappa$ the set a_α defined by

$$a_\alpha = \begin{cases} \{\xi < \kappa \mid \mathrm{in}(\xi) \wedge \alpha \leq \xi \wedge < V_\xi, \in > \prec_{\mathcal{L}_{\alpha\alpha}} < V_\kappa, \in > \} & \text{if } \mathrm{reg}(\alpha), \\ \kappa & \text{otherwise}, \end{cases}$$

is stationary in κ . Hence by Π^1_1-indescribability (cp. Theorem 3.6 of GLOEDE 1971):

$$a = \{\xi < \kappa \mid \mathrm{in}(\xi) \wedge \forall \eta < \xi \quad a_\eta \cap \xi \text{ is stationary in } \xi\}$$

is stationary in κ . Since $\mathfrak{a}$ is a subset of

$$\{\xi<\kappa \mid in(\xi) \wedge \forall\eta<\xi \;\; \xi\in \mathfrak{a}_\eta\} \subseteq \{\xi<\kappa \mid in(\xi) \wedge \langle V_\xi, \in\rangle \prec_{\mathcal{L}_{\xi\xi}} \langle V_\kappa, \in\rangle\},$$

the latter set is stationary in κ , too. □

We conclude this section with the following remark: Theorem 8.15 suggests to investigate an infinitary version of a strengthening of ACKERMANN's set theory A (called ZA by REINHARDT 197o and Sb by LEVY 196o) by adding to A the following priciple of reflection

$$\forall \vec{v} \varepsilon V \; (\quad \varphi(\vec{v}) \longleftrightarrow \varphi^V(\vec{v}))$$

for every formula φ of $\mathcal{L}_{VV}$ with free variables as indicated which does not contain the constant V (the class V referred to in the definition of $\mathcal{L}_{VV}$ clearly is the class of all sets in the corresponding metatheory).

9 *Foundation and choice*

In the preceeding sections we were mainly concerned with a generalization of the usual set theoretical axioms to the infinitary case but the results still applied to the finitary case except in some applications. We shall now consider some consequences of infinitary axioms which are peculiar to the infinitary case.

A fundamental concept of set theory which cannot be characterized by a finitary formula (and even not in $\mathcal{L}_{\omega_1\omega}$) but by a formula of $\mathcal{L}_{\omega_1\omega_1}$ is the notion of well-foundedness (and similarly the notion of a well-ordering). A finitary definition of " R is a well-founded relation " is usually given as follows:

$$\mathrm{Fund}(R): \longleftrightarrow \; \mathrm{Rel}(R) \wedge \forall u \,(u \neq 0 \rightarrow \exists v \varepsilon u \; \forall w \,(w R v \rightarrow \neg\, w \varepsilon u)$$

(As usual, we write $u R v$ in place of $\langle u,v\rangle \varepsilon R$.)

Occasionally it is more convenient to introduce a stronger concept by requiring that there does not exist an infinitely long descending chain with respect to the relation R :

$$\mathrm{NDS}(R): \longleftrightarrow \; \mathrm{Rel}(R) \wedge \; \neg \exists u \; (\mathrm{Ft}(u) \wedge \mathrm{dom}(u) \equiv \omega \;\; \wedge \forall n{<}\omega \;\; u(n+1) \, R \, u(n)).$$

However, the equivalence

$$\mathrm{Fund}(R) \longleftrightarrow \; \mathrm{NDS}(R)$$

can be proved in finitary set theory (e.g. ZF) only if the axiom of dependent choices, DC , is assumed. Instead we shall now make use of sufficiently strong infinitary languages in order to prove for every relation R :

$$\mathrm{Fund}(R) \longleftrightarrow \quad \neg \exists \vec{v} \bigwedge_{n<\omega} v_{n+1} \, R \, v_n \qquad (\vec{v} = \langle v_n \mid n<\omega\rangle)$$

using a logical axiom which corresponds to the (set theoretical) axiom of dependent choices (which, however, will now be required for the metatheory in order to prove that the infinitary logical axiom of dependent choices is true in every model).

9.1 THEOREM

If $\omega \in B$, then

$$\mathcal{L}^*_{AB}\text{-FundS} \vdash \neg \exists \vec{v} \bigwedge_{n<\omega} v_{n+1} \varepsilon v_n \qquad (\vec{v} = \langle v_n \mid n<\omega \rangle).$$

Proof: Suppose $\exists \vec{v} \bigwedge_{n<\omega} v_{n+1} \varepsilon v$. Consider the formula

$$\varphi(u) := \exists \vec{v} \bigwedge_{n<\omega} [v_{n+1} \varepsilon v_n \wedge v_0 \equiv u].$$

Then $\exists u\ \varphi(u)$ by our assumption. Therefore, by the schema of foundation, there is a set u such that

$\varphi(u) \wedge \forall v \varepsilon u\ \neg \varphi(v)$, i.e.

(i) $\exists \vec{v} \bigwedge_{n<\omega} [v_{n+1} \varepsilon v_n \wedge v_0 \equiv u]$ and

(ii) $\forall v \varepsilon u\ \neg \exists \vec{v} \bigwedge_{n<\omega} [v_{n+1} \varepsilon v \wedge v_0 \equiv v]$.

By (i) there is a sequence $\vec{v}$ such that $\bigwedge_{n<\omega} \vec{v}(n+1) \varepsilon \vec{v}(n) \wedge \vec{v}(0) \equiv u$.

Let $\vec{w}$ be the sequence of length ω such that $\vec{w}(n) = \vec{v}(n+1)$ for $n<\omega$.

Then

$$\vec{w}(0) \varepsilon u \wedge \bigwedge_{n<\omega} \vec{w}(n+1) \varepsilon \vec{w}(n) \wedge \vec{w}(0) \equiv \vec{w}(0)$$

contrary to (ii). □

9.2 THEOREM

If $\omega \in B$, then

$$KP^*_{AB} \vdash \text{Fund}(R) \rightarrow \neg \exists \vec{v} \bigwedge_{n<\omega} \vec{v}(n+1) R \vec{v}(n).$$

Proof: Suppose $\bigwedge_{n<\omega} \vec{v}(n+1) R \vec{v}(n)$ for some sequence $\vec{v}$, $d(\vec{v}) = \omega$.

By Pair^B there is a set u such that

$$\forall w\ (w \varepsilon u \leftrightarrow \bigvee_{n<\omega} w \equiv \vec{v}(n)).$$

Obviously we have $u \not\equiv 0 \wedge \forall w \varepsilon u\ \exists v \varepsilon u\ v R w$, which implies $\neg \text{Fund}(R)$. □

Following KARP 1964 (p.12o) we introduce the following schema of dependent choices (as logical axioms):

$$DC_{\mathcal{L}}: \quad \exists v_0\ \varphi_0 \wedge \bigwedge_{n<\omega} \forall v_n \exists v_{n+1}\ \varphi_{n+1} \rightarrow \exists \vec{v} \bigwedge_{n<\omega} \varphi_n$$

where $\vec{v} = \langle v_n \mid n<\omega \rangle$ and $\langle \varphi_n \mid n<\omega \rangle$ is any sequence of $\mathcal{L}$-formulas such that $\bigwedge_{n<\omega} \varphi_n$ is a formula of $\mathcal{L}$ and v_n does not occur in $\bigwedge_{m<n} \varphi_m$.

(The proper generalization of KARP's axiom schema would allow sequences of variables $\vec{v}_n$ in place of the single variables v_n ; however, for our purposes the above schema will be sufficient.)

9.3 Corollary

If $\omega \in \mathcal{B}$ and if we assume the schema of dependent choices, $DC_{\mathcal{L}}$, then:

$$KP_{\mathcal{L}} \vdash \text{Fund}(R) \leftrightarrow \text{Rel}(R) \wedge \neg \exists \vec{v} \bigwedge_{n<\omega} \vec{v}(n+1) R \vec{v}(n) \leftrightarrow \text{NDS}(R)$$

for every $\mathcal{L}$ -definable class term R .

(The proof is similar to the proof of Theorem 9.4 below, and the last equivalence follows from Theorem 9.6 below.) □

9.4 THEOREM

Denote by $KP^{o}_{\mathcal{L}}$ the theory obtained from $KP_{\mathcal{L}}$ by deleting the axiomschema of foundation, $\mathcal{L}$ -FundS. If $\omega \in \mathcal{B}$, and if we assume $DC_{\mathcal{L}}$, then

$$KP^{o}_{\mathcal{L}} + \neg \exists \vec{v} \bigwedge_{n<\omega} \vec{v}(n+1) \varepsilon \vec{v}(n) \vdash \mathcal{L}\text{-FundS} .$$

Proof: Suppose $\varphi(u,w)$ is a formula of $\mathcal{L}$ with free variables as indicated and

$$\exists u \quad \varphi(u,\vec{w}) \wedge \forall u [\varphi(u,\vec{w}) \rightarrow \exists w (\varphi(w,\vec{w}) \wedge w \varepsilon u)] .$$

Then we have:

$$\exists v_0 \quad \varphi(v_0,\vec{w}) \wedge \bigwedge_{n<\omega} \forall v_n \exists v_{n+1} [\varphi(v_n,\vec{w}) \rightarrow \varphi(v_{n+1},\vec{w}) \wedge v_{n+1} \varepsilon v_n].$$

Applying the law of dependent choices we have:

$$\exists \vec{v} \bigwedge_{n<\omega} \varphi_n(v_n, v_{n+1},\vec{w}) \quad , \quad \text{where}$$

$$\varphi_0(v_0,v_1,\vec{w}) := \varphi(v_0,\vec{w}) \quad ,$$

$$\varphi_n(v_n,v_{n+1},\vec{w}) := [\varphi(v_n,\vec{w}) \rightarrow \varphi(v_{n+1},\vec{w}) \wedge v_{n+1} \varepsilon v_n] \quad \text{if} \quad n \neq 0$$

Hence $\exists \vec{v} \bigwedge_{n<\omega} \vec{v}(n+1) \varepsilon \vec{v}(n)$. □

9.5 Remark

If $\omega \in \mathcal{B}$ and if we assume the axioms $DC_{\mathcal{L}}$ (as logical axioms), then by Theorem 9.1 and 9.5 the axiomschema of foundation, $\mathcal{L}$ -FundS, can be replaced in $KP_{\mathcal{L}}$ by the single axiom

$$(+) \quad \neg \exists \vec{v} \bigwedge_{n<\omega} \vec{v}(n+1) \varepsilon \vec{v}(n) \quad .$$

The same result holds for $ZF_{\mathcal{L}}$; moreover, in $ZF_{\mathcal{L}}$ the axiom of foundation,

Fund, already implies the axiomschema of foundation, $\mathcal{L}$-FundS (mainly by using the $\mathcal{L}$-AusS), and the axiom of foundation can be shown to be equivalent to (+) in $ZF^{o}_{\mathcal{L}}$ (= $ZF_{\mathcal{L}}$ without Fund) by using the single instance

$$\exists v_o (v_o \varepsilon u) \wedge \bigwedge_{n<\omega} \forall v_n \exists v_{n+1} (v_{n+1} \varepsilon v_n) \rightarrow$$
$$\rightarrow \exists \vec{v} \; [\vec{v}(0) \varepsilon u \wedge \bigwedge_{n<\omega} \vec{v}(n+1) \varepsilon \vec{v}(n)]$$

of the laws of dependent choices.
(+) in turn may be replaced by the finitary sentence

(++) $\neg \exists u [Ft(u) \wedge dom(u) \equiv \omega \wedge \forall n < \omega \; u(n+1) \varepsilon u(n)]$

(i.e. $NDS(\varepsilon)$) using only $Pair^{B}$:

9.6 <u>THEOREM</u>

If $\omega \in B$, then (without using $DC_{\mathcal{L}}$):

$$KP_{\mathcal{L}} \vdash Rel(R) \,.\rightarrow.\, \exists \vec{v} \bigwedge_{n<\omega} \vec{v}(n+1) R \vec{v}(n)$$
$$\leftrightarrow \exists u [Ft(u) \wedge dom(u) \equiv \omega \wedge \forall n < \omega \; u(n+1) R u(n)].$$

Proof: Suppose that R is a relation and for some ω-sequence $\vec{v}$:

$$\bigwedge_{n<\omega} \vec{v}(n+1) R \vec{v}(n) \quad .$$

By Lemma 5.4 there is a set u such that $Rp_{\omega}(u,\vec{v})$, i.e.

$Ft(u) \wedge dom(u) \equiv \overline{\omega} \wedge \forall n < \overline{\omega} \; u(n+1) R u(n)$ ((2) of Lemma 5.4).

Since $\omega \equiv \overline{\omega}$ (cp. Lemma 4.7), we have proved " $\rightarrow$ " .
In order to show the converse, suppose

$$Ft(u) \wedge dom(u) \equiv \omega \wedge \forall n < \omega \; u(n+1) R u(n) \quad .$$

Again, since $\omega = \overline{\omega}$ and by Lemma 4.5 we have:

$$\bigwedge_{n<\omega} u(\overline{n+1}) R u(\overline{n}) \; ,$$

and the conclusion is now immediate. □

9.7 <u>Application</u>

Let DC denote the (set theoretical) axiom of dependent choices of BERNAYS (see e.g. FELGNER 1971, pp.146ff). In Meta-(ZF+DC) one can prove:

If B is an admissible set, $\omega \in B$, then every model $< M, E >$ of the theory $Ext + Pair^{B} + Fund$ (where M is a set) is isomorphic to a standard model, i.e. a model of the form $< M_o, \in >$ for some transitive set M_o.

In fact, E is extensional. Hence, in order to abe able to apply MOSTOWSKI's collapsing theroem (MOSTOWSKI 1949, Theorem 3) it suffices to show that E is well-founded. Suppose not; then by DC there is a function such that

$$Ft(f) \wedge dom(f) = \omega \quad \wedge rng(f) \subseteq M \quad \wedge \forall n<\omega \quad f(n+1)\, E\, f(n).$$

this means that

$$\langle M,E\rangle \models \bigwedge_{n<\omega} v_{n+1} \varepsilon v_n [f] \quad \text{, hence}$$

$$\langle M,E\rangle \models \exists \vec{v} \bigwedge_{n<\omega} \vec{v}(n+1) \varepsilon \vec{v}(n) \quad \text{and hence by Theorem 9.2}$$

$$\langle M,E\rangle \models \neg \text{ Fund} \quad \text{, a contradiction.}$$

Using Theorem 9.1 instead of Theorem 9.2, the same result holds if $\langle M,E\rangle$ is a model of the schema $\mathcal{L}_{BB}$-FundS. In particular, if $\omega \in B$, then every model of KP^*_{AB} (a fortiori every model of ZF^*_{AB}) is isomorphic to a standard model (using DC in the metatheory).

REFERENCES

BARWISE, J.

1968a Implicit definability and compactness in infinitary languages, in: *The syntax and semantics of infinitary languages* (ed. by J. Barwise), 1-35. Lecture Notes in Mathematics, Vol. 72, Springer-Verlag, Berlin - Heidelberg - New York (1968)

1969a Infinitary logic and admissible sets, *Journ. Symb. Logic* 34 (1969), 226-252

1969b Applications of strict Π^1_1 predicates to infinitary logic, *Journ. Symb. Logic* 34 (1969), 4o9-423

BARWISE, J., GANDY, R.O. and MOSCHOWAKIS, Y.N.

1971 The next admissible set, *Journ. Symb. Logic* 36 (1971), 1o8-12o

BOWEN, K.A.

1973 Infinite forcing and natural models of set theory, *Bull. Acad. Polon. Sci.* XXI,3 (1973), 195-199

CHANG, C.C.

1971 Sets constructible using $L_{\kappa\kappa}$, in: SCOTT 1971, 1-8

FELGNER, U.

1971 *Models of ZF-set theory*, Lecture Notes in Mathematics, Vol. 223, Springer-Verlag, Berlin - Heidelberg - New York (1971), VI + 173 pp.

GLOEDE, K.

1971 Filters closed under Mahlo's and Gaifman's operation, in: *Proc. Cambridge Summer School in Mathematical Logic* (ed. by A.R.D. Mathias and H. Rogers), 495-53o. Lecture Notes in Maths, Vol. 337. Springer-Verlag, Berlin - Heidelberg - New York

1974 Mengenlehre in infinitären Sprachen, Universität Heidelberg (1974), 188 pp.

GÖDEL, K.

194o The consistency of the axiom of choice and the generalized continuum hypothesis, *Annals of Math. Studies 3*, Princeton, N. J. (2nd printing 1951, 74 pp.)

JENSEN, R.B. and KARP, C.

1971 Primitive recursive set functions, in: SCOTT 1971, 143 -167

KARP, C.

1964 *Languages with expressions of infinite length,* North - Holland Publ. Co., Amsterdam (1964), IX + 183 pp.

KEISLER, J.H.

1971 *Model theory for infinitary logic,* North - Holland Publ. Co., Amsterdam (1971), X + 2o8 pp.

LEVY, A.

196o Axiom schemata of strong infinity in axiomatic set theory, *Pacific J. Math.* 1o (196o), 223-238

1965 *A hierarchy of formulas in set theory,* Memoirs Am. Math. Soc. 57, Providence, R.I., 76 pp.

MOSTOWSKI, A.

1949 An undecidable arithmetical statement, *Fund. Math.* 36 (1949), 143-164

REINHARDT, W.N.

197o Ackermann's set theory equals ZF, *Annals of Math. Logic* 2 (197o), 189-249

SCOTT, D.

1971 (editor) *Axiomatic set theory.* Proc. of Symposia in Pure Mathematics 13, Providence, R.I.

SCOTT, D. and MYHILL, J.

1971 Ordinal definability, in SCOTT 1971, 271-278

TARSKI, A., MOSTOWSKI, A. and ROBINSON, R.M.

1953 *Undecidable theories.* North - Holland Publ. Co., Amsterdam (1953), 98 pp.

SUR LA MÉTHODE EN HISTOIRE DE LA LOGIQUE

C. Imbert (Paris)

Jusqu'à une date récente, l'histoire de la logique s'est donné pour objet la description des formes logiques et pour principe la logistique, selon un programme défini par H. Scholz : "La logique formelle est la seule logique qui, sous la forme de la logistique, soit aujourd'hui assez élaborée pour servir de point d'appui à un examen rétrospectif de l'histoire de la logique, quelles que soient les limitations imposées"[(1)]. Le précepte fut remarquablement servi par J. Lukasiewicz et I.M. Bochenski, pour ne citer que les plus illustres. De fait, l'école logistique a imposé une méthode issue d'une lecture partielle, et du fait de cette limitation partiale, de travaux de G. Frege. H. Scholz semble tenir compte des seuls premiers paragraphes de la Begriffsschrift et de la description rudimentaire de la syntaxe logique qu'on en peut tirer[(2)]. Or la logique de Frege tient sa nouveauté et sa puissance, voire ses mécomptes, des hypothèses sémantiques successivement adoptées par le logicien d'Iéna. On se propose de montrer qu'il serait souhaitable de reconsidérer l'histoire de la logique en privilègiant les déterminations sémantiques des systèmes logiques. Leur rôle fut toujours déterminant bien qu'ils soient, c'est trop évident, moins aisément assignables que ne l'est la forme logique.

Pour respecter les limites d'une communication, on s'en tiendra aux points suivants :

I Un bilan des résultats obtenus par la méthode logiciste, peut être établi sur le cas qui lui semble le plus favorable : la "restitution" de la dialectique stoïcienne et son opposition à l'Analytique aristotélicienne.

II On indiquera ensuite quelques présupposés de la méthode logiciste.

III Après avoir résumé la leçon des trois systèmes logiques successivement adoptés par G. Frege, on empruntera à cette oeuvre, aporétique autant que novatrice, les raisons du déplacement des questions sémantiques léguées par la tradition grecque.

IV La connaissance des clivages sémantiques pourrait offrir un nouveau principe pour une division raisonnée de l'histoire de la logique, alors capable de prendre en compte des doctrines jusqu'à présent dépréciées (telles par exemple les recherches logiques du XVIIè siècle).

I

Soit, pour pierre d'épreuve, la restitution moderne de la dialectique stoïcienne. L'histoire en est bien connue. Après une longue période d'oubli, puis de mépris (auXIXè siècle), les textes stoïciens ont été reconsidérés après que leur intérêt eut été mis en évidence par un article de J. Lukasiewicz, daté de 1934. Ce regain d'attention est lié à des circonstances très particulières. D'une part le logicien polonais avait su distinguer dans la <u>Begriffschrift</u> une partie propre correspondant au calcul des propositions (3). D'autre part, J. Lukasiewicz, parallèlement à une analyse philosophique du déterminisme élaborait alors une logique à plusieurs valeurs de vérité (4). Soumise aux deux critères de l'adhésion au principe de bivalence et de l'affinité avec le système logistique le plus simple (lecalcul des propositions), la dialectique stoïcienne put être opposée à l'Analytique aristotélicienne. En premier lieu, les tropes stoïciens pouvaient être directement confrontées aux règles de la logistique ; en second lieu, Aristote n'avait fait qu'un usage tacite et limité du principe de bivalence, dès lors qu'il avait voulu inclure dans l'Analytique une théorie du syllogisme modal (5). J. Lukasiewicz a repris, trois fois pour le moins (6), l'étude de la dialectique stoïcienne ajoutant à

1 - La thèse de bivalence

les caractéristiques suivantes :

2 - Les ordinaux dont usent les Stoïciens (le premier, le second etc..) sont des variables propositionnelles tandis que les variables aristotéliciennes représentent des termes.

3 - Les stoïciens ont fait usage de connecteurs binaires dont la signification est celle des fonctions de vérité des modernes

4 - Un syllogisme aristotélicien est une thèse, un syllogisme stoïcien est une règle d'inférence.

Ces résultats ont été confirmés par H. Scholz et I. M. Bochenski. Les travaux de Benson Mates (<u>Stoic Logic</u>, 1953) et de W. et M. Kneale (<u>The Development of Logic,</u> 1962) ont apporté les précisions suivantes :

5 - Les Stoïciens connaissaient le théorème de la déduction (voir B. Mates, opus cit V, 3 : <u>The principle of conditionalization</u>)

6 - Plutôt que le calcul des propositions, les règles d'inférence stoïciennes anticipent la déduction naturelle, entendue au sens de Gentzen (W. and M. Kneale et O. Becker : <u>Zwei Untersuchungen zur Antiken Logik</u>).

7 - Il y a entre la dialectique du Portique et l'Analytique un rapport qui

mérite d'être comparé à celui qui vaut entre le calcul des propositions et ce que l'on appelle communément "calcul des prédicats".

8 - Enfin, on peut comparer la sémantique stoïcienne à celle de G. Frege (par quoi on entend celle de l'article Sens et Dénotation).

L'interprétation moderne était achevée ; elle semblait même confirmée par l'étendue et la richesse du parallèle. Toutefois, une confrontation avec les textes originaux montre que chacune de ces thèses est contestable. De cette critique on ne peut donner ici qu'une ébauche soutenue par quelques contre-exemples. Posons en principe que, loin d'opposer des arguments philologiques disjoints à telle ou telle des interprétations modernes, on voudrait confronter le système grec dont Aristote et les Stoïciens ont donné deux variantes, à la conception moderne de la logique. On montrera que l'interprétation inspirée de la "logistique" est victime d'une double erreur quand elle inscrit les systèmes grecs dans quelque partie des systèmes modernes et quand elle oppose entre elles les syllogistiques grecques sur des points où elle ne furent pas en conflit. Seule la considération exclusive de "formes" mal comprises en est responsable.

Soit d'abord les trois premières thèses qui concernent la bivalence, la nature des variables et celle des connecteurs. On leur opposera la définition Stoïcienne de la dialectique. "Elle est la science de ce qui est vrai, de ce qui est faux et de ce qui n'est ni l'un ni l'autre. Elle a trait à ce qui signifie et à ce qui est signifié" (7). La formule, inspirée de Chrysippe, indique excellemment trois traits déterminants de la discipline dialectique. En premier lieu elle est une critique (8) (on en précisera plus bas l'intention), et les Stoïciens ont voulu donner les critères d'une orthologie ; en second lieu elle est l'art de discerner, dans le continuum donné de la langue naturelle, ce qui peut être asserté ou nié de plein droit et ce qui est soumis à une dépendance contextuelle ; en troisième lieu le signifiant y est examiné eu égard à ses valeurs sémantiques. La définition apporte donc une restriction à la bivalence, qui n'est pas attachée de plein droit aux unités apparentes du discours (les énoncés ou lekta). Une proposition susceptible d'être vraie ou fausse est ou bien celle qui comporte un élément deictique (adverbe, forme verbale déterminée, particule démonstrative) ou l'ensemble d'une proposition antécédente et d'une proposition conséquente reliées par un anaphorique. Il est donc exclu d'associer au conditionnel "si... alors" une fonction de vérité, et tout aussi impossible de voir dans les ordinaux dont usaient les Stoïciens des variables propositionnelles. Ce n'est pas le lieu de discuter la notion de variable qui était assurément étrangère à l'horizon conceptuel des Grecs (9) ; il suffit ici d'apporter des contre-exemples. Si la définition des connecteurs était celle des fonctions de vérité, Chrysippe n'aurait pas refusé l'équivalence entre l'implication et la négation d'une conjonction comme il est rapporté dans un célèbre argument de De Fato. Quant aux expressions le premier, le second, elles renvoient au contexte ; elles en sont la citation abrégée. Or ce contexte introduit toujours un rapport de dépendance quant au sens et aux conditions de vérité, entre les substitués.

Ex : Si Platon marche, Platon bouge.

Il est en outre précisé qu'un conditionnel où ce rapport manquerait, qu'un conditionnel composé de propositions réellement indépendantes, est faux. Dioclès cite le suivant :

S'il fait jour, Dion se promène (VII, 73)

Ces exemples suffisent à invalider l'hypothèse d'extensionalité qui soutient les trois premières thèses de notre liste. Ils ruinent également toute tentative visant à identifier la sémantique stoïcienne et celle de Frege, soit la thèse 8 de la liste. L'article Sens et Dénotation, auquel Mates emprunte, relève du deuxième système frégéen (voir infra III). L'auteur met en jeu trois facteurs : le signe, le sens, et la dénotation alors que la dialectique stoïcienne n'en considéra jamais que deux : les signifiants et les signifiés. En outre, l'intention de G. Frege n'y est pas de capturer la sémantique de l'usage "naturel" du discours, mais bien au contraire de l'hypothèquer afin de soutenir par un argument de vraisemblance la théorie, toute nouvelle, des fonctions et valeurs de vérité [10].

Il reste à considérer les thèses 4 à 7. Elles empruntent à la syntaxe des systèmes modernes plusieurs notions qui, employées d'ailleurs d'une manière aléatoire, n'en ont pas moins pour intention de projeter les "formes" grecques dans les systèmes modernes et d'utiliser le spectre ainsi obtenu pour opposer entre elles (mais toujours par référence à l'interprétation moderne) l'Analytique et la dialectique stoïcienne. L'école logistique a ainsi obtenu un parallèle par effets contrastés assurément brillant, mais aussi sûrement artificiel. On a contesté que le syllogisme aristotélicien soit une thèse [11]. En outre, il semble évident que la formule :

Si A se dit de tout ce dont B se dit
et si B se dit de tout ce dont Γ se dit
alors A se dit de tout ce dont Γ se dit

ne peut être entendue comme un segment homogène de la langue objet. De manière générale, le souci de reconstituer les fragments de la logique grecque avec les critères et l'appareil de la logique post-frégéenne ont établi une preuve de consistance relative. Mais les logisticiens voulaient en outre démontrer l'identité à soi-même de la logique formelle, d'Aristote à Leibniz, et de Leibniz aux Principia russelliens. Bref à sauver du naufrage de l'histoire une logica perennis, capable d'ailleurs de protéger la substance, plus précieuse encore, d'une philosophia perennis [12] comme "science rigoureuse". Si on laisse de côté ce propos trop évidemment daté, il demeure que le programme logisticien s'est divisé par le fait en deux tentatives de succès inégal. La première fut de confirmer les syllogistiques grecques en en donnant un modèle satisfaisant aux critères contemporains, Elle fut conduite avec toute la rigueur que l'on peut demander au propos de formaliser une doctrine "naïve". La seconde tentative, moins heureuse, fut d'utiliser l'image des logiques grecques dans la systèmatique moderne (et on peut en concevoir plusieurs) pour juger de leur nature et de leurs rapports. Ce faisant, on les traitait comme des secteurs disjoints de la doctrine moderne sinon comme une anticipation de celle-ci. Ici les résultats - et particulièrement les thèses 4 à 7 - sont contradictoires, et ils sont arbitraires puisqu'il n'existe aucun moyen de les contrôler : ce qui est thèse pour l'un est règle pour l'autre. Et comment donner un sens au théorème de la déduction dans un système qui répugne à la distinction entre théorie syntaxique et théorie sémantique de la déduction ? Sans préjuger si les Grecs avaient ou non notre concept de déduction : on sait qu'Aristote identifie toujours le lien syllogistique à une causalité. L'incertitude des résultats est celle de la méthode.

Il est cependant possible de tracer la différence qui prévaut entre les systèmes aristotélicien et stoïcien. Elle apparaît fort clairement dès que l'on veut bien connaître le problème auquel l'une et l'autre apportent une solution. Deux conditions déterminèrent la naissance et le cours des recherches logiques des Grecs. La première fut l'usage des livres Sur la Nature pour l'enseignement de la physique et comme véhicule du savoir entre la Grèce d'outre-mer (qui fut son point d'origine) et l'Attique. De là vient la nécessité d'une méthode capable de valider une connaissance non directement portée par la perception. C'est l'origine de "l'examen des énoncés" que Platon lia ouvertement à la parution du livre d'Anaxagore [13]. La seconde condition déterminante fut que les procédures d'analyse ou de lecture [14] ont eu pour but non seulement d'identifier les unités de langue pertinentes et leur composition, mais encore de situer dans l'inscription résiduelle des perceptions (ou mémoire) le paradigme de contrôle des énoncés. Ces analyses ont pris diverses formes, dont les plus notoires sont l'aristotélicienne et la stoïcienne. Leur stratégie de lecture s'oppose principalement sous les trois chefs :

a) de la nature des unités linguistiques jugées pertinentes

b) de leur mode d'association

c) de la nature et de l'organisation des données empiriques auxquelles sont rapportées les unités distinguées par la lecture

Le tableau ci-dessous rend brièvement compte des choix respectifs et dessine entre les deux logiques un rapport d'enveloppement, nullement d'opposition :

Lycée	le terme	prédication	sensibles abstraits distribués dans l'intellect en dix séries de catégorèmes.
Portique	l'énoncé, qui peut être incomplet (et limité à un terme prédicatif)	prédication, complémentation (directe ou indirecte) subordination	prolepses déposés dans la mémoire par l'expérience ou composées à partir d'elle

L'analyse du Portique est donc armée pour traiter (lire) toutes les inférences validées par la réduction aux termes comme le prescrit la méthode d'Aristote; mais aussi quelques autres, dont les tournures de subordination, qui échappaient à la technique des Analytiques. Car les figures de l'Analytique, réductibles aux modes parfaits de la première, sont à leur tour lues par les deux premiers tropes anapodictiques stoïciens. Mais si ces deux analyses méritent de plein droit le nom d'exercice sémantique, elles ne visent nullement à confronter directement un texte et une donnée physique, perçue ou idéale. La première intention a été invalidée dès que fut établie, dans le Cratyle, la preuve de l'arbitraire (ou plutôt de la fragilité) du signifiant : la seconde fut invalidée par l'échec du Sophiste (15), qui supposait une organisation physique strictement isomorphe à la syntaxe. Les tropes ou modes syllogistiques transmis par la tradition ne sont donc pas des formes immanentes aux unités de langage et dont on a vraiment cherché la trace dans les textes aristotéliciens ou stoïciens ; ce sont des protocoles de lecture indiquant comment une suite argumentative doit être validée c'est à dire confrontée avec les traces mêmes que la nature a déposées dans le sens, dans l'imagination, enfin sous une forme resserée et ordonnée dans la mémoire.

Ce recours à la mémoire n'introduit cependant aucune instance psychologique, en aucun des sens modernes où on pourrait l'entendre. Car le réalisme de la mémoire ne fut que la reprise à peine masquée, mais en des termes naturalistes, des méthodes en usage dans les écoles athéniennes et alexandrines. D'une part, l'analyse du texte soumis à l'examen (qu'il s'agisse d'un théorème de géométrie, de la conclusion d'un syllogisme physique ou d'un jugement de téléologie morale) est régie par une série protocolaire de questions qui déploient l'information ressérée dans l'ordonnance du discours et qui en est exigible. Ces questions sollicitent de l'énoncé l'une ou l'autre des déterminations catégoriales (qu'il s'agisse des dix catégories aristotéliciennes ou des quatre stoïciennes), assurant par là un couplage entre le texte et un système de traits déterminants, non soumis à l'ordre linéaire du discours et empruntés aux réalités naturelles qu'ils schématisent - Modèle réduit en quelque sorte, intermédiaire entre le foisonnement de l'empirie et l'économie hétérogène du discours. Dans les deux systèmes que nous confrontons, le modèle a pris la forme soit d'une hiérarchie de termes (Lycée) soit de séries rayonnantes de prolepses à partir d'une détermination donnée (Portique). Le diagramme de la mémoire est donc l'apriori sémantique responsable du choix des tropes syllogistiques. D'autre part, cette organisation mémorielle réplique les diagrammes scolaires : tableaux de définition, arbres de classification naturelle, analyse lexicale, bref l'une ou l'autre des formules adoptées pour la rédaction des recueils d'éléments dont on peut voir la forme primitive dans les lexiques de termes homériques et la forme élaborée dans les livres d'enseignement alexandrins - dont les Eléments euclidiens [15 bis]. Leur intention commune fut de fixer les liens d'association privilégiée entre deux concepts, qu'elle soit d'essence ou d'usage. Remarquons que ces diagrammes ne sont pas des modèles extensionnels. L'assimilation des répertoires d'analyse à la mémoire résume en fait deux thèses épistémologiques. L'une est l'antériorité culturelle des schématisations figuratives par rapport aux déductions linguistiques que les premières doivent valider. L'autre est d'attribuer à l'action secrète de la nature cette antériorité.

Si l'on néglige cette dernière thèse, constitutive de la philosophie grecque, et le fait historique des bibliothèques athéniennes, on retiendra la solution singulière proposée aux apories de la lecture. Ici la longue expérience qu'avaient les Grecs des arts représentatifs a infléchi l'analyse logique en lui donnant toutes les ressources de l'exegèse picturale [16] En particulier on savait à la fois jouir d'une image et profiter de son instruction sans cependant que le modèle en soit donné, ni même puisse jamais en être donné [17]; l'art dominant a prêté à l'autre. Mais dans les deux cas il fut admis qu'il s'agissait de reconnaître l'archétype de la nature dans la perspective picturale ou dans le profil de l'énoncé. Aussi la logique grecque a légué à la tradition en même temps qu'une méthode de lecture un poids égal de préjugés. Dont l'identification du concept et de la représentation, une impossibilité de principe à reconnaître tant l'autonomie de la syntaxe que les limitations liées au caractère linéaire du langage. Ces deux préjugés, dont le premier fut souvent dénoncé, n'ont jamais pu être évités tant que la langue prise en compte par les logiciens fut une langue naturelle et que l'instance sémantique fut de fait entièrement ou partiellement confondue avec l'acte de perception (sensible ou intellectuelle) ; on leur doit les trois thèses suivantes qui ont affecté les premiers systèmes de logique symbolique. Frege fut le premier à les énoncer clairement et pour deux d'entre-elles à s'en libérer. Mais ce fut au prix d'une critique dirigée contre lui-même et poursuivie pendant plus de quarante ans.

a) - La première est qu'une langue, artificielle sinon naturelle, peut cumuler les propriétés d'un calcul et d'une caractéristique universelle, unir les deux finalités d'une langue descriptive et d'une logique déductive. Dans cette hypothèse, deux langues scientifiques données ne sauraient être ni arbitraires ni hétérogènes. C'était là reproduire à l'intérieur des systèmes formels l'idéal grec d'une histoire naturelle qui aurait l'envergure d'une Encyclopédie. En outre, la même thèse demande que les fonctions descriptives soient de même nature que les fonctions de décision, qu'un caractère soit en même temps un critère [18].

b) - La deuxième thèse est la conséquence directe du naturalisme. Elle identifie l'analyse logique, assignable et inscrite dans un ensemble fini de règles, avec le processus obscur, mais naturel et supposé infrangible, de la connaissance. Elle conduit à assimiler des concepts à certaines unités syntaxiques (prédicats ou fonctions). Elle est encore la cause de la répugnance qu'ont montrée les logiciens à distinguer les différents niveaux de structuration logique dissimulés dans l'usage spontané, ou stylistiquement étendu, de la langue naturelle. Il s'agit moins ici des logiques non-aristotéliciennes, au sens de J. Lukasiewicz (Logiques à plusieurs valeurs de vérité), que des systèmes relativement indépendants constitués par le calcul des propositions, la théorie de la quantification, la logique des prédicats avec égalité etc...

c) - La troisième fut d'importer dans les langues symboliques la propriété d'évidence ou de transparence attachée, comme un impératif biologique, à la langue vernaculaire. Le langage, d'abord crédité des mêmes vertus représentatives que la perception, reçut ensuite la clarté à titre de qualité rhétorique et de caractère normatif. L'analyse intentionnelle du discours interdit, parce qu'elle en est l'alternative, une caractérisation des énoncés eu égard à des critères syntaxiques et sémantiques dénombrés dans la métalangue et relatifs à elle seule. La métalangue ne fut que tardivement substituée au point de vue absolu du lecteur grec. Il ne servirait à rien d'objecter que les scolastiques opposaient la suppositio materialis à la suppositio formalis. La doctrine ne fait que repéter la distinction stoïcienne des signifiants et des signifiés ; elle n'a pas d'autre rôle qu'une médecine contre les sophismes (Ex : mus edit caseum, mus est syllaba, ergo syllaba edit caseum). Il ne servirait pas non plus de dire que l'Arsgrammatica des Alexandrins, que déjà même les Analytiques premiers ou encore le résumé de Dioclès de Magnésie sont, de fait, parcourus de propositions énoncés en métalangue : ils n'ont pas été compris comme tels. Leur titre général d'Eisagogé (Introduction) indique leur fonction, qui est de mettre le disciple dans la position d'un lecteur absolu, intrônisé par la nature même des données épistémologiques. A l'inverse la distinction d'une langue objet et d'une métalangue est le signe certain d'une ruine du naturalisme logique.

Ces trois thèses, qui jouent encore dans les deux premières idéographies frégéennes ont été discutées par le logicien allemand dans des textes demeurés inédits de son vivant, et abandonnées pour l'essentiel dans les Recherches logiques. On notera qu'elles ont toutes une implication sémantique. Ce fait explique que, malgré les recherches des mathématiciens au XVIIè et XVIIIè siècle, et par le fait d'une réflexion mal centrée sur la forme, la logique n'ait pas éprouvé de réelles modifications avant Frege (19).

II

La méthode des logisticiens a donc peu de prise sur les raisons qui ont déterminé la logique grecque. On peut estimer que l'inventaire des traités grecs et scolastiques, l'examen de leur consistance et la distribution des mérites et des blâmes était sa fin, et qu'elle l'a atteinte. Mais on s'inquiètera plutôt de savoir si elle pouvait viser à autre chose, et si le catalogue des résultats n'était pas fermé d'avance, par les limites mêmes du point de vue adopté.

Deux motifs ont conduit H. Scholz au programme de l'Esquisse. Le premier fut la critique de la logique transcendantale kantienne ; le second fut d'ériger en méthode quelques thèmes de la Begriffsschrift. Or ces deux motifs n'en font qu'un. H. Scholz reconnaît que Kant fut le premier à exposer le concept de logique formelle et à reconnaître son empire, puisqu'il tenta d'établir entre celle-ci et la logique dite transcendantale "un lien très problèmatique" qui "ne résiste pas à un examen rigoureux". Mais cette doctrine transcendantale est inutile dès lors que la logistique offre "une logique formelle parfaite" suffisant à une construction déductive de la mathématique moderne (p. 36 et 99). Quant à la juste notion de la forme, elle fut exposée dans l'opuscule de Frege.

Or les conséquences de l'argument en contredisent les prémisses. H. Scholz inscrit en effet l'histoire de la logique dans la perspective ménagée par Kant : elle n'est que le développement sans révolution du "noyau aristotélicien". Et cet asservissement au "fil conducteur" kantien se double d'une cécité quant au véritable apport de la Begriffsschrift que Frege n'a pas manqué de signaler dans sa préface. Ce fut la construction de la fonction successeur, obtenue grâce à la quantification (entre autres), donc grâce à une altération fondamentale de l'analyse aristotélicienne de la proposition. H. Scholz a manqué l'invention ici pertinente, et sa notion de "forme parfaite" ne suffit pas à porter les conséquences qu'il en espère. Il suffit de citer l'Esquisse : "l'exemple le plus simple d'une telle forme est, dans la symbolique élémentaire d'Aristote, l'expression "Tous les S sont des P". Est-il besoin de rappeler que la Begriffsschrift est dans le détail de son idéographie, une révision polémique de cette expression. Si jamais on en doutait Frege s'en est expliqué dans sa réponse au compte-rendu de Schröder [20]. Scholz commet la même erreur que l'auteur des Vorlesungen über die Algebra der Logik.

Si l'on prend en considération le commentaire que Frege lui-même fit de la Begriffsschrift et le développement qu'il lui donna, il conviendrait de repartir les mises à l'inverse. Car Frege fut directement guidé par Kant dans sa réflexion sur la quantification et la construction des concepts [21]. En outre, on aurait tort de confondre une logique formelle et la langue formulaire que conçut Frege. Aussi bien a-t-il pris position contre le formalisme tout au long de sa carrière [22]. Sans aucun doute son analyse logique a servi les formalismes modernes, mais dans la mesure où désormais une logique qui s'en tiendrait au seul volet de la syntaxe serait jugée incomplète.

III

Il reste à établir comment l'oeuvre de Frege, si elle n'offre pas une notion univoque et canonique de la "forme" peut donner les moyens de reconnaître les clivages diachroniques pertinents. Le logicien d'Iéna a proposé successivement trois systèmes, et chacun d'entre eux est la correction d'une version antérieure. Cette remise en question, qui pour les deux derniers systèmes est une autocritique, ne fut jamais subordonnée à une recherche de la forme logique. L'opposition de la forme et du contenu, vestige de l'aristotélicisme, n'a pas survécu à la première idéographie où elle jouait d'ailleurs un rôle inverse à celui que Scholz suppose. Frege mit en effet l'accent sur la représentation des "contenus", et prenant parti sur la représentation des rapports logiques proposés par Leibniz et remaniée récemment par Boole, St. Jevons, E. Schröder et d'autres, il constate : "ce qui manque, c'est le contenu" (23). Encore faut-il prendre garde à un déplacement des termes de l'opposition, puisque dans les années 80 Frege entend par <u>forme</u> logique les rapports de déduction inscrits dans la dimension verticale de la page d'écriture et par <u>contenu</u> ce que prend en charge l'écriture horizontale. Cette distinction arbitraire, vite abandonnée par Frege, montre combien le couple forme/contenu était inadéquat à traduire les intentions de l'idéographie. Lorsque Frege répondit au compte-rendu malveillant de Schröder, s'il accepte de distinguer le calcul de la représentations des contenus, il précise que "le calcul de la déduction est partie obligée d'une idéographie" (24). Ce qui est nier l'opposition.

Pour ne pas outrepasser les limites d'une communication, on considera les trois systèmes frégéens sous le seul chef de la <u>caractéristique</u>. On verra que la notion, d'abord entendue au sens leibnizien, à peu à peu cédé la place à un nouvel ensemble de concepts sémantiques, aujourd'hui déterminants. D'ailleurs Frege a pris soin d'en indiquer le poids de cette recherche. Pour la <u>Begriffsschrift</u>, nous venons de citer les textes. Dans l'introduction aux <u>Grundgesetze</u>, l'auteur rapporte à la distinction du sens et de la dénotation des symboles le progrès de la deuxième idéographie et voit l'origine de toutes les modifications apportées à la <u>Begriffsschrift</u> (25). On n'en constatera pas la nature sémantique. Enfin la première des <u>Recherches Logiques</u> (<u>La pensée</u>, 1918) est en son entier une réflexion sur le concept de vérité. On est donc en droit de mettre en place les trois sémantiques frégéennes (26).

1) En 1879, Frege ne se propose rien de plus que de compléter le symbolisme de l'arithmétique de manière à représenter idéographiquement les parties de la déduction jusqu'alors énoncées en langue naturelle, ou même passées sous silence. Quel qu'ait été le nombre et l'importance des inventions logiques présentées dans cet opuscule, Frege ne voulut d'abord y voir rien d'autre qu'un exercice particulièrement subtil de traduction. On y trouve représentés idéographiquement aussi bien le quaterne des propositions aristotéliciennes, que la subordination causale (§ 12) et la fonction <u>successeur</u> (§ 27). La seule norme sémantique clairement posée est celle de l'univocité, et Frege stigmatise la faiblesse de la logique

de Leibniz et de Boole. Puisque les signes + et y représentent des opérations logiques, il est impossible de les employer dans le contexte d'une preuve arithmétique où ces mêmes symboles ont leur signification usuelle. Mais si Frege a parfait le système Leibnizien en évitant qu'un même signe soit crédité de deux significations, il n'est jamais dit quelle est la nature de ces significations. Que représentent les symboles : les pensées ou les choses ? Les formules de l'arithmétique, ou les objets de l'arithmétique et les opérations dont ils sont le domaine ? Les deux cas sont envisagés, sans que Frege voie difficulté, dans l'Introduction (27). Apparemment, il s'agit plutôt de reproduire adéquatement les raisonnements arithmétiques, c'est à dire de capturer dans l'idéographie un autre langage, supposé sain. Cependant l'homogénéité sémantique recherchée entre les deux aspects de l'idéographie, la déduction et l'expression d'un "contenu conceptuel", entraîne le système vers une interprétation où les formules seraient référées non à des significations (ou à des énoncés d'une autre langue) mais à un ensemble d'objets et aux opérations définies sur ces objets. L'incohérence sémantique qui en résulte est particulièrement sensible dans l'analyse du vrai et dans celle de l'égalité, dont on montrera la modification dans les systèmes ultérieurs.

En 1879, l'introduction du signe d'égalité (§ 8) donne à Frege l'occasion de mettre en relief deux particularités de son emploi. Les signes qui figurent de part et d'autre du symbole d'égalité doivent, écrit Frege, être pris pour eux-mêmes et non pour ce qu'ils représentent. L'égalité exprime la possibilité de substituer les noms des choses mais non les choses mêmes, ce qui n'aurait aucun sens : on ne substitue pas une chose à elle-même. En outre, Frege en tire une propriété sémantique générale, en vertu de laquelle l'identité introduit un dédoublement de signification (Zwiespaltigkeit) affectant tous les symboles. Frege est ici plus rigoureux que Leibniz (28) et incline en faveur d'une interprétation extensionnelle de l'égalité : mais au prix d'une équivocité affectant tous les signes, et d'une curieuse intrusion d'une règle métalinguistique dans la langue objet.

Quant à la propriété d'être vraie, Frege la représente par le signe $\vdash$, juxtaposé au jugement et qu'il interprète comme "le prédicat commun" de tous les jugements (§ 3). Il n'a donc pas la notion de valeur de vérité bien que le conditionnel soit défini par la conjonction de ses cas de vérité (§ 5).

On voit que la Begriffsschrift a en partage quelques uns des traits par lesquels nous avons caractérisé la logique grecque. En premier elle assimile au sein d'une langue caractéristique la déduction et la description. Et si Frege a pris garde de représenter les unes et les autres par des symboles distincts ce fut afin d'assurer leur intégration dans un langage homogène, et de remédier à la division introduite au XVIIè siècle avec les langues "spécieuses". Ainsi l'égalité et la fonction successeur ont le même statut que les opérateurs logiques et les quantificateurs. En outre, Frege ne distingue aucunement les deux niveaux logiques conjoints dans l'idéographie (calcul des propositions et quantification). Bien plus, il se loue du passage aisé de l'un à l'autre, de l'implication matérielle à l'implication formelle [28]. Enfin, la description de l'égalité autorise une substitution *ad hoc* du signe, pris en tant que tel, à ce qu'il signifie, comme si c'était là un jeu linguistique normal et sans craindre les paradoxes que ne manque pas d'entraîner l'absence de distinction entre langue et métalangue.

2) La deuxième idéographie tient son originalité de l'opposition qui organise maintenant la sémantique du système, entre le sens et la dénotation des symboles. Elle est bien connue, ainsi que ses difficultés, et le paradoxe des extensions de concepts : aussi peut-on faire l'économie de son exposé. On se contentera de résumer les avantages et les inconvénients de cette innovation sémantique.

En posant que deux symboles distribués de part et d'autre du signe d'égalité doivent avoir la même dénotation malgré la différence des sens, Frege met fin à l'énigme de l'égalité et à la solution qu'il lui avait donnée dans la *Begriffsschrift* ("division de la signification," cf. (1) ci-dessus). En outre, puisque tout symbole ou ensemble bien construit de symboles a un sens et une dénotation, on dira que la proposition a pour dénotation le vrai (cf. l'article *Fonction et concept*, 1891). Ces deux thèses fondent conjointement le calcul des propositions en tant qu'il étudie des fonctions particulières ayant le vrai (le faux) pour argument et (ou) pour valeur. Ainsi Frege peut-il éliminer de la deuxième idéographie le signe d'équivalence ($\equiv$). Son rôle est parfaitement tenu par le signe d'égalité, dès lors que l'équivalence entre propositions n'est rien d'autre que l'identité de leur valeur de vérité [29].

Mais la notion, nouvelle, des fonctions de vérité entraîne avec elle une difficulté majeure. Conçues dans une analogie stricte avec les fonctions appelées aujourd'hui *prédicats*, Frege est conduit à penser que leur argument (le vrai) est un objet, et que leur extension (comme les autres extensions du concept) est un objet. Ce dernier point est examiné dans l'obscur § 10 de la *Darlegung der Begriffsschrift* (*Grundgesetze*, tome I). Se demandant si une valeur de vérité peut être un "parcours de valeur" (une extension de concept), Frege répond par l'affirmative, et pose que le vrai est l'extension de la fonction : — (der Wagerechte) dont on sait qu'elle intervient dans la composition de toutes les fonctions propositionnelles [30]. Du même coup Frege introduit dans l'idéographie une fonction qui a pour nature de prendre pour argument la valeur de vérité de ce qui suit, et pour résultat paradoxal d'admettre pour argument sa propre extention. On vérifiera que cette même fonction intervient, avec sa propriété paradoxale, dans la transcription idéographique que Frege a donnée de l'antinomie russellienne (Nachwort, p. 256 sq.) Grundgesetze t. II

Le progrès et la faiblesse, aussi notoires l'un que l'autre, de la seconde idéographie sont donc l'effet d'une innovation sémantique qui n'affecte en rien la forme ; celle-ci est passée de la première à la seconde idéographie sans altération sensible [31].

On remarquera enfin une oscillation entre le point de vue syntaxique et le point de vue sémantique dans la présentation tant de l'égalité que de la quantification. Ces deux fonctions (la quantification étant dite fonction de second niveau) sont d'une part données avec les règles de remplacement et d'instanciation (Lois fondamentales III et II a). Mais, d'autre part, elles sont décrites dans leur rôle sémantique. C'est même leur justification explicite : l'égalité admet pour arguments ceux dont la dénotation est identique, tandis que la quantification s'applique à une fonction argument dont l'extension n'est pas vide [32].

En subordonnant l'idéographie à une théorie générale des fonctions, Frege menait à son terme (cf. Fonction et concept) l'analyse du quantificateur existenciel et de la construction des concepts (Aufbau der Begriffe) engagée dans les Fondements. Reprenant l'examen de la preuve ontologique et de quelques préceptes de la Logique Kantienne il avait alors montré combien le philosophe était resté soumis à la logique classique. C'est donc simultanément que Frege emprunte à la logique transcendantale et entend en libérer les règles. On voit les effets de cette indépendance récemment acquise dans la réforme que Frege impose maintenant à la langue arithmétique quand il réinterprète la formule :

$$y = a x$$

au moyen du triplet de termes : fonction, argument et valeur de la fonction pour un argument donné. Les rapports entre logique et arithmétique sont inversés, la seconde reçoit de la première les règles de son symbolisme.

3) La troisième logique frégéenne resta inachevée par la mort de l'auteur. On en connaît les trois premières Recherches logiques publiées entre 1918 et 1925 ; la première page d'une quatrième Recherche, dont on verra l'intérêt, a été publiée dans le volume d'Inédits - Quatre traits majeurs montrent les modifications apportées au premier projet de la caractéristique.

a) - La première Recherche logique ébauche une définition sémantique (au sens moderne, celui de Tarski) du vrai, Sous le couvert d'une critique du psychologisme et de l'idéalisme, Frege examine les conditions d'objectivité d'une pensée qui n'est pas dite vraie à la manière dont l'est une représentation, et dont les conditions de vérité ne sauraient être extrinsèques à son énoncé. Le pseudo prédicat"... est vrai" n'est qu'une redondance propre à la langue naturelle et dont avait été victime la première idéographie et à certains égard la seconde. Frege exclut donc la fonction :—— qui prenait pour argument sa propre extension. Enfin si le vrai est toujours désigné comme une valeur de vérité, il n'est plus conçu comme un objet.

b) - Dans les deuxième et troisième Recherches (La Négation, la composition des pensées) Frege expose une théorie équivalente à une restriction du calcul des propositions. Toute proposition composée est fonction de vérité des propositions élémentaires qui y figurent ; de telles compositions sont dites "mathématiques". Frege y associe des règles de déduction analogues à celle de Gentzen. L'équivalence, pouvant être définie à partir des compositions primitives n'est plus un signe primitif, elle est par là à nouveau distinguée de l'identité [33].

c) - La théorie de la quantification, ébauchée dans les premiers paradoxes d'une quatrième Recherche est distinguée des fonctions logiques précédentes. La méthode mise en oeuvre dans la troisième Recherche laisse supposer avec vraisemblance que Frege se proposait d'associer aux formules de quantification de nouvelles règles de déduction. Les quantificateurs n'auraient alors d'autre contenu que leurs règles d'usage ; ils cessent d'exprimer une propriété des prédicats auxquels ils sont préposés.

d) - Enfin Frege introduit une distinction qui résout bon nombre des obscurités sémantiques des idéographies précédentes en opposant le Darlegungsprache (l'allemand dans lequel est exposé le système logique) et le Hilfsprache (le système logique lui-même, au service de l'exposé mathématique). Sans lui être identique, cette distinction se rapproche de l'opposition (due à Carnap) entre langue objet et métalangue, Elle éclaire en tout cas la pénible discussion où Frege s'épuisait, quelques quinze ans plus tôt, à convaincre Kerry que "le concept cheval" n'est pas un concept mais un objet [34].

Sans vouloir forcer le parallèle, la parenté philosophique la plus claire serait ici celle qui unirait les Recherches Logiques aux écrits contemporains de Wittgenstein. On sait le rôle que Frege a joué dans la publication du Tractatus bien que le contenu lui en ait paru obscur. Faute de connaître le propos des conversations que tinrent les deux logiciens, et dont le procès-verbal semble irrémédiablement perdu, il serait sans doute fructueux de comparer la manière du Tractatus et les 17 Kernsätze zur Logik, publiées dans le Nachlass. On remarquera encore que Post, vers la même époque, établissait l'indépendance d'une partie propre des Principia correspondant au calcul des propositions et en donnait une traduction algébrique. L'extensionnalité prenait la place du logicisme.

Partant de ce résumé des trois logiques frégéennes il semble que l'on puisse conclure à une irrémédiable insuffisance des seules considérations formelles, au sens où le concept de forme fut mis à l'honneur par l'école des logisticiens et sous le patronage usurpé de Frege. Bien que la forme ait peu varié de la Begriffsschrift aux Recherches Logiques (on voudrait ajouter : aux Principia) les logiques associées à des schèmes syntaxiques pourtant comparables sont fondamentalement différentes. La forme n'offre donc qu'une caractérisation débile tant que les règles de transformation et les assignations sémantiques associées n'ont pas été soigneusement définies. Or cette définition prit un double aspect, dogmatique et critique, dont le second fut dominant. On a vu qu'il s'est agi d'analyser et de restreindre les pouvoirs spontanément attribués à la langue naturelle, ceux que les langages artificiels ont d'abord tenté de perfectionner sans rien vouloir perdre des propriétés ressenties comme évidentes. Un des résultats bien connu de l'analyse frégéenne fut d'exclure d'une logique extensionnelle nombre de moyens stylistiques dont est pourvue la langue commune (tels que la déixis, la modalité et toutes les nuances de la subordination des grammairiens : causalité, concession, hypothèse irréelle), et dont la syntaxe méritait aussi bien le titre de forme.

Confrontées au programme d'une histoire de la logique, ces remarques suffisent à invalider l'hypothèse de l'héritage aristotélicien et d'une tradition logique continue, qu'on l'entende au sens d'un progrès ou d'une interprétation rétrospective. En conviendrait-on, qu'il faut admettre le caractère singulier de la formalisation des Grecs, opérée sous le truchement d'une théorie des facultés. On a vu qu'il fallait y reconnaître le souci d'indexer la logique d'une histoire naturelle par des figures empruntées aux schémas représentatifs non extensionnels qu'avaient développé l'art pictural et les exercices dichotomiques de l'école. Des Grecs à Frege rien n'est demeuré semblable, ni les éléments syntaxiques jugés pertinents, ni les modèles auxquels ils sont confrontés ni la manière de la confrontation. Si l'on tient compte enfin des disciplines au service desquelles fut conçue l'analyse logique -

l'histoire naturelle pour les Anciens, les mathématiques pour les modernes ; on peut penser que la comparaison de ces systèmes doit prendre en principe le contre-pied du précepte de Wittgenstein : que la logique doit prendre soin d'elle-même. Car si une logique détermine elle-même ses normes, l'arbitraire de celles-ci lui échappe.

IV

Si l'on abandonne aussi bien le naturalisme kantien [35] que le programme des logisticiens, plus fidèles à Kant qu'il ne paraît [36], il reste à proposer - à titre d'hypothèses - les révolutions qui ont marqué le développement de la logique occidentale. On préférera à cette métaphore l'emprunt, méthodologiquement plus certain, du concept de modification diachronique. Aussi bien, l'histoire de la logique ne peut pas être désolidarisée de celle du langage vernaculaire, même si maintenant les prestations de service se font parfois à l'inverse. La diachronie logique affecterait non seulement la syntaxe et la sémantique, mais encore la manière de concevoir l'une et l'autre et de concevoir leurs rapports.

Dans cette hypothèse on pourrait distinguer quatre états diachroniques. L'analyse aristotélicienne résout l'énoncé en termes directement confrontés aux représentations et portant, seuls, la fonction sémantique. La syntaxe du discours n'est qu'un apparat spécieux, et on craindrait d'appeler logique une méthode qui fait si peu de cas des contraintes et des ressources liées à la linéarité du discours. D'ailleurs Aristote n'a pas manqué de condamner le nom en même temps que les rudiments de la logique platonicienne. A l'inverse l'analyse stoïcienne a préparé une logique autonome en examinant les propositions en tant que telles et selon les dépendances de la syntaxe interne et externe. Les règles de compatibilité et d'incompatibilité ont pris le pas sur la hiérarchie des termes ; le discours étant un équivalent précieux de la représentation est doté d'une fonction sémantique de plein droit ; et quand l'iconoclasme a proclamé la vanité des images, il a pris place dans les conséquences du stoïcisme. Il suppose en effet une technique achevée d'interprétation des énoncés ; et la crise signale à l'historien le moment où la cohérence du texte fut préférée à l'évidence des images. La logique scolastique des consequentiae, bien adaptée à la langue naturelle fut à son tour frappée d'archaïsme quand celle-là fut confrontée au paradigme des langues spécieuses. Les chaînes de raisons imitées de la géométrie furent préférées aux implications de la scolastique; la logique dit de Port-Royal et les essais de combinatoire leibniziens témoignent de cette substitution. Vient, enfin, un dernier état caractérisé par la juridiction de principes sémantiques extensionnels.

On ne craindra pas de soutenir le projet, involontairement paradoxal, d'une histoire diachronique qui situerait la première révolution non-aristotélicienne dans les conséquences iconoclastes du stoïcisme, la deuxième dans la substitution des combinaisons d'idées simples à la transivité de la prédication (37), et la troisième moins dans la Begriffsschrift, système transitoire, que dans l'analyse concertée de la quantification et des fonctions de vérité. En plus de l'extensionnalité, il faut retenir cette différence majeure qui oppose tous les systèmes modernes à l'aristotélisme : avec la Begriffsschrift la logique a cessé d'être une règle critique de lecture ; elle détermine jusqu'à ses règles d'écriture, prenant à cet égard "soin d'elle-même". Ce en quoi Wittgenstein à vu une déontologie philosophique.

Claude IMBERT
ECOLE NORMALE SUPERIEURE de JEUNES FILLES
PARIS

NOTES

(1) Heinrich Scholz, Esquisse d'une histoire de la logique, tr. fese p. 440. Voir aussi les deux articles de J. Lukasiewicz : Logistic and Philosophy (1936) et In defence of Logistic (1937) réunis dans Selected Works.

(2) "Nous entendons par forme en général, une expression où apparaît au moins une variable et qui est telle qu'elle se transforme en une proposition vraie ou fausse lorsque nous substituons quelque chose à cette variable, ou, en bref, lorsque nous donnons à cette variable un contenu approprié". H. Scholz, ibidem, p. 22 - Disposant d'une technique logique beaucoup plus sûre, J. Lukasiewicz ne prend pas moins pour critère d'identification des systèmes logiques le type des variables figurant dans une forme logique.

(3) "Je pense qu'il est important d'établir ce qui ne semble pas être communément reconnu, même en Allemagne, à savoir que le fondateur de la logique propositionnelle moderne est Gottlob Frege". Sur l'histoire de la logique des propositions, Selected Works, p. 198, trad. fese par J. Largeault dans Logique mathématique. Voir aussi la communication de J. Lukasiewicz et A. Tarski à la Société savante de Varsovie : Recherches sur le calcul des propositions (1930) publié dans A. Tarski Logic, Semantics, Metamathematies p. 38, trad. fese sous la direction de G. Granger.

(4) Voir les trois articles : On three-valued Logic (1920), Two-valued Logic (1921) et On Determinism (1922) dans Selected Works.

(5) Voir la Syllogistique d'Aristote dans la perspective de la logique formelle moderne paragraphes 23 et 62.

(6) En plus des deux articles cités : Two-valued Logic et sur l'histoire de la logique des propositions, voir La Syllogistique d'Aristote dans la perspective de la logique formelle moderne (1ère éd. 1951) chap. I en particulier, trad. fese par F. Zaslawsky Caujolle.

(7) Définition rapportée par Dioclès, cité par Diogène Laërce, Vies illustres, VII 62. Le résumé de Dioclès est le seul exposé cohérent que nous ayons de la logique stoïcienne. Les citations de Sextus Empiricus ne sont pas utilisables directement, elles relèvent elles-mêmes d'une autre économie logique qui est l'argumentation sceptique.

(8) Rappelons seulement que le terme kritikos désignait le lecteur et l'amateur d'art. La nature de cette activité critique fait l'objet d'un travail en cours de rédaction sur Logique et Langage dans l'ancien stoïcisme.

(9) Il est vain de parler de variables sans définir un domaine où ces variables prennent leur valeur. En outre, il conviendrait de préciser la fonction dont cette variable est l'argument. Enfin, on sait combien la notion est, pour l'analyse logique, imparfaite, tant qu'on n'y a pas joint la distinction entre variable libre et variable liée.

(10) Voir notre Introduction aux Ecrits logiques et philosophiques de G. Frege, p. 31 sq.

(11) Voir G. Granger - Le syllogisme catégorique d'Aristote, dans l'Age de la Science, oct. déc. 1970.
Voir aussi l'étude originale de Lynn E. Rose, Aristotle's Syllogistic, Springfield 1968.

(12) Voir I. M. Bochenski : La philosophie contemporaine en Europe et F. Caujolle Zaslawsky, Logique et philosophie chez Ian Lukasiewicz dans l'Age de la Science janv. mars 1970. Le lien de l'école logiciste et de la phénoménologie n'est plus à établir.

(13) Voir Phédon 99 c

(14) Le préfixe ana (analusis, anamnésis, anagnosis) indique le mouvement de référer des segments linguistiques à des entités réelles, par l'intermédiaire de leur trace mémorielle.

(15) Voir la critique d'Aristote Met A 9, M 4, N 2 et Catégories chap. 1 à 3

(15 bis) On sait que les termes euclidiens de : définition, demande, axiome relèvent du vocabulaire stoïcien. Placés en tête du traité, ils constituent la mémoire nécessaire à son intelligence.

(16) Cette subordination apparaît encore dans la formule d'Horace : Ut pictura poesis où les pouvoirs et règles du langage sont placés dans la dépendance des capacités représentatives de la peinture.

(17) Le glissement des méthodes est établi dans l'étude citée plus haut, consacrée à la logique stoïcienne.

(18) Critère de classement et critère d'existence. Le § 89 des Fondements de l'arithmétique invalide la seconde fonction en distinguant entre les Merkmale d'un concept et ses Eingenschaften. L'appendice au deuxième tour des Grundgesetze discute la première : "doit-on admettre que la loi du tiers exclu ne vaut pas pour les classes" ? (p. 254)

(19) Il peut être nécessaire de rappeler que l'adjectif formalis des scolastiques désigne le contenu, le signifié d'un énoncé. Et le formalisme leibnizien, dans la mesure où il a pour méthode la combinatoire, est l'art de lier l'inférence au contenu, en toute indépendance de la syntaxe logique.

(20) Sur le but de l'idéographie, 1882

(21) Voir Fondements de l'arithmétique §§ 59, 88, 89, et notre Introduction p. 77 sq. - Egalement Dialog mit Punjer uber Existenz, Nachlass p. 60 sq.

(22) Fondements, § 28, Grundgesetze II, p. 80 à 153, et Antwort auf die Ferienplauderei des Herrn Thomae (1906)

(23) La Science justifie le recours à une idéographie (1882), trad. fese p. 68. Voir aussi la préface à la Begriffsschrift.

(24) Sur le but de l'idéographie, trad. fese p. 71

(25) Vorwort p. IX et X : "Je peux, dans une certaine mesure être juge de la résistance que ces idées nouvelles ne manqueront pas de faire naître car il me fallut vaincre en moi-même un tel sentiment avant de m'y arrêter".

(26) Cf. notre article : Sur la sémantique de Frege, Annales de la Faculté des Lettres de Lille, 1972.

(27) Comparer Vorwort p. IX : "une méthode de déduction qui fasse abstraction de la nature particulière des choses" et s'en tienne aux lois de la pensée" à la p. XI : "une méthode de désignation qui adhère aux choses mêmes".

(28) Les signes représentent, sans que Leibniz ait perçu la difficulté, les choses et les pensées : "tout raisonnement humain s'accomplit par le moyen de signes ou de caractères. Ce ne sont pas seulement les choses elles-mêmes mais aussi les idées des choses qui ne doivent pas être observées distinctement par l'esprit, et c'est pourquoi afin d'abréger on les remplace par des signes". Voir Couturat, la Logique de Leibniz, chap. II et III.

(29) Grundgesetze I, Vorwort p. IX

(30) Voir Begriffsschrift §§ 5 et 7 et Grundgesetze 5

(31) "Les anciens signes fondamentaux qui ont été maintenus sans modification d'aspect et dont l'algorithme n'a pas non plus été véritablement altéré, ont cependant été pourvus d'autres définitions (Erklärungen) " Grundgesetze, Vorwort p. 10

(32) Fondements de l'arithmétique § 53 et Grundgesetze § 8

(33) Cf. Nachgel. Schriften p. 40, et notre introduction aux Ecrits logiques et philosophiques p. 57

(34) Concept et Objet (1892)

(35) Naturalisme limité à la logique, qui assure le lien entre l'usage spontané et l'usage transcendental des facultés. Sur l'analogie entre la logique et la grammaire générale, voir Logique, Introduction, I et J. Cavaillès Sur la logique et la théorie de la science, p. 1 et 2

(36) On comparera le programme de H. Scholz avec l'Abrégé d'une histoire de la logique dans Logique, Introduction II, trad. fese p. 20. Voir aussi le reproche de Kant, qui marque la fin d'une logique purement synchronique et dogmatique : "De façon générale, les logiciens sont historiquement ignorants" Ibidem, Introduction VI, trad. fese p. 49

(37) Cf. parmi d'autres exemples, la réduction des figures du syllogisme par les diagrammes d'Euler. Cette technique élabore une méthode proposée par Port-Royal (Logique, Livre III). Kant en a tiré les conclusions ; dans l'opuscule Sur la fausse subtilité des quatre figures dessyllogismes. La syllogistique n'est plus qu'un "colosse aux pieds d'argile".

THE MODEL THEORY OF LOCAL FIELDS

Simon Kochen

Dedicated to the memory of Abraham Robinson

1. INTRODUCTION

In 1965 Ax and Kochen [1] gave a metamathematical connection between the p-adic fields $\mathbb{Q}_p$ of algebraic number theory and the formal power series fields $\mathbb{Z}_p((t))$ of algebraic geometry. This principle states that an elementary statement about valued fields is valid in $\mathbb{Q}_p$ if and only if it is valid in $\mathbb{Z}_p((t))$, for all but a finite number of primes p. The principle was applied to solve particular Diophantine conjectures of E. Artin and Lang by transferring these problems from $\mathbb{Q}_p$ to $\mathbb{Z}_p((t))$. The principle itself was an immediate consequence of a theorem on ultraproducts, namely that for any non-principal ultrafilter D on the set of primes, $\Pi_p\mathbb{Q}_p/D \cong \Pi_p\mathbb{Z}_p((t))/D$ (assuming the continuum hypothesis).

The basic method in that paper was subsequently modified and generalized in Ax and Kochen [2] and [3] and also in Ersov [4] and [5] and Ax [6] to prove, among other results, that $\mathbb{Q}_p$ is a decidable field, and that the decidability and elementary equivalence type of $F((t))$ is determined by that of the field F (for char. F = 0). In Ax and Kochen [3] it was also shown that an elimination of quantifiers was valid for these fields using a test of A. Robinson. In [7] P. Cohen outlined such an actual procedure for eliminating quantifiers, leading again to a proof of the original metamathematical principle. A. Robinson [8] proved the principle by the methods of non-standard analysis by showing that $\mathbb{Q}_\pi$ is elementarily equivalent to $\mathbb{Z}_\pi((t))$ for π a non-standard prime number.

In these notes we prove a single theorem, the Isomorphism Theorem, which gives a unified generalization of the previous results.

At the same time, the proof is substantially simpler than the original proof. The Isomorphism Theorem leads to a classification of Hensel fields of finite ramification index e under elementary equivalence. The case $e = 0$ includes power series fields $F((t))$ of characteristic 0 as well as the generalized power series fields $F((t^G))$ of Hahn; the case $e = 1$ includes the field $\mathbb{Q}_p$ of p-adic numbers, and more generally the Witt vector field $W(\mathfrak{K})(W(\mathbb{Z}_p) = \mathbb{Q}_p)$, as well as unramified extensions such as the cyclotomic extension of $\mathbb{Q}_p$. In these cases the elementary equivalence type of the field is completely determined by the elementary type of the residue class field and the value group. We present the cases $e = 0$ and $e = 1$ (the "unramified" case) in these notes. The case of ramified fields, with $e > 1$, requires some more algebraic information and will be treated in a subsequent paper.

In our present simpler proof we have been able to dispense with much of the algebraic machinery of the original proof. This has enabled us to write this paper in a self-contained manner with the non-specialist in mind, and in particular for those without a number-theoretic background. The basic approach remains the same as in the original papers: to prove elementary equivalence results via the isomorphism of ultraproducts. We do this by selecting the salient properties of the ultraproducts (ω-pseudo-complete $\aleph_1$-Hensel fields with cross-sections) which uniquely determine the field in terms of the residue class field and value group. The organization of the proof also has common features with the non-standard proof in A. Robinson [8].

We shall assume familiarity with the definition and basic properties of ultraproducts. For background the reader may consult a standard text such as Bell and Slomson [9]. (The first few pages of Ax and Kochen [1] actually suffice.) To prove the ω-pseudo-completeness we have used the ω_1-saturation property of the ultraproduct V:

a countable set $\{\varphi_i(x)\}$ of elementary formulas is satisfiable by an element of V if every finite subset is. Those who do not wish to rely on this fact may easily prove the ω-pseudo-completeness directly or refer to the proof in Ax and Kochen [1], Lemma 9.

For number-theoretic applications the focus of these results remains the metamathematical connection between the p-adic (now, more generally, Witt vector) fields and the power series fields. We stress that this connection is not surprising but rather formalizes the natural analogy between these two classes of local fields, which stems from the similarity between the global algebraic number fields and algebraic function fields in one variable. Indeed, Hensel introduced the local fields of number theory precisely in analogy with the power series fields over $\mathbb{C}$ in order to explain and complete Kummer's work on the decomposition of rational primes in algebraic number fields. For Hensel the rational primes correspond to points z_0 in $\mathbb{C}$ considered as primes $z - z_0$ in the ring $\mathbb{C}[z]$. The decomposition of the prime p in an algebraic number field corresponds to the resolution of a point z_0 into the separate points lying above z_0 on the Riemann surface of the algebraic function field.

From this point of view the results presented in this and the subsequent paper on finitely ramified Hensel fields forms a natural completion of the earlier papers, since we now consider all points of the compact Riemann surface (with finitely many sheets) the branch points as well as the unramified points. It is possible to include a restricted class of Hensel fields of infinite ramification in our classification, those satisfying Kaplansky's Hypothesis A (see Schilling [10], Chapter 7, p. 220). The modifications necessary to include this class are routine and we leave them to the reader.

2. FINITELY RAMIFIED HENSEL FIELDS

Let F be a field and G and ordered abelian group. Denote by F^* the multiplicative group of non-zero elements of F. A <u>valuation</u> ord: $F^* \to G$ is a group epimorphism (ord (ab) = ord a + ord b) satisfying the inequality

$$\text{ord}\,(a+b) \geq \min\,(\text{ord}\,a, \text{ord}\,b)\ .$$

We follow the convention of writing ord $0 = \infty$. We shall frequently use the easily proved fact that ord $(a+b) = \min\,(\text{ord}\,a, \text{ord}\,b)$ if ord $a \neq$ ord b. The ring $O_F = \{a \mid \text{ord}\,a \geq 0\}$ is called the <u>valuation ring</u> of the valuation ord. The set $\mathcal{P} = \{a \mid \text{ord}\,a > 0\}$ forms a (unique) maximal ideal in O_F. We call the field $\bar{F} = O_F/\mathcal{P}$ the <u>residue class field</u> of F.

<u>Examples</u>. Let $\mathfrak{K}$ be a field and G an ordered abelian group.

(1) $F = \mathfrak{K}(t)$, the field of rational functions in one variable over $\mathfrak{K}$.

$$\text{ord}\,a(t) = \begin{cases} \text{order of zero of } a(t) \text{ at } 0 \\ -\text{ order of pole of } a(t) \text{ at } 0 \end{cases}$$

$\bar{F} = \mathfrak{K}$, ord $F = \mathbb{Z}$.

(2) $F = \mathfrak{K}((t))$, the field of formal power series

$$= \left\{ \sum_{i=m}^{\infty} a_i t^i \mid a_i \in \mathfrak{K},\ m \in \mathbb{Z} \right\}.$$

$$\text{ord} \sum_{i=m}^{\infty} a_i t^i = m, \text{ assuming } a_m \neq 0.$$

$\bar{F} = \mathfrak{K}$, ord $F = \mathbb{Z}$.

(3) $F = \mathfrak{K}((t^G))_\beta$, the field of generalized power series of Hahn $(\beta \geq \aleph_0)$

$$= \left\{ \sum_{\alpha\in S} a_\alpha t^\alpha \mid S \text{ well ordered} \subset G,\ \#S = \beta \right\}$$

$$\text{ord} \sum_{\alpha \in S} a_\alpha t^\alpha = \alpha_0, \text{ the smallest element of } S \text{ such that } a_{\alpha_0} \neq 0.$$

$\bar{F} = \mathfrak{K}$, ord $F = G$.

If $\#G = \beta$ we write $F = \mathfrak{K}((t^G))$.

(4) A further generalization of power series fields. A 2 co-cycle $f: G \times G \to \mathfrak{K}^*$ is a map such that

(a) $f(\alpha,\beta) = f(\beta,\alpha)$;

(b) $f(0,0) = f(0,\alpha) = f(\alpha,0)$; and

(c) $f(\alpha,\beta + \gamma)f(\beta,\gamma) = f(\alpha + \beta,\gamma)f(\alpha,\beta)$.

Let

$$F = \mathfrak{K}((t^G;f))_\beta$$
$$= \left\{ \sum_{\alpha \in S} a_\alpha t^\alpha \,|\, S \text{ well ordered} \subset G,\ \#S = \beta \right\}$$

Multiplication of power series is modified by requiring $t^\alpha t^\beta = f(\alpha,\beta)t^{\alpha+\beta}$.

$$\text{ord} \sum_{\alpha \in S} a_\alpha t^\alpha = \alpha_0, \text{ the smallest element of } S \text{ such that } a_{\alpha_0} \neq 0.$$

$\bar{F} = \mathfrak{K}$, ord $F = G$.

The relevance of these fields and the significance of the 2 co-cycle will become apparent in Section 9.

(5) $F = \mathbb{Q}$, the field of rational numbers, p a fixed rational prime. ord $a = m$ where $a = (r/s)p$, $r,s \in \mathbb{Z}$, $p \nmid rs$. $\bar{F} = \mathbb{Z}_p (= \mathbb{Z}/p)$) ord $F = \mathbb{Z}$. This is the <u>p-adic valuation</u> of $\mathbb{Q}$.

Let F be valued in $\mathbb{Z}$. A sequence $\{a_n\}$ in F is a Cauchy sequence if for every $0 < k \in \mathbb{Z}$ there exists n_0 such that for all $n > m > n_0$ $\text{ord}(a_n - a_m) > k$. This accords with notion of a Cauchy sequence in a metric space if we define a norm $|\ |$ by $|a| = r^{-\text{ord}\, a}$ for $r > 1$, $r \in \mathbb{R}$. We may now form the completion of F with respect to Cauchy sequences. The completion of $\mathfrak{K}(t)$ is then $\mathfrak{K}((t))$. The completion of $\mathbb{Q}$ with respect to the p-adic valuation is the field $\mathbb{Q}_p$ of p-adic numbers. We then have

$\mathrm{ord}\colon \mathbb{Q}_p \to \mathbb{Z}$ and $\bar{\mathbb{Q}}_p = \mathbb{Z}_p$. Every element a of $\mathbb{Q}_p$ may be uniquely written in the form $\Sigma_{i=m}^{\infty} a_i p^i$ with $0 \leq a_i < p$, and $\mathrm{ord}\, a = m$ where $a_m \neq 0$. Addition and multiplication are performed componentwise with carrying as in the decimal expansion of real numbers.

The construction of the complete field $\mathbb{Q}_p$ may be generalized. If $\mathfrak{K}$ is any field, then there exists a unique (up to isomorphism) complete field F valued in $\mathbb{Z}$ with $\bar{F} = \mathfrak{K}$. If char F = char $\mathfrak{K}$, then $F \cong \mathfrak{K}((t))$. If char $F = 0$ and char $\mathfrak{K} = p$ with $\mathfrak{K}$ a perfect field ($\mathfrak{K}^p = \mathfrak{K}$), then $F \cong W(\mathfrak{K})$, the field of Witt vectors (see Greenberg [11] Chapter 6). For $\mathfrak{K}$ not perfect Witt's construction has been generalized in Teichmüller [12] (see Schilling [10], Chapter 7). In this case also we write $W(\mathfrak{K})$ for the unique field F.

Definition. A valued field F is a Hensel field if it has the following property. Let $f(X) \in O_F[X]$ be such that the image of $f(X)$ in $\bar{F}$, $\bar{f}(X)$, has a non-singular root α in $\bar{F}$, i.e., $\bar{f}(\alpha) = 0$ and $\bar{f}'(\alpha) \neq 0$. Then $f(X)$ has a root a in O_F with $\bar{a} = \alpha$.

For power series fields this property is simply a statement of the Implicit Function Theorem for polynomials. A (weak) form of Hensel's Lemma state that complete fields are Hensel fields. To prove this we form the Newton sequence for approximating a root: let a_0 be any element of O_F with $\bar{a}_0 = \alpha$ and $a_{n+1} = a_n - f(a_n)/f'(a_n)$. We claim that $\mathrm{ord}(a_{n+1}-a_n) > n$ and $\mathrm{ord}\, f(a_n) > n$. For $n = 0$ this is clear. Assuming it for n, we have $\mathrm{ord}(a_n-a_0) > 0$, so that $\mathrm{ord}(f'(a_n)-f'(a_0)) > 0$. Hence, $\mathrm{ord}\, f'(a_n) = 0$. The Taylor expansion

$$f(a_{n+1}) = f(a_n) - \frac{f(a_n)}{f'(a_n)} f'(a_n) + \frac{f(a_n)^2}{f'(a_n)^2} \cdot b, \text{ with } b \in O_F$$

shows that $\operatorname{ord} f(a_{n+1}) > n+1$. Also $\operatorname{ord} f'(a_{n+1}) = 0$, so that $\operatorname{ord}(a_{n+2}-a_{n+1}) > n+1$, completing the induction. Let $a = \lim a_n$; then clearly $f(a) = 0$ and $\bar{a} = \alpha$.

It is not difficult to check that the generalized power series fields also form Hensel fields.

Let K be a valued field. We define the (absolute) ramification index e of K to be

$$e = \begin{cases} \#\{\gamma \mid 0 < \gamma \leq \operatorname{ord} p\} & \text{if } \operatorname{char} \bar{K} = p > 0 \\ 0 & \text{if } \operatorname{char} \bar{K} = 0 \end{cases}.$$

If $e < \infty$ we say that K is (absolutely) finitely ramified. In other words K is finitely ramified if $\{\gamma \mid 0 < \gamma \leq \operatorname{ord} n\}$ is a finite set for all $n \in Z$. Clearly this implies that K is of characteristic 0. If $e = 0$ or 1 we say that K is unramified.

Definition. A cross-section (x-section) $\pi : \operatorname{ord} K \to K^*$ of a valued field K is a group homomorphism $(\pi(\alpha+\beta) = \pi(\alpha)\pi(\beta))$ such that $\operatorname{ord} \pi(\alpha) = \alpha$ for all $\alpha \in \operatorname{ord} K$. If K has finite ramification index $e > 0$, then π is a normalized cross-section if in addition $\pi(e) = p$. If $e = 0$, then a normalized x-section is simply a x-section.

We shall often write π^α for $\pi(\alpha)$ and, if $H \subseteq \operatorname{ord} K$, π^H for $\pi(H)$.

Examples

(1) The field $K((t^G))_\beta$ has the x-section $\pi(\alpha) = t^\alpha$.

(2) The field $K((t^G;f))_\beta$ has in general no x-section. We shall study this situation in Section 9.

(3) The field Q_p and, more generally the Witt vector field $W(\mathfrak{K}_p)$, has the normalized x-section $\pi(n) = p^n$.

(4) If $\{K_i \mid i \in I\}$ is a family of valued fields with (normalized) x-sections π_i, then the ultraproduct $\Pi_{i\in I}K_i/D$ has the

(normalized) x-section π defined by $\pi^{\gamma^*} = (\pi^{\gamma(i)})$. Here * denotes the equivalence class defined by the ultrafilter D.

The rest of the paper is devoted to the study of finitely ramified Hensel fields with a x-section. Our aim is the proof of the Isomorphism Theorem of Section 6, which states that two such fields V and V′ of cardinality $\aleph_1$ which are ω-pseudo-complete are isomorphic if they have the same residue class field $\bar{V} = \bar{V}' = \mathfrak{K}$ and value group ord V = ord V′ = G. We now indicate the plan of the proof. We shall build up the isomorphism of V and V′ step by step from subfields E and E′. The steps fall into three categories:

(i) unramified extensions, where the residue class field $\bar{E}$ extends but not the value group ord E (Section 5);

(ii) totally ramified extensions, where the value group extends but not the residue class field (Section 3); and

(iii) immediate extensions, where neither $\bar{E}$ nor ord E extends (Section 4).

The unramified case is handled at one blow by showing that one can start the isomorphism with subfields E,E′ such that $\bar{E} = \bar{E}'$ is the full residue class field $\mathfrak{K}$. We take care of the totally ramified case by reducing it to the case of immediate extensions, using here the existence of a x-section. The main body of the proof lies in Section 4, which deals with the immediate extension case, using the Hensel property and the ω-pseudo-completeness of V and V′.

We now make this plan more precise. The instrument we use to handle the extension of isomorphisms is the notion of a pure map, which we now define.

Let K,K′ be two valued fields with x-sections π, π'. An <u>analytic isomorphism</u> $\varphi: K \to K'$ is a field isomorphism which induces an isomorphism of ord K onto ord K′. If ord K = ord K′ we require that the induced map be the identity. The analytic isomorphism φ is <u>x-analytic</u> if the diagram

$$\begin{array}{ccc} K & \xrightarrow{\ \varphi\ } & K' \\ \pi \uparrow & & \uparrow \pi' \\ \text{ord } K & \longrightarrow & \text{ord } K' \end{array}$$

commutes.

If E is a subfield of V we let $\tilde{E}$ denote the algebraic closure of E in V. The field E is <u>algebraically closed in</u> V if $\tilde{E} = E$. If H is a subgroup of the abelian group G, we let $\tilde{H}$ denote the divisible hull of H in G, i.e., $\tilde{H} = \{\alpha | \alpha \in G, \exists n \in Z, n\alpha \in H\}$. The subgroup H is <u>pure</u> in G if $\tilde{H} = H$. Let E be a subfield of V for which π is a x-section. Then $\text{ord } \tilde{E} = \widetilde{\text{ord}}\, E$. (We shall write $\widetilde{\text{ord}}\, E$ for $\widetilde{\text{ord } E}$.) For if $\gamma \in \widetilde{\text{ord}}\, E$, then $n\gamma \in \text{ord } E$ for some $n \in Z$, so that $\pi^{\gamma} = (\pi^{n\gamma})^{1/n} \in \tilde{E}$, and $\gamma \in \text{ord } \tilde{E}$. On the other hand, if $c \in \tilde{E}$, then $\Sigma_{i<n} a_i c^i = 0$ with $a_0, \ldots, a_n \in E$. Hence $\text{ord } \Sigma a_i c^i = \infty$, so that $\text{ord } a_i c^i = \text{ord } a_j c^j$ for some $0 \leq i \neq j \leq n$. Thus, $\text{ord } c = (\text{ord } a_i - \text{ord } a_j)/(j-i) \in \widetilde{\text{ord}}\, E$.

For the definition of a pure map and the reduction procedure of Section 3 we shall require only that V and V' are valued fields with x-sections and common value group $\mathfrak{G}$ and residue class field $\mathfrak{R}$. In Sections 4 and 5 we add the requirements that V and V' be finitely ramified ω-pseudo-complete Hensel fields with normalized x-sections.

<u>Definition</u>. A <u>pure</u> map $\varphi: E \to E'$ is a x-analytic isomorphism of subfields E, E' of V, V' with $\text{ord } E = \text{ord } E'$ a pure countable subgroup of G and $\bar{E} = \bar{E}' = \mathfrak{R}$.

We shall show that pure maps exist (Section 5, Proposition 4) and that every pure map can be extended to a pure map which includes in its domain an arbitrary new element a of V (Section 4, Proposition 3). A simple transfinite induction then yields the main theorem. Proposition 3 is proved by reducing the case where $\text{ord } E(a) \neq \text{ord } E$ (Section 3, Proposition 1) to the case where $\text{ord } E(a) = \text{ord } E$, which is handled by Proposition 2, Section 4.

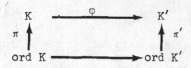

3. TOTALLY RAMIFIED EXTENSIONS

In this section we prove that if $\varphi: E \to E'$ is a pure map and $a \in V$, then there exists a pure map $\hat{\varphi}: \hat{E} \to \hat{E}'$ extending φ, with $\operatorname{ord} \hat{E}(a) = \operatorname{ord} \hat{E}$.

We now motivate our method for proving this result. To construct the field $\hat{E}$ such that $\operatorname{ord} \hat{E}(a) = \operatorname{ord} \hat{E}$, we certainly require that $\operatorname{ord} \hat{E} \supset \operatorname{ord} E(a)$. Thus, we first extend the group $\operatorname{ord} E$ to $\operatorname{ord} E(a)$ and we extend the corresponding field E via the x-section π. We do this stepwise by adding one new element α_n at a time from $\operatorname{ord} E(a) = \{\alpha_1, \alpha_2, \ldots, \alpha_n, \ldots\}$. Thus, as a first step we add π^{α_1} to E and extend the isomorphism $E \to E'$ to $E(\pi^{\alpha_1}) \to E'(\pi'^{\alpha_1})$ (Lemma 1). However to do this we need the purity of $\operatorname{ord} E$. We cannot continue directly to add α_2 since $\operatorname{ord} E(\pi^{\alpha_1})$ is not necessarily pure. We must first extend the isomorphism to the field

$$E(\pi^{\beta} \mid \beta \in \widetilde{\operatorname{ord}}\, E(\pi^{\alpha_1}))$$

the value group of which is the pure group $\widetilde{\operatorname{ord}}\, E(\pi^{\alpha_1})$. Lemma 3, which is proved via Lemma 2, achieves this.

<u>Lemma 1</u>. Let $\varphi: E \to E'$ be a pure map and $\alpha \in G - \operatorname{ord} E$. There exists a x-analytic map $\psi: E(\pi^{\alpha}) \to E'(\pi'^{\alpha})$ extending φ.

<u>Proof</u>. The element π^{α} is transcendental over E, for otherwise $n\alpha \in \operatorname{ord} E$ for some $n \in \mathbb{Z}$, so that $\alpha \in \operatorname{ord} E$, by the purity of $\operatorname{ord} E$ in G. Similarly π'^{α} is transcendental over E'. Hence $\pi^{\alpha} \mapsto \pi'^{\alpha}$ defines an algebraic isomorphism $\psi: E(\pi^{\alpha}) \to E'(\pi'^{\alpha})$. To see that ψ is analytic, let $x = \Sigma\, a_i \pi^{i\alpha} \in E[\pi^{\alpha}]$. If $\operatorname{ord} a_i \pi^{i\alpha} = \operatorname{ord} a_j \pi^{j\alpha}$ for some $i \neq j$, then $(i-j)\alpha = \operatorname{ord} a_j - \operatorname{ord} a_i \in \operatorname{ord} E$, so that $\alpha \in \operatorname{ord} E$, a contradiction. Hence $\operatorname{ord} x = \min_i(\operatorname{ord} a_i + i\alpha) = \min_i(\operatorname{ord} \varphi(a_i) + i\alpha) = \operatorname{ord} \psi(x)$, so that ψ is analytic. Also this shows that $\operatorname{ord} E(\pi^{\alpha}) = \operatorname{ord} E \oplus \mathbb{Z}\alpha$, and, for $h \in \operatorname{ord} E$, $n \in \mathbb{Z}$, $\psi(\pi^{h+n\alpha}) = \psi(\pi^{h})\psi(\pi^{\alpha})^{n} = \pi'^{h+n\alpha}$, so that ψ is x-analytic.

Lemma 2. Let $\varphi : E \to E'$ be a x-analytic isomorphism, $p\beta \in \operatorname{ord} E$, $\beta \in G - \operatorname{ord} E$, with p a rational prime. There exists a x-analytic isomorphism $\psi : E(\pi^{\beta}) \to E'(\pi'^{\beta})$ extending φ.

Proof. We claim that the polynomial $X^p - \pi^{p\beta}$ is irreducible over E. If not, then for some $\Sigma a_i X^i \in E[X]$ of positive degree less than p, $\Sigma a_i \pi^{i\beta} = 0$, so that $\operatorname{ord} a_i \pi^{i\beta} = \operatorname{ord} a_j \pi^{j\beta}$ for some $i \neq j$. It follows that $(i-j)\beta = \operatorname{ord} a_j - \operatorname{ord} a_i \in \operatorname{ord} E$. Since $(i-j,p) = 1$ we can find $\ell, m \in Z$ such that $\ell(i-j) + mp = 1$. Hence, $\beta = \ell(i-j)\beta + mp\beta \in \operatorname{ord} E$, a contradiction. Since $X^p - \pi^{p\beta}$ is irreducible, the map $\pi^{\beta} \mapsto \pi'^{\beta}$ defines an algebraic isomorphism $\psi : E(\pi^{\beta}) \to E'(\pi'^{\beta})$. Now let $x = \Sigma_{i<p} a_i \pi^{i\beta}$ be in $E(\pi^{\beta})$. If $\operatorname{ord} a_i \pi^{i\beta} = \operatorname{ord} a_j \pi^{j\beta}$ for some $i \neq j$, then as before $\beta \in \operatorname{ord} E$, a contradiction. Thus $\operatorname{ord} x = \min_i(\operatorname{ord} a_i + i\beta) = \min_i(\operatorname{ord} \varphi(a_i) + i\beta) = \operatorname{ord} \psi(x)$. Finally, for $h \in \operatorname{ord} E$, $i < p$, $\psi(\pi^{h+i\beta}) = \psi(\pi^h)\psi(\pi^{\beta})^i = (\pi')^{h+i\beta}$. Hence ψ is a x-analytic isomorphism.

Lemma 3. Let $\varphi : E \to E'$ be a pure map and $\alpha \in G - \operatorname{ord} E$. Let $H = \operatorname{ord} E(\pi^{\alpha})$, $H' = \operatorname{ord} E'(\pi'^{\alpha})$. There exists a pure map $\psi : E(\pi^{\tilde{H}}) \to E'(\pi'^{\tilde{H}'})$ extending φ.

Proof. We can write $\tilde{H} = \bigcup_{n=1}^{\infty} H_n$, where $H_1 = H$, and H_{n+1} is the subgroup of G generated by H_n and an element β_n in $G - H_n$ with $p\beta_n \in H_n$, for some prime p. Let $E_1 = E(\pi^{\alpha})$ and $E_{n+1} = E_n(\pi^{\beta_n})$. Then clearly $\operatorname{ord} E_n = H_n$, $E(\pi^{\tilde{H}}) = \bigcup_{n=1}^{\infty} E_n$, and $\operatorname{ord} E(\pi^{\tilde{H}}) = \tilde{H}$. Define E_n' in a similar fashion. By Lemma 1, φ extends to a x-analytic isomorphism $\varphi_1 : E_1 \to E_1'$. By Lemma 2, every x-analytic $\varphi_n : E_n \to E_n'$ extends to a x-analytic $\varphi_{n+1} : E_{n+1} \to E_{n+1}'$. Then $\psi = \bigcup_{n=1}^{\infty} \varphi_n$ is the required pure map

Proposition 1. Let $\varphi : E \to E'$ be a pure map and $a \in V$. There exists a pure map $\hat{\varphi} : \hat{E} \to \hat{E}'$ extending φ with $\operatorname{ord} \hat{E}(a) = \operatorname{ord} \hat{E}$.

Proof. First note that ord $E(a)$ is countable. For if ord $E(a) \neq$ ord E, then ord $c \notin$ ord E for some $c \in E(a)$. Clearly, ord $E(c)$ is the (countable) group generated by ord E and ord c, viz., $\{\gamma + n \text{ ord } c \mid \gamma \in \text{ord } E, n \in \mathbb{Z}\}$. Since every element of the field $E(a)$ is algebraic over $E(c)$, it follows that ord $E(a) \subset \widetilde{\text{ord}}\, E(c)$, which is countable.

Order ord $E(a)$ in a sequence $\{\alpha_1, \alpha_2, \ldots\}$. Let

$$H_0 = \text{ord } E \qquad E_0 = E$$

$$H_i = \widetilde{\text{ord}}\, E_{i-1}(\pi^{\alpha_i}) \qquad E_i = E(\pi^{H_i})$$

$$H_\omega = \bigcup_{i=0}^{\infty} H_i \qquad E_\omega = \bigcup_{i=0}^{\infty} E_i .$$

Clearly $H_\omega = \widetilde{\text{ord}}\, E(a)$, which is pure and countable and $E_\omega = E(\pi^{H_\omega})$. Define E'_ω and H'_ω similarly. Then an induction using Lemma 3 shows immediately that there is a pure map $\varphi_\omega : E_\omega \to E'_\omega$ extending φ.

At this point we have that ord $E_\omega = \widetilde{\text{ord}}\, E(a)$. However, it may still happen that ord $E_\omega(a) \neq$ ord E_ω so we continue the above procedure to enforce the equality. Let

$$\mathcal{H}_0 = H_\omega \qquad \mathcal{E}_0 = E_\omega$$

$$\mathcal{H}_i = \widetilde{\text{ord}}\, \mathcal{E}_{i-1}(a) \qquad \mathcal{E}_i = E(\pi^{\mathcal{H}_i})$$

$$\hat{H} = \bigcup_{i=0}^{\infty} \mathcal{H}_i \qquad \hat{E} = \bigcup_{i=0}^{\infty} \mathcal{E}_i .$$

Clearly ord $\hat{E} = \hat{H}$, which is pure and countable and $\hat{E} = E(\pi^{\hat{H}})$. If $\gamma \in$ ord $\hat{E}(a)$, then for some i, $\gamma \in \mathcal{H}_i$, so that $\gamma \in$ ord $\mathcal{E}_i \subset$ ord $\hat{E}$. Thus ord $\hat{E}(a) =$ ord $\hat{E}$. Defining $\hat{E}'$ and $\hat{H}'$ similarly, we have by induction the existence of a pure map $\hat{\varphi} : \hat{E} \to \hat{E}'$ extending φ.

4. IMMEDIATE EXTENSIONS

Let $\varphi: E \to E'$ be a pure map and $a \in V$ such that $\underset{\cdot}{E}(a)$ is an immediate extension of E. Our aim in this section is to show that we can extend φ to a pure map which includes a in its domain. To achieve this we shall use the machinery of pseudo-convergent sequences. These sequences were introduced in Ostrowski [13] specifically to study immediate extensions.

We begin our discussion with a short account of the original motivation for considering pseudo-convergent sequences. For a field K valued in $\mathbb{Z}$ the Cauchy completeness of K is equivalent to the maximal completeness of K, i.e., the non-existence of proper immediate extensions. For a field valued in an arbitrary group the two notions of completeness no longer coincide. In the case studied by Ostrowski the field K is $\mathbb{C}((t^{1/n} \mid n = 2,3,4,\ldots))$, the algebraic closure of $C((t))$, which is valued in Q. The Cauchy completeness $\check{K}$ of K consists of the field of power series $\Sigma_{i=0}^{\infty} c_{\alpha_i} t^{\alpha_i}$ with $c_{\alpha_i} \in \mathbb{C}$ and $\alpha_i \in \mathbb{Q}$ such that $\alpha_i \to \infty$. The field $\check{K}$ is not maximally complete. It has the immediate extension $\mathbb{C}((t^{\mathbb{Q}}))$, which is maximally complete. From the point of view of sequences the essential difference of the two completions lies in the following. If we form the partial sums $a_n = \Sigma_{i=0}^{n} c_{\alpha_i} t^{\alpha_i}$ of an element $\Sigma\, c_{\alpha_i} t^{\alpha_i}$ then for any $m > n$, $\mathrm{ord}(a_m - a_n) = \alpha_m$ increases, but not necessarily to ∞, as is the case in $\check{K}$.

These considerations led Ostrowski to introduce the concepts of the pseudo-convergence and pseudo-limit of a sequence. He then showed that the completion of a valued field with respect to pseudo-convergent sequences yields the maximal completion. Kaplansky [14] proved uniqueness theorems for maximal completions via Ostrowski's methods. Thus for one class of maximally complete fields K Kaplansky showed that $K \cong \bar{K}((t^{\mathrm{ord}\, K}))$. If ultraproducts of valued fields were maximally complete we could apply Kaplansky's results to prove the relevant isomorphism theorem. Although ultraproducts are not maximally complete, the ω_1-saturation property implies that they are ω-pseudo-complete, i.e., every countable pseudo-convergent sequence has a pseudo-limit. It is this property in the presence of Hensel's Lemma and the existence of a x-section that enables us to still obtain an isomorphism theorem. It may be helpful in our discussion of pseudo-convergence to keep in mind the ω-pseudo-complete field $\mathfrak{K}((t^G))_\beta$ of generalized power series. In fact to anticipate the final structure theorem (Section 9) the field V is isomorphic to $\mathfrak{K}((t^G))_\omega$ in the case of ramification index $e = 0$.

We now treat pseudo-convergence in the context of our problem. Let E be a subfield of V and let $\{a_n\}$ be a sequence in E.

Definition

(a) The sequence $\{a_n\}$ is pseudo-convergent if for some integer n_0, $\mathrm{ord}(a_m-a_n) > \mathrm{ord}(a_n-a_k)$ for all $m > n > k > n_0$. If this is so, we write $\mathrm{ord}(a_m-a_n)\uparrow$eventually or, more briefly, $\mathrm{ord}(a_m-a_n)\uparrow$ev.

(b) An element a in V is a pseudo-limit of the sequence $\{a_n\}$, in symbols $a_n \to a$ or $a \in \mathrm{ps.\,lim}\, a_n$, if for some integer n_0, $\mathrm{ord}(a-a_n) > \mathrm{ord}(a-a_k)$ for all $n > k > n_0$ (i.e., $\mathrm{ord}(a-a_n)\uparrow$ev.).

The field V is called ω-pseudo-complete if every pseudo-convergent sequence in V has a pseudo-limit in V. We now add the requirement that V and V' are finitely ramified ω-pseudo-complete Hensel fields with normalized x-sections.

If $\{a_n\}$ is pseudo-convergent, then $\mathrm{ord}(a_m-a_n) = \mathrm{ord}((a_m-a_{n+1}) + (a_{n+1}-a_n)) = \mathrm{ord}(a_{n+1}-a_n)$ev.. We write $\gamma_n = \mathrm{ord}(a_{n+1}-a_n)$, so that $\gamma_n\uparrow$ev.. If a is an element such that $\mathrm{ord}(a-a_n) = \gamma_n$ ev., then clearly $a_n \to a$. The converse is also true. Thus, if $\{a_n\}$ is any sequence such that $a_n \to a$, then for any $m > n > k > n_0$,

$\operatorname{ord}(a_m-a_n) = \operatorname{ord}((a-a_n) - (a-a_m)) = \operatorname{ord}(a-a_n) > \operatorname{ord}(a-a_k) =$ $\operatorname{ord}(a-a_k) - (a-a_n)) = \operatorname{ord}(a_n-a_k)$, so that $\{a_n\}$ is pseudo-convergent. Note that if we set $n = k + 1$ in the above argument we have that if $a_n \to a$ then $\operatorname{ord}(a-a_k) = \gamma_k$ ev..

It is far from being the case that a pseudo-convergent sequence $\{a_n\}$ has a unique pseudo-limit. In fact, if $a_n \to a$ and a' is any element such that $\operatorname{ord}(a'-a) > \gamma_k$ ev., then for $n > k$ $\operatorname{ord}(a'-a_n) = \min(\operatorname{ord}(a'-a), \operatorname{ord}(a-a_n)) = \gamma_n > \gamma_k = \operatorname{ord}(a'-a_k)$ ev. so that $a_n \to a'$. Conversely, if $a_n \to a$ and $a_n \to a'$, then $\operatorname{ord}(a'-a) \geq \min(\operatorname{ord}(a-a_{k+1}), \operatorname{ord}(a'-a_{k+1})) = \gamma_{k+1} > \gamma_k$ ev..

Pseudo-convergent sequences fall into two classes depending upon whether $a_n \to 0$ or $a_n \not\to 0$. Assume first that $a_n \not\to 0$. We claim that $\operatorname{ord} a_n$ is eventually constant. If not, then for all n there is an $m > n$ such that $\operatorname{ord} a_m \neq \operatorname{ord} a_n$. On the other hand, since $a_n \not\to 0$, we have that for every k, there is an $n > k$ such that $\operatorname{ord} a_n \leq \operatorname{ord} a_k$. Hence $\operatorname{ord}(a_m-a_n) = \min(\operatorname{ord} a_m, \operatorname{ord} a_n)$ $\leq \operatorname{ord} a_n = \min(\operatorname{ord} a_n, \operatorname{ord} a_k) \leq \operatorname{ord}(a_n-a_k)$, a contradiction. If now $a_n \to a$, then $\operatorname{ord} a = \operatorname{ord} a_n$ ev.. For otherwise $\gamma_n = \operatorname{ord}(a-a_n) = \min(\operatorname{ord} a, \operatorname{ord} a_n)$ ev., a constant, which contradicts $\gamma_n\uparrow$ev.. Finally, note that $\gamma_n = \operatorname{ord}(a-a_n) \geq$ $\min(\operatorname{ord} a, \operatorname{ord} a_n) = \operatorname{ord} a$.

Next assume that $a_n \to 0$. Then $\operatorname{ord} a_n\uparrow$ev., so that $\operatorname{ord} a \neq \operatorname{ord} a_n$ ev.. If $\operatorname{ord} a < \operatorname{ord} a_n$ ev., then $\gamma_n = \operatorname{ord}(a-a_n) = \operatorname{ord} a$ ev., contradicting $\gamma_n\uparrow$ev.. Thus, $\operatorname{ord} a > \operatorname{ord} a_n$ ev., and $\gamma_n = \operatorname{ord}(a-a_n) = \operatorname{ord} a_n$ ev.. We collect these facts together in the following

Lemma 4. Let $\{a_n\}$ be a pseudo-convergent sequence and $a_n \to a$. Then

(a) $a_n \not\to 0$ if and only if $\operatorname{ord} a = \operatorname{ord} a_n < \gamma_n$ ev.

(b) $a_n \to 0$ if and only if $\operatorname{ord} a > \operatorname{ord} a_n = \gamma_n$ ev..

The more fundamental division of pseudo-convergent sequences than the above is given by the following.

Definition. The pseudo-convergent sequence $\{a_n\}$ in E is algebraic if, for some polynomial $p(X) \in E[X]$, $p(a_n) \to 0$. Otherwise, $\{a_n\}$ is a transcendental sequence.

If $\{a_n\}$ is an algebraic sequence, we call a polynomial $q(X)$ over E of least degree such that $q(a_n) \to 0$ a minimal polynomial of $\{a_n\}$ in E. If $a_n \to c \in V$ and $q(c) = 0$ for a minimal polynomial $q(X)$ of $\{a_n\}$ in E, we say that c is a minimal algebraic pseudo-limit of $\{a_n\}$.

It is clear that in order to study algebraic and transcendental sequences we shall need information on the value of a polynomial $p(X)$ of degree $k > 0$ at a pseudo-limit a of the sequence $\{a_n\}$. For this the natural tool to use is the Taylor expansion of $p(X)$ about the point a:

$$p(a_n) - p(a) = \sum_{j=1}^{k} \frac{p^j(a)}{j!} (a_n - a)^j$$

where $p^j(X)$ is the jth derivative of the polynomial $p(X)$. Let $\delta_j = \text{ord}(p^j(a)/j!)$. Note that, for a non-constant polynomial $p(X)$, $\delta_j < \infty$ for some $1 \leq j \leq k$. For if $p^j(a) = 0$ for all $1 \leq j \leq k$, then $p(a_n) = p(a)$ for every n, contradicting the fact that a polynomial takes the same value only finitely often.

Since $\gamma_n \uparrow \text{ev.}$ we have for fixed $i \neq j$, either $(j-i)\gamma_n < \delta_i - \delta_j$ ev. or $(j-i)\gamma_n > \delta_i - \delta_j$ ev.. In other words, either $\delta_j + j\gamma_n < \delta_i + i\gamma_n$ ev. or $\delta_j + j\gamma_n > \delta_i + i\gamma_n$ ev..

Thus, eventually there is a single term in the Taylor expansion with a strictly minimum ord value $\delta_j + j\gamma_n$. Hence

$$\text{ord}(p(a_n) - p(a)) = \delta_j + j\gamma_n \text{ ev.} \; .$$

Since $\delta_j + j\gamma_n \uparrow \text{ev.}$, we have proved that $p(X)$ is a continuous function for pseudo-convergence in the sense that $a_n \to a$

implies that $p(a_n) \to p(a)$. This means in particular that $\{p(a_n)\}$ is a pseudo-convergent sequence. Applying Lemma 4 to $\{p(a_n)\}$, we have that

$$\left.\begin{array}{l} p(a_n) \not\to 0 \text{ if and only if } \operatorname{ord} p(a) - \operatorname{ord} p(a_n) < \delta_j + j\gamma_n \text{ ev.} \\ p(a_n) \to 0 \text{ if and only if } \operatorname{ord} p(a) > \operatorname{ord} p(a_n) = \delta_j + j\gamma_n \text{ ev.} \end{array}\right\} \quad (*)$$

This discussion suffices to treat the case of transcendental sequences, but to deal with algebraic sequences we need more precise information about the approximation of a polynomial by the terms of its Taylor expansion. In particular, we require that the first term $p'(a)(a_n-a)$ is the closest approximation to $p(a_n) - p(a)$. To achieve this we need to assume that $p'(a_n) \not\to 0$.* We shall now use for the first time the assumption that V is finitely ramified. Since we need to know about the value of $p'(X)$ at the point a, we consider the Taylor expansion of $p'(X)$ about a:

$$p'(a_n) - p'(a) = \sum_{j=2}^{k} \frac{p^j(a)}{j!} j(a_n-a)^{j-1} \quad .$$

By a similar argument as before we see that for some integer n_0 there is a single term with a strictly minimum ord value $\delta_j + (j-1)\gamma_n + \operatorname{ord} j$ for all $n \geq n_0$. By taking n_0 large enough we can also ensure that $\gamma_n \uparrow$ for $n \geq n_0$.

Let $m = \#\{\gamma | 0 < \gamma \leq \operatorname{ord} j\} < \infty$. Then $0 < \gamma_{n+1} - \gamma_n < \gamma_{n+2} - \gamma_n < \dots < \gamma_{n+m} - \gamma_n$ for $n \geq n_0$ so that $\gamma_{n+m} - \gamma_n > \operatorname{ord} j$. Hence $(j-1)\gamma_{n+m} > (j-1)\gamma_n + \operatorname{ord} j$, for $j > 1$.

Since $p'(a_n) \not\to 0$ we have by (*) $\operatorname{ord} p'(a) = \operatorname{ord} p'(a_n)$ ev.. Thus, $\delta_1 = \operatorname{ord} p'(a) \leq \operatorname{ord}(p'(a_n)-p'(a)) = \delta_j + (j-1)\gamma_n + \operatorname{ord} j$.

---

* To see that such an assumption is necessary, let $p(X) = X^2 + \pi^{\gamma}X + \pi^{\gamma}$, and $\gamma_n \uparrow$ with $\gamma_n < \gamma$ for all n. Then $\{\pi^{\gamma_n}\}$ is an algebraic sequence with $\pi^{\gamma_n} \to 0$ and $\operatorname{ord}(p(\pi^{\gamma_n}) - p(0)) = \operatorname{ord} 1/2\, p''(0)(\pi^{\gamma_n})^2 < \operatorname{ord} p'(0)\pi^{\gamma_n}$.

Hence $\delta_1 < \delta_j + (j-1)\gamma_n$ for $n \geq n_0 + m$ or $\delta_1 + \gamma_n < \delta_j + j\gamma_n$ ev..
We have thus proved that if $a_n \to a$ and $p'(a_n) \not\to 0$, then
$\mathrm{ord}(p(a_n)-p(a)) = \delta_1 + \gamma_n$ ev..

It follows that in this case (*) holds with $j = 1$, i.e., if $a_n \to a$ and $p'(a_n) \not\to 0$, then

$$\left.\begin{array}{l} p(a_n) \not\to 0 \text{ if and only if } \mathrm{ord}\, p(a) = \mathrm{ord}\, p(a_n) < \delta_1 + \gamma_n \text{ ev.} \\ p(a_n) \to 0 \text{ if and only if } \mathrm{ord}\, p(a) > \mathrm{ord}\, p(a_n) = \delta_1 + \gamma_n \text{ ev.} \end{array}\right\} \quad (\dagger)$$

We are now in a position to apply pseudo-convergent sequences to the study of immediate extensions.

<u>Lemma 5</u>. Let $\{a_n\}$ be a transcendental sequence in E and $a_n \to a \in V$. Then

(i) a is a transcendental element over E

(ii) Let $\varphi : E \to E'$ be an analytic isomorphism and $a' \in V'$ be such that $\varphi(a_n) \to a'$. There exists an analytic isomorphism $\psi : E(a) \to E'(a')$ extending φ.

<u>Proof</u>.

(i) If $p(X) \in E(X)$, then $\mathrm{ord}\, p(a) = \mathrm{ord}\, p(a_n) < \gamma_n$ ev. by (*). Therefore $\mathrm{ord}\, p(a) \neq \infty$, i.e., $p(a) \neq 0$. Hence a is transcendental over E.

(ii) The map $a \mapsto a'$ thus gives an algebraic isomorphism $\psi : E(a) \to E'(a')$. Let $p_\varphi(X)$ be the image of $p(X)$ under the map φ. Since $\mathrm{ord}\, p(a) = \mathrm{ord}\, p(a_n) = \mathrm{ord}\, p_\varphi(\varphi(a_n)) = \mathrm{ord}\, p(a')$ ev., the map ψ is analytic.

<u>Lemma 6</u>. Let $\{a_n\}$ be an algebraic sequence in E.

(i) If $p(a_n) \to 0$ and $p'(a_n) \not\to 0$, then there is $a \in V$ such that $a_n \to a$ and $p(a) = 0$. In particular, there exists a minimal algebraic ps. lim. $a \in V$ of $\{a_n\}$.

(ii) Let $\varphi : E \to E'$ be an analytic isomorphism and $a \in V$ a minimal algebraic ps. lim. of $\{a_n\}$. There exists a minimal algebraic ps. lim. a' of $\{\varphi(a_n)\}$ and an analytic isomorphism $\psi: E(a) \to E'(a')$ extending φ.

Proof.

(i) Let $a \in$ ps. lim a_n. If $p(a) = 0$, then we are done. Suppose $p(a) \neq 0$. Since $\operatorname{ord} p(a) > \operatorname{ord} p'(a) + \gamma_n$ ev., $p'(a) \neq 0$. Let

$$q(X) = p\left(a + \frac{p(a)}{p'(a)} X\right) \Big/ p(a) .$$

Then

$$q(X) = 1 + X + \sum_{j>1} \frac{p^j(a)}{j!} \frac{p(a)^{j-1}}{p'(a)^j} X^j .$$

Now

$$\begin{aligned} \operatorname{ord} \frac{p^j(a)}{j!} \frac{p(a)^{j-1}}{p'(a)^j} &= \delta_j + (j-1)\operatorname{ord} p(a) - j\delta_1 \\ &> (\delta_j + j\gamma_n) - (\delta_1 + \gamma_n) \text{ ev., by } (\dagger) \\ &> 0 \text{ ev..} \end{aligned}$$

Thus $\bar{q}(X) = 1 + X$, which has the root $X = -1$ in $\mathfrak{K}$. Since $\bar{q}'(X) = 1 \neq 0$, we may use the Hensel property of V to obtain $u \in O_V$ such that $q(u) = 0$ and $\bar{u} = -1$. From $\bar{u} = -1$ it follows that $\operatorname{ord} u = 0$. Hence, $p(X)$ has a root $b = a + \frac{p(a)}{p'(a)} u$ in V. Now $\operatorname{ord}(b-a) = \operatorname{ord} p(a) - \delta_1 > \gamma_n$ ev. by $(\dagger)$. Thus $a_n \to b$ and we are done. Finally, if $p(X)$ is a minimal polynomial for $\{a_n\}$ then $p(a_n) \to 0$ and $p'(a_n) \not\to 0$, so that b is a minimal algebraic ps. lim. of $\{a_n\}$.

(ii) If $p(X)$ is a minimal polynomial for $\{a_n\}$, then $p(X)$ is irreducible over E. For if $p(X) = q_1(X)q_2(X)$ with $\deg q_i < \operatorname{ord} p$, then $q_i(a_n) \not\to 0$ so that $\operatorname{ord} q_i(a_n) = \operatorname{ord} q_i(a)$ ev. by (*). Hence $\operatorname{ord} p(a_n) = \operatorname{ord} p(a)$ ev., contradicting $p(a_n) \to 0$.

Clearly $p_\varphi(X)$ is a minimal polynomial for $\{\varphi(a_n)\}$. Let $a' \in V'$ be a ps. lim. of $\{\varphi(a_n)\}$ such that $p_\varphi(a') = 0$. Then $a \mapsto a'$ defines an algebraic isomorphism $\psi: E(a) \to E'(a')$ extending φ. Every element of $E(a)$ has the form $q(a)$, where $\deg q(X) < \deg p(X)$

Hence $q(a_n) \neq 0$, so that $\text{ord}\ q(a) = \text{ord}\ q(a_n) = \text{ord}\ q_\varphi(\varphi(a_n)) = \text{ord}\ q_\varphi(a')$ ev., proving that ψ is analytic.

(We note that in the case of each of these two lemmas the field $E(a)$ is an immediate extension of E. The proof of the lemmas shows that $\text{ord}\ E(a) = \text{ord}\ E$. To see that $\overline{E(a)} = \bar{E}$, let $q(X) \in E[X]$ (with $\deg q(X) < \deg p(X)$ in the second case). Assume $\text{ord}\ q(a) \geq 0$. Since $q(a_n) \neq 0$ we have $\text{ord}\ q\ (a) = \text{ord}\ q(a_n)$ ev.. Thus, $\text{ord}(q(a)-q(a_n)) \geq 0$. But $q(a_n) \to q(a)$, so that $\text{ord}\ (q(a)-q(a_n))\uparrow$ ev.. It follows that $\text{ord}(q(a)-q(a_n)) > 0$. In other words $\overline{q(a)} = \overline{q(a_n)} \in \bar{E}$. Of course, in our case $\bar{\bar{E}} = \mathfrak{K} = \bar{V}$, so that $\overline{E(a)} = \bar{E}$, trivially.)

We now prove a converse to these results.

Lemma 7. Let E be a subfield V with $\text{ord}\ E$ countable and let F be an immediate extension of E. If $a \in F - E$, then there exists a sequence $\{a_n\}$ in E with a as a pseudo-limit but with no pseudo-limit in E.

Proof. Let $H = \{\text{ord}(a-x) \mid x \in E\}$. We show that H has no greatest element. Let $\text{ord}(a-b) \in H$; then there is $c \in E$ such that $\text{ord}\ c = \text{ord}(a-b)$, since $\text{ord}\ F = \text{ord}\ E$. Also since $\bar{F} = \bar{E}$, there is $d \in E$ such that $\bar{d} = \overline{(a-b)/c}$. Thus, $\text{ord}\ (1 - \frac{cd}{a-b}) > 0$, or $\text{ord}(a-(b+cd)) > \text{ord}(a-b)$, giving a greater element of H.

Since then H is countably infinite, there exists an increasing sequence $\{\gamma_n\}$ of elements of H cofinal with H. In other words, there is a sequence $\{a_n\}$ in E with $\text{ord}(a-a_n) = \gamma_n\uparrow$. This shows that $a_n \to a$.

If $a_n \to b \in E$, then $\text{ord}(a-b) > \gamma_n$ ev., contradicting the fact that $\{\gamma_n\}$ is cofinal with H. Thus, $\{a_n\}$ has no pseudo-limit in E.

We are now ready to prove Proposition 2, extending an analytic isomorphism from a subfield E of V to an immediate extension

E(a). First let us motivate the method of proof. The proof divides naturally into two cases. We first suppose that a is algebraic over E. We wish to apply Lemma 6; however the element a may not be accessible directly via an algebraic sequence in E, since a may not be a minimal algebraic ps. lim. of a sequence in E. Nevertheless, we shall see that a may be reached by a (transfinite) succession of minimal algebraic extensions, so that Lemma 6 applies. In fact, the proof shows that the isomorphism extends to the algebraic closure $\tilde{E}$ of E in V.

If a is transcendental over E, then we need to know that a is a ps. lim. of a transcendental sequence. This need not be the case unless the field E is algebraically closed in V. This necessitates our first extending the isomorphism to $\tilde{E}$ (by the above). However we may now have lost the immediacy of the extension, i.e., ord $\tilde{E}(a) \neq$ ord $\tilde{E}$. We may deal with this problem by using Proposition 1 to form $\hat{\tilde{E}}$, so that ord $\hat{\tilde{E}}(a) =$ ord $\hat{\tilde{E}}$, but now once again $\hat{\tilde{E}}$ may not be algebraically closed in V. We take care of both problems by alternating the two constructions countably often.

Proposition 2. Let $\varphi : E \to E'$ be a pure map and $a \in V$ with E(a) an immediate extension of E. There exists a pure map $\psi : F \to F'$ extending φ with $a \in F$.

Proof.

(a) First suppose that a is algebraic over E. Consider the set P of all pure maps $\rho : K \to K'$ with $E \subset K \subset \tilde{E}$. P is a non-empty set partially ordered under inclusion with every chain having an upper bound, the union. By Zorn's lemma, P has a maxi- element $\psi : F \to F'$. We claim that $F = \tilde{E}$, so that $a \in F$ and we are done. If $F \neq \tilde{E}$, there is $b \in V - F$ algebraic over F. By Lemma 7 there exists a sequence $\{a_n\}$ in F with b as a ps. lim. and no ps. lim. in F. The sequence $\{a_n\}$ is algebraic by Lemma 5

(i) and has a minimal algebraic ps. lim. $c \in V$ by Lemma 6 (i). From Lemma 6 (ii) it follows that there is a minimal algebraic ps. lim. $c' \in V$ of the sequence $\{\psi(a_n)\}$ in F' such that ψ has a pure extension $\xi: F(c) \to F'(c')$, contradicting the maximality of ψ.*

(b) Now suppose that a is transcendental over E. Let $E_0 = E$, $E_{2i} = \tilde{E}_{2i-1}$, $E_{2i+1} = \hat{E}_{2i}$, and $E_\omega = \bigcup_{n=0}^{\infty} E_n$. Then the field E_ω is clearly algebraically closed in V and $E_\omega(a)$ is an immediate extension of E. By Proposition 1 and Part (a) of this proof, φ can be extended to a pure map φ_ω with domain E_ω. If $a \in E_\omega$ we are done. If $a \notin E_\omega$, then by Lemma 7 there exists a sequence $\{a_n\}$ in E_ω with a as a ps. lim. and no ps. lim. in E_ω. If $\{a_n\}$ is an algebraic sequence, then Lemma 6 yields a proper algebraic extension of E_ω. Hence $\{a_n\}$ is a transcendental sequence. Let a' be a ps. lim. in V' of the sequence $\{\varphi_\omega(a_n)\}$. Then Lemma 5 (ii) gives us the required pure map $\psi: E_\omega(a) \to E'_\omega(a')$.

Putting Propositions 1 and 2 together we have

Proposition 3. For every pure map $\varphi: E \to E'$ and $a \in V$ there exists a pure map $\psi: F \to F'$ extending φ with $a \in F$.

---

* To obtain a pure extension of φ to $\psi: E(a) \to E'(a')$ it suffices to take a maximal element $\psi: F \to F'$ of the set of pure maps $\rho: K \to K'$ with $E \subset K \subset E(a)$. Then $F = E(a)$. We have constructed the larger field $F = \tilde{E}$ for later use in the proof. The reader may find it instructive to construct ψ alternatively by transfinite induction. The field $\tilde{E}$ (resp. $E(a)$) appears then as the closure of E under minimal algebraic ps. lims. of algebraic sequences in E (resp. with a as ps. lim.).

5. UNRAMIFIED EXTENSIONS

In this section we prove that pure maps exist. Up to this point our proof has been uniform, not depending upon the ramification index e. Now our proof divides into the case $e = 0$, and $e = 1$. This is not surprising since even in the classical case the existence and uniqueness of a complete field K valued in $\mathbb{Z}$ with given residue class field $\mathfrak{K}$ is treated separately in the cases of equal and unequal characteristic of K and $\mathfrak{K}$. A basic difference that appears here is that for $e > 0$ there is a smallest positive element 1 of the value group G, and hence the ordered group $\mathbb{Z}$ of integers lies as a convex subgroup inside G. For $e = 0$ this need not be the case.

(a) We first treat the case $e = 0$. Here char V = char $\mathfrak{K} = 0$. The prototype for this case is the formal power series field $F((t))$ with char. $F = 0$. Here $\overline{F((t))} \cong F$, but F also occurs as a subfield of $F((t))$. We shall now see that this property is also valid for V. To motivate the proof, we note that in the case of the field $F((t))$ ord is trivial on F, and F is a maximal field with this property.

Lemma 8. Let F be a Hensel field with char. $\bar{F} = 0$. Let F_0 be a maximal subfield of F on which the valuation ord is trivial. Then $F_0 \cong \bar{F}_0 = \bar{F}$.

Proof. The field F_0 exists by Zorn's Lemma. Since F_0 is a field, the residue class map is a monomorphism on F_0. Suppose now that $\bar{F}_0 \neq \bar{F}$, so that there is $\alpha \in \bar{F} - \bar{F}_0$.

(i) α is algebraic over $\bar{F}_0$.

There is an irreducible $f(X) \in \bar{F}_0[X]$ with $f(\alpha) = 0$. Let $q(X) \in F_0[X]$ be such that $\bar{q}(X) = f(X)$. Since $f(X)$ is irreducible, $f'(\alpha) \neq 0$, so that we may apply Hensel's Lemma to obtain $a \in O_F$ such that $q(a) = 0$ and $\bar{a} = \alpha$. Now $q(X)$ is irreducible over F_0 since $f(X)$ is irreducible over $\bar{F}_0$. Hence the residue class map gives an isomorphism $F_0(a) \to \bar{F}_0(\alpha)$. Since $g(a) \in F_0(a)$ with $\deg g(X) < \deg q(X)$ and $\operatorname{ord} g(a) > 0$ yields $\bar{g}(\alpha) = 0$, a contradiction, we have that ord is trivial on $F_0(a)$. This contradicts the maximality property of F_0.

(ii) α is transcendental over $\bar{F}_0$.

If $a \in F - F_0$ with $\bar{a} = \alpha$, then clearly a is transcendental over F_0. Hence the residue class map $F_0(a) \to \bar{F}_0(\alpha)$ is an isomorphism. As in (i), ord is trivial on $F_0(a)$, contradicting the maximality of F_0.

<u>Proposition 4. (a)</u> In the case $e = 0$, there exists a pure map $\varphi_0 : E_0 \to E_0'$.

<u>Proof</u>. Let E_0, E_0' be the subfields of V, V' provided by Lemma 8 with $E_0 \cong \mathfrak{K} \cong E_0'$. The isomorphism thus defined is clearly a pure map.

(b) Next we treat the case $e = 1$. Here $\operatorname{char} V = 0$ and $\operatorname{char} \mathfrak{K} = p$. Let 1 denote the smallest positive element of G. Then $\operatorname{ord} p = 1$.

We now need a construction from general valuation theory. This construction allows us to decompose valuations in terms of convex subgroups of the valuation group. A <u>convex</u> subgroup Δ of an ordered abelian group G is defined by the condition that if $\delta \in \Delta$ and $\gamma \in G$ such that $|\gamma| < |\delta|$, then $\gamma \in \Delta$. Then $\Gamma = G/\Delta$ forms an ordered abelian group with the inherited ordering. Conversely, if $h : G \to \Gamma$ is an order-preserving homomorphism of ordered abelian groups then the kernel of h is a convex subgroup of G.

Let $\operatorname{ord} : K \to G$ be a valuation and let Δ be a convex subgroup of G. Let $h : G \to G/\Delta (= \Gamma)$ be the canonical homomorphism. We define $v : K \to \Gamma$ to be the composition $v = h \circ \operatorname{ord}$. Then v

is itself a valuation. Let K_v be the residue class field of K under the valuation v. Define the map $u: K_v \to \Delta$ by $u(\alpha) = \mathrm{ord}(a)$ where a is any element of K such that α is the image of a under the residue class map of the valuation v. Clearly u is a valuation with residue class $\bar{K}$. The valuation ord is called the <u>composition</u> of the valuations u and v. We summarize by the diagram:

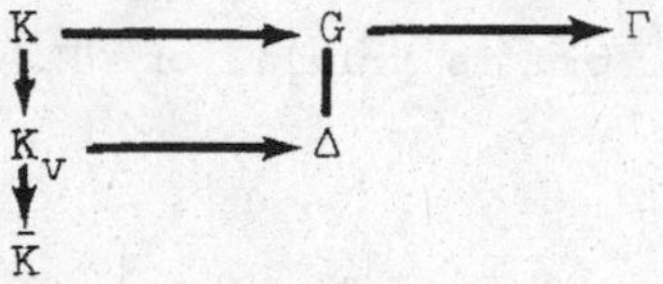

In the case at hand the element 1 generates the convex subgroup $\mathbb{Z}(=\mathbb{Z}1)$ in G. Decomposing the valuation ord according to the previous paragraph we have

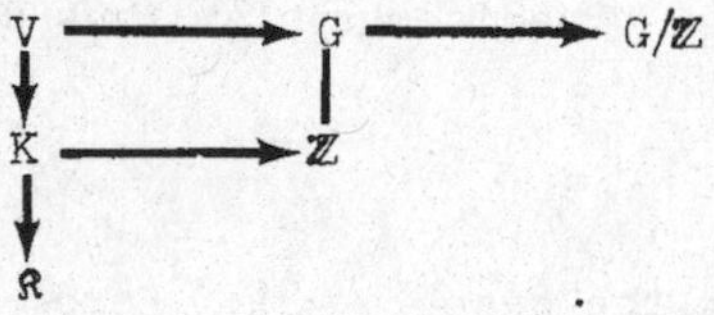

Thus K is an unramified field valued in $\mathbb{Z}$. Since the field V is ω-pseudo-complete, then clearly so is K. But since K is valued in $\mathbb{Z}$ this means that K is (Cauchy) complete. As stated in Section 2 this implies that $K \cong W(\mathfrak{K})$ under an analytic isomorphism.

Now the field V valued in $G/\mathbb{Z}$ forms a Hensel field with residue class field K of characteristic 0. By Lemma 8, there is a subfield E_0 of V valued in $\mathbb{Z}$ which is isomorphic to K. A similar consideration for V' yields a subfield E_0'. Now $E_0 \cong W(\mathfrak{K}) \cong E_0'$ gives us an analytic isomorphism $\varphi_0: E_0 \to E_0'$. The fact that the x-sections π, π' are normalized means that $\pi(n) = p^n = \pi'(n)$ so that φ_0 is x-analytic and hence pure. We have thereby established

<u>Proposition 4 (b)</u>. In the case $e = 1$, there exists a pure map $\varphi_0: E_0 \to E_0'$.

6. THE ISOMORPHISM THEOREM

<u>Theorem 1</u>. Let V and V' be unramified ω-pseudo-complete Hensel fields of cardinality $\aleph_1$ with normalized x-sections π, π'. Then there is a x-analytic isomorphism $\varphi: V \to V'$ if and only if $\bar{V} \simeq \bar{V}'$ and $\text{ord } V \simeq \text{ord } V'$.

<u>Proof</u>. To simplify notation we assume that $\bar{V} = \mathfrak{K} = \bar{V}'$ and $\text{ord } V = G = \text{ord } V'$. We define the map φ by transfinite induction. We well order V, V' by ordinals $< \omega_1$. Let $\varphi_0: V_0 \to V_0'$ be the pure map provided by Proposition 4 (a) and (b). For $\alpha < \omega_1$ assume inductively that $\varphi_\alpha: V_\alpha \to V_\alpha'$ is a pure map. If α is an even ordinal (i.e., $\alpha = \lambda + n$, λ a limit ordinal and n a positive even integer) let a be the first element in $V - V_\alpha$. Let $\varphi_{\alpha+1}: V_{\alpha+1} \to V_{\alpha+1}'$ be the pure map provided by Proposition 3 (with $V_\alpha = E$, $V_\alpha' = E'$, $\varphi_\alpha = \varphi$). If α is an odd ordinal, let a' be the first element of $V' - V_\alpha'$. Let $\varphi_{\alpha+1}: V_{\alpha+1} \to V_{\alpha+1}'$ be the inverse of the pure map provided by Proposition 3 (with $V_\alpha' = E$, $V_\alpha = E'$, $\varphi_\alpha^{-1} = \varphi$). For α a limit ordinal, let $V_\alpha = \bigcup_{\beta<\alpha} V_\beta$, $V_\alpha' = \bigcup_{\beta<\alpha} V_\beta'$, and $\varphi_\alpha = \bigcup_{\beta<\alpha} \varphi_\beta$. It is clear that $V = \bigcup_{\alpha<\omega_1} V_\alpha$ and $V' = \bigcup_{\alpha<\omega_1} V_\alpha'$, and that $\varphi = \bigcup_{\alpha<\omega_1} \varphi_\alpha$ defines the required x-analytic isomorphism of V onto V'.

By introducing transfinite pseudo-convergent sequences (see Schilling [10], Chapter 2) we may generalize the theorem to apply to unramified Hensel fields of cardinality $\aleph_\alpha$ with normalized x-sections which are pseudo-complete for transfinite sequences of cardinality $< \aleph_\alpha$. However, no new ideas are involved, only added notation.

7. APPLICATIONS

This section contains some of the model-theoretic and number-theoretic consequences of the theorem we have just proved. Our first and most important application is the metamathematical principle which connects power series fields with Witt vector fields.

Theorem 2. For each prime p let $\mathfrak{K}_p$ be a field of cardinality $\leq 2^{\aleph_0}$ and characteristic p. Let D be a non-principal ultrafilter on the set of primes. Then, assuming $2^{\aleph_0} = \aleph_1$, we have

$$\prod_p W(\mathfrak{K}_p)/D \cong \prod_p \mathfrak{K}_p((t))/D$$

via a x-analytic isomorphism. In particular,

$$\prod_p \mathbb{Q}_p/D \cong \prod_p \mathbb{Z}_p((t))/D$$

Proof. That the fields $\Pi_p W(\mathfrak{K}_p)/D$ and $\Pi_p \mathfrak{K}_p((t))/D$ are unramified Hensel fields of char. 0, cardinality $2^{\aleph_0}$, and $e = 0$ is clear. Example 4 of Section 2 shows that they have x-sections. It remains to show that these fields are ω-pseudo-complete. We prove this by using the fact that these ultraproducts are ω_1-saturated.

Let $\{a_n\}$ be a pseudo-convergent sequence (in either field). Then for some integer n_0, $ord(a_m - a_n) > ord(a_n - a_k)$ for all $m > n > k > n_0$. Let $\varphi_k(x)$ be the elementary formula

$$ord(x - a_{n_0+k}) = ord(a_{n_0+k+1} - a_{n_0+k}) \quad .$$

Then $\bigwedge_{k=1}^{m} \varphi_k(x)$ is satisfied by the element $x = a_{n_0+m+1}$. By the ω_1-saturation property there exists an element $x = a$ in the field satisfying all the formulas $\varphi_k(x)$ simultaneously. Hence,

$$ord(a - a_n) = ord(a_{n+1} - a_n)$$

for all $n \geq n_0$, proving that $a_n \to a$.

Corollary 1. Let Δ be an elementary statement about valued fields (with x-section). There exists a finite set U_Δ of primes such that Δ is valid in $W(\mathfrak{K}_p)$ if and only if Δ is valid in $\mathfrak{K}_p((t))$ for all $p \notin U_\Delta$.

Proof. If Δ differed in validity on an infinite set T of primes for $W(\mathfrak{K}_p)$ and $\mathfrak{K}_p((t))$, then we could include T as an element of a non-principal ultrafilter D. In that case Δ would be valid in one of $\Pi_p W(\mathfrak{K}_p)/D$ and $\Pi_p \mathfrak{K}_p((t))/D$ but not the other, contradicting the theorem above. This assumes $2^{\aleph_0} = \aleph_1$, but this assumption may be dropped by a standard argument via absoluteness using Gödel [15].

This corollary has been applied to settling specific number-theoretic conjectures. A field K has the property $C_i(d)$ if every form (i.e., homogeneous polynomial) over K of degree d with $d^i + 1$ variables has a non-trivial zero in K. Chevalley proved that every finite field has the property $C_1(d)$ for all d. Lang [16] used this result to prove that the field $\mathbb{Z}_p((t))$ has the property $C_2(d)$ for all d. E. Artin conjectured that $\mathbb{Q}_p$ is $C_2(d)$ for all d. This was known (Meyer) for $d = 2$ and later proved (Lewis, Demyanov) for $d = 3$. Since $C_2(d)$ is an elementary statement which is valid in $\mathbb{Z}_p((t))$ for all p, we have

Corollary 2. For each d, there exists a finite set U_d of primes such that $\mathbb{Q}_p$ has the property $C_2(d)$ for all $p \notin U_d$.

Subsequently Terjanian showed that this result was best possible and that Artin's original conjecture is false for $d = 4$ and $p = 2$. Counterexamples for other d and p were later found by S. Schanuel, P. Samuel, and J. Browkin.

Lang conjectured that if a form of degree d has coefficients in $\mathbb{Z}$ then it suffices that it have $d + 1$ variables in order to have a non-trivial zero in $\mathbb{Q}_p$ for all but a finite number of p.

If the coefficients are interpreted modulo p then this property holds for finite fields by Chevalley's result. Since $\mathbb{Z}_p \subset \mathbb{Z}_p((t))$ it also holds for $\mathbb{Z}_p((t))$. It follows from Corollary 1 that Lang's conjecture is true. A purely algebraic proof of Lang's conjecture was given by Greenleaf independently. No purely algebraic proof of Corollary 2 is known to date. Note that Lang's conjecture does not require the full Isomorphism Theorem, only the fact that $\Pi_p \mathbb{Z}_p/D$ is imbeddable in $\Pi_p \mathbb{Q}_p/D$, which follows from Lemma 8. This is because Lang's conjecture is an existential statement whereas Artin's conjecture is universal existential.

We now turn to some other applications of our isomorphism theorem.

Theorem 3. Let char $\mathfrak{K} = 0$. Then

(a) $\mathfrak{K} \equiv \mathfrak{K}'$ and $G \equiv G'$ if and only if $\mathfrak{K}((t^G)) \equiv \mathfrak{K}'((t^{G'}))$. In particular,

$$\mathfrak{K} \equiv \mathfrak{K}' \text{ if and only if } \mathfrak{K}((t)) \equiv \mathfrak{K}'((t)) .$$

(b) $\mathfrak{K}$ and G are decidable if and only if $\mathfrak{K}((t^G))$ is decidable. In particular, $\mathfrak{K}$ is decidable if and only if $\mathfrak{K}((t))$ is decidable.

Proof. We apply the Löwenheim-Skolem Theorem if necessary to replace the fields $\mathfrak{K}((t^G))$ and $\mathfrak{K}'((t^{G'}))$ by elementarily equivalent fields K and K' of cardinality $\leq 2^{\aleph_0}$. An application of the Isomorphism Theorem to the countably indexed non-principal ultrapowers K^I/D and K'^I/D yields the result. To prove the second half of (b) note that the ordered abelian group $\mathbb{Z}$ is decidable. (A complete axiomization of the theory of $\mathbb{Z}$ is given by the statement that the group G is a Z-group: G has a smallest positive element and the Euclidean algorithm holds for G, i.e., $\#(G/nG) = n$, for all positive integers n. This result is due essentially to Presburger.)

A complete axiomatization of $\mathfrak{K}((t^G))$ is as follows: V is a Hensel field with x-section such that $Th(\bar{V}) = Th(\mathfrak{K})$ and $Th(\text{ord } V) = Th(G)$. We shall see in Section 8 that we may drop the x-section from this axiomatization.

Let $\mathfrak{M}$ be the valued field of germs of meromorphic functions in the complex plane. Then $\mathfrak{M}$ forms an unramified Hensel field with value group $\mathbb{Z}$ and residue class field $\mathbb{C}$. Hence $\mathfrak{M} \equiv \mathbb{C}((t))$ and $\mathfrak{M}$ is decidable. In fact it is easy to extend our results to show that $\mathfrak{M}$ is an elementary subsystem of $\mathbb{C}((t))$. An algebraic consequence is that every system of polynomials over $\mathfrak{M}$ which has a common zero in $\mathbb{C}((t))$ (i.e., a formal zero) has already a zero in $\mathfrak{M}$ (i.e., a convergent zero).

Theorem 4. Let char $\mathfrak{K} = p$. Then

(a) $\mathfrak{K} \equiv \mathfrak{K}'$ if and only if $W(\mathfrak{K}) \equiv W(\mathfrak{K}')$.

(b) $\mathfrak{K}$ is decidable if and only if $W(\mathfrak{K})$ is decidable. In particular, $\mathbb{Q}_p$ is decidable.

The proof is similar to the proof of the previous theorem.

A complete axiomatization of $W(\mathfrak{K})$ is as follows: V is a Hensel field with normalized x-section ord V a Z-group, ord p = 1, and $Th(\bar{V}) = Th(\mathfrak{K})$. For $\mathbb{Q}_p (= W(\mathbb{Z}_p))$ we may of course replace $Th(\bar{V}) = Th(\mathfrak{K})$ by $\bar{V} = \mathbb{Z}_p$. Again Section 8 will show that we may drop the x-section function.

As another application, let $\check{\mathbb{Q}}_p$ be the cyclotomic extension of $\mathbb{Q}_p$ obtained by adjoining all roots of unity to $\mathbb{Q}_p$. The field $\check{\mathbb{Q}}_p$ is the maximal unramified algebraic extension of $\mathbb{Q}_p$. The residue class field of $\check{\mathbb{Q}}_p$ is $\tilde{\mathbb{Z}}_p$ the algebraic closure of $\mathbb{Z}_p$ and the value group of $\check{\mathbb{Q}}_p$ is $\mathbb{Z}$. Since $\tilde{\mathbb{Z}}_p$ is a decidable field, we conclude by the previous theorem that $\check{\mathbb{Q}}_p$ is a decidable field. Also $\check{\mathbb{Q}}_p \equiv W(\tilde{\mathbb{Z}}_p)$. We draw a number-theoretic consequence from this. Lang [16] showed that $W(\tilde{\mathbb{Z}}_p)$ is a $C_1(d)$ field for all d.

It now follows that $\check{\mathbb{Q}}_p$ is also a $C_1(d)$ field for all d. It is possible to give a purely algebraic proof of the fact that $\check{\mathbb{Q}}_p$ is $C_1(d)$, using the generalized Hensel's Lemma in Greenberg [17]. However, that is not the whole story. Greenberg's theorem is the result of an analysis of the algebraic content of Cohen's elimination of quantifiers for $\mathbb{Q}_p$. Cohen's method in turn analysed the constructive content of the model-theoretic proof of the decidability of $\mathbb{Q}_p$ in Ax and Kochen [2]. In this way, the circle of logical and algebraic ideas is closed. Incidentally, it is an unsolved conjecture of E. Artin that the corresponding global field, the cyclotomic field of $\mathbb{Q}$ is a $C_1(d)$ field.

Other number-theoretic consequences of the above theorems are given in Kochen [18].

We indicate finally how the elementary equivalence of real closed fields may be subsumed under the Isomorphism Theorem. This may seem surprising since the Theorem refers to fields with non-Archimedean valuation whereas the ordering of the field R of reals corresponds to the Archimedean valuation of absolute value. The answer is that ultraproducts of real closed fields admit non-Archimedean valuations. Thus let $\mathcal{R} = \mathbb{R}^I/D$ be a non-principal countably indexed ultrapower of $\mathbb{R}$ (or of any real closed field of cardinality $\leq 2^{\aleph_0}$). Then the ring F of elements a of $\mathcal{R}$ which are bounded by $\mathbb{R}$, i.e., $|a| < r$ for some $r \in \mathbb{R}$, forms a valuation ring. This ring then defines a valuation ord on $\mathcal{R}$. The unique maximal ideal $\mathcal{I}$ consists of the infinitesimal elements, i.e., $\mathcal{I} = \{a \mid \forall r \in \mathbb{R}, |a| < |r|\}$. We have $\bar{\mathcal{R}} \cong \mathbb{R}$ and ord $\mathcal{R}$ is a divisible group (in fact, ord $\mathcal{R} \cong (\mathbb{Z}^I/D)/\mathbb{Z})$. Moreover, under this valuation $\mathcal{R}$ forms a ω-pseudo-complete Hensel field with x-section. This allows us via the Isomorphism Theory to reduce the elementary equivalence of real closed fields to that of divisible ordered abelian groups. The elementary equivalence of these groups is easily proved (see for example A. Robinson [19]).

8. THE CROSS-SECTION

All our results on the elementary equivalence and decidability of valued fields have assumed the existence of a x-section $\pi: G \to V$. Since the standard fields which motivated this study all have a x-section this may be considered a reasonable requirement. Nonetheless, this function is not among the usual functions and relations in terms of which a valued field is defined, namely the field operations, the value group addition and inequality relation, and the valuation function ord.

We shall show in this section that the results obtained for fields with x-section may also be obtained for valued fields without x-section. These results are neither stronger nor weaker than the previous ones. On the one hand, allowing a x-section π among the functions enriches the class of elementary statements and so strengthens decidability results. On the other hand, these results do not apply to fields without x-section.

For an instance of such a field, consider the ultrapower field $\mathcal{Q}_p \to \mathcal{Z}$ ($\mathcal{Q}_p = \mathbb{Q}_p^I/D$, $\mathcal{Z} = \mathbb{Z}^I/D$). The group $\mathbb{Z}$ is a convex subgroup of $\mathcal{Z}$. This allows us to define a new valuation on $\mathcal{Q}_p$ namely, $\mathcal{Q}_p \overset{v}{\to} \mathcal{Z}/\mathbb{Z}$ with $\bar{\mathcal{Q}}_p = \mathbb{Q}_p$. Now as we have seen $\mathbb{Q}_p$ is a decidable field of characteristic 0 and $\mathcal{Z}/\mathbb{Z}$, being a divisible ordered abelian group, is also decidable (see Robinson [19]). One can show however that this valuation v has no x-section. Thus the previous results do not allow us to conclude that $\mathcal{Q}_p \overset{v}{\to} \mathcal{Z}/\mathbb{Z}$ is a decidable valued field. The results of this section will show that this is indeed the case.

Now all our results followed from applying the Isomorphism Theorem to appropriate ultraproducts. This in particular required that a x-section exist for these ultraproducts. We assured this by simply adding a x-section function to the first order language. We shall now show that these ultraproducts have a x-section automatically, even when the factors do not. This is a consequence of the ω_1-saturation property of the ultraproducts. We have so far used the ω_1-saturation property to show that the ultraproducts are ω-pseudo-complete. This uses only the additive properties of the field. We shall now use the saturation property on the multiplicative group of the field. This will take the form of proving that the group is complete in a certain topology, the Z-topology.

We are concerned with the existence of a x-section for the valuation $V \xrightarrow{\text{ord}} G$. Now this is in fact just a problem about ord as a homomorphism of the abelian group V^* on G. The fact that V is a field and that G is ordered will only enter indirectly via the fact that V^* has meager torsion and that G is torsion-free.

Let us now switch to additive notation and assume that A is an abelian group. The Z-topology on A is defined by letting the family $\{nA \mid n = 1,2,\ldots\}$ be a base of a system of neighborhoods of the identity 0 of A. Note that the smaller family $\{n!A \mid n = 1,2,\ldots\}$ is also of a base of a system of neighborhoods of 0. The latter has the advantage of being a nested sequence. The group A is Z-complete if it is complete in the Z-topology on A.

Lemma 9. An ω_1-saturated abelian group A is Z-complete. In particular, a countably indexed non-principal ultraproduct of abelian groups is Z-complete.

The proof is entirely similar to the proof of the ω-pseudo-completeness of ultraproducts given in Theorem 2 and so will be omitted.

Clearly an abelian group A is Hausdorff in the Z-topology if and only if the subgroup $A_\omega = \bigcap_{n=1}^{\infty} nA$ is the trivial group 0. For the class of groups in which we are interested we shall now see that A_ω is the largest divisible subgroup of A. This will enable us to reduce our problem to the case of Hausdorff groups.

Definition. An abelian group A has meager torsion if for each integer $n > 0$, the subgroup $\{a|na = 0\}$ is finite.

As our principal example, note that a subgroup of the multiplicative group of a field has meager torsion.

Lemma 10. Let A be an abelian group with meager torsion. Then the group $A_\omega = \bigcap_{n=1}^{\infty} nA$ is the largest divisible subgroup of A.

Proof. It clearly suffices to show that A_ω is divisible. Let $a \in A_\omega$. By hypothesis, the set $S_n = \{s \in A|ns = a\}$ is a non-empty finite set, for each integer $n > 0$. For each $j|i$ define the map $\theta_i^j : S_i \to S_j$ by $\theta_i^j(b) = (i/j)b$. Then $\{S_i, \theta_i^j\}$ forms an inverse system of non-empty finite sets, so that $S = \varprojlim S_i \neq \emptyset$. Let $\beta \in S$. We have the canonical maps $\theta_k : S \to S_k$, with $\theta_k^i \cdot \theta_k = \theta_i$ if $i|k$. Let $\beta_k = \theta_k(\beta)$. Then $\beta_i = j\beta_{ij}$, so that $\beta_i \in A_\omega$. Thus for each integer $n > 0$, $a = \beta_1 = n\beta_n$, with $\beta_n \in A_\omega$. This proves that A_ω is divisible.

The existence of a x-section π to a homomorphism $H \xrightarrow{h} G$ of abelian groups (i.e., a homomorphism $\pi : G \to H$ such that $h(\pi(\alpha)) = \alpha$ for all $\alpha \in G$) is equivalent to the splitting of the exact sequence

$$0 \longrightarrow A \longrightarrow H \xrightarrow{h} G \longrightarrow 0$$

where A is the kernel of H. Now, given G and A, the splitting of the above exact sequence for every abelian group H is equivalent to the homological condition $\mathrm{Ext}_{\mathbb{Z}}^1(G,A) = 0$. It is therefore natural to make use of homological techniques in proving the existence of a x-section. We recall here those facts from homological algebra which we shall need. We shall work in the category of abelian groups (considered as $\mathbb{Z}$-modules). The only special property of $\mathbb{Z}$-modules we use is that the global dimension of $\mathbb{Z}$ is one, so that $\mathrm{Ext}^2_{\mathbb{Z}}(G,A) = 0$. The reader may consult e.g., Northcott [20], Chapter 7 for background material.

As usual $\mathrm{Hom}(G,A)$ denotes the group of homomorphisms from G into A. $\mathrm{Ext}(G,A)$ $(= \mathrm{Ext}^1_{\mathbb{Z}}(G,A))$ denotes the group of (abelian) extensions of A by G.

We shall make use of the following properties of the functor Ext.

(1) The exact sequence

$$0 \longrightarrow R \longrightarrow S \longrightarrow T \longrightarrow 0$$

induces the exact sequence

$$0 \longrightarrow \mathrm{Hom}(T,A) \longrightarrow \mathrm{Hom}(S,A) \longrightarrow \mathrm{Hom}(R,A)$$

$$\longrightarrow \mathrm{Ext}(T,A) \longrightarrow \mathrm{Ext}(S,A) \longrightarrow \mathrm{Ext}(R,A) \longrightarrow 0$$

(2) $\mathrm{Ext}(G,A_1 \oplus A_2) \cong \mathrm{Ext}(G,A_1) \oplus \mathrm{Ext}(G,A_2)$.

(3) $\mathrm{Ext}(G,A) = 0$, for A a divisible group (since a divisible subgroup of a group is a direct summand).

<u>Lemma 11</u>. Let A be a Z-complete abelian group with meager torsion. Then $\mathrm{Ext}(G,A) = 0$, for every torsion-free abelian group G.

<u>Proof</u>.

(a) Since A_ω is divisible it is a direct summand of A. Writing $A = A_\omega \oplus B$ we have by Properties (2) and (3)

$$\mathrm{Ext}(G,A) = \mathrm{Ext}(G,A_\omega) \oplus \mathrm{Ext}(G,B) = \mathrm{Ext}(G,B) .$$

Also, since B is a direct summand of A, B is Z-complete; and by the previous lemma, B is Hausdorff in the Z-topology.

(b) Let

$$0 \longrightarrow G \longrightarrow S \longrightarrow T \longrightarrow 0$$

be an exact sequence. Then by Property (1) we have the exact sequence

$$\text{Ext}(S,A) \longrightarrow \text{Ext}(G,A) \longrightarrow 0 .$$

Thus, to prove that $\text{Ext}(G,A) = 0$ it suffices to prove $\text{Ext}(S,A) = 0$ for some extension S of G. In particular, we may take for S a divisible group, say the divisible hull $G \otimes_{\mathbb{Z}} Q$ of G.

(c) Parts (a) and (b) show that it suffices to prove $\text{Ext}(G,A) = 0$ for A Hausdorff and Z-complete and for G divisible and torsion-free. In other words, we must show under these conditions that the exact sequence

$$0 \longrightarrow A \longrightarrow H \xrightarrow{\varphi} G \longrightarrow 0$$

splits. We effect the splitting by showing that there is a x-section $\psi : G \to H$ of φ.

Let $\gamma \in G$. Choose $c_n \in \varphi^{-1}(\gamma/n!)$ for all integers $n > 0$. Let $a_n = n!c_n - c_1$. Then $a_n \in A$, since $\varphi(a_n) = 0$. Now $\{a_n\}$ forms a Cauchy sequence in the Z-topology of A, since $a_n - a_k = k!\,(\frac{n!}{k!} c_n - c_k) \in k!A$. Hence, A being Z-complete and Hausdorff, $\{a_n\}$ has a unique limit a in A. Define the map $\psi : G \to H$ by

$$\psi(\gamma) = a + c_1 .$$

Then ψ is well-defined for if $c_n' \in \varphi^{-1}(\gamma/n!)$ for all $n > 0$, and $a_n' = n!c_n' - c_1'$, then

$$(a_n + c_1) - (a_n' + c_1') = n!c_n - n!c_n' \in n!A ,$$

so that $a + c_1 = a' + c_1'$.

Clearly ψ is a homomorphism, for let $\gamma_1, \gamma_2 \in G$, $c_{ni} \in \varphi^{-1}(\gamma_i/n!)$, $a_{ni} = n!c_{ni} - c_{1i}$ for $n > 0$, $i = 1,2$. Choose $d_n \in \varphi^{-1}((\gamma_1 + \gamma_2)/n!)$ to be $d_n = c_{n1} + c_{n2}$. Then

$$\psi(\gamma_1 + \gamma_2) = a_1 + a_2 + c_{11} + c_{12} = \psi(\gamma_1) + \psi(\gamma_2) ,$$

where $a_i = \lim_{n \to \infty} a_{ni}$, $i = 1,2$.

Finally, $\varphi(\psi(\gamma)) = \varphi(a + c_1) = \gamma$.

Proposition 5 (a). Let $V_i \xrightarrow{\text{ord}} G_i$ be a countable family of valued fields. Let $V \xrightarrow{\text{ord}} G$ be a non-principal ultraproduct of this family. Then there exists a x-section $\pi: G \to V$.

Proof. If $V = \Pi V_i/D$, and U_i is the group of units of V_i, then the group of units U of V is $\Pi U_i/D$. Hence U is Z-complete group. Since U is a subgroup of V^*, U has meager torsion. Also, G, being ordered, is torsion-free. Thus we may apply Lemma 11 to obtain $\mathrm{Ext}(G,U) = 0$ from which the existence of a x-section follows.

For the case of ramification index $e = 1$ we require the existence of a normalized x-section. This will follow from the following homological lemma.

Lemma 12. Let

$$\begin{array}{ccccccccc} 0 & \longrightarrow & A & \longrightarrow & H & \longrightarrow & G & \longrightarrow & 0 \\ & & \uparrow & & \uparrow & & \uparrow & & \\ 0 & \longrightarrow & A_1 & \longrightarrow & H_1 & \longrightarrow & G_1 & \longrightarrow & 0 \end{array}$$

be split exact sequences. Assume that $\mathrm{Ext}(G/G_1, A) = 0$. Then every x-section $\pi_1: G_1 \to H_1$ can be extended to a x-section $\pi: G \to H$.

Proof. By Property 1, the exact sequence

$$0 \longrightarrow G_1 \longrightarrow G \longrightarrow G/G_1 \longrightarrow 0$$

induces the exact sequence

$$\mathrm{Hom}(G_1, A) \longrightarrow \mathrm{Hom}(G, A) \longrightarrow \mathrm{Ext}(G/G_1, A) = 0 .$$

Hence, every homomorphism from G_1 into A extends to one from G into A.

Now by hypothesis there exists a x-section $\pi_0: G \to H$. Define the homomorphism $h_1: G_1 \to A$ by $h_1 = \pi_1 - \pi_0$. Extend h_1 to a homomorphism $h: G \to A$. Let the map $\pi: G \to H$ be defined by $\pi = \pi_0 + h$. Then clearly π is a x-section extending π_1.

<u>Proposition 5 (b)</u>. Let $V_i \xrightarrow{\text{ord}} G_i$ be a countable family of Hensel fields. Let $V \xrightarrow{\text{ord}} G$ be a non-principal ultraproduct of this family. Assume that V has ramification index $e = 1$. Then there exists a normalized x-section $\pi : G \to V$.

<u>Proof</u>. As we have seen in the proof of Proposition 4 (b), we have the exact sequences

$$\begin{array}{ccccccccc} 0 & \longrightarrow & U & \longrightarrow & V & \longrightarrow & G & \longrightarrow & 0 \\ & & \uparrow & & \uparrow & & \uparrow & & \\ 0 & \longrightarrow & U_0 & \longrightarrow & W(\bar{V}) & \longrightarrow & \mathbb{Z} & \longrightarrow & 0 \end{array}$$

where U_0 is the group of units of $W(\bar{V})$. The upper sequence splits by Proposition 5 (a); the lower one splits because there is a x-section $\pi_1 : Z \to W(\bar{V})$ given by $\pi_1(n) = p^n$. Since $\mathbb{Z}$ is a convex subgroup of G, $G/\mathbb{Z}$ is ordered and hence torsion-free. As before U is Z-complete and with meager torsion. Hence $\mathrm{Ext}(G/Z, U) = 0$ by Lemma 11. It now follows from Lemma 12 that there is a x-section $\pi : G \to V$ extending π_0, so that $\pi(1) = p$.

As a consequence of Proposition 5 (a) and (b) all the results of Section 7 apply without assuming the existence of a x-section function for the valued fields considered there.

9. STRUCTURE THEOREM

The Isomorphism Theorem states that an unramified ω-pseudo-complete $\aleph_1$-Hensel field V with x-section is uniquely determined by its value group G and residue class field $\mathfrak{K}$. This naturally calls for the identification of this uniquely determined field. In other words, what is required is a canonical construction of V from G and $\mathfrak{K}$. In this section we give such a construction.

For the case of ramification index $e = 0$, this is straightforward. The generalized power series field $\mathfrak{K}((t^G))_\omega$ is a Hensel field of cardinality $2^{\aleph_0}$, with a x-section $\pi(\alpha) = t^\alpha$. It is easily checked that this field is also ω-pseudo-complete. For the case $e = 1$, the situation is more complicated. Since the group $\mathbb{Z}$ is a convex subgroup of G in this case we know that the valuation $\mathrm{ord}: V \to G$ may be decomposed into u and $v = h \circ \mathrm{ord}$.

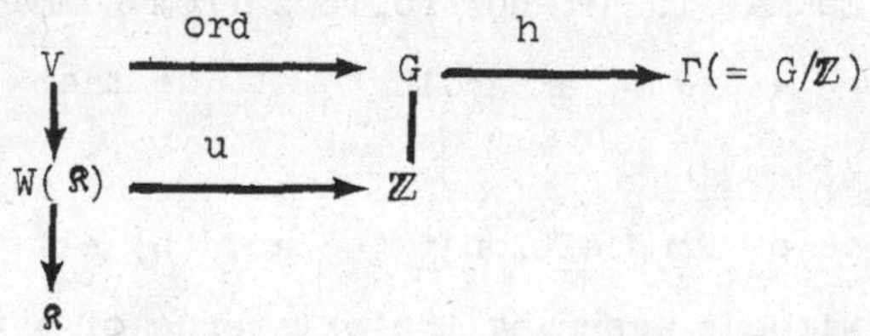

Under the valuation v the field V remains an ω-pseudo-complete Hensel field. If v had a x-section, then we could conclude by the Isomorphism Theorem that $V \cong W(\mathfrak{K})((t^{\Gamma}))_\omega$. However, the valuation v does not have a x-section. The obstruction is that $\mathrm{Ext}(\Gamma, \mathbb{Z}) \neq 0$ so that $0 \to \mathbb{Z} \to G \to \Gamma \to 0$ does not split. For if the canonical homomorphism $G \xrightarrow{h} \Gamma$ had a x-section, i.e., a homomorphism $\Gamma \xrightarrow{g} G$ such that $hg(\alpha) = \alpha$, then $\pi \circ g$ would be a x-section for v. In any case, the extension G of $\mathbb{Z}$ by Γ is given by a 2 co-cycle $m: \Gamma \times \Gamma \to \mathbb{Z}$; namely, let $g: \Gamma \to G$ be any map (not necessarily a homomorphism) such that $h(g(\alpha)) = \alpha$. Then $m(\alpha, \beta) = g(\alpha) + g(\beta) - g(\alpha+\beta)$ is such a

2 co-cycle. Then $\pi \circ m$ is a 2 co-cycle of Γ in $W(\mathfrak{K})$. Note that $\pi \circ m(\alpha,\beta) = \pi(m(\alpha,\beta)) = p^{m(\alpha,\beta)}$, so we may write $\pi \circ m = p^m$.

Now, the construction of an ω-pseudo-complete Hensel field without x-section via a 2 co-cycle $f:\Gamma \times \Gamma \to K$ was described in Example 4 of Section 2. This is the field $K((t^\Gamma;f))_\omega$. (Our present discussion should make the significance of the 2 co-cycle f in Example 4 clear; and also why such fields admit no x-section in general.) This suggests that we construct the field $W(\mathfrak{K})((t^\Gamma;p^m))_\omega$ to describe the field V valued in Γ. Finally we must compose the valuation v with u to regain the valuation $\mathrm{ord}:V \to G$. This results finally in the following canonical construction. Let $0 \to \Delta \to G \to \Gamma \to 0$ be an exact sequence of ordered abelian groups. Let $m:\Gamma \times \Gamma \to \Delta$ be a 2 co-cycle for this extension. We assume that $u:K \to \Delta$ is a valuation of a field K with x-section π_0. We define the valued field $K((t^\Gamma;m))_\beta \xrightarrow{\mathrm{ord}} G$ as follows. The elements of the field are power series $\Sigma_{\alpha \in S} c_\alpha t^\alpha$ where S is a well-ordered subset of Γ of cardinality β. Multiplication in the field is defined formally, with the proviso that $t^\alpha t^\beta = \pi_0^{m(\alpha,\beta)} t^{\alpha+\beta}$. An element of G may be considered as an ordered pair (γ,δ) with $\gamma \in \Gamma$, $\delta \in \Delta$, where $(\gamma,\delta) + (\gamma',\delta') = (\gamma + \gamma', \delta + \delta' + m(\gamma,\gamma'))$. The ordering on the pairs (γ,δ) is defined lexicographically. We then define $\mathrm{ord}(\Sigma_{\alpha \in S} c_\alpha t^\alpha) = (\alpha_0, u(c_{\alpha_0}))$ where α_0 is the smallest element of S with $c_{\alpha_0} \neq 0$. The field $K((t^\Gamma;m))_\beta$ is an ω-pseudo-complete Hensel field with x-section $\pi((\gamma,\delta)) = \pi_0^\delta t^\gamma$. It has residue class field $\bar{K}$ and value group G. This completes our construction. We now summarize the results of this section.

Theorem 5. Let V be an unramified ω-pseudo-complete $\aleph_1$-Hensel field with x-section π, residue class field $\mathfrak{K}$, and value group G. Then, assuming $2^{\aleph_0} = \aleph_1$,

(a) if $e = 0$, $V \cong \mathfrak{K}((t^G))_\omega$;

(b) if $e = 1$, $V \cong W(\mathfrak{K})((t^{G/\mathbb{Z}};m))_\omega$ where m is any 2 co-cycle from $G/\mathbb{Z}$ to $\mathbb{Z}$.

Moreover, the isomorphism is x-analytic.

BIBLIOGRAPHY

[1] J. Ax and S. Kochen, Diophantine problems over local fields, I, Am. J. of Math. 87 (65) 605-630.

[2] ——————, Diophantine problems over local fields, II, Am. J. of Math. 87 (65) 631-648.

[3] ——————, Diophantine problems over local fields, III, Ann. Math. 83 (66) 437-456.

[4] J. Ersov, On elementary theories of local fields, <u>Alg. and Log. Sem.</u> 4 (65) 5-30.

[5] ———, On the elementary theory of maximal normed fields, Doklady 165 (65) 21-23.

[6] J. Ax, A metamathematical approach to some problems in number theory, AMS Symposium (73) 161-190.

[7] P. Cohen, Decision problems for real and p-adic fields, Comm. on Pure and Appl. M. 22 (69) 131-153.

[8] A. Robinson, Problems and methods in model theory, Lecture Notes, Varenna CIME (68) 183-266.

[9] J.Bell and A. Slomson, Models and ultraproducts, North-Holland 1971.

[10] O. F. G. Schilling, The theory of valuations, AMS Mathematical Survey, 1950.

[11] M. Greenberg, Lectures on forms in many variables, Benjamin, 1969.

[12] O. Teichmüller, Diskret bewertete perfekte Körper mit unvolkommenem Restklassen körper, J. für Math. 176 (36) 141-152.

[13] A. Ostrowski, Untersuchungen zur arithmetischen Theorie der Körper, Math. Zeit. 39 (34) 269-404.

[14] I. Kaplansky, Maximal fields with valuations, I and II, Duke J. 9 (42) 303-321, Duke J. 12 (45) 243-248.

[15] K Gödel, The consistency of the generalized continuum hypothesis, Princeton University Press, 1940.

[16] S. Lang, On quasi algebraic closure, Ann. Math. 55 (52) 373-390.

[17] M. Greenberg, Rational points in Henselian discrete valuation rings, Publ. Math. IHES 31 (67) 59-64.

[18] S. Kochen, Integer-valued rational functions over the p-adic numbers, AMS Proc. Symp. in Pure M. 12 (70) 57-73.

[19] A. Robinson, Complete theories, North-Holland, 1956.

[20] D. Northcott, An introduction to homological algebra, Cambridge U.P., 1960.

QUANTIFIER ELIMINATION

Peter Krauss
State University College
New Paltz,NY 12561
USA

Although quantifier elimination plays an important role in mathematical logic, thus far no general definition of this notion has been given in the literature. This is not accidental because the technical difficulties inherent in such a task are quite puzzling. Frequently applied is the method of quantifier elimination. Given a theory Σ , the aim of this method is to determine a set of "simple" formulas such that every formula is equivalent in Σ to a simple formula. For all practical purposes this "definition" is useless because we do not see how to give the word "simple" a precise meaning so that all known applications are covered. A moment of reflection reveals that such explications of "simple as "finite" or "recursive" are absurd. As the term "quantifier elimination" suggests, simple formulas usually have "few" quantifiers. But again, an explication in terms of prefix classes appears to be hopeless in view of known examples. Also often equivalence to simple formulas in the theory Σ turns out to be effective and, moreover, frequently it happens to be decidable whether a simple formula holds in Σ or not. This then yields a decision procedure for the theory Σ . However, tieing such requirements into a definition of quantifier elimination would eliminate many useful and interesting applications, in particular all those which we are going to discuss in this paper. Indeed, if we take the term "simple formula" to mean "quantifierfree formula", then we come up with a very fruitful theory of quantifier elimination, although the expression "simple formula" can now be rather misleading as some examples show. Moreover, the effectiveness of quantifier elimination usually gets lost, although in some cases it can be recaptured by complimentary techniques. In fact, the most striking feature of this approach is that syntactical and combinatorial techniques can be replaced by model theoretic and algebraic methods. Of course, the most direct approach to quantifier elimination is syntactical. By some combinatorial or inductive procedure one shows that every formula is equivalent in the theory Σ to a simple formula. Such arguments are sustained by astounding combinatorial ingenuity although usually they are tedious and exasperating in detail. Several famous results have been established by this method. To mention a few examples, Tarski [12] eliminates quantifiers in the theory of Boolean algebras, Mostowski and Tarski [5] in the theory of well-orderings, Szmielew [11] in the theory of abelian groups and finally Tarski [13] in the theories of algebraically closed fields and real closed fields. Subsequently it was discovered that some of these arguments could be substantially simplified by model theoretic and algebraic methods. Again we only mention a few well-known examples. Robinson [7] uses the notion of model completeness, Kochen [4] uses ultra products, Shoenfield [9] uses the substructure and isomorphism conditions and finally Shoenfield [10] , Blum [1] and Sacks [8] use saturated structures. In all of these cases simple formulas are quantifierfree formulas. We shall add another variant of this approach which is a refinement of the method of saturated structures. By combining this method with the method of extending mappings "an element at a time" (essentially due to Fraissé) we are able to give "almost" purely algebraic necessary and sufficient conditions for elimination of quantifiers which have strikingly simple applications. The algebraic facts we appeal to in these applications are usually quite elementary and the model theory involved is clearly delineated. In fact, our approach reveals how much model theory is indispensible and where exactly it enters into the picture.

Throughout this paper we shall use standard notation and terminology so that we may restrict ourselves to a few preliminary remarks. Given is a basic denumerable similarity type t determined by a countable set Rl of relation symbols and a countable set Op of operation symbols. t is called finitely based if both Rl and Op are finite sets. A t-structure $\mathfrak{A}$ has universe $|\mathfrak{A}|$, and for each n-ary relation symbol $R \in Rl$ has an n-ary relation $R^{\mathfrak{A}}$, and for each n-ary operation symbol $f \in Op$ has an n-ary operation $f^{\mathfrak{A}}$. $\mathfrak{A}$ is called an algebra if $Rl=\emptyset$ and $\mathfrak{A}$ is called a relational structure if $Op=\emptyset$. $\mathfrak{A}$ is called locally finite if every non-empty finite subset of $|\mathfrak{A}|$ generates a finite substructure of $\mathfrak{A}$. Notice that relational structures are locally finite. We write

$$\mathfrak{A} \subseteq \mathfrak{B} \quad , \quad \mathfrak{A} \prec \mathfrak{B} \quad , \quad \mathfrak{A} \equiv \mathfrak{B}$$

in case $\mathfrak{A}$ is a substructure, an elementary substructure, and elementarily equivalent to $\mathfrak{B}$ respectively. If $\mathfrak{M}$ is a class of structures then we define

$$S\mathfrak{M} = \{\mathfrak{B} \mid \mathfrak{B} \subseteq \mathfrak{A} \text{ and } \mathfrak{A} \in \mathfrak{M}\}$$

If $\mathfrak{A}$ is a t-structure and $\emptyset \neq X \subseteq |\mathfrak{A}|$ then we form the similarity type t(X) by adjoining for each $x \in X$ an individual constant (0-ary operation symbol) $\dot{x}$. Correspondingly $\mathfrak{A}' = (\mathfrak{A}, x)_{x \in X}$ is a t(X)-structure, where we take

$$\dot{x}^{\mathfrak{A}'} = x \quad \text{for all} \quad x \in X .$$

More generally, a t(X)-structure is of the form $\mathfrak{B}' = (\mathfrak{B}, f(x))_{x \in X}$, where $\mathfrak{B}$ is a t-structure, $f: X \to |\mathfrak{B}|$ and

$$\dot{x}^{\mathfrak{B}'} = f(x) \quad \text{for all} \quad x \in X .$$

We shall only be interested in the case where X is non-empty and finite. Of course, most of our definitions have obvious and well-known generalizations, however we shall attempt to drive to the point as directly as possible rather than clutter our presentation with extraneous generalities.

Suppose $\emptyset \neq X \subseteq |\mathfrak{A}|$, where X is finite, and $f: X \to |\mathfrak{B}|$. f is called a <u>local isomorphism</u> from $\mathfrak{A}$ into $\mathfrak{B}$, in symbols

$$(\mathfrak{A}, x)_{x \in X} \equiv_0 (\mathfrak{B}, f(x))_{x \in X} ,$$

if there exists an isomorphism from the substructure of $\mathfrak{A}$ generated by X onto the substructure of $\mathfrak{B}$ generated by f(X) which extends f. (Notice that in this case the isomorphism extending f is uniquely determined by f). Now, following Fraissé, we define inductively for each $n < \omega$

$$(\mathfrak{A}, x)_{x \in X} \equiv_{n+1} (\mathfrak{B}, f(x))_{x \in X}$$

if for every $a \in |\mathfrak{A}|$ there exists $b \in |\mathfrak{B}|$ such that

$$(\mathfrak{A}, x, a)_{x \in X} \equiv_n (\mathfrak{B}, f(x), b)_{x \in X}$$

and conversely, for each $b \in |\mathfrak{B}|$ there exists $a \in |\mathfrak{A}|$ such that

$$(\mathfrak{A}, x, a)_{x \in X} \equiv_n (\mathfrak{B}, f(x), b)_{x \in X} .$$

A local isomorphism f from $\mathfrak{A}$ into $\mathfrak{B}$ is called <u>immediately extendible</u> if

$$(\mathfrak{A}, x)_{x \in X} \equiv_1 (\mathfrak{B}, f(x))_{x \in X} .$$

f is <u>called elementary</u> if

$$(\mathfrak{A}, x)_{x \in X} \equiv (\mathfrak{B}, f(x))_{x \in X} .$$

Next we consider the (finitary) first-order language of similarity t. A set Σ of t-formulas is called <u>substructure complete</u> if every local isomorphism between models of Σ is elementary. Notice that this definition is <u>purely model theoretic</u>. This notion has been vagrant in model theory for a long time, however the first formal definition apparently is due to Sacks [8]. Our first lemma is well-known and will only be put to auxiliary use in this paper.

<u>Lemma</u> 1 Σ is substructure complete if and only if Σ is model complete and SMod Σ has the amalgamation property.

We say that Σ <u>admits quantifier elimination</u> if for every t-formula φ there exists a quantifierfree t-formula ψ such that

$$\Sigma \vdash \varphi \leftrightarrow \psi$$

Notice that this definition may be interpreted as being <u>purely syntactical</u>. As we pointed out in our introductory remarks, this "definition" of quantifier elimination has no claim to universality. However many <u>examples</u> of quantifier elimination can be successfully treated with this definition. Moreover, this is the only definition known thus far which gives rise to a fruitful <u>theory</u> of quantifier elimination.

The next theorem converts the syntactical notion of quantifier elimination into the model theoretic notion of substructure completeness. Again it appears that the first explicit statement and proof of this important result are due to Sacks [8].

<u>Theorem</u> 2 Σ admits quantifier elimination if and only if Σ is substructure complete.

The model theoretic-algebraic approach to quantifier elimination now is characterized by an attempt to establish substructure completeness by "purely" algebraic methods. As to be expected, with the help of ultraproducts this goal can always be fully attained. However, we shall develop a method which is less restrictive at the expense of being "almost" algebraic. On the other hand we shall see that the model theoretic residue of this method is easy to handle once it is fully exposed. Our basic tool is an important result which is due to Fraissé [2].

Theorem 3 Suppose $\emptyset \neq X \subseteq |\mathfrak{A}|$, where X is finite, and $f:X \to |\mathfrak{B}|$. If for every $n < \omega$,

$$(\mathfrak{A},x)_{x \in X} \equiv_n (\mathfrak{B},f(x))_{x \in X}$$

then

$$(\mathfrak{A},x)_{x \in X} \equiv (\mathfrak{B},f(x))_{x \in X} .$$

Remark 4 Notice that the hypothesis of Theorem 3 is purely algebraic although it involves a rather complicated inductive procedure. The converse of Theorem 3 is not true in general. In fact, it is true only in case both $\mathfrak{A}$ and $\mathfrak{B}$ are locally finite of finitely based similarity type. Since Fraissé's Theorem has been proved originally for relational structures with finitely many relations (for which the converse is true), the first applications in the literature have been restricted to this case. We shall soon see that our method of establishing substructure completeness will fail to be "purely" algebraic by exactly the same margin as the converse of Theorem 3 fails to be true. In fact, in order to establish that a local isomorphism f from $\mathfrak{A}$ into $\mathfrak{B}$ is elementary we shall show purely algebraically that for every $n < \omega$,

$$(\mathfrak{A},x)_{x \in X} \equiv_n (\mathfrak{B},f(x))_{x \in X} .$$

Since the converse of Theorem 3 is not true, in general we are attempting to show something too strong. Therefore the models $\mathfrak{A}$ and $\mathfrak{B}$ have to be suitably chosen. This choice constitutes the model theoretic residue of our method. First we reduce the inductive procedure involved in the hypothesis of Theorem 3 to immediate extensions of local isomorphisms. Of course, for practical applications this reduction is quite crucial.

Lemma 5 Suppose every local isomorphism from $\mathfrak{A}$ into $\mathfrak{B}$ is immediately extendible, let $\emptyset \neq X \subseteq |\mathfrak{A}|$, where X is finite, and suppose $f:X \to |\mathfrak{B}|$ is a local isomorphism. Then for every $n < \omega$,

$$(\mathfrak{A},x)_{x \in X} \equiv_n (\mathfrak{B},f(x))_{x \in X} .$$

Proof: By induction on n. Assume the assertion is true for n and let $a \in |\mathfrak{A}|$. By hypothesis there exists $b \in |\mathfrak{B}|$ such that

$$(\mathfrak{A}, x, a)_{x \in X} \equiv_0 (\mathfrak{B}, f(x), b)_{x \in X} .$$

By induction hypothesis,

$$(\mathfrak{A}, x, a)_{x \in X} \equiv_n (\mathfrak{B}, f(x), b)_{x \in X},$$

and by symmetry,

$$(\mathfrak{A}, x)_{x \in X} \equiv_{n+1} (\mathfrak{B}, f(x))_{x \in X}.$$

Corollary 6 If every local isomorphism between models of Σ is immediately extendible then Σ is substructure complete.

Proof: Use Theorem 3 and Lemma 5 .

Remark 7 Notice that the hypothesis of Corollary 6 is purely algebraic and does not involve any inductive procedure any more. Again, the converse of Corollary 6 is not true. In the remainder of this paper we shall be concerned with the margin by which Corollary 6 fails to be true. This will reveal the model theoretic residue in our method of eliminating quantifiers. First we need a few more definitions.

$\mathfrak{A}$ is called locally homogeneous if every local automorphism of $\mathfrak{A}$ is immediately extendible. Notice that a countable locally homogeneous structure is homogeneous.

Corollary 8 If $\mathfrak{A}$ is locally homogeneous then every local automorphism of $\mathfrak{A}$ is elementary.

Proof: Use Lemma 5 .

Remark 9 Again the converse of Corollary8 is not true. Although an investigation of the converse of Corollary 8 leads to interesting results we shall not be sidetracked by such a pursuit.

$\mathfrak{A}$ is called locally saturated if for every finite $X \subseteq |\mathfrak{A}|$, every 1-type for $Th(\mathfrak{A}, x)_{x \in X}$ is realizable in $(\mathfrak{A}, x)_{x \in X}$. Notice that a finite structure is locally saturated and a countable locally saturated structure is saturated. Now we can see to what extent the converse of Theorem 3 is true.

Theorem 10 Suppose $\mathfrak{A}$ and $\mathfrak{B}$ are locally saturated. Then $\mathfrak{A} \equiv \mathfrak{B}$ if and only if for every $n < \omega$, $\mathfrak{A} \equiv_n \mathfrak{B}$.

Proof: The assertion is trivial in case $\mathfrak{A}$ and $\mathfrak{B}$ are finite. So assume that $\mathfrak{A}$ and $\mathfrak{B}$ are infinite. We shall prove by induction on n that for every finite $X \subseteq |\mathfrak{A}|$ and every $f: X \to |\mathfrak{B}|$, if

$$(\mathfrak{A}, x)_{x \in X} \equiv (\mathfrak{B}, f(x))_{x \in X}$$

then

$$(\mathfrak{A}, x)_{x \in X} \equiv_n (\mathfrak{B}, f(x))_{x \in X} .$$

Clearly the assertion holds for $n = 0$. Now assume that the assertion is true for n, let $X \subseteq |\mathfrak{A}|$ be finite and let $f: X \to |\mathfrak{B}|$, where

$$(\mathfrak{A}, x)_{x \in X} \equiv (\mathfrak{B}, f(x))_{x \in X} .$$

Consider $a \in |\mathfrak{A}|$ and let Σ be the type of a in $(\mathfrak{A}, x)_{x \in X}$. Then Σ is a 1-type for $\mathrm{Th}(\mathfrak{B}, f(x))_{x \in X}$. Since $\mathfrak{B}$ is locally saturated, there exists $b \in |\mathfrak{B}|$ such that b realizes Σ in $(\mathfrak{B}, f(x))_{x \in X}$. Thus

$$(\mathfrak{A}, x, a)_{x \in X} \equiv (\mathfrak{B}, f(x), b)_{x \in X} .$$

By induction hypothesis,

$$(\mathfrak{A}, x, a)_{x \in X} \equiv_n (\mathfrak{B}, f(x), b)_{x \in X} ,$$

and by symmetry,

$$(\mathfrak{A}, x)_{x \in X} \equiv_{n+1} (\mathfrak{B}, f(x))_{x \in X} .$$

This completes the inductive proof and the converse of Theorem 3 is established.

Denumerable (locally) saturated models do not always exist. In fact, necessary and sufficient conditions for the existence of a denumerable saturated model of a theory Σ are well-known (see Vaught [14]). Nevertheless, locally saturated models always exist in abundance. The next theorem tells us where to look for them (see Sacks [8]).

Theorem 11 If $\mathfrak{A}$ is infinite then there exists a locally saturated $\mathfrak{B} \succ \mathfrak{A}$ such that

$$\mathrm{card}\,\mathfrak{B} \leq \max(2^{\aleph_0}, \mathrm{card}\,\mathfrak{A}) .$$

Notice that in Theorem 11 we may not be able to obtain $\mathfrak{B}$ denumerable in case $\mathfrak{A}$ is denumerable. Now we can state the main result of this paper.

<u>Theorem</u> 12 The following are equivalent :

(i) Σ is substructure complete.

(ii) Every local isomorphism between locally saturated models of Σ is immediately extendible.

(iii) For every $\mathfrak{A}, \mathfrak{B} \in \mathrm{Mod}\,\Sigma$ there exist $\mathfrak{C} \succ \mathfrak{A}$ and $\mathfrak{D} \succ \mathfrak{B}$ such that every local isomorphism from $\mathfrak{C}$ into $\mathfrak{D}$ is immediately extendible.

<u>Proof</u>: Assume (i) and suppose $\mathfrak{A}, \mathfrak{B} \in \mathrm{Mod}\,\Sigma$ are locally saturated. Let $\emptyset \neq X \subseteq |\mathfrak{A}|$, where X is finite, and let $f: X \to |\mathfrak{B}|$, where

$$(\mathfrak{A}, x)_{x \in X} \equiv_0 (\mathfrak{B}, f(x))_{x \in X} .$$

By hypothesis,

$$(\mathfrak{A}, x)_{x \in X} \equiv (\mathfrak{B}, f(x))_{x \in X} .$$

Consider any $a \in |\mathfrak{A}|$ and let Σ be the type of a in $(\mathfrak{A}, x)_{x \in X}$. Then Σ is a 1-type for $\mathrm{Th}(\mathfrak{B}, f(x))_{x \in X}$. Since $\mathfrak{B}$ is locally saturated, there exists $b \in |\mathfrak{B}|$ realizing Σ in $(\mathfrak{B}, f(x))_{x \in X}$. Thus

$$(\mathfrak{A}, x, a)_{x \in X} \equiv (\mathfrak{B}, f(x), b)_{x \in X} ,$$

and therefore

$$(\mathfrak{A}, x, a)_{x \in X} \equiv_0 (\mathfrak{B}, f(x), b)_{x \in X} .$$

By symmetry,

$$(\mathfrak{A}, x)_{x \in X} \equiv_1 (\mathfrak{B}, f(x))_{x \in X} .$$

Next, assume (ii) and suppose $\mathfrak{A}, \mathfrak{B} \in \mathrm{Mod}\,\Sigma$. By Theorem 11, there exist locally saturated $\mathfrak{C} \succ \mathfrak{A}$ and $\mathfrak{D} \succ \mathfrak{B}$ and (iii) follows at once. Finally, assume (iii). Suppose $\mathfrak{A}, \mathfrak{B} \in \mathrm{Mod}\,\Sigma$, let $\emptyset \neq X \subseteq |\mathfrak{A}|$, where X is finite, and suppose $f: X \to |\mathfrak{B}|$, where

$$(\mathfrak{A}, x)_{x \in X} \equiv_0 (\mathfrak{B}, f(x))_{x \in X} .$$

Choose $\mathfrak{C} \succ \mathfrak{A}$ and $\mathfrak{D} \succ \mathfrak{B}$ according to the hypothesis. Then

$$(\mathfrak{C}, x)_{x \in X} \equiv_0 (\mathfrak{D}, f(x))_{x \in X} ,$$

and by Lemma 5 , for all $n<\omega$,

$$(\mathcal{C},x)_{x\in X} \equiv_n (\mathcal{D},f(x))_{x\in X} .$$

By Theorem 3,

$$(\mathcal{C},x)_{x\in X} \equiv (\mathcal{D},f(x))_{x\in X} ,$$

and therefore

$$(\mathfrak{A},x)_{x\in X} \equiv (\mathfrak{B},f(x))_{x\in X} .$$

This establishes (i).

Remark 13 Theorem 12 clearly reveals the model theoretic component in our method of quantifier elimination. It gives us two options to apply model theory. Upon first view it appears that in applications we still may get rather deeply involved in model theory. In (ii) it appears that we have to determine the locally saturated models of the theory Σ . In (iii) it is not clear which elementary extensions of models of Σ to choose (unless we go back to (ii) !). However, in all applications we have investigated thus far the task of characterizing the locally saturated models of Σ turns out to be surprisingly simple. Once this is accomplished the application of (ii) only requires some well-known algebraic facts. Applying (iii) we have discovered that some well-known algebraic facts together with a direct appeal to the upward Löwenheim-Skolem Theorem are often successful. We shall amply illustrate these remarks with examples. First we shall further refine our results.

It is often possible to establish substructure completeness by considering locally saturated structures internally.

Corollary 14 Suppose that either Σ is complete or $\mathrm{SMod}\,\Sigma$ has the amalgamation property. Then Σ is substructure complete if and only if all locally saturated models of Σ are locally homogeneous.

Proof: We first consider the case where Σ is complete. Assume all locally saturated models of Σ are locally homogeneous and suppose $\mathfrak{A}$, $\mathfrak{B}\in \mathrm{Mod}\,\Sigma$ are locally saturated. Let $\emptyset \neq X\subseteq |\mathfrak{A}|$, where X is finite, and let $f:X\to |\mathfrak{B}|$, where

$$(\mathfrak{A},x)_{x\in X} \equiv_0 (\mathfrak{B},f(x))_{x\in X} .$$

Since Σ is complete, $\mathfrak{A}\equiv\mathfrak{B}$. Consider any $a\in|\mathfrak{A}|$. Since $\mathfrak{B}$ is locally saturated it follows by a well-known argument that there exist $g: X \to |\mathfrak{B}|$ and $c \in |\mathfrak{B}|$ such that

$$(\mathfrak{A}, x, a)_{x \in X} \equiv_0 (\mathfrak{B}, g(x), c)_{x \in X} .$$

Therefore

$$(\mathfrak{B}, g(x))_{x \in X} \equiv_0 (\mathfrak{B}, f(x))_{x \in X} ,$$

and by hypothesis there exists $b \in |\mathfrak{B}|$ such that

$$(\mathfrak{B}, g(x), c)_{x \in X} \equiv_0 (\mathfrak{B}, f(x), b)_{x \in X} .$$

Thus

$$(\mathfrak{A}, x, a)_{x \in X} \equiv_0 (\mathfrak{B}, f(x), b)_{x \in X} ,$$

and by symmetry,

$$(\mathfrak{A}, x)_{x \in X} \equiv_1 (\mathfrak{B}, f(x))_{x \in X} .$$

By Theorem 12, Σ is substructure complete. The converse follows directly from Theorem 12.

The case where SModΣ has the amalgamation property follows from the first case because in this case Σ is substructure complete if and only if all complete extensions of Σ are substructure complete.

Remark 15 There are again two options to apply Corollary 14. Often we wish to eliminate quantifiers in order to establish completeness. In this case the first option is not available. Then it is sometimes known from algebra that SModΣ has the amalgamation property. On the other hand it is also known from algebra that it is usually rather difficult to establish the amalgamation property.

Next we shall discuss the case where model theory can be completely eliminated from our method of quantifier elimination. From Remark 4 it is clear when to expect this case.

Corollary 16 Suppose the similarity type is finitely based and every model of Σ is locally finite. If either Σ is complete or SModΣ has the amalgamation property then the following are equivalent :

(i) Σ is substructure complete.

(ii) Every model of Σ is locally homogeneous.

(iii) Every denumerable model of Σ is homogeneous.

Proof: We again first consider the case where Σ is complete. Assume (i) and suppose $\mathfrak{A} \in \mathrm{Mod}\Sigma$. By Theorem 11 there exists locally saturated $\mathfrak{B} \succ \mathfrak{A}$, and by Corollary 14, $\mathfrak{B}$ is locally homogeneous. Now it is well-known (see Morley and Vaught [6]) that we actually can give a set Γ of formulas such that for any locally finite structure $\mathfrak{C}$, $\mathfrak{C} \in \mathrm{Mod}\Sigma$ if and only if $\mathfrak{C}$ is locally homogeneous. It follows that $\mathfrak{A}$ is locally homogeneous. Thus (i) implies (ii). Conversely, (i) follows from (ii) by Corollary 14, and the equivalence of (ii) and (iii) is obvious. The case where $\mathrm{SMod}\Sigma$ has the amalgamation property can be treated as before in the proof of Corollary 14.

Remark 17 Suppose, under the hypothesis of Corollary 16, that Σ is substructure complete. If, moreover, Σ is complete then it follows from Corollary 16 and a theorem of Morley and Vaught [6] that Σ is $\aleph_0$-categorical. Thus in either case all complete extensions of Σ are $\aleph_0$-categorical, and therefore every denumerable model of Σ is saturated. It follows that every model of Σ is locally saturated. This explains why in this case model theory can be completely eliminated from our method of quantifier elimination (Compare also Theorem 10).

Now it is time to give some examples. Since these examples are well-known from the literature we do not have to go into much detail. We begin with the most direct purely algebraic cases where Corollary 16 is applicable.

Examples 18 (i) Let (DNO) be the theory of dense linear orderings $\langle A, \leq \rangle$ without endpoints. By definition, every dense linear ordering without endpoints is locally homogeneous. Since SMod(DNO) is the class of linear orderings, which has the amalgamation property, (DNO) is substructure complete.

(ii) Let (ALBA) be the theory of atomless Boolean algebras $\langle A, \wedge, \vee, ^- \rangle$. Again it is a well-known fact that every atomless Boolean algebra is locally homogeneous. Indeed, atomless Boolean algebras may be defined as the locally homogeneous Boolean algebras. Since SMod(ALBA) is the class of Boolean algebras, which has the amalgamation property, (ALBA) is substructure complete.

(iii) For each prime number p, let (AG_p) be the theory of infinite elementary abelian p-groups $\langle A, +, -, 0 \rangle$. Since an infinite elementary abelian p-group can be considered as an infinite dimensional vector space over the integers modulo p, it follows at once that every infinite elementary abelian p-group is locally homogeneous. Since $SMod(AG_p)$ is the class of elementary abelian p-groups, which has the amalgamation property, (AG_p) is substructure complete.

<u>Remarks</u> 19 In all Examples 18 we have chosen the second option in the hypothesis of Corollary 16 using the amalgamation property. In two cases we just as well could have chosen the first option using the completeness of (DNO) and (AG_p) respectively. However, by essentially the same argument we can also directly establish condition (ii) of Theorem 12 for <u>all</u> models $\mathfrak{A}$ and $\mathfrak{B}$. In this case we obtain the amalgamation property and completeness as immediate consequences of Lemma 1.

Next we give some examples where an almost blindfolded application of the upward Löwenheim-Skolem Theorem together with condition (iii) of Theorem 12 are successful.

<u>Examples</u> 20 (i) Let (ACF) be the theory of <u>algebraically closed fields</u> $\langle A,+,-,0,\cdot,^{-1},1\rangle$. Suppose $\mathfrak{A},\mathfrak{B}\in Mod(ACF)$. Then $\mathfrak{A}$ and $\mathfrak{B}$ are infinite and therefore there exist uncountable $\mathfrak{C}\succ\mathfrak{A}$ and $\mathfrak{D}\succ\mathfrak{B}$. Let $\emptyset \neq X \subseteq |\mathfrak{C}|$, where X is finite, and suppose $f: X \to |\mathfrak{D}|$, where

$$(\mathfrak{C},x)_{x\in X} \equiv_0 (\mathfrak{D},f(x))_{x\in X}.$$

Let $\mathfrak{C}_0$ be the algebraic closure of the subfield of $\mathfrak{C}$ generated by X. Then there exists an embedding $g:\mathfrak{C}_0 \to \mathfrak{D}$ such that $f \subseteq g$. Now consider any $a \in |\mathfrak{C}|$. If $a \in |\mathfrak{C}_0|$ then

$$(\mathfrak{C},x,a)_{x\in X} \equiv_0 (\mathfrak{D},f(x),g(a))_{x\in X}.$$

Otherwise a is transcendental over $\mathfrak{C}_0$. Since $\mathfrak{D}$ is uncountable it has infinite transcendence degree over $g(\mathfrak{C}_0)$. It follows at once that there exists $b\in|\mathfrak{D}|$ such that

$$(\mathfrak{C},x,a)_{x\in X} \equiv_0 (\mathfrak{D},f(x),b)_{x\in X}.$$

By symmetry,

$$(\mathfrak{C},x)_{x\in X} \equiv_1 (\mathfrak{D},f(x))_{x\in X},$$

and (ACF) is substructure complete.

(ii) Let (DTFA) be the theory of <u>infinite divisible torsionfree</u>

abelian groups $\langle A,+,-,0\rangle$. Suppose $\mathfrak{A},\mathfrak{B} \in \mathrm{Mod(DTFA)}$. Then there exist uncountable $\mathfrak{C} \succ \mathfrak{A}$ and $\mathfrak{D} \succ \mathfrak{B}$. Let $\emptyset \neq X \subseteq |\mathfrak{C}|$,where X is finite, and suppose $f:X \to |\mathfrak{D}|$, where

$$(\mathfrak{C},x)_{x\in X} \equiv_0 (\mathfrak{D},f(x))_{x\in X} .$$

Let $\mathfrak{C}_0$ be the divisible hull of the subgroup of $\mathfrak{C}$ generated by X. Then there exists an embedding $g: \mathfrak{C}_0 \to \mathfrak{D}$ such that $f \subseteq g$. Now consider any $a \in |\mathfrak{C}|$. If $a \in |\mathfrak{C}_0|$ then

$$(\mathfrak{C},x,a)_{x\in X} \equiv_0 (\mathfrak{D},f(x),g(a))_{x\in X} .$$

Otherwise we consider $\mathfrak{C}$ and $\mathfrak{D}$ as infinite dimensional vector spaces over the rationals. Then $\mathfrak{C}_0$ may be identified with the subspace of $\mathfrak{C}$ generated by X and it follows at once that there exists $b \in |\mathfrak{D}|$ such that

$$(\mathfrak{C},x,a)_{x\in X} \equiv_0 (\mathfrak{D},f(x),b)_{x\in X} .$$

By symmetry,

$$(\mathfrak{C},x)_{x\in X} \equiv_1 (\mathfrak{D},f(x))_{x\in X} ,$$

and (DTFA) is substructure complete.

Remark 21 In both Examples 20 we did apply the upward Löwenheim-Skolem Theorem just to obtain uncountable elementary extensions. However, this weak appeal to the Löwenheim-Skolem Theorem is somewhat misleading. In both cases the theory is $\aleph_1$-categorical and therefore all uncountable models are saturated.

Finally we give some examples where a more subtle argument together with condition (ii) of Theorem 12 are required.

Example 22 Let (RCF) be the theory of real closed fields $\langle A,+,-,0,\cdot,^{-1},1,\leq\rangle$. We first separate the property of locally saturated real closed fields which will enable us to eliminate quantifiers. Later we shall see that this property actually characterizes these fields. Let $\mathfrak{A} \in \mathrm{Mod(RCF)}$ and let $\mathfrak{B} \subseteq \mathfrak{A}$. (X,Y) is called a cut of $\mathfrak{B}$ if

(i) $X \cup Y = |\mathfrak{B}|$

(ii) $X \cap Y = \emptyset$

(iii) if $x \in X$ and $y \in Y$ then $x < y$.

Let $a \in |\mathfrak{A}|$. We say that a fills (X,Y) if for all $x \in X$ and all $y \in Y$,

$$x < a < y .$$

We say that (X,Y) <u>can be filled in</u> $\mathfrak{A}$ if there exists $a \in |\mathfrak{A}|$ which fills (X,Y) .

<u>Lemma</u> 23 If $\mathfrak{A}$ is a locally saturated real closed field then every cut of the real closure of a finitely generated subfield of $\mathfrak{A}$ can be filled in $\mathfrak{A}$.

<u>Proof</u>: Suppose $\mathfrak{B}$ is a finitely generated subfield of $\mathfrak{A}$ and let $\bar{\mathfrak{B}}$ be the real closure of $\mathfrak{B}$ in $\mathfrak{A}$. Then every element of $|\bar{\mathfrak{B}}|$ can be characterized as the k-th root of some polynomial with coefficients in $|\mathfrak{B}|$. Thus for each $x \in |\bar{\mathfrak{B}}|$ there exists a $t(\mathfrak{B})$-formula φ_x such that for all $y \in |\mathfrak{A}|$,

$$(\mathfrak{A},b)_{b \in |\mathfrak{B}|} \models \varphi_x[y] \text{ if and only if } y = x .$$

Now suppose (X,Y) is a cut of $\bar{\mathfrak{B}}$ and let Σ be the set of following formulas :

$$\forall v[\varphi_x[v] \rightarrow v < u] \quad \text{if } x \in X ,$$

$$\forall v[\varphi_x[v] \rightarrow u < v] \quad \text{if } x \in Y .$$

Since $\mathfrak{A}$ is a real closed field, $(|\mathfrak{A}|, \leqslant)$ is a dense linear ordering without endpoints. Thus every finite subset of Σ is realizable in $(\mathfrak{A},b)_{b \in |\mathfrak{B}|}$. Since $\mathfrak{A}$ is locally saturated, there exists $a \in |\mathfrak{A}|$ realizing Σ in $(\mathfrak{A},b)_{b \in |\mathfrak{B}|}$. It follows at once that a fills (X,Y).

Now we continue Example 22 and suppose $\mathfrak{A}, \mathfrak{B} \in \text{Mod(RCF)}$ are locally saturated. Let $\emptyset \neq X \subseteq |\mathfrak{A}|$, where X is finite, and suppose $f:X \rightarrow |\mathfrak{B}|$, where

$$(\mathfrak{A},x)_{x \in X} \equiv_0 (\mathfrak{B},f(x))_{x \in X} .$$

Let $\mathfrak{A}_0$ be the real closure of the subfield of $\mathfrak{A}$ generated by X. Then there exists an embedding $g: \mathfrak{A}_0 \rightarrow \mathfrak{B}$ such that $f \subseteq g$. Finally consider any $a \in |\mathfrak{A}|$. If $a \in |\mathfrak{A}_0|$ then

$$(\mathfrak{A},x,a)_{x \in X} \equiv_0 (\mathfrak{B},f(x),g(a))_{x \in X} .$$

Otherwise a determines a cut of $g(\mathfrak{A}_0)$. By Lemma 23 there exists $b \in |\mathfrak{B}|$ which fills this cut. It follows at once from field theory that

$$(\mathfrak{A},x,a)_{x\in X} \equiv_0 (\mathfrak{B},f(x),b)_{x\in X} .$$

By symmetry,

$$(\mathfrak{A},x)_{x\in X} \equiv_1 (\mathfrak{B},f(x))_{x\in X} ,$$

and (RCF) is substructure complete.

We also notice now that to eliminate quantifiers we used a property of locally saturated real closed fields which actually characterizes these fields.

Corollary 24 A real closed field $\mathfrak{A}$ is locally saturated if and only if every cut of the real closure of a finitely generated subfield of $\mathfrak{A}$ can be filled in $\mathfrak{A}$.

Proof: Assume the right-hand side of the assertion. Suppose $\mathfrak{B}$ is a finitely generated subfield of $\mathfrak{A}$ and let Σ be a 1-type for $\mathrm{Th}(\mathfrak{A},b)_{b\in|\mathfrak{B}|}$. Then there exist $\mathfrak{C} \succ \mathfrak{A}$ and $y \in |\mathfrak{C}|$ such that y realizes Σ in $(\mathfrak{C},b)_{b\in|\mathfrak{B}|}$. Let $\bar{\mathfrak{B}}$ be the real closure of $\mathfrak{B}$ in $\mathfrak{A}$. If $y \notin |\bar{\mathfrak{B}}|$ then y determines a cut of $\bar{\mathfrak{B}}$. By hypothesis there exists $x \in |\mathfrak{A}|$ which fills this cut. It follows from field theory that

$$(\mathfrak{A},b,x)_{b\in|\mathfrak{B}|} \equiv_0 (\mathfrak{C},b,y)_{b\in|\mathfrak{B}|} .$$

From Example 22 we now obtain that

$$(\mathfrak{A},b,x)_{b\in|\mathfrak{B}|} \equiv (\mathfrak{C},b,y)_{b\in|\mathfrak{B}|} ,$$

and therefore x realizes Σ in $(\mathfrak{A},b)_{b\in|\mathfrak{B}|}$ .We have shown that $\mathfrak{A}$ is locally saturated.

Example 25 Let (PDA) be the theory of prüferized divisible abelian groups $\langle A,+,-,0\rangle$ (that is divisible abelian groups which have, for each prime p, infinitely many elements of order p).We first review a few facts from group theory.

Every infinite divisible abelian group $\mathfrak{A}$ is a direct sum

$$\Sigma\langle \mathfrak{A}_i \mid i \in I\rangle ,$$

where each summand $\mathfrak{A}_i$ is either isomorphic to the rationals $\langle \mathbb{Q},+,-,0\rangle$

or, for some prime p, is isomorphic to the p-Prüfer group $\langle \mathbb{Z}_{p^\infty},+,-,0\rangle$. Let $\iota_0^{\mathfrak{A}}$ be the number of summands $\mathfrak{A}_i$ in this direct decomposition which are isomorphic to the rationals. Similarly, for each prime p, let $\iota_p^{\mathfrak{A}}$ be the number of summands which are isomorphic to the p-Prüfer group. Then $\iota_0^{\mathfrak{A}}$ together with $\iota_p^{\mathfrak{A}}$, for all primes p, determine $\mathfrak{A}$ up to isomorphism and

$$\mathfrak{A} \in \mathrm{Mod(PDA)} \text{ if and only if } \iota_p^{\mathfrak{A}} \geq \aleph_0 \text{ for all primes } p .$$

We again first separate the property of locally saturated prüferized divisible abelian groups which we need to eliminate quantifiers.

Lemma 26 If $\mathfrak{A}$ is a locally saturated prüferized divisible abelian group then $\iota_0^{\mathfrak{A}} \geq \aleph_0$.

Proof: Suppose $\iota_0^{\mathfrak{A}} = n < \aleph_0$. Then there exists a sequence of n independent elements of infinite order, and any sequence of independent elements of infinite order has at most n elements. If $n = 0$, let $\mathfrak{B}$ be the trivial subgroup of $\mathfrak{A}$. Otherwise let $x \in |\mathfrak{A}|^n$ be a sequence of independent elements of infinite order, and let $\mathfrak{B}$ be the subgroup of $\mathfrak{A}$ generated by $\{x_i \mid i < n\}$. Then

$$\mathfrak{B} \cong \langle \mathbb{Z},+,-,0\rangle^n .$$

Now let Σ be the set of following formulas :

$$\neg\, ku = 0 \quad , \text{ where } 0 < k < \omega ,$$

$$\dot{b} + ku = 0 \rightarrow [\, \dot{b} = 0 \wedge ku = 0 \,] \quad , \text{ where } b \in |\mathfrak{B}| \text{ and } k \in \mathbb{Z} .$$

Since $\mathfrak{A}$ is a prüferized divisible abelian group, every finite subset of Σ is realizable in $(\mathfrak{A},b)_{b\in|\mathfrak{B}|}$. Since $\mathfrak{A}$ is locally saturated, there exists $a \in |\mathfrak{A}|$ realizing Σ in $(\mathfrak{A},b)_{b\in|\mathfrak{B}|}$. It follows that a has infinite order and $\langle x_0,\dots,x_{n-1},a\rangle$ is a sequence of independent elements. This is a contradiction.

Now we continue Example 25 and suppose $\mathfrak{A},\mathfrak{B} \in \mathrm{Mod(PDA)}$ are locally saturated. Let $\emptyset \neq X \subseteq |\mathfrak{A}|$, where X is finite, and suppose $f: X \rightarrow |\mathfrak{B}|$, where

$$(\mathfrak{A},x)_{x\in X} \equiv_0 (\mathfrak{B},f(x))_{x\in X} .$$

Let $\mathfrak{A}_o$ be a divisible hull of the subgroup of $\mathfrak{A}$ generated by X. Then there exists an embedding $g: \mathfrak{A}_o \to \mathfrak{B}$ such that $f \subseteq g$. Moreover, there exists divisible $\mathfrak{A}_1 \subseteq \mathfrak{A}$ such that

$$\mathfrak{A} = \mathfrak{A}_o \oplus \mathfrak{A}_1 ,$$

and similarly there exists divisible $\mathfrak{B}_1 \subseteq \mathfrak{B}$ such that

$$\mathfrak{B} = g(\mathfrak{A}_o) \oplus \mathfrak{B}_1 .$$

Finally consider any $a \in |\mathfrak{A}|$. Then there exist unique $a_o \in |\mathfrak{A}_o|$ and $a_1 \in |\mathfrak{A}_1|$ such that

$$a = a_o + a_1$$

By Lemma 26, there exists $b_1 \in |\mathfrak{B}_1|$ which is disjoint from $g(\mathfrak{A}_o)$ and has the same order as a_1. Let

$$b = g(a_o) + b_1 .$$

Then

$$(\mathfrak{A},x,a)_{x \in X} \equiv_o (\mathfrak{B},f(x),b)_{x \in X} .$$

By symmetry,

$$(\mathfrak{A},x)_{x \in X} \equiv_1 (\mathfrak{B},f(x))_{x \in X} ,$$

and (PDA) is substructure complete.

Again we notice that we used a property of locally saturated prüferized divisible abelian groups which actually characterize these groups.

Corollary 27 A prüferized divisible abelian group $\mathfrak{A}$ is locally saturated if and only if $\iota_o^{\mathfrak{A}} \geq \aleph_o$.

The proof is simple and may be left to the reader.

Our method of quantifier elimination is not "purely" algebraic because we have to test local isomorphisms between locally saturated ($\aleph_o$-saturated) models of the theory. We now notice that we can further narrow the choice of test models. First, it suffices to test local isomorphisms between $\aleph_1$-saturated models. However, this still leaves us with the task of investigating the $\aleph_1$-saturated models of the theory. To completely avoid model theory we have to take a second step and use a result due to Keisler [3].

Theorem 29 If F is an ω-incomplete ultrafilter on I then $\prod_F \langle \mathfrak{A}_i \mid i \in I \rangle$ is $\aleph_1$-saturated.

With the help of Theorem 29 it is easy to see that we only have to test local isomorphisms between ultrapowers of models of the theory. We collect these observations and leave the proof to the reader.

Corollary 30 Let F be an ω-incomplete ultrafilter on ω. Then the following are equivalent :

(i) Σ is substructure complete.

(ii) Every local isomorphism between $\aleph_1$-saturated models of Σ is immediately extendible.

(iii) For every $\mathfrak{A}, \mathfrak{B} \in \mathrm{Mod}\,\Sigma$, every local isomorphism between $\mathfrak{A}^\omega_F$ and $\mathfrak{B}^\omega_F$ is immediately extendible.

We know of no striking examples where the purely algebraic approach of Corollary 30(iii) yields a marked advantage over an application of Theorem 12(ii). However, Corollary 30(ii) sometimes simplifies the argument a little, and we give an example.

Example 31 Suppose $\mathfrak{A}, \mathfrak{B} \in \mathrm{Mod(RCF)}$ are $\aleph_1$-saturated. Let $\emptyset \neq X \subseteq |\mathfrak{A}|$, where X is finite, and suppose $f: X \to |\mathfrak{B}|$, where

$$(\mathfrak{A}, x)_{x \in X} \equiv_0 (\mathfrak{B}, f(x))_{x \in X} .$$

We continue to use the notation introduced in Example 22 and obtain

$$(\mathfrak{A}, x)_{x \in |\mathfrak{A}_0|} \equiv_0 (\mathfrak{B}, g(x))_{x \in |\mathfrak{A}_0|} ,$$

where card $\mathfrak{A}_0 < \aleph_1$. If $a \notin |\mathfrak{A}_0|$, let Σ be the set of following formulas :

$$u < \dot{x} \quad \text{if} \quad a < x$$

$$\dot{x} < u \quad \text{if} \quad x < a$$

Then every finite subset of Σ is realizable in $(\mathfrak{B}, g(x))_{x \in |\mathfrak{A}_0|}$. Since $\mathfrak{B}$ is $\aleph_1$-saturated, there exists $b \in |\mathfrak{B}|$ realizing Σ in $(\mathfrak{B}, g(x))_{x \in |\mathfrak{A}_0|}$. It follows that

$$(\mathfrak{A}, x, a)_{x \in |\mathfrak{A}_0|} \equiv (\mathfrak{B}, g(x), b)_{x \in |\mathfrak{A}_0|} .$$

Although the argument presented in Example 31 appears to be somewhat simpler than the one in Example 22, this simplification is rather accidental. Indeed, we are using the fact that the ordering of an $\aleph_1$-saturated real closed field is $\aleph_1$-dense, and this property actually characterizes $\aleph_1$-saturated real closed fields. It just so happens that it is a little easier to say what the $\aleph_1$-saturated real closed fields are than what the $\aleph_0$-saturated ones are.

REFERENCES

1. Blum, L. , Ph.D. Thesis, Massachusetts Institute of Technology, 1968
2. Fraissé, R. , Sur quelques classifications des relations, basées sur des isomorphismes restreints. I.Étude générale. II.Application aux relations d'ordres. Alger-Mathématiques 2 (1955), 16-60, 273-295
3. Keisler, H.J. , Utraproducts and saturated models, Indag.Math. 26 (1964), 178-186.
4. Kochen, S. , Ultraproducts in the theory of models, Ann.of Math. 74 (1961), 221-261.
5. Mostowski, A. ,and Tarski, A. , Arithmetical classes and types of well-ordered systems, Bull.Amer.Math.Soc. 55 (1949), 65.
6. Morley, M. ,and Vaught, R. , Homogeneous universal models, Math. Scand. 11 (1962), 37-57.
7. Robinson, A. , Complete Theories, Amsterdam, 1956, 129pp.
8. Sacks, G.E. , Saturated Model Theory, Reading, Mass., 1972, xii + 335pp.
9. Shoenfield, J.R. , Mathematical Logic, Reading, Mass., 1967, vii + 344pp.
10. Shoenfield, J.R. , A theorem on quantifier elimination, Symposia Mathematica 5 (1970), 173-176.
11. Szmielew, W. , Elementary properties of abelian groups, Fund.Math. 41 (1955), 203-271.
12. Tarski, A. , A Decision Method for Elementary Algebra and Geometry, Berkeley and Los Angeles, 1951, iii + 63pp.
13. Tarski, A. , Arithmetical classes and types of Boolean algebras, Bull.Amer.Math.Soc. 55 (1949), 64.
14. Vaught, R. , Denumerable models of complete theories, Infinitistic Methods, Wardzawa, 1961, 303-321.

Intensional Semantics for Natural Language

Franz von Kutschera

In this paper I shall try to give a survey of the connections between intensional semantics and semantics for natural languages, i.e. between a logical and a linguistic discipline. Since these connections are the result of a long and still very active development, this survey can only be concerned with the general outlines and so is not primarily adressed to the specialists in the field. When I was asked to give such a survey I accepted the offer as an opportunity to make a little bit of propaganda among logicians, whose interest is concentrated on mathematics, for another promising field of application of logic that may in the future become equally important as that of mathematics.

I

First let me briefly sketch the development in theoretical linguistics that has been leading up to today's close cooperation with logic.

Logic first gained influence in linguistics when its standards of preciseness for the syntactical description of languages were taken over by linguists. It is, among others, the merit of Y.Bar-Hillel and of N.Chomsky to have firmly implanted this idea in modern grammar. Modern logic from its beginning - essentially since Frege's "Begriffsschrift" (1879) - has been using artificial languages that are syntactically and, since Tarski's paper on the concept of truth of 1931, also semantically built up in a rigorous manner. The, so to speak, idealized experimental conditions under which such artificial languages are constructed allow an exactness of their grammatical rules and therefore of linguistic analysis that contrasts very positively with the vague concepts and the assertions of doubtful generality in traditional grammar. Clearly natural languages, evolving from long historical developments are much more complex and difficult to describe by exact rules than constructed languages. But if the property of well-formedness of the sentences of a natural language L is decidable, as it should be as a precondition to them being easily understandable, then on Church's thesis on the mathematical definability of the concept of decidability and in view of the development of general systems for generating decidable sets of expressions in metamathematics, there must be such systems for generating the sentences of L. Generative grammar mostly uses Semi-Thue-systems. If "S" (for "sentence"), "NP" (for "noun phrase"), "VP" (for "verb phrase"), "A" (for "article"), "N" (for "noun"), "VT" (for "transitive verb"), etc. are (grammatical) symbols, and the expressions from the lexicon of L provide the terminal vocabulary, the well-formed sentences of L can (in a first approximation) be described as the expressions derivable from the symbol S by applications of the rules of the system. These rules are of the form $X\sigma Y \rightarrow X\tau Y$, where σ is a grammatical symbol and τ such a symbol or a terminal expression. We obtain for instance this derivation of the sentence "The man hits the dog":

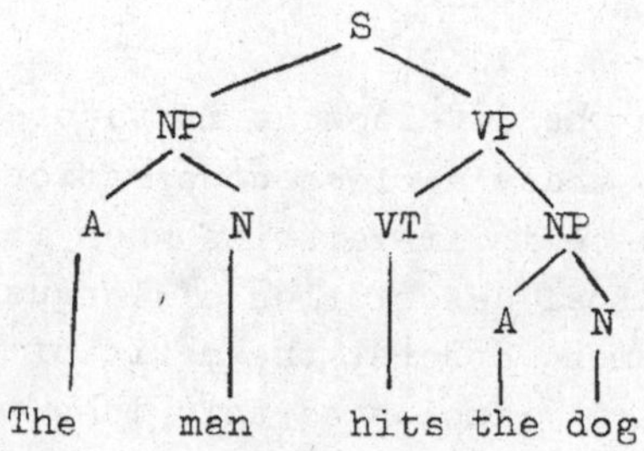

This model has the advantage of being familiar for linguists: the sentences of a language are analysed into a linearly concatenated sequence of constituents and this parsing operation can be performed at various levels of generality to yield a hierarchical branching-diagram.

There are many complications involved in this grammatical model that I shall not discuss here. Let me just say that the end-expressions of such derivations represent only the deep-structure of the sentences of L which in many cases do not coincide with their surface structure, i.e. their normal form, which then has to be derived from its deep structure by transformation rules which rearrange the expressions, take care of congruence, mode, number etc.

But even if you count the theory of Semi-Thue-Systems as a logical theory, this is not a syntactical analysis of the sentences of L that could be termed "logical", since it is based on the categories "verb phrase" etc. of traditional grammar. So this was a step in the right direction but it did not carry very far.

The first attempt at a generative semantics as made by Fodor and Katz in (63) was even less sucessful. They tried to coordinate semantical rules to the syntactical ones, but since the basic type of their projection rules was only that of forming a conjunction of one-place attributes, this attempt ended in failure.

The failure, however, of these projects to integrate logical ideas into the framework of traditional grammar cleared the way to linguistic analyses that are logical in a deeper sense. The idea seemed more and more attractive to depart from the categories of traditional grammar and use logical categories instead, as developed by K. Ajdukiewicz, St. Leśniewski, Y. Bar-Hillel, H.B. Curry and others, and to represent the deep-structure of the sentences by formulae of a logical language. Syntactically this idea was not very revolutionary since the complications of natural languages were already deferred to the transformational part of the grammar, which now could be left essentially unchanged. The only syntactical problem was not to make the deep structure too different from the surface structure of a sentence which it will be if the usual logical representation is used.

Semantics, however, at first presented the difficulty that natural languages are full of non-extensional contexts, while logic, till about 15 years ago, had only extensional semantics to offer and then till about the end of the sixties only intensional semantics for elementary types of language.

II

W.V.Quine in his paper "The Problem of Meaning in Linguistics" (51) and in other papers since has argued that, while the theory of reference, i.e. of the extensions of expressions, is, thanks to the work of Tarski and others, a sound and rigorous discipline, the theory of meaning is still in a desolate state since it has not even been able to define its basic notions, as those of proposition, attribute, synonymity, analyticity etc. Neither, according to Quine, was it ever likely to attain the state of a sound discipline since these concepts cannot be rigorously defined. To vary a Wittgensteinean dictum, Quine thought that all that can be said clearly can be said in an extensional language, and whereof we cannot speak clearly, we should be silent.

In his "Meaning and Necessity" (47), however, R.Carnap had already shown the way to a rigorous definition of these concepts in the same set-theoretical framework extensional semantics uses. His idea was roughly this: If we know the meaning of a sentence A, then we know under which conditions it is true. We can express this by saying: If we know the meaning of A we know in which possible worlds it is true. The inversion of this principle is not so obvious: Do we know the meaning of a sentence if we know under which conditions it would be true? But we can at least define a concept of intension as a first approximation to that of meaning by postulating that this inversion holds. Then we have for two sentences A and B: The intension of A is identical with that of B iff they have the same truth value in all possible worlds.

And we can define the intension of A by abstraction to be that function f, s. t. for every world i f(i) is the truth value of A in i.

This can be generalized for other types of expressions: The intension of an expression E is that function which assigns to every world i the extension of E in i.

A (possible) world is no distant cosmos on whose existence we speculate, but, as our world can be defined, according to Wittgenstein, as the set of all facts, a (possible) world can be defined as a set of propositions that is consistent and maximal, i.e. as a "complete novel."

As two logically equvalent sentences like "2+2=4" and "$dx^2/dx=2x$" have identical intensions but different meanings - meanings are to be defined so that two expressions, that are identical in meaning, may be substituted for each other in all contexts salva veritate - intensions are but approximations to meanings. They are, however, good approximations since it is possible, as we shall see, to define meanings with the help of intensions.

III

Carnap's ideas were first put to use in modal logic by S.Kripke and others, although with a slight modification of the basic idea: instead of sets of worlds they used sets of interpretations. The language L is that of propositional or of first-order predicate logic with an additional sentential operator N for necessity, and a model of L is a set of functions Φ_i $i \in I$ that have the properties of the usual extensional interpretations while $\Phi_i(NA)$ depends not only on $\Phi_i(A)$ but also on the values $\Phi_j(A)$ with $j \neq i$. A model for propositional modal logic for instance is a triple $\langle I,S,\Phi \rangle$, so that

a) I is a non-empty set of worlds (or of indices for interpretations).
b) For all $i \in I$ S_i is a subset of I with $i \in S$.
c) For all $i \in I$ Φ_i is a function from the set of sentences into the set $\{t,f\}$ of truth-values so that
   c1) Φ_i satisfies the conditions for extensional propositional interpretations, and
   c2) $\Phi_i(NA)=t$ iff $S_i \subseteq \lfloor A]$,
   where $\lfloor A]$ is the set $\{j \in I: \Phi_j(A)=t\}$ of A-worlds.

Such intensional models made it possible for the first time to define the formal properties of the intuitive notions of necessity exactly and to prove the soundness and completeness of systems of modal logics with respect to such notions. Up to Kripke's work there was a host of competing axiomatic systems of modal logic, while nobody could justify his intuition that his axioms should make up an adequate system, nor say how his notion of necessity compared with others.

There has been a lot of fruitful research in modal logic in the wider sense since, including for instance deontic, epistemic and conditional logic. Instead of sets S_i families of sets or families of sets of sets were used. But all this did not give the general framework for the application of this sort of semantics to natural languages. What was needed was a richer language than that of first-order predicate logic, and a simple and general characterization for the different types of intensional functors.

IV

This was provided at the end of the sixties in several papers, foremost in R.Montague's "Universal Grammar" (70). Let me briefly sketch his language, call it M, in an extensional and an intensional interpretation, so that we get a better notion of what intensional semantics is like.

First we define categories:

D1: a) σ and ν are categories (of sentences and proper names).

b) If τ and ρ are categories, $\tau(\rho)$ is a category (of functors which applied to expressions of category ρ produce expressions of category τ).

M is to contain the symbols λ (for functional abstraction), $\equiv$ (for identity), brackets and an infinite supply of constants and variables for each category.

The well-formed expressions of M are called terms of M:

D2: a) All constants of M of category τ are terms of category τ.

b) If F is a term of category $\tau(\rho)$ and t a term of category ρ F(t) is a term of category τ.

c) If A[b] is a term of category τ and b a constant and x a variable (not occuring in A[b]) of category ρ, then $\lambda x A[x]$ is a term of category $\tau(\rho)$.

d) If s and t are terms of the same category, $(s\equiv t)$ is a term of category σ.

For the interpretation of M we first define the sets of possible extensions of terms of category τ relative to the universe of discourse U:

D3: $E_{\nu,U}=U$

$E_{\sigma,U}=\{t,f\}$

$E_{\tau(\rho),U} = E_{\tau,U}{}^{E_{\rho,U}}$

where A^B is the set of functions from B into A.

D4: An extensional interpretation of M over U is a function Φ such that

a) $\Phi(a) \in E_{\tau,U}$ for all constants a of category τ.

b) $\Phi(F(t)) = \Phi(F)(\Phi(t))$.

c) $\Phi(\lambda x A[x])$ is that function $f \in E_{\tau(\rho),U}$ so that for all Φ' with $\Phi' \overline{\overline{b}} \Phi$ $f(\Phi'(b)) = \Phi'(A[b])$ (where the constant b does not occur in $\lambda x A[x]$ and $\Phi' \overline{\overline{b}} \Phi$ says that Φ' and Φ coincide with the possible exception of the values $\Phi(b)$, $\Phi'(b)$).

d) $\Phi(s\equiv t) = t$ iff $\Phi(s) = \Phi(t)$.

M is a type-theoretical language with predicates treated as truth-value functions as Frege proposed in "Funktion und Begriff" (1891) and two and more-place functions treated as one-place functions as in combinatory logic. As Tarski has shown we can define the usual logical operators, $\neg, \wedge, \Lambda, \epsilon$ in M.

Intensional interpretations of M may then be defined thus: We supplement the alphabet of M by two new symbols μ and δ. μt is to be an expression whose extension is the intension of t. μt occurring instead of t signifies that t stands in an <u>indirect</u> or non-extensional context, where its extension, according to Frege, is its usual intension. We need then new categories for such expressions and incorporate into D1 the condition:

<u>D1c)</u> If τ is a category then $\iota(\tau)$ is a category (of expressions of the form μt).

and into D3 the definition

$E_{\iota(\tau),U} = E_{\tau,U}{}^{I}$,

so that extensions of expressions of category $\iota(\tau)$ are intensions of expressions of category τ.

δ is to be an operator such that $\delta\mu t \equiv t$. D2 is then supplemented by two stipulations:

c) If t is a term of category τ, μt is a term of category $\iota(\tau)$.

f) If t is a term of category $\iota(\tau)$, δt is a term of category τ.

<u>D5:</u> An <u>intensional interpretation</u> of M over U and I (a non-empty set of worlds) is a function Φ such that for all $i \epsilon I$:

a) Φ_i satisfies the conditions for extensional interpetations of M over U according to D4.

b) $\Phi_i(\mu t) = \lambda^* j\Phi_j(t)$ (where λ^* is a metalinguistic symbol for functional abstraction).

c) $\Phi_i(\delta a) = \Phi_i(a)(i)$.

Condition (c) of D4 now is to be modified so that Φ' is an interpretation with $\Phi_i'(b) = \Phi_j'(b)$ for all $j \epsilon I$: We want to quantify over $E_{\rho,U}$ and since there are more functions in $E_{\rho,U}{}^{I}$ than objects of $E_{\rho,U}$, and since $\Phi_i'(A[b])$ may depend on values $\Phi_j'(b)$ for $j \neq i$, we must restrict the Φ's accordingly. If $\Phi_i'(A[b])$ does not depend on values $\Phi_j(b)$ for $j \neq i$, then the nature of the restriction does not matter; if it does, then $\lambda x A[x]$ may make no sense - that was Quine's argument against quantifying into modal contexts - and in that case again any restriction will do. If we interpret individual constants b as <u>standard names</u>, however, so that $\Phi_i(a) = \Phi_j(a)$ for all $j \epsilon I$ - and S.Kripke has given good reasons for that in "Naming and Necessity" (72) - then quantification over individuals into modal contexts makes sense, the same sense as our interpretation of expressions of the form $\lambda x A[x]$.

A word may be in order on the much discussed problem whether all the worlds in I should contain the same individuals, as we have stipulated, following Montague, or not, and how transworld-identity is to be understood, or if there can only be correspondences, counterpart-relations as D.Lewis suggests in (68) e.g. but no identities.

First the objects in U are to be possible objects. For each $i \in I$ we may introduce sets $U_i \subseteq U$ of objects existing in i and these sets may be different for different i's. If E is a constant of category $\sigma(\nu)$ and $\Phi_i(E)=U_i$ we may define quantification over existing instead of possible objects in the manner of Free Logic by $\Lambda.xA[x] :=\Lambda x(E(x) \supset A[x])$. Second we can take the identity of objects as a basic notion that need not be defined for each world by the Leibniz-principle of coincidence of properties, or for different worlds by a restricted Leibniz-principle of coincidence of "essential" properties or something of that sort. Introducing counterpart-relations in the sense of Lewis certainly makes for higher generality, but I know of no cases where this increase in generality is fruitful and therefore I prefer simplicity.

Since non-extensional contexts are very frequent in natural languages the use of the μ-operator is somewhat tedious. Therefore we might either treat all functors as correlating extensions to intensions, or assign intensions to the expressions directly. But as we want to distinguish, for instance, between quantification over extensions and that over intensions, between quantification over individuals and quantification over individual concepts, we have to mark the difference syntactically in any way so that we cannot hope to get off much cheaper by such approaches than in languages of the Montague-type.

IV

If L is a natural language and M an interpreted Montague-language then a logical grammar for L is defined by an analysing relation R(A,B) on T(M) $\times$ T(L), where T(M) is the set of wellformed expressions of M and T(L) this set for L, such that

1) For all $B \in T(L)$ there is an A with R(A,B).
2) If R(A,B) then the meaning of A is a possible meaning of B.

If R(A,B), A is called an analysing expression for B.

If R is explicitly defined, all essential grammatical concepts for L can be defined from this relation.

If R(A,B), then the expression A represents the deep-structure of B with constants of M in place of words or morphemes of L. There is no need now to supply analyses of deep-structures in the form of their derivations, since the structure of the terms of M is unambiguous, R may be taken to contain the rules of substitution of the terminal vocabulary of L for grammatical symbols in Generative Grammar as well as its transformational part.

V

Analysing relations have been given only for very small fragments of natural languages. There are numerous difficulties to overcome if they are to be defined for larger and more interesting parts of language. I shall only mention some to convey an impression of the complexity of a logical analysis of natural language:

1) First there is the syntactical problem that logical deep structure, i.e. the structure of the terms of M, is often very different from the surface structure of the terms of L. This makes for very complicated transformations, and therefore is an inventive to change the usual logical representation. Take the following two examples:
a) Quantifiers like "everybody", "somebody", "nobody" are treated in English like proper names in the sentences Joe sings, Everybody sings, Nobody sings. Instead of representing those sentences in the usual form $G(a)$, $\wedge xG(x)$ and $\neg \vee xG(x)$, there have been attempts therefore, to assimilate proper names to quantifiers by treating them as functors of category $\sigma(\sigma(\nu))$, or by treating quantifiers ("a man", "all men", "no man"), as well as proper names, as names for bundels of properties (the "universal-generic man" having those properties that all men have, the "existential-generic man" having the properties that some man has etc.). Cf. Lewis (70), e.g.
b) In the German sentences

α) Fritz singt laut (Fritz sings loudly)
β) Fritz singt gern (Fritz likes to sing)
γ) Fritz singt wahrscheinlich (Probably Fritz sings)

the adverbs have the same function in surface structure though logically they are to be treated quite differently: "wahrscheinlich" is applied to the proposition that Fritz sings, "laut" characterizes the verb, and "gern" has itself the function of a verb, as becomes apparent in the English translations. The usual logical representations of the three sentences would look something like this $\vee f(S(f) \wedge f(a) \wedge L(f))$ ("There is an action of singing that Fritz performs and that has the property of being loud"), $F(a,g)$, and $P(f(a))$. "singt" occurs in (α) as a 2nd-order predicate, in (β) and (γ) as a 1st-order predicate.

These two examples show that we should look for non-standard logical representations of ordinary language sentences closer to their syntactical structure.

2) Generally speaking, there is a variability and plasticity of the terms of natural languages quite unparalleled in logic. The same term of L often has to be coordinated by the analyzing relation R to many categorially and semantically different terms of M. The task of getting along with a minimum of morphemes without ending up with ambiguity in too many cases is solved much better by natural languages, it seems , than by logic. It is quite an interesting problem whether we could not do better in logic even if we hold on, as we should, to the principle of unambiguity in <u>all</u> cases.

3) Besides the syntactical problems of natural language analysis there are semantic problems which call for generalizations of the concept of an interpretation of M defined in D5. While we usually only consider eternal sentences in logic, many sentences of L contain index-expressions like "I", "you", "here", "now", "yesterday", "this" etc., whose extensions vary for different utterances of the same sentence. Therefore extensions and intensions must be defined for <u>utterances</u>, i.e. pairs $\langle A,j\rangle$ of a sentence A and an occurrence of A. If I is a set of n-tuples of parameters, specifying speaker, audience, time, place, indicated things etc., i.e. a set of <u>points of reference</u>, then we may introduce in D5 besides i another index j for Φ so that $\Phi_{i,j}(A)$ is the extension, $\lambda^* i\Phi_{i,j}(A)$ the intension of the <u>utterance</u> $\langle A,j\rangle$ of A, while $\lambda^* j\Phi_{i,j}(A)$ is the extension and $\lambda^* ij\Phi_{i,j}(A)$ the intension of the sentence A.

There is, however, no obvious limitation of the parameters in j, so that we must perhaps take j as an index for a space-time-point in i where A was uttered, as suggested by D. Lewis in (69). The meaning of an utterance may depend, for instance, on the facts obvious for speaker and audience in the situation of its occurrence as in the sentence "I shall now go (which may mean: walk, drive, go by train, fly) to Boston".

4) In ordinary language there are wellformed but meaningless expressions as "17 laughs", "The king of Bavaria is sitting in the audience" "If we were alive, we could read this paper", etc. Most empirical predicates are not defined for all syntactically admissible argu-

ments and many sentences for being meaningful presuppose that something is the case which in fact may not be the case at all. Therefore we should, following D.Scott in (70), define the sets of possible extensions for the non-basic categories by

$E_{\tau(\rho),U}=E_{\tau,U}{}^{(E_{\rho,U})}$ and $E_{\iota(\tau),U}=E_{\tau,U}{}^{(I)}$,

where $A^{(B)}$ is the set of functions from <u>subsets</u> of B into A.

5) Besides syntactical ambiguity (as "Flying planes can be dangerous") there is also semantic ambiguity (as in "Peter is going to the bank") and pragmatic ambiguity (as in "The problem I mentioned above was first noted by Quine"). As semantic ambiguity is often eliminated by the context ("Peter is going to the bank to cash a cheque"), we should not represent all ambiguous words by different constants of M. Instead we might assign <u>classes of extensions</u> to expressions and formulate the conditions in D5 thus:

a) $\Phi_i(a) \subseteq E_{\tau,U}$ for all constants a of category τ.

b) $\Phi_i(F(t)) = \{\gamma\in E_{\tau,U}: \bigvee\alpha\beta(\alpha\in\Phi_i(F) \wedge \beta\in\Phi_i(t) \wedge \alpha(\beta) = \gamma)\}$.

c) $\Phi_i(\lambda xA[x])$ is that class of functions $f\in E_{\tau(\rho),U}$ such that for all Φ' with $\Phi' \overset{b}{=} \Phi, \Phi'_i(b) = \{\alpha\}$ and $\Phi'_j(b) = \Phi'_i(b)$ for all $j\in I$ there is a $\beta\in\Phi'_i(A[b])$ with $f(\alpha)=\beta$.

d) $\Phi_i(s\equiv s') = \{\gamma\in E_{\sigma,U}: \bigvee\alpha\beta(\alpha\in\Phi_i(s) \wedge \beta\in\Phi_i(s') \wedge (\alpha=\beta \wedge \gamma=t . \vee . \alpha\neq\beta \wedge \gamma=f))\}$.

e) $\Phi_i(\mu t) = \{f\in I_{\tau,U}: \bigwedge j\bigvee\alpha(\alpha\in\Phi_j(t) \wedge f(j) = \alpha)\}$.

Then an expression t is unambiguous in i iff $\Phi_i(t)$ is a unit-class. We may then also abandon partial interpretations as considered under (4), since we can represent a function $f = \Phi_i(F) \in E_{\tau,U}{}^{(E_{\rho,U})}$ which is defined on the subset $E'\subseteq E_{\rho,U}$ by the set of functions from $E_{\tau,U}{}^{E_{\rho,U}}$ coinciding on E' with f.

6) Not all differences in meaning can be represented by differences in intension. The two sentences "Jack believes, that 2+2=4" and "Jack believes, that $dx^2/dx = 2x$" may have different truth-values though "2+2=4" and "$dx^2/dx = 2x$" have the same intensions, as we saw. There is one approach to meaning, first taken by S.Kripke in his completeness proofs for the modal systems S1, S2 and S3, envisaging abnormal worlds in which not all logically true sentences hold. This has the advantage of formal simplicity but there is no way of determining what sort of absurd worlds we should assume to account for the logical incapabilities of all possible people in all our possible worlds.

Another approach is this: We introduce indices $k \in K$ for the terms of M. Let k(A) be the index of the term A. Then we define $\Phi_{i,k}$ as in D5 and introduce an operator $\varkappa$ such that $\Phi_{i,k}(\varkappa t)=\lambda^{*}i\Phi_{i,k(t)}(t)$. This way we assign a term t an intension for every context A, represented by k(A), in which it occurs.

$\Phi_{i,k}$ can, for instance, be defined so that $\Phi_{i,k}(\varkappa s) = \Phi_{i,k}(\varkappa t)$ iff t is obtained from s by substituting constants with the same intensions. Then this concept of meaning coincides with Carnap's notion of <u>intensional isomorphism</u> in (47).

7) Besides descriptive sentences natural languages also contain questions, imperatives, exclamations, guesses, suggestions etc. As has been emphasized especially by J.L.Austin in (55) and J.R. Searle in (70) a semantics of natural language has also to account for these <u>illocutionary modes</u> of sentences or utterances.

We may, however, assign the question "Is Tom coming?", adressed by John to Jack the (descriptive) meaning of the assertion "John asks Jack, whether Tom is coming". And the question "Is Tom coming?", as a <u>sentence</u>, can be assigned the (descriptive) meaning of the predicate "to ask, whether Tom is coming". In this way, which is essentially identical with what D.Lewis proposed in (70), we can, with the help of illocutionary verbs like "order", "ask", "promise" etc., define the semantics for other illocutionary modes in the framework of a semantics for assertions.

V

So the attempt at a logical analysis of natural languages suggests quite a few syntactical and semantical modifications of the language M. Besides the specific difficulties encountered in logical grammar we should also mention some fundamental objections that have been raised against the whole project:

1) Natural languages are vague in many respects, syntactical and semantical. Analysing such languages, it has been said, by assigning them exact logical descriptions is therefore inadequate in principle since it projects on them a higher degree of precision than they actually have and is therefore a modification rather than a description. It is not the task of a grammar of a language L to transform L into a precise language in the sense of logic, but to mirror faithfully the properties L actually has.

This is not just the difficulty of how to derive the properties of

L from observations of how L is used, as D.Lewis suggests in (69), pp.200seq, but L as a natural language itself is not something precise but fuzzy all over. Instead of a well-defined class of wellformed expressions there are degrees of grammaticalness; instead of predicates with well-defined domains there are predicates more or less welldefined for different arguments; instead of a well defined class of possible interpretations of a term t there is a class of more or less possible or natural interpretations of t.

In view of this John R.Ross in (73) gave the advice to grammarians "You have to get yourself thinking the fuzzy way!" Now, for logicians at least, this cannot mean thinking the vague or unprecise way, but only thinking the comparative instead of the classificatory way. This means that, after the more fundamental difficulties of logical grammar are overcome, we should think of defining notions like "Expression s is more wellformed than expression t", "Φ is a more typical (or normal) interpretation of t than Φ'" and "s is less vague than t". In that way we may also define comparative concepts of synonymy and analyticity, as advocated by Quine. If, just to give an example, we have a relation of comparative similarity of worlds, as employed for instance by R.Stalnaker in (68) and D.Lewis in (73) in their analyses of conditionals, we might say that sentence A is at most as analytical as B iff ¬A-worlds are at least as similar to the real world as ¬B-worlds. Such comparative concepts certainly make for higher complexity, but I see no a priori reasons why logic should not be able to mirror the fuzziness of natural languages this way.

2) Accounting for vagueness in this way would also solve another fundamental problem, pointed out by Quine: The interpretation of M - and if we analyse a natural language L by M also that of L - depends on the set I of possible worlds. Now we cannot take I to be the set of all logically possible worlds, since the (analysing expressions of the) analytic sentences of L are to hold in all worlds of I. If, on the other hand, we determine I as the set of worlds in which all analytic sentences of L hold, then I is not well-defined since, as Quine has convincingly shown, the set of analytic sentences is not well defined. There is no firm boundary between analytic and synthetic truths, and with a little ingenuity you can always think of bizarre words, where the validity of supposedly analytic statements becomes doubtful. But if we admit partial interpretations, vagueness and a comparative concept of analyticity, we can take I to be the set of all logically possible worlds, 5-dimensional ones

and those with married bachelors included, but with the non-logical terms (almost) undefined there.

3) The most fundamental objection against intensional semantics, at last, comes to this: The whole approach of this semantics is based on the realistic idea, that we confer extensions, intensions and meanings on linguistic expressions by coordinating extra-linguistic entities, concrete things, attributes, propositions etc. to them. That way we can abstract semantics from pragmatics, semantic coordination from the use of the expressions in accordance with these correlations. But this idea has been questioned with, as I believe, very sound arguments from Peirce onward. The slogan of today's Philosophy of Language is: "The meaning of a word is determined by its use". Use, therefore, comes before, not after meaning, and therefore pragmatics, not semantics, is the fundamental discipline. Though we can certainly distinguish and identify many properties and facts without the use of language, a large and important class of concepts and propositions is defined only with the help of linguistic distinctions. In this sense Wittgenstein said: "How do I know that this color is red?" - An answer would be: "I have learned English" ((53),381). Semantics, therefore, is not a theory of correlations of words with meanings, defined independently of language, but it has to be based on a theory of linguistic behavior.

In his introduction to "Word and Object" ((60), p.IX) Quine said: "Language is a social art. In acquiring it we have to depend entirely on intersubjectively available cues as what to say and when. Hence there is no justification for collating linguistic meanings, unless in terms of men's dispositions to respond overtly to socially observable stimulations."

His "hence", however, is a non sequitur: Every semantics that is useful for the analysis of linguistic phenomena is thereby practically justified, no matter what theoretical constructs it employs, if it makes no pretense of being able to explain the fundamental facts of language; that, however, has never been the aim of intensional semantics. A deeper, philosophical analysis of meaning has to start from linguistic conventions in the sense of D.Lewis in (69). It can also be shown, how the descriptions of meanings in the framework of intensional semantics may be based upon descriptions of such conventions. But that is another story.

To sum up this brief survey we can say then that intensional semantics for natural languages, though still facing a lot of problems, has proved to be a very effective instrument for linguistic analyses. From a logical point of view, on the other hand, its interest lies in the fact that a closer look at the phenomena of natural languages is giving new stimulations to logical developments.

References

Austin, J.L. (55): How to Do Things with Words, (ed. by J.O.Urmson), London 1962.

Carnap, R. (47): Meaning and Necessity, Chicago 1956.

Fodor, J.A. and Katz, J.J. (63): "The Structure of a Semantic Theory", Language 39 (1963), 170-210.

Kripke, S. (70): "Naming and Necessity", in Harman and Davidson (eds): Semantics of Natural Language, Dordrecht 1972, 253-355, 763-769.

Lewis, D. (68): "Counterpart Theory and Quantified Modal Logic", The Journal of Philosophy 65(1968), 113-126.

Lewis, D.(69): Convention, Cambridge, Mass. 1969.

Lewis, D.(70): "General Semantics", Synthese 22(1970), 18-67.

Lewis, D. (73): Counterfactuals, Oxford 1973.

Montague, R. (70): "Universal Grammar", Theoria 36(1970), 373-398.

Quine, W.V. (51): "The Problem of Meaning in Linguistics", reprinted in Quine: From a Logical Point of View, Cambridge/Mass. [1]1953.

Quine, W.V. (60): Word and Object, Cambridge/Mass. 1960.

Ross, J.R. (73): "Clause-Matiness", to appear in E.Keenan (ed.): Formal Semantics for Natural Language, Cambridge 1975.

Scott, D. (70): "Advice on Modal Logic", in K.Lambert (ed.): Philosophical Problems in Logic, Dordrecht 1970.

Searle, J.R. (70): Speech Acts, Cambridge 1970.

Stalnaker, R. (68): "A Theory of Conditionals", in N.Rescher (ed.): Studies in Logical Theory, Oxford 1968.

Wittgenstein, L. (53): Philosophical Investigations, ed. by G.E.M. Anscombe and R.Rhees, Oxford 1953.

On extendability of models of ZF set theory to the models of Kelley-Morse theory of classes

By W.Marek and A.Mostowski (Warszawa)

§-1. Introduction

The standard axiomatisation of set theory due to Zermelo, Fraenkel and others was extended by von Neumann, Bernays and Gödel to an axiomatisation in which there appears, apart from the basic notion of a "set", the notion of a "class". Intuitively, classes are properties of sets, it being understood that we identify properties with the same extensions. This intuition derives from Cantor, who spoke of consistent and inconsistent classes.

The introduction of classes also extends the notion of "function": some classes which are not necessarily sets are functions. But the basic intuition of set theory that the image of a set is again a set is preserved.

If we restrict the scheme of class existence:

$$(EX)(x)(x \in X \iff \Phi(x))$$

(where "X" does not appear in Φ) to formulae in which no quantifier binds class variables, we obtain the so-called "predicative class theory" of Bernays and Gödel (or "GB"). If, however, no restriction of this sort is imposed, i.e.if Φ may contain quantifiers with variables ranging over all classes, the theory which results is called the Kelley-Morse theory of classes (or "KM"). In both cases, GB and KM, we accept the class form of the replacement axiom.

The system GB corresponds to the intuition that classes do not form an acceptable totality, although some operations on classes are acceptable, by the composition of appropriate operations we get classes $\{ x : \Phi(x) \}$ for predicative Φ's. The system KM corresponds to the conviction that the aggregate of all classes does form an acceptable totality and is a legitimate mathematical object.

The authors' standpoint is the following. They agree with the opinion that there is no reason to assume the properties of sets form an aggregate which is a legitimate mathematical object; but they think that the extensions of properties do form such an acceptable totality, and therefore that the system KM has as strong an intuitive basis as the system ZF. We claim, in fact, that KM is a formalisation of "second order ZF set theory", and that, in particular, the form of the replacement axiom which is accepted in KM is in accordance with this claim. Thus we believe that KM is a very good system for the formalisation and development of mathematics.

This conviction leads to a comparison of the relationships between Peano arithmetic and second order arithmetic on the one hand and between ZF and KM on the other.

The intermediate systems (i.e. systems lying between GB and KM, for instance KM_n, in which the scheme of class existence is restricted to Σ^1_n formulas) are not considered here, although a substantial number of results (in particular §3) may be extended to that case. An interesting interpretation of the Δ^1_1 class existence scheme is given in [11].

It is well-known that the theory KM is stronger than ZF, and that there are statements in the language of ZF set theory provable in KM but not (under assumption of the consistency of ZF) in ZF. Sentences which assert the existence of transitive models of ZF set theory may serve as examples; another example is a sentence asserting the existence of a model of CB and the Σ^1_n-class existence scheme. Let ZF^{KM} be the set of formulas Φ of the language of ZF set theory such that the relativisation Φ^V of Φ to the universe of sets is provable in KM. The system ZF^{KM} is an extension of ZF, and is axiomatisable, but no axiomatisation (in the language of set theory) is known. Our feeling is that ZF^{KM} consists of sentences as true as those of ZF set theory. (Clearly the consistency of ZF^{KM} is formally equivalent to that of KM).

Consider a model $\langle M,E \rangle$ of ZF^{KM}. For general model-theoretic reasons $\langle M, E \rangle$ is elementarily equivalent to a model $\langle N,E' \rangle$ obtained from a model $\langle R,E'' \rangle$ of KM by restricting it to sets of the model (with restricted membership relation). Thus it seems that it is most natural to begin studies of models of ZF^{KM} by considering models which are restrictions (in the above sense) of the models of KM. Such models are called "extendable".

In the present paper we study extendable models.

The paper is organized as follows: §0 deals with preliminaries. In Part One we give a number of extendable and non-extendable models. We introduce the important notion of β - extendability (corresponding to β - models of second order arithmetic), which is a restriction of the notion of extendability. We show that every elementary class consisting of models of signature $\langle 1, 2 \rangle$ contains an element not extendable to a model of KM. For the ω - models we have a stronger result, which, in view of recent work of Krajewski, is optimal. We show that extendability is a PC and β - extendability is PCPC property. We establish the Σ^1_1 - reflection for the theory KM. We discuss the connections between the extendability phenomenon and the height of the model.

In Part Two we deal mainly with β - extendability. For a given transitive model $\langle M, \in \rangle$ of ZF set theory, we define a class $R.A.^M \subseteq \mathcal{P}(M)$ and show that $\langle M, \in \rangle$ is a β - extendable model iff $\langle R.A.^M, M, \in \rangle$ is a model of KM. Since the class $R.A.^M$ is defined by a constructive (although transfinite) process, this result may be considered as a criterion of β - extendability : to see whether $\langle M, \in \rangle$ is a β - extendable model, we have only to check whether the images of sets in M by functions from $R.A.^M$ belong to M or not. In the former case M is β - extendable, in the latter not. The class $R.A.^M$ is called ramified analysis over M; its construction closely follows work of Gandy and Putnam, who proved similar results in the case of second order arithmetic. (Gandy's results were unfortunately never published.). Transposing other unpublished ideas of Gandy to set theory, we prove that $\langle R.A.^{M}, M, \in \rangle$ is the smallest β - extension of the model M. Considerations used on this proof allow us to get inner interpretations of KM in itself satisfying in addition various forms of choice. In the course of this argument, we discover an interesting difference between second order arithmetic and the theory KM, namely that the latter has minimal transitive models; the former has not, as was shown by Friedman in [3]. This solves a problem in [3].

The results of Part Three were proved by W.Marek. The most important result is a proof that the notions of extendability and of β - extendability of transitive models of ZF are different. (This is not surprising, on analogy with second order arithmetic, but it has to be proved). Using methods of Barwise and Wilmers, we show that the least ordinal α for which $\langle L_{\alpha}, \in \rangle$ is extendable is smaller than the least ordinal α' for which $\langle L_{\alpha'}, \in \rangle$ is β - extendable. Under reasonable assumptions (namely that α' exists), both α and α' are denumerable. Moreover, α is denumerable in $\langle L_{\alpha'}, \in \rangle$, and $\langle L_{\alpha}, \in \rangle$ is extendable in $\langle L_{\alpha'}, \in \rangle$.

We give sufficient conditions of extendability and β - extendability. They are entirely constructive; they appeal to the next admissible set, in the case of extendability, and to the next Σ^0_∞ - admissible set, in the case of β - extendability.

The subject of this paper has been studied by both of us for quite a long time. We started systematic research on it in 1970 and discussed it over many hours; hence our common authorship of the paper. Our earlier related results appeared separately in [12], [6], [9], [8], [10], [14] (arranged in order of appearance).

In our opinion, the notion of extendability merits further research. The most important problem seems to us to be the axiomatisation of the ZF^{KM} set theory. In particular, we are interested in mathematically interesting consequences of ZF^{KM} which are not consequences of ZF.

We are grateful to our colleagues from Warsaw: W.Guzicki, St.Krajewski, M.Srebrny and P.Zbierski for many valuable discussions and remarks. We owe a lot to R.O. Gandy whose ideas are so clearly seen in the part two.

§ 0. Preliminaries

Four basic theories we deal with in the paper are ZFC, ZF^{KM}, KM and KMC. They all are formulated in the same language L_{ST} ZFC is the usual set theory of Zermelo and Fraenkel (with choice). ZF^{KM} was defined in the introduction. In order to introduce KM we proceed as follows: We change the language L_{ST} into two-sorted language defining a predicate $\mathfrak{Z}(X) \Leftrightarrow (EY)(X \in Y)$ and using small Latin letters for variables ranging over sets i.e. classes with the property $\mathfrak{Z}(.)$. KM is the theory based on the following axioms:

1) Extensionality
2) Pairing for sets
3) Sum for sets
4) Powerset axiom
5) Infinity axiom
6) Foundation axiom
7) Choice axiom
8) Class existence scheme: $(EX)(x)(x \in X \Longleftrightarrow \Phi(x))$
   X not free in $\Phi(.)$
9) Replacement axiom (class form)
   If in addition we add the scheme 10)
10) Choice scheme
   $(x)(EY)\,\Phi(x,Y) \Rightarrow (EY)(x)\,\Phi(x, Y^{(x)})$

where $Y^{(x)} = \{ y : \langle x,y \rangle \in Y \}$,

the resulting theory is called KMC.

Models of KM (KMC) may always be represented in the form $\langle \mathcal{F} \cup M, M, E \rangle$ where $\mathcal{F} \subseteq \mathcal{P}(M)$: namely it follows from the axiom of extensionality and the definition of $\mathfrak{z}(\cdot)$ that proper classes may be uniquely represented by subsets of M where M is the set of all sets of the model.

The height of a model $\langle M,E \rangle$ is the supremum of ordinals represented in $\langle M, E \rangle$. In case when $\langle M, E \rangle$ is a transitive model of ZF, the height h(M) is equal to $M \cap On$. But when $\langle M, E \rangle$ is a model of KM then the height of it is usually much larger since there are wellorderings in M of length bigger than On.

If $\langle N,E \rangle$ is a structure and $x \in N$ then we put $x^* = \{y : N \models y \in x\}$. In case when N is transitive and $E = \in \restriction N$ then $x^* = x$.

Throughout the paper we use standard model theoretic and set-theoretic terminology. If X is a class then $X^{(x)}$ is called the $x^{\underline{th}}$ section of X. We write $X \eta Y$ instead of $(Ex)(X = Y^{(x)})$. In this way - intuitively - Y codes a collection $\{Y^{(x)} : x \in \mathrm{Dom}\, Y\}$. In part two we use the following property of the theory KM: wellorderings are comparable i.e. If X and Y are wellorderings then either X is similar to a (unique) initial segment of Y or conversly. This in turn implies that any two non-standard wellorderings which are wellorderings in the sense of a given model have the same initial wellordered type (it is in fact the height of the model). In part three we use-standard by now-facts from the theory of so called admissible sets. In particular we assume the working knowledge of classical results of Barwise on Σ_1 - compactness of denumerable admissible sets. We assume also some knowledge of constructible hierarchy.

§ 1. Extendability

Definition: Let $\langle M,E \rangle$ be a model of ZFC set theory and T a theory. $\langle M,E \rangle$ is called T - extendable iff there is $\mathcal{F} \subseteq \mathcal{P}(M)$ such that $\langle \mathcal{F} \cup M, M, E' \rangle \models T$ where $E' = E \cup [(M \times \mathcal{F}) \cap \in]$ We shall consider models which are KM- or KMC - extendable. When from the context it is clear what is T we call $\langle M,E \rangle$ just extendable.

In case when M is a transitive set and $E = \in \cap M^2$ then - if $\langle M, \in \rangle$ is extendable - we may find $\mathcal{F}$ such that $M \subseteq \mathcal{F} \subseteq \mathcal{P}(M)$ and $\langle \mathcal{F}, M, \in \rangle \models$ KM. In fact we shall be mainly interested in such structures.

The simplest property of the theory KM is that for every formula of L_{ST}, if ZFC $\vdash \varphi$ then KM $\vdash (\varphi)^V$. It follows in particular that if $\langle M, V^M, E \rangle \models$ KM then $\langle V^M, E \restriction V^M \rangle \models$ ZFC. Let $ZF^{KM} = \{\varphi : KM \vdash (\varphi)^V\}$. Then ZF^{KM} is a recursively enumerable set of sentences and therefore it has a recursive axiomatization Moreover it is easy to see that the theory KM is consistent. Unfortunately we do not know any axiomatization of ZF^{KM}. The theory ZF^{KM} is much richer than ZFC. In particular $ZF^{KM} \vdash$ "There exists a transitive model of ZFC". A much stronger statement provable in ZF^{KM} is the following

(*) "For every ordinal α there exists a sequence f, defined on α and such that f is an elementary tower of natural models of ZFC".

Indeed, using the classical reasoning of Montague Vaught (which we present below under the name "over-and-over-and-over-again") one can prove that V is the union of an elementary tower $\{R_{\xi_\alpha}\}_{\alpha \in On}$

The last statement is not expressible in L_{ST} as a statement about sets since it has the form: "There exists an increasing function $\xi : On \longrightarrow On$ such that $\alpha < \beta$ implies $\langle R_{\xi_\alpha}, \in \rangle \prec \langle R_{\xi_\beta}, \in \rangle$". The quantifier "there exists ξ" binds a class variable and not a set variable. A (very unsatisfactory) interpretation of this statement is the formula (*). It should be noted that the above statement is provable already in quite weak subsystems of KM. Although not provable in GB theory of classes it is derivable already from the Σ^1_1 class existence scheme. Another type of statement provable in KM is the following

"There exists an increasing function $\rho : On \longrightarrow On$ such that $\alpha < \beta$ implies $\langle R_{\rho_\alpha}, \in \rangle \prec \langle R_{\rho_\beta}, \in \rangle$ and such that

$(\alpha)_{On} (EX)(X \subseteq R_{\rho_\alpha + 1}$ & "$\langle X, R_{\rho_\alpha}, \in \rangle$ satisfies GB + Σ^1_n - comprehension")".

A typical reasoning which we call "over-and-over-and-over-again" and use at least four times in this paper is the following one:

Theorem: There are arbitrarily large α such that $\langle R_\alpha, \in \rangle \prec \langle V, \in \rangle$.

Proof: Using the scheme 8 we are able to prove full scheme of induction and so we are able to prove that for every class X /and in particular for the class V there exists the class/ $Stsf_X$ consisting of pairs $\langle \varphi, \vec{x} \rangle$ such that all terms of $\vec{x}$ belong to X and $\langle X, \in \rangle \models \varphi[\vec{x}]$. Applying the class form of the Skolem Löwenheim theorem (it is provable in KM, cf [14]) we get a set s_0 such that $\langle s_0, \in \rangle \prec \langle V, \in \rangle$

We may assume that $u \subseteq s_0$ (where u is a fixed set).

Notice that there exists α_0 such that $s_0 \subseteq R_{\alpha_0}$.

Now we define inductively two sequences $\{\alpha_n\}_{n\in\omega}$, $\{s_n\}_{n\in\omega}$ such that:

I $\langle s_n, \in\rangle \prec \langle V, \in\rangle$

II $s_n \subseteq R_{\alpha_n} \subseteq s_{n+1}$

It is clear that $\bigcup_{n\in\omega} s_n = \bigcup_{n\in\omega} R_{\alpha_n} = R_{\bigcup_{n\in\omega}\alpha_n}$.

Put $\alpha = \bigcup_{n\in\omega}\alpha_n$. Since $\langle \bigcup_{n\in\omega} s_n, \in\rangle \prec \langle V, \in\rangle$ therefore we have

$$\langle R_\alpha, \in\rangle \prec \langle V, \in\rangle$$

By the construction $u \subseteq R_\alpha$ so α may be chosen arbitrarily big. The proof as we presented it needs global form of choice, the Gödel's axiom E. Considering s_n's of least possible rank we are able to use only the local form of choice. ∎

The sequence $\{R_{\xi_\alpha}\}_{\alpha\in On}$ of the consecutive natural elementary submodels of V may be characterized as follows: ξ_0 is the least ordinal majorizing all ordinals definable in V (i.e. definable by formulas of the form $\varphi^V(.)$). Similarly $\xi_{\alpha+1}$ is the least ordinal bigger than all ordinals definable in V by formulas with the parameter R_{ξ_α}.

In the language L_{ST} we are able to express the notion of wellordering, namely :

Definition: $W.O.(X) \Longleftrightarrow X \subseteq V^2 \& (x,y)(x \in Dom\, X \& y \in Dom\, X \& \langle x,y \rangle \in X$

$\& \langle y,x \rangle \in X \Rightarrow y = x) \& (Z)(Z \subseteq Dom\, X \& Z \neq \emptyset \Rightarrow (Ez)(z \in Z \& (w)(w \in Z \Rightarrow$

$\langle z,w \rangle \in X)))$.

Similarly as in case of models of the second order arithmetic we introduce the notion of β - models.

Definition: The structure $\langle \mathcal{F}, M, \in \rangle$ (where $\mathcal{F} \subseteq \mathcal{P}(M)$) is called β - model (or - equivalently - is said to possess β - property) iff for all $X \in \mathcal{F}$, $\langle \mathcal{F}, M, \in \rangle \models W.O.[X]$ implies that X is a wellordering.

This leads to the following natural definition:

Definition: The model $\langle M, \in \rangle$ is called β - KM - extendable (β - KMC - extendable) iff there is $\mathcal{F} \subseteq \mathcal{P}(M)$ such that $\langle \mathcal{F}, M, \in \rangle$ is a β - model of KM (β - model of KMC).

Transitive models of ZF set theory are necessarily β - models. *) This follows from the fact that

$ZF \vdash (X)(W.O.(X) \Longrightarrow (E\alpha)(Ord(\alpha) \& X \cong \langle \alpha, \in \rangle))$ and

additionally from the fact that the formula "(.) is an ordinal" is an absolute formula with respect to transitive structures. In case of the theory KM the situation is different (we show this in § 3, although earlier it was proved in [9]). A curiosity with respect to this is the following lemma:

---

*) Indeed much weaker theories have this property; it is sufficient to assume Δ_0 - collection and Σ_1 - comprehension. The fact that powerset axiom is not used we employ later.

Lemma: Formula W.O.(.) is equivalent (in KM) to a predicative one.

Proof: Consider the following formula w.o.(X): (x,y) $(x \in \mathrm{Dom}\, X \,\&\, y \in \mathrm{Dom}\, X \,\&\, \langle x,y\rangle \in X \,\&\, \langle y,x\rangle \in X \Rightarrow y = x) \,\&\, (z)(z \subseteq \mathrm{Dom}\, X \,\&\, z \neq \emptyset \Rightarrow (Eu)(u \in z \,\&\, (v)(v \in z \Rightarrow \langle u,v\rangle \in X)$. Clearly $W.O.(X) \Rightarrow w.o.(X)$. Assume now $\neg W.O.(X)$; Consider x_0 of least rank such that the Class $X \cap \{t : \langle t, x_0\rangle \in X\}^2$ is not wellfounded. Now inductively define x_{n+1} as an element of $\{t : \langle t, x_n\rangle \in X\} - \{x_n\}$ of least possible rank such that $X \cap \{t : \langle t, x_{n+1}\rangle \in X\}^2$ is not wellfounded. $\{x_n\}_{n\in\omega}$ is a set not wellfounded in X. Again we can eliminate global choice from the proof. ∎

The notions of extendability and β - extendability coincide on some classes of models; as shown in [9], if $\langle M, \in\rangle$ is a transitive model of ZFC and $\mathrm{cf}(M \cap On) \geq \omega_1$ then every extension of $\langle M, \in\rangle$ is necessarily a β - extension. Indeed if $\mathcal{F}$ is an extension and $\langle \mathcal{F}, M, \in\rangle \models W.O.[X]$ then, if $\{x_n\}_{n\in\omega}$ is an X - descending sequence then by our assumption on the cofinality character of $On \cap M$, $\{x_n\}_{n\in\omega} \subseteq R^M_\alpha$ for some $\alpha \in On \cap M$. But then, $X \upharpoonright R^M_\alpha$ is not a wellordering; since $X \upharpoonright R^M_\alpha \in M$ therefore $\langle M, \in\rangle \models \neg W.O.[X \upharpoonright R^M_\alpha]$ contradicting $\langle \mathcal{F}, M, \in\rangle \models W.O.[X]$.

We investigate now the extendability of some models of ZFC.

Proposition: If α is a strongly inaccessible cardinal then $\langle R_\alpha , \in \rangle$ is β - KMC extendable.

Proof: $\langle R_{\alpha+1} , R_\alpha , \in \rangle$ is a desired extension. ∎

Notice that by Skolem-Löwenhein theorem, $\langle R_\alpha , \in \rangle$ has also other extensions. It has even more than α of them. Under assumption of regularity of 2^α it has even at least 2^α of them.

Proposition: If α is a weakly inaccessible cardinal then $\langle L_\alpha , \in \rangle$ is β - KMC - extendable.

Proof: If α is weakly inaccessible then $\langle L , \in \rangle \models$ "α is a strongly inaccessible cardinal". Moreover $\langle L, \in \rangle \models$ "$L_\alpha = R_\alpha$" thus $\langle L, \in \rangle \models$ "L_α is β - KMC - extendable". However the latter statement is absolute because L is a transitive model of ZF and therefore it is a β - model. Note that one of β - extensions of $\langle L_\alpha , \in \rangle$ is $L_{\alpha^+} \cap \mathcal{P}(L_\alpha)$. ∎

Proposition: The least transitive model of ZFC is not extendable.

Proof: The formula "(.) is a transitive model of ZFC" is Δ_1^{ZF} and thus absolute with respect to transitive models of ZF. If the least transitive model were extendable then it would satisfy the statement "there exists a transitive model of ZFC". But it does not since it is the smallest one.*) ∎

---

*) Once again we do not use the full power of KM; Σ_1^1 class existence scheme is enough.

Let $\{\Theta_\xi\}_{\xi \in On}$ be the consecutive enumeration of heights of transitive models of ZFC.

Proposition: Let φ be a formula such that ZFC $\vdash (E!\alpha)\varphi$ and φ is absolute with respect to transitive models of ZFC.

Let α_φ be the unique object satisfying φ . Then $\langle L_{\Theta_{\alpha_\varphi}}, \in \rangle$ is not extendable.

Proof: If $\langle M, \in \rangle$ is extendable then, since $\alpha_\varphi \in M$ therefore also $\Theta_{\alpha_\varphi} \in M$ and $L_{\Theta_{\alpha_\varphi}} \in M$. Thus $M \neq L_{\Theta_{\alpha_\varphi}}$. ∎

The above proposition can be generalized to the ordinal numbers definable in theories stronger than ZFC. Indeed it is enough that T is a recursive theory such that KM $\vdash$ "$\langle V, \in \rangle \models T$"

For instance $ZF^{KM_n} = \{\varphi : GB + \Sigma^1_n$ class existence scheme $\vdash \varphi\}$

Definition: If Ξ is a strongly inaccessible cardinal then α_Ξ is the least β such that $\langle R_\beta, \in \rangle \prec \langle R_\Xi, \in \rangle$

Proposition: If Ξ is a strongly inaccessible cardinal then $\langle R_{\alpha_\Xi}, \in \rangle$ is a non extendable transitive model of ZF^{KM}.

Proof: Since $\langle R_\Xi, \in \rangle$ is extendable therefore it satisfies ZF^{KM}. Thus $\langle R_{\alpha_\Xi}, \in \rangle$ also satisfies ZF^{KM} .

By absoluteness of the notion of rank with respect to transitive models of ZFC we get, for $\eta < \alpha_\Xi$ $(R_\eta)^{\langle R_{\alpha_\Xi}, \in \rangle} = R_\eta \cap R_{\alpha_\Xi} = R_\eta$.

If $\langle R_{\alpha_\Xi}, \in \rangle$ were extendable then there would exist $\eta < \alpha_\Xi$

such that in the extension $\langle \mathcal{F}, R_{\alpha_\Xi}, \in \rangle$ we would have

$$\langle \mathcal{F}, R_{\alpha_\Xi}, \in \rangle \models \text{"}\langle R_\gamma, \in \rangle \prec \langle V, \in \rangle\text{"}$$

$$\langle R_\Xi, \in \rangle$$

But then $\langle R_\gamma, \in \rangle \prec \langle R_{\alpha_\Xi}, \in \rangle$ i.e.

$$\langle R_\gamma, \in \rangle \prec \langle R_{\alpha_\Xi}, \in \rangle$$

contradicting minimality of α_Ξ. ∎

The analysis of the proof shows that $\langle R_{\alpha_\Xi}, \in \rangle$ is in fact not extendable to a model of GB + $\sum_1^1$ class existence scheme.

Theorem (Krajewski): If $\langle M, E \rangle$ is a model of ZFC then there is $\langle N, E' \rangle$ such that $\langle M, E \rangle \equiv \langle N, E' \rangle$ and such that $\langle N, E' \rangle$ is not extendable.

Proof: By the main result of [17] there is a model $\langle N, E' \rangle$ such that $\langle N, E' \rangle \equiv \langle M, E \rangle$ and such that all ordinals of $\langle N, E' \rangle$ are definable in $\langle N, E' \rangle$. We claim that $\langle N, E' \rangle$ is the desired model.

Indeed, if it were extendable then, in the extension $\langle \mathcal{F}, N, E'' \rangle$ we would have $\langle \mathcal{F}, N, E'' \rangle \models$ "$\langle R_\alpha, \in \rangle \prec \langle V, \in \rangle$" for some ordinal α of $\langle N, E' \rangle$. Consider the replica, $(R_\alpha^{\langle N,E' \rangle})^*$ of the object $R_\alpha^{\langle N,E' \rangle}$. Then in particular

$$\langle (R_\alpha^{\langle N,E' \rangle})^*, E' \upharpoonright (R_\alpha^{\langle N,E' \rangle})^* \rangle$$ is a proper elementary subsystem of $\langle N, E' \rangle$. Under this condition all definable elements of $\langle N, E' \rangle$ must be in $(R_\alpha^{\langle N,E' \rangle})^*$. But α is

not there, which gives the desired contradiction. ∎

Corollary: If $\mathcal{E}$ is the class of all extendable models and $\mathcal{K}$ an arbitrary elementary class in the language of L_{ST} then $\mathcal{K}-\mathcal{E} \neq \emptyset$

In the case of ω - models i.e. models with natural numbers ordered in the type ω, we get a slightly stronger result: *)

Theorem: If $\langle M,E \rangle$ is an ω - model of ZFC then either $\langle M,E \rangle$ is not extendable or there is an ordinal α of the model $\langle M,E \rangle$ such that $\langle (R_\alpha^{\langle M,E \rangle})^*, E \restriction (R_\alpha^{\langle M,E \rangle})^* \rangle \prec \langle M,E \rangle$ and $\langle (R_\alpha^{\langle M,E \rangle})^*, E \restriction (R_\alpha^{\langle M,E \rangle})^* \rangle$ is not extendable.

Proof: The key fact is the following tedious lemma:

Lemma: If x,y are two elements of M such that $\langle M,E \rangle \models$ " x is a pair $\langle x_0, E' \rangle$ and y is a pair $\langle y_0, E'' \rangle$ and $E' \subseteq x_0^2$ and $E'' \subseteq y_0^2$ " then ($\langle M,E \rangle \models$ "$x < y$" iff $\langle x_0^*, (E')^0 \rangle \prec \langle y_0^*, (E'')^0 \rangle$).

Proof of the lemma: We show that the satisfaction class for x inside of the model $\langle M,E \rangle$ and the satisfaction class for $\langle x_0^*, E^0 \rangle$ in V are isomorphic. Indeed consider the object z which is the satisfaction class for x in the model $\langle M,E \rangle$. Then z^* consists of objects being pairs $\langle \varphi, \vec{s} \rangle$ (in the sense of M) where $\langle M,E \rangle \models$ " φ is a formula" and

---

*) We first define:

Definition: If $x \in M$ then $x^0 = \{ \langle z, t \rangle : \langle M,E \rangle \models$ " $\langle z, t \rangle \in x$" $\}$

$\langle M, E\rangle \models$ " s is a finite sequence of elements of x_0 " and $\langle M, E\rangle \models$ " $x \models \varphi[\vec{s}]$ ". Now we use the fact that $\langle M, E\rangle$ is an ω - model and thus the notion of a formula is absolute with respect to $\langle M, E\rangle$. Also the notion of finiteness is absolute i.e. $\langle M,E\rangle \models$ " s is finite " implies that $(s)^*$ is actually finite. Now we show by induction (which is allowed since $\langle M, E\rangle$ is an ω - model) that $\langle M, E\rangle \models$ " $x \models \varphi[\vec{s}]$" iff $\langle x_0^*, E^o\rangle \models \varphi[((s)^*)^o]$. Thus we had shown the isomorphism. Finally let us notice that $\langle x_0^*, E^o\rangle < \langle y_0^*, (E')^o\rangle$ is equivalent to the fact $\mathrm{Stsf}_{\langle (x_0)^*, E^o\rangle} = \mathrm{Stsf}_{\langle (y_0)^*, (E')^o\rangle} \cap (\mathrm{Form} \times \bigcup_{n\in\omega} {}^n(x_0^*))$.

Making the calculations inside and outside of the model and taking into account that $(x \cap y)^* = x^* \cap y^*$. (Where the symbol $\cap$ on the left hand side denotes an operation in the model and on the right hand side a set theoretic operation) an finally using once more absolutness of a finite sequence, we get the result.

With the lemma proved we prove the theorem as follows.

Let α be the least ordinal - in the sense of $\langle M, E\rangle$ - such that in the extension $\langle \mathcal{T}, M, E'\rangle \models$ "$\langle R_\alpha, \epsilon\rangle < \langle V, \epsilon\rangle$" (Clearly under the assumption of extendability there must be α with this property).

We claim that $\langle (R_\alpha^{\langle M,E\rangle})^*, E\restriction(R_\alpha^{\langle M,E\rangle})^*\rangle$ is not extendable. Suppose it were. Then there must be an ordinal β – in the sense of $\langle (R_\alpha^{\langle M,E\rangle})^*, E\restriction(R_\alpha^{\langle M,E\rangle})^*\rangle$ – such that in the extension $\langle \mathfrak{S}, (R_\alpha^{\langle M,E\rangle})^*, (E\restriction(R_\alpha^{\langle M,E\rangle})^*)'\rangle$ the formula "$\langle R_\beta, \in\rangle \prec \langle V, \in\rangle$" holds.

Thus $\langle (R_\beta^{\langle (R_\alpha^{\langle M,E\rangle})^*, E\restriction(R_\alpha^{\langle M,E\rangle})^*\rangle})^*, E\restriction(R_\beta^{\langle (R_\alpha^{\langle M,E\rangle})^*, E\restriction(R_\alpha^{\langle M,E\rangle})^*\rangle})^*\rangle$ is an elementary subsystem of $\langle (R_\alpha^{\langle M,E\rangle})^*, E\restriction(R_\alpha^{\langle M,E\rangle})^*\rangle$

But $\langle M,E\rangle$ is an rank extension of $\langle (R_\alpha^{\langle M,E\rangle})^*, E\restriction(R^{\langle M,E\rangle})^*\rangle$ (cf [17]) and therefore $(R_\beta^{\langle (R_\alpha^{\langle M,E\rangle})^*, E\restriction(R_\alpha^{\langle M,E\rangle})^*\rangle})^* =$ $= (R_\beta^{\langle M,E\rangle})^*$.

Thus $\langle (R_\beta^{\langle M,E\rangle})^*, E\restriction(R_\beta^{\langle M,E\rangle})^*\rangle \prec \langle (R_\alpha^{\langle M,E\rangle})^*, E\restriction(R_\alpha^{\langle M,E\rangle})^*\rangle$

and so, using the lemma we have $\langle M,E\rangle \models$ "$\langle R_\beta, \in\rangle \prec \langle R_\alpha, \in\rangle$" contradicting the choice of α. *) ∎

The extendable models always satisfy ZF^{KM}. The compactness theorem implies the following theorem:

---

*) As shown by St.Krajewski the assumption that $\langle M,E\rangle$ is an ω - model cannot be omitted. Indeed he shows the following theorem

Theorem: If $\mathfrak{M} = \langle M,E\rangle$ is an extendable model then there exists a cardinal λ and an ultra filter D on λ such that the ultrapower $\mathfrak{N} = \mathfrak{M}^\lambda/D$ is extendable and for every ordinal α of $\mathfrak{N}$, if in the extension $\langle \mathfrak{F}, \mathfrak{N}\rangle \models$ "$\langle R_\alpha, \in\rangle \prec \langle V, \in\rangle$" holds then $\langle (R_\alpha^{\mathfrak{N}})^*, E'\restriction(R_\alpha^{\mathfrak{N}})^*\rangle$ is extendable.

Theorem: $\langle M,E \rangle$ is a model of ZF^{KM} iff there is model $\langle N,E' \rangle$ such that $\langle M,E \rangle \equiv \langle N,E' \rangle$ and $\langle N,E' \rangle$ is extendable model.

Proof: Implication from the right hand side to the left hand side is obvious. Assume $\langle M,E \rangle \models ZF^{KM}$. It is enough to show that $KM + (Th(\langle M,E \rangle))^{\vee}$ is consistent. Otherwise $KM \vdash (\neg\varphi)^{\vee}$ for some $\varphi \in Th(\langle M,E \rangle)$ thus $\neg\varphi \in ZF^{KM}$. But $ZF^{KM} \subseteq Th(\langle M,E \rangle)$, contradiction. ∎

The ultrapower $\mathfrak{M}^{\lambda}/D$ of an extendable model is again extendable. Thus applying the theorem of Frayne we get the following result:

Proposition (St.Krajewski): If $\langle M,E \rangle$ is a model of ZF^{KM} then there is an elementary extension of it $\langle N,E' \rangle$ which is an extendable model.

We come back now to the discussion of the ordinal α_{Ξ}

Proposition: α_{Ξ} is a cardinal.

Proof: Since $\langle V, \in \rangle$ is a rank extension of $\langle R_{\alpha_{\Xi}}, \in \rangle$ therefore the notion of a cardinal is absolute with respect to $\langle R_{\alpha_{\Xi}}, \in \rangle$. Since $\langle R_{\alpha_{\Xi}}, \in \rangle$ is a model of ZFC therefore it is a limit of its own cardinals. Thus α_{Ξ} is a limit of cardinals and so is itself a cardinal. ∎

Notice that the cofinality character of α_{Ξ} is always ω. As is well known, if $\mathfrak{M}$ is a natural model of the theory KM (and even of the theory GB) i.e. a model of the form $\langle R_{\alpha+1}, R_{\alpha}, \in \rangle$ then α is a strongly inaccessible cardinal. If however we consider models of the form $\langle \mathcal{F}, R_{\alpha}, \in \rangle$ without stipulation that $\mathcal{F} = R_{\alpha+1}$, then, under the assumption that inaccessible cardinals exist we may find extendable models of the form $\langle R_\alpha, \in\rangle$ Indeed we have the following theorem:

Theorem: If Ξ is an inaccessible cardinal then there are arbitrarity large $\eta < \Xi$ such that $\langle R_\eta, \in\rangle$ is extendable, $\langle R_\eta, \in\rangle \prec \langle R_\Xi, \in\rangle$, $\operatorname{cf} \eta = \omega$.

Proof: We use the "over-and-over-and-over-again" method.

Let $u \in R_\Xi$. Consider the system $\langle R_{\Xi+1}, R_\Xi, \in\rangle$ and its subsystem $\langle A^0, A_1^0, \in\rangle$ such that $u \subseteq A_1^0$. The objects in A_1^0 are elements of R_Ξ, objects in $A^0 - A_1^0$ are elements of $R_{\Xi+1} - R_\Xi$. We define as before sequences $\{A^n\}_{n\in\omega}$, $\{A_1^n\}_{n\in\omega}$ $\{\eta_n\}_{n\in\omega}$ such that:

$$\langle A^j, A_1^j, \in\rangle \prec \langle R_{\Xi+1}, R_\Xi, \in\rangle$$

$$A_1^j \subseteq R_{\eta_j} \subseteq A_1^{j+1},$$

$$\eta_j < \Xi, \quad \overline{\overline{A^{j+1}}} = \beth_{\eta_j}$$

As before $A_1^j \subseteq R_\Xi$, $A^j - A_1^j \subseteq R_{\Xi+1} - R_\Xi$

Now set $$A = \bigcup_{j\in\omega} A^j, \quad A_1 = \bigcup_{j\in\omega} A_1^j, \quad \eta = \bigcup_{j\in\omega} \eta_j$$

Then $\langle A, A_1, \in\rangle \prec \langle R_{\Xi+1}, R_\Xi, \in\rangle$

Let us note firstly that A_1 is transitive (though A is not) and $A_1 = R_\eta$. This follows from the fact that $A_1^j \subseteq R_{\eta_j} \subseteq A_1^{j+1}$.

Thus $\langle A, R_\eta, \in\rangle \prec \langle R_{\Xi+1}, R_\Xi, \in\rangle$. For each $X \in A$ put

$X^+ = X \cap R_\eta$. Let $\mathcal{F} = \{X^+ : X \in A\}$. Clearly $\mathcal{F}$ arises simply from A by the contraction procedure and so $\langle A, R_\eta, \in\rangle \cong \langle \mathcal{F}, R_\eta, \in\rangle$. Thus $\langle \mathcal{F}, R_\eta, \in\rangle \models KM$. By our construction $\langle R_\eta, \in\rangle \prec \langle R_\Xi, \in\rangle$ and of $\eta = \omega$. ∎

Notice that the system $\langle \mathcal{F}, R_\eta, \in\rangle$ is a β - model. Indeed let $\langle \mathcal{F}, R_\eta, \in\rangle \models W.O.[X]$. Then for some $Z \in A$, $X = Z^+$. By our construction $\langle A, R_\eta, \in\rangle \models W.O.[Z]$. The analysis of the form of Z shows that X is a restriction of Z to R_η. Now $\langle A, R_\eta, \in\rangle \prec \langle R_{\Xi+1}, R_\Xi, \in\rangle$. Thus $\langle R_{\Xi+1}, R_\Xi, \in\rangle \models W.O.[Z]$ and so Z is a wellordering. Since Z is a wellordering, therefore all its restrictions are and so X is a wellordering.

When we look closely to the proof we find that we did'nt use all power of inaccesability. This leads to the following definition.

Definition: A model $\langle M, \in\rangle$ is called 2-extendable iff there exist two extensions $\mathcal{F}_1$ and $\mathcal{F}_2$ of $\langle M, \in\rangle$ such that: $\langle \mathcal{F}_i, M, \in\rangle \models KM$, $i = 1,2$ and $(EX)(X \in \mathcal{F}_2 \,\&\, (Y)(Y \in \mathcal{F}_1 \Longleftrightarrow Y_\eta X))$ i.e the extension $\mathcal{F}_1$ is codable within $\mathcal{F}_2$; in particular $\mathcal{F}_1 \subseteq \mathcal{F}_2$.

By virtually the same reasoning as above we have the following theorem:

Theorem: If $\langle M, \in\rangle$ is 2-extendable then there exists $\alpha \in On \cap M$ such that $\langle R^M_\alpha, \in\rangle \prec \langle M, \in\rangle$ and $\langle R^M_\alpha, \in\rangle$

is extendable. If in particular the smaller extension $\mathcal{F}_1$ is a β-extension, then $\langle R^M_\alpha, \in\rangle$ may be chosen β-extendable.

Analogous hierarchies of n-extendability and α-extendability may be introduced, with analogous results. However, we shall not pursue the matter here.

Since we had shown that extendability is not an elementary property of models, it seems reasonable to investigate whether this property is connected with the height of model of ZF.

Theorem: If there exists a strongly inaccessible cardinal and $0^{\#}$ exists then there are transitive models M_1 and M_2 of ZF^{KM} such that $On \cap M_1 = On \cap M_2$, M_1 is extendable (even β-extendable) and M_2 is not. M_1 and M_2 can be chosen denumerable.

Proof: We first produce uncountable models M_1 and M_2 with the desired property as follows. If κ is an inaccessible cardinal then consider α_κ. As we proved before $\langle R_{\alpha_\kappa}, \in\rangle$ is not extendable. But α_κ is a cardinal and since $0^{\#}$ exists it is strongly inaccessible in L. Thus $\langle L, \in\rangle \models$ "$\langle L_{\alpha_\kappa}, \in\rangle$ is β-extendable". Since $\langle L, \in\rangle$ is transitive therefore $\langle L_{\alpha_\kappa}, \in\rangle$ is actually β-extendable. Set $M_1 = \langle L_{\alpha_\kappa}, \in\rangle$ $M_2 = \langle R_{\alpha_\kappa}, \in\rangle$. Clearly both of them satisfy ZF^{KM}. We construct denumerable models M_1 and M_2 with the above properties as follows. Let ζ be the least ordinal β such that $\alpha_\kappa \in \beta$ and $\langle R_\beta, \in\rangle \prec \langle R_\kappa, \in\rangle$. Consider the structure $\langle R_\zeta, \in, \prec_\kappa\rangle$. Clearly $\langle R_\zeta, \in, \prec_\kappa\rangle$ satisfies "$\langle R_{\alpha_\kappa}, \in\rangle$ has no elementary submodel of the form $\langle R_\eta, \in\rangle$ & $\langle L_{\alpha_\kappa}, \in\rangle$ is β-extendable".

Pick denumerable transitive model $\langle M, \in, \lambda \rangle$ elementarily equivalent to $\langle R_\zeta, \in, \alpha_\kappa \rangle$. We claim that $\langle R_\lambda^{\langle M, \in \rangle}, \in \rangle$ is not extendable. If $\langle R_\lambda^{\langle M, \in \rangle}, \in \rangle$ were extendable then for some $\eta < \lambda$

$$\langle R_\eta^{\langle R_\lambda^{\langle M, \in \rangle}, \in \rangle}, \in \rangle \prec \langle R_\lambda^{\langle M, \in \rangle}, \in \rangle$$

By a reasoning we used twice, $R_\eta^{\langle R_\lambda^{\langle M, \in \rangle}, \in \rangle} = R_\lambda^{\langle M, \in \rangle}$

Thus $\langle R_\eta^{\langle M, \in \rangle}, \in \rangle \prec \langle R_\lambda^{\langle M, \in \rangle}, \in \rangle$ and so

$$\langle M, \in \rangle \models \text{"} \langle R_\eta, \in \rangle \prec \langle R_\lambda, \in \rangle \; \& \; \eta \in \lambda \text{"}$$

This however contradicts the fact that $\langle M, \in, \lambda \rangle \equiv \langle R_\zeta, \in, \alpha_\kappa \rangle$

Since $\langle M, \in \rangle$ is transitive and $\langle M, \in \rangle \models$ "L_λ is β - extendable" therefore $\langle L_\lambda^{\langle M, \in \rangle}, \in \rangle$ is indeed β- extendable i.e $\langle L_\lambda, \in \rangle$ is β -extendable. ∎

Thus the height of the model does not determine the extendability property.

There is positive result concerning Cohen extensions of extendable models.

Theorem: If $\langle M, \in \rangle$ is a denumerable transitive extendable model, $\langle P, \leq \rangle \in M$ is a notion of forcing, G any M-generic ultrafilter in $\langle P, \in \rangle$ then $\langle M[G], \in \rangle$ is again extendable.

Proof: Following [2] we find that if $\langle N, V^N, \in \rangle$ is a denumerable transitive model of KM, $\langle P, \leq \rangle \in V^N$ then $\langle N[G], V^{N[G]}, \in \rangle$ is a model of KM (Actually, Chuaqui proves this for a larger class of notions of forcing, some of them being proper classes of N). Thus we only need to show that, if $M = V^N$ then $M[G] = V^{N[G]}$. This follows from the fact that if G is M-generic then (under assumption $\langle P, \in \rangle \in M$) it is necessarily N-generic, and the fact that if $K_G(X) \in V^{N[G]}$ then for some set x, $K_G(X) = K_G(x)$. ∎

We show now a strong form of the reflexion principle for the theory KM.

Let X be a class. We define a relation $Sat(X, \varphi, \vec{x})$ between formulas of L_{ST} and finite sequences of elements of Dom X which satisfies the following conditions

$$Sat(X, \ulcorner v_i \in v_j \urcorner, \vec{x}) \Longleftrightarrow X^{(x_i)} \in X^{(x_j)}$$

$$Sat(X, \ulcorner v_i = v_j \urcorner, \vec{x}) \Longleftrightarrow X^{(x_i)} = X^{(x_j)}$$

$$Sat(X, \ulcorner \neg \varphi \urcorner, \vec{x}) \Longleftrightarrow \neg Sat(X, \varphi, \vec{x})$$

$$Sat(X, \ulcorner \varphi \& \psi \urcorner, x) \Longleftrightarrow Sat(X, \varphi, \vec{x}) \& Sat(X, \psi, \vec{x})$$

$$Sat(X, \ulcorner (Ev_i)\varphi \urcorner, \vec{x}) \Longleftrightarrow (Ex)_{DomX}\, Sat(X, \varphi, \vec{x}\left(\tfrac{i}{x}\right))$$

where $\vec{x}\left(\frac{i}{x}\right) = (\vec{x} - \{i\} \times V) \cup \{\langle i, x \rangle\}$

We define Sat(.,.,.) as the smallest relation satisfying the above. In case when X is a set, $Sat(X, \ulcorner\varphi\urcorner, \langle x_1 \ldots, x_k \rangle)$ is equivalent to the following: $\langle \{y : y \eta X\}, \in \rangle \models \varphi[X^{(x_1)}, \ldots, X^{(x_k)}]$

We have the following lemma:

Lemma: If φ is a predicative formula and X a class such that

$(x)(x \in V \Rightarrow x \eta X)$ and $X_1 = X^{(x_1)}, \ldots, X_k = X^{(x_k)}$ then

$$\varphi(X_1, \ldots, X_k) \iff Sat(X, \varphi, x)$$

Proof: By induction on the complexity of formulas. For atomic formulas and boolean connectives the proof is obvious. In the case of the existential quantifier we use the fact that $(x)(x \eta X)$.

Lemma: If φ is a Σ^1_1 formula then $(X_1) \ldots (X_n)(\varphi(X_1, \ldots, X_n)$

$$\iff (EX)(Ex_1) \ldots (Ex_n)\big[X_1 = X^{(x_1)} \& \ldots \& X_n = X^{(x_n)} \& Sat(X, \ulcorner\varphi\urcorner, x)$$

$$\& (x)(x \eta X)\big])$$

Proof: Let $\ulcorner\varphi\urcorner = \ulcorner(E v_i)\Psi\urcorner$ where Ψ is a predicative formula. Let $X_1, \ldots, X_n, Z$ be given such that $\Psi(Z, X_1, \ldots, X_n)$. We form the class X as follows: $X = \{0\} \times Z \cup \bigcup_{x \in V} \{\{x\}\} \times x \cup \bigcup_{i=1}^{n} \{i+1\} \times X_i$.

Then by the preceding lemma $Sat(X, \ulcorner\Psi\urcorner, \langle 0, 2, \ldots, n+1 \rangle)$

thus $Sat(X, \ulcorner\varphi\urcorner, \langle 2, \ldots, n+1 \rangle)$. Since $X^{(2)} = X_1, \ldots, X^{(n+1)} = X_n$

therefore $(Ex_1) \ldots (Ex_n)(X_1 = X^{(x_1)} \& \ldots \& X_n = X^{(x_n)} \& Sat(X, \ulcorner\varphi\urcorner, x)$.

Conversely, assume $\mathrm{Sat}(X, \ulcorner\varphi\urcorner, \vec{x})$ & $X_1 = X^{(x_1)}$ &...& $X_n = X^{(x_n)}$.

Then $\mathrm{Sat}(X, \ulcorner(Ev_i)\psi\urcorner, \vec{x})$. So for some $z \in \mathrm{Dom}X$, $\mathrm{Sat}(X, \ulcorner\psi\urcorner, \vec{x}(\begin{smallmatrix} i \\ z \end{smallmatrix}))$. Consider $X^{(z)}$. By the preceding lemma again $\psi(X^{(x)}, X_1, \ldots, X_n)$ and thus $(EZ)\psi(X_1, \ldots, X_n)$ ∎

Theorem (Σ^1_1 reflection principle): If $\varphi \in \Sigma^1_1$ then for every $X_1, \ldots, X_n$ there are arbitrarily large α's such that:

(a) $\varphi(X_1, \ldots, X_n) \Longrightarrow \langle R_{\alpha+1}, R_\alpha, \in\rangle \models \varphi[X_1 \cap R_\alpha, \ldots, X_n \cap R_\alpha]$

(b) $\langle R_\alpha, \in\rangle \prec \langle V, \in\rangle$.

Proof: Let $X_1, \ldots, X_n$ be given. If $\neg\varphi(X_1, \ldots, X_n)$ then let α be appropriately large such that $\langle R_\alpha, \in\rangle \prec \langle V, \in\rangle$.

Suppose now $\varphi(X_1, \ldots, X_n)$. By the preceding lemma there is a class X and a sequence $\vec{x}$ such that $\mathrm{Sat}(X, \varphi, \vec{x})$ where, for each i, $1 \leq i \leq n$, $X^{(x_i)} = X_i$. Moreover, $(x)(x \,\eta\, X)$. Now we use the "over-and-over-and-over-again" method once more (using Skolem-Löwenheim, class version). We define four sequences $\{z^n\}_{n\in\omega}, \{z^n_o\}_{n\in\omega}, \{X_n\}_{n\in\omega}, \{t_n\}_{n\in\omega}$ inductively as follows: Let u be a given set. There is $w \subseteq \mathrm{Dom}\,X$ such that $u = \{X^{(s)} : s \in w\}$.

z^o is any subset of $\mathrm{Dom}\,X$ such that, for all $\vec{x} \in \bigcup_{n\in\omega} {}^n(z^o)$, for all φ's $\mathrm{Sat}(X_o, \varphi, \vec{x}) \Longleftrightarrow \mathrm{Sat}(X, \varphi, \vec{x})$ where $X_o = X \cap (z^o \times V)$, $z^o_o = \{z \in z^o : X^{(z)} \in V\}$;

t_o is any set containing z_o^o such that for some η

$\{ X^{(z)} : z \in t_o \} = R_\eta$ (Since z_o^o is a set, t_o can be found)

Now assume z^n, z_o^n, X_n, t_n are known.

z^{n+1} is a subset of Dom X such that $t_n \subseteq z^{n+1}$, $\overline{\overline{z^{n+1}}} = \overline{\overline{t_n}}$

and such that for all $\vec{x} \in \bigcup_{k \in \omega} {}^k(z^{n+1})$, all $\varphi \in L_{sr}$

$$\mathrm{Sat}(X_{n+1}, \ulcorner\varphi\urcorner, \vec{x}) \Longleftrightarrow \mathrm{Sat}(X, \ulcorner\varphi\urcorner, \vec{x})$$

where $X_{n+1} = X \cap (z^{n+1} \times V)$.

$z_o^{n+1} = \{ v \in z^{n+1} : X^{(v)} \in V \}$

t_{n+1} is a set containing z_o^{n+1} such that for some η,

$\{ X^{(z)} : z \in t_{n+1} \} = R_\eta$. Now form $z^\omega = \bigcup_{n \in \omega} z^n$,

$X_\omega = X \cap (z^\omega \times V)$, $z_o^\omega = \bigcup_{n \in \omega} z_o^n$ and $t_\omega = \bigcup_{n \in \omega} t_n$

We find that $z_o^\omega = \{ v \in z^\omega : X^{(v)} \in V \}$

By our construction there is η such that $\{ X^{(v)} : v \in t_\omega \} = R_\eta$

The following holds:

(I) $(\vec{z})(\vec{z} \in \bigcup_{n \in \omega} ({}^n(z^\omega)) \Longrightarrow (\mathrm{Sat}(X_\omega, \ulcorner\varphi\urcorner, \vec{z}) \Longleftrightarrow \mathrm{Sat}(X, \ulcorner\varphi\urcorner, \vec{z}))$

(II) $z_o^\omega = t_\omega$

Since z^ω is a set we may assume that for $z_1, z_2 \in \mathrm{Dom}\, X_\omega$,

$z_1 \neq z_2 \Longrightarrow X_\omega^{(z_1)} \neq X_\omega^{(z_2)}$.

Define now on Dom X_ω a relation E as follows

$$\langle x, y\rangle \in E \iff X_\omega^{(x)} \in X_\omega^{(y)}$$

The relation E is wellfounded and extensional (for the last fact we need, for $z_1 \neq z_2$, $X_\omega^{(z_1)} \neq X_\omega^{(z_2)}$). Thus $\langle \mathrm{Dom} X_\omega, E\rangle \simeq \langle N, \in\rangle$ for some transitive N.

We find that

(III) $(x)(\mathrm{Sat}(X_\omega,\ v_i \in V, \{\langle i,x\rangle\})) \iff X_\omega^{(x)} \in R_\eta$

Thus $N \subseteq R_{\eta+1}$, $V^N = R_\eta$

The analysis of the isomorphism $i : \langle \mathrm{Dom} X_\omega, E\rangle \longrightarrow \langle N, \in\rangle$ gives the following : $i(x) = X_\omega^{(x)} \cap R_\eta = X^{(x)} \cap R_\eta$

Since $\mathrm{Sat}(X_\omega, \varphi, \vec{x})$ therefore $\langle N, V^N, \in\rangle \models \varphi[X^{(x_1)} \cap R_\eta, \ldots, X^{(x_n)} \cap R_\eta]$

Finally for $\varphi \in \Sigma_1^1$ $\langle N, R_\eta, \in\rangle \models \varphi[z_1, \ldots, z_n]$ implies $\langle R_{\eta+1}, R_\eta, \in\rangle \models \varphi[z_1, \ldots, z_n]$. Thus $\mathrm{Sat}(X, \varphi, \vec{x})$ implies $\langle R_{\eta+1}, R_\eta, \in\rangle \models \varphi[X^{(x_1)} \cap R_\eta, \ldots, X^{(x_n)} \cap R_\eta]$ (whenever $\varphi \in \Sigma_1^1$) Clearly $\langle R_\eta, \in\rangle \prec \langle V, \in\rangle$.

Considering z^n's, and t_n's of least possible rank we may eliminate the usage of the global form of the axiom of choice. ∎

Definition: (a) A class $\mathcal{K}$ of models is a PC class with respect to the class $\mathcal{L}$ iff $\mathcal{K} \subseteq \mathcal{L}$ and there is a set of sentences S in the language $L_{ST}(A)$ (arising from L_{ST} by adding unary predicate A) such that

$$(\mathfrak{M})\ (\mathfrak{M} \in \mathcal{L} \Rightarrow (\mathfrak{M} \in \mathcal{K} \Leftrightarrow (EX)(X \subseteq |\mathfrak{M}| \,\&\, (\varphi)(\varphi \in S \Rightarrow \langle \mathfrak{M}, X \rangle \models \varphi))))$$

(b) A class $\mathcal{K} \subseteq \mathcal{L}$ is CPC with respect to $\mathcal{L}$ iff $\mathcal{L} - \mathcal{K}$ is PC class. Analogously we define PCPC, CPCPC classes etc.

Theorem: The class of extendable models is a PC class with respect to the class of all models of ZFC.

Proof: If $\mathfrak{M}$ is extendable then - by virtue of Skolem-Löwenheim theorem there is $C \subseteq \mathcal{P}(|\mathfrak{M}|)$ which extends $\mathfrak{M}$ and such that $\bar{\bar{C}} = \overline{\overline{|\mathfrak{M}|}}$.

Let f be an enumeration of C with elements of $|\mathfrak{M}|$. Finally let $X = \{ \langle x,y \rangle^{\mathfrak{m}} : y \in f(x) \}$.

We have the following lemma:

Lemma: For every formula φ of L_{ST} there is a formula Ψ_φ of $L_{ST}(A)$ such that

$$\langle C, M, E' \rangle \models \varphi\, [x_1, \ldots, x_k, f(y_1), \ldots, f(y_k)]$$

$$\langle M, E, X \rangle \models \Psi_\varphi\, [x_1 \ldots x_k, y_1 \ldots y_k]$$

Moreover the mapping $\varphi \mapsto \Psi_\varphi$ is effective.

We leave the proof to the willingful reader.

Now: $\mathfrak{M}$ is extendable $\Longleftrightarrow$

$$(EC)(C \subseteq \mathcal{P}(|\mathfrak{M}|) \cup M \;\&\; (\varphi)(\varphi \in KM \Rightarrow \langle C, M, E' \rangle \models \varphi))$$

$$\Longleftrightarrow (EX)(X \subseteq |\mathfrak{M}| \;\&\; (\varphi)(\varphi \in KM \Rightarrow \langle M, E, X \rangle \models \Psi_\varphi)$$

thus the class of extendable models is a PC class. ∎

One shows (just by appropriate modification of the above definition) that the class of β- extendable models is a PCPC class. By the existence of the least β - extension (cf § 2) it is also a CPCPC class. We do not know whether it is a CPC class (with respect to the class of all models of ZFC).

§ 2. Ramified analysis and β -extendability

We present here a construction of the least β -extension of a model $\langle M,E \rangle$ (provided $\langle M,E \rangle$ is β -extendable). The construction follows closely the one of Gandy used in his proof of existence of the least β -model of analysis. However we have to change some details since not every model has a definable wellordering.

We use instead another interesting property of transitive models of ZF; Every transitive model of ZF has a definable prewellordering (according to the rank of elements) such that every equivalence class of this prewellordering is a set. This fact will be used to show that ramified analysis has a prewell-ordering such that every equivalence class of it is codable as a class with the domain being a set.

Let M be a transitive set, $U \subseteq \mathcal{P}(M)$ family of subsets of M, we consider a structure $\langle U, M, \in \rangle$.

Definition: $\mathcal{D}(\langle U,M,\in\rangle)$ is the family of subsets of M, parametrically definable over $\langle U,M,\in\rangle$ i.e. of the form

$$\{ x \in M : \langle U,M,\in\rangle \models \varphi[x, \vec{z}] \}$$

where φ is a formula and $\vec{z}$ a sequence of parameters.

Let us note that $\langle U,M,\in\rangle \models \zeta[A]$ is equivalent with $A \in M$.

Let $X \subseteq M$.

We define $R.A_0^{M,X} = M \cup \{X\}$

$$R.A_{\alpha+1}^{M,X} = \mathcal{D}(\langle R.A_\alpha^{M,X}, M, \in\rangle)$$

$$R.A_\lambda^{M,X} = \bigcup_{\xi<\lambda} R.A_\xi^{M,X}$$

and finally $R.A_\cdot^{M,X} = \bigcup_{\xi\in On} R.A._\xi^{M,X}$

Since $R.A_\alpha^{M,X} = R.A_{\alpha+1}^{M,X}$ implies that $R.A_\alpha^{M,X} = R.A_\cdot^{M,X}$ therefore by cardinality argument there must be ρ such that $R.A._\rho^{M,X} = R.A_\cdot^{M,X}$.

Definition: Let $K \subseteq M$. K is called M-amenable iff either K is not function or K is a function and $(x)(x\in M \Rightarrow K * x \in M)$.

Theorem 2.1. (Ramified analysis theorem). If $\langle M,\in\rangle$ is a transitive model of ZFC and if every element of $R.A_\cdot^{M,X}$ is M-amenable then $\langle R.A_\cdot^{M,X}, M, \in\rangle$ is a β-model of KM.

Proof. The family $R.A.^{M,X}$ is – by virtue of construction – closed under the scheme of class existence (Indeed, if $R.A.^{M,X} = R.A._{\eta}^{M,X}$ then $R.A_{\eta+1}^{M,X} \subseteq R.A.^{M,X} = R.A_{\eta}^{M,X}$ i.e. $\mathcal{D}(\langle R.A._{\eta}^{M,X}, M, \in \rangle) \subseteq R.A._{\eta}^{M,X}$).

The axiom of substitution holds by the amenability property of $R.A.^{M,X}$ Since $V^{R.A.^{M,X}} = M$ (by our previous remark) therefore the axiom of power set holds too.

The proof of the fact that $R.A.^{M,X}$ is a β – model we defer until we get appropriate technique. ∎

Lemma 2.1: If $\langle R.A._{\xi}^{M,X}, M, \in \rangle \models KM$ then $R.A.^{M,X} = R.A._{\xi}^{M,X}$.

Proof: If $\langle R.A._{\xi}^{M,X}, M, \in \rangle \models KM$ therefore $\mathcal{D}(\langle R.A._{\xi}^{M,X}, M, \in \rangle) \subseteq R.A._{\xi}^{M,X}$. By induction $R.A._{\eta}^{M,X} = R.A._{\xi}^{M,X}$ for all $\eta \geqslant \xi$. ∎

Our task – in fact for all the rest of this chapter – will be to prove that $R.A.^{M,X}$ is definable in every β – model $\langle \mathcal{F}, M, \in \rangle$ of KM such that $X \in \mathcal{F}$.

Before we go into the proof we need certain extension of the language. We add to the language of L_{ST} predicates $\mathcal{M}(.)$ and $\mathfrak{X}(.)$ and assume the following axioms:

1) $(X)(\mathcal{M}(X) \longrightarrow \mathfrak{Z}(X))$

2) Trans $(\mathcal{M}(.))$

3) $(Z\ F\ C)^{\mathcal{M}(.)}$

4) $(Y)(\mathfrak{X}(Y) \Rightarrow \mathfrak{M}(Y))$

Since there is no reason to assume that any of objects M,X is definable in KM we have to start at the language level. The theory $KM_{\mathfrak{M},\mathfrak{X}}$ is the theory KM + 1) ... 4) where in the class existence scheme we allow $\mathfrak{M}$ and $\mathfrak{X}$ to appear.

It is clear that $KM_{\mathfrak{M},\mathfrak{X}}$ is a conservative extension of KM. Let us notice that if $\langle \mathcal{F}, M, \in \rangle \models KM$, $X \in \mathcal{F}$ then the structure $\langle \mathcal{F}, M, \in, M, X \rangle$ is a model of $KM_{\mathfrak{M},\mathfrak{X}}$. There are models of $KM_{\mathfrak{M},\mathfrak{X}}$ of different form. For instance $\langle \mathcal{F}, M, \in, L^M, X \rangle$ where $X \in \mathcal{F}$ and $X \subseteq L^M$. In the sequel proofs will be done in $KM_{\mathfrak{M},\mathfrak{X}}$.

For a moment we are going to study prewellorderings

Definition: A prewellordering (p.w.o) is any relation reflexive, transitive and satisfying the wellfoundedness condition

$(Z)(Z \neq \emptyset \;\&\; Z \subseteq \mathrm{Dom}(\prec) \Longrightarrow (Ez)(z \in Z \;\&\; (t)(t \in Z \Longrightarrow z \prec t)$

If $\prec$ is a p.w.o we define $\sim_\prec$ as follows: $x \sim_\prec y \Longleftrightarrow (x \prec y \;\&\; y \prec x)$. $\sim_\prec$ is an equivalence; let $Cl_\prec(x)$ be an equivalence class of x in $\sim_\prec$

Definition: A p.w.o $\prec$ is a good p.w.o. (gpwo) iff

$(x)(x \in \mathrm{Dom} \prec \Longrightarrow Cl_\prec(x) \in V)$

If $\prec$ is a gpwo then $\prec$ determines a wellordering $\tilde{\prec}$ on the class $\mathrm{Dom} \prec / \sim_\prec$ as follows: $Cl(x) \tilde{\prec} Cl(y) \Longleftrightarrow x \prec y$.

Conversly if $<$ is a wellordering and $F : \mathrm{Dom}(<) \longrightarrow V$ satisfies conditions: a) $x \neq y \implies F(x) \cap F(y) = \emptyset$

b) $F(x) \neq \emptyset$

then $<$ and F determine natural p.w.o $\precsim$ on $\bigcup_{x \in \mathrm{Dom}(<)} F(x)$

namely $x_1 \precsim x_2 \iff (Ey_1)(Ey_2)(x_1 \in F(y_1)\ \&\ x_2 \in F(y_2)\ \&\ y_1 < y_2)$

Operations $\tilde{\prec}$ and $\precsim$ commute (up to isomorphism).

In the sequel we will need one more operation:

Let Y be a class such that Dom Y is set and

a) $(y)(y \in \mathrm{Dom}\ Y \implies \mathrm{W.O.}(Y^{(y)}))$ and

b) $(y_1)(y_2)(y_1, y_2 \in \mathrm{Dom}\ Y \implies Y^{(y_1)} \cong Y^{(y_2)})$.

We call Y mixable iff it satisfies a) and b).

The ordering Y^{mix} is defined as follows:

$$\mathrm{Dom}\ Y^{mix} = \{ f \in {}^{(\mathrm{Dom}\ Y)}V : (x_1)(x_2)(x_1, x_2 \in \mathrm{Dom}\ Y \implies Y^{(x_1)} \restriction f(x_1) \cong Y^{(x_1)} \restriction f(x_2) \}$$

$$f_1 <_{Y^{mix}} f_2 \iff (Ex)(x \in \mathrm{Dom}\ Y\ \&\ f_1(x) \prec_{Y(x)} f_2(x))$$

Lemma 2.2. If Y is mixable then for all $x \in \mathrm{Dom}\ Y$, $Y^{mix} \cong Y^{(x)}$

Definition: (a) If Y_1 and Y_2 are classes then

$\{0\} \times Y_1 \cup \{1\} \times Y_2$ is called ordered pair of Y_1, Y_2 and is denoted by $\langle Y_1, Y_2 \rangle$

(b) If Y is a class, $Y \subseteq V \times V$. $\prec$ is a gpwo of Dom Y then the pair $\langle Y, \prec \rangle$ is called a gpwo family.

(c) If $\prec$ happens to be a wellordering of Dom Y then $\langle Y, \prec \rangle$ is called a wellordered family.

Definition: A proper formula is a formula Φ such that

(a) $0 \in Fr\,\Phi$ (b) $i \in Fr\,\Phi \Longrightarrow i \dot{-} 1 \in Fr\,\Phi$

($Fr\,\Phi$ is the set of indices of free variables in Φ)

Since we identify formulas with their Gödel numbers, the set of proper formulas is a set of numbers; we denote it by Pform. Usage of proper formulas allows us not to bother about which are the free variables of the formula, thus simplyfying the formalization of the operation $\mathcal{D}(\cdot)$.

If $\prec$ is a pwo then $\bigcup_{n \in \omega}{}^{n}\mathrm{Dom}(\prec)$ has a natural pwo.

We denote it by $\prec_{alex}$. It is the following ordering:

$$lh(\vec{x}) < lh(\vec{y}) \vee \big[lh(\vec{x}) = lh(\vec{y}) \;\&\; (Ek)(j)(j<k \Longrightarrow \vec{x}(j) \sim \vec{y}(j) \;\&$$
$$\vec{x}(k) \prec \vec{y}(k) \;\& \neg\, \vec{y}(k) \prec \vec{x}(k)) \vee (lh(\vec{x}) = lh(\vec{y}) \;\&\; (j)(j \in \mathrm{Dom}\, x \Rightarrow x(j) \sim y(j))$$

Lemma 2.3. If $\prec$ is a gpwo then $\prec_{alex}$ is a gpwo.

Proof: Assume $\vec{s}_1 \prec_{alex} \vec{s}_2 \prec_{alex} \vec{s}_1$: Then it is obvious that $lh(\vec{s}_1) = lh(\vec{s}_2)$. Let $\vec{s}_1 = \langle z_1, \dots, z_k \rangle$,

$\vec{s}_2 = \langle t_1, \dots, t_k \rangle$. We show by induction that

$z_1 \sim t_1 \quad \dots \quad z_k \sim t_k$. This however shows how the classes of $\sim_{\prec_{alex}}$ look like:

$Cl_{\prec_{alex}}(\langle z_1 \ldots z_k \rangle) = \{ \langle t_1 \ldots t_k \rangle : z_1 \sim t_1 \,\&\ldots\&\, z_k \sim t_k = Cl(z_1) \times \ldots \times Cl(z_k)$. The latter class is a set. ∎

If $\prec$ is a pwo then in the class $\bigcup_{\varphi \in Pform} \{\varphi\} \times ({}^{(Fr(\varphi)-\{0\})}Dom(\prec))$ there is a special pwo called the derived ordering of X and denoted by X' namely

$$\langle \varphi, \vec{x} \rangle \prec' \langle \psi, \vec{y} \rangle \Leftrightarrow (\varphi \prec \psi) \vee (\varphi = \psi \,\&\, \vec{x} \prec_{alex} \vec{y})$$

One proves that: If $\prec$ is a gpwo then $\prec'$ is a gpwo and in particular that if $\prec$ is a wellordering then $\prec'$ is also a wellordering.

We recall that the formula $\mathfrak{z}(.)$ served as formalization of the predicate "$x \in V$"

Definition: Let Y be a class such that

(i) $(x)(\mathcal{M}(x) \Longrightarrow x \,\eta\, Y)$

(ii) $Sat(Y, \mathfrak{z}(.), t) \Longrightarrow \mathcal{M}(Y^{(t)})$

then we define $\mathcal{D}_{\mathcal{M}}(Y) = \{\langle\langle \varphi, \vec{z} \rangle, x \rangle : \varphi \in Pform \,\&$

$(\vec{z} \in {}^{(Fr\ -\{0\})}Dom\ Y) \,\&\, (Et)(t \in Dom Y \,\&\, Y^{(t)} = x$

$\& Sat(Y, \varphi, \vec{z}(\overset{o}{t})))$.

Let Y be a set, then we say that the family of sets $\{X : X \,\eta\, Y\}$ i.e. $\{Y^{(t)} : t \in Dom\ Y\}$ is codable by Y (or - equivalently - that Y is a code for $\{Y^{(t)} : t \in \mathrm{Dom}\, Y\}$.

In order to explain the meaning of the operation $\mathcal{D}_{\mathcal{R}}(\cdot)$, let us remark that if Y,M are sets $Y^{(t)} \subseteq M$ for all $t \in \mathrm{Dom}\, Y$, $\mathcal{R}(x) \iff x \in M$, then $\mathcal{D}_{\mathcal{R}}(Y)$ is nothing else but a certain code for the family $\mathcal{D}(\langle\{Z : Z \,\eta\, Y\}, M, \in\rangle)$ as it was defined on page 491.

Let us note that the operation $\mathcal{D}_{\mathcal{R}}(\cdot)$ makes sense also in case when Y is a proper class and $\mathcal{R}(x) \iff x \in M$ where M is a proper class; In this case however $\mathcal{D}(\langle\{Z : Z \,\eta\, Y\}, M, \in\rangle)$ was not defined.

Lemma 2.4.: If $\langle Y, \prec\rangle$ is a gpwo family (i.e. $\prec$ is a gpwo) then $\langle \mathcal{D}_{\mathcal{R}}(Y), \prec'\rangle$ is also a gpwo family.

Lemma 2.5.: If $\langle Y_1, \prec_1\rangle$ is a gpwo family then there is a unique gpwo family $\langle Y_2, \prec_2\rangle$ (called the concentration of $\langle Y_1, \prec_1\rangle$) satisfying the following conditions

a) $(W)(W \,\eta\, Y_1 \iff W \,\eta\, Y_2)$

b) $(x)(x \in \mathrm{Dom}\, Y_2 \Rightarrow Y_1^{(x)} = Y_2^{(x)})$

c) $\prec_2 = \prec_1 \restriction \mathrm{Dom}\, Y_2$

d) $(x)(y)(x \in \mathrm{Dom}\, Y_1 \,\&\, y \in \mathrm{Dom}\, Y_2 \,\&\, x \prec y \,\&\, \neg\, y \prec x \Rightarrow Y_1^{(x)} \neq Y_1^{(y)})$

e) $(x)(y)(x \sim_{\prec_1} y \,\&\, Y_1^{(x)} = Y_1^{(y)} \Rightarrow (x \in \mathrm{Dom} Y_2 \iff y \in \mathrm{Dom} Y_2))$

Proof: Let us describe how $\langle Y_2, \prec_2 \rangle$ arises from $\langle Y_1, \prec_1 \rangle$.

Let $Z \eta Y_1$, $Z = Y_1^{(z)}$. We pick $\prec_1$-least z_1 with this property. Since $\prec_1$ is a gpwo z_1 needs not to be unique. We leave in the $\mathrm{Dom}(\prec_2)$ all these $\prec_1$ - least z_1's but erase all other z's which have the property that $Y_1^{(z)} = Z$. This procedure determines both Y_2 and $\prec_2$. The conditions a) - e) were determined to give this procedure. ∎

The unique pair constructed in the concentration procedure is called concentration of $\langle Y, \prec \rangle$ and denoted $\langle Y^-, \prec^- \rangle$. Now let M be the class $\{x : \mathcal{M}(x)\}$, $\prec$ the class $\{\langle x,y \rangle : x \in M \& y \in M \& \rho(x) \leq \rho(y)\}$, $X = \{x : \mathcal{X}(x)\}$, define $\widetilde{E}_M = \{\langle \{y\}, x \rangle : x \in y\}$, $E'_M = E_M \cup \{\langle o, x \rangle : x \in X\}$

In the case when M is a set then $\widetilde{E}_M$ is a code for M and E'_M is a code for $M \cup \{X\}$.

Notice that $\mathrm{Dom}\, E'_M$ has a special gpwo $\prec^+$ defined as follows

$$x \prec^+ y \iff (x = o) \vee (Et)(Eu)(x = \{t\} \& y = \{u\} \& t \prec u)$$

Lemma 2.6.: If T is a wellordering then there are unique classes U_T and ϑ_T satisfying the following conditions:

1) $\mathrm{Dom}\, U_T = \mathrm{Dom}\, T = \mathrm{Dom}\, \vartheta_T$

2) $(x)(x \in \mathrm{Dom}\, T \Rightarrow \langle U_T^{(x)}, \vartheta_T^{(x)} \rangle$ is a gpwo family)

3) If x_0 is the first element of T then

$$U_T^{(x_0)} = E_M \qquad \vartheta_T^{(x_0)} = \prec^+$$

4) If y is a successor of x in the wellordering T then

$$U_T^{(y)} = \mathcal{D}_{\mathfrak{r}}(U_T^{(x)}) \qquad \vartheta_T^{(y)} = (\vartheta_T^{(x)})'$$

5) If y is limit in T then

$$U_T^{(y)} = \bigcup_{x <_T y} \; \bigcup_{z \in \mathrm{Dom}\, U_T^{(x)}} \{\langle x, z\rangle\} \times (U_T^{(x)})^{(z)}$$

$$\vartheta_T^{(y)} = \{\langle\langle x,z\rangle , \langle x_1, z_1\rangle\rangle : x <_T y \;\&\; x_1 <_T y$$
$$\& \; z \in \mathrm{Dom}\, U_T^{(x)} \;\&\; z_1 \in \mathrm{Dom}\, U_T^{(x_1)} \;\&$$
$$(x <_T x_1 \;\&\; x \neq x_1) \vee (x = x_1 \;\&\; z \prec_{\vartheta_T^{(x)}} z_1)\}$$

Proof: We make use of the theorem on inductive definitions by transfinite induction on elements of wellorderings. We pick inductive clauses to correspond to the construction described on the lemma. One point which needs some explanation is that $\vartheta_T^{(y)}$ is a gpwo when y is limit. Notice however that this is obvious by the method we produce $\vartheta_T^{(y)}$. It is a direct union (according to T) of disjoint copies of $\vartheta_T^{(x)}$ for $x <_T y$.

The reader who had enough patience to come to this point deserves an explication.

Intuitively $U_T^{(x)}$ is a code for $R.A_{\alpha}^{M,X}$ where $\alpha = \overline{T \restriction x}$ ($T \restriction x$ is an initial segment of T determined by x) and U_T is a code for the sequence $\{\langle \alpha , R.A_{\alpha}^{M,X} \rangle : \alpha < \beta\}$ where $\beta = \overline{T}$. $\vartheta_T^{(x)}$ is a code for certain uniformly definable prewellordering of $R.A_{\alpha}^{M,X}$. This prewellordering is "thin" in the following sense: each of its equivalence classes is codable as a subclass Z of M such that Dom $Z \in M$.

The main point of this construction is that U_T and ϑ_T depend very loosely on T. If T_1 and T_2 are similar wellorderings then the unique similarity function F of T_1 and T_2 generates a sort of similarity between U_{T_1} and U_{T_2}. Similarly for ϑ_{T_1} and ϑ_{T_2}. U_T is called a diagram of construction of the ramified analysis along the wellordering or simply a diagram. ϑ_T is called the diagram of prewellordering of the ramified analysis along the wellordering T or simply pwo diagram. Notice that the complications we came into the clause 5 of the preceding lema arose from the fact that in order to avoid use of choice scheme we had to pass the limit points in a uniform vay.

Definition: Let $Z \subseteq M$, Od (Z, T) is an abbreviaton of the following formula: $W.O.(T) \& (E\, t)$ ("t is the last element of T" $\& \; Z \eta U_T^{(t)} \& (u)(u \in (\mathrm{Dom}(T) - \{t\}) \Rightarrow \neg Z \eta U_T^{(u)})$)

Intuitively Od (Z, T) means that Z belongs to

$R.A._{\alpha}^{M,X} - \bigcup_{\beta<\alpha} R.A._{\beta}^{M,X}$ and $\alpha = \overline{T}\upharpoonright t$ where t is last element of T.

Lemma 2.7. If T_1, T_2 are wellorderings, $T_1 \cong T_2$ and F establishes their isomorphism then: $(x)(x \in \mathrm{Dom}\, T_1 \implies (Y)$ $(Y \,\eta\, U_{T_1}^{(x)} \iff Y \,\eta\, U_{T_2}^{(Fx)}$).

Proof: By induction on the length of the wellordering T_1. For the first elements of T_1 and T_2 the equivalence is obvious. All the rest follows from the following: $(Z)(Z \,\eta\, X_1 \iff Z \,\eta\, X_2) \Rightarrow$ $(Z)(Z \,\eta\, \mathcal{D}_{\mathcal{R}}(X_1) \iff Z \,\eta\, \mathcal{D}_{\mathcal{R}}(X_2))$. Similar fact may be proved for ϑ_T. ∎

Using the lemma 2.7. we show the lemma 2.8., formalising the remark preceding lemma 2.7.

Lemma 2.8. Od (Z, T) & Od (Z, T_1) $\implies T \cong T_1$

Definition: Let $\langle Y, \prec \rangle$ be a gpwo family and $Z \,\eta\, Y$ We define $\langle Y, \prec \rangle (Z) = \{ s : Y^{(s)} = Z$ & s is $\prec$ - minimal with this property $\}$.

Notice that $\langle Y, \prec \rangle (Z) = \langle Y^{-}, \prec^{-} \rangle$ (Z)

Definition: (a) r.a. (T,Z) is an abbreviation of the formula:

$(Et)(t \in \mathrm{Dom}\, T$ & $Z \,\eta\, U_T^{(t)}$)

(b) r.a. (Z) is an abbreviation of the formula:

(ET)(W.O. (T) & r.a. (T,Z))

(c) $Z_1 \prec_{r.a.} Z_2$ is an abbreviation of the formula:

$$(T_1)(T_2)\left[\text{Od}(Z_1, T_1) \,\&\, \text{Od}(Z_2, T_2) \Longrightarrow (Ez)(z \in \text{Dom}(T_2) \,\&\, T_1 \cong T_2 \restriction z)\right] \vee (ET)\{ \text{Od}(Z_1, T) \,\&\, \text{Od}(Z_2, T) \,\&$$

$$(t)(t \in \text{Dom}(T) \,\&\, Z_1 \eta\, U_T^{(t)} \,\&\, Z_2 \eta\, U_T^{(t)} \Longrightarrow$$

$$(w_1)(w_2)(w_1 \in \langle U_T^{(t)}, \vartheta_T^{(t)} \rangle (Z_1)$$

$$\& w_2 \in \langle U_T^{(t)}, \vartheta_T^{(t)} \rangle (Z_2) \Longrightarrow w_1 \prec_{(\vartheta_T^{(t)})} w_2))$$

Intuitively r.a. (T,Z) means $Z \in R.A_{\overline{T}}^{M,\chi}$, r.a.$(Z)$ means $Z \in R.A^{M,\chi}$ and $Z \prec_{r.a.} Y$ means, Z is constructed in the process of construction of $R.A^{M,\chi}$ earlier than Y i.e. either the order of construction of Z is smaller than that of Y or (if their order is the same then either Z was defined by a formula whose Gödel number was smaller than that used to define Y or alternatively if it is the same formula then the parameters used to define Z are lexicographically earlier than that used to define Y).

Let us note that we could use U_T and ϑ_T as terms since indeed they were unique by the lemma 2.2. Formally we should use formulas $U(.,.)$ and $\vartheta(.,.)$ such that:

(a) $U(T, Y) \Longleftrightarrow W.O.(T) \,\&\, Y = U_T$

(b) $\vartheta(T,Y) \Longleftrightarrow W.O(T) \,\&\, Y = \vartheta_T$.

Thus while speaking on absoluteness of U_T and ϑ_T we mean the absoluteness of the formulas U and ϑ.

Lemma 2.8. : $\prec_{r.a.}$ defines a gpwo of r.a.

Proof: The precise meaning of the lemma is that:

$$(E\,Z)(r.a\,(Z)\,\&\,\Phi(Z)) \Rightarrow (E\,Z)(r.a.(Z)\,\&\,\Phi(Z)\,\&\,(Y)(\Phi(Y)\,\&\,r.a(Y) \Rightarrow Z \prec_{r.a} Y))$$

Pick firstly least T such that $(EY)(Od\,(T,Y)\,\&\,\Phi(Y))$. Consider now any $\vartheta_T^{(t)}$ minimal $z \in \mathrm{Dom}\, U_T^{(t)}$ such that $\Phi((U_T^{(t)})^{(z)})$ (where t is the last element of T). As $(\vartheta_T^{(t)})^-$ is an initial segment of $\prec_{r.a.}$ we are done. ∎

Now let M be a β - extendable transitive model of ZFC.

Let $\mathcal{F} \subseteq \mathcal{P}(M)$ be a β - extension of M, $X \in \mathcal{F}$ is fixed subset of M. Then the structure $\langle \mathcal{F}, M, \in, M, X \rangle$ is a β - model of $KM_{\Omega, \mathcal{X}}$.

Recall that $h(\mathcal{F})$ is a supremum of the types of well-orderings in $\mathcal{F}$.

The construction of the relativized ramified analysis was conducted above two times.

In the first definition we constructed a family of subsets of a transitive set, in the second we defined a predicate in the theory KM. . The next lemma connects these two definitions and allows us - while interpreting Ω as M - to use interchangingly $R.A^{M,X}_{\alpha}$ and the family defined by the predicate r.a.(T,.) (where $\overline{T} = \alpha$). $\mathcal{F}$, M, X are fixed for the time being.

Lemma 2.9. If $T \in \mathcal{F}$, $\overline{T} = \alpha$ and $\langle \mathcal{F}, M, \in \rangle$ is a β - model, $X \in \mathcal{F}$ then, for all $Z \in \mathcal{F}$

a) $\langle \mathcal{F}, M, \in, M, X \rangle \models$ r.a. $[T,Z] \iff Z \in R.A.^{M,X}_{\alpha}$

and so, for all $Z \in \mathcal{F}$

$\langle \mathcal{F}, M, \in, M, X \rangle \models$ r.a. $[Z] \iff Z \in R.A^{M,X}_{\eta}$ where $\eta = h(\mathcal{F})$.

b) $R.A^{M,X}_{\eta+1} \subsetneq \mathcal{F}$

c) $R.A^{M,X}_{\eta+1}$ does not contain a wellordering of type η

Proof: Clearly a) implies b) and b) implies c).

To prove a) we have to show the absoluteness - with respect to $\langle \mathcal{F}, M, \in, M, X \rangle$ of U_T which follows from the fact that $\mathcal{F}$ is a β - model. ∎

Lemma 2.10. If $\alpha < \eta$ then the structure $\langle R.A^{M,X}_{\alpha}, M, \in \rangle$ has the β - property i.e. $(Y)(Y \in R.A^{M,X}_{\alpha} \Rightarrow (\langle R.A^{M,X}_{\alpha}, M, \in \rangle \models W.O[Y] \Rightarrow Y$ is a wellordering)).

Proof: By lemma 2.9. $Y \in \mathcal{F}$ and so, if Y is not a wellordering then $\langle \mathcal{F}, M, \in \rangle \models \neg W.O.[Y]$ (here we use β - property of $\langle \mathcal{F}, M, \in \rangle$). Thus there is $x \in M$ such that $\langle \mathcal{F}, M, \in \rangle \models$ "$x \subseteq \mathrm{Dom}\, Y$ & $x \neq \emptyset$ & "x has no Y-first element". The formula in " "'s is predicative and all the parameters are in $R.A^{M,X}_{\alpha}$. Thus $\langle R.A^{M,X}_{\alpha}, M, \in \rangle \models$ "$x \subseteq \mathrm{Dom}\, Y$ & $x \neq \emptyset$ & "x has no Y-first element". So $\langle R.A^{M,X}_{\alpha}, M, \in \rangle \models \neg W.O.[Y]$. ∎

Definition: γ_0 is the first ordinal γ such that $R.A^{M,X}_{\gamma+1}$ does not contain a wellordering of type $\geqslant \gamma$.

From the lemma 2.9. (c) it follows that $\gamma_0 \leqslant \eta$.

We are going to prove that $\langle R.A^{M,X}_{\gamma_0}, M, \in \rangle$ has a definable prewellordering; one such ordering is $\prec_{r.a.}$ restricted to this set (which is absolute with respect to $\langle R.A^{M,X}_{\gamma_0}, M, \in \rangle$).

Lemma 2.11. γ_0 is a limit ordinal.

Proof: from a wellordering of type γ one can-putting the first element to the end - produce a wellordering of type $\gamma_0 + 1$. This construction does not lead outside of $R.A.^{M,X}_{\gamma_0}$. ∎

Lemma 2.12. If $\alpha < \gamma_0$, $\beta < \gamma_0$ then $\alpha + \beta$, $\alpha \cdot \beta < \gamma_0$

Proof: If T_1, T_2 are wellorderings of types α and β

respectively and if they belong to some $R.A^{M,X}_{\xi}$ then orderings of type $\alpha+\beta$, $\alpha\cdot\beta$ belong to $R.A^{M,X}_{\xi+1}$ (as they are definable over $R.A^{M,X}_{\xi}$). Since ζ_0 is limit we get the result. ∎

Lemma 2.13. Each of the structures $\langle R.A^{M,X}_{\alpha}, M, \in\rangle, (\alpha \leq \zeta_0)$ is a model of Gödel Bernays theory of classes.

Proof: for $\alpha = 1$ the statement is well known. The union of an ordered family of models of Gödel Bernays theory of classes (with fixed V) is a model of Gödel Bernays theory of classes. This fixes the limit case. So what we need to prove is the successor case.

This is shown as usual by proving the closure under operations corresponding to the axioms of group B. Note that the fact that $R.A.^{M,X}_{\zeta_0} \subseteq \mathcal{F}$ implies the validity of the axiom of replacement. ∎

Generally, in the theory GB we are not able to prove the comparability of wellorderings (this needs Σ^1_1 class existence scheme). But the structure $\langle R.A^{M,X}_{\zeta_0}, M, \in \rangle$ does have the comparability property.

Lemma 2.14. If T_1, T_2 are wellorderings, $T_1, T_2 \in R.A^{M,X}_{\alpha}$, $T_1 \cong T_2$ Then the similarity function may be found in $R.A.^{M,X}_{\alpha+\overline{T}_1+1}$.

Proof: We show that the similarity function is definable over

$\langle R.A^{M,X}_{\alpha+\bar{T}_1}, M, \in\rangle$. We prove it inductively. Let F be a similarity function for T_1 and T_2. Then $F\restriction O_{T_1}(x)$ is a similarity function of $T_1\restriction x$ and $T_2\restriction F(x)$. By inductive assumption $F\restriction O_{T_1}(x)$ belongs to $R.A^{M,X}_{\alpha+\bar{T}_1}$ (for all $x \in \mathrm{Dom}T_1$). By their uniqueness it follows that $G = \underbrace{\quad}_{x \in \mathrm{Dom}T_1} (F\restriction O_{T_1}(x))$ belongs to $R.A^{M,X}_{\alpha+\bar{T}_1+1}$. If T_1 has no last element then this union is the desired similarity function. If T_1 has last element-say t_0-then T_2 has also last element-say u_0- and $G \cup \{\langle t_0, u_0\rangle\} \in R.A^{M,X}_{\alpha+\bar{T}_1+1}$ since the latter is a model of Gödel-Bernays theory of classes. Since $F = G \cup \{\langle t_0, u_0\rangle\}$ we are done. ∎

Lemma 2.15. Relations of similarity and of "less then" for the wellorderings are absolute with respect to $\langle R.A^{M,X}_{\gamma_0}, M, \in\rangle$

Proof: By 2.12. and 2.14. ∎

Lemma 2.16. a) If $Y \in R.A^{M,X}_{\alpha}$ then $\mathcal{D}_{\mathcal{M}}(Y) \in R.A^{M,X}_{\alpha+1}$

b) If $T \in R.A^{M,X}_{\alpha}$ then $T' \in R.A^{M,X}_{\alpha}$

Proof: Since $\langle R.A^{M,X}_{\alpha}, M, \in\rangle$ is a model of Gödel-Bernays theory of classes therefore for each φ and t $\{\langle\langle\varphi, t\rangle, x\rangle : (Ey)(Y^{(y)} = x \;\&\; \mathrm{Sat}(Y, \varphi, y^\frown t))\}$ belongs to $R.A^{M,X}_{\alpha}$. Since $\langle R.A^{M}_{\alpha}, M, \in\rangle$ is an ω - model, the notion of the formula is absolute and thus $\mathcal{D}_{\mathcal{M}}(Y)$ is definable over $\langle R.A^{M,X}_{\alpha}, M, \in \rangle$. Thus $\mathcal{D}_{\Omega}(Y)$ belongs to $R.A.^{M,X}_{\alpha+1}$.

b) T' is definable from T by predicative formula. ∎

Lemma 2.17. (a) $R.A^{M,X}_{\gamma_0}$ is closed with respect to the operation $\mathcal{D}_{\Omega}(.)$

(b) The formula defining the operation $\mathcal{D}_{\Omega}(.)$ is absolute with respect to $\langle R.A^{M,X}_{\gamma_0}, M, \in \rangle$.

Proof: a) Follows from 2.16.a and 2.11.

(b) We establish firstly that the formula Sat(.,.,.) is absolute with respect to $\langle R.A^{M,X}_{\gamma_0}, M, \in \rangle$. This in turn implies absoluteness of $\mathcal{D}_{\Omega}(.)$. ∎

Lemma 2.18. If $T \in R.A^{M,X}_{\alpha}$, T is a wellordering then U_T and ϑ_T belong to $R.A.^{M,X}_{\alpha+\bar{T}+1}$

Proof: Analogous to the proof of the lemma 2.14 using the lemma 2.16. ∎

Lemma 2.19. (a) $R.A^{M,X}_{\gamma_0}$ is closed with respect to the operations U_T and ϑ_T. (b) (Formulas defining) The operations U_T and ϑ_T are absolute with respect to $\langle R.A^{M,X}_{\gamma_0}, M, \in \rangle$

Proof: (a) From lemma 2.18 and 2.11.

(b) As before by the analysis of defining formulas. ∎

Lemma 2.20. (a) The concentration operation $(\cdot)^{-}, (\cdot)^{-}$ is absolute with respect to $\langle R.A^{M,X}_{\gamma_0}, M, \in \rangle$.

(b) The value operation $\langle \cdot, \cdot \rangle (Y)$ is absolute with respect to $\langle R.A.^{M,X}_{\gamma_0}, M, \in \rangle$

Lemma 2.21. (Analogue of Gödel's $\langle L, \in \rangle \models V = L$)
$\langle R.A.^{M,X}_{\gamma_0}, M, \in \rangle \models (X)$ r.a. (X)

Proof: We need to prove that in $\langle R.A^{M,X}_{\gamma_0}, M, \in \rangle$ the following formula is satisfied:

$(Y)(ET)(W.O.(T) \& (Et)(t \in \mathrm{Dom}T \& Y_{\eta} U_T^{(t)})$. Let Y be given, $Y \in R.A^{M,X}_{\gamma_0}$. Thus for some $\alpha < \gamma_0$, $Y \in R.A.^{M,X}_{\alpha}$. By the definition of γ_0 there must be a wellordering T in $R.A.^{M,X}_{\alpha+1}$ such that the type of T, $\bar{T}$ is bigger than or equal to α

Thus $U_T \in R.A^{M,X}_{\alpha+\alpha+2}$ (by the lemma 2.18). Once again by 2.12 $U_T \in R.A^{M,X}_{\gamma_0}$. As order of construction of Y is at most α, we are done.*)

---

*) It is clear from the reasoning that the consideration of γ_0 (This trick is due to Gandy) is basic to the success of our construction. Because it may well happen that $\gamma_0 < \eta$ and then we would have inside of $R.A.^{M,X}$ too few wellorderings to reach η (η is $h(\mathcal{F})$). In case when $\langle \mathcal{F}, M, \in \rangle \models$ KMC we can show directly - in $\langle \mathcal{F}, M, \in \rangle$ - that $\langle R.A^{M,X}_{\eta}, M, \in \rangle$ satisfies KM. But this reasoning does not lead to basic lemmas 2.21 and 2.22.

Lemma 2.22. The formula $\prec_{r.a.}$ is absolute with respect to $\langle R.A._{\gamma_0}^{M,X}, M, \in \rangle$

Proof: Using previous lemmas it is enough to prove absoluteness of the formula $Od(.,.)$ which we leave to the reader. ∎

To prove the reflection principle (and thus that $\langle R.A^{M,X}_{\gamma_0}, M, \in \rangle$ is a model of KM) we follow the classical proof of Levy of the reflection principle ZF.

Theorem 2.2.: (Reflection Principle for $\langle R.A._{\gamma_0}^{M,X}, M, \in \rangle$

For every formula Φ of L_{ST} there are arbitrarily large $\alpha < \gamma_0$ such that for all $X_1 \ldots X_n \in R.A^{M,X}_{\alpha}$

$$\langle R.A._{\alpha}^{M,X}, M, \in \rangle \models \Phi [X_1, \ldots, X_n] \iff$$

$$\iff \langle R.A^{M,X}_{\gamma_0}, M, \in \rangle \models \Phi [X_1, \ldots, X_n]$$

To show this we need three facts:

1^0 The possibility of bounding the places where examples for existential formulas appear.

2^0 Every definable functional on $\langle R.A._{\gamma_0}^{M,X}, M, \in \rangle$ which takes as values wellorderings, is invariant under similarity of wellorderings and is continuous is majorized by a functional of the same sort which is in addition increasing.

3^0 Every definable, increasing, invariant and continuous functional has arbitrarily large critical points.

We show 1° leaving 2° and 3° to the reader. In both cases the idea of proof is similar to that of 1°. Namely in showing that appropriate supremum of wellorderings exist.

Proof of 1° Let $U_T^{(x)}$ be given (i.e. a code for $R.A\frac{M,X}{T\upharpoonright x}$).

Assume that for every $Z \,\eta\, U_T^{(x)}$ there is $W \in R.A^{M,X}_{\gamma_0}$ such that $\langle R.A^{M,X}_{\gamma_0}, M, \in\rangle \models \Phi[Z, W]$. We show that there is $\rho < \gamma_0$ such that for every $Z \in U_T^{(x)}$ there is $W \in R.A^{M,X}_{\rho}$ such that $\langle R.A^{M,X}_{\gamma_0}, M, \in\rangle \models \Phi[Z, W]$.

For every $Z \,\eta\, U_T^{(x)}$ i.e. for every $z \in \mathrm{Dom}\, U_T^{(x)}$ we may find $\prec_{r.a.}$ - minimal wellordering T_z such that appropriate W may be found in $R.A.\frac{M,X}{T_z}$. Unfortunately T_z is not unique. Consider the shortest T_z's. Still we are not able to claim T_z unique. However we shall find a new wellordering similar to T_z and which is uniquely determined by z.

Definition: $S \subseteq V^2$ is called small class ordinal (s.c.o) iff

a) $\mathrm{Dom}\, S \in V$

b) $(Y)(Y \,\eta\, S \Rightarrow W.O.(Y))$

c) $(Y_1)(Y_2)(Y_1 \,\eta\, S \;\&\; Y_2 \,\eta\, S \Rightarrow Y_1 \cong Y_2)$

Definition: Classes Z_1, $Z_2 \subseteq V^2$ are almost equal (a.e) iff $(Y)(Y \,\eta\, Z_1 \Longleftrightarrow Y \,\eta\, Z_2)$

Lemma: There is a predicate Sel(.,.) such that:

a) $\mathrm{Sel}(Z, Y) \Rightarrow \mathrm{W.O.}(Y)\ \&\ \mathrm{s.c.o.}(Z)$

b) Z_1 a.e. $Z_2 \Rightarrow (\mathrm{Sel}(Z_1, Y) \Leftrightarrow \mathrm{Sel}(Z_2, Y))$

c) $\mathrm{s.c.o.}(Z) \Rightarrow (E!\,Y)\ \mathrm{Sel}(Z, Y)$

d) $(x)(Y)(x \in \mathrm{Dom} Z\ \&\ \mathrm{Sel}(Z, Y) \Rightarrow Y \cong Z^{(x)})$

The most natural idea would be, to consider instead of (elements just set of those) functions on Dom Z taking as values elements on the same level (viz the picture). Unfortunately sets of elements on different levels may be identical. Let us notice however that elements f of Dom Z^{mix} such that for given z, $\mathcal{R}$ f = z, form necessarily an element of V.

So preceeding formally call a level of Z an $\mathcal{R}$ f for some f $\in$ Dom Z^{mix}

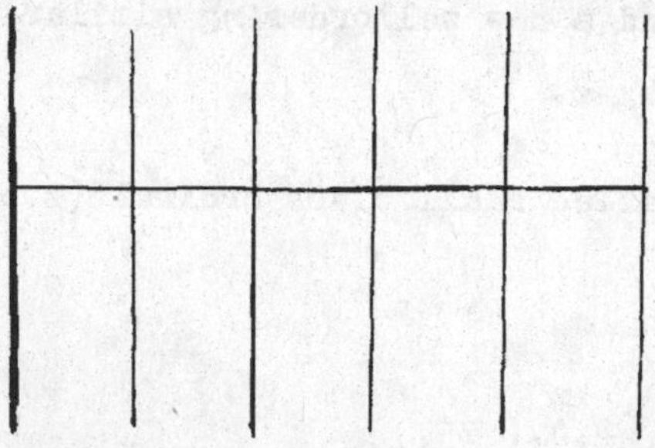

Let us consider an order type of these f $\in$ Dom Z^{mix} such that $\mathcal{R}$ f = z (z fixed). This type is an ordinal and does not depend at all on Z in the sense that if Z_1 and Z_2 are a.e. then the appropriate types in Z_1^{mix} and Z_2^{mix} are the same.

Define now, for z being the set of all elements on the same level α_z to be the type of the set of all f's such that $\mathcal{R} f = z$.

Form now the class $H_Z = \bigcup\{\alpha_z \times \{z\} : z$ is the set of all elements on some level of Z $\}$.

Notice again that if Z_1 a.e. Z_2 then $H_{Z_1} = H_{Z_2}$.

Order now H_Z as follows: $\langle \alpha, s\rangle \prec^* \langle \beta, t\rangle$ iff "The initial segment of Z^{mix} determined by the $\beta^{\underline{th}}$ function f(in Z^{mix}) such that $\mathcal{R} f = t$ contains a subset ordered by Z^{mix} in type $\geq \alpha$ of functions g such that $\mathcal{R} g = s$".

The predicate Sel(.,.) is a description of $\prec^*$ from Z as constructed above. ∎

Lemma: $\langle R.A^{M.X}_{\gamma_0}, M, \in\rangle$ is closed with respect to the operation determined by Sel, moreover Sel(.,.) is absolute with respect to $\langle R.A^{M,X}_{\gamma_0}, M, \in\rangle$.

Using the above lemma we are able to prove certain uniformization principle for $R.A^{M.X}$

Lemma: Let $\mathcal{H}$ (.,.) be a predicate such that:

$$\langle R.A^{M,X}_{\gamma_0}, M, \in\rangle \models (\mathcal{H}(Z,Y) \Rightarrow W.O.(Y))\ \&$$

$$(\mathcal{H}(Z,Y)\ \&\ Y \cong Y_1 \Rightarrow \mathcal{H}(Z, Y_1))$$

Then there is a predicate $\mathcal{H}'(.,.)$ such that

$$\langle R.A._{\rceil_0}^{M,X}, M, \in \rangle \models (\mathcal{H}'(Z, Y) \Rightarrow \mathcal{H}(Z,Y)) \& [(EY) \mathcal{H}(Z,Y) \Rightarrow (E!\, Y) \mathcal{H}'(Z,Y)$$

Proof: We describe a construction of $\mathcal{H}'$. Given Z consider all $\prec_{r.a}$ minimal and shortest Y's such that $\mathcal{H}(Z,Y)$. This collection may be coded as an s.c.o. Any two s.c.o 's coding it are almost equal. Using Sel we get the appropriate wellordering. ∎

Now, to finish the proof of 1°:

Let T_z be the wellordering obtained when the uniformization principle was applied to the predicate:

$$\mathcal{H}(z,S) \Leftrightarrow (E\,W)(Ey)(y \in \mathrm{Dom}\, S \;\&\; W \,\eta\, U_S^{(y)} \;\&\; \Phi((U_T^{(x)})^{(z)}, W))$$

Form the class K as follows:

$$K = \bigcup_{z \in \mathrm{Dom}\, \mathcal{U}_T^{(x)}} \{z\} \times \mathrm{Dom}\, T_z$$

Define a relation $\sim$ on K as follows:

$$\langle z_1, x\rangle \sim \langle z_2, y\rangle \Leftrightarrow T_{z_1} \upharpoonright x \cong T_{z_2} \upharpoonright y$$

From every equivalence class of $\sim$ pick elements of the smallest rank; Let L be a class of these sets. Define now

$$[\langle z_1, x\rangle] \prec^s [\langle z_2, y\rangle] \text{ iff } (Et)(t\, T_{z_2}\, y \;\&\; T_{z_1} \upharpoonright x \cong T_{z_2} \upharpoonright t)$$

$\prec^S$ is a wellordering majorizing all T_z's. $\prec^S$ is definable over $\langle R.A.^{M,X}_{\gamma_0}, M, \in \rangle$ and so, belongs to $R.A.^{M,X}_{\gamma_0+1}$. Thus the type of $\prec^S$ must be less than γ_0 and so we have shown that the appropriate supremum exists below γ_0. ∎

The functional which we adjoin now to the formula Φ is the following (we use - as before - the symbols $R.A.^{M,X}_{\prec}$ to make it more readable)

$$R'_{\Phi}(T_1,T_2) \Leftrightarrow (Z)(Z \in R.A.^{M,X}_{T_1} \& (EY)\Phi(X,Y) \Rightarrow (EY)(Y \in R.A.^{M,X}_{T_2} \& \Phi(Z,Y)$$

& " T_2 is a shortest wellordering with this property")).

The functional R' is definable, continuous and invariant with respect to the similarity of wellorderings. In order to get critical point used to reflect Φ we have to majorize it by a definable functional with the same properties and in addition increasing.

This is the reason why we prove 2° and 3°.

We leave the details to the experienced reader.

Since $\langle R.A.^{M,X}_{\gamma_0}, M, \in \rangle$ has the reflection property therefore it is a model of KM. From the existence of a definable gpwo we derive:

Theorem 2.3. $\langle R.A.^{M,X}_{\gamma_0}, M, \in \rangle$ satisfies the following collection scheme:

$$(x)(EY)\Phi(x,Y) \Rightarrow (EY)(x)(Ey)(y \in \text{Dom } Y \& \Phi(x, Y^{(y)}))$$

Proof: We pick $\prec_{r.a}$ - minimal Y's good for X and give them together. ▮

Let Coll $_\Phi$ and $\mathbb{C}_\Phi$ be collection scheme and choice scheme instance for Φ respectively

Theorem 2.4. KM + Coll $_\Phi$ + Global Choice $\vdash \mathbb{C}_\Phi$

Proof: Assume Coll $_\Phi$ and global choice i.e. let $\prec$ be a wellordering of the whole class V. Assume $(x)(EY)\Phi(x, Y)$. Then by Coll $_\Phi$, $(EY)(x)(Ey)(y \in \mathrm{Dom}\, Y \;\&\; \Phi(x, Y^{(x)})$.

Let R_x be a subset of Dom Y consisting of y's such that $\Phi(x, Y^{(y)})$. Let z_x be a $\prec$ - first element of Dom Y. Form $Y_1 = \bigcup_{x \in V} \{x\} \times Y^{(z_x)}$. Y_1 makes $\mathbb{C}_\Phi$ true. ▮

Thus we see that, if $\langle R.A._{\gamma_0}^{M,X}, M, \in \rangle$ satisfies the global choice then it automatically satisfies the choice scheme. This happens for instance when M has a wellordering definable in $\langle R.A^{M,X}_{\gamma_0}, M, \in \rangle$

We have a much nicer situation when $\langle M, \in \rangle$ has a definable wellordering, say $\prec$. Applying the whole construction to $\prec$ (i.e. letting $\mathcal{U}_T^{(t_0)} = \prec$) we get a definable wellordering of the whole $\langle R.A^{M,X}_{\gamma_0}, M, \in \rangle$

Since the existence of definable wellordering in the presence of choice scheme implies the scheme of dependent choices we sum up the situation as follows:

Theorem 2.5. (a) If $\langle M, \in \rangle$ is a transitive β-extendable model of ZFC then there is the smallest β-extension of $\langle M, \in \rangle$. This extension has a definable without parameters good prewellordering and, apart of the axioms of KM satisfies additionally the collection scheme.

(b) If $\langle M, \in \rangle$ is a transitive β-extendable model of ZFC, $\mathcal{F}$ is any β-extension of $\langle M, \in \rangle$, $X \subseteq M$, $X \in \mathcal{F}$ then there is the smallest β-extension of $\langle M, \in \rangle$ containing X. As before this extension has a good prewellordering definable with the parameter X and satisfies additionally the collection scheme.

(c) If $\langle M, \in \rangle$ is a transitive β-extendable model of ZFC and has a definable wellordering then the smallest β-extension of $\langle M, \in \rangle$ has a wellordering definable without parameters, satisfies the choice scheme and the scheme of dependent choices.

(d) If $\langle M, \in \rangle$ is a transitive β-extendable model of ZFC and has a definable wellordering, and if $\mathcal{F}$ is a β-extension of $\langle M, \in \rangle$, $X \in \mathcal{F}$ then the smallest β-extension of $\langle M, \in \rangle$ containing X has a definable wellordering (with a parameter X) and satisfies the choice scheme and the scheme of dependent choices.

Careful inspection shows that $\prec_{r.a.}$ is Δ^1_1 and r.a. is Σ^1_1.

The reasoning used in the proof of the theorem 2.5. may be

applied to a proof that V = L is relatively consistent with KM. Indeed when M is interpreted as L the formula r.a. defines an inner model of KM + V = L in KM. More precisely let $r.a.^{L}(.)$ be this formula (i.e. $\mathcal{M}(x) \Longleftrightarrow x \in L$, $X = \emptyset$.)

Definition: If T is a wellordering, T + 1 is the class arising from T by putting the first element of T to the end.

Let $r.a_{\square}^{L}$ be $r.a^{L}$ if there is no wellordering X such that $\neg(EY)\, Y \cong X \,\&\, r.a^{L}(X + 1, Y)$ and let $r.a_{\square}^{L}(.)$ be r.a. (Z,.) if Z is the shorlest wellordering with this property (Intuitively we consider $R.A.^{L}$ if there is no ξ such that $R.A^{L}_{\xi+1}$ does not contain a wellordering of type ξ or $R.A.^{L}_{\zeta_0}$ for the least ζ_0 such that $R.A^{L}_{\zeta_0+1}$ does not contain a wellordering of type ζ_0). By similar reasoning as in the proof of the theorem 2.5. we show that the formula $r.a._{\square}^{L}$ is an inner interpretation of KM + V = L in KM (the trick with $\square$ is again due to Gandy).

There is an important modification. We need to show that the classes satisfying $r.a_{\square}^{L}$ are L - amenable i.e. that if X is an $r.a_{\square}^{L}$ - class which is a function, $x \in L$ then $X * x \in L$. This needs a form of the condensation Lemma of Gödel proved as in the ZF case (In fact this was the beginning of investigations of the second author on the problems of this paper). Note that when we knew that M was

β - extendable then the property of M - amenability of $R.A.^M$ classes was automatic.

The syntactic contents of the reasoning leading to the theorem 2.5 may be summed up in the following

Metatheorem: a) There is a formula $\Phi(.)$ such that

1) $KM \vdash \Phi(V)$

2) $KM \vdash (x)(x \in V \Rightarrow \Phi(x))$

3) For every Ψ being an axiom of KM or an instance of the collection scheme

$KM \vdash (\Psi)^{\Phi}$

b) There is a formula $\Theta(.)$ such that

4) $KM \vdash (V = L)^{\Theta}$

5) $KM \vdash \Theta(L)$

6) $KM \vdash (x)(x \in L \Rightarrow \Theta(x))$

7) For every Ψ being an axiom of KM or an instance of the scheme of choice, $KM \vdash (\Psi)^{\Theta}$

Proof: In case a) take as Φ the formula $r.a_{\square}(.)$ with $M(x) \Longleftrightarrow \mathfrak{z}(x)$.

Incase b) take as Θ the formula $r.a._{\square}^{L}$. ∎

Now we are finally able to complete

The proof of theorem 2.1.

Assume that $\langle R.A.^{M,X}, M, \in \rangle$ is not a β - model (though it is a model). By the comparability of wellorderings all false (or as we say below nonstandard) wellorderings are longer than all standard (i.e. true) wellorderings in $R.A.^{M,X}$ and so all these nonstandard wellorderings have the same type of the maximal wellordered initial segment. Call the type of this segment α. Clearly $R.A.^{M,X}$ does not contain a wellordering of type α. Since $R.A^{M,X}_{\alpha+1} \subseteq R.A.^{MX}$ therefore also $R.A^{M,X}_{\alpha+1}$ does not contain a wellordering of type α. Let γ_0 - as before - be least γ such that $R.A^{M.X}_{\gamma+1}$ does not contain a wellordering of type γ.

<u>Case A:</u> $\gamma_0 < \alpha$. We claim that $\langle R.A^{M,X}_{\gamma_0}, M, \in \rangle$ has the property β.

First we remark that if $\langle R.A^{M,X}_{\gamma_0}, M, \in \rangle \models W.O.[T]$ then $\langle R.A^{M,X}, M, \in \rangle \models W.O.[T]$. Otherwise, since $\langle R.A.^{M,X}, M, \in \rangle$ is a model of KM, there would be a set (i.e. an element of M) not wellfounded in T. Since $M \subsetneq R.A^{M,X}_{\gamma_0}$ therefore $\langle R.A^{M,X}_{\gamma_0}, M, \in \rangle \models \neg W.O[T]$ contrary to the assumption. Now assume again $\langle R.A^{M,X}_{\gamma_0}, M, \in \rangle \models W.O.[T]$ but T is not a wellordering. By the above $\langle R.A.^{M,X}, M, \in \rangle \models W.O.[T]$ and so the initial wellordered segment of T has a type α which is bigger than γ_0 But then there is a initial segment of T of type $> \gamma_0$ in

$R.A^{M,X}_{\gamma_0}$ and thus also in $R.A^{M,X}_{\gamma_0+1}$ contradicting the choice of γ_0.

Moreover every element of $R.A^{M.X}_{\gamma_0}$ is M - amenable (since $R.A^{M,X}_{\gamma_0} \subseteq R.A^{M,X}$).

Now we know that $\langle R.A^{M,X}_{\gamma_0}, M, \in \rangle$ has the β - property and as before - by the property of γ_0 we prove the reflection property of $\langle R.A^{M,X}_{\gamma_0}, M, \in \rangle$. Thus it happens that $\langle R.A^{M,X}_{\gamma_0}, M, \in \rangle \models KM$.

By the lemma 2.1. $R.A^{M,X}_{\gamma_0} = R.A^{M,X}$ and so $\langle R.A^{M,X}, M, \in \rangle$ is a β - model of KM contradicting our assumption.

Case B: $\gamma_0 = \alpha$. As before we show that $\langle R.A^{M,X}_{\alpha}, M, \in \rangle$ is a model of GB theory of classes.

We prove now that:

1) For standard wellorderings $T \in R.A^{M,X}_{\alpha}$, $U_T \in R.A^{M,X}_{\alpha}$

2) For nonstandard wellorderings $T \in R.A^{M,X}_{\alpha}$, $U_T \notin R.A^{M,X}_{\alpha}$

(The point 2) has to be understood as follows: If T is not a standard wellordering then $\langle R.A^{M,X}_{\alpha}, M, \in \rangle \models \neg (EX)U(X,T)$)

Point 1) is proved by the same reasoning as the proof of theorem 2.5.

Point 2) we prove as follows: Since $\langle R.A^{M,X}_{\alpha}, M, \in \rangle$ is a model of GB therefore together with U_T for nonstandard T we get $U_T^{(x)}$ for some x such that $T \upharpoonright x$ is nonstandard.

By 1) $U_T^{(x)}$ contains all the classes belonging to $R.A_\alpha^{M,X}$. Now we construct the diagonal class for $U_T^{(x)}$ (i.e. the class $\{z \in \mathrm{Dom}\, U_T^{(x)} : z \notin (U_T^{(x)})^{(z)}\}$. This class being predicative in U_T belongs to $R.A_\alpha^{M,X}$ but is different from all the $(U_T^{(x)})^{(z)}$ which contradicts the fact that $U_T^{(x)}$ contains all classes of $R.A_\alpha^{M,X}$.

The points 1) and 2) allow us to discern the well-orderings among the objects satisfying in $\langle R.A_\alpha^{M,X}, M, \in \rangle$ the formula W.O. (We still do not know that $\langle R.A_\alpha^{M,X}, M, \in \rangle$ is a β - structure!) namely these are the objects for which U_T exists. As before we check the absoluteness of $\mathrm{Od}(\cdot,\cdot)$, $U(\cdot,\cdot)$ and $\vartheta(\cdot,\cdot)$ and $\prec_{r.a.\eta}(\cdot,\cdot)$. Now as before we show the reflection principle using instead of all objects satisfying in $\langle R.A_\alpha^{M,X}, M, \in \rangle$ the formula W.O. only those for which U_T exists.

So $\langle R.A_\alpha^{M,X}, M, \in \rangle$ is a model, and since $KM \vdash (T)(W.O.(T) \Rightarrow (EX)(U(X,T))$ we get the desired contradiction with the presence of nonstandard wellorderings in $R.A_\alpha^{M,X}$. Indeed we proved that for the nonstandard T's there is no U_T in $R.A_\alpha^{M,X}$. Thus also in this case $R.A^{M,X}$ (i.e. $R.A_\alpha^{M,X}$) is a β - model. ∎

Lemma 2.22.: If $\langle \mathcal{F}, M, \in \rangle$ is a β - model for KM and $\mathcal{G} \subseteq \mathcal{F}$ is a family of subsets of M such that $\langle \mathcal{G}, M, \in \rangle \models$ KM then $\langle \mathcal{G}, M, \in \rangle$ is also a β - model.

Proof: If $X \in \mathcal{G}$ is not a wellordering therefore, since $\mathcal{G} \subseteq \mathcal{F}$, $\langle \mathcal{F}, M, \in \rangle \models \neg\, W.O.[X]$ ($\mathcal{F}$ is a β -model).

Thus there is $x \in M$ such that x is not wellfounded in X.

Thus $\langle \mathcal{G}, M, \in \rangle \models \neg\, W.O.[X]$

Corollary: If there exists a β - model of KM then there is a β - model $\langle \mathcal{F}, M, \in \rangle$ of KM such that, for all $\mathcal{G} \subseteq \mathcal{F}$ if $\langle \mathcal{G}, M, \in \rangle$ is a model then $\mathcal{G} = \mathcal{F}$

Proof: Let $\langle \mathcal{F}, M, \in \rangle$ be a β - model; consider $\langle R.A.^M, M, \in \rangle$. It is the least β - model for KM (with M as the universe of sets). This together with the lemma 2.22, completes the proof. ∎

The corollary shows an important difference between ω - models of second order arithmetic and transitive models of KM. In the case of the former system there is no minimal ω - model as shown by H.Friedman in [3]. In the case of models of KM there are - under suitable assumptions - minimal transitive models. This answers the question of H.Friedman from the Introduction of the aforenamed paper.

If the constructed family $R.A.^M$ is not a model (thus not a β - model) then in the process of construction there must appear a class which is not M - amenable. Thus there must be a least α such that $R.A^{M,X}_{\alpha}$ contains non - M - amenable class. Clearly $R.A._0^{M,\emptyset}$ does not contain such an animal. Similarly $R.A._1^{M,\emptyset}$ - since this class is the least model for GB which has M as the universe of sets.

Theorem 2.7.: Let $\mathcal{N}$ be the least transitive model of ZF. Then $R.A._2^{\mathcal{N}}$ contains a class which is not $\mathcal{N}$ - amenable.

Proof: As shown in [13], $\mathcal{N} = L_\alpha$ where α is the least β such that $\langle L_\beta, \in \rangle$ models ZF. In [7] it is shown that $(L_{\alpha+2} - L_\alpha) \cap \mathcal{P}(\omega) \neq \emptyset$. It is easy to show that $L_{\alpha+2} \cap \mathcal{P}(\omega) \subseteq R.A._2^{\mathcal{N}}$. Thus there is a subclass of ω which is not a set and so non $\mathcal{N}$ - amenable class. ∎

One can however prove the following

Theorem 5: If M is a transitive model of ZFC, and M is extendable, then all classes in $R.A.^{M,\emptyset}_{h(M)^+}$ are M - amenable (where $h(M)^+$ is the least admissible ordinal bigger than $h(M)$'

Sketch of the proof: All the ordinals less than $(h(M))^+$ are representable in every extension of the model M (it follows from the fact that the standard part of any possibly nonstandard admissible set containing M has to contain α^+, as a subset, for every standard α). Thus also classes U_T for

$T < \alpha^+$ are in every extension and so the classes constructed before α^+ are M - amenable. ∎

Finally let us notice that in case when $\langle M, \in \rangle$ is not β - extendable but is a model of ZFC set theory we can show by slightly modified reasoning that $\langle R.A.^M, M, \in \rangle$ is a least β - model of KM - {replacement axiom}. This gives the following theorem:

Theorem 2.1'. If $\langle M, \in \rangle$ is a transitive model of ZFC then $\langle R.A.^M, M, \in \rangle$ is the least β - model of (KM - {replacement axiom}) with the class of sets equal M (i.e. it can process semisets in sense of Vopenka & Hajek). It is a model of KM just in case when $\langle M, \in \rangle$ is β - extendable.

§ 3. Extendability vs. β - extendability

We shall deal now with models of KMC. Indeed we remember that every transitive model $\langle \mathcal{F}, M, \in \rangle$ of KM has a transitive submodel $\langle \mathcal{G}, L^M, \in \rangle$ which satisfies KM + scheme of choice.

Thus while considering the heights of transitive extendable models we may restrict ourselves to the models extendable to the models of KMC. (Notice that $h(M) = h(L^M)$ and $L^M = L_{h(M)}$)

Let T denote the following sentence of L_{ST} :

$(Ex)(\omega \in x \;\&\; (z)(z \in x \Rightarrow \mathcal{P}(z) \in x) \;\&\; (f)(y)(y \in x \;\&\; Func(f) \;\&$

$f \subseteq x \Rightarrow f * y \in x))$

Thus the sentence T means that "There is an inaccessible family of sets". We notice that T is Σ_2 formula and the formula "(.) is an inaccessible family of sets" is Π_1.

In [6] the following fact is proved:

Proposition 3.1.: The theory $ZFC^- + T$ is interpretable in KMC by means of wellfounded trees.

Let us look more carefully at this interpretation. The inaccessible family is representable by a tree coding V. Trees of rank less than On represent elements of the maximal inaccessible family.

The trees of the rank less than or equal to On have realizations; in case of trees of rank less than On the realization is a set. In case of trees of rank On the realization is a proper class.

The proposition 3.1. Leads to the following: If $\langle \mathcal{F}, M, \in \rangle \models KMC$ then $\langle \text{Trees}^{\langle \mathcal{F}, M, \in \rangle}, \text{Eps}, \text{Eq} \rangle \models ZFC + T$ (where Eps and Eq are appropriate relations interpreting $\in$ and $=$). Every tree has a rank which is a wellordering. Assume $\langle \mathcal{F}, M, \in \rangle \models$ "T is a tree" & "U is a rank of T".

Then T is a tree iff U is wellordering. This immediately implies that the notion of a tree is absolute exactly for β - models.

Putting all these together we get a semantic version of the results from [6]:

Proposition 3.2. $\langle \mathcal{F}, M, \in \rangle$ is a β - model of KMC iff There is a transitive model $\langle N, \in \rangle$ of $ZFC^- + T$ such that

1) $M \in N$

2) $\langle N, \in \rangle \models$ " M is an inaccessible family of sets"

3) $\mathcal{F} = \mathcal{P}(M) \cap N$

The proof $\Rightarrow$ is roughly the following. We take all well-founded trees (without nontrivial automorphisms) and take as Eq and Eps isomorphism and membership of trees relations. Then $\langle \mathrm{Trees}^{\langle \mathcal{F}, M, \in \rangle}, \mathrm{Eq}, \mathrm{Eps} \rangle$ is a model (without absolute equality) of $ZFC^- + T$. Thus the structure $\langle \mathrm{Trees}^{\langle \mathcal{F}, M, \in \rangle}, \mathrm{Eps} \rangle / \mathrm{Eq} \models ZFC^- + T$. We take now realizations of trees from $\mathrm{Trees}^{\langle \mathcal{F}, M, \in \rangle}$. Since $\langle \mathcal{F}, M, \in \rangle$ was a β - model they are really trees and so they indeed have realizations. (The process of realization is similar to contraction procedure). We get an isomorphic model $\langle N, \in \rangle$. The equivalence class of a tree coding M is a desired inaccessible family. By class existence in $\langle \mathcal{F}, M, \in \rangle$ the subsets of M being in $\mathcal{F}$ and only them are in $N \cap \mathcal{P}(M)$. The proof of $\Leftarrow$ is obvious. ∎

Corollary: $\langle M, \in \rangle$ is β - KMC - extendable iff there is transitive model $\langle N, \in \rangle$ of $ZFC^- + T$ such that

1) $M \in N$

2) $\langle N, \in \rangle \models$ "M is an inaccessible family of sets".

If $\langle M,E \rangle$ is a relational structure, $E \subseteq M \times M$ then Sp M is the set of those $m \in M$ which are wellfounded i.e. those for which there is no infinite E - descending sequence beginning with m. If $\langle M,E \rangle$ satisfies extensionality then $\langle \mathrm{Sp}\, M, E \restriction \mathrm{Sp}\, M \rangle$ is isomorphic to a transitive structure $\langle A, \in \rangle$. Thus we may simply assume that Sp M is transitive (when $\langle M,E \rangle$ satisfies extensionality).

Further analysis of the notion of the tree allows us to give an analogue of the proposition 3.2. for extendable but not necessarily β - extendable transitive models.

Proposition 3.3. $\langle \mathcal{F}, M, \in \rangle$ is a transitive model of KMC iff there is a model $\langle N,E \rangle$ of $ZFC^- + T$ such that

1) $M \in \mathrm{Sp}\, N$

2) $\langle N, E \rangle \models$ "M is an inaccessible family of sets"

3) $\mathcal{F} = \mathcal{P}(M) \cap N$

Proof: Again $\Leftarrow$ is obvious (we tacitly assume that the objects in N - Sp N are not subsets of M)

$\Rightarrow$ Once more consider (wellfounded trees) $\langle \mathcal{T}, M, \in \rangle$ i.e. objects which satisfy in $\langle \mathcal{T}, M, \in \rangle$ the formula "(.) is a tree". All real trees which are in $\mathcal{F}$ are there but there may be also some "nonstandard trees".

When we make the model $\mathfrak{M} = \langle (\text{Trees})^{\langle \mathcal{F},M,\in\rangle}, \text{Eps}^{\langle \mathcal{F},M,\in\rangle} \rangle / \text{Eq}^{\langle \mathcal{F},M,\in\rangle}$

then: The standard part of the model $\mathfrak{M}$ will consist of equivalence classes of wellfounded trees. Nonstandard trees (if there are any) give nonstandard elements of $\mathfrak{M}$. But the tree representing M has rank On and so is standard; thus its realization exists and is in Sp $\mathfrak{M}$. By class existence in $\langle \mathcal{F}, M, \in \rangle$ the subsets of M being in $\mathcal{F}$, and only them, are the subsets of M in $\mathfrak{M}$. ∎

Corollary: $\langle M, \in \rangle$ is KMC -extendable iff there is $\langle N,E \rangle$ a model of $ZFC^- + T$ such that:

1) $M \in \text{Sp}\,\mathcal{N}$

2) $\langle N,E \rangle \models$ "M is an inaccessible family of sets".

As we noted, if $\langle M, \in \rangle$ is extendable then $\langle L^M, \in \rangle$ is extendable. In case when $\langle M, \in \rangle$ is β - extendable and we take constructibility interpretation Θ (cf §2) within the β - extension $\langle \mathcal{F}, M, \in \rangle$ we get a structure $\langle \mathcal{G}, L^M, \in \rangle$ which is a β - model. Therefore, if $\langle M, \in \rangle$ is β - extendable then $\langle L^M, \in \rangle$ is also β - extendable. This leads us to the following definition:

Definition: a) α is extendable ordinal iff $\langle L_\alpha, \in \rangle$ is an extendable model

b) α is β - extendable ordinal iff $\langle L_\alpha, \in \rangle$ is a β - extendable model.

Let us notice that - by our results in the §2 - KM and KMC -

extendability of L_α coincide (since it posesses a definable wellordering).

The same fact holds for β - extendability.

Our criterions of extendability and β - extendability were highly ineffective in the sense that it was not clear where to look for the extensions. For the models of the form L_α and β - extendability we have quite nice criterion; For other models and weaker form of extendability we show later some criterions.

Definition: If α is an ordinal then α^* is the least ordinal β such that: 1) $\alpha \in \beta$
2) $\langle L_\beta, \in \rangle \models ZF^-$.

Theorem 3.1. α is β - extendable iff $\langle L_{\alpha^*}, \in \rangle \models$ "L_α is an inaccessible family".

Proof.: $\Rightarrow$ By the corollary after the proposition 3.1. there is $\langle N, \in \rangle$ such that $\langle N, \in \rangle \models$ "L_α is an inaccessible family of sets", $\langle N, \in \rangle \models ZFC^-$.

Then $\langle L^N, \in \rangle \models ZFC^-$ and since "(.) is an inaccessible family" is a Π_1 formula therefore $\langle L^N, \in \rangle \models$ "L_α is an inaccessible family of sets". Since $L^N = L_{h(N)}$ we have $h(N) \geq \alpha^*$ and so using again the fact that "(.) is an inaccessible family of sets" is Π_1 we get $\langle L_{\alpha^*}, \in \rangle \models$ "L_α is an inaccessible family of sets".

$\Leftarrow$ Immediate by the same corollary and the fact that $\langle L_\alpha, \in\rangle \models ZF^-$ implies $\langle L_\alpha, \in\rangle \models ZFC^-$. ∎

Corollary: If α is β - extendable than $R.A.^{L_\alpha} = L_{\alpha^*} \cap \mathcal{P}(L_\alpha)$ *)

Proof: Both $\langle R.A^{L_\alpha}, L_\alpha, \in\rangle$ and $\langle L_{\alpha^*} \cap \mathcal{P}(L_\alpha), L_\alpha, \in\rangle$ are the smallest β - extensions of $\langle L_\alpha, \in\rangle$ ∎

For the rest of this paper we assume that there are β - extendable ordinals.

Definition: a) α^0 is the least extendable ordinal

b) α^1 is the least β - extendable ordinal.

Lemma 3.1. Both α^0 and α^1 are denumerable.

Proof: Obvious by Skolem-Löwenheim. ∎

Theorem 3.2. (On difference) a) $\alpha^0 < \alpha^1$

b) $\langle L_{\alpha^1}, \in\rangle \models$ " α^0 is denumerable"

c) $\langle L_{\alpha^1}, \in\rangle \models$ " α^0 is extendable".

We will prove our theorem using Barwise' Σ_1 compactness theorem. The basic fact is that Barwise theorem is provable in ZFC and thus valid in every model of ZFC. This fact was first noted by Barwise [1] and then by Wilmers [18]. We assume that the reader is familiar with the theorem of Barwise and some of its standard applications.

---

*) As pointed to us by M.Srebrny, R.B.Jensen in his Habilitationschrift (unpublished) proves this equality for all α's.

We need the following lemma:

Lemma 3.2.: α^{+}, the least admissible ordinal greater than α is smaller than α^{*}.

Proof: It is known that there is a single sentence Φ such that Φ +"scheme of foundation" is equivalent to KP (we were informed about this by G.Kreisel and C.Smorynski). By the reflection principle in $\langle L_{\alpha^*}, \in \rangle$ there is $\beta < \alpha^*$ such that Φ holds in $\langle L_{\beta}, \in \rangle$ (Since Φ holds in $\langle L_{\alpha^*}, \in \rangle$). Thus L_{β} is admissible and since $\alpha^{+} \leq \beta$ we are done.

Proof of the theorem 3.2. ("on difference")

Consider the system $\langle L_{(\alpha^1)^+}, \in \rangle$ and the theory $\mathcal{T}$ in the infinitary language $\mathcal{L}_{L_{(\alpha^1)^+}}$ based on 3 groups of axioms:

a) $\bigwedge$ ZFC$^-$

b) $\in$ - diagram of $L_{(\alpha^1)^+}$

c) "L_{α^1} is an inaccessible family of sets"

The theory $\mathcal{T}$ is definable over $\langle L_{(\alpha^1)^+}, \in, \{\alpha^1\} \rangle$ by a Σ_1 formula Θ and is consistent since it has a model (for instance $\langle L_{(\alpha^1)^*}, \in \rangle$ is a model of $\mathcal{T}$). Therefore the structure $\langle L_{(\alpha^1)^+}, \in, \{\alpha^1\} \rangle$ satisfies the formula Consis_{Θ}, where Consis_{Θ} is a finitary sentence of L_{ST} expressing consistency of $\mathcal{T}$.

Since $L_{(\alpha^1)^+} \in L_{(\alpha^1)^*}$ therefore since $\langle L_{(\alpha^1)^*}, \in \rangle$ satisfies the full scheme of choice and has a definable well-ordering we have <u>inside</u> $L_{(\alpha^1)^*}$ a denumerable (within $L_{(\alpha^1)^*}$) elementary substructure $\langle A, \in \restriction A, \{\alpha^1\} \rangle$ of $\langle L_{(\alpha^1)^+}, \in, \{\alpha^1\} \rangle$

The structure $\langle A, \in \restriction A, \{\alpha^1_1\} \rangle$ is isomorphic again within $L_{(\alpha^1)^*}$ to a structure $\langle B, \in, \{\delta\} \rangle$ where B is transitive. By standard reasoning $B = L_\gamma$ for some γ. γ is denumerable within $\langle L_{(\alpha^1)^*}, \in \rangle$ and so it is denumerable within $\langle L_{\alpha^1}, \in \rangle$ since $\langle L_{(\alpha^1)^*}, \in \rangle \models$ "L_{α^1} is an inaccessible family of sets". Consider now $\langle L_\gamma, \in, \{\delta\} \rangle$. First of all we notice that $\gamma = \delta^+$.

Moreover $\langle L_\gamma, \in, \{\delta\} \rangle \models \mathrm{Consis}_\Theta$

Now let us look what the formula Θ defines over $\langle L_\gamma, \in, \{\delta\} \rangle$.

It is clear that it defines the following theory:

a') $\bigwedge\!\!\bigwedge$ ZFC$^-$

b') $\in$ diagram of L_γ

c') "L_δ is an inaccessible family of sets"

As this theory is Σ_1 definable and $\langle L_\gamma, \in, \{\delta\} \rangle \models \mathrm{Consis}_\Theta$ we apply now Barwise compactness theorem within $\langle L_{\alpha^1}, \in \rangle$. Since γ is denumerable in $\langle L_{\alpha^1}, \in \rangle$, $\langle L_\gamma, \in, \{\delta\} \rangle$ is Σ_1 - complete in $\langle L_{\alpha^1}, \in \rangle$.

Thus we get within $\langle L_{\alpha'}, \in \rangle$ a denumerable model of the theory definable over $\langle L_{\gamma}, \in, \{\delta\} \rangle$ by Θ, i.e. of the axiom groups a'), b'), c').

Let $\langle N, E \rangle$ be a model of this theory. By the condiction b') $\langle N, E \rangle$ is an end extension of $\langle L_{\gamma}, \in \rangle$ (within $\langle L_{\alpha'}, \in \rangle$ but this is an absolute statement). Since $\delta < \gamma$ therefore δ belongs to the standard part of $\langle N, E \rangle$. We apply now the corollary of the proposition 3.3. So $\langle L_{\delta}, \in \rangle$ is an extendable model. Clearly $\alpha^{\circ} \leq \delta$ and so both a) and b) of the theorem hold.

To show c) we apply within $\langle L_{\alpha'}, \in \rangle$ Skolem Löwenheim result of Nadel [16], since α° is denumerable in $\langle L_{\alpha'}, \in \rangle$ and thus between α° and $\omega_1^{\langle L_{\alpha'}, \in \rangle}$ there are recursively inaccessible ordinals.

Definition: We call an admissible set A Σ_1- complete iff for every Σ_1 definable theory $\mathcal{T}$,

$$\langle A, \in \rangle \models \mathrm{Consis}_{\Theta} \text{ iff } \mathcal{T} \text{ has a model}$$

where Θ is a Σ_1 formula defining $\mathcal{T}$.

By the Barwise compactness theorem together with completeness theorem for languages $\mathcal{L}_M$ (M denumerable) we find that all denumerable admissible sets are Σ_1-complete.

Analyzing the proof of the theorem 3.2. we get

Theorem 3.3. There is a formula Φ such that whenever $\langle M^+, \in \rangle$ is Σ_1 complete then :

$$\langle M, \in \rangle \text{ is KMC - extendable iff } \langle M^+, \in, M \rangle \models \Phi$$

Proof: Φ is a Π_1 sentence stating the consistency of the following theory $\mathcal{T}$:

a) $\bigwedge$ ZFC$^-$

b) " $\in$ - diagram of the world (it is called EE in [5])

c) "M is inaccessible family of sets"

Let us notice that b) uniformly defines an $\in$ diagram of admissible set over itself.

We use the following fact :

If $\langle N, E \rangle \models$ KP then $M \in \mathrm{Sp}\, N$ iff $M^+ \subseteq \mathrm{Sp}\, N$

To prove the theorem assume firstly that $\langle M, \in \rangle$ is KMC extendable. By the corollary to the proposition 3.3. we find that there is a model $\langle N, E \rangle$ of a) and c) and such that $M \in \mathrm{Sp}\, N$. Thus $M^+ \subseteq \mathrm{Sp}\, N$ and so $\langle N, E \rangle$ satisfies an $\in$-diagram of M^+. Thus $\langle N, E \rangle$ is a model of $\mathcal{T}$ (more precisely of the theory defined over $\langle M^+, \in \rangle$ by Θ).

Conversly, if $\langle M^+, \in, M \rangle \models \mathrm{Consis}_{\Theta}$ then, by Σ_1 completeness of $\langle M^+, \in \rangle$ and by the fact that $\mathcal{T}$ is Σ_1

definable we get a model $\langle N, E \rangle$ of $\mathcal{T}$.

$M^+ \subseteq \mathrm{Sp}\, N$ and so $M \in \mathrm{Sp}\, N$. Using once more the corollary to the proposition 3.3. we are done. ∎

Corollary: If M is denumerable then

$$\langle M, \in \rangle \text{ is extendable iff } \langle M^+, \in, M \rangle \models \Phi$$

We come back to the proof of the theorem 3.2. It was definitely not economic for the following two reasons.

1) Remark that $\langle L_\alpha, \in \rangle$ need not be β - extendable in order to make our reasoning work. What we need is that there is an extension $\mathcal{F}$ of $\langle L_\alpha, \in \rangle$ such that $h(\mathcal{F}) > \alpha^+$.

2) We did not use the following fact: Every ω - model of ZFC which is extendable contains its own theory.

We define: $\alpha^{(0)} = \alpha \qquad \alpha^{(\xi+1)} = (\alpha^{(\xi)})^+$

$\alpha^{(\lambda)} = \bigcup_{\xi<\lambda} \alpha^{(\xi)}$ if this ordinal is admissible or $(\bigcup_{\xi<\lambda} \alpha^{(\xi)})^+$ otherwise.

Definition: An extendable model $\langle L_\alpha, \in \rangle$ is "ξ - good" iff it has an extension $\langle \mathcal{F}, L_\alpha, \in \rangle$ such that $\alpha^{(\mathcal{F})} < h(\mathcal{F})$

Using the reasoning of the proof of the theorem 3.2. we get

Theorem 3.4. a) Every 1 - good model contains as an element 0-good (i.e. transitive extendable) model.

b) If $k \in \omega$ then every $(k+1)$-good model contains as an element k - good model.

The theorem 3.4. may be extended to all recursive ordinals.

Following the line of 2) we find that in the proof of the theorem 3.2. we could add the following clause d) to the a),b),c): $\left(Th(L_{\alpha^1}, \in)\right)^{L_{\alpha^1}}$ as the latter is L_{α^1} - finite. Therefore we have the following:

Theorem 3.5. If $\langle M, \in \rangle$ is β - KMC - extendable then there is $N \in M$ such that $\langle N, \in \rangle \equiv \langle M, \in \rangle$ and

$\langle M, \in \rangle \models \bar{\bar{N}} = \aleph_0$ & $\langle N, \in \rangle$ is extendable".

(Thus $\langle N, \in \rangle$ is indeed extendable).

The proof of 3.5. needs a subtler considerations of β - extendable models.

Namely in the proposition 3.2. one may add 4) "Every set is equipollent to an ordinal". The model produced from trees satisfies Skolem-Löwenheim theorem and so we work as in 3.2.

Additionally we must prove that $L_{\alpha^+}[M] \in L_{\alpha^*}[M]$ which is again obvious. We close the paper with the informations on the number of extensions of $\langle M, E \rangle$.

In [9] the following is proved:

Proposition 3.4. If $\langle \mathcal{F}, M, \in \rangle$ is a denumerable model of KM then there is a proper extension $\mathcal{G}$ of $\mathcal{F}$ such that:

1) $\langle \mathcal{F}, M, \in \rangle \prec \langle \mathcal{G}, M, \in \rangle$

2) $\langle \mathcal{G}, M, \in \rangle$ is not a β - model

Moreover there is $2^{\aleph_0}$ $\mathcal{G}$'s of power $\aleph_0$ and $2^{\aleph_1}$ of power $\aleph_1$.

We do not know any necessary and sufficient condition under which a β - model $\langle \mathcal{F}, M, \in \rangle$ has a proper elementary extension $\langle \mathcal{G}, M, \in \rangle$ also being a β - model.

There are however some necessary and some sufficient conditions:

Some of them are due to Guzicki [4]

1) If we want to get a model of the same height as $\mathcal{F}$ then $\langle \mathcal{F}, M, \in \rangle$ must satisfy the negation of the class form of relative constructibility.

2) Sufficient: The ones given in [4]. They give stronger results than those of our proposition 3.5. (although they go in

different direction)

Guzicki's models are forcing models-quite exeptional fact since they are also elementary extensions.

Under assumption of Martin's axiom Guzicki's construction gives $2^{2^{\aleph_0}}$ β- models of power $2^{\aleph_0}$

Definition: A model $\langle \mathcal{F}, M, \in \rangle$ of KM satisfies condition (β) iff there is a model $\langle N,E \rangle$ of $ZFC^- + T$ such that:

1) $\mathcal{F} \in Sp\, N$

2) $\langle N,E \rangle \models \mathcal{F} = \mathcal{P}(M)$

3) $\langle N,E \rangle \models$ "M is an inaccessible family of sets"

4) $\langle N,E \rangle \models$ " $\overline{\overline{\mathcal{P}(M)}} > \bar{\bar{M}}^+$ "

5) $\langle N,E \rangle$ is M^+ - standard

(here M^+ denotes next cardinal in $\langle N,E \rangle$)

Proposition 3.5. ([10]) If $\langle \mathcal{F}, M, \in \rangle$ is a denumerable model of KMC satisfying condition (β) then there are $2^{\aleph_0}$ proper elementary denumerable extensions $\langle \mathcal{G}, M, \in \rangle$ satisfying conditions (β) and $2^{\aleph_1}$ of such extensions of power $\aleph_1$, all these extensions can be chosen to have the same height as $\mathcal{F}$.

Proof: Using the quantifier "there is more than M^+ " in $\langle N, E \rangle$ ∎

We have the following lemma:

Lemma 3.2. If $\langle \mathcal{F}, M, \in \rangle$ has a property (β) then it is a β - model.

By this lemma, a countable model satisfying (β) has $2^{\aleph_0}$ proper elementary denumerable extensions each of which is a β- model.

For the non-denumerable models almost nothing is known. If α is a strongly inaccessible cardinal then $\langle R_\alpha, \in \rangle$ has 2^α extensions of power α. There are even 2^α extensions being elementary subsystem of $\langle R_{\alpha+1}, R_\alpha, \in \rangle$

If $V = L$ then the elementary subsystems of $\langle R_{\alpha+1}, R_\alpha, \in \rangle$ are linearly ordered by inclusion. In the same time it is relatively consistent to assume that they are not linearly ordered by inclusion; even under the assumption that $\langle R_\alpha, \in \rangle \models V = L$.

Mathematical Institute of the Polish Academy of Sciences
Institute of Mathematics, University of Warsaw.

References

[1] J.Barwise: Infinitary methods in the model theory of set theory. In : Logic Colloquium 69, Editors R.O.Gandy and C.M.E. Yates, Amsterdam 1971, pp 53 - 66.

[2] R.Chuaqui: Forcing for the impredicative theory of classes, Journal of Symb.Logic 37(1972), pp 1 - 18.

[3] H.Friedman: Countable models of set theories, in Springer Lecture Notes 337.

[4] W.Guzicki: Ph.D. thesis, Warsaw, 1973

[5] J.L. Krivine, K. McAloon: Some true unprovable formulas for set theory
In: Proceedings of the Bertrand Russel memorial Logic conference, Leeds 1973, pp 332-41

[6] W.Marek: On the metamathematics of impredicative set theory. Diss.Math. XCVII.

[7] W.Marek, M.Srebrny: Gaps in constructible universe, Annals of Math. Logic 6(1974), pp 359-394

[8] W.Marek, M.Srebrny: There is no minimal transitive model of Z^-, to appear in Zeitschrift für Math. Logik.

[9] W.Marek, P.Zbierski: On higher order set theories Bull.Acad.Pol.Sci. XXI(1973), pp 97-101

[10] W.Marek, P.Zbierski: A lemma on quantifier with applications, to appears

[11] Y.Moschovakis: Predicative classes, in the Proceeding of Symposia AMS XIII 1. Editor D.Scott, Providence 1971, pp 247-264

[12] A.Mostowski: Some impredicative definitions in set theory, Fundamenta Math. XXXIV(1947), pp

[13] A.Mostowski: Models of ZF set theory satisfying..., Acta Phil. Fennica 18(1965), pp 135-144

[14] A.Mostowski: Constructible sets with applications, Amsterdam-Warszawa 1970.

[15] A.Mostowski: Remarks on models of Gödel-Bernays set theory to appear in: Sets and Classes; Bernays memorial volume.

[16] M.Nadel: Some Skolem-Löwenheim results for admissible sets. Israel Journal of Math.12(1972), pp 427-432.

[17] J.B.Paris: Minimal models of ZF. In: Proceedings of the Bertrand Russel memorial Logic conference, Leeds 1973, pp 327-331

[18] G.Wilmers: An $\aleph_1$ - standard model of ZF set theory which is an element.... In Proceedings of the Bertrand Russel memorial Logic conference, Leeds 1973, pp 315-326

MANY-VALUED ALGORITHMIC LOGIC

H. Rasiowa
Institute of Mathematics
University of Warsaw
PKiN, 00-901 Warsaw, Poland

Attempts to systematize theoretical research concerning programs have led to the application of various ideas, methods and approaches. For instance methods of graphs associated with programs, algebraic treatments, axiomatic methods, an abstract approach using lattice theory and λ-calculus [30], have all been applied.

One of the research methods is to develop the theory of programs on the basis of formalized logical systems. The attempt to find simple logical systems, which would serve as a basis for programming theory and be sufficiently rich to allow sophisticated investigations, caused the creation of algorithmic logic and its various extensions.

Algorithmic logic was formulated by A. Salwicki in his Ph. D. thesis ([26], [27], [28]) and developed in several papers by L. Banachowski ([1]-[5]), A. Kreczmar ([8]-[11]), G. Mirkowska ([14]-[16]), A. Salwicki [29], and others. Formalized systems of algorithmic logic contain in their languages expressions interpreted as programs and formulas describing properties of programs. For instance the stop property, correctness and partial correctness, various equivalence relations between programs, etc., are expressible by means of these formulas. This approach permits one to formulate most of the important laws on computational processes in the form of logical tautologies. Also, it turned out that methodological investigations dealing with problems which occur in programming can be carried out within the framework of algorithmic logic. Research which could improve programming is a further aim.

The formulation of many-valued extensions of algorithmic logic was carried out for the following two reasons. In programming practice there are situations in which one of n programs $\pi_1, \ldots, \pi_n$ should be performed according to which one of n conditions $\alpha_1, \ldots, \alpha_n$ is satisfied. If condition α_i is satisfied, then the program π_i $(i = 1, \ldots, n)$ should be realized. The instruction CASE, which occurs in certain programming languages, is obviously appropriate in this situation. The application of this instruction for arbitrary $n \geq 2$, considerably simplifies programming. In order to have a logical tool to investigate programs with case instructions ω^+-valued algorithmic logic was formulated [18] and examined ([19], [20]). The second aim in constructing many-valued extensions of algorithmic logic has been as follows. In formalized languages of algorithmic logic there are no expressions representing recursive procedures. However, investigations concerning recursive procedures may be carried out within algorithmic logic using a certain kind of implicit definition [29]. This approach is rather complicated. Moreover, it is not possible on the basis of algorithmic logic to investigate programs containing the instruction go to. The attempt to construct a logical tool to examine programs with labels as well as the recursive procedures--understood as certain expressions realized as modified Mazurkiewicz's pushdown algorithms (see [6])--led to the formulation of various versions of extended ω^+-valued algorithmic logic ([21], [22]).

ω^+-Valued algorithmic logic and its extensions are closely related to the theory of Post algebras. The notion of a Post algebra of any finite order $m \geq 2$ was introduced by P. C. Rosenbloom in 1942. The paper [7] by G. Epstein, in which the definition of Post algebra was formulated in a much simpler way, initiated research in this field. Over the last 14 years Post algebras have been investigated from a number of points of view and in increasing generality by various authors (e.g., G. Epstein, T. Traczyk, Ph. Dwinger, C. C. Chang and A. Horn, A. Malcev, V. Kirin, G. Rousseau, E. Włodarska, H. Sawicka, Cat-Ho Nguyen, Z. Saloni, B. Dahn, L. Maksimowa and D. Wakarelov, T. P. Speed, and the present author).

Generalized Post algebras of order ω^+, as formulated in [17], and those which satisfy a finite representability condition (see [20], [23]), play for extended ω^+-valued algorithmic logic and for ω^+-valued algorithmic logic, respectively, a role analogous to that of Boolean algebras for classical logic. On the other hand, ω^+-valued predicate calculi [17] and mixed-valued predicate calculi [23] constitute a starting point for the construction of formalized systems of extended ω^+-valued algorithmic logic and of ω^+-valued algorithmic logic, respectively. These predicate calculi as well as generalized Post algebras of order ω^+ have been formulated and examined from the point of view of their applications in a logical approach to programming theory.

This paper is a brief survey of results concerning generalized Post algebras of order ω^+, mixed-valued and ω^+-valued predicate calculi, algorithmic logic and its many-valued extensions.

1. GENERALIZED POST ALGEBRAS OF ORDER ω^+, MIXED-VALUED AND ω^+-VALUED PREDICATE CALCULI

A generalized Post algebra of order ω^+ (or briefly Post algebra of order ω^+) is an abstract algebra.

(1) $\mathfrak{P} = (P, \vee, \cup, \cap, \Rightarrow, \neg, (d_n)_{n \in N}, (e_i)_{0 \le i \le \omega})$, where N is the set of positive integers, and for all $n \in N$, $k \in N$, $0 \le i \le \omega$, $a, b \in P$, the following conditions are satisfied:

(p_0) $(P, \vee, \cup, \cap, \Rightarrow, \neg)$ is a pseudo-Boolean algebra with a unit element $\vee$ and a zero element $\wedge = \neg \vee$,

(p_1) $d_n(a \cup b) = d_n a \cup d_n b$, (p_2) $d_n(a \cap b) = d_n a \cap d_n b$,

(p_3) $d_n(a \Rightarrow b) = (d_1 a \Rightarrow d_1 b) \cap \ldots \cap (d_n a \Rightarrow d_n b)$, (p_4) $d_n \neg a = \neg d_1 a$,

(p_5) $d_n d_k a = d_k a$

(p_6) $d_n(e_i) = \begin{cases} \vee & \text{if } n \le i \\ \wedge & \text{if } n > i \end{cases}$

(p_7) $d_1 a \cup \neg d_1 a = \vee$ (p_8) $a = \bigcup_{n=1}^{\infty} (d_n a \cap e_n)$

(p_9) $d_{n+1} a \le d_n a$ (p_{10}) $e_\omega = \vee$.

The following definitions are adopted in $\mathfrak{P}$:

(2) $j_0 a = \neg d_1 a$, $j_n a = \neg d_{n+1} a \cap d_n a$, $n \in N$.

It follows that

(3) $j_n e_i = \begin{cases} \vee & \text{if } n = i \\ \wedge & \text{if } n \ne i \end{cases}$ $n \in N_0$, $0 \le i \le \omega$,

where N_0 is the set of non-negative integers.

In every Post algebra $\mathfrak{P}$ of order ω^+

(4) $(P, \vee, \cup, \cap)$ is a distributive lattice with a unit element $\vee$ and a zero element $\wedge = \neg \vee$,

(5) $\wedge = e_0 \leq e_1 \leq \ldots \leq e_\omega = \vee$

(6) if $a \leq b$, then $d_n a \leq d_n b$ for each $n \in N$,

(7) if $\mathfrak{P}$ is nondegenerate, then $i_1 \neq i_2$ implies $e_{i_1} \neq e_{i_2}$, $0 \leq i_1, i_2 \leq \omega$,

(8) $a = b$ iff $d_n a = d_n b$ for each $n \in N$,

(9) the set $B_{\mathfrak{P}} = \{d_n a : n \in N$ and $a \in P\}$ coincides with the set of all complemented elements in $(P, \vee, \cup, \cap)$ and $\mathfrak{B}_{\mathfrak{P}} = (B_{\mathfrak{P}}, \vee, \cup, \cap, \Rightarrow, \neg)$ is a Boolean algebra which is said to correspond to $\mathfrak{P}$.

It follows from (9) that every nondegenerate Post algebra $\mathfrak{P}$ of order ω^+ determines a nondegenerate Boolean algebra $\mathfrak{A}_{\mathfrak{P}}$ and a chain (5) of the type ω^+.

The simplest example of a Post algebra of order ω^+ is offered by the following algebra $\mathfrak{P}_\omega$ which plays a role analogous to that of the two-element Boolean algebra in the class of all Boolean algebras.

(10) $\mathfrak{P}_\omega = (P_\omega, \vee, \cup, \cap, \Rightarrow, \neg, (d_n)_{n \in N}, (e_i)_{0 \leq i \leq \omega})$,

where $P_\omega = \{e_i\}_{0 \leq i \leq \omega}$ the elements e_i, $0 \leq i \leq \omega$, form a chain $\wedge = \neg \vee = e_0 \leq e_1 \leq \ldots \leq e_\omega = \vee$ of the type ω^+, $(P_\omega, \vee, \cup, \cap, \Rightarrow, \neg)$ is a linear pseudo-Boolean algebra with the lattice ordering $\leq$, i.e., for all $0 \leq i \leq \omega$, $0 \leq k \leq \omega$

(11) $e_i \cup e_k = e_{\max(i,k)}$, $\qquad e_i \cap e_k = e_{\min(i,k)}$,

$$e_i \Rightarrow e_k = \begin{cases} \vee & \text{if } i \leq k \\ e_k & \text{if } i > k, \end{cases} \qquad \neg e_i = e_i \Rightarrow e_0 = \begin{cases} \vee & \text{if } i = 0 \\ \wedge & \text{if } i \neq 0, \end{cases}$$

and the operations d_n, $n \in N$, are defined by means of (p_6).

Notice that the subalgebra $\mathfrak{B}_0 = (\{e_0, e_\omega\}, \vee, \cup, \cap, \Rightarrow, \neg)$ of the corresponding reduct $(P_\omega, \vee, \cup, \cap, \Rightarrow, \neg)$ of $\mathfrak{P}_\omega$ is the two-element Boolean algebra. Moreover, for each $m \geq 2$, the subalgebra

$$\mathfrak{P}_m = (P_m, \vee, \cup, \cap, \Rightarrow, \neg, d_1, \ldots, d_{m-1}, e_0, \ldots, e_{m-2}, e_\omega),$$

where $P_m = \{e_0, \ldots, e_{m-2}, e_\omega\}$, of the corresponding reduct $(P_\omega, \vee, \cup, \cap, \Rightarrow, \neg, d_1, \ldots, d_{m-1}, e_0, \ldots, e_{m-2}, e_\omega)$ of $\mathfrak{P}_\omega$ is the m-element Post algebra of order m.

Other examples of Post algebras of order ω^+ may be obtained by the application of the following method. Let $\mathfrak{B} = (B,\vee,\cup,\cap,\Rightarrow,\neg)$ be a Boolean algebra and let P be the set of all decreasing sequences $b = (b_1,b_2,\ldots)$, $b_1 \geq b_2 \geq \ldots$, of elements in B. Define the operations $\vee,\cup,\cap,\Rightarrow,\neg,d_n,n \in N,e_i$ for $0 \leq i \leq \omega$, on P as follows:

(12) $\vee = (\vee,\vee,\ldots)$,

(13) $b \cup c = (b_1 \cup c_1, b_2 \cup c_2, \ldots)$,

(14) $b \cap c = (b_1 \cap c_1, b_2 \cap c_2, \ldots)$,

(15) $b \Rightarrow c = (b_1 \Rightarrow c_1, (b_1 \Rightarrow c_1) \cap (b_2 \Rightarrow c_2), \ldots)$,

(16) $\neg b = (\neg b_1, \neg b_1, \ldots)$,

(17) $d_n b = (b_n, b_n, \ldots)$,

(18) $e_i = (\underbrace{\vee,\ldots,\vee}_{i\text{-times}},\wedge,\wedge,\ldots)$.

Then $\mathfrak{P}(\mathfrak{B}) = (P,\vee,\cup,\cap,\Rightarrow,\neg,(d_n)_{n\in N},(e_i)_{0\leq i\leq\omega})$ and all its subalgebras are Post algebras of order ω^+. It can be shown that for each Post algebra $\mathfrak{P}$ of order ω^+ there is a Boolean algebra $\mathfrak{B}$, such that $\mathfrak{P}$ is isomorphic either to $\mathfrak{P}(\mathfrak{B})$ or to a subalgebra of $\mathfrak{P}(\mathfrak{B})$.

Another representation theorem for Post algebras of order ω^+, viz. as algebras of subsets of certain quasi-ordered sets, has been proved by L. Maksimowa and D. Vakarelov [12]. They also considered representations preserving some infinite joins and meets. A topological representation of the algebras under consideration has been given by Z. Saloni [25].

Among Post algebras of order ω^+ we single out those which satisfy the following finite representability condition:

(fr) for each element a there is $m \geq 2$ such that

$$a = (d_1 a \cap e_1) \cup \ldots \cup (d_{m-2} a \cap e_{m-2}) \cup d_{m-1} a.$$

They constitute a special case of those as examined by Speed [31]. In particular $\mathfrak{P}_\omega$ satisfies condition (fr). A representation (fr) for a given element a is not unique. Because if (fr) holds, then for each $n \geq m-1$, $d_n a = d_{m-1} a$ and hence $a = (d_1 a \cap e_1) \cup \ldots \cup (d_{n-2} a \cap e_{n-2}) \cup d_{n-1} a$. By the order of a, in symbols $\mathrm{ord}(a)$, we mean the least $m > 2$ such that

$a = (d_1 e \cap e_1) \cup \ldots \cup (d_{m-2} a \cap e_{m-2}) \cup d_{m-1} a$. For instance, $\mathrm{ord}(e_\omega) = 2$, $\mathrm{ord}(e_i) = i + 2$ for $0 \leq i < \omega$. The set of all elements of orders not greater than m forms a Post algebra of order m.

The class of all Post algebras of order ω^+ which satisfy condition (fr) is characterized by the axioms (p_0)-(p_7) and (fr). Every nondegenerate Post algebra $\mathfrak{P}$ of order ω^+, which satisfies condition (fr) is a coproduct of a nondegenerate Boolean algebra $\mathfrak{B}$ and a chain of type ω^+. $\mathfrak{P}$ is isomorphic to the subalgebra of $\mathfrak{P}(\mathfrak{B})$ formed by all decreasing sequences $b = (b_1, b_2, \ldots)$ of elements in $\mathfrak{B}$ which are constant from some point on. In other words, for each $b = (b_1, b_2, \ldots)$ there is $m \geq 2$ such that $b_{m-1} = b_{m+k}$ for $k \in N_0$.

Notice that for each $m \geq 2$, Post algebras of order m are abstract algebras

(19) $\mathfrak{P} = (P, \vee, \cup, \cap, \Rightarrow, \neg, d_1, \ldots, d_{m-1}, e_0, \ldots, e_{m-2}, e_\omega)$

satisfying for all $1 \leq n \leq m-1$, $1 \leq k \leq m-1$, $i \in \{0, \ldots, m-2, \omega\}$ and $a, b \in P$ the axioms (p_0)-(p_7) and moreover

(p_m) $a = (d_1 a \cap e_1) \cup \ldots \cup (d_{m-2} a \cap e_{m-2}) \cup d_{m-1} a$.

Thus for each $m \geq 2$, the class of all Post algebras of order m is equationally definable. Every such algebra is a coproduct of a Boolean algebra $\mathfrak{B}$ and an m-element chain $\wedge = e_0 \leq \ldots \leq e_{m-2} \leq e_\omega = \vee$. It is then isomorphic to the algebra of all decreasing $(m-1)$-element sequences $b = (b_1, \ldots, b_{m-1})$, $b_1 \geq \ldots \geq b_{m-1}$, of elements in $\mathfrak{B}$, the operations $\vee, \cup, \cap, \Rightarrow, \neg, d_1, \ldots, d_{m-1}$, $e_0, \ldots, e_{m-2}, e_\omega$ being defined in a way similar to that specified by equations (12)-(18).

In Post algebras of order ω^+ (of orders $m \geq 2$) d-filters, i.e., filters ∇ satisfying the condition

(20) $a \in \nabla$ iff $d_n a \in \nabla$ for each $n \in N$ (for $n = 1, \ldots, m-1$)

play the role analogous to that of filters in Boolean algebras. If ∇ is a prime d-filter in a Post algebra $\mathfrak{P}$ of order ω^+ (of order m), then $\mathfrak{P}/\nabla$ is isomorphic to $\mathfrak{P}_\omega$ (to $\mathfrak{P}_m$). Moreover, if $\mathfrak{P}$ is a Post algebra of order ω^+ (order m), a is an element different from $\vee$ and S is a countable set of infinite joins and infinite meets in $\mathfrak{P}$, then there exists a prime d-filter ∇ in $\mathfrak{P}$ such that

$a \notin \nabla$ and ∇ preserves all infinite joins and meets in S (see [24],[17]).

The following generalization of Epstein's lemma [7] also holds for Post algebras of order ω^+ (see [17]): for any Post algebra $\mathfrak{P}$ of order ω^+ (of order m) and any elements a, a_t, $t \in T$, in $\mathfrak{P}$

$$a = \bigcup_{t\in T} a_t \text{ iff } d_n a = \bigcup_{t\in T} d_n a_t \text{ for each } n \in N \ (n = 1, \ldots, m-1),$$

$$a = \bigcap_{t\in T} a_t \text{ iff } d_n a = \bigcap_{t\in T} d_n a_t \text{ for each } n \in N \ (n = 1, \ldots, m-1).$$

The theorems formulated above are useful in metamathematical investigations concerning ω^+-valued and mixed-valued predicate calculi.

ω^+-Valued predicate calculi contain in their formalized languages predicates realized as k-argument $(k \in N)$ mappings from the universe of a realization into P_ω, and logical connectives $\vee, \wedge, \rightarrow, \sim, D_n, n \in N$, as well as propositional constants E_i, $0 \le i \le \omega$, realized as operations $\cup, \cap, \Rightarrow, \neg, d_n$, $n \in N$, e_i, $0 \le i \le \omega$, in $\mathfrak{P}_\omega$, respectively. The realization of quantifiers is by infinite joins and meets in $\mathfrak{P}_\omega$. Thus the Post algebra $\mathfrak{P}_\omega$ is adopted as a semantic basis. The class of all Post algebras of order ω^+ is applied as an algebraic tool in metamathematical investigations. A Hilbert-style formalization of ω^+-valued predicate calculi (see [17]) needs an inference rule of ω-type: $\frac{D_n\alpha,\ n \in N}{\alpha}$. The notion of ultraproducts of ω^+-valued realizations has also been introduced and an analogue of Łoś' theorem has been proved and applied to a proof of the theorem on the existence of ω^+-valued models for consistent uncountable theories based on ω^+-valued predicate calculi. The theorem on a prenex form of formulas, the compactness theorem, the first ε-theorem and a modification of the second ε-theorem also hold.
It is worth mentioning that there is a way of interpreting ω^+-valued predicate calculi and theories in corresponding elementary theories of classical logic. The formalized languages in which such interpretations are given have to be much richer. With every predicate of ω^+-valued predicate calculus or of a theory one associates a sequence of two-valued predicates. The provability of a formula in ω^+-valued predicate calculus or theory is equivalent to the provability of a set of formulas in a corresponding classical theory.

A Kripke-style semantics for ω^+-valued predicate calculi and the completeness theorem with respect to this semantics has been given by L. Maksimowa and D. Vakarelov [13].

Mixed-valued predicate calculi have been formulated and examined in [23]. Their formalized languages contain m-valued predicates and m-valued propositional variables for arbitrary $m \geq 2$, but there occur neither infinitely many-valued predicates nor infinitely many-valued propositional variables. Any m-valued k-argument predicate is realized as a k-argument mapping from the universe of a realization into $P_m = \{e_0, \ldots, e_{m-2}, e_\omega\}$, and any valuation assigns to each m-valued propositional variable an element in P_m. Propositional connectives and propositional constants being the same as in ω^+-valued predicate calculi, are realized as corresponding algebraic operations in $\mathfrak{P}_\omega$. The quantifiers are realized as infinite joins and meets in $\mathfrak{P}_\omega$. A Hilbert-style formalization [23] does not need any infinitistic rule of inference. Post algebras of order ω^+ which satisfy condition (fr) are applied as an algebraic tool for metamathematical investigations.

Mixed-valued predicate calculi have properties analogous to those which hold for the classical ones. A great part of metamathematics can easily be proved using algebraic methods. For instance, a theorem on a prenex form of formulas, an analogue of the deduction theorem, a theorem on diagrams of formulas and a Gentzen-style formalization, the compactness theorem, a theorem on the ultraproducts of ω^+-valued models, the theorem on the existence of ω^+-valued models for uncountable consistent theories, both ε-theorems, the Herbrand theorem, the Craig theorem, and others.

The following remark is worth making. Suppose that a k-argument m-valued predicate ρ^m is realized as $\rho^m_R : U^k \to \{e_0, \ldots, e_{m-2}, e_\omega\}$. Then ρ^m_R determines characteristic functions of $m-1$ k-argument relations on U as follows:

$$d_n\rho^m_R(u_1, \ldots, u_k) = \begin{cases} \vee & \text{if } \rho^m_R(u_1, \ldots, u_k) \geq e_n \\ \wedge & \text{otherwise} \end{cases}, \quad n = 1, \ldots, m-1.$$

Obviously, $d_1\rho^m_R(u_1, \ldots, u_k) \geq \ldots \geq d_{m-1}\rho^m_R(u_1, \ldots, u_k)$. These characteristic functions may be treated as coordinates of ρ^m_R. On the other hand ρ^m_R also determines characteristic functions of $m-1$ k-argument relations on U as follows:

$$j_n\rho^m_R(u_1, \ldots, u_k) = \begin{cases} \vee & \text{if } \rho^m_R(u_1, \ldots, u_k) = e_n \\ \wedge & \text{otherwise} \end{cases}, \quad n = 0, \ldots, m-2.$$

The first assignment suggests an interpretation of mixed-valued predicate calculi and theories in corresponding elementary theories of classical logic, whose formalized languages are obtained by assigning to each m-valued predicate ρ^m an $(m - 1)$-element sequence $\rho^m_1, \ldots, \rho^m_{m-1}$ of two-valued predicates, and to each m-valued propositional variable p^m an $(m - 1)$-element sequence $p^m_1, \ldots, p^m_{m-1}$ of two-valued propositional variables. The second assignment is important with respect to applications in programming theory.

Restricting a formalized language L_{mix} of a mixed-valued predicate calculus to n-valued predicates and n-valued propositional variables for $2 \leq n \leq m$, where m is a fixed integer, and adopting among D_n, $n \in N$, and E_i, $0 \leq i \leq \omega$, only $D_1, \ldots, D_{m-1}$ and $E_0, \ldots, E_{m-2}, E_\omega$, we obtain a language L_m of mixed-valued predicate calculus in which there are m possible truth-values, i.e., $e_0, \ldots, e_{m-2}, e_\omega$. Formalization of these predicate calculi (see [23]) gives a weak form of separation theorem for mixed-valued predicate calculi [23]. By $\mathrm{ord}(\alpha)$, for any formula α of a mixed-valued predicate calculus we mean the least m, such that $\alpha \in L_m$. We also distinguish, with respect to the syntax, Boolean formulas in any mixed-valued predicate calculus. Their orders may be arbitrarily high. If $\mathrm{ord}(\alpha) = m$, then for any realization R and valuation v, $\alpha_R(v) \in \{e_0, \ldots, e_{m-2}, e_\omega\}$. If α is a Boolean formula then $\alpha_R(v) \in \{e_0, e_\omega\}$. The weak separation theorem mentioned above asserts that for any set A of formulas and any formula α, if $\mathrm{ord}(\alpha) \leq m$ and for each $\beta \in A$, $\mathrm{ord}(\beta) \leq m$, then α is derivable from A in the mixed-valued predicate calculus under consideration iff α is derivable from A in the mixed-valued predicate calculus restricted to m possible truth-values.

2. ALGORITHMIC LOGIC

Formalized systems of algorithmic logic are extensions of first-order predicate calculi without quantifiers. Their languages contain certain expressions called programs, generalized terms, and generalized formulas describing properties of programs. In generalized formulas may occur iteration quantifiers which are infinite disjunctions and infinite conjunctions of a special kind.

More exactly, let $L_0 = (A_0, T, F_0)$ be an enumerable first-order predicate language without quantifiers, where A_0 is its alphabet, T the set of terms and F_0 the set of formulas. Assume that countable sets V and V_0 of individual variables and of propositional variables, respectively, are contained in A_0, and that propositional constants E_0 and E_ω, corresponding to any false statement and to any true statement, respectively, belong to A_0. Extend A_0 to A by adjoining three program operations signs: $\circ$ (composition sign), $\underline{\vee}$ (branching sign) and $*$ (iteration sign), and moreover iteration quantifiers $\bigcup$, $\bigcap$ and auxiliary signs [,], /.

Let R be a realization of predicates and functors of L_0 in a set $U \neq \phi$ and let W_u be the set of all valuations of individual variables in U and of propositional variables in $\{\wedge, \vee\} = \{e_0, e_\omega\}$. The valuations are considered as memory states (state vectors).

Programs are realized as partial mappings from the set W_u into itself.

Atomic programs are substitutions, i.e., expressions

(1) $[x_1/\tau_1 \ldots x_n/\tau_n \; p_1/\alpha_1 \ldots p_k/\alpha_k]$, $n, k \in N_0$,

where $x_1, \ldots, x_n$ are different individual variables, $p_1, \ldots, p_k$ are different propositional variables, $\tau_1, \ldots, \tau_n$ are any terms and $\alpha_1, \ldots, \alpha_k$ are any formulas in F_0. The set of all substitutions will be denoted by S. If $s \in S$ and has form (1), then its realization for a state vector $v \in W_u$ is defined thus:

(2) $s_R(v) = v' \in W_u$, where $v'(x_i) = \tau_{iR}(v)$ for $i = 1, \ldots, n$,

$v'(p_i) = \alpha_{iR}(v)$, $i = 1, \ldots, k$,

and $v'(x) = v(x)$ for $x \neq x_1, \ldots, x_n$, $v'(p) = v(p)$ for $p \neq p_1, \ldots, p_k$, $x \in V$, $p \in V_0$.

The set FS of programs is the least set containing S and satisfying

(fs) if $K, M \in FS$ and $\alpha \in F_0$, then $\circ[KM]$, $\underline{\vee}[\alpha KM]$, $*[\alpha K] \in FS$.

In order to extend the realization R to FS we adopt the following equations:

(fsr1) $\circ\,[KM]_R(v) = \begin{cases} M_R(K_R(v)) & \text{if this is defined} \\ \text{undefined otherwise} \end{cases}$

(fsr2) $\underline{\vee}\,[\alpha KM]_R(v) = \begin{cases} K_R(v) & \text{if this is defined and } \alpha_R(v) = e_\omega \\ M_R(v) & \text{if this is defined and } \alpha_R(v) = e_0 \\ \text{undefined otherwise} \end{cases}$

(fsr3) $*\,[\alpha K]_R(v) = \begin{cases} K_R^i(v), \text{ where } i \text{ is the least non-negative integer such that} \\ \qquad \alpha_R(K_R^i(v)) = e_0 \text{ and } K_R^i(v) \text{ is defined} \\ \text{undefined if such } i \in N_0 \text{ does not exist} \end{cases}$

where $K_R^0(v) = v$, $K_R^{n+1}(v) = K_R(K_R^n(v))$, for $n \in N_0$.

Programs in FS may be translated into an ALGOL-like language as follows. Substitution (1) should be read:

$x_1 := \tau_1$ <u>and</u> . . . <u>and</u> $x_n := \tau_n$ <u>and</u> $p_1 := \alpha_1$ <u>and</u> . . . <u>and</u> $p_k := \alpha_k$.

Programs $\circ[KM]$, $\underline{\vee}\,[\alpha KM]$, $*\,[\alpha K]$ correspond respectively to

<u>begin</u> , <u>if</u> α <u>then</u> K <u>else</u> M, <u>while</u> α <u>do</u> K;
K;
M
<u>end</u>

From terms and programs expressions of a new kind are constructed to be called generalized terms. The set FST of generalized terms is the least set containing T and satisfying the conditions:

(fst1) if $\tau \in FST$ and $K \in FS$, then $K\tau \in FST$,

(fst2) if φ is an n-argument functor and $\tau_1, \ldots, \tau_n \in FST$, then $\varphi(\tau_1 \ldots \tau_n) \in FST$.

Generalized terms are realized as partial functions from W_u into U. More exactly, in order to extend realization R on FST we adopt the following equations

(fstr1) $K\tau_R(v) = \begin{cases} \tau_R(K_R(v)) & \text{if this is defined} \\ \text{undefined otherwise} \end{cases}$

(fstr2) $\varphi(\tau_1 \ldots \tau_n)_R(v) = \begin{cases} \varphi_R(\tau_{1R}(v), \ldots, \tau_{nR}(v)) & \text{if } \tau_{iR}(v) \text{ are defined, } i = 1, \ldots, n \\ \text{undefined otherwise} \end{cases}$

The set FSF of generalized formulas is the least set containing F_0, and all expressions $\rho(\tau_1 \ldots \tau_n)$, where ρ is a predicate and

$\tau_1, \ldots, \tau_n \in$ FST, closed with respect to the propositional connectives $\vee$, $\wedge$, $\rightarrow$, $\sim$, and satisfying the condition

(fsf) if $\alpha \in$ FSF and $K \in$ FS, then $K\alpha$, $\bigcup K\alpha$, $\bigcap K\alpha \in$ FSF.

In order to extend the realization R to FSF the following additional equations are adopted

$$\text{(fsfr1)}\quad \rho(\tau_1 \ldots \tau_n)_R(v) = \begin{cases} \rho_R(\tau_{1R}(v) \ldots \tau_{nR}(v)) & \text{if all } \tau_{iR}(v) \text{ are defined,} \\ & i = 1, \ldots, n \\ e_0 & \text{otherwise} \end{cases}$$

$$\text{(fsfr2)}\quad K\alpha_R(v) = \begin{cases} \alpha_R(K_R(v)) & \text{if this is defined} \\ e_0 & \text{otherwise} \end{cases}$$

$$\text{(fsfr3)}\quad \bigcup K\alpha_R(v) = \bigcup_{i=0}^{\infty} (K^i\alpha)_R(v),\quad \bigcap K\alpha_R(v) = \bigcap_{i=0}^{\infty} (K^i\alpha)_R(v),$$

where $K^0\alpha = \alpha$, $K^{i+1}\alpha = KK^i\alpha$ for $i \in N_0$, and $\bigcup$, $\bigcap$ on the right-hand sides of these equations denote infinite joins and meets in the two-element Boolean algebra.

Properties of programs are expressible by means of generalized formulas. For instance, for any $K \in$ FS, KE_ω describes the stop property of K. Indeed, $KE_{\omega R}(v) = e_\omega$ if and only if $K_R(v)$ is defined. Formulas $(\alpha \rightarrow K\beta)$ describe a correctness of $K \in$ FS with respect to an initial condition α for input state vectors and a terminal condition β for output state vectors. Similarly $((\alpha \wedge KE_\omega) \rightarrow K\beta)$ expresses the partial correctness of K with respect to α and β. Thus the examination of properties of programs can be reduced to examining satisfiability and validity of corresponding generalized formulas in certain or in all realizations.

Systems of algorithmic logic have been investigated by G. Mirkowska ([14], [15], [16]), who obtained several metamathematical results (e.g., a Hilbert-style formalization with infinitistic rules of inference and a Gentzen-style formalization, an analogue of Löwenheim-Skolem-Gödel theorem, an analogue of Herbrand's theorem for certain generalized formulas, a theorem on a normal form of a program and others).

Effectivity problems in algorithmic logic have been examined by A. Kreczmar ([8], [9], [10], [11]). He proved that the set of all valid generalized formulas of algorithmic logic is recursively isomorphic to the set of all formulas true in the standard model of arithmetic and that the set of all consequences of a set A of generalized formulas is hyperarithmetical with respect to A. Other results concerned the degrees of unsolvability of fundamental properties of programs in various classes of realizations. Using algebraic and metamathematical methods he obtained new simple proofs, eliminating Gödel enumerations and Turing machines, of known theorems and certain new results.

Problems of the definability and programmability of functions, relations and relational systems in algorithmic logic have been investigated by A. Salwicki [29] who presented a theory of programmability and its relationship with the theory of recursive functions.

Problems concerning correctness of programs and modular properties of programs within the framework of algorithmic logic extended by the usual quantifiers, as well as metamathematical problems dealing with this logic have been examined by L. Banachowski ([1], [3], [4], [5]). Moreover he has applied algorithmic logic to investigations of data structures [1].

An approach to recursive procedures by means of a special kind of implicit definitions in algorithmic logic has been presented by A. Salwicki [29].

3. ω^+-VALUED ALGORITHMIC LOGIC

In certain programming languages the instruction <u>CASE</u> occurs. It is a generalization of <u>if then else</u> and corresponds to m-ary branchings, for all $m \geq 2$. In order to have a logical tool to investigate programs with this instruction, which greatly simplifies programming, ω^+-valued algorithmic logic was invented ([18], [19], [20]). It is an extension of algorithmic logic.

Formalized languages of ω^+-valued algorithmic logic are constructed in a way similar to that in which those of algorithmic logic were constructed. But here one begins with a mixed-valued predicate language $L_{mix} = (A_{mix}, T, F_{mix})$ without quantifiers instead of a usual first-order predicate language without quantifiers as in Section 2.

Let R be any realization of predicates and functors of L_{mix} in a set $U \neq 0$ and let W_u be the set of all valuations of individual and propositional variables of L_{mix}.

Atomic programs are generalized substitutions, i.e., expressions

(1) $[x_1/\tau_1 \ldots x_n/\tau_n\ p_1^{m_1}/\alpha_1 \ldots p_k^{m_k}/\alpha_k]$, $n,k \in N_0$,

in which $x_1, \ldots, x_n$, $\tau_1, \ldots, \tau_n$ are as in the case of substitutions, different individual variables and arbitrary terms, respectively, $p_i^{m_i}$, for $i = 1, \ldots, k$, are different m_i-valued propositional variables, and α_i, for $i = 1, \ldots, k$, are any formulas such that $\mathrm{ord}(\alpha_i) \leq m_i$. The set of all generalized substitutions will be denoted by S_ω. The realization R is extended to S_ω by equations analogous to (2) in Section 2.

The set $F_\omega S$ of programs is the least set containing S_ω and satisfying the following conditions:

(f_ωs1) if $K,M \in F_\omega S$ then $\circ[KM] \in F_\omega S$,

(f_ωs2) if $\mathrm{ord}(\alpha) = m$ and $K_0, \ldots, K_{m-2}, K_\omega \in F_\omega S$, then $\underline{\vee}\ [\alpha K_\omega K_{m-2} \ldots K_0] \in F_\omega S$,

(f_ωs3) if α is a Boolean formula and $K \in F_\omega S$, then $*\ [\alpha K] \in F_\omega S$.

The realization R is extended on $F_\omega S$ by adopting the equations (fsr1), (fsr3) in Section 2 and also

$$(f_\omega sr)\quad \underline{\vee}\ [\alpha K_\omega K_{m-2} \ldots K_0]_R(v) = \begin{cases} K_{iR}(v) & \text{if this is defined and } \alpha_R(v) = e_i, \\ & i = 0, \ldots, m-2, \omega \\ \text{undefined otherwise} & \end{cases}$$

Consider the following example. Let K be the following program in $F_\omega S$

$$K = \underline{\vee}\ [\rho^3(x)[x/0][x/y][y/1]]$$

and let R be the standard realization in the set N_0, by the assumption that

$$\rho_R^3(n) = \begin{cases} e_0 & \text{if } n < 0 \\ e_1 & \text{if } n = 0 \\ e_\omega & \text{if } n > 0. \end{cases}$$

Then K_R may be translated into the program

<u>CASE</u> x > 0; x = 0; x < 0 <u>of</u> <u>begin</u> x : = 0; x : = y; y : = 1 <u>end</u>.

The set $F_\omega ST$ of generalized terms is defined analogously to FST in Section 2 and likewise for their realizations by a given realization R.

The set $F_\omega SF$ of generalized formulas is the least set containing formulas in L_{mix} and all expressions $\rho^m(\tau_1 \ldots \tau_n)$, where ρ^m is an m-valued n-argument predicate and $\tau_1, \ldots, \tau_n \in F_\omega ST$, is closed under all connectives in L_{mix} and satisfies the condition

($f_\omega sf$) if $K \in F_\omega S$ and $\alpha \in F_\omega SF$, then $K\alpha$, $\bigcup K\alpha$, $\bigcap K\alpha \in F_\omega SF$.

In order to extend a realization R to $F_\omega SF$ the equations (fsfr1), (fsfr2), (fsfr3) in Section 2 are adopted, where $\bigcup$ and $\bigcap$ on the right-hand sides of (fsfr3) denote infinite joins and infinite meets in $\mathfrak{P}_\omega$, respectively.

Each ω^+-valued algorithmic language L_ω uniquely determines for every $m \geq 2$ a mixed-valued algorithmic language L_m with m possible truth-values: $e_0, \ldots, e_{m-2}, e_\omega$. The language L_m was obtained by restricting the alphabet of L_ω to n-valued predicates and n-valued propositional variables, $2 \leq n \leq m$, and by adopting among D_n, $n \in N$, and E_i, $0 \leq i \leq \omega$, only $D_1, \ldots, D_{m-1}$ and $E_0, \ldots, E_{m-2}, E_\omega$. The sets $F_\omega S$, $F_\omega ST$ and $F_\omega SF$ are then restricted to $F_m S$, $F_m ST$ and $F_m SF$, respectively. Realizations of L_ω restricted to L_m are realizations of L_m. The same is true of valuations.

Hilbert-style formalizations with completeness theorems for systems of ω^+-valued algorithmic logic and for mixed-valued algorithmic logics with logical values restricted to m were given in [20]. A weak form of separation theorem for ω^+-valued algorithmic logic also holds just as it does for mixed-valued predicate calculi. Metamathematical results concerning algorithmic logic may be extended to ω^+-valued algorithmic logic. Moreover it can serve as a tool for research analogous to that carried out on the basis of algorithmic logic.

4. EXTENDED ω^+-VALUED ALGORITHMIC LOGIC

The attempt to construct a logical tool to investigate programs with labels and with recursive procedures caused the formulating of two versions of extended ω^+-valued algorithmic logic. The first has been proposed in [21], the second in [22]. Other modifications are also considered in order to investigate programs with coroutines. The version to be presented here is an extension of that in [22].

The main idea of the construction of formalized languages of extended ω^+-valued logic and their realizations is connected with a modification of the notion of a deterministic pushdown algorithm (see [6]).

A deterministic pushdown algorithm is a system

$$A\ell = [W, L^*, \iota, I], \text{ where}$$

(1) W is a set (of objects of $A\ell$),

(2) L is a finite set and L^* is the set of all words under L including the empty word e_0,

(3) $\iota \in L$ is an initial label of $A\ell$,

(4) I is a finite set of instructions,

(5) every $I \in I$ is an ordered pair (f_I, r_I) of partial functions, $f_I \subset L^* \times L^*$ (a control function) and $r_I \subset W \times W$ (an action),

(6) with every $I \in I$ there is associated a label e_I; one of the instructions has as its label ι, and all instructions have different labels,

(7) dom $f_I = \{e_I w\}_{w \in L^*}$, i.e., the set of all words beginning with the label e_I of I,

(8) for each f_I there is $u \in L^*$ such that

$$f_I(e_I w) = uw, \text{ for each } w \in L^*.$$

The ordered pairs $(w,v) \in L^* \times W$ are said to be states of $A\ell$.

A computation of an algorithm $A\ell$ is a finite sequence of states

$$(\iota, v_0), (u_1, v_1), \ldots, (u_n, e_0),$$

such that $(u_{k+1}, v_{k+1}) = (f_I(u_k), r_I(v_k))$ for some $I \in I$, $k = 0, \ldots, n-1$.

Let L_ω be an ω^+-valued algorithmic language based on an initial mixed-valued predicate language $L_{mix} = (A_{mix}, T, F_{mix})$ without quantifiers as presented in Section 3.

Suppose that in the alphabet of L_ω we replace the iteration sign $*$ by a procedure operation sign $\circ *$, the iteration quantifiers $\bigcup, \bigcap$ by the infinite disjunction and the infinite conjunction signs $\bigvee, \bigwedge$, respectively. Moreover, let us adjoin a set $V_L = \{a_n\}_{n \in N}$ of label variables. In such a way we obtain a new alphabet A. Let R be a realization of functors and predicates occurring in A_{mix} in a set $U \neq \phi$ and let W_u be the set of all valuations (state vectors). Now we introduce valuations of a new kind to be called label valuations or label vectors. They are mappings $v_L : V_L \to \{e_i\}_{0 \leq i < \omega}$ satisfying the conditions:

(9) for each v_L there is $n \in N$ such that $v_L(a_n) = e_0$,

(10) for each v_L and $n \in N$, if $v_L(a_n) = e_0$, then $v_L(a_{n+1}) = e_0$.

The set of all label valuations will be denoted by W_L. It follows from (9) and (10) that identifying each $v_L \in W_L$ with the sequence $(v_L(a_1), v_L(a_2), \ldots)$, every label vector is either $(e_0, e_0, \ldots)$ or $(e_{k_1}, \ldots, e_{k_n}, e_0, e_0, \ldots)$, where $k_1, \ldots, k_n \neq 0$. The first label vector may be interpreted as the empty word over the set $\{e_n\}_{n\in N}$ of labels, and the second one, as the word $e_{k_1} \ldots e_{k_n}$ over $\{e_n\}_{n\in N}$. The ordered pairs $(v_L, v) \in W_L \times W_u$ will be called states.

The set F_LS of programs is now defined in another way. Programs will be realized as partial functions from $W_L \times W_u$ into itself.

Three kinds of atomic programs are adopted: generalized substitutions, label substitutions and label supervisors.

For any generalized substitution $s \in S_\omega$ in the form (1) of Section 3,

(11) $s_R(v_L, v) = (v_L, v')$, where v' is defined as previously.

We shall also use the notation $s_R(v) = v'$. The following expressions are called label substitutions:

(S_L1) $[a_1/E_{k_1}\ a_2/a_1]$, $k_1 \in N$,

(S_L2) $[a_1/E_{k_1} \ldots a_n/E_{k_n}\ a_{n+1}/a_2]$, $k_1, \ldots, k_n \in N$, $n \in N$,

(S_L3) $[a_1/a_2]$,

(S_L4) $[a_1/E_0]$,

(S_L5) $[\quad]$.

The set of all label substitutions will be denoted by S_L^0. The set S_L consists of all label substitutions with the exception of $[a_1/E_0]$.

In order to extend a realization R to S_L^0 we adopt the following definition: for each $S^* \in S_L^0$

(12) $s_R^*(v_L, v) = (v_L', v)$, where

(S_Lr1) $v_L' = (e_{k_1}, v_L(a_1), v_L(a_2), \ldots)$ in the case of (S_L1),

(S_Lr2) $v_L' = (e_{k_1}, \ldots, e_{k_n}, v_L(a_2), v_L(a_3), \ldots)$ in the case of (S_L2),

(S_Lr3) $v_L' = (v_L(a_2), v_L(a_3), \ldots)$ in the case of (S_L3),

(S_Lr4) $v_L' = (e_0, e_0, \ldots)$ in the case of (S_L4),

(S_Lr5) $v_L' = v_L$ in the case of (S_L5).

We shall also use the notation: $s_R^*(v_L) = v_L'$.

The following expressions

(sp) $[J_k a_n]$, $k \in N_0$, $n \in N$,

where $J_k a_n \overset{df}{=} (\sim D_{k+1} a_n \wedge D_k a_n)$, $k \in N_0$, are called label supervisors. In order to extend the realization R to the set Sp of all label supervisors we adopt the following definition:

(spr) $$[J_k a_n]_R (v_L, v) = \begin{cases} (v_L, v) & \text{if } j_k(v_L(a_n)) = e_\omega, \text{ i.e., if } v_L(a_n) = e_k \\ \text{undefined otherwise.} \end{cases}$$

Thus in any realization R this program tests whether the n-th coordinate of a label vector v_L is equal to e_k, or not.

The set $F_L S$ has also the following properties:

(f_Ls1) if $H_1, H_2 \in F_L S$, then $\circ[H_1 H_2] \in F_L S$,

(f_Ls2) if $\mathrm{ord}(\alpha) = m$, $\alpha \in F_{mix}$ and $H_0, \ldots, H_{m-2}, H_\omega \in F_L S$, then $\underline{\vee}\, [\alpha H_\omega H_{m-2} \ldots H_0] \in F_L S$.

For further extension of the realization R we adopt

($f_L$sr1) $$\circ[H_1 H_2]_R (v_L, v) = \begin{cases} H_{2R}(H_{1R}(v_L, v)) & \text{if this is defined} \\ \text{undefined otherwise} \end{cases}$$

($f_L$sr2) $$\underline{\vee}\, [\alpha H_\omega H_{m-2} \ldots H_0]_R (v_L, v) = \begin{cases} H_{iR}(v_L, v) & \text{if this is defined and} \\ & \alpha_R(v) = e_i,\ i = 0, \ldots, m-2, \omega \\ \text{undefined otherwise.} \end{cases}$$

The following expressions are called instructions of order 1.

(i1) $\circ[\circ[[J_k a_1] s^*] s]$, where $k \in N$, $s^* \in S_L$, $s \in S_\omega$,

(i2) $\circ[[J_k a_1]\ \underline{\vee}\, [\alpha \circ [s_\omega^* s_\omega] \circ [s_{m-2}^* s_{m-2}] \ldots \circ [s_0^* s_0]]]$, where $k \in N$, $\alpha \in F_{mix}$, $\mathrm{ord}(\alpha) = m$ and $s_i^* \in S_L$, $s_i \in S_\omega$ for $i = 0, \ldots, m-2, \omega$.

The set of all instructions of order 1 will be denoted by I_1. It follows from (spr), ($f_L$sr1), ($f_L$sr2) that if $H \in I_1$ and has form (i1) then

(ir1) $$H_R(v_L, v) = \begin{cases} (s_R^*(v_L), s_R(v)) & \text{if } v_L(a_1) = e_k \\ \text{undefined otherwise.} \end{cases}$$

If $H \in I_1$ and has form (i2) then

$$\text{(ir2)}\quad H_R(v_L,v) = \begin{cases} (s^*_{iR}(v_L),\ s_{iR}(v)) & \text{if } v_L(a_1) = e_k \text{ and } \alpha_R(v) = e_i, \\ \quad i = 0, \ldots, m-2, & \\ \text{undefined otherwise.} & \end{cases}$$

Since in both cases the definability of $H_R(v_L,v)$ depends on the satisfaction of $v_L(a_1) = e_k$, it is natural to call e_k the label of instructions (i1) and (i2). Observe that the realizations of instructions (i1), (i2) correspond to instructions in pushdown algorithms.

Now we are going to define new expressions to be called procedures of order 1. The set of all procedures of order 1 will be denoted by $\mathcal{P}_1$. The set $\mathcal{P}_1$ consists of the following expressions:

(p1) $\circ^* [H_{k_1 t} H_{k_1} \ldots H_{k_n} H_t]$, $n \in N$, where

1° $H_{k_1}, \ldots, H_{k_n}$ are instructions of order 1 with different labels $e_{k_1}, \ldots, e_{k_n}$, and e_{k_1} is adopted as the label of this procedure,

2° E_t does not occur in $H_{k_1}, \ldots, H_{k_n}$,

3° $H_{k_1 t} = \circ[[J_{k_1} a_1][a_1/E_{k_1}\ a_2/E_t\ a_3/a_2]]$ and is called a preparatory instruction,

4° $H_t = \circ[[J_t a_1][a_1/a_2]]$ and is called a terminal instruction.

In order to extend the realization R to the set $\mathcal{P}_1$ we introduce the notion of a computation of $H \in \mathcal{P}_1$ by R for a state (v_L,v). This is a finite sequence of states

(c) $(v_L^0,v^0), (v_L^1,v^1), \ldots, (v_L^m,v^m), (v_L^{m+1},v^{m+1})$

such that the following conditions are satisfied:

(c1) $(v_L^0,v^0) = H_{k_1 tR}(v_L,v)$,

(c2) for each $i = 0, \ldots, m-1$, $(v_L^{i+1},v^{i+1}) = H_{k_j R}(v_L^i,v^i)$ for some $j = 1, \ldots, n$,

(c3) $(v_L^{m+1},v^{m+1}) = H_{tR}(v_L^m,v^m)$,

(c4) all states in (c) are defined.

Observe that (v_L^0,v^0) for a procedure H in form (p1) is defined iff $v_L(a_1) = e_{k_1}$, and in that case $(v_L^0,v^0) = ((e_{k_1},e_t,v_L(a_2), \ldots),v)$. Thus the preparatory instruction, if it can be performed, separates the first label in v_L from $v_L(a_2), v_L(a_3), \ldots$, by the terminal label e_t. During a computation this label e_t plays a part analogous to that of the empty word in computations of pushdown algorithms. The terminal instruction, if it can be performed, cancels the terminal label e_t. It can be shown that if (c) is a computation of a procedure (pl), then $v_L^{m+1} = (v_L(a_2), v_L(a_3), \ldots)$.

Now for any procedure $H \in P_1$, if H is in form (pl), then we define

$$\text{(prl)} \quad H_R(v_L, v) = \begin{cases} (v_L^{m+1}, v^{m+1}) & \text{if (c) is a computation of } H \text{ by } R \text{ for } (v_L, v) \\ \text{undefined} & \text{if a computation of } H \text{ by } R \text{ for } (v_L, v) \text{ does not exist.} \end{cases}$$

The following procedure $H \in P_1$ is an implementation of a recursive program $P : F(x) \Leftarrow \underline{\text{if}}\ x = 0\ \underline{\text{then}}\ 1\ \underline{\text{else}}\ x \cdot F(x-1)$, over N_0.

$H = \circ^* [H_{13}H_1H_2H_3]$, where

$H_{13} = \circ[[J_1a_1][a_1/E_1\ a_2/a_3\ a_3/a_2]]$,

$H_1 = \circ[[J_1a_1]\ \underline{\vee}\ [x = 0\ \circ[[a_1/a_2][y/1]]\circ[[a_1/E_1\ a_2/E_2\ a_3/a_2][x/x-1]]]]$,

$H_2 = \circ[\circ[[J_2a_1][a_1/a_2]][x/x+1\ y/(x+1)\cdot y]]$,

$H_3 = \circ[[J_3(a_1)][a_1/a_2]]$.

We adopt as R the standard realization of functors and of the equality predicate in the set N_0. The following sequence of states is an example of a computation of H by R for $(v_L, v) \in W_L \times W_{N_0}$, where $v_L = (e_1, e_5, e_0, \ldots)$ and $v(x) = 3$, $v(y) \in N_0$, $v(z) \in N_0$ for $z \in V$. In writing state vectors in this computation we shall only give the values of the variables which occur in H.

$v_L^0 = (e_1, e_3, e_5, e_0, \ldots)$	$v^0(x) = 3$	$v^0(y) = v(y)$
$v_L^1 = (e_1, e_2, e_3, e_5, e_0, \ldots)$	$v^1(x) = 2$	$v^1(y) = v(y)$
$v_L^2 = (e_1, e_2, e_2, e_3, e_5, e_0, \ldots)$	$v^2(x) = 1$	$v^2(y) = v(y)$
$v_L^3 = (e_1, e_2, e_2, e_2, e_3, e_5, e_0, \ldots)$	$v^3(x) = 0$	$v^3(y) = v(y)$
$v_L^4 = (e_2, e_2, e_2, e_3, e_5, e_0, \ldots)$	$v^4(x) = 0$	$v^4(y) = 1$
$v_L^5 = (e_2, e_2, e_3, e_5, e_0, \ldots)$	$v^5(x) = 1$	$v^5(y) = 1$
$v_L^6 = (e_2, e_3, e_5, e_0, \ldots)$	$v^6(x) = 2$	$v^6(y) = 2$
$v_L^7 = (e_3, e_5, e_0, \ldots)$	$v^7(x) = 3$	$v^7(y) = 6$
$v_L^8 = (e_5, e_0, \ldots)$	$v^8(x) = 3$	$v^8(y) = 6$

It can be proved that $H_R(v_L, v)$ is defined for each state $(v_L, v) \in W_L \times W_{N_0}$ such that $v_L(a_1) = e_1$. Moreover, if $H_R(v_L, v) = (\overline{v}_L, \overline{v})$, then $\overline{v}(y) = v(x)!$

Suppose that we have defined instructions and procedures of all orders $\leq m$. Then we define instructions of order $m+1$ in a similar way as those of order 1, admitting in (i1) and (i2) procedures of orders $\leq m$, at least one of them being of order m, instead of generalized substitutions. In that case the label substitutions which occur in instructions must have special forms. Procedures of order $m+1$ are defined as in the case of order 1, but instructions occurring in these procedures may be of orders $\leq m+1$, at least one of them being of order $m+1$.

A realization R is extended to the set I_{m+1} of instructions of order $m+1$ and to the set $\mathcal{P}_{m+1}$ of procedures of order $m+1$ in a manner similar to that in which R is extended to I_1 and $\mathcal{P}_1$.

Let $\mathcal{P} = \bigcup_{m=1}^{\infty} \mathcal{P}_m$. The set F_LS is the least set containing $S_\omega \cup S_L^0 \cup Sp \cup \mathcal{P}$ and satisfying the conditions (f_Ls1) and (f_Ls2).

The following theorem concerning procedures is worth mentioning. If $H \in \mathcal{P}$, then for every realization R in a set $U \neq \phi$ and any two states (v_L, v), (w_L, w) such that $v = w$ and $v_L(a_1) = w_L(a_1)$, either both $H_R(v_L, v)$ and $H_R(w_L, w)$ are defined or both are undefined. Moreover, if $H_R(v_L, v) = (\overline{v}_L, \overline{v})$ and $H_R(w_L, w) = (\overline{w}_L, \overline{w})$, then $\overline{v} = \overline{w}$ and $\overline{v}_L = (v_L(a_2), v_L(a_3), \ldots)$, $\overline{w}_L = (w_L(a_2), w_L(a_3), \ldots)$. Thus the resulting state vectors are equal.

The following theorem on a normal form for programs in F_LS is also worth mentioning. For each $H \in F_LS$ there is nor $H \in F_LS$, effectively defined, such that for each realization R in $U \neq 0$ and for any state (v_L, v), $H_R(v_L, v)$ is defined iff nor $H_R(v_L, v)$ is defined. Moreover, if $H_R(v_L, v) = (\overline{v}_L, \overline{v})$, then nor $H_R(v_L, v) = (v_L, \overline{v})$. Thus after performing nor H_R the label vector is not changed.

The set T of terms is extended to the set F_LST of generalized terms. F_LST is the least set containing T and satisfying the conditions analogous to (fst1) and (fst2) in Section 2, viz.

(f_Lst1) if $\tau \in F_LST$ and $H \in F_LS$, then $H\tau \in F_LST$,

(f_Lst2) if φ is n-argument functor and $H_1, \ldots, H_n \in F_LST$, then $\varphi(\tau_1 \ldots \tau_n) \in F_LST$.

The realization R is extended to F_LST by adopting the following equations:

$$(f_L sr)\quad \begin{cases} \tau_R(v_L,v) = \tau_R(v) \quad \text{for each } \tau \in T, \\ H\tau_R(v_L,v) = \begin{cases} \tau_R(H_R(v_L,v)) & \text{if this is defined} \\ \text{undefined otherwise} \end{cases} \\ \varphi(\tau_1 \ldots \tau_n)_R(v_L,v) = \begin{cases} \varphi_R(\tau_{1R}(v_L,v), \ldots, \tau_{nR}(v_L,v)) & \text{if all } \tau_{iR}(v_L,v) \text{ are defined for } i = 1, \ldots, n \\ \text{undefined in the opposite case} \end{cases} \end{cases}$$

The next step is to extend the set F_{mix} of formulas in the initial language to the set F_LSF of generalized formulas. The set F_LSF is the least set containing all propositional variables, propositional constants E_i, $0 \le i \le \omega$, label variables a_i, $i \in N$, formulas $\rho^m(\tau_1 \ldots \tau_n)$--where ρ^m is any m-valued n-argument predicate and $\tau_1, \ldots, \tau_n \in F_LST$--, closed under the logical connectives $\vee, \wedge, \rightarrow, \sim, D_n$, $n \in N$, and satisfying the following conditions:

$(f_L sf1)$ if $\alpha \in F_LSF$ and $H \in F_LS$, then $H\alpha \in F_LSF$,

$(f_L sf2)$ if $(\alpha_n)_{n \in N}$ is a sequence of generalized formulas and the set of all individual variables and of all propositional variables which occur in these generalized formulas is finite, then $\bigvee(\alpha_1\alpha_2 \ldots)$ and $\bigwedge(\alpha_1\alpha_2 \ldots)$ are in F_LSF.

In order to extend the realization R to F_LSF the following additional equations are adopted:

$$(f_L sfr)\quad \begin{cases} \alpha_R(v_L,v) = \alpha_R(v) \quad \text{for each formula } \alpha \in F_{mix} \\ a_{iR}(v_L,v) = v_L(a_i) \quad \text{for } i \in N \\ \rho^m(\tau_1 \ldots \tau_n)_R(v_L,v) = \begin{cases} \rho^m_R(\tau_{1R}(v_L,v), \ldots, \tau_{nR}(v_L,v)) & \text{if all } \tau_{iR}(v_L,v) \text{ for } i = 1, \ldots, n \text{ are defined} \\ e_0 & \text{otherwise} \end{cases} \\ H\alpha_R(v_L,v) = \begin{cases} \alpha_R(H_R(v_L,v)) & \text{if this is defined} \\ e_0 & \text{in the opposite case} \end{cases} \\ \bigvee(\alpha_1\alpha_2 \ldots)_R(v_L,v) = \bigcup_{i=1}^{\infty} \alpha_{iR}(v_L,v) \\ \bigwedge(\alpha_1\alpha_2 \ldots)_R(v_L,v) = \bigcap_{i=1}^{\infty} \alpha_{iR}(v_L,v) \end{cases}$$

By means of realizations we introduce in the usual way a semantic consequence operation. A formalization in the style of Hilbert, using infinitistic inference rules, for systems of extended ω^+-valued algorithmic logic can be given, and the completeness theorem holds. The set of formulas derivable from logical axioms by means of the inference rules coincides with the set of all valid formulas.

Observe that properties of programs in $F_L S$ may be described by means of generalized formulas. For instance for each $H \in F_L S$, HE_ω express the stop property of H. For every procedure $H \in P$, whose label is e_k, formulas $(J_k a_1 \to (\alpha \to H\beta))$, and $((J_k a_1 \to ((\alpha \wedge HE_\omega) \to H\beta))$ describe correctness and partial correctness of H, respectively, with respect to an input formula α and an output formula β.

Extended ω^+-valued algorithmic logic has been formulated in a simplified form in [22]. The approach differs from that presented above in that all m-valued predicates and m-valued propositional variables for $m > 2$ are eliminated. On the other hand it is also possible to construct formalized languages of extended ω^+-valued algorithmic logic that include those of ω^+-valued algorithmic logic and in particular of algorithmic logic.

Various systems of extended ω^+-valued algorithmic logic may be applied to research analogous to that carried out on the basis of algorithmic logic and concerning programs with instructions <u>go to</u>, <u>CASE</u> and with recursive procedures.

REFERENCES

[1] Banachowski, L. Modular approach to the logical theory of programs, Proc. Intern. Symp. Math. Found. Comp. Sci., Warsaw-Jadwisin, 1974, Springer.

[2] Banachowski, L. An axiomatic approach to the theory of data structures, Bull. Acad. Pol. Sci., Ser. Math. Astron. Phys. to appear.

[3] Banachowski, L. Extended algorithmic logic and properties of programs, ibid. to appear.

[4] Banachowski, L. Modular properties of programs, ibid. to appear.

[5] Banachowski, L. Investigations of properties of programs by means of the extended algorithmic logic, Ph. D. Thesis, Faculty of Mathematics and Mechanics, University of Warsaw, 1975.

[6] Blikle, A.; Mazurkiewicz, A. An algebraic approach to the theory of programs, algorithms, languages and recursiveness, Proc. Intern. Symp. and Summer School Math. Found. Comp. Sci., Warsaw-Jabłonna, 1972, CCPAS Reports, 1972.

[7] Epstein, G. The lattice theory of Post algebras, Trans. Amer. Math. Soc., 95 (1960), 300-317.

[8] Kreczmar, A. The set of all tautologies of algorithmic logic is hyperarithmetical, Bull. Acad. Pol. Sci., Ser. Math. Astron. Phys., 21 (1971), 781-783.

[9] Kreczmar, A. Degree of recursive unsolvability of algorithmic logic, ibid. 20 (1972), 615-617.

[10] Kreczmar, A. Effectivity problems of algorithmic logic, Automata, Languages and Programming, Lec. Not. Comp. Sci., 14, Springer, 1974, 584-600.

[11] Kreczmar, A. Effectivity problems of algorithmic logic (in Polish), Ph. D. Thesis, Faculty of Mathematics and Mechanics, University of Warsaw, 1973.

[12] Maksimowa, L.; Vakarelov, D. Representation theorems for generalized Post algebras of order ω^+, Bull. Acad. Pol. Sci., Ser. Math. Astron. Phys., 22 (1974), 757-764.

[13] Maksimowa, L.; Vakarelov, D. Semantics for ω^+-valued predicate calculi, ibid. 765-771.

[14] Mirkowska, G. On formalized systems of algorithmic logic, ibid., 18 (1971),

421-428.

[15] Mirkowska, G. Herbrand theorem in algorithmic logic, ibid., 22 (1974), 539-543.

[16] Mirkowska, G. Algorithmic logic and its applications in program theory (in Polish), Ph. D. Thesis, Faculty of Mathematics and Mechanics, University of Warsaw, 1972.

[17] Rasiowa, H. On generalized Post algebras of order ω^+ and ω^+-valued predicate calculi, Bull. Acad. Pol. Sci., Ser. Math. Astron. Phys., 21 (1973), 209-219.

[18] Rasiowa, H. On logical structures of mixed-valued programs and ω^+-valued algorithmic logic, ibid., 451-458.

[19] Rasiowa, H. Formalized ω^+-valued algorithmic systems, ibid., 559-565.

[20] Rasiowa, H. A simplified formalization of ω^+-valued algorithmic logic, ibid. 22 (1974), 595-603.

[21] Rasiowa, H. Extended ω^+-valued algorithmic logic, ibid., 605-610.

[22] Rasiowa, H. ω^+-Valued algorithmic logic as a tool to investigate procedures, Proc. Intern. Symp. Math. Found. Comp. Sci., Warsaw-Jadwisin, 1974, Springer.

[23] Rasiowa, H. Mixed-valued predicate calculi, Studia Logica, to appear.

[24] Rasiowa, H. Post algebras as a semantic foundation of many-valued logic, MAA Studies in Mathematics, 1975.

[25] Saloni, Z. A topological representation of generalized Post algebras of order ω^+, Bull. Acad. Pol. Sci., Ser. Math. Astron. Phys., to appear.

[26] Salwicki, A. Formalized algorithmic languages, ibid., 18 (1970), 227-232.

[27] Salwicki, A. On the equivalence of FS-expressions and programs, ibid., 275-278.

[28] Salwicki, A. On the predicate calculi with the iteration quantifiers, ibid., 279-285.

[29] Salwicki, A. Programmability and recursiveness (an application of algorithmic logic to procedures), Dissertationes Mathematicae, to appear.

[30] Scott, D. Outline of a mathematical theory of computation, Oxford mon. PRG-2, Oxford University, 1970.

[31] Speed, T. P. A note on Post algebras, Coll. Math., 24 (1971), 37-44.

THE LEAST Σ_2^1 AND Π_2^1 REFLECTING ORDINALS

Wayne Richter[1]

University of Minnesota

1. Introduction

In our lectures at the 1974 Kiel Summer Institute we gave an exposition of the general theory of inductive definitions. With the recent publications of Moschovakis [6] and [7] and the earlier papers of Aanderaa [1], Richter [8], and Richter-Aczel [9] most of this material is now available. For this reason the present paper is concerned with an application of the general theory.

Recall from [9] that σ_m^n is the least Σ_m^n -reflecting ordinal and π_m^n is the least Π_m^n -reflecting ordinal. [Definitions appear below]. The main results of [9] establish a connection between these reflecting ordinals and the closure ordinals of certain sets of operators. An operator $\Phi : P(\omega) \to P(\omega)$ determines a transfinite sequence $\langle \Phi^\xi : \xi \in ON \rangle$ of subsets of ω, where $\Phi^\lambda = \bigcup\{\Phi(\Phi^\xi) : \xi < \lambda\}$. The closure ordinal $|\Phi|$ of Φ is the least ordinal λ such that $\Phi^{\lambda+1} = \Phi^\lambda$. Let $\Phi^\infty = \Phi^{|\Phi|}$. For $\mathcal{F}$ a set of second order relations on ω, $\Phi \in \mathcal{F}$ means the second order relation $\lambda n, X[n \in \Phi(X)]$ belongs to $\mathcal{F}$. Let

$$|\mathcal{F}| = \sup\{|\Phi| : \Phi : P(\omega) \to P(\omega) \ \& \ \Phi \in \mathcal{F}\} .$$

$\mathcal{F}$-IND is the set of (first order) relations X on ω such that for some $\Phi \in \mathcal{F}$ and some $m_1, \ldots, m_j$,

$$X = \{(n_1, \ldots, n_k) : \langle m_1, \ldots, m_j , n_1, \ldots, n_k\rangle \in \Phi^\infty\} ,$$

where $\langle \ \rangle$ is the usual coding function. $\mathcal{F}$-HYP is the set of relations X such that both X and its complement are in $\mathcal{F}$-IND.

Let δ_m^n be the supremum of the order types of well-orderings on ω which are Δ_m^n definable on the structure $\langle \omega, \in \restriction \omega\rangle$. In the statement of Theorem 1.1 (which is part of Theorem E of [9]) Σ_m^n is the set of Σ_m^n second order relations on $\langle \omega, \in \restriction \omega\rangle$; similarly for Π_m^n.

<u>Theorem 1.1.</u>(i) $\delta_m^n \le \sigma_m^n \le |\Sigma_m^n|$;

[1]Research supported in part by the U.S. National Science Foundation under Grant GP-20846.

(ii) $\delta^n_m \leq \pi^n_m \leq |\Pi^n_m|$.

In the special case $m = n = 1$, Theorem B of [9] implies

(1) $\delta^1_1 < \sigma^1_1 = |\Sigma^1_1|$ and $\delta^1_1 < \pi^1_1 = |\Pi^1_1|$.

The following main result of the present paper answers questions left open in [9].

<u>Theorem A.</u> (i) $\sigma^1_2 = \delta^1_2$;

(ii) $\pi^1_2 = |\Pi^1_2|$;

(iii) A relation on ω belongs to $L_{\pi^1_2}$ iff it belongs to Π^1_2 -HYP .

Thus in terms of the inequalities in Theorem 1.1, σ^1_2 is as small as possible and π^1_2 is as large as possible. [(i) corrects an error in remarks on p. 306 of [9]]. For $n \geq 1$, $m \geq 1$, Theorem 1.1, (1), and Theorem A are probably the best results in this direction without further assumptions such as $V = L$. Other information about $|\Pi^1_2|$ may be found in Cenzer [3].

2. Σ^1_2 -reflection

α, β, γ are used to denote ordinals. L_α is the set of constructible sets of order less than α . All structures we consider are of one of the forms $\langle L_\alpha, \in \restriction \alpha, \ldots \rangle$ or $\langle \omega, \in \restriction \omega, \ldots \rangle$. To simplify notation we abbreviate by omitting $\in \restriction \alpha$, $\in \restriction \omega$. And we occassionally write L_α and ω for $\langle L_\alpha, \in \restriction \alpha \rangle$ and $\langle \omega, \in \restriction \omega \rangle$, respectively. We frequently omit mention of the language of specific formulas when this is clear from the context. We use the Lévy hierarchy [4] where formulas with all quantifiers restricted are $\Sigma_0 (= \Sigma^0_0)$. For the most part terminology follows that of [6], [7], or [9].

<u>Definition</u>. (i) Let $A \subseteq ON$. The structure $\langle L_\alpha, R_1, \ldots \rangle$ is Σ^1_2 <u>-reflecting on</u> A if for every Σ^1_2 sentence θ of the language for the structure, $\langle L_\alpha, R_1, \ldots \rangle \models \theta \Rightarrow \langle L_\beta, R_1 \restriction \beta, \ldots \rangle \models \theta$ for some $\beta \in A \cap \alpha$.

(ii) $\langle L_\alpha, R_1, \ldots \rangle$ is Σ^1_2 <u>-reflecting</u> if it is Σ^1_2 -reflecting on ON .

(iii) α is Σ^1_2 <u>-reflecting</u> if L_α is Σ^1_2 -reflecting.

(iv) The definitions of a Π^1_2 -reflecting structure and Π^1_2 -reflecting ordinal are obtained from (i)-(iii) by replacing Σ^1_2 by Π^1_2 .

(v) σ_2^1 is the least Σ_2^1 -reflecting ordinal and π_2^1 is the least Π_2^1 -reflecting ordinal.

Recall that a structure of the form $\langle A, \in \restriction A, R_1, \ldots \rangle$, where A is transitive and closed under pairing and union, is <u>admissible</u> if it satisfies Σ_o -Collection and Σ_o -Separation. If such a structure is admissible it also satisfies Δ_1 -Collection and Δ_1 -Separation. An ordinal α is <u>admissible</u> if L_α is admissible. α^+ is the smallest admissible ordinal greater than α , and for a structure $\langle A, R_1, \ldots \rangle$, $\langle A, R_1, \ldots \rangle^+$ is the smallest admissible set having $A, R_1, \ldots$ as members.

The following lemma is Theorem 6.2 (i) of [9]. It is a uniform version of the basic theorem of Barwise-Gandy-Moschovakis [2] (cf. [6, p. 202]) .

<u>Lemma 2.1</u>. If $\varphi(v_1,\ldots,v_n)$ is a Π_1^1 formula then there is a Σ_1 formula $\varphi^+(v_o,v_1,\ldots,v_n)$ having the same constants as $\varphi(v_1,\ldots,v_n)$ such that for every non-empty countable transitive set A and every admissible set B such that $A \in B$, if $a_1,\ldots,a_n \in A$ then

$$A \models \varphi(a_1,\ldots,a_n) \text{ iff } B \models \varphi^+(A, a_1,\ldots,a_n) \ .$$

<u>Definition</u>. α is <u>stable</u> if for every Σ_1 formula $\varphi(v_1,\ldots,v_n)$ and $a_1,\ldots,a_n \in L_\alpha$,

$$L_\alpha \models \varphi(a_1,\ldots,a_n) \text{ iff } \varphi(a_1,\ldots,a_n) \ .$$

<u>Theorem 2.2</u>. If α is stable and countable then α is Σ_2^1 -reflecting.

<u>Proof</u>. Let $\varphi(U)$ be a Π_1^1 formula with U a free n-ary relation (variable) such that $L_\alpha \models \exists U\varphi(U)$. Then

$$\exists V \exists U \in V[V \text{ is countable } \& \ V \text{ is admissible } \& \ \alpha \in V \ \& \ L_\alpha \models \varphi(U)] \ ;$$

$$(1) \quad \exists V \exists U \in V \exists \beta \in V[V \text{ is admissible } \& \ U \subseteq L_\beta^n \ \& \ V \models \varphi^+(L_\beta, U)] \ .$$

The statement "V is admissible" may be expressed by a Π_3 sentence, <u>V is admissible</u> (cf. Theorem 2.4 of [9]) with quantifiers restricted by V . The function $\beta \to L_\beta$ is uniformly Δ_1 -definable in any admissible V and hence since φ^+ is Σ_1 , $V \models \varphi^+(L_\beta,U)$ may be expressed by a Σ_1 formula, <u>$V \models \varphi^+(L_\beta,U)$</u> . Thus (1) may be expressed by a Σ_1 sentence with constants from L_α . Since α is stable, there is a $V \in L_\alpha$ such that

$$L_\alpha \models \exists U \in V \exists \beta \in V[\underline{V \text{ is admissible}} \ \& \ U \subseteq L_\beta^n \ \& \ \underline{V \models \varphi^+(L_\beta, U)}] \ .$$

By the absoluteness of Σ_0 formulas in transitive structures, for some $V, U, \beta \in L_\alpha$, $U, \beta \in V$ and V is admissible and $U \subseteq L_\beta^n$ and $V \models \varphi^+(L_\beta, U)$; i.e. $U, \beta \in L_\alpha$ and $L_\beta \models \varphi(U)$.

Corollary 2.3. $\sigma_2^1 = \delta_2^1 < |\Sigma_2^1|$.

Proof. $\sigma_2^1 \geq \delta_2^1$ by Theorem E of [9]. Kripke and Platek observed that δ_2^1 is stable (for a proof see Lemma 6.2 of [5]). Hence by Theorem 2.2, $\sigma_2^1 \leq \delta_2^1$. To see that $\delta_2^1 < |\Sigma_2^1|$ let $\Phi : P(\omega) \to P(\omega)$ be a complete Σ_2^1 operator. Let $\mathcal{F}$ be the set of Σ_2^1 second order relations on ω . Then $\Phi(\emptyset) \in \mathcal{F}$-HYP . $\Phi(\emptyset)$ is a complete Σ_2^1 set. Hence there is a well-ordering elementary in $\Phi(\emptyset)$, and hence in $\mathcal{F}$-HYP , of order type greater than δ_2^1 . Hence by Theorem 8 of [7] $|\Sigma_2^1| = |\mathcal{F}| = \sup\{\text{rank } (<) : < \in \mathcal{F}\text{-HYP} \ \& < \text{ is well-founded}\} > \delta_2^1$.

3. Elementary basis properties

We reserve W, X, Y, Z, $Z_1, \ldots$ to denote relations or variables on ω , and i, j, k, m, n, $m_1, n_1, \ldots$ to denote members of ω . The following well known basis result is an immediate corollary of a relativization of the Novikoff-Kondo-Addison uniformization theorem.

Basis Theorem. If φ is Π_1^1 on ω then

$$\exists X \varphi(X, Z) \Rightarrow \exists X \in \Delta_2^1(Z) \varphi(X, Z) \ .$$

For a proof (of the non-relativized version) see Shoenfield [11, p. 188-9] .

We also need the following relativized version of Shoenfield's important theorem ([10], [11, p. 319]). Let $\delta_2(Z)$ be the supremum of the order types of well-orderings on ω which are Δ_2^1 on $\langle \omega, Z \rangle$.

Shoenfield's Theorem. If $Z \in L$ then X is Δ_2^1 on $\langle \omega, Z \rangle$ iff $X \in L_{\delta_2(Z)}$.

Definition. $\alpha \in B$ iff α is countable, a limit of admissible ordinals, and for every Σ_0 second order relation $\varphi(X, Y, Z)$ on ω and every $Z \in L_\alpha$,

$$\forall X \exists Y \varphi(X, Y, Z) \Longleftrightarrow \forall X \in L_\alpha \exists Y \in L_\alpha \varphi(X, Y, Z) \ .$$

Let $\text{Lim}(B) = \{\alpha : \alpha \text{ is countable} \ \& \ \alpha \text{ is a limit of members of } B\}$.

Lemma 3.1. $X \in L \Rightarrow \delta_2(X) \in B$.

Proof. Let $X \in L$ and $\alpha = \delta_2(X)$. Then α is stable and hence is a limit of admissible ordinals. Let $W \in L_\alpha$ and φ be Σ_0 . Then by the Basis Theorem and Shoenfield's Theorem

$$\forall Y \exists Z \varphi(Y,Z,W) \iff \forall Y \in \Delta^1_2(W) \exists Z \in \Delta^1_2(W) \varphi(Y,Z,W)$$
$$\iff \forall Y \in L_\alpha \exists Z \in L_\alpha \varphi(Y,Z,W) \quad .$$

Lemma 3.2. Let β be countable and a limit of admissible ordinals. Then $\beta \in B$ iff for every Σ^1_1 second order relation $\varphi(X,Z)$ on ω and every $Z \in L_\beta$,

(1) $\forall X \in L_\beta \varphi(X,Z) \Rightarrow \forall X \varphi(X,Z)$.

Proof. We first observe that if $\psi(X,Y,Z)$ is Σ_0 on ω and $X , Z \in L_\beta$ then

(2) $\exists Y \psi(X,Y,Z) \Rightarrow \exists Y \in L_\beta \psi(X,Y,Z)$.

For suppose $\exists Y \psi(X,Y,Z)$, where $X, Z \in L_\gamma$, $\gamma < \beta$. By the Kleene Basis Theorem (cf. [11, p. 148]) relativized to X , Z there is a W which is Π^1_1 on $\langle \omega,X,Z \rangle$ and a Y arithmetical in W such that $\psi(X,Y,Z)$. By the main theorem of Barwise-Gandy-Moschovakis [6, p. 202] , W is Σ_1 on $\langle \omega,X,Z \rangle^+$. Since $X,Z \in L_\gamma$, $\langle \omega,X,Z \rangle^+ \subseteq L_{\gamma^+}$. Either $\langle \omega,X,Z \rangle^+ = L_{\gamma^+}$ or $\langle \omega,X,Z \rangle^+ \in L_{\gamma^+}$. In either case W is Σ_1 on L_{γ^+} . Since Y is arithmetical in W , $Y \in L_{\gamma^+ +1} \subseteq L_\beta$ by definition of L .

To prove the lemma assume the hypothesis. Let $Z \in L_\beta$ and ψ be Σ_0 on ω and $\varphi(X,Z) \iff \exists Y \psi(X,Y,Z)$. Let $\beta \in B$. Then

$$\forall X \in L_\beta \varphi(X,Z) \Rightarrow \forall X \in L_\beta \exists Y \psi(X,Y,Z)$$
$$\Rightarrow \forall X \in L_\beta \exists Y \in L_\beta \psi(X,Y,Z) \quad \text{by (2)}$$
$$\Rightarrow \forall X \varphi(X,Z) \quad \text{, since } \beta \in B \text{ .}$$

Now suppose (1) holds for every Σ^1_1 φ . Then for ψ Σ_0 on ω ,

$$\forall X \exists Y \psi(X,Y,Z) \Rightarrow \forall X \in L_\beta \exists Y \in L_\beta \psi(X,Y,Z)$$

by (2). Also,

$$\forall X \in L_\beta \exists Y \in L_\beta \psi(X,Y,Z) \Rightarrow \forall X \in L_\beta \varphi(X,Z)$$
$$\Rightarrow \forall X \varphi(X,Z)$$

by (1). Thus $\beta \in B$.

Lemma 3.3. $\mathrm{Lim}(B) \subseteq B$.

Proof. Let $\beta \in \mathrm{Lim}(B)$. Clearly β is a limit of admissible ordinals. Let $\varphi(X,Z)$ be Σ^1_1 on ω and $Z \in L_\beta$ such that $\forall X \in L_\beta \varphi(X,Z)$. For some $\gamma < \beta$,

$\gamma \in B$ and $Z \in L_\gamma$. Then $\forall X \in L_\gamma \varphi(X,Z)$. Since $\gamma \in B$, $\forall X\varphi(X,Z)$ by Lemma 3.2. Hence $\beta \in B$.

Lemma 3.4. There is a Π^1_2 sentence $\underline{B}$ such that $L_\beta \models \underline{B}$ iff $\beta \in B$.

Proof. By Lemma 3.2, $\beta \in B$ iff

(3) [β is a limit of admissible ordinals and for every Σ^1_1 second order relation $\varphi(X,Z)$ on ω and every $Z \in L_\beta$

$$\forall X \in L_\beta \varphi(X,Z) \Rightarrow \forall X\varphi(X,Z)] \quad .$$

It is easy to find a Π^1_2 sentence $\underline{B}$ such that (3) is equivalent to $L_\beta \models \underline{B}$.

Lemma 3.5. There is an elementary formula θ such that if $\beta \in B$ then $\alpha \in B \cap \beta \Leftrightarrow L_\beta \models \theta(\alpha)$.

Proof. Let $\beta \in B$. Then for $Z \in L_\beta$ and Σ_0 φ ,

$$\forall X \exists Y \varphi(X,Y,Z) \Leftrightarrow \forall X \in L_\beta \exists Y \in L_\beta \varphi(X,Y,Z) \quad .$$

Hence $\alpha \in B \cap \beta$ iff

(4) [$\alpha < \beta$ & α is a limit of admissible ordinals & for every Σ_0 second order relation $\varphi(X,Y,Z)$ on ω and every $Z \in L_\alpha$

$$\forall X \in L_\beta \exists Y \in L_\beta \varphi(X,Y,Z) \Leftrightarrow \forall X \in L_\alpha \exists Y \in L_\alpha \varphi(X,Y,Z)] \quad .$$

It is easy to find an elementary formula θ such that the right side, (4), is equivalent to $L_\beta \models \theta(\alpha)$ for $\beta \in B$.

4. Π^1_2 -reflection

We turn now to the proof that $\pi^1_2 = |\Pi^1_2|$. Let $\kappa = \pi^1_2$. The crucial part of the proof, Lemmas 4.3 and 4.4, consists in showing that $\langle L_\kappa, B \cap \kappa\rangle$ is admissible and Π^1_2 -reflecting on $\mathrm{Lim}(B)$.

Lemma 4.1. $\kappa \in B$.

Proof. κ is recursively inaccessible and hence is a limit of admissible ordinals. In [9] we observed that $\delta^1_2 \leq \kappa$. An easy relativization of this argument shows that for each $Z \in L_\kappa$, $\delta_2(Z) \leq \kappa$. Let φ be a Σ^1_1 second order relation on ω and $Z \in L_\kappa$. Suppose $\forall X \in L_\kappa \varphi(X,Z)$ but $\exists X \neg \varphi(X,Z)$. Then by the Basis Theorem, $\exists X \in \Delta^1_2(Z) \neg \varphi(X,Z)$; hence by Shoenfield's Theorem, $\exists X \in L_{\delta_2(Z)} \neg \varphi(X,Z)$. But this implies $\exists X \in L_\kappa \neg \varphi(X,Z)$ which contradicts our assumption.

<u>Lemma 4.2</u>. κ is Π^1_2 -reflecting on $\mathrm{Lim}(B)$.

<u>Proof</u>. Since $\kappa \in B$, $L_\kappa \models \underline{B}$, and hence for each $\alpha < \kappa$, $L_\kappa \models \underline{B}$ & $\alpha = \alpha$. Since κ is Π^1_2 -reflecting there is some $\beta < \kappa$ such that $L_\beta \models \underline{B}$ & $\alpha = \alpha$; i.e. $\forall \alpha < \kappa\ \exists \beta < \kappa[\alpha < \beta\ \&\ L_\beta \models \underline{B}]$. Thus $\kappa \in \mathrm{Lim}(B)$. It is easy to find a Π^1_2 sentence ψ such that for all γ ,

$$\forall \alpha < \gamma\ \exists \beta < \gamma[\alpha < \beta\ \&\ L_\beta \models \underline{B}] \text{ iff } L_\gamma \models \psi ,$$

i.e. $\gamma \in \mathrm{Lim}(B)$ iff $L_\gamma \models \psi$. Since $\kappa \in \mathrm{Lim}(B)$, $L_\kappa \models \psi$. Now let φ be a Π^1_2 sentence such that $L_\kappa \models \varphi$. Then $L_\kappa \models \varphi\ \&\ \psi$. Since κ is Π^1_2 -reflecting there is some $\beta < \kappa$ such that $L_\beta \models \varphi\ \&\ \psi$. Hence $\beta \in \mathrm{Lim}(B)$ and $L_\beta \models \varphi$.

<u>Lemma 4.3</u>. $\langle L_\kappa, B \cap \kappa\rangle$ is Π^1_2 -reflecting on $\mathrm{Lim}(B)$.

<u>Proof</u>. Let $\varphi(U,V)$ be a Σ_0 formula of the language for $\langle L,B\rangle$ such that $\langle L_\kappa, B \cap \kappa\rangle \models \forall U \exists V \varphi(U,V)$. Let φ^* be obtained from φ be replacing each occurence of the form $B(\alpha)$ by $\theta(\alpha)$. Then φ^* is an elementary formula of the language of set theory such that if $\beta \in \mathrm{Lim}(B)$ then for all U,V,

$$\langle L_\beta, B \cap \beta\rangle \models \varphi(U,V) \text{ iff } L_\beta \models \varphi^*(U,V) .$$

Since $\langle L_\kappa, B \cap \kappa\rangle \models \forall U \exists V \varphi(U,V)$, $L_\kappa \models \forall U \exists V \varphi^*(U,V)$. Since κ is Π^1_2 -reflecting on $\mathrm{Lim}(B)$ there is some $\beta \in \mathrm{Lim}(B) \cap \kappa$ such that $L_\beta \models \forall U \exists V \varphi^*(U,V)$; hence $\langle L_\beta, B \cap \beta\rangle \models \forall U \exists V \varphi(U,V)$.

<u>Lemma 4.4</u>. $\langle L_\kappa, B \cap \kappa\rangle$ is admissible.

<u>Proof</u>. We first show $\langle L_\kappa, B \cap \kappa\rangle$ satisfies Σ_0 -Collection. Let φ be a Σ_0 formula and $a \in L_\kappa$. Suppose $\langle L_\kappa, B \cap \kappa\rangle \models \forall u \in a \exists v \varphi$. Let φ^* be obtained from φ as in the proof of Lemma 4.3. Then for all $\beta \in \mathrm{Lim}(B)$ and $u,v \in L_\beta$, $\langle L_\beta, B \cap \beta\rangle \models \varphi(u,v)$ iff $L_\beta \models \varphi^*(u,v)$. Since $\kappa \in \mathrm{Lim}(B)$, $L_\kappa \models \forall u \in a \exists v \varphi^*(u,v)$. Since φ^* is elementary and hence Π^1_2 , so is the formula $\forall u \in a \exists v \varphi^*(u,v)$. Since κ is Π^1_2 -reflecting on $\mathrm{Lim}(B)$ there is a $\beta \in \mathrm{Lim}(B)$ such that $L_\beta \models \forall u \in a \exists v \varphi^*(u,v)$. Since $\beta \in \mathrm{Lim}(B)$, for all $u,v \in L_\beta$, $L_\beta \models \varphi^*(u,v)$ iff $\langle L_\beta, B \cap \beta\rangle \models \varphi(u,v)$. Hence $\langle L_\beta, B \cap \beta\rangle \models \forall u \in a \exists v \varphi(u,v)$, i.e. $\forall u \in a \exists v \in L_\beta[\langle L_\beta, B \cap \beta\rangle \models \varphi(u,v)]$. Since φ is Σ_0 ,

$\langle L_\beta, B\cap B\rangle \models \varphi(u,v)$ iff $\langle L_\varkappa, B\cap\varkappa\rangle \models \varphi(u,v)$. Hence
$\langle L_\varkappa, B\cap\varkappa\rangle \models \forall u\in a\exists v\in L_\beta\varphi(u,v)$.

To show Σ_o-Separation suppose φ is Σ_o and $a\in L_\varkappa$. Choose $\beta\in \mathrm{Lim}(B)\cap\varkappa$ so that $a\in L_\beta$ and all constants in φ belong to L_β . Since φ is Σ_o ,

$$\{u:u\in a \;\&\; \langle L_\varkappa, B\cap\varkappa\rangle \models \varphi(u)\} = \{u:u\in a \;\&\; \langle L_\beta, B\cap\beta\rangle \models \varphi(u)\}$$

$$= \{u:u\in a \;\&\; L_\beta \models \varphi^*(u)\}\in L_{\beta+1}\subseteq L_\varkappa ,$$

since φ^* is elementary.

At this point we may complete the proof that $\pi_2^1 = |\Pi_2^1|$ by applying either Theorem 10.3 of [9] or Theorem 24 of [7]. Since the approach of Moschovakis is more general we use the latter.

Let $\mathcal{F}$ be the set of second order relations on ω which are Π_2^1 on ω . Given $\varphi\in\mathcal{F}$ and a set M , let $\varphi^M = \varphi\upharpoonright M$, i.e.

$$\varphi^M(n_1,\dots,n_k,Z_1,\dots,Z_\ell) \iff Z_1,\dots,Z_\ell\in M \;\&\; \varphi(n_1,\dots,n_k,Z_1,\dots,Z_\ell) .$$

We abbreviate φ^{L_α} by φ^α and let $\mathcal{F}^\alpha = \{\varphi^\alpha:\varphi\in\mathcal{F}\}$. Following [7] we say L_α is <u>$\mathcal{F}$-admissible</u> if the structure $\langle L_\alpha,\mathcal{F}^\alpha\rangle$ is admissible, where the distinguished relations of $\langle L_\alpha,\mathcal{F}^\alpha\rangle$ are $\in\upharpoonright L_\alpha$ and the members of $\mathcal{F}^\alpha$. L_α is said to be <u>$\Pi_2^1(\mathcal{F})$-reflecting</u> if $\langle L_\alpha,\mathcal{F}^\alpha\rangle$ is Π_2^1 -reflecting.

<u>Lemma 4.5</u>. $L_\varkappa$ is $\mathcal{F}$-admissible and $\Pi_2^1(\mathcal{F})$-reflecting.

<u>Proof</u>. In order to show that $L_\varkappa$ is $\mathcal{F}$-admissible it suffices to show that $\varphi\in\mathcal{F}$ implies $\varphi^\varkappa$ is Δ_1 on $\langle L_\varkappa, B\cap\varkappa\rangle$ for then every relation Σ_o on $\langle L_\varkappa,\mathcal{F}^\varkappa\rangle$ is Δ_1 on $\langle L_\varkappa, B\cap\varkappa\rangle$ and hence Σ_o-Collection and Σ_o-Separation for $\langle L_\varkappa,\mathcal{F}^\varkappa\rangle$ are implied by Δ_1-Collection and Δ_1-Separation for $\langle L_\varkappa, B\cap\varkappa\rangle$. Suppose $\varphi\in\mathcal{F}$, say $\varphi(n,Z)\iff\forall X\exists Y\psi(n,X,Y,Z)$, where ψ is Σ_o . Then there is a Σ_1 formula φ' of the language for $\langle L,B\rangle$ such that for $\beta\in\mathrm{Lim}(B)$,

$$\varphi^\beta(n,Z) \iff Z\in L_\beta \;\&\; \forall X\exists Y\psi(n,X,Y,Z)$$

$$\iff \exists\gamma<\beta[\gamma\in B\cap\beta \;\&\; Z\in L_\gamma \;\&\; \forall X\in L_\gamma\exists Y\in L_\gamma\psi(n,X,Y,Z)]$$

$$\iff \langle L_\beta, B\cap\beta\rangle \models \varphi'(n,Z) .$$

Similarly, there is a Σ_1 formula $(\neg\varphi)'$ such that

$$\neg\varphi^\beta(n,Z) \iff \langle L_\beta, B\cap\beta\rangle \models (\neg\varphi)'(n,Z) .$$

In particular, we have shown that if $\varphi \in \mathcal{F}$ then φ^{κ} is Δ_1 on $\langle L_\kappa, B \cap \kappa\rangle$.

Now suppose φ is a $\Pi^1_2(\mathcal{F})$ sentence such that $\langle L_\kappa, \mathcal{F}^\kappa\rangle \models \varphi$. We find $\beta \in \kappa$ such that $\langle L_\beta, \mathcal{F}^\beta\rangle \models \varphi$. Let $\varphi_1,\ldots,\varphi_k$ be the (names of) members of $\mathcal{F}$ appearing in φ . Let φ^* be the sentence of the language of $\langle L, B\rangle$ obtained by replacing each occurrence of the form $\varphi_i(n_1,\ldots,n_j, Z_1,\ldots,Z_k)$ by $\varphi_i'(n_1,\ldots,n_j, Z_1,\ldots,Z_k)$. Then φ^* is Π^1_2 and for $\beta \in \mathrm{Lim}(B)$,

$$\langle L_\beta, \mathcal{F}^\beta\rangle \models \varphi \text{ iff } \langle L_\beta, B \cap \beta\rangle \models \varphi^* .$$

Hence $\langle L_\kappa, B \cap \kappa\rangle \models \varphi^*$ and since $\langle L_\kappa, B \cap \kappa\rangle$ is Π^1_2 -reflecting on $\mathrm{Lim}(B)$ there is a $\beta \in \mathrm{Lim}(B) \cap \kappa$ such that $\langle L_\beta, B \cap \beta\rangle \models \varphi^*$; hence $\langle L_\beta, \mathcal{F}^\beta\rangle \models \varphi$.

Given a set M let $o(M) = \sup\{\alpha: \alpha \in M\}$. Summarizing the relevant portions of Moschovakis [6, Ch. 9], [7, p. 75] on companions we have:

Lemma 4.6. Let

$$\omega^+(\Pi^1_2) = \cap\{M : M \text{ is } \mathcal{F}\text{-admissible, } \Pi^1_2(\mathcal{F})\text{-reflecting, and } \omega \in M\} .$$

Then: (i) $\omega^+(\Pi^1_2)$ is $\mathcal{F}$-admissible and $\Pi^1_2(\mathcal{F})$-reflecting;

(ii) $o(\omega^+(\Pi^1_2)) = |\Pi^1_2|$;

(iii) for any X ,

$$X \in \mathcal{F}\text{-IND iff } X \text{ is } \Sigma_1 \text{ on } \langle \omega^+(\Pi^1_2), \mathcal{F}^\alpha\rangle ,$$

where $\alpha = o(\omega^+(\Pi^1_2))$.

Theorem 4.7 (i) $|\Pi^1_2| = \pi^1_2$;

(ii) for any X

$$X \in \Pi^1_2\text{-IND iff } X \text{ is } \Sigma_1 \text{ on } \langle L_\kappa, B \cap \kappa\rangle, \text{ where } \kappa = \pi^1_2 ;$$

(iii) in particular, for any X

$$X \in \Pi^1_2\text{-HYP iff } X \in L_\kappa .$$

Proof. (i) $|\Pi^1_2| \geq \pi^1_2$ by Theorem 1.1. By Lemma 4.5, $\omega^+(\Pi^1_2) \subseteq L_\kappa$; hence by Lemma 4.6,

$$|\Pi^1_2| = o(\omega^+(\Pi^1_2)) \leq o(L_\kappa) = \pi^1_2 .$$

(ii) Since $\omega^+(\Pi^1_2)$ is admissible and $o(\omega^+(\Pi^1_2)) = \kappa$ it follows easily that $L_\varkappa \subseteq \omega^+(\Pi^1_2)$; hence $L_\varkappa = \omega^+(\Pi^1_2)$. Then by Lemma 4.6,

$$X \in \Pi^1_2\text{-IND iff } X \text{ is } \Sigma_1 \text{ on } \langle L_\varkappa, \mathcal{F}^\varkappa\rangle .$$

From the proof of Lemma 4.5, if $\varphi \in \mathcal{F}$ then $\varphi^\varkappa$ is Δ_1 on $\langle L_\varkappa, B \cap \varkappa\rangle$ and hence every relation Σ_1 on $\langle L_\varkappa, \mathcal{F}^\varkappa\rangle$ is Σ_1 on $\langle L_\varkappa, B \cap \varkappa\rangle$. For the other direction, let $\varphi \in \mathcal{F}$ parametrize the relations in $\mathcal{F}$ of signature (1,1), say $\varphi(m,n,Z) \iff \forall X \exists Y \psi(m,n,X,Y,Z)$, where ψ is Σ_o on ω . Then

$$\alpha \in B \cap \varkappa \iff \alpha < \varkappa \ \& \ \alpha \text{ is a limit of admissible ordinals}$$

$$\& \ \forall m,n \in \omega \forall Z \in L_\alpha[\varphi^\varkappa(m,n,Z) \iff \forall X \in L_\alpha \exists Y \in L_\alpha \psi(m,n,X,Y,Z)]$$

and hence $B \cap \varkappa$ is Δ_1 on $\langle L_\varkappa, \mathcal{F}^\varkappa\rangle$. It follows that every relation Σ_1 on $\langle L_\varkappa, B \cap \varkappa\rangle$ is Σ_1 on $\langle L_\varkappa, \mathcal{F}^\varkappa\rangle$.

(iii) By (ii) and Δ_1-Separation, for any X

$$X \in \Pi^1_2\text{-HYP iff } X \text{ is } \Delta_1 \text{ on } \langle L_\varkappa, B \cap \varkappa\rangle$$

$$\text{iff } X \in L_\varkappa .$$

REFERENCES

[1] S. Aanderaa, Inductive definitions and their closure ordinals, in Generalized Recursion Theory, J.E. Fenstad, P.G. Hinman (eds.), North-Holland (1974), 207-220.

[2] K.J. Barwise, R.O. Gandy and Y.N. Moschovakis, The next admissible set, J. Symb. Logic 36 (1971), 108-120.

[3] D. Cenzer, Ordinal recursion and inductive definitions, in Generalized Recursion Theory, J.E. Fenstad, P.G. Hinman (eds.), North-Holland (1974), 221-264.

[4] A. Lévy, A hierarchy of formulas in set theory, Memoirs of the Amer. Math. Soc. 57, Amer. Math. Soc., Providence, (1965).

[5] W. Marek and M. Srebrny, Gaps in the constructible universe, Ann. of Math. Logic 6 (1974), 359-394.

[6] Y.N. Moschovakis, Elementary Induction on Abstract Structures, North-Holland (1974).

[7] Y.N. Moschovakis, On nonmonotone inductive definability, Fund. Math. 82 (1974), 39-83.

[8] W. Richter, Recursively Mahlo ordinals and inductive definitions, in Logic Colloquium '69, R.O. Gandy and C.E.M. Yates (eds.), North-Holland (1971), 273-288.

[9] W. Richter and P. Aczel, Inductive definitions and reflecting properties of admissible ordinals, in Generalized Recursion Theory, J.E. Fenstad, P.G. Hinman (eds.), North-Holland (1974), 301-381.

[10] J.R. Shoenfield, The problem of predicativity, Essays on the Foundations of Mathematics, Y. Bar-Hillel (ed.), Magnes Press (1961) and North-Holland (1962), 132-139.

[11] J.R. Shoenfield, Mathematical Logic, Addison-Wesley (1967).

DATA TYPES AS LATTICES

by

Dana Scott
Merton College, Oxford

To the memory of
Christopher Strachey
1916-1975

0. INTRODUCTION

Investigations begun in 1969 with Christopher Strachey led to the idea that the denotations of many kinds of expressions in programming languages could be taken as elements of certain kinds of spaces of "partial" objects. As these spaces could be treated as function spaces, their structure at first seemed excessively complicated - even impossible. But then the author discovered that there were many more spaces than we had first imagined - even wanted. They could be presented as lattices (or as some prefer, semilattices), and the main technique was to employ topological ideas, in particular the notion of a continuous function. This approach and its applications have been presented in a number of publications, but that part of the foundation concerned with *computability* (in the sense of recursion theory) was never before adequately exposed. The purpose of the present paper is to provide such a foundation and to simplify the whole presentation by a de-emphasis of abstract ideas. An appendix and the bibliography[1] provide a partial guide to the literature and an indication of connections with other work.

The main innovation in this report is to model everything within one "universal" domain $P\omega = \{x | x \subseteq \omega\}$, the domain of all subsets of the set ω of non-negative integers. The advantages are many: by the most elementary considerations $P\omega$ is recognised to be a lattice and a topological space. In fact, $P\omega$ is a *continuous lattice*, even an *algebraic lattice*, but in the beginning we do not even need to define such an "advanced" concept; we can save these ideas for an *analysis* of what has been done here in a more direct way. Next by taking the set of integers as the basis of the construction, the connection with the ordinary theory of number-theoretic (especially, general recursive) functions can be made clear and straight forward.

The model $P\omega$ can be intuitively viewed as the domain of *multiple-valued* integers; what is new in the presentation is that functions are not only multiple valued but also "multiple argumented". This remark is given a precise sense in Section 2 below, but the upshot of the idea is that multiple-valued integers are regarded as objects in themselves — possibly infinite — and as something more than just the collection of single integers contained in them. This combination of the finite with the infinite into a single domain, together with the idea that a continuous function can be reduced to its *graph* (in the end, a set of integers), makes it possible to view an $x \in P\omega$ at one time as a value, at another as an argument, then as an integer, then as a function, and still later as a functional (or combinator). The "paradox" of self-application (as in $x(x)$) is solved by allowing the *same* x to be used in two *different* ways. This is done in ordinary recursion theory *via* Gödel numbers (as in $\{e\}(e)$), but the advantage of the present theory is that not only is the function concept the *extensional* one, but it includes *arbitrary* continuous functions and not just the computable ones.

Section 1 introduces the elementary ideas on the topology of $P\omega$ and the continuous functions including the fixed-point theorem. Section 2 has to do with computability and definability. The language LAMBDA is introduced as an extension of the pure λ-calculus by four arithmetical combinators; in fact, it is indicated in Section 3 how the whole system could be based on *one* combinator. What is shown is that computability in $P\omega$ according to the natural definition (which assumes that we already know what a recursively enumerable set of integers is) is equivalent to LAMBDA definability. The main tool is, not surprisingly, the First Recursion Theorem formulated with the aid of the so-called *paradoxical combinator* Y. The plan is hardly original, but the point is to work out what it all means in the model.

Along the way we have to show how to give every λ-term a denotation in $P\omega$; the resulting principles of λ-calculus that are thereby verified are summarized in Table 1. Of these the first three (α), (β), and (ξ) are indeed valid in the model; however, rule (η), which is a stronger version of extensionality, *fails* in the $P\omega$ model. This should not be regarded as a disadvantage since the import of (η) is to suppose every object is a function. A quick construction of these special models is indicated at the end of Section 5. Since $P\omega$ is partially ordered by $\subseteq$, there are also laws involving this relation. Law (ξ*) is an improvement of (ξ); while (μ) is a form of monotonicity for application.

Section 3 has to do with enumeration and degrees. Gödel numbers for LAMBDA are defined in a very easy way which takes advantage of the notation of combinators. This leads to the Second

(α)	$\lambda x.\tau = \lambda y.\tau[y/x]$
(β)	$(\lambda x.\tau)(y) = \tau[y/x]$
(ξ)	$\lambda x.\tau = \lambda x.\sigma$ iff $\forall x.\tau = \sigma$
(η)	$y = \lambda x.y(x)$
(ξ*)	$\lambda x.\tau \subseteq \lambda x.\sigma$ iff $\forall x.\tau \subseteq \sigma$
(μ)	$x \subseteq y$ and $u \subseteq v$ imply $u(x) \subseteq v(y)$

Table 1. Some Laws of λ-Calculus

Recursion Theorem, and results on incompleteness and undecidability follow along standard lines. *Relative recursiveness* is also very easy to define in the system, and we make the tie-in with *enumeration degrees* which correspond to finitely generated combinatory subalgebras of $P\omega$. Finally a theorem of Myhill and Shepherdson is interpreted as a most satisfactory completeness property for definability in the system.

Sections 4 and 5 show how a calculus of *retracts* leads to quite simple definitions of a host of useful domains (as lattices). Section 6 investigates the classification of other subsets (non-lattices) of $P\omega$; while Section 7 contrasts partial (multiple-valued) functions with total functions, and interprets various theories of functionality. Connections with category theory are mentioned.

What is demonstrated in this work is how the language LAMBDA, together with its interpretation in $P\omega$, is an extremely convenient vehicle for *definitions* of computable functions on complex structures (all taken as subdomains of $P\omega$). It is a "high-level" programming language for recursion theory. It is *applied* combinatory logic, which in usefulness goes far beyond anything envisioned in the standard literature. What has been shown is how many interesting predicates can be *expressed* as equations between continuous functions. What is needed next is a development of the *proof theory* of the system along the lines of the work of Milner, which incorporates the author's extension of McCarthy's rule of Recursion Induction to this high-level language. Then we will have a flexible and practical "mathematical" theory of computation.

1. CONTINUOUS FUNCTIONS

The domain $P\omega$ of all subsets of the set ω of non-negative integers is a complete lattice under the partial ordering $\subseteq$ of set inclusion, as is well known. We use the usual symbols $\cup$, $\cap$, $\bigcup$, $\bigcap$ for the finite and infinite lattice operations of union and intersection. $P\omega$ is of course also a Boolean algebra; and for complements we write

$$\sim x = \{n \mid n \notin x\}$$

where it is understood that such variables as i, j, k, l, m, n range over integers *in* ω, while u, v, w, x, y, z range over *subsets of* ω.

The domain $P\omega$ can also be made into a topological space — in many ways. A common method is to match each $x \subseteq \omega$ with the corresponding characteristic function in $\{0,1\}^\omega$, and to take the induced product topology. In this way $P\omega$ is a totally disconnected compact Hausdorff space homeomorphic to the Cantor "middle-third" set. This is *not* the topology we want; it is a *positive-and-negative* topology which makes the function $\sim x$ continuous. We want a weaker topology: the topology of positive "information", which has the advantage that all continuous functions possess fixed points. (The equation $x = \sim x$ is impossible.) The topology that we do want is exactly that appropriate to considering $P\omega$ to be a continuous lattice. But all this terminology of abstract mathematics is quite unnecessary, since the required definitions can be given in very elementary terms.

To make the topology "visible", we introduce a standard enumeration $\{e_n | n \in \omega\}$ of all *finite* subsets of ω. Specifically we set

$$e_n = \{k_0, k_1, \ldots, k_{m-1}\},$$

provided that $k_0 < k_1 < \ldots < k_{m-1}$ and $n = \sum_{i<m} 2^{k_i}$. Thus n is the code number for e_n, and the elements of e_n are the exponents in the binary expansion of the integer n. This is a one-one enumeration of finite subsets, where $k \in e_n$ always implies $k < n$, the function $\max(e_n)$ is (primitive) recursive in n, and the relations $k \in e_n$, $e_n \subseteq e_m$, $e_n = e_m \cup e_k$ are all (primitive) recursive in k, m, n.

Topologically speaking the finite sets e_n are *dense* in the space $P\omega$, for each $x \in P\omega$ is the "limit" of its finite subsets in the sense that

$$x = \bigcup\{e_n | e_n \subseteq x\}.$$

To make this precise we need a rigorous definition of *open subset* of $P\omega$.

DEFINITION. A *basis* for the neighbourhoods of $P\omega$ consists of those sets of the form:

$$\{x \in P\omega \mid e_n \subseteq x\},$$

for a given e_n. An arbitrary *open subset* is then a union of basic neighbourhoods.

It is easy to prove that an open subset $U \subseteq P\omega$ is just a set of "finite character"; that is, a set such that for all $x \in P\omega$ we have $x \in U$ *if and only if* some finite subset of x also belongs to U. An alternate approach would define directly what we mean by a continuous function using the idea that such functions must preserve limits.

DEFINITION. A function $f : P\omega \to P\omega$ is *continuous* iff for all $x \in P\omega$ we have:

$$f(x) = \bigcup\{f(e_n) \mid e_n \subseteq x\}.$$

Again it is an easy exercise to prove that a function is continuous in the sense of this definition iff it is continuous in the usual topological sense (namely: inverse images of open sets are open). For giving proofs it is even more convenient to have the usual ε-δ formulation of continuity.

THEOREM 1.1. (The Characterization Theorem). *A function $f : P\omega \to P\omega$ is continuous iff for all $x \in P\omega$ and all e_m we have:*

$$e_m \subseteq f(x) \text{ iff } \exists e_n \subseteq x.\ e_m \subseteq f(e_n).$$

Note that open sets and continuous functions have a *monotonicity property*:

whenever $x \subseteq y$ and $x \in U$, then $y \in U$; and
whenever $x \subseteq y$, then $f(x) \subseteq f(y)$.

This gives a precise expression to the "positive" character of our topology. However, note too that openness and continuity mean rather more than just monotonicity. In particular, a continuous function is completely determined by the pairs of integers such that $m \in f(e_n)$, as can be seen from the definition. (Hence, there are only a continuum number of continuous functions, but more than a continuum number of monotonic functions.) This brings us to the definition of the *graph* of a continuous function.

To formulate the definition, we introduce a standard enumeration (n,m) of all pairs of integers. Specifically we set

$$(n,m) = \tfrac{1}{2}(n+m)(n+m+1)+m.$$

This is the enumeration along the "little diagonals" going from left to right, and it produces the ordering:

$$(0,0),(1,0),(0,1),(2,0),(1,1),(0,2),(3,0),(2,1),\ldots\ .$$

Note that $n \leq (n,m)$ and $m \leq (n,m)$ with equality possible only in the cases of (0,0) and (1,0). This is a one-one enumeration, and the inverse functions are (primitive) recursive — but we do not require at the present any notation for them.

DEFINITION. The *graph* of a continuous function $f : \mathcal{P}\omega \to \mathcal{P}\omega$ is defined by the equation:

$$\mathrm{graph}(f) = \{(n,m) \mid m \in f(e_n)\};$$

while the *function* determined by any set $u \subseteq \omega$ is defined by the equation:

$$\mathrm{fun}(u)(x) = \{m \mid \exists e_n \subseteq x.\,(n,m) \in u\}\ .$$

THEOREM 1.2. (The Graph Theorem). *Every continuous function is uniquely determined by its graph in the sense that:*

(i) $\qquad \mathrm{fun}(\mathrm{graph}(f)) = f.$

Conversely, every set of integers determines a continuous function and we have:

(ii) $\qquad u \subseteq \mathrm{graph}(\mathrm{fun}(u)),$

where equality holds just in case u satisfies:

(iii) whenever $(k,m) \in u$ *and* $e_k \subseteq e_n$, *then* $(n,m) \in u$.

Besides functions of one variable we need to consider also functions of several variables. The official definition for one variable given above can be extended simply by saying $f(x,y,\ldots)$ is continuous iff it is continuous in *each* of $x,y,\ldots$. Those familiar with the product topology can prove that for our special positive topology on $\mathcal{P}\omega$ this is equivalent to being continuous on the product space (continuous in the variables *jointly*). Those interested only in elementary proofs can calculate out directly from the definition (with the aid of 1.1) that continuity behaves under combinations by substitution [as in: $f(g(x,y),h(y,x,y))$].

THEOREM 1.3. (The Substitution Theorem). *Continuous functions of several variables on* $\mathcal{P}\omega$ *are closed under substitution.*

The other general fact about continuous functions that we shall use constantly concerns fixed points, whose existence can be proved using a well-known method.

THEOREM 1.4. (The Fixed-Point Theorem). *Every continuous function* $f : P\omega \to P\omega$ *has a least fixed point given by the formula:*

$$\mathrm{fix}(f) = \bigcup\{f^n(\emptyset) \mid n \in \omega\},$$

where $\emptyset$ *is the empty set and* f^n *is the n-fold composition of* f *with itself.*

Actually fix is a *functional* with continuity properties of its own. We shall not give the required definitions here because they can be more easily derived from the construction of the model given in the next section.

For those familiar with the abstract theory of topological spaces we give in conclusion two general facts about continuous functions with values in $P\omega$ which indicate the scope and generality of our method.

THEOREM 1.5. (The Extension Theorem). *Let* X *and* Y *be arbitrary topological spaces where* $X \subseteq Y$ *as a subspace. Then every continuous function* $f : X \to P\omega$ *can be extended to a continuous function* $\overline{f} : Y \to P\omega$ *defined by the equation:*

$$\overline{f}(y) = \bigcup\{\bigcap\{f(x) \mid x \in X \cap U\} \mid y \in U\},$$

where $y \in Y$, *and* U *ranges over the open subsets of* Y.

THEOREM 1.6. (The Embedding Theorem). *Every* T_0*-space* X *with a countable basis* $\{U_n \mid n \in \omega\}$ *for its topology can be embedded in* $P\omega$ *by the continuous function* $\varepsilon : X \to P\omega$ *defined by the equation:*

$$\varepsilon(x) = \{n \mid x \in U_n\}.$$

Technically the T_0-hypothesis is what is needed to show that ε is one-one. The upshot of these two theorems is that in looking for (reasonable) topological structures we can confine attention to the subspaces of $P\omega$ and to continuous functions defined on *all* of $P\omega$. Thus the emphasis on a single space is justified structurally. What we shall see in the remainder of this work is that the use of a single space is also justified practically because the required subspaces and functions can be *defined* in very simple ways by a natural method of equations.

In order to make the plan of the work clearer, the proofs of the theorems have been placed in an appendix when they are more than simple exercises.

2. COMPUTABILITY AND DEFINABILITY.

The purpose of the first section was to introduce in a simple-minded way the basic notions about the topology of $P\omega$ and its continuous functions. In this section we wish to present the details of a powerful language for defining *particular* functions — especially computable functions — and initiate the study of the *use* of these functions. This study is then extended in different ways in the following sections.

Before looking at the language, a short discussion of the "meaning" of the elements $x \in P\omega$ will be helpful from the point of view of motivation. Now in itself $x \in P\omega$ is a *set*, but this does not reveal its *meaning*. Actually x has no "fixed" meaning, because it can be *used* in strikingly different ways; we look for meaning here solely in terms of use. Nevertheless it is possible to give some coherent guidelines.

In the first place it is convenient to let the singleton subsets $\{n\} \in P\omega$ stand for the corresponding integers. In fact, we shall enforce by convention the *equation* $n = \{n\}$ as a way of simplifying notation. In this way $\omega \subseteq P\omega$ as a subset. (Note that our convention conflicts with such set-theoretical equations as $5 = \{0,1,2,3,4\}$. What we have done is to abandon the usual set-theoretical conventions in favour of a slight redefinition of *set of integers* which produces a more helpful convention for present purposes.) So, if we choose, a singleton "means" a single integer. The next question is what a "large" set $x \in P\omega$ could mean. Here is an answer: if we write:

$$x = \{0,5,17\} = 0 \cup 5 \cup 17,$$

we are thinking of x as a *multiple* integer. This is especially useful in the case of multiple-valued function where we can write:

$$f(a) = 0 \cup 5 \cup 17.$$

Then "$m \in f(a)$" can be interpreted as "m is *one* value of f at a." Now $a \in P\omega$, too, and so it is a multiple integer also. This brings us to an important point.

A multiple integer is (often) *more* than just the (random) collection of its elements. From the definition of continuity, $m \in f(a)$ is equivalent to $m \in f(e_n)$ with $e_n \subseteq a$. We may not be able to reduce this to $m \in f(\{n\})$ with $n \in a$ without additional assumptions on f. Indeed we shall take advantage of the feature of continuous functions whereby the elements of an argument a can join in *cooperation* in determining $f(a)$. Needless to say, continuity implies that the cooperation cannot extend beyond *finite* configurations, and so we can say that a *is* the union (or limit) of its finite subsets. However, finitary cooperation will be found to be quite a powerful notion.

Where does this interpretation leave the empty set $\emptyset$? When we write "$f(a) = \emptyset$" we can read this as "f has *no* value at a", or "f is *undefined* at a". In this case $f(a)$ exists (as a set), but it is not "defined" as an *integer*. Single- (or singleton) valued functions are "well defined", but multiple-valued functions are rather "over defined".

How does this interpretation fit in monotonicity? In case $a \subseteq b$ and $m \in f(a)$, then we must have $m \in f(b)$. We can read "$a \subseteq b$" as "b is an *improvement* of a" or roughly "b is *better-defined* than a". The point of monotonicity is that the better we define an argument, the better we define a value. "Better" does not imply "well" (that is, single-valuedness), and overdefinedness may well creep in. This is not the fault of the function; it is our fault for not choosing a different function.

As a subspace $\omega \subseteq P\omega$ is *discrete*. This implies that *arbitrary* functions $p : \omega \to \omega$ are continuous. Note that $p : \omega \to P\omega$ as well, because $\omega \subseteq P\omega$. By 1.5 we can extend p continuously to $\overline{p} : P\omega \to P\omega$. The formula given produces this function:

$$(1) \qquad \begin{aligned} \overline{p}(x) &= \bigcap\{p(n) \mid n \in \omega\}, && \text{if } x = \emptyset; \\ &= p(n), && \text{if } x = n \in \omega; \\ &= \omega, && \text{otherwise.} \end{aligned}$$

This is a rather abrupt extension of p (the *maximal* extension); a more gradual, continuous extension (the *minimal* extension) is determined by this equation:

$$(2) \qquad \hat{p}(x) = \bigcup\{p(n) \mid n \in x\}.$$

The same formulae work for *all* multiple-valued functions $p : \omega \to P\omega$. Functions like $f = \hat{p}$ are exactly characterized as being those continuous functions $f : P\omega \to P\omega$ which in addition are *distributive* in the sense of these equations:

$$f(x \cup y) = f(x) \cup f(y) \quad \text{and} \quad f(\emptyset) = \emptyset.$$

The sets $\emptyset$ and ω play special roles. When we consider them as *elements* of $P\omega$ we shall employ the notation:

$$\bot = \emptyset \qquad \text{and} \qquad \top = \omega.$$

The element $\bot$ is the most "undefined" integer, and $\top$ is the most "overdefined". All others are in between.

One last general point on meaning: suppose $x \in P\omega$ and $k \in x$. Then $k = (n,m)$ for suitable (uniquely determined) integers n and m. That is to say, every element of x can be regarded as an *ordered pair*; thus, x can be used as a *relation*. Such an operation as

(3) $$x;y = \{(n,l) \mid \exists m.(n,m) \in x, (m,l) \in y\}$$

is then a continuous function of two variables that treats both x and y as relations. On the other hand we could define a quite different continuous function such as

(4) $$x + y = \{n + m \mid n \in x, m \in y\}$$

which treats both x and y *arithmetically*. The only reason we shall probably never write $(x+y);x$ again is that the values of this perfectly well-defined continuous function are, for the most part, quite uninteresting. There is, however, no theoretical reason why we cannot use the same set with several different "meanings" in the same formula. Of course if we do so, it is to be expected that we will show the point of doing this in the special case. We turn now to the definition of the general language for defining all such functions.

The *syntax* and *semantics* of the language LAMBDA are set out in Table 2. The syntax is indicated on the left, and the meanings of the combinations are shown on the right as subsets of $P\omega$. This is the basic language and could have been given (less understandably) in terms of combinators (see 2.4). It is, however, a very primitive language, and we shall require many definitions before we can see why such functions as in (3) and (4) are themselves definable.

The definition has been left somewhat informal in hopes that it will be more understandable. In the above, τ is any term of the language. LAMBDA is *type-free* and allows any combination to be made by substitution into the given functions. There is one primitive constant (0); there are two unary functions ($x+1, x-1$); there is one binary function ($u(x)$) and one ternary function ($z \supset x,y$); finally there is one variable binding operator ($\lambda x.\tau$). The first three equations have obvious sense. In the fourth, $z \supset x,y$ is McCarthy's *conditional expression* (a test for *zero*). Next $u(x)$ defines *application* (u is treated as a graph and x as a set), and $\lambda x.\tau$ is *functional abstraction* (compare the definition of fun). In defining $\lambda x.\tau$, we use $\tau[e_n/x]$ as a shorthand for *evaluating* the term τ when the variable x is given the value e_n.

Note that the functions are all multiple-valued. Thus we

0	$=$	$\{0\}$
$x+1$	$=$	$\{n+1 \mid n \in x\}$
$x-1$	$=$	$\{n \mid n+1 \in x\}$
$z \supset x,y$	$=$	$\{n \in x \mid 0 \in z\} \cup \{m \in y \mid \exists\, k.k+1 \in z\}$
$u(x)$	$=$	$\{m \mid \exists\, e_n \subseteq x.(n,m) \in u\}$
$\lambda x.\tau$	$=$	$\{(n,m) \mid m \in \tau[e_n/x]\}$

Table 2. The Language LAMBDA

have such a result as:

(5) $(6 \cup 10) + 1 = 7 \cup 11.$

The partial character of subtraction has expression as:

(6) $0 - 1 = \bot.$

We shall see how to define + and - in LAMBDA later. The conditional could also have been defined by cases:

(7)
$$
\begin{aligned}
z \supset x, y &= \bot, && \text{if } z = \bot;\\
&= x, && \text{if } z = 0;\\
&= y, && \text{if } 0 \notin z \neq \bot;\\
&= x \cup y, && \text{if } 0 \in z \neq 0.
\end{aligned}
$$

We say that a LAMBDA-term τ defines a function of its free variables (at least). Other results depend on this fundamental proposition:

THEOREM 2.1. (The Continuity Theorem). *All* LAMBDA *definable functions are continuous.*

Once that is proved, we can use 1.2. to establish:

THEOREM 2.2. (The Conversion Theorem). *The three basic principles* (α), (β), (ξ) *of* λ*-conversion are all valid in the model.*

By "model" here we of course understand the interpretation of the language where the semantics gives terms denotations in $P\omega$ according to the stated definition. Through this interpretation, more properly speaking, $P\omega$ becomes a model for the axioms (α), (β), (ξ). Two well-known results of the calculus of λ-conversion allow the reduction of functions of several variables to those of *one*, and the reduction of all the primitives to *combinators* (constants) — all this with the aid of the binary operation of application.

THEOREM 2.3. (The Reduction Theorem). *Any continuous function of k-variables can be written as*

$$f(x_0, x_1, \ldots, x_{k-1}) = u(x_0)(x_1)\ldots(x_{k-1}),$$

where u is a suitably chosen element of $P\omega$.

THEOREM 2.4. (The Combinator Theorem). *The* LAMBDA*-definable functions can be generated (from variables) by iterated application with the aid of these six constants:*

$$
\begin{aligned}
\mathsf{0} &= 0\\
\mathsf{suc} &= \lambda x.x+1\\
\mathsf{pred} &= \lambda x.x-1\\
\mathsf{cond} &= \lambda x \lambda y \lambda z.\, z \supset x, y
\end{aligned}
$$

$$K = \lambda x \lambda y. x$$
$$S = \lambda u \lambda v \lambda x. u(x)(v(x))$$

But the result that makes all this model building and combinatory logic *mathematically* interesting concerns the so-called *paradoxical combinator* defined by the equation:

$$(8) \quad Y = \lambda u. (\lambda x. u(x(x)))(\lambda x. u(x(x))).$$

THEOREM 2.5. (The First Recursion Theorem). *If u is the graph of a continuous function f, then $Y(u) = \mathrm{fix}(f)$, the least fixed point of f.*

There are two points to note here: the fixed point is LAMBDA-definable if f is; and Y defines a *continuous* operator. The word "recursion" is attached to the theorem because fixed points are employed to solve recursion equations. It would not be correct to call the Fixed-Point Theorem 1.4 the Recursion Theorem since it only shows that fixed points *exist* and not how they are *definable* in a language. The *Second* Recursion Theorem (in Kleene's terminology) is related, but it involves Gödel numbers as introduced in Section 3.

From this point on we see no need to distinguish continuous functions from elements of $P\omega$; a continuous function will be *identified* with its graph. Note that u is a graph iff $u = \lambda x. u(x)$, which is equivalent to 1.2 (iii). For this reason (functions are graphs) we propose the name *Graph Model* for this model of the λ-calculus. (There is more to LAMBDA than just λ, however.)

The identification of functions with graphs entails that the function *space* of all continuous functions from $P\omega$ into $P\omega$ is to be identified (one-one) with the subspace

$$\mathbf{FUN} = \{u \mid u = \lambda x. u(x)\} \subseteq P\omega.$$

The identification is *topological* in that the subspace topology agree with the product topology on the function space. This is the topology of pointwise convergence and is closely connected with the lattice structure on the function space which is also defined pointwise (that is, argumentwise). In the notation of λ-abstraction we can express this as the extension of the axiom of extensionality called (ξ^*) in Table 1 of the Introduction. The laws in Table 1 are not the only ones valid in the model, however. We may also note such argumentwise distributive laws as:

$$(9) \quad (f \cup g)(x) = f(x) \cup g(x)$$

$$(10) \quad (\lambda x. \tau) \cup (\lambda x. \sigma) = \lambda x. (\tau \cup \sigma)$$

(11) $(f \cap g)(x) = f(x) \cap g(x)$

(12) $(\lambda x.\tau) \cap (\lambda x.\sigma) = \lambda x.(\tau \cap \sigma)$

In the above f and g must be graphs. It is also true that if $\mathscr{F} \subseteq \mathsf{P}\omega$, then

(13) $\bigcup\{f \mid f \in \mathscr{F}\}(x) = \bigcup\{f(x) \mid f \in \mathscr{F}\}$,

but the same does not hold for $\bigcap$.

We state now a sequence of minor results which show why some simple functions and constants are LAMBDA-definable.

(14) $\bot = (\lambda x.x(x))(\lambda x.x(x))$

(15) $x \cup y = (\lambda z.0) \supset x,y$ (HINT: $0,1 \in \lambda z.0$)

(16) $\top = \mathsf{Y}(\lambda x.0 \cup (x+1))$

(17) $x \cap y = \mathsf{Y}(\lambda f \lambda x \lambda y.x \supset (y \supset 0,\bot), f(x-1)(y-1)+1)(x)(y)$

The elements $\bot$ and $\top$ are graphs, by the way, and we can characterize them as the only fixed points of the combinator K:

(18) $a = \lambda x.a$ iff $a = \bot$ or $a = \top$.

Next we use the notation $\langle x_0, x_1, \ldots, x_{n-1} \rangle$ for the function $\hat{p}$ where $p : \omega \to \mathsf{P}\omega$ is defined by:

$$p(i) = x_i \quad, \text{ if } i < n;$$
$$= \bot \quad, \text{ if } i \geq n.$$

(19) $\langle\rangle = \bot$

(20) $\langle x \rangle = \lambda z.z \supset x, \bot$

(21) $\langle x,y \rangle = \lambda z.z \supset x, (z-1 \supset y, \bot)$

(22) $\langle x_0, x_1, \ldots, x_n \rangle = \lambda z.z \supset x_0, \langle x_1, \ldots, x_n \rangle (z-1)$

Obviously we should formalize the subscript notation so that $u_x = u(x)$; then we find:

(23) $\langle x_0, x_1, \ldots, x_{n-1} \rangle_i = x_i \quad, \text{ if } i < n;$
$= \bot \quad, \text{ if } i \geq n.$

This gives us the method of LAMBDA-defining *finite* sequences (in a quite natural way), and the next step is to consider *infinite* sequences. But these are just the functions $\hat{p}$ where $p : \omega \to \mathcal{P}\omega$ is arbitrary. What we need then is a condition expressible in the language equivalent to saying $u = \hat{p}$ for some p. This is the same as

$$u = \lambda x. \bigcup \{u_i | i \in x\},$$

but the $\bigcup$- and set-notation is not part of LAMBDA. We are forced into a recursive definition:

(24) $\quad \$ = \mathsf{Y}(\lambda s \lambda u \lambda z. z \supset u_0, s(\lambda t. u_{t+1})(z-1))$

This equation generalises (22) and we have:

(25) $\quad \$(u) = \lambda x. \bigcup \{u_i | i \in x\}$

Thus the combinator \$ "revalues" an element as a distributive function. This suggests introducing the λ-notation for such functions by the equation:

(26) $\quad \lambda n \in \omega . \tau = \$(\lambda z. \tau[z/n])$

With all these conventions LAMBDA-notation becomes very much like ordinary mathematical notation without too much strain.

Suppose that f is any continuous function and $a \in \mathcal{P}\omega$. We can define $p : \omega \to \mathcal{P}\omega$ in the ordinary way by *primitive recursion* where:

$$p(0) = a \; ;$$

$$p(n+1) = f(n)(p(n)).$$

The question is: can we give a LAMBDA-definition for $\hat{p}$ (in terms of f and a as constants, say)? The answer is clear, for we can prove:

(27) $\quad \hat{p} = \mathsf{Y}(\lambda u \lambda n \in \omega . n \supset a, f(n-1)(u(n-1)))$

This already shows that a large part of the definitions of recursion theory can be given in this special language. Of course, *simultaneous* (primitive) recursions can be transcribed into LAMBDA with the aid of the ordered tuples of (22), (23) above. But we can go further and connect with partial (and general) recursive functions. We state first a definition.

<u>DEFINITION</u>. A continuous function f of k-variables is *computable* iff the relationship

$$m \in f(e_{n_0})(e_{n_1}) \ldots (e_{n_{k-1}})$$

is recursively enumerable in the integer variables $m, n_0, n_1, \ldots, n_{k-1}$.

If q is a *partial* recursive function in the usual sense, then we can regard it as a mapping $q : \omega \to \omega \cup \{\bot\}$, where $q(n) = \bot$ means that q is undefined at n. Saying that q is partial *recursive* is just to say that $m \in q(n)$ is r.e. as a relation in n and m. It is easy to see that this is in turn equivalent to the recursive enumerability of the relationship $m \in \hat{q}(e_n)$; and so our definition is formally a generalization of the usual one. But it is also intuitively reasonable. To "compute" $y = f(x)$, in the one-variable case, we proceed by enumeration. First we begin the enumeration of all finite subsets $e_n \subseteq x$. For each of these f starts up an enumeration of the set $f(e_n)$; so we sit back and observe which $m \in f(e_n)$ by enumeration. The totality of all such m for all $e_n \subseteq x$ forms in the end the set y.

THEOREM 2.6. (The Definability Theorem) *For a k-ary continuous function f, the following are equivalent:*

(i) f *is computable;*

(ii) $\lambda x_0 \lambda x_1 \ldots \lambda x_{k-1} . f(x_0)(x_1)\ldots(x_{k-1})$ *as a set is r.e.;*

(iii) $\lambda x_0 \lambda x_1 \ldots \lambda x_{k-1} . f(x_0)(x_1)\ldots(x_{k-1})$ *is* LAMBDA-*definable.*

As a hint for the proof we may note that the method of equation (27) shows that all primitive recursive functions p have the corresponding $\hat{p}$ LAMBDA-definable. Next we remark that a non-empty r.e. set is the range of a primitive recursive function; but the range of p is $\hat{p}(\top)$, which is clearly LAMBDA-definable. That any LAMBDA-definable set (graph) is r.e. is obvious from the definition of the language itself. More details are given in the appendix.

We may draw some interesting conclusions from the Definability Theorem. In the first place, we see that the countable collection $\mathrm{RE} \subseteq P\omega$ of r.e. sets is *closed* under application and LAMBDA definability. Indeed it forms a model for the λ-calculus (axioms (α), (β), (ξ^*) at least) and it also contains the arithmetical combinators. (Clearly there will be *many* intermediate submodels.) In the second place, we can see now how very easy it is to interpret λ-calculus in ordinary arithmetical recursion theory by means of quite elementary operations on r.e. sets. Thus the equivalence of λ-definability with partial recursiveness seems not to be all that good a piece of evidence for Church's Thesis. In his 1936 paper (a footnote on p. 346) Church says about λ-definability:

> The fact, however, that two such widely different and (in the opinion of the author) equally natural definitions of effective calculability turn out to be equivalent adds to the strength of the reasons adduced below for believing that they constitute as general a characterization of this notion as is consistent with the usual intuitive understanding of it.

The point never struck the present author as an especially telling one, and the reduction of λ-calculus to r.e. theory shows that the divergence between the theories is not at all wide. Of course it is a pleasan surprise to see how many complicated things can be defined in *pure* λ-calculus (without arithmetical combinators), but this fact cuts the wrong way as evidence for the thesis (we want stronger theories, not weaker ones). Post systems (or even first-order theories) are much better to mention in this connection, since they are obviously *more* inclusive in giving enumerations than Turing machines or Herbrand-Gödel recursion equations. But the equivalence proofs are all so easy! What one would like to see is a "natural" definition where the equivalence with r.e. is not just a mechanical exercise involving a few tricks of coding.

In the course of the development in this section we have stated many equations which are not found in Table 1, and which involve new combinators. In conclusion we would like to mention an equation about Y which holds in the model, which can be stated in pure λ-calculus, and which cannot be proved by ordinary reduction (though we shall not try to justify this last statement here). In order to shorten the calculations, we note from definition (8) that $Y(u) = Y(\lambda y.u(y))$; so by 2.5 this also equals $u(Y(u))$.

(28) $$Y(\lambda f \lambda x.g(x)(f(x))) = \lambda x.Y(g(x)).$$

Call the left side f' and the right f''. Now

$$f'' = \lambda x.g(x)(Y(g(x))) = \lambda x.g(x)(f''(x))$$

thus $f' \subseteq f''$, because f' is a least fixed point. On the other hand $f' = \lambda x.g(x)(f'(x))$, so $f'(x) = g(x)(f'(x))$. Thus $f''(x) \subseteq f'(x)$, because $f''(x)$ is a least fixed point. As this holds for all x, we see that $f'' \subseteq f'$; and so they are equal. There must be many other such equations.

3. ENUMERATION AND DEGREES.

A great advantage of the combinators from the formal point of view is that (bound) variables are eliminated in favour of "algebraic" combinations. The disadvantage is that the algebra is not all that pretty, as the combinations tend to get rather long and general laws are rather few. Nevertheless as a technical device it is mildly remarkable that we can have a notation for all r.e. sets requiring so few primitives. In the model defined here the reduction to one combinator rests on a lemma about conditionals:

(1) $\mathrm{cond}(x)(y)(\mathrm{cond}(x)(y)) = y$

Recall that cond (or $\supset$) is a test for zero, so that:

(2) $\mathrm{cond}(x)(y)(0) = x$

This suggests that we lump all combinators of 2.4 into this one:

(3) $\mathrm{G} = \mathrm{cond}(\langle \mathrm{suc},\mathrm{pred},\mathrm{cond},\mathrm{K},\mathrm{S}\rangle)(0)$

We can then readily prove:

THEOREM 3.1. (The Generator Theorem). *All* LAMBDA-*definable elements can be obtained from* G *by iterated application.*

A distributive function f is said to be *total* iff $f(n) \in \omega$ for all $n \in \omega$. As they come from obvious primitive recursive functions, we do not stop to write out LAMBDA-definitions of these three total functions:

(4) $\mathrm{apply} = \lambda n\in\omega.\lambda m\in\omega.(n,m)+1$

(5) $\mathrm{op}((n,m)) = n$

(6) $\mathrm{arg}((n,m)) = m$

The point of these auxiliary combinators concerns our Gödel numbering of the r.e. sets. The number 0 will correspond to the generator G; while $(n,m)+1$ will correspond to the application of the nth set to the mth. This is formalized in the combinator val which is defined as the least fixed point of the equation:

(7) $\mathrm{val} = \lambda k\in\omega.k\supset\mathrm{G},\mathrm{val}(\mathrm{op}(k-1))(\mathrm{val}(\mathrm{arg}(k-1)))$

This function accomplishes the enumeration as follows:

THEOREM 3.2. (The Enumeration Theorem). *The combinator* val *enumerates the* LAMBDA-*definable elements in that* $\mathsf{RE} = \{\mathrm{val}(n) \mid n\in\omega\}$. *Further* :

(i) $\mathrm{val}(0) = \mathrm{G}$

(ii) $\mathrm{val}(\mathrm{apply}(n)(m)) = \mathrm{val}(n)(\mathrm{val}(m))$

As a principal application of the Enumeration Theorem we may mention the following: suppose u is *given* as LAMBDA-definable. We look at its definition and rewrite it in terms of combinators — eventually in terms of G alone. Then using 0 and apply we write down the name of an integer corresponding to the combination — say, n. By 3.2 we see that we have *effectively found* from the definition an integer such that val $n = u$. This remark can be strengthened by some numerology.

(8) $\mathrm{apply}(0)(0) = 1$ and $\mathrm{val}(1) = 0$;

(9) $\mathrm{apply}(0)(1) = 3$ and $\mathrm{val}(3) = \langle \mathrm{suc}, \ldots \rangle$;

(10) $\mathrm{apply}(3)(1) = 12$ and $\mathrm{val}(12) = \mathrm{suc}$.

Thus, define as the least fixed point:

(11) $\mathrm{num} = \lambda n \in \omega . n \supset 1, \mathrm{apply}(12)(\mathrm{num}(n-1))$,

and derive the equation for all $n \in \omega$:

(12) $\mathrm{val}(\mathrm{num}(n)) = n$.

We note that num is a primitive recursive (total) function. The combinator num allows us now to effectively find a LAMBDA-definition, corresponding to a given LAMBDA-definable element u, of an element v such that uniformly in the integer variable n we have $\mathrm{val}(v(n)) = u(n)$. Further, v is a primitive recursive (total) function. This is the technique involved in the proof of Kleene's well-known result:

<u>THEOREM 3.3. (The Second Recursion Theorem).</u> *Take a* LAMBDA-*definable element* v *such that:*

(i) $\mathrm{val}(v(n)) = \lambda m \in \omega . \mathrm{val}(n)(\mathrm{apply}(m)(\mathrm{num}(m)))$,

and then define a combinator by:

(ii) $\mathrm{rec} = \lambda n \in \omega . \mathrm{apply}(v(n))(\mathrm{num}(v(n)))$.

Then we have a primitive recursive function with this fixed-point property:

(iii) $\mathrm{val}(\mathrm{rec}(n)) = \mathrm{val}(n)(\mathrm{rec}(n))$.

Note that if u is LAMBDA-definable, then we find first an n such that $\mathrm{val}(n) = u$. Next we calculate $k = \mathrm{rec}(n)$. This effectively gives us an integer such that $\mathrm{val}(k) = u(k)$. Gödel numbers represent expressions (combinations in G), and val maps the numbers to the values denoted by the expressions in the model. The k just found thus represents an expression whose value is defined in terms of its *own* Gödel number. In recursion theory there are many applications of this result. Another familiar argument shows:

<u>THEOREM 3.4. (The Incompleteness Theorem).</u> *The set of integers* n *such that* $\mathrm{val}(n) = \bot$ *is not r.e.; hence, there can be no effectively given formal system for enumerating all true equations between* LAMBDA *terms.*

(A critic may sense here an application of Church's Thesis in stating the metatheoretic consequence of the non-result.) A few details of the proof can be given to see how the notation works. First let v be a (total) primitive recursive function such that:

$$\mathrm{val}(v(n)) = n \cap \mathrm{val}(n),$$

and note that:

$$n \cap \mathrm{val}(n) = \bot \text{ iff } n \notin \mathrm{val}(n).$$

Call the set in question in 3.4 the set b. If it were r.e., then so would be:

$$\{n \in \omega \mid v(n) \in b\} = (\lambda n \in \omega.\, v(n) \cap b \supset n, n)(\top).$$

That would mean having an integer k such that:

$$\mathrm{val}(k) = \{n \in \omega \mid v(n) \in b\}.$$

But then:

$$\begin{aligned} k \in \mathrm{val}(k) &\text{ iff } v(k) \in b \\ &\text{ iff } \mathrm{val}(v(k)) = \bot \\ &\text{ iff } k \notin \mathrm{val}(k), \end{aligned}$$

which gives us a contradiction. This is the usual diagonal argument.

The relationship $\mathrm{val}(n) = \mathrm{val}(m)$ means that the expressions with Gödel numbers n and m have the *same* value in the model. (This is not only not r.e. but is a complete Π^0_2-predicate.) A total mapping can be regarded as a *syntactical* transformation on expressions defined *via* Godel numbers. Such a mapping p is called *extensional* if it has the property:

$$\mathrm{val}(p(n)) = \mathrm{val}(p(m)) \textit{ whenever } \mathrm{val}(n) = \mathrm{val}(m).$$

The Myhill-Shepherdson Theorem shows that extensional, syntactical mappings really depend only on the *values* of the expressions. Precisely we have:

<u>THEOREM 3.5. (The Completeness Theorem for Definability).</u> *If a (total) extensional mapping* p *is* LAMBDA-*definable, then there is a* LAMBDA-*definable* q *such that* $\mathrm{val}(p(n)) = q(\mathrm{val}(n))$ *for all* $n \in \omega$.

Of course q is uniquely determined (because the values of q are given at least on the finite sets). Thus any attempt to define something *new* by means of some strange mapping on Gödel numbers is bound to fail as long as it is *effective* and *extensional*. The main part of the argument is concentrated on showing these mappings to be *continuous*; that is why q exists.

The preceding results indicate that the expected results on r.e. sets are forthcoming in a smooth and unified manner in this setting. Some knowledge of r.e. theory was presupposed, but analysis shows that the knowledge required is slight. The notion of primitive recursive functions should certainly be well understood together with standard examples. Partial functions need not be introduced separately since they are naturally incorporated into LAMBDA (the theory of multiple-valued functions). As a working definition of r.e. one can take either "empty or the range of a primitive recursive function" or, more uniformly, "a set of the form $\{m \mid \exists n. m+1 = p(n)\}$ where p is primitive recursive." A few obvious closure properties of r.e. sets should then be proved, and then an adequate foundation for the discussion of LAMBDA will have been provided. The point of introducing LAMBDA is that further closure properties are more easily expressed in a theory where equations can be variously interpreted as involving numbers, functions, functionals, etc., without becoming too heavily involved in intricate Gödel numbering and encodings. Another useful feature of the present theory concerns the ease with which we can introduce *relative recursiveness*.

As we have seen $\{\mathrm{val}(n) \mid n \in \omega\}$ is an enumeration of all r.e. sets. Suppose we add a new set a as a new *constant*. What are the sets enumerable in a? Answer: $\{\mathrm{val}(n)(a) \mid n \in \omega\}$, since in combinatory logic a parameter can always be factored out as an extra argument. Another way to put the point is this: for b to be *enumeration reducible* to a it is necessary and sufficient that $b = u(a)$ where $u \in \mathsf{RE}$. This is *word for word* the definition given by Rogers (1967) pp.146-7. What we have done is to put the theory of enumeration operators (Friedberg-Rogers and Myhill-Shepherdson) into a general setting in which the language LAMBDA not only provides definitions but also the basis of a calculus for demonstrating properties of the operators defined. The algebraic style of this language throws a little light on the notion of enumeration degree. In the first place we can identify the degree of an element with the set of all objects reducible to it (rather than just those equivalent to it) and write

$$\mathrm{Deg}(a) = \{u(a) \mid u \in \mathsf{RE}\}.$$

The set-theoretical inclusion is then the same as the partial ordering of degrees. What kind of a partially ordered set do we have?

THEOREM 3.6. (The Subalgebra Theorem). *The enumeration degrees are exactly the finitely generated combinatory subalgebras of* $P\omega$.

By "subalgebras" here we of course mean subsets containing G and closed under application (hence, they contain all of RE, the least subalgebra). Part of the assertion is that every *finitely* generated subalgebra has a *single* generator (under application). This fact is an easy extension of 3.1. Not very much seems to be known about enumeration degrees. *Joins* can obviously be formed using the pairing function $\langle x,y\rangle$ on sets. Each degree is a countable set; hence, it is trivial to obtain the existence of a sequence of degrees whose *infinite join* is not a degree (not finitely generated). The intersection of subalgebras is a subalgebra — but it may not be a degree even starting with degrees. There are no minimal degrees above RE, but there are minimal *pairs* of degrees. Also for a given degree there are only countably many degrees minimal over it; but the question of whether the partial ordering of enumeration degrees is dense seems still to be open.

Theorem 3.6 shows that the semilattice of enumeration degrees is naturally extendable to a *complete* lattice (the lattice of all subalgebras of $P\omega$), but whether there is anything interesting to say about this complete lattice from the point of view of structure is not at all clear. Rogers has shown ((1967)pp.151-3) that Turing degrees can be defined in terms of enumeration degrees by restricting to special elements. In our style of notation we would define the space:

$$\mathsf{TOT} = \{u \mid u = \$(u) \text{ and } \forall n\in\omega . u(n)\in\omega\},$$

the space of all graphs of total functions. Then the system $\{\mathrm{Deg}(u) \mid u\in \mathsf{TOT}\}$ is isomorphic to the system of Turing degrees. Now there are many other interesting subsets of $P\omega$. Whether the degree structure of these various subsets is worth investigation is a question whose answer awaits some new ideas.

Among the subsets of $P\omega$ with natural mathematical structure, we of course have FUN, which is a semigroup under $\circ = \lambda u \lambda v \lambda x . u(v(x))$. It is, however, a rather complicated semigroup. We introduce for its study three new combinators:

$$\text{(13)} \qquad \mathsf{R} = \lambda x . \langle 0, x\rangle ;$$

$$\text{(14)} \qquad \mathsf{L} = \lambda x . x_1(x_2) ;$$

$$\text{(15)} \qquad \overline{u} = \lambda x . x_0 \supset \langle 1, u, x_1\rangle , \overline{u(x_1)}(x_2) .$$

<u>THEOREM 3.7. (The Semigroup Theorem).</u> *The countable semigroup* RE∩FUN *of computable enumeration operators is finitely generated by* R, L *and* $\overline{G}$.

The proof rests on the verification of two equations which permit an application of 3.1:

$$\text{(16)} \qquad L \circ \overline{u} \circ R = \lambda x.u(x)$$

$$\text{(17)} \qquad \overline{u} \circ \overline{v} \circ R = u(v).$$

Certainly the word problem for RE∩FUN is unsolvable: indeed, not even recursively enumerable. Can the semigroup be generated by *two* generators by the way?

4. RETRACTS AND DATA TYPES

Data can be structured in many ways: ordered tuples, lists, arrays, trees, streams, and even operations and functions. The last point becomes clear if one thinks of *parameters*. We would normally hardly consider the pairing function $\lambda x \lambda y.\langle x,y\rangle$ as being in itself a piece of data. But if we treat the first variable as a parameter, then it can be specialized to a *fixed value*, say the element a, producing the function $\lambda y.\langle a,y\rangle$. This function is more likely to be the output of some process and *in itself* can be considered as a datum. It is rather like one *whole row* of a matrix. If we were to regard a two argument function f as being a matrix, then its ath row would be exactly $\lambda y.f(a)(y)$. If s were a selection function, then, for example, $\lambda y.f(s(y))(y)$ would represent the selection of one element out of each column of the matrix. This selection could be taken as a specialization of parameters in the operator $\lambda u \lambda v \lambda y.u(v(y))(y)$. We have not been very definite here about the exact nature of the fixed a, f, or s, or the range of the variable y or the range of values of the function f. The point is only to recall a few elements of structure and to suggest an abstract view of data going beyond the usual iterated arrays and trees.

What then is a *data type*? Answer: a type of data. That is to say a collection of data that have been grouped together for reasons of similarity of structure or perhaps mere convenience. Thus the collection may very well be a mixed bag, but more often than not canons of taste or demands of simplicity dictate an adherence to regularity. The grouping may be formed to eliminate irrelevant objects and focus the attention in other ways. It is frequently a matter of good organization that aids the understanding of complex definitions. In programming languages one of the major reasons for making an explicit declaration of a data type (that is, the restriction of certain variables to certain "modes") is that the computed objects of that type can enjoy a special *representation* in the machine that allows the manipulation of these objects *via* the chosen representation to be reasonably efficient. This is a very critical matter for good language design and good compiler writing. In this report, however, we cannot discuss the problems of representation, important as they may be. Our objective here is conceptual organization, and we wish to show how such ideas, in the language for computable functions used here, can find the proper expression.

Which are the data types that can be defined in LAMBDA? No final answer can be given since the number is infinite and inexhaustible.

From one point of view, however, there is only one: $P\omega$ itself. It is the universal type and all other types are subtypes of it; so $P\omega$ plays a primary role in this exposition. But in a way it is too big, or at least too complex, since each of its elements can be used in so many different ways. When we specify a subtype the intention is to restrict attention to a special use. But even then the various subtypes overlap, and so the same elements still get different uses. Style in writing definitions will usually make the differentiation clear though. The main innovation to be described in this section is the use of LAMBDA expressions to define types *as well as* elements. Certain expressions define *retracts* (or better: retraction mappings), and it is the *ranges* (or as we shall see: *sets of fixed points*) of such retracts that form the groupings into types. Thus LAMBDA provides a calculus of type definitions including *recursive* type definitions. Examples will be explained both here and in the following sections. Note that types as retracts turn out to be types as *lattices*, that is types of partial and many-valued objects. The problem of cutting these lattice types down to the perfect or complete objects is discussed in Section 6. Another view of types and functionality of mappings is presented in Section 7.

The notion of a retract comes from (analytic) topology, but it seems almost an accident that the idea can be applied in the present context. The word is employed not because there is some deep tie-up with topology but because it is short and rather descriptive. Three easy examples will motivate the general plan:

(1) $$\mathrm{fun} = \lambda u \lambda x . u(x);$$

(2) $$\mathrm{pair} = \lambda u . \langle u_0, u_1 \rangle;$$

(3) $$\mathrm{bool} = \lambda u . u \,\overline{\supset}\, 0, 1.$$

Here $\overline{\supset}$ is the *doubly strict* conditional defined by

(4) $$z \,\overline{\supset}\, x, y = z \supset (z \supset x, \top), (z \supset \top, y),$$

which has the property that if z is both zero and positive, then it takes the value $\top$ instead of the value $x \cup y$.

<u>DEFINITION.</u> An element $a \in P\omega$ is called a *retract* iff it satisfies the equation $a = a \circ a$.

Of course the $\circ$-notation is used for functional composition in the standard way:

(5) $$u \circ v = \lambda x . u(v(x)).$$

And it is quite simple to prove that each of the three combinators in (1)-(3) is a retract according to the definition. But what is the point?

Consider fun. No matter what $u \in P\omega$ we take, fun(u) is (the graph of) a *function*. And if u already is (the graph of) a function, then u = fun(u). That is to say, the *range* of fun is the same as the set of *fixed points* of fun is the same as the set of all (graphs of) functions. Any mapping a whose range and fixed-point set coincide satisfies $a = a \circ a$, and conversely. A retract is a mapping which "retracts" the whole space onto its range *and* which is the identity mapping on its range. That is the import of the equation $a = a \circ a$. Strictly speaking the range is the retract and the mapping is the *retraction*, but for us the mapping is more important. (Note, however, that distinct retracts can have the same range.) We let the mapping stand in for the range.

Thus the combinator fun represents in itself the concept of a *function* (continuous function on $P\omega$ into $P\omega$). Similarly pair represents the idea of a *pair* and bool the idea of being a boolean value as an element of $\{\bot, 0, 1, \top\}$, since we must think in the multiple-valued mode. What is curious (and, as we shall see, useful) is that all these retracts which are defining *subspaces* are at the same time *elements* of $P\omega$.

<u>DEFINITION.</u> If a is a retract, we write $u{:}a$ for $u = a(u)$ and $\lambda u{:}a.\tau$ for $\lambda u.\tau[a(u)/u]$.

Since retracts are *sets* in $P\omega$, we cannot use the ordinary membership symbol to signify that u *belongs* to the range of a; so we write $u{:}a$. The other notation with the λ-operator *restricts* a function to the range of a. For f to be so restricted simply means $f = f \circ a$. For the *range* of f to be *contained in* that of the retract a means $f = a \circ f$. These algebraic equations will be found to be quite handy. We are going to have a calculus of retracts *and* mappings between them involving many operators on retracts yet to be discovered. Before we turn to this calculus, we recall the well-known connection between lattices and fixed points.

<u>THEOREM 4.1. (The Lattice Theorem).</u> *The fixed points of any continuous function form a complete lattice (under $\subseteq$); while those of a retract form a continuous lattice.*

We note further that by The Embedding Theorem 1.6, it follows that any separable (by which we mean countably-based) continuous lattice is a retract of $P\omega$; hence, our universal space is indeed rich in retracts. A very odd point is that $a = a \circ a$ is a fixed-point equation itself ($\lambda u. u \circ u$ is obviously continuous). Thus the retraction mappings form a complete lattice. Is this a continuous lattice? (Ershov has proved it is not; see appendix for a sketch.) A related question is solved positively in the next section. Actually the ordering of retracts under $\subseteq$ does not seem to be all that interesting; a more algebraic ordering is given by:

DEFINITION. For retracts a and b we write $a \lll b$ for $a = a \circ b = b \circ a$.

The idea here should be clear : $a \lll b$ means that a is a retract of b. It is easy to prove the:

THEOREM 4.2. (The Partial Ordering Theorem). *The retracts are partially ordered by* $\lll$.

There do not seem to be any lattice properties of $\lll$ of a general nature. Note, however, that if retracts *commute*, $a \circ b = b \circ a$, then $a \circ b$ is the greatest lower bound under $\lll$ of a and b. Also if we have a sequence where both $a_n \lll a_{n+1}$ and $a_n \subseteq a_{n+1}$ for all $n \in \omega$, then $\bigcup \{a_n | n \in \omega\}$ is the upper bound for the a_n under $\lll$, as can easily be argued from the definition by continuity of $\circ$.

Certainly there is no "least" retract under $\lll$. One has $\perp = \perp \circ a$ (recall: $\perp = \lambda x. \perp$), but not $a \circ \perp = \perp$. This last equation means more simply that $a(\perp) = \perp$; that is, a is *strict*. For retracts strictness is thus equivalent to $\perp \lll a$, so we can say that there is a least strict retract. The combinator $\mathrm{I} = \lambda u. u$ clearly represents the largest retract (the *whole* space), and it is strict also. In a certain sense strictness can be assumed without loss of generality. For if a is not strict, let

$$b = \lambda x. \{n | a(x) \neq a(\perp)\}.$$

This function takes values in $\{\perp, \top\}$ and is continuous because $\{x \in \mathcal{P}\omega | a(x) \neq a(\perp)\}$ is *open*. Next define:

$$a^* = \lambda u. a(u) \cap b(u).$$

This is a strict retract whose range is homeomorphic (and lattice isomorphic) to that of a. Note, however, that the mapping from a to a^* is not continuous (or even monotonic).

To have a more uniform notation for retracts we shall often write nil, id, and seq for the combinators $\perp$, I, \$. Two further retracts of interest are

(6) $$\mathrm{open} = \lambda u. \{m | \exists e_n \subseteq e_m . n \in u\}$$

(7) $$\mathrm{int} = \lambda u. u \supset 0, \mathrm{int}(u-1) \supset u, u.$$

The range of open is lattice isomorphic to the lattice of open subsets of $P\omega$; definition (6) is not a LAMBDA-definition of the retract, but such can be given.

In (7) we intend int to be the least fixed point of the equation. By induction on the least element of u (if any) one proves that:

$$\begin{aligned} \mathrm{int}(u) &= \bot, \quad \textit{if } u = \bot; \\ &= u, \quad \textit{if } u \in \omega; \\ &= \top, \quad \textit{otherwise}. \end{aligned}$$

This retract wipes out the distinctions between multiple values, moving all above the singletons up to $\top$; its range thus has a very simple structure. The retract int clearly generalizes bool. The range of fun is homeomorphic to the space of all continuous functions from $P\omega$ into $P\omega$; the range of pair, to the space of all ordered pairs; the range of seq, to the space of all infinite sequences. A combination like $\lambda u.\mathrm{int} \circ \mathrm{seq}(u)$ is a retract whose range is homeomorphic to the space of infinite sequences of elements from the range of int.

We now wish to introduce some operators that provide systematic ways of forming new combinations of retracts. There are three principal ones:

(8) $\quad a \hookrightarrow b = \lambda u.b \circ u \circ a;$

(9) $\quad a \otimes b = \lambda u.\langle a(u_0), b(u_1)\rangle;$

(10) $\quad a \oplus b = \lambda u.u_0 \supset \langle 0, a(u_1)\rangle, \langle 1, b(u_1)\rangle.$

These equations clearly generalize (1) - (3). Before we explain our operators, note these three equations which hold for *arbitrary* $a, b, a', b' \in P\omega$:

(11) $\quad (a \hookrightarrow b) \circ (a' \hookrightarrow b') = (a' \circ a) \hookrightarrow (b \circ b');$

(12) $\quad (a \otimes b) \circ (a' \otimes b') = (a \circ a') \otimes (b \circ b');$

(13) $\quad (a \oplus b) \circ (a' \oplus b') = (a \circ a') \oplus (b \circ b').$

The reversal of order $(a' \circ a)$ on the right-hand side of (11) should be remarked. These equations will be used not only for properties of types (ranges of retracts) but also for the mappings between the types.

<u>THEOREM 4.3. (The Function Space Theorem).</u> *Suppose a, b, a', b', c are retracts. Then we have:*

(i) $\quad$ *$a \hookrightarrow b$ is a retract, and it is strict if b is;*

(ii) $\quad$ *$u : a \hookrightarrow b$ iff $u = \lambda x : a.u(x)$ and $\forall x : a.u(x) : b$;*

(iii) if $a \lessdot a'$ and $b \lessdot b'$, then $a \multimap b \lessdot a' \multimap b'$;

(iv) if $f: a \multimap b$ and $f': a' \multimap b'$, then $f \multimap f': (b \multimap a') \multimap (a \multimap b')$;

(v) if $f: a \multimap b$ and $f': b \multimap c$, then $f' \circ f: a \multimap c$.

Parts *(i)*, *(iii)*, *(iv)*, and *(v)* can be proved using equations (8) and (11) in an algebraic (formal) fashion. It is *(ii)* that tells us what it all means: the range of $a \multimap b$ consists exactly of those functions which are restricted to (the range of) a and which have values in b. So we can read $u: a \multimap b$ in the normal way: u is a (continuous) mapping from a into b. In technical jargon, we can say that the (strict) retracts and continuous functions form a *category*. In fact, it is equivalent to the category of separable continuous lattices and continuous maps. In this context, *(iv)* shows that $\multimap$ operates not only on spaces (retracts) but also on maps: it is a *functor* contravariant in the first argument and covariant in the second. Further categorical properties will emerge.

THEOREM 4.4. (The Product Theorem). *Suppose a, b, a', b' are retracts. Then we have:*

(i) $a \otimes b$ *is a retract, and it is strict if a and b are;*

(ii) $u : a \otimes b$ *iff* $u = \langle u_0, u_1 \rangle$ *and* $u_0 : a$ *and* $u_1 : b$;

(iii) *if* $a \lessdot a'$ *and* $b \lessdot b'$, *then* $a \otimes b \lessdot a' \otimes b'$;

(iv) *if* $f : a \multimap b$ *and* $f' : a' \multimap b'$, *then* $f \otimes f' : a \otimes a' \multimap b \otimes b'$.

Again the operator proves to be a functor, but what is stated in 4.4 is not quite enough for the standard identification of $\otimes$ as the categorical product. For this we need some additional combinators:

(14) $$\mathrm{fst} = \lambda u.u_0;$$

(15) $$\mathrm{snd} = \lambda u.u_1;$$

(16) $$\mathrm{diag} = \lambda u.\langle u, u \rangle.$$

Then we have these properties:

(17) $$\mathrm{fst} \circ (a \otimes b) : (a \otimes b) \multimap a;$$

(18) $$\mathrm{snd} \circ (a \otimes b) : (a \otimes b) \multimap b;$$

(19) $$\mathrm{diag} \circ a : a \multimap a \otimes a;$$

(20) $$\mathrm{fst} \circ (f \otimes f') = f \circ \mathrm{fst};$$

(21) $$\mathrm{snd} \circ (f \otimes f') = f' \circ \mathrm{snd}.$$

Here a and b are retracts and f and f' are functions. Now suppose

a, b, and c are retracts and $f : c \multimap a$ and $g : c \to b$. Let

$$h = (f \otimes g) \circ \mathrm{diag} \circ c.$$

we can readily prove that:

$$h : c \multimap (a \otimes b),$$

and

$$\mathrm{fst} \circ h = f \quad \text{and} \quad \mathrm{snd} \circ h = g.$$

Furthermore, h is the *unique* such function. It is this uniqueness and existence property of functions into $a \otimes b$ that identifies the construct as a product.

There are important connections between $\multimap$ and $\otimes$. To state these we require some additional combinators:

(22) $$\mathrm{eval} = \lambda u. u_0(u_1)$$

(23) $$\mathrm{curry} = \lambda u \lambda x \lambda y. u(\langle x, y \rangle)$$

If a, b, and c are retracts, the mapping properties are:

(24) $$\mathrm{eval} \circ ((b \multimap c) \otimes b) : ((b \multimap c) \otimes b) \multimap c$$

(25) $$\mathrm{curry} \circ ((a \otimes b) \multimap c) : ((a \otimes b) \multimap c) \multimap (a \multimap (b \multimap c))$$

Suppose next that $f : (a \otimes b) \multimap c$ and $g : a \multimap (b \multimap c)$. We find that

$$\mathrm{eval} \circ (\mathrm{curry}(f) \otimes b) = f$$

and

$$\mathrm{curry}(\mathrm{eval} \circ (g \otimes b)) = g.$$

This shows that our category of retracts is a *cartesian closed category*, which means roughly that product spaces and function spaces within the category interact harmoniously.

<u>THEOREM 4.5. (The Sum Theorem)</u>. *Suppose a, b, a', b' are retracts. Then we have*

(i) *$a \oplus b$ is a retract, and it is always strict;*

(ii) *$u : a \oplus b$ iff $u = \bot$ or $u = \top$ or*

$$u = \langle 0, u_1 \rangle \text{ and } u_1 : a \text{ or}$$
$$u = \langle 1, u_1 \rangle \text{ and } u_1 : b;$$

(iii) *if $a \lll a'$ and $b \lll b'$, then $a \oplus b \lll a' \oplus b'$;*

(iv) *if $f : a \multimap b$ and $f' : a' \multimap b'$, then $f \oplus f' : a \oplus a' \multimap b \oplus b'$.*

There are several combinators associated with $\oplus$:

(26) $$\mathrm{inleft} = \lambda x. \langle 0, x \rangle;$$

(27) $$\text{inright} = \lambda x.\langle 1,x\rangle ;$$

(28) $$\text{outleft} = \lambda u.u_0 \supset u_1,\bot ;$$

(29) $$\text{outright} = \lambda u.u_0 \supset \bot,u_1 ;$$

(30) $$\text{which} = \lambda u.u_0;$$

(31) $$\text{out} = \lambda u.u_1.$$

(The last two are the same as fst and snd, but they will be used differently.) We find:

(32) $$(a\oplus b)\circ\text{inleft}\circ a : a \hookrightarrow (a\oplus b);$$

(33) $$(a\oplus b)\circ\text{inright}\circ b : b \hookrightarrow (a\oplus b);$$

(34) $$a\circ\text{outleft}\circ(a\oplus b) : (a\oplus b) \hookrightarrow a;$$

(35) $$b\circ\text{outright}\circ(a\oplus b) : (a\oplus b) \hookrightarrow b;$$

(36) $$\text{which}\circ(a\oplus b) : (a\oplus b) \hookrightarrow \text{bool};$$

(37) $$a\circ\text{out}\circ(a\oplus a) : (a\oplus a) \hookrightarrow a;$$

where a and b are retracts. Most of these facts as they stand are trivial until one sets down the relations between all these maps; but there are too many to put them down here. Note, however, if a, b, and c are retracts and $f : a \hookrightarrow c$ and $g : b \hookrightarrow c$, then if we let

$$h = c\circ\text{out}\circ(f\oplus g),$$

we have:

$$h : (a\oplus b) \hookrightarrow c,$$

and

$$h\circ\text{inleft} = f \quad\text{and}\quad h\circ\text{inright} = g.$$

But, though h exists, it is *not* unique. So $a\oplus b$ is not the categorical sum (coproduct). The author does not know a neat categorical characterization of this operator.

There would be no difficulty in extending $\otimes$ and $\oplus$ to more factors by expanding the range of indices from 0,1 to 0,1,...,n-1. The explicit formulae need not be given; but if we write $a_0 \otimes a_1 \otimes \ldots \otimes a_{n-1}$, we intend this expanded meaning rather than the iterated binary product.

To understand sums and other facts about retracts, consider the least fixed point of this equation:

(38) $$\text{tree} = \text{nil} \oplus (\text{tree} \otimes \text{tree}).$$

To be certain that tree *is* a retract, we need a general theorem:

THEOREM 4.6. (The Limit Theorem). *Suppose F is a continuous function that maps retracts to retracts and let* $c = \mathsf{Y}(F)$. *Then c is also a retract. If in addition F maps strict retracts to strict retracts and is monotone in the sense that* $a \lll b$ *implies* $F(a) \lll F(b)$ *for all (strict) retracts a and b, then the range of c is homeomorphic to the inverse limit of the ranges of the strict retracts* $F^n(\bot)$ *for* $n \in \omega$.

This can be applied in the case of (38) where $F = \lambda z.\mathtt{nil} \oplus (z \otimes z)$. Thus we can analyze tree as an inverse limit. This approach has the great advantage over the earlier method of the author where limits were required in showing that tree *exists*. Here we use Y to give existence at once, and then apply 4.3 - 4.5 to figure out the nature of the retract.

In 4.6, the fact that c is a retract can be reasoned as follows: $\bot$ is a retract. Thus each $F^n(\bot)$ is a retract. We compute:

$$\begin{aligned} c \circ c &= \bigcup\{F^n(\bot)|n\in\omega\} \circ \bigcup\{F^n(\bot)|n\in\omega\} \\ &= \bigcup\{F^n(\bot) \circ F^n(\bot)|n\in\omega\} \qquad \text{(Note: same } n.\text{)} \\ &= \bigcup\{F^n(\bot)|n\in\omega\} = c. \end{aligned}$$

In case F is monotone and preserves strictness, then we can argue that each $F^n(\bot) \lll c$. The retracts $F^n(\bot)$ are the *projections* of c onto the terms of the limit. Of course $F^n(\bot) \lll F^m(\bot)$ if $n \le m$. The $u:c$ can be put into a one-one correspondence (homeomorphism, lattice isomorphism) with the infinite sequences $\langle v_0, v_1, \ldots, v_n, \ldots \rangle$, where $v_n : F^n(\bot)$ and $v_n = F^n(\bot)(v_{n+1})$. Indeed $v_n = F^n(\bot)(u)$ and $u = \bigcup\{v_n|n\in\omega\}$. This is exactly the inverse limit construction.

Retreating from generalities back to the example of tree, we can grant that it exists and is provably a retract. Two things in its range are $\bot$ and $\top$ by 4.5 *(ii)*, but they are not so interesting. Now $\bot$: nil, so by 4.5 *(ii)* we have $\langle 0 \rangle = \langle 0, \bot \rangle$: tree. Let us think of this as *the* atom. What else can we have? If x,y : tree, then $\langle x,y \rangle$: tree$\otimes$tree and so $\langle 1, \langle x,y \rangle \rangle$: tree. Thus (the range of) tree contains an atom and is closed under a binary operation. Note that the atomic and non-atomic trees are distinguished by which and that suitable constructor and destructor functions are definable on tree. But the space also contains infinite trees since we can solve for the least fixed point of:

$$t = \langle 1, \langle \langle 0 \rangle, t \rangle \rangle$$

and t : tree. (Why?) And there are many other examples of infinite elements in tree.

A point to stress in this construction is that tree being LAMBDA definable is *computable*, and there are many computable functions definable on or to (the range of) tree. All the "structural" functions, for example, are computable. These are functions which in other languages would be called isatom or construct or node, and they are all easily LAMBDA definable. Just as with $\oplus$, $\otimes$, $\multimap$ they are not explicit in the *notation*, but they are definable nevertheless. In the case of node, we could use finite sequences of Boolean values to pick out or name nodes. Thus solve for name = nil $\oplus$ bool $\otimes$ name , and then give a recursive definition of:

$$\text{node} : \text{name} \multimap (\text{tree} \multimap \text{tree}).$$

Any combination of retract preserving functors can be used in this game. For example:

(39) $$\text{lamb} = \text{int} \oplus (\text{lamb} \multimap \text{lamb})$$

This looks innocent, but the range of lamb would give a quite different and not unattractive model for the λ-calculus (plus arithmetic). What we do to investigate this model is to modify LAMBDA slightly by replacing the ternary conditional $z \supset x,y$ by a *quarternary* one $w \supset x,y,z$; otherwise the syntax of the language remains the same. The semantics, however, is a little more complex.

Let us use τ, σ, ρ, θ as syntactical variables for expressions in the modified language. The semantics is provided by a function $\mathcal{H}$ that maps the expressions of the language to their values in (the range of) lamb. To be completely rigorous we also have to confront the question of free and bound variables. For simplicity let us index the variables of the language by integers, and let us take the variables to be $v_0, v_1, v_2, \ldots, v_n, \ldots$. We cannot simply evaluate out an expression τ to its value $\mathcal{H}[\![\tau]\!]$ until we know the values of the free variables in τ. The values of these variables will be given by an "environment" t which can be construed as a *sequence* of values in lamb. We can restrict these environments to the retract:

(40) $$\text{env} = \lambda t.\text{lamb}\circ\text{seq}(t).$$

When t : env, then t_n : lamb is the value that the environment gives to the variable v_n. We also need to employ a transformation on environments as follows:

(41) $$t[x/n] = \lambda m\in\omega.\,\text{eq}(n)(m) \supset x, t_m.$$

Here eq is the primitive recursive function that is 0, if n,m are equal, and is 1, otherwise, for $n,m\in\omega$. The effect of $t[x/n]$ is to replace the nth term of the sequence t by the value of x, otherwise to leave the rest of the sequence unchanged. To correspond with our use of very simple variables we have selected a simple notion of environment: in the semantics of more general languages it is customary to regard an environment as a function from the set of variables into the domain of denotable values.

The correct way to evaluate a term τ given an environment t is to find $\mathcal{H}[\![\tau]\!](t)$. We use the brackets $[\![$ and $]\!]$ here simply as an aid to the eye in keeping the syntactical part separated from the rest. The environment enters as a function-argument in the usual way; thus we shall have:

(42) $\mathcal{H}[\![\tau]\!] : \mathrm{env} \rightarrow \mathrm{lamb}.$

(43) $\mathcal{H}[\![v_n]\!](t) = t_n$

$\mathcal{H}[\![0]\!](t) = \mathrm{inleft}(0)$

$\mathcal{H}[\![\tau+1]\!](t) = \mathrm{which}(\mathcal{H}[\![\tau]\!](t)) \,\bar{\supset}\, \mathrm{inleft}(\mathrm{out}(\mathcal{H}[\![\tau]\!](t))+1)),\bot$

$\mathcal{H}[\![\tau-1]\!](t) = \mathrm{which}(\mathcal{H}[\![\tau]\!](t)) \,\bar{\supset}\, \mathrm{inleft}(\mathrm{out}(\mathcal{H}[\![\tau]\!](t))-1)),\bot$

$\mathcal{H}[\![\theta\supset\tau,\sigma,\rho]\!](t) = \mathrm{lamb}(\mathrm{which}(\mathcal{H}[\![\theta]\!](t)) \supset$

$(\mathrm{out}(\mathcal{H}[\![\theta]\!](t)) \supset \mathcal{H}[\![\tau]\!](t),\mathcal{H}[\![\sigma]\!](t)),\mathcal{H}[\![\rho]\!](t))$

$\mathcal{H}[\![\tau(\sigma)]\!](t) = \mathrm{which}(\mathcal{H}[\![\tau]\!](t)) \,\bar{\supset}\, \bot,\mathrm{out}(\mathcal{H}[\![\tau]\!](t))(\mathcal{H}[\![\sigma]\!](t))$

$\mathcal{H}[\![\lambda v_n.\tau]\!](t) = \mathrm{inright}(\lambda x{:}\mathrm{lamb}.\mathcal{H}[\![\tau]\!](t[x/n]))$.

A good question is: why does $\mathcal{H}$ exist? The answer is: because of the Fixed-point Theorem.

If we rewrite the semantic equations $\mathcal{H}[\![\tau]\!](t) = (\ldots)$ in (43) by the equation $\mathcal{H}[\![\tau]\!] = \lambda t{:}\mathrm{env}(\ldots)$, then $\mathcal{H}$ is seen to be a function from expressions to values in lamb. As the range of lamb is contained in $P\omega$, we can say more broadly that $\mathcal{H} \in P\omega^{\mathrm{Exp}}$, where Exp is the syntactical set of expressions and the exponential notation designates the set of *all* functions from Exp into $P\omega$. This function set is a complete lattice because $P\omega$ is. Therefore if we read (43) as a definition by cases on Exp, then we can find $\mathcal{H}$ as a suitable fixed point in the complete lattice $P\omega^{\mathrm{Exp}}$. Indeed it is the fixed point of a continuous operator.

Actually we can regard Exp as being a *subset* of $P\omega$ to avoid dragging in other lattices. What we need is another recursive definition of a data type:

(44) $\mathrm{exp} = \mathrm{int} \oplus \mathrm{nil} \oplus \mathrm{exp} \oplus \mathrm{exp} \oplus (\mathrm{exp} \otimes \mathrm{exp} \otimes \mathrm{exp} \otimes \mathrm{exp})$

$\oplus (\mathrm{exp} \otimes \mathrm{exp}) \oplus (\mathrm{int} \otimes \mathrm{exp})$

Note that there are as many summands in (44) as there are clauses in (43). We can think of exp as giving the "abstract" syntax of the language. We use the integers to index the variables and the nil element to stand for the individual constant. Read (44) as saying that every expression is *either* a variable *or* a constant or the successor of an expression *or* the predecessor of an expression *or* the conditional formed from a tuple of expressions *or* the abstraction formed from a pair of a variable and an expression. We do not need in (44) to introduce special "symbols" for the successor, application, *etc.*, because the separation by cases given by the ⊕ operation is sufficient to make the distinctions. (That is why the syntax is "abstract".) The point is that for recursive definitions it does not matter how we make the distinctions as long as they can be made. From this new point of view, we could rewrite (43) so as to show:

$$\text{(45)} \qquad \mathcal{K} : \mathsf{exp} \hookrightarrow (\mathsf{env} \hookrightarrow \mathsf{lamb})$$

which is clearly more satisfactory — especially as it is now clear that $\mathcal{K}$ is *computable*. And this is a method that can be generalized to many other languages. The method also shows why it is useful to allow *function spaces* as particular data types.

Another example of this method can be illustrated, if the reader will recall the Gödel numbering of Section 3. It will be seen that there are similarities with the tree construction : instead of 0 and apply$(n)(m)$, tree uses $\langle 0\rangle$ and $\langle 1,\langle x,y\rangle\rangle$. Note, however, that Gödel numbers are *finite* while tree has *infinite* objects. But the infinite objects are always *limits* of finite objects, so there *are* connections. (We discuss this again in Section 6.) In particular recursive definitions on Godel numbers, like that of val, have analogues on tree. Here is the companion of equation (7) of Section 3:

$$\text{(46)} \quad \mathsf{vaal} = \lambda x{:}\mathsf{tree}.\mathsf{which}(x) \supset \mathsf{G},\ \mathsf{vaal}(\mathsf{fst}(\mathsf{out}(x)))(\mathsf{vaal}(\mathsf{snd}(\mathsf{out}(x))))$$

We have vaal : tree $\hookrightarrow$ id, where of course (46) is taken as defining vaal as the least fixed point. This is an example of a computable function between effectively given retracts. The LAMBDA-definable elements of $P\omega$ are the computable elements in the range of vaal.

We have discussed the category of retracts and continuous maps, but if they are all LAMBDA-definable then they fall within the countable model RE. Thus there is *another* category of effectively given retracts and effectively given continuous maps.

(Examples: tree, id, vaal, and all those retracts and maps generated by ⊕, ⊗, and ↪.) This category seems to deserve the status of a generalised recursion theory; though this is not to say that as yet very much is known about it. In fact, the proper formulation may require an enriched category rather than a restricted one. Thus instead of confining attention to the computable retracts and computable maps, it might be better to use the full category with all maps and to single out the computable ones (also maybe the finite ones) by special predicates. In effect we have avoided any methodological decisions by working in the universal space $P\omega$ and by *defining* a notion when required – if possible with the aid of LAMBDA. This makes it possible to give all the necessary definitions and to prove the theorems without at first having to worry about axiomatic problems.

5. CLOSURE OPERATIONS AND ALGEBRAIC LATTICES

Given any family of (finitary) operations on a set (say, ω) there is a closure operation defined on the subsets of that set obtained by forming the *least subset* including the given elements and *closed* under the operations. Examples are very familiar from algebra: the subgroup generated by a set of elements, the subspace spanned by a set of vectors, the convex hull of a set of geometric points. We simplify matters here by restricting attention to closures operating on sets in $P\omega$, but the idea is quite general. The main point about these "algebraic" closure operations – as distinguished from topological closure operations – is that they are *continuous*. Thus, in the case of subgroups, if an element belongs to the subgroup generated by some elements, then it also belongs to the subgroup generated by *finitely* many of them. In the context of $P\omega$ we can state the characteristic condition very simply.

DEFINITION. An element $a \in P\omega$ is called a *closure operation* iff it satisfies: $I \subseteq a = a \circ a$.

We see by definition that a closure operation is not only continuous, but it is also a retract. This is reasonable since the closure of the closure of a subset must be equal to the closure. To say of a function that $I \subseteq a$, means that $x \subseteq a(x)$ for all $x \in P\omega$. In other words, every set is contained in its closure. (Note that closures are opposite to the "projections", those retracts where $a \subseteq I$.) Among examples of closure operations we find I and $\top$; the first has the most closed sets (fixed points), the second has the least. (Note that $\top = \omega$ always is a fixed point of a closure operation; $\top = \lambda x.\top$ is thus the most trivial closure operation.) The examples fun, open, int of Section 4 are all closure operations (cf. equations (1), (6), (7) of the last section). We remarked that fun is a retract, but the reader should prove in addition:

(1) $$u \subseteq \lambda x.u(x),$$

for all $u \in P\omega$ (cf. Theorem 1.2). We note that this fact can be rewritten in the language of retracts as:

(2) $$I \subseteq I \multimap I,$$

the significance of which will emerge after we develop a bit of the theory of closure operations.

Unfortunately the natural definition of the retract bool does not yield a closure operation. In this section we adopt this modification:

$$\text{(3)} \qquad \text{boool} = \lambda u.\ u \supset 0, \tau + 1$$

The closed sets of boool are $\perp$, 0, $\tau + 1$, and τ. Note that with any closure operation a, the function value $a(x)$ is the least closed set (fixed point of a) including as a subset the given set x. Thus given any family $\mathcal{C} \subseteq P\omega$ of "closed" sets which is closed under the intersection of subfamilies, if we define

$$\text{(4)} \qquad a(x) = \bigcap\{y \in \mathcal{C} \mid x \subseteq y\},$$

then this will be a closure operation *provided* it is continuous. This remark makes it easy to check that certain functions are closure operations if we can spot easily the family $\mathcal{C}$ of fixed points.

Alas, the "natural" definition of ordered pairs (cf. equation (21) of Section 2) leads to projections rather than closures. Here we must choose another:

$$\text{(5)} \qquad [x,y] = \{2n \mid n \in x\} \cup \{2m+1 \mid m \in y\},$$

with these inverse functions:

$$\text{(6)} \qquad [u]_0 = \{n \mid 2n \in u\}$$

$$\text{(7)} \qquad [u]_1 = \{m \mid 2m+1 \in u\}.$$

We shall find that the main advantage of these equations lies in the obvious equation:

$$\text{(8)} \qquad u = [[u]_0, [u]_1],$$

which is not true for the other pairing functions. Of course we have:

$$\text{(9)} \qquad [[x,y]]_0 = x$$

$$\text{(10)} \qquad [[x,y]]_1 = y$$

We shall not extend the idea of these new functions to triples and sequences, though it is clear what to do.

Abstractly, an *algebraic lattice* is a complete lattice in which the isolated points are dense. An isolated (sometimes called: *compact*) point in a lattice is one that is not the limit (sup or lub) of any directed family of its proper subelements. This definition works in continuous lattices, but more generally it is better to say that if the isolated point is *contained in* a sup, then it is also contained in a finite subsup (a sup of a finite selection of elements out of the given sup). In the case of the lattice of subgroups of a group, the isolated ones are the finitely generated subgroups. The isolated points of $P\omega$ are the finite sets e_n. To say that isolated points are *dense* means that every element in the lattice is the sup of the isolated points it contains. The sequel to Theorem 4.1 for closure operations relates them to algebraic lattices.

THEOREM 5.1. (The Algebraic Lattice Theorem). *The fixed points of any closure operation form an algebraic lattice.*

The proof is very easy if one notes that the isolated points of $\{x \mid x = a(x)\}$, where a is a closure operation, are exactly the images $a(e_n)$ of the finite sets in $P\omega$. What makes Theorem 5.1 more interesting is the converse.

THEOREM 5.2. (The Representation Theorem for Algebraic Lattices). *Every algebraic lattice with a countable number of isolated points is isomorphic to the range of some closure operation.*

By Theorem 1.6 we know that the algebraic lattice is a retract, but a more direct argument makes the closure property clear. Thus, let D be the algebraic lattice with $\{d_n \mid n \in \omega\}$ as the set of all isolated points with the indicated enumeration. We shall use the square notation with symbols $\sqsubseteq$ and $\sqcup$ for the lattice ordering and sup. The desired closure operation is defined by:

$$a(x) = \{m \mid d_m \sqsubseteq \sqcup\{d_n \mid n \in x\}\}.$$

It is an easy exercise to show that from the definition of "isolated" it follows that a is continuous; and from density, it follows that D is in a one-one order preserving correspondence with the fixed points of a.

In the last section we introduced an algebra of retracts, much of which carries over to closure operations given the proper definitions. Without any change we can use Theorem 4.3 on function spaces, provided we check that the required retracts are closures.

THEOREM 5.3. (The Function-Space Theorem for Algebraic Lattices). *Suppose that a and b are closure operations, then so is $a \hookrightarrow b$.*

The proof comes down to showing that:

$$(11) \qquad u(x) \subseteq b(u(a(x))),$$

whenever a and b are closure operations. But this is easy by monotonicity. Note that (1) is needed.

For those interested in topology, one can give a construction of the isolated points of the function space which is much more direct than just taking the functions $b \circ e_n \circ a$, which on the face of it do not tell us too much. But we shall not need this explicit construction here.

The reason for changing the pairing functions is to be able to form products and sums of closure operations. In the case of products, the analogue of $\otimes$ is straight forward:

$$(12) \qquad a \boxtimes b = \lambda u.[a([u]_0), b([u]_1)] \; ;$$

while for sums using $a' = \lambda x. 0 \cup a(x-1)+1$ and similarly for b' we write:

$$a \boxplus b = \lambda u.([u_0]_0 \supset 0,0) \cup ([u]_1 \supset 1,1) \overline{\supset} [a'([u]_0),\bot],[\bot,b'([u]_1)].$$

We can then establish with the aid of (8) - (10):

<u>THEOREM 5.4. (The Product and Sum Theorem for Algebraic Lattices).</u> *Suppose that a and b are closure operations; then so are $a \boxtimes b$ and $a \boxplus b$. Analogues of the results in Theorems 4.4 and 4.5 carry over.*

Following the discussion in Section 4, we can also show that the closure operations form a cartesian closed category, which in some ways is better than the category of *all* retracts. What makes it better is the existence of a "universe".

Every continuous operation generates a closure operation by just closing up the sets under the continuous function (as a set operation). We can institutionalize this thought by means of this definition:

$$(14) \qquad \mathrm{V} = \lambda a \lambda x. \mathrm{Y}(\lambda y.\ x \cup a(y))$$

Clearly V is LAMBDA-definable, continuous, *etc.* A more understandable characterization would define $\mathrm{V}(a)(x)$ by this equation:

$$(15) \qquad \mathrm{V}(a)(x) = \bigcap \{y \mid x \subseteq y \text{ and } a(y) \subseteq y\}$$

These two definitions are easily seen to be equivalent. What is unexpected is the discovery (due in a different form to Peter Hancock and Per Martin-Löf) that V itself is a closure operation.

<u>THEOREM 5.5. (The Universe Theorem for Algebraic Lattices).</u> *The function V is a closure operation and its fixed points comprise the set of all closure operations.*

Thus to say a is a closure operation, write $a : \mathrm{V}$. To have a mapping on closure operations, write $f : \mathrm{V} \rightarrow \mathrm{V}$. Remark that 5.5 allows us to write V:V. It all seems rather circular, but it is quite consistent. The category of separable algebraic lattices "contains itself" - if we are careful to work through retracts of $P\omega$.

The proof of 5.5 requires a few steps. We note first that for all $x, a \in P\omega$:

$$(16) \qquad x \subseteq \mathrm{V}(a)(x)$$

Let $y = V(a)(x)$. This is the least y with $x \cup a(y) \subseteq y$. What is the least z with $y \cup a(z) \subseteq z$? The answer is of course y, which proves:

(17) $$V(a)(V(a)(x)) = V(a)(x).$$

Thus $V(a)$ is always a closure operation. If a is already a closure operation, then clearly $V(a)(x) = a(x)$. Therefore we have shown:

(18) $$a = V(a) \text{ iff } a \text{ is a closure operation.}$$

But then by (16) and (17) we have by (18):

(19) $$V(a) = V(V(a)).$$

From (16) by monotonicity we see:

(20) $$a(x) \subseteq a(V(a)(x)) \subseteq V(a)(x).$$

Hence by (1) we can derive:

(21) $$a \subseteq \lambda x.a(x) \subseteq \lambda x.V(a)(x) = V(a).$$

From (19) and (21) it follows that V itself is a closure operation.

The operation V forms the least closure operation containing a given element, and it shows that the lattice of closure operations is not only a retract of $P\omega$ but also an algebraic lattice.

Since we can now use V as a retract, the earlier results become formulas:

(23) $$(\lambda a{:}V.\ \lambda b{:}V.\ a \boxtimes b) : V \circ\!\!\rightarrow (V \circ\!\!\rightarrow V);$$

(24) $$(\lambda a{:}V.\ \lambda b{:}V.\ a \boxplus b) : V \circ\!\!\rightarrow (V \circ\!\!\rightarrow V);$$

We can also state such functorial properties as:

(25) $$(\lambda a{:}V.\ \lambda b{:}V.\ a \circ\!\!\rightarrow b) : V \circ\!\!\rightarrow (V \circ\!\!\rightarrow V).$$

Using this style of notation we have:

<u>THEOREM 5.6. (The Limit Theorem for Algebraic Lattices)</u>.
$(\lambda f : V \circ\!\!\rightarrow V.Y(f)) : (V \circ\!\!\rightarrow V) \circ\!\!\rightarrow V$

In words: if f is a mapping on closure operations, then its least fixed point is also a closure operation. The proof of course holds with *any* retract in place of V, but we are more interested in applications to V. For example, note that $V(\bot) = I$. Now let $f = \lambda a{:}V.a \circ\!\!\rightarrow a$. The least fixed point of this f is the limit of the sequence:

$$\bot,\ I,\ I \circ\!\!\rightarrow I,\ (I \circ\!\!\rightarrow I) \circ\!\!\rightarrow (I \circ\!\!\rightarrow I),\ ((I \circ\!\!\rightarrow I) \circ\!\!\rightarrow (I \circ\!\!\rightarrow I)) \circ\!\!\rightarrow ((I \circ\!\!\rightarrow I) \circ\!\!\rightarrow (I \circ\!\!\rightarrow I)), \ldots$$

and we see that all these retracts are *strict*. This means $Y(f)$ is non-trivial in that it has at least *two* fixed points (*viz.* $\bot$ and $\top$). But $d = Y(f)$ must be the least closure operation satisfying

$$d = d \multimap d, \tag{26}$$

and we have thus proved that there are *non-trivial* algebraic lattices isomorphic to their own function spaces. This construction (which rests on hardly more that (2), since we could take $d = \mathrm{Y}(\lambda a.\, \mathrm{I} \cup (a \multimap a)))$ is much quicker than the inverse limit construction originally found by the author to give λ-calculus models satisfying (η). There are many other fixed points of (26) besides this least closure operation, but their connection with inverse limits is not fully investigated.

We note in conclusion that most constructions by fixed points give algebraic lattices (like lamb in Section 4), and so we could just as well do them in V if we remember to use $\boxtimes$ and $\boxplus$. The one-point space is $\top$ (*not* nil), and so the connection with inverse limits *via* 4.6 is not as clear when non-strict functions are used. For many purposes, this may not make any difference.

6. SUBSETS AND THEIR CLASSIFICATION.

Retracts produce very special subsets of $P\omega$: a retract always has a nonempty range which forms a lattice under $\subseteq$. For example, the range of int is $\{\bot,\top\}\cup\omega$. We often wish to eliminate $\bot$ and $\top$; and with a retract like tree the situation is more complex, since combinations like $\langle 1,\langle\langle 1,\langle\bot,\langle 0\rangle\rangle\rangle,\top\rangle\rangle$ might require elimination. In these two cases the method is simple.

Consider these two functions:

(1) $$\mathrm{mid} = \lambda x : \mathrm{int}.\ x \supset 0,0$$

(2) $$\mathrm{perf} = \lambda u{:}\mathrm{tree}.\mathrm{which}(u) \supset 0,\Delta(\mathrm{perf}(\mathrm{fst}(\mathrm{out}(u))))(\mathrm{perf}(\mathrm{snd}(\mathrm{out}(u))))$$

where Δ is a special combinator:

(3) $$\Delta = \lambda x\lambda y.(x\supset(y\supset 0,\top),\top) \cup (y\supset(x\supset 0,\top),\top).$$

We find that $\omega = \{x{:}\mathrm{int}\,|\,\mathrm{mid}(x) = 0\}$. In the case of trees, note first this behaviour of Δ:

Δ	$\bot$	0	$\top$
$\bot$	$\bot$	$\bot$	$\top$
0	$\bot$	0	$\top$
$\top$	$\top$	$\top$	$\top$

The question is: what subset is $\{u{:}\mathrm{tree}\,|\,\mathrm{perf}(u) = 0\}$?

Now perf is defined recursively. We can see that

$$\mathrm{perf}(\bot) = \bot,\ \mathrm{perf}(\top) = \top,\ \mathrm{perf}(\langle 0\rangle) = 0,$$

and

$$\mathrm{perf}(\langle 1,\langle x,y\rangle\rangle) = \Delta(\mathrm{perf}(x))(\mathrm{perf}(y))$$

when x,y : tree. Every tree, aside from $\top$ or $\bot$, is either atomic or a pair of trees. The atomic tree is "perfect" (that is, $\mathrm{perf}(\langle 0\rangle) = 0$). A *finite* tree which does not contain $\bot$ or $\top$ is perfect — as we can see inductively using the table above for Δ. An infinite tree is never perfect: either some branch ends in $\top$ and perf maps it to $\top$, or $\top$ is never reached and perf maps it to $\bot$. Thus the subset in question is then seen to be the set of *finite* trees generated from the atom by pairing. This is clearly a desirable subset, and it is sorted out by a function with a simple recursive definition. The general question is: what subsets can be characterized by equations? The answer can be given by reference to the *topology* of $P\omega$.

<u>DEFINITION</u>. Let $\mathfrak{G}$ be the class of *open* subsets of $P\omega$, and $\mathfrak{F}$ be the class of *closed* subsets. Further let $\mathfrak{B}$ be the class of all (finite) *Boolean combinations* of open sets.

We recall from Section 1 that $U \in \mathfrak{G}$ just in case for all $x \in P\omega$, we have $x \in U$ if and only if some finite subset of x is in U. The class of open sets contains ϕ and $P\omega$ and is closed under finite intersection and arbitrary union; in fact, it can be generated by these two closure conditions from subsets of the special form $\{x \in P\omega | n \in x\}$ for the various $n \in \omega$. An open set is always *monotonic* (whenever $x \in U$ and $x \subseteq y$, then $y \in U$), so that every non-empty $U \in \mathfrak{G}$ has $\top \in U$.

Another characterization of openness can be given by continuous functions. Suppose $U \in \mathfrak{G}$. Define $f : P\omega \to \{\bot,\top\}$ so that

$$U = \{x | f(x) = \top\},$$

then f is continuous. Conversely, if such an f is continuous, then U is open. But if we do not assume the range of f is included in $\{\bot,\top\}$, this is not true. For the case of general functions we know that f is continuous if and only if $\{x | f(x) \in V\}$ is open for all open V. This defines continuity in terms of openness, but we can turn it the other way around:

<u>THEOREM 6.1. (The $\mathfrak{G}$ Theorem).</u> *The open subsets of $P\omega$ are exactly the sets of the form :*

$$\{x | f(x) \supseteq 0\},$$

where $f : P\omega \to P\omega$ is continuous.

We could have written $0 \in f(x)$ or the equation $f(x) \cap 0 = 0$ instead of $f(x) \supseteq 0$. Note that in case $f : P\omega \to \{\bot,\top\}$, then $f(x) \supseteq 0$ is equivalent to $f(x) = \top$. Also any other integer could have been used in place of 0.

We can say that $\{x | 0 \in x\}$ is the *typical* open set, and that every other open set can be obtained as an inverse image of the typical set by a continuous function. We shall extend this pattern to other classes, especially looking for equations. In the case of openness an inequality could also be used, giving as the typical set $\{x | x \neq \bot\}$. But since closed sets are just the complements of open sets, this remark gives us:

<u>THEOREM 6.2. (The $\mathfrak{F}$ Theorem).</u> *The closed subsets of $P\omega$ are exactly the sets of the form:*

$$\{x \mid f(x) = \bot\},$$

where $f : P\omega \to P\omega$ *is continuous.*

Aside from $\{x | x = \bot\}$, we could have used $\{x | x \subseteq a\}$ as the typical closed set where $a \in P\omega$ is any element whatsoever aside from $\top$. This $\top$ has, by the way, a special character. We note:

$$\{\top\} = \bigcap\{\{x | n \in x\} | n \in \omega\}.$$

Thus $\{\top\}$ is a *countable intersection* of open sets, otherwise called a $\mathfrak{G}_\delta$-set. There are of course many other $\mathfrak{G}_\delta$-sets, but $\{\top\}$ is the typical one:

<u>THEOREM 6.3. (The $\mathfrak{G}_\delta$ Theorem)</u>. *The countable intersections of open subsets of* $P\omega$ *are exactly the sets of the form:*

$$\{x \mid f(x) = \top\},$$

where $f : P\omega \to P\omega$ *is continuous.*

It may not be obvious that every $\mathfrak{G}_\delta$-set has this form. Certainly, as we have remarked, every $\mathfrak{G}$-set has this form. Thus if W is a $\mathfrak{G}_\delta$, we have:

$$W = \bigcap\{U_n | n \in \omega\}$$

and further

$$U_n = \{x \mid f_n(x) = \top\},$$

where the f_n are suitably chosen continuous functions. Define the function g by the equation:

$$g(x) = \{(n,m) | m \in f_n(x)\},$$

Clearly g is continuous, and we have:

$$W = \{x | g(x) = \top\},$$

as desired.

We let $\mathfrak{F} \dot{\cap} \mathfrak{G}$ denote the class of all sets of the form $C \cap U$, where $C \in \mathfrak{F}$ and $U \in \mathfrak{G}$. Similarly for $\mathfrak{F} \dot{\cap} \mathfrak{G}_\delta$. Now $\{x | x \subseteq 0\}$ is closed and $\{x | x \supseteq 0\}$ is open. Thus $\{0\} \in \mathfrak{F} \dot{\cap} \mathfrak{G}$. This set is typical.

<u>THEOREM 6.4. (The $\mathfrak{F} \dot{\cap} \mathfrak{G}$ Theorem).</u> *The sets that are intersections of closed sets with open sets are exactly the sets of the form:*

$$\{x \mid f(x) = 0\},$$

where $f : P\omega \to P\omega$ *is continuous.*

Again it may not be obvious that every $\mathfrak{F} \dot{\cap} \mathfrak{G}$ set has this form. We can write:

$$C = \{x \mid f(x) = \bot\},$$

and

$$U = \{x \mid g(x) = 0\},$$

where $C \in \mathfrak{F}$ and $U \in \mathfrak{G}$ and the continuous f and g are suitably chosen. Define

$$h(x) = \{2n+1 \mid n \in f(x)\} \cup \{2n \mid n \in g(x)\},$$

and remark that h is continuous. We have:

$$C \cap U = \{x \mid h(x) = 0\},$$

as desired.

It is easy to see that $\{e\} \in \mathfrak{F} \dot{\cap} \mathfrak{G}$ if e is finite, but in general $\{a\} \in \mathfrak{F} \dot{\cap} \mathfrak{G}_\delta$. In case a is infinite but not equal to $\top$ (say, $a = \{n \mid n > 0\} = \top+1$), then $\{a\}$ is typical in its class.

<u>THEOREM 6.5. (The $\mathfrak{F} \dot{\cap} \mathfrak{G}_\delta$ Theorem).</u> *The sets that are intersections of closed sets with countable intersections of open sets are exactly the sets of the form:*

$$\{x \mid f(x) = a\},$$

where $f : P\omega \to P\omega$ is continuous and a is a fixed infinite set not equal to $\top$.

Note that $(\mathfrak{F} \dot{\cap} \mathfrak{G})_\delta$ is the same class as $\mathfrak{F} \dot{\cap} \mathfrak{G}_\delta$, so we see by 6.4 that a good choice of a is $\lambda n \in \omega . 0$.

There is no single subset of $P\omega$ typical for $\mathfrak{B}$, which can be viewed as the *finite unions* of sets from the class $\mathfrak{F} \dot{\cap} \mathfrak{G}$.

<u>THEOREM 6.6. (The $\mathfrak{B}$ Theorem).</u> *The sets that are Boolean combinations of open sets are exactly the sets of the form:*

$$\{x \mid f(x) \in \mathscr{L}\},$$

where $f : P\omega \to P\omega$ is continuous and $\mathscr{L}$ is a finite set of finite elements of $P\omega$.

To see that every $\mathfrak{B}$ set has this form, suppose that

$$V = W_0 \cup W_1 \cup \ldots \cup W_{n-1},$$

where each $W_i \in \mathfrak{F} \dot{\cap} \mathfrak{G}$. We can write:

$$W_i = \{x \mid f_i(x) = 0\}$$

where $f_i : P\omega \to \{\bot, 0, \top\}$ is continuous. Then define:

$$g(x) = \{2i + j \mid j \in f_i(x) \cap \{0,1\},\ i < n\},$$

and note that g is continuous. Let:

$$\mathscr{L} = \{y \subseteq \{m \mid m < 2n\} \mid \exists i < n.\, 2i \in y, 2i+1 \notin y\}$$

we have:

$$V = \{x \mid g(x) \in \mathscr{G}\}$$

as desired.

<u>THEOREM 6.7. (The $\mathfrak{B}_\delta$ Theorem).</u> *The sets that are countable intersections of Boolean combinations of open sets are exactly the sets of the form:*

$$\{x \mid f(x) = g(x)\},$$

where $f, g : \mathrm{P}\omega \to \mathrm{P}\omega$ *are continuous.*

This is clearly the most interesting of these characterization theorems, because equations like $f(x) = g(x)$ turn up all the time and the collection is a very rich totality of subsets of $\mathrm{P}\omega$. It includes all the retracts, since they are of the form $\{x|x=a(x)\}$. And much more. That every such set in 6.7 is $\mathfrak{B}_\delta$ follows from these logical transformations:

$$\begin{aligned}\{x|f(x)=g(x)\} &= \{x \mid \forall n\in\omega[n\in f(x) \leftrightarrow n\in g(x)]\}\\ &= \bigcap_{n=0}^{\infty} (\{x|n\in f(x), n\in g(x)\} \cup \{x|n\notin f(x), n\notin g(x)\})\end{aligned}$$

That puts the set in the class $(\mathfrak{F} \dot{\cup} \mathfrak{G})_\delta \subseteq \mathfrak{B}_\delta$

On the other hand we can see that $(\mathfrak{F} \dot{\cup} \mathfrak{G})_\delta$ is exactly $\mathfrak{B}_\delta$. Because, in view of 6.6, $\mathfrak{B}_\sigma$, the class of *countable* unions of $\mathfrak{B}$-sets, is exactly $(\mathfrak{F} \dot{\cap} \mathfrak{G})_\sigma$. The remark we want to make then follows by taking complements.

Now let S be an arbitrary $\mathfrak{B}_\delta$-set. We can write:

$$S = \bigcap_{n=0}^{\infty} (\{x|f_n(x)=0\} \cup \{x|g_n(x)=0\}),$$

where $f_n, g_n : \mathrm{P}\omega \to \{\bot, 0, \top\}$ are continuous. Now let u, v be continuous functions which on $\{\bot, 0, \top\}$ realize these two tables:

u	$\bot$	0	$\top$
$\bot$	$\bot$	0	0
0	0	0	$0'$
$\top$	0	$0'$	$0'$

v	$\bot$	0	$\top$
$\bot$	0	0	$0'$
0	0	0	$0'$
$\top$	$0'$	$0'$	$\top$

where $0' = 0 \cup 1$. This is an exercise in many-valued logic, and we find for $x, y \in \{\bot, 0, \top\}$:

$$u(x)(y) = v(x)(y) \quad \text{iff} \quad x = 0 \quad \text{or} \quad y = 0$$

Thus define continuous functions f' and g' such

$$f' = \lambda x \lambda n\in\omega.\, u(f_n(x))(g_n(x))$$

$$g' = \lambda x \lambda n\in\omega.\, v(f_n(x))(g_n(x)),$$

and we find:

$$S = \{x \mid f'(x) = g'(x)\}$$

as desired.

This is as far as we can go with *equations*. More complicated sets can be defined using *quantifiers*. For example the Σ_1^1 or *analytic* sets can be put in the form:

$$\{x \mid \exists y.f(x)(y) = g(x)(y)\},$$

and their complements, the Π_1^1 sets, in the form

$$\{x \mid \forall y \, \exists z.h(x)(y)(z) = 0\},$$

with continuous f, g, h. For the three classes we then have as "typical" sets those shown in the Table 3.

It should be remarked that $\mathfrak{B}_\delta$ contains all the closed sets in the *Cantor Space* topology on $P\omega$ (that is, the topology obtained when it is regarded as the infinite product of *discrete* two-point spaces). Therefore the Σ_1^1 sets for the two topologies on $P\omega$ are the *same*. Hence, since we know for Cantor Space that $\Delta_1^1 = \Sigma_1^1 \cap \Pi_1^1$ is the class of *Borel Sets*, we can conclude that the two topologies on $P\omega$ have the *same* Borel sets. (That is, in both cases Δ_1^1 is the Boolean σ-algebra generated from the open sets.)

Returning now to the example involving trees mentioned at the beginning of this section, we see that the set of perfect (finite) trees can be written in the form:

$$\{x \mid x = \mathrm{tree}(x),\ \mathrm{perf}(x) = 0\} = \{x \mid \langle x, \mathrm{perf}(x)\rangle = \langle \mathrm{tree}(x), 0\rangle\},$$

thus it is a $\mathfrak{B}_\delta$-set. (Note that $\mathfrak{B}_\delta$ are obviously closed under finite intersection by the ordered pair method just illustrated; that they are closed under finite union is a little messier to make explicit, but the essential idea is contained in the proof of 6.7.)

As another example, we might wish to allow infinite trees but not the strange tree $\top$. Consider the following function:

(4) $\mathrm{top} = \lambda u{:}\mathrm{tree}.\mathrm{which}(u) \supset \bot, \mathrm{top}(\mathrm{fst}(\mathrm{out}(u))) \cup \mathrm{top}(\mathrm{snd}(\mathrm{out}(u)))$

We can show that $\mathrm{top} : P\omega \to \{\bot, \top\}$. For a tree u the equation $\mathrm{top}(u) = \bot$ means that it *does not* contain $\top$, or as we might say: it is *topless*. The topless trees form a closed subset of the subspace of trees. (An interesting retract is the function $\lambda u.\mathrm{tree}(u) \cup \mathrm{top}(u)$ whose range consists exactly of the topless trees plus *one* exceptional tree $\top$.) Such a closed subset of (the range of) a retract is a kind of *semilattice*. (We shall not introduce a precise definition here.) Every directed subset has a limit (least upper bound) and every pair with an

Classes	Typical Sets
$\mathfrak{G}$	$\{x \mid 0 \in x\}$
$\mathfrak{F}$	$\{\bot\}$
$\mathfrak{G}_\delta$	$\{\top\}$
$\mathfrak{F} \dot{\cap} \mathfrak{G}$	$\{0\}$
$\mathfrak{F} \dot{\cap} \mathfrak{G}_\delta$	$\{\top+1\}$
$\mathfrak{B}_\delta$	$\{u \mid u_0 = u_1\}$
Σ^1_1	$\{u \mid \exists y. u_0(y) = u_1(y)\}$
Π^1_1	$\{u \mid \forall y\, \exists z. u(y)(z) = 0\}$

Table 3. Classes and Typical Sets

upper bound has a least upper bound. But generally least upper bounds do not have to exist within the semilattice. The type of domains that interest us *become* continuous lattices with the *addition* of a top element ⊤ larger than all the other elements. The elimination of ⊤ is done with a function like top of our example. This is convincing evidence to the author that an independent theory of semilattices is quite unnecessary: they can all be *derived* from lattices. The problem is simply to define the top-cutting operation, then restriction to the "topless" elements is indicated by an equation (like top(u) = ⊥). In this way all the constructions are kept within the control of a smooth-running theory based on LAMBDA. This point seems to be important if one wants to keep track of which functions are computable.

An aspect of the problem of classification treated in this section which has not been given close enough attention is the explicitly constructive way of verifying the closure properties of the classes. Consider the class $\mathfrak{B}_\delta$ for example. Let B be the typical set as shown in Table 3. Then whatever $f \in P\omega$ we choose, the set

$$\{x \mid f(x) \in B\}$$

is a $\mathfrak{B}_\delta$-set and every such set has this form. Thus the f's *index* the elements of the class. Suppose $f,g \in P\omega$. What we should look for are two LAMBDA-definable combinators such that union(f)(g) and inter(f)(g) give the functions that index the union and intersection of the sets determined by f and g. That is we want:

$$\{x \mid \text{union}(f)(g)(x) \in B\} = \{x \mid f(x) \in B\} \cup \{x \mid g(x) \in B\}$$

It should be possible to extract the precise definition from the outline of the proofs given above, but in general this matter needs more investigation. There may very well be certain classes where such operations are not constructive, even though the classes are simply defined.

7. TOTAL FUNCTIONS AND FUNCTIONALITY.

There is an inevitable conflict between the concepts of *total* and *partial* functions: we desire the former, but it is the latter we usually get. Total functions are better because they are "well-defined" at all their arguments, but the rub is that there is no general way of deciding when a *definition* is going to be well defined in all its uses. In analysis we have singularities, and in recursion theory we have endless, non-finishing computations. In the present theory we have in effect evaded the question in two ways. First we have embraced the partial function as the norm. But secondly, and possibly confusingly, the multiple-valued functions are normal, total functions from $P\omega$ into $P\omega$. The point, of course, is that we are making a *model* of the partial functions in terms of ordinary mathematical functions. But note that the success of the model lies in *not* using arbitrary functions: it is only the continuous functions that correspond to the kind of partial functions we wanted to study. It would be a mistake to think of the variables in λ-calculus as ranging over arbitrary functions — and this mistake was made by both Church and Curry. The fixed-point operator Y shows that we must restrict attention to functions which do have fixed points. It is certainly the case that $P\omega$ is not the only model for the λ-calculus, but it is a very satisfactory model and is rich enough to illustrate what can and what cannot be done with partial functions.

Whatever the pleasures of partial functions (and the multiple-valued ones, too), the desire for total functions remains. Take the integers. We are more interested in ω than $\omega \cup \{\bot, \top\}$. Since the multiple values $\bot$ and $\top$ are but two in number, it is easy to avoid them. The problem becomes tiresome in considering functions, however. The lattice represented by the retract $\mathrm{int} \hookrightarrow \mathrm{int}$ is much too large, in that there are as many non-total functions in this domain as total ones. The aim of the present section is to introduce an interpretation of a theory of functionality in the model $P\omega$ that provides a convenient way of restricting attention to the functions (or other objects) that are total in the desired sense. The theory of functionality is rather like proposals of Curry, but not quite the same for important reasons as we will see.

In the theory of retracts of Sections 5 and 6, the plan of "restricting attention" was the very simple one of restricting to a *subset*. It was made notationally simple as the subsets in question could be parameterized by continuous functions. The retraction mappings stand in for their ranges. Even better, certain continuous functions act on these retractions as space-forming functors (such as $\oplus$ and $\leftrightarrow$), which gives greater notational simplicity because one language is able to serve for several tasks. When we pass to the theory of total functions, this same kind of simplicity is no longer possible owing to an increase in quantifier complexity in the necessary definitions. (This remark is made definite below.) Another point where there is some loss of simplicity concerns the representation of entities in $P\omega$: subsets will no longer be enough, since we will need *quotients* of subsets. This is not a very startling point. Many constructions are affected in a natural way *via* equivalence classes. An equivalence relation makes you blind to certain distinctions. It may be easier also to remain a bit blind than to search for the most beautiful representative of an equivalence class: there may be nothing to choose between several candidates, and it can cost too much effort to attempt a choice. Thus our first agreement is that for many purposes a *kind* of object can be taken as a set of equivalence classes for an equivalence relation on a subset of $P\omega$.

Because $P\omega$ is closed under the pairing function $\lambda x \lambda y.\langle x,y\rangle$, we shall construe relations on subsets of $P\omega$ *as* subsets of $P\omega$ all of whose elements are ordered pairs. That is a relation A satisfies this inclusion:

(1) $$A \subseteq \{\langle x,y\rangle \mid x,y \in P\omega\}.$$

<u>DEFINITION.</u> A (restricted) *equivalence relation* on $P\omega$ is a symmetric and transitive relation on $P\omega$.

Such relations are restricted because they are only reflexive on their domains — which are the same as their ranges — and these are the subsets with which the relations are concerned. We shall write $x \mathrel{A} y$ for $\langle x,y\rangle \in A$ and $x : A$ for $x \mathrel{A} x$. What we assume about these relations is the following:

(2) $$x \mathrel{A} y \text{ implies } y \mathrel{A} x.$$

(3) $$x \mathrel{A} y \text{ and } y \mathrel{A} z \text{ imply } x \mathrel{A} z.$$

In case a is a retract, we introduce an equivalence relation to correspond:

(4) $$E_a = \{\langle x,x\rangle \mid x{:}a\}.$$

This is the identity relation restricted to the range of a. Such relations (for obvious reasons) and many others satisfy an additional *intersection property*:

(5) $x \mathrm{A} y$ and $x \mathrm{A} z$ imply $x \mathrm{A} (y \cap z)$

We shall not generally assume (5) in this short discussion, but it is often convenient.

Each equivalence relation represents a *space*: the space of all its equivalence classes. Such spaces form a category more extensive than the category of retracts studied above. The familiar functors can be extended to this larger category by these definitions:

(6) $\mathrm{A} \to \mathrm{B} = \{\langle \lambda x.u(x), \lambda x.v(x)\rangle \mid u(x) \mathrm{B} v(y) \text{ whenever } x \mathrm{A} y\}$

(7) $\mathrm{A} \times \mathrm{B} = \{\langle\langle x,x'\rangle, \langle y,y'\rangle\rangle \mid x \mathrm{A} y \text{ and } x' \mathrm{B} y'\}$

(8) $\mathrm{A} + \mathrm{B} = \{\langle\langle 0,x\rangle, \langle 0,y\rangle\rangle \mid x\mathrm{A}y\} \cup \{\langle\langle 1,x'\rangle, \langle 1,y'\rangle\rangle \mid x'\mathrm{B}y'\}$

THEOREM 7.1. (The Closure Theorem). *If* A *and* B *are restricted equivalence relations, then so are* A → B, A × B *and* A + B. *We find:*

(i) $f : \mathrm{A} \to \mathrm{B}$ *iff* $f = \lambda x.f(x)$ *and whenever* $x \mathrm{A} y$, *then* $f(x) \mathrm{B} f(y)$, *in particular:*

(ii) *if* $f : \mathrm{A} \to \mathrm{B}$ *and* $x : \mathrm{A}$, *then* $f(x) : \mathrm{B}$; *furthermore,*

(iii) $u : \mathrm{A} \times \mathrm{B}$ *iff* $u = \langle u_0, u_1\rangle$ *and* $u_0 : \mathrm{A}$ *and* $u_1 : \mathrm{B}$;

(iv) $u : \mathrm{A} + \mathrm{B}$ *iff either* $u = \langle 0, u_1\rangle$ *and* $u_1 : \mathrm{A}$,
or $u = \langle 1, u_1\rangle$ *and* $u_1 : \mathrm{B}$.

It follows easily from 7.1 that the restricted equivalence relations form a cartesian closed category which – in distinction to the category of retracts – has disjoint sums (or *coproducts* as they are usually called in category theory). This result is probably a special case of a more general theorem. The point is that $P\omega$ itself is a space in a cartesian closed category (that of continuous lattices and continuous maps) and it contains as subspaces the Boolean space and especially its own function space and cartesian square. In this circumstance any such rich space must be such that its restricted equivalence relations again form a good category. Our construction is not strictly categorical in nature, as we have used the *elements* of $P\omega$ and have relied on being able to form arbitrary subsets (arbitrary relations). But a more abstract formulation must be possible. The connection with the category of retracts is indicated in the next theorem.

THEOREM 7.2. (The Isomorphism Theorem). *If* a *and* b *are retracts we have the following isomorphisms and identities relating the spaces:*

(i) $\mathrm{E}_a \cong \{\langle x,y\rangle \mid a(x) = a(y)\}$;

(ii) $\mathrm{E}_{a \leftrightarrow b} \cong \mathrm{E}_a \to \mathrm{E}_b$;

(iii) $E_{a\otimes b} = E_a \times E_b$;

(iv) $E_{a\oplus b} = E_a + E_b \cup \{\langle \bot,\bot\rangle, \langle \top,\top\rangle\}$.

Part *(iv)* is not categorical in nature as it stands, but and *(iii)* indicate that E is a functor from the category of retracts into the category of equivalence relations that shows that the former is a full subcartesian-closed category of the latter. We cannot pursue the categorical questions here, but note that there are many subcategories that might be of interest; for example, the equivalence relations with the intersection property are closed under $\to$, $\times$, and $+$.

Returning to the question of total functions we introduce this notation:

(9) $$N = \{\langle n,n\rangle \mid n \in \omega\} .$$

This is the type of the integers *without* $\bot$ and $\top$, *i.e.* the total integers. We note that:

(10) $$N = \{u \mid u = \langle u_0,u_0\rangle,\ u_0 = \mathrm{int}(u_0),\ \mathrm{mid}(u_0) = 0\}.$$

Thus N is a $\mathfrak{B}_\delta$-set. What is $N \to N$? We see:

(11) $f : N \to N$ iff f : fun and $f(n) \in \omega$ whenever $n \in \omega$.

This $N \to N$ is indeed the type of *all* total functions from ω into ω. It can be shown that $N \to N$ is also a $\mathfrak{B}_\delta$-set. Good. But what is $(N \to N) \to N$? This is no longer a $\mathfrak{B}_\delta$-set, the best we can say is Π^1_1.

By 7.1 it corresponds to the type of all (extensional) *continuous* total functions from $N \to N$ into N. (The condition on A and B on the right side of (6) makes the concept of function of function embodied in $A \to B$ extensional, since the functions are meant to preserve the equivalence relations.)

A more precise discussion identifies $N \to N$ as a topological space, usually called the *Baire Space*. If we introduce the finite discrete spaces by:

(12) $$N_k = \{\langle n,n\rangle \mid n < k\},$$

then $N \to N_2$ can also be identified with a topological space, usually called the Cantor Space. In this identification we find at the next type, say either $(N \to N) \to N$ or $(N \to N_2) \to N_2$, that elements correspond to the usual notion of continuous function defined in topological terms. However, these higher type spaces are not at all conveniently taken as *topological* spaces. Certain of them can be identified as *limit spaces* according to the work of Hyland, and for these $\to$, $\times$, and $+$ have the natural interpretation. We cannot enter into these details here, but we can remark that the higher type spaces become ever more complicated. Thus $((N \to N) \to N) \to N$ is a Π^1_2-set and each $\to$ will add another quantifier to the definition. This is reasonable, because to say that a function is total is to say that *all* its values are well behaved. But if its domain is a complex space, this statement of totality is even more complex. Despite this complexity, however, it is possible to sort out what kind of mapping properties many functions have. We shall mention a few of the combinators.

THEOREM 7.3. (The Functionality Theorem). *The combinators* I, K, *and* S *enjoy the following functionality properties which hold for all equivalence relations* A, B, C:

(i) $\quad I : A \to A;$

(ii) $\quad K : A \to (B \to A);$

(iii) $\quad S : (A \to (B \to C)) \to ((A \to B) \to (A \to C)).$

Furthermore these combinators are uniquely determined by these properties.

Let us check that S satisfies *(iii)*. Suppose that:

$$f(A \to (B \to C))f'$$

We must show that:

$$S(f)((A \to B) \to (A \to C))S(f').$$

To this end suppose that:

$$g(A \to B)g'.$$

We must show that:

$$S(f)(g)(A \to C)S(f')(g').$$

To this end suppose that:

$$x \; A \; x'$$

We must show that:

$$S(f)(g)(x) \; C \; S(f')(g')(x').$$

Now by definition of the combinator S we have:

$$S(f)(g)(x) = f(x)(g(x))$$

$$S(f')(g')(x') = f'(x')(g'(x')).$$

By assumptions on g, g' and on x, x', we know:

$$g(x) \; B \; g'(x')$$

By assumptions on f, f' and on x, x', we know:

$$f(x)(B \to C)f'(x')$$

The desired conclusion now follows when we note such combinations as $S(f)$ and $S(f)(g)$ are indeed functions. (We are using 7.1*(i)* several times in this case.)

In the case of the converse, let us suppose by way of example that $k \in P\omega$ is such that

$$k : A \to (B \to A)$$

holds for all equivalence relations A and B. By specializing to, say, the identity relation we see that whatever $a \in P\omega$ we take, both k and $k(a)$ are functions. To establish that $k = K$ we need to show that the equation:

$$k(a)(b) = a$$

holds for all $a,b \in P\omega$. This is easy to prove, for we have only to set:

$$A = \{\langle a,a\rangle\} \quad \text{and}$$
$$B = \{\langle b,b\rangle\},$$

and the equation follows at once. Not all proofs are quite so easy, however.

In the case of the combinator S it is not strictly true to say that 7.3 *(iii)* determines it outright. The exact formulation is this: *if* $s \in P\omega$ is such that:

$$s(f) = s(\lambda x \lambda y. f(x)(y)) \quad \text{and}$$
$$s(f)(g) = s(f)(\lambda x. g(x))$$

for all $f,g \in P\omega$; and *if*

$$s : (A \to (B \to C)) \to ((A \to B) \to (A \to C))$$

for all A, B, C, *then* $s = S$. In other words we need to know that s converts its first two arguments into functions with the right number of places before we can say that its explicit functionality identifies as being the combinator S.

In Hindley, Lercher and Seldin (1972) they show that the functionality property:

(13) $\lambda f \lambda g. f \circ g : (B \to C) \to ((A \to B) \to (A \to C)),$

follows from (7.1) *(ii)* and (7.3) *(ii)* and *(iii)* in view of the identity:

(14) $\lambda f \lambda g. f \circ g = S(K(S))(K)$

(see pp. 77-80).

A more interesting result concerns the *iterators* defined as follows:

(15) $$Z_0 = \lambda f \lambda x.x$$

(16) $$Z_{n+1} = \lambda f \lambda x.f(Z_n(f)(x)).$$

In other words, $Z_n(f)(x) = f^n(x)$. These natural combinators can be typed very easily, but Gordon Plotkin has shown that the obvious typing actually characterizes them.

<u>THEOREM 7.4. (The Iterator Theorem).</u> *The combinators* Z_n *enjoy the following functionality property which holds for all equivalence relations* A:

(i) $$Z_n : (A \to A) \to (A \to A)$$

Further if any element $z \in P\omega$ *satisfies (i) for all* A, *then it must be one of the iterators, provided that* $z(f) = z(\lambda x.f(x))$ *holds for all* $f \in P\omega$.

That each of the Z_n satisfies 7.4 *(i)* is obvious. Suppose z were another such element. Then clearly:

$$z = \lambda f \lambda x.z(f)(x).$$

Suppose f and x are fixed for the moment. Let:

$$A = \{\langle f^n(x), f^n(x)\rangle \mid n \in \omega\},$$

where we can suppose in addition that:

$$f = \lambda x.f(x).$$

Then $f : A \to A$ is clear, and so $z(f) : A \to A$ also. But $x : A$, therefore $z(f)(x) = f^n(x)$, for some $n \in \omega$, because $z(f)(x) : A$. The trouble with this easy part of the argument is that the integer n *depends* on f and x. What we must show is that it is independent of f and x, then $z = Z_n$ will follow.

Plotkin's method for this case is to introduce some independent successor functions:

(17) $$\sigma_j = \lambda x.\{(j,k+1) \mid (j,k) \in x\}$$

Note that:

(18) $$\sigma_j^m((j',0)) = (j,m)\ , \text{ if } j = j';$$
$$= \bot\ , \text{ if } j \neq j'.$$

It then follows that:

(19) $$(\sigma_j \cup \sigma_{j'})^m((j,0) \cup (j',0)) = (j,m) \cup (j',m).$$

Having these identities, we return to the argument.

From what we saw before, given $j \in \omega$, there is an n_j such that:

$$z(\sigma_j)((j,0)) = \sigma_j^{n_j}(j,0) = (j,n_j).$$

Take any two $j,j' \in \omega$. We also know there is an $n \in \omega$ where:

$$z(\sigma_j \cup \sigma_{j'})((j,0) \cup (j',0)) = (j,n) \cup (j',n),$$

in view of (19). But since $\sigma_j \subseteq \sigma_j \cup \sigma_{j'}$, and $\sigma_{j'} \subseteq \sigma_j \cup \sigma_{j'}$, we have:

$$(j,n_j) \cup (j',n_{j'}) \subseteq (j,n) \cup (j',n).$$

It follows that

$$n_j = n = n_{j'}$$

and so they are all equal. This determines the fixed $n \in \omega$ we want.

Suppose that both f and x are finite sets in $P\omega$. Choose $j > max(f \cup x)$. Let A this time be the least equivalence relation such that:

$$f^m(x) \text{ A } f^m(x) \cup (j,m)$$

holds for all $m \in \omega$. We then check that:

$$\lambda x.f(x)(\text{A} \to \text{A})(\lambda x.f(x)) \cup \sigma_j$$

Therefore, we have:

$$z(\lambda x.f(x))(\text{A} \to \text{A})z((\lambda x.f(x)) \cup \sigma_j),$$

and since $x \text{ A } x \cup (j,0)$, we get:

$$z(\lambda x.f(x))(x) \text{ A } z((\lambda x.f(x)) \cup \sigma_j)(x \cup (j,0)).$$

Now there is an integer $m \in \omega$ such that:

$$z((\lambda x.f(x)) \cup \sigma_j)(x \cup (j,0)) = f^m(x) \cup \sigma_j^m((j,0))$$
$$= f^m(x) \cup (j,m),$$

where we have been able to separate f and σ because j is so large. But the right-hand side must contain $z(\sigma_j)(j,0) = (j,n)$. Thus m and our fixed n are the same. The other element

$$z(\lambda x.f(x))(x) = f^q(x)$$

for some $q \in \omega$. Thus we have:

$$f^q(x) \text{ A } f^n(x) \cup (j,n).$$

Again since j is so large, $(j,n) \notin f^q(x)$. Thus by our choice of A we must have $f^q(x) = f^n(x)$. This means then, since n is fixed, that for *all finite* f,x:

$$z(\lambda x.f(x))(x) = f^n(x).$$

But then by continuity this equation holds for *all* f, x. It follows now that $z = Z_n$, by the proviso of the theorem.

These results bring up many questions which we leave unanswered here. For example, which combinators (*i.e.* pure λ-terms) have functionality as in the examples above? and can we decide when a term is one such? In particular can the diagonal combinator $\lambda x.x(x)$ be typed? (The argument of Hindley, et al., p.81, is purely formal and does not apparently apply to the model.) What about terms in LAMBDA beyond the pure λ-calculus?

APPENDIX 1. PROOFS AND TECHNICAL REMARKS.

For Section 1. If we give the two-point space $\{\bot,\top\}$ the weak T_0-topology with just three open sets: $\emptyset$, $\{\top\}$, $\{\bot,\top\}$, we have what is called the Sierpinski Space and its infinite product $\{\bot,\top\}^\omega$ with the product topology is the same as $P\omega$. The finite sets $e_n \in P\omega$ correspond exactly to the usual basic open sets for the product. For those familiar with such notions, this well-known observation makes many of the facts mentioned in this section fairly obvious. From any point of view, 1.1 and the remarks in the following paragraph are simple exercises.

Proof of 1.2. Equation *(i)* as a functional equation comes down to

$$\{m \mid \exists e_n \subseteq x . m \in f(e_n)\} = f(x),$$

which is just another way of writing the definition of continuity. Thus it is indeed true for all x. Next, inclusion *(ii)* means that if $(n,m) \in u$, then $\exists e_k \subseteq e_n . (k,m) \in u$. Clearly all we need to do is take $k=n$. If we also want the converse inclusion to hold, then what we need is condition *(iii)*.

Proof of 1.3. Substitution is generalized composition of functions of many variables with all possible identifications and permutations of the variables; however, as we are able to define continuity by separating the variables, the argument reduces to a few special cases. The first trick is to take advantage of monotonicity. Thus, suppose $f(x,y)$ is continuous in each of its variables. What can we say of $f(x,x)$, a very special case of substitution? We calculate

$$\begin{aligned} f(x,x) &= \bigcup\{f(e_n,x) \mid e_n \subseteq x\} \\ &= \bigcup\{f(e_n,e_m) \mid e_n \subseteq x, e_m \subseteq x\}. \end{aligned}$$

Then if we think of $e_k = e_n \cup e_m$ and realize that $f(e_n,e_m) \subseteq f(e_k,e_k)$, we see that

$$f(x,x) = \bigcup\{f(e_k,e_k) \mid e_k \subseteq x\}.$$

This means that $f(x,x)$ is continuous in x. This same argument works if other variables are present, as in the passage from $f(x,y,z,w)$ to $f(x,x,z,w)$. When an identification of more than two variables is required, as from $f(x,y,z,w)$ to $f(x,x,x,x)$, the principle is just applied several times.

Finally to show that $f(g(x,y),h(y,x,y))$ is continuous, it is sufficient to show that $f(g(x,y),h(z,u,v))$ is continuous in each of its variables <u>separately</u>. By simply overlooking the remaining variables, this comes down to showing that $f(g(x))$ is continuous if f and g are.

But the proof for ordinary composition is very easy with the aid of the Characterization Theorem 1.1.

Proof of 1.4. This well-known fact holds for continuous functions on many kinds of chain-complete partial orderings; but $P\omega$ illustrates the idea well enough. Suppose f had a fixed point $x = f(x)$. Then since $\emptyset \subseteq x$ and f is monotonic, we see that $f(\emptyset) \subseteq f(x) = x$. But then again, $f(f(\emptyset)) \subseteq f(x) = x$; and so by induction, $f^n(\emptyset) \subseteq x$. This proves that $\mathrm{fix}(f) \subseteq x$; and thus if $\mathrm{fix}(f)$ is a fixed point, it must be the least one. To prove that it is a fixed point, we need a fact that will often be useful:

LEMMA. If $x_n \subseteq x_{n+1}$ for all n, and if f is continuous, then

$$f(\bigcup\{x_n | n \in \omega\}) = \bigcup\{f(x_n) | n \in \omega\}.$$

Proof. By monotonicity, the inclusion holds in one direction. Suppose $e_m \subseteq f(\bigcup\{x_n | n\in\omega\})$. Then by 1.1 we have $e_m \subseteq f(e_k)$ for some $e_k \subseteq \bigcup\{x_n | n\in\omega\}$. Because e_k is finite and the sequence is increasing, we can argue that $e_k \subseteq x_n$ for some n. But then $f(e_k) \subseteq f(x_n)$. This shows that $e_m \subseteq \bigcup\{f(x_n) | n\in\omega\}$ and proves the inclusion in the other direction. (Exercise: Does this property characterize continuous functions?)

Proof of 1.4 concluded. Noting that $f^n(\emptyset) \subseteq f^{n+1}(\emptyset)$ holds for all n, we can calculate:

$$\begin{aligned} f(\mathrm{fix}(f)) &= \bigcup\{f(f^n(\emptyset)) | n\in\omega\} \\ &= \bigcup\{f^{n+1}(\emptyset) | n\in\omega\}. \end{aligned}$$

But this is just $\mathrm{fix}(n)$, since the only term left out is $f^0(\emptyset) = \emptyset$.

Proof of 1.5. The function $\overline{f}$ is clearly well defined even when $y \in Y$ has a neighbourhood U where $X \cap U = \emptyset$: in that case $\overline{f}(y) = \omega$ by convention on the meaning of $\bigcap$ in $P\omega$. In case $x \in X$, it is obvious that $\overline{f}(x) \subseteq f(x)$. For the opposite inclusion, suppose that $m \in f(x)$. Because f is continuous and $\{z | m \in z\}$ is open in $P\omega$, there is an open subset V of X such that $x' \in V$ always implies that $m \in f(x')$. But X is a subspace of Y, so $V = X \cap U$ for some open subset U of Y. Thus we can see why $m \in \overline{f}(x)$. It remains to show that $\overline{f}$ is itself continuous.

We must show that the inverse image under $\overline{f}$ of every open subset of $P\omega$ is open in Y. But the open subsets of $P\omega$ are unions of finite intersections of sets of the form $\{z | m \in z\}$. Thus it is enough to show that $\{y | m \in \overline{f}(y)\}$ is always open in Y. But this set equals $\bigcup\{U | m \in \bigcap\{f(x) | x \in X \cap U\}\}$, which being a union of open sets is open. Note that what we have proved is that $\overline{f}$ is continuous no matter what function f is given; however, if f is not continuous, then $\overline{f}$ cannot be an extension of f.

For readers not as familiar with general topology we note that the idea of 1.5 can be turned into a definition. Suppose $\mathsf{X} \subseteq P\omega$ is a subset of $P\omega$. It becomes a subspace with the relative topology. What are the continuous functions $f : \mathsf{X} \to P\omega$? From 1.5 we see that a necessary and sufficient condition is that the $\overline{f} : P\omega \to P\omega$ be an extension of f. Thus for $x \in \mathsf{X}$ we can write the equation $f(x) = \overline{f}(x)$ as a biconditional:

$$m \in f(x) \text{ iff } \exists e_n \subseteq x\ \forall x' \in \mathsf{X}[e_n \subseteq x' \text{ implies } m \in f(x')],$$

which is to hold for all $m \in \omega$. This form of the definition of continuity on a subspace is more complicated than the original definition, because in general $e_n \notin \mathsf{X}$ and we cannot write $f(e_n)$.

Proof of 1.6. What T_0 means is that every point of X is uniquely determined by its neighbourhoods. Now $\varepsilon(x)$ just tells you the set of indices of the (basic) neighbourhoods of x. Thus it is clear that ε is one-one. To prove that it is continuous, we need only note:

$$\{x | n \in \varepsilon(x)\} = U_n ,$$

which is always open. To show that ε is an embedding, we must finally check that the *images* of the open sets U_n are open in $\varepsilon(X)$. This comes down to showing:

$$\varepsilon(U_n) = \varepsilon(X) \cap \{z | n \in z\},$$

which is clear.

For Section 2. Equation (1) defines a continuous function because it is a special case of 1.5, where we have been able to simplify the definition into cases because ω is a very elementary subset of $P\omega$. Equation (2) gives a continuous function since the definition makes $\hat{p}$ distributive, as remarked in the text for finite unions, but it is just as easy to show that $\hat{p}$ distributes over arbitrary unions. The difference between a continuous f and a distributive $\hat{p}$ is this: to find $m \in f(x)$ we need a finite subset $e_n \subseteq x$ with $m \in f(e_n)$; however, to find $m \in \hat{p}(x)$ we need only one element $n \in x$ with $m \in \hat{p}(\{n\}) = \hat{p}(n) = p(n)$. Continuous functions are generalizations of distributive functions. The generality is necessary. For example in equation (3) we see another function $x;y$ distributive in each of its variables; but take care: the function $x;x$ is not distributive in x — there is no closure under substitution. This is just one reason why continuous functions are better. Another good example comes from (4) if you compare the functions x, $x+x$, $x+x+x$, *etc.*

Equations (5) - (7) are very elementary. Note that $z \supset x,y$

is distributive in each of its variables. We could write: $z \supset x,y = \hat{p}(z)$, where $p(0) = x$ and $p(n+1) = y$, to show that it is distributive in z.

Proof of 2.1. If we did not use the λ-notation, then all LAMBDA-definable functions would be obtained by substitution from the first five (*cf.* Table 2). Since they are all seen to be continuous, the result would then follow by the Substitution Theorem 1.3. Bringing in λ-abstraction means that we have to combine 1.3 with this fact:

LEMMA. If $f(x,y,z,\ldots)$ is a continuous function of all its variables, then $\lambda x.f(x,y,z,\ldots)$ is a continuous function of the remaining variables.

Proof. It is enough to consider one extra variable. We compute from the definition of λ in Table 2 as follows:

$$\begin{aligned}\lambda x.f(x,y) &= \{(n,m) \mid m \in f(e_n,y)\}\\ &= \{(n,m) \mid \exists e_k \subseteq y.\, m \in f(e_n,e_k)\}\\ &= \bigcup\{\{(n,m) \mid m \in f(e_n,e_k)\} \mid e_k \subseteq y\}\\ &= \bigcup\{\lambda x.f(x,e_k) \mid e_k \subseteq y\}.\end{aligned}$$

Thus $\lambda x.f(x,y)$ is continuous in y.

Proof of 2.2. The reason behind this result is the restriction to continuous functions. Theorem 2.1 shows that we cannot violate the restriction by giving definitions in LAMBDA, and the Graph Theorem 1.2 shows that continuous functions correspond perfectly with their graphs.

The verification of (α) of Table 1 is obvious as the 'x' in '$\lambda x.\tau$' is a bound variable. (Care should be taken in making the proviso that 'x' is not otherwise free in τ.) The same would of course hold for any other pair of variables. We do not bother very much about alphabetic questions.

The verification of (β) is just a restatement of 1.2*(i)*. Let τ define a function f (of x). Then by definition $f(x) = \tau$ and $\lambda x.\tau = \mathrm{graph}(f)$. Also $\mathrm{fun}(u)(x)$ in the notation of 1.2 is the same as the binary operation $u(x)$ in the notation of LAMBDA. Thus in 1.2*(i)* if we apply both sides to y we get nothing else than (β).

Half of property (ξ) is already implied by (β): the implication from left to right. (Just apply both sides to x.) In the other direction, $\forall x.\tau = \sigma$ means that τ and σ define the *same* function of x; thus, the two graphs must be equal.

Remarks on other laws. The failure of (η) simply means that not every set in $P\omega$ *is* the graph of a function. Condition 1.2*(iii)* is equivalent to saying that $u = \lambda x.u(x)$, in other words, u is the graph of some function if and only if it is the graph of the function determined by u.

Law (μ) is the monotone property of application (in both variables); therefore, (μ) and (ξ) together imply (ξ*) from left to right. Suppose that $\forall x.\tau \subseteq \sigma$, then clearly:

$$\{(n,m) | m \in \tau[e_n/x]\} \subseteq \{(n,m) | m \in \sigma[e_n/x]\},$$

which gives (ξ*) from right to left.

There are, by the way, other laws valid in the model, as explained in the later results.

Proof of 2.3. This is a standard result combinatory logic. We have only to put:

$$u = \lambda x_0 \lambda x_1 \ldots \lambda x_{n-1}.f(x_0, x_1, \ldots, x_{n-1}).$$

That is, we use the iteration of the process of forming the graph of a continuous function. As each step (from the inside out) keeps everything continuous, we are sure that the equation of 2.3 will hold for iterated application.

Proof of 2.4. This can be found in almost any reference on combinatory logic or λ-conversion. The main idea is to eliminate the λ in favour of the combinators. The fact that we have a few other kinds of terms causes no problem if we introduce the corresponding combinators. The method of proof is to show, for any LAMBDA-term τ with free variables among $x_0, x_1, \ldots, x_{n-1}$, that there is a combination γ of combinators such that:

$$\tau = \gamma(x_0)(x_1) \ldots (x_{k-1}).$$

This can be done by induction on the complexity of τ.

Proof of 2.5. The well-known calculation shows that we have from (β):

$$Y(u) = (\lambda x.u(x(x)))(\lambda x.u(x(x)))$$
$$= u(Y(u)).$$

Thus $Y(u)$ *is* a fixed point of the function $u(x)$. What is needed is the proof to show that it is the least one.

Let $d = \lambda x.u(x(x))$ and let a be any other fixed point of $u(x)$. To show, as we must, that $d(d) \subseteq a$, it is enough to show that $e_\ell \subseteq d$ always implies $e_\ell(e_\ell) \subseteq a$; because by continuity we have:

$$d(d) = \bigcup\{e_\ell(e_\ell) \mid e_\ell \subseteq d\}.$$

By way of induction suppose that this implication holds for all $n < \ell$. Assume that $e_\ell \subseteq d$ and that $m \in e_\ell(e_\ell)$. We will want to use the induction hypothesis to show that $m \in a$. By the definition of application, there exists an integer n such that $(n,m) \in e_\ell$ and $e_n \subseteq e_\ell$. But $n \leq (n,m) < \ell$, and $e_n \subseteq d$. By the hypothesis, we have $e_n(e_n) \subseteq a$. Note that $(n,m) \in d$ also, and that d is defined by λ-abstraction; thus, $m \in d(e_n)$ by definition. By monotonicity $u(e_n(e_n)) \subseteq u(a) = a$; therefore $m \in a$. This shows that $e_\ell(e_\ell) \subseteq a$, and the inductive proof is complete.

Remark. Note that we did not actually use the Fixed-Point Theorem in the proof, but we did use rather special properties of the pairing (n,m) and the finite sets e_ℓ.

Equation (9) is proved easily from the definition of application; indeed $u(x)$ is distributive in u. Equation (10) is proved even more easily from the definition of λ-abstraction. For equation (11), we see that the inclusion from left to right would hold in general by monotonicity. In the other direction, suppose $m \in f(x) \cap g(x)$. Then for suitable k and ℓ we have $(k,m) \in f$ and $e_k \subseteq x$, also $(\ell,m) \in g$ and $e_\ell \subseteq x$. Let $e_n = e_k \cup e_\ell \subseteq x$. Because f and g are *graphs*, we can say $(n,m) \in f \cap g$; and thus $m \in (f \cap g)(x)$. This is the only point where we require the assumption on graphs. Equation (12) follows directly from the definition of abstraction. For equation (13), which generalizes (9), we can also argue directly from the definition of application. In the case of intersection is is easy to find u_n such that $0 \in u_n(\top)$ for all n, but $\bigcap\{u_n \mid n \in \omega\} = \bot$.

Equation (14) is obvious because the least fixed point of the identity function must be $\bot$. A less mysterious definition would be $\bot = 0 - 1$, but the chosen one is more "logical".

For (15) we note that by definition:

$$\lambda z.0 = \{(n,m) \mid m \in 0\} = \{(n,0) \mid n \in \omega\}$$

Because $0 = (0,0)$ and $1 = (1,0)$, we get the hint. Equation (16) makes use of $\cup$ for iteration. If $x = 0 \cup (x+1)$, then x must contain all integers; hence $x = \top$. The iteration for $\cap$ in (17) is more complex. The fundamental equation we need is:

$$x \cap y = x \supset (y \supset 0, \bot), ((x-1) \cap (y-1))+1.$$

This says to compute the intersection of two sets x and y, we first test whether $0 \in x$. If so, then test whether $0 \in y$. If so, then we know $0 \in x \cap y$. In the meantime we begin testing x for positive elements. If we could compute (by the same program) the intersection $(x-1) \cap (y-1)$, then we would get the positive elements of the intersection $x \cap y$ by adding one. This is a very slow program, but we can argue by induction that it gives us all the desired elements. Of course, $\cap$ is the least function satisfying this equation.

In the case of (18) it is clear that we have:

$$\lambda x.\bot = \{(n,m)|m \in \bot\} = \bot;$$

$$\lambda x.\top = \{(n,m)|m \in \top\} = \top;$$

because in the last every integer is a (number of a) pair. Suppose now that $a = \lambda x.a$ and $a \neq \top$. Let k be the least integer where $k \notin a$. Now $k = (n,m)$ for some n and m. If $m \in a$, then $(n,m) \in a = \lambda x.a$; hence $m \notin a$. But $m \leq k$ and k is minimal; therefore, $m = k$. But this is only possible if $k = m = n = 0$. Suppose further $a \neq \bot$ and that ℓ is the least integer where $\ell \in a$. Now $\ell = (i,j)$ with $j \in a$ and $j \leq \ell$. So $j = \ell$ and $\ell = j = i = 0$. This contradiction proves that $a = \bot$ or $a = \top$.

Equations (19) - (22) are definitions, and (23) is proved easily by induction on i. Equation (24) is also a definition. To prove (25) we note that $\{u_i | i \in x\}$ is continuous (even: distributive) in u and x. Thus, there is a continuous function $\mathrm{seq}(u)(x)$ giving this value. What is required is to prove that it is LAMBDA-definable. We see:

$$\begin{aligned}\mathrm{seq}(u)(x) &= \{n \in u_0 | 0 \in x\} \cup \{m \in \bigcup\{u_{i+1} | i+1 \in x\} | \exists k. k+1 \in x\}\\ &= x \supset u_0, \mathrm{seq}(\lambda t.u_{t+1})(x-1).\end{aligned}$$

that is, seq satisfies the fixed-point equation for \$. Thus $\$ \subseteq \mathrm{seq}$. To establish the other inclusion we argue by induction on i for:

$$\forall x,u[i \in x \Rightarrow u_i \subseteq \$(u)(x)]$$

This is easy by cases using what we are given about \$ in (24), and it implies that $\mathrm{seq} \subseteq \$$. Note that:

$$\lambda n{\in}\omega.\tau = \lambda n{\in}\omega.\sigma \text{ iff } \forall n{\in}\omega.\tau = \sigma.$$

For primitive recursive functions, even of several variables, there is no trouble in transcribing into LAMBDA-notation any standard definition — especially as we can use the abstraction operator $\lambda n{\in}\omega$. If we recall that every r.e. set a has the form:

$$a = \{m | \exists n.\ p(n) = m+1\},$$

where p is primitive recursive, we then see that $a = \hat{p}(\top)-1$. This means that every r.e. set is LAMBDA-definable.

Proof of 2.6. In case of a function of several variables, we remark:

$$\lambda x_0 \lambda x_1 \ldots \lambda x_{k-1}.f(x_0)(x_1)\ldots(x_{k-1}) = \{(n_0,(n_1,(\ldots,(n_{k-1},m)\ldots))) \mid m \in f(e_{n_0})(e_{n_1})\ldots(e_{n_{k-1}})\}$$

This makes the implication from (*i*) to (*ii*) obvious. Conversely, if a *graph* u is r.e., then from the definition of application we have:

$$m \in u(e_{n_0})(e_{n_1})\ldots(e_{n_k}) \text{ iff } (n_0,(n_1,(\ldots,(n_{k-1},m)\ldots))) \in u,$$

which is r.e. in m, n_0, $n_1,\ldots,n_{k-1}$. Therefore (*i*) and (*ii*) are equivalent.

We have already proved that (*ii*) implies (*iii*). For the converse we have only to show that all LAMBDA-definable sets are r.e. For this argument we could take advantage of the Combinator Theorem 2.4. Each of the six combinators are r.e., and there is no problem of showing that if u and x are r.e., then so is $u(x)$; because it is defined in such an elementary way with existential and bounded universal number quantifiers and with membership in u and in x occurring positively. Explicitly we have:

$$m \in u(x) \text{ iff } \exists e_n \subseteq x.(n,m) \in u$$

$$\text{iff } \exists n\, \forall m < n[[m \in e_n \text{ implies } m \in x] \text{ and } (n,m) \in u].$$

For Section 3. For the proof of (1) we distinguish cases. In case $x = y = \bot$, we note that $\mathrm{cond}(\bot)(\bot) = \bot$ and $\bot(\bot) = \bot$, so the equation checks in this case. Recall:

$$\mathrm{cond}(x)(y) = \lambda z.z \supset x\, y$$

$$= \{(n,m) \mid m \in (e_n \supset x,y)\}.$$

We can show $0 \notin \mathrm{cond}(x)(y)$. Note first $0 = (n,m)$ iff $n = 0 = m$; furthermore, $e_0 = \bot$ and $\bot \supset x,y = \bot$; but $0 \notin \bot$. Also we have:

$$\mathrm{cond}(x)(y)(0) = x \quad\text{and}\quad \mathrm{cond}(x)(y)(1) = y;$$

so if either $x \neq \bot$ or $y \neq \bot$, then $\mathrm{cond}(x)(y) \neq \bot$. In this case, $\mathrm{cond}(x)(y)$ must contain *positive* elements. The result now follows.

Theorem 3.1 is obvious from the construction of G, because $G(G) = 0$ and $G(0)(0) = \mathrm{suc}$, and so the $G(0)(i)$ give us all the other combinators.

The primitive recursive functions needed for (4) - (6) are standard. Equation (7) is a definition — if we rewrote it using the Y-operator — and the proof of 3.2 is easy by induction. There is also no difficulty with (8) - (12). The idea of the proof of 3.3 is contained in the statement of the theorem itself. The proof of 3.4 is already outlined in the text.

Proof of 3.5. The argument is essentially the original one of Myhill-Shepherdson. Suppose p is computable, total and extensional. Define:

$$q = \{(j,m) | m \in \mathrm{val}(p(\mathrm{fin}(j)))\},$$

where fin is primitive recursive, and for all $j \in \omega$:

$$\mathrm{val}(\mathrm{fin}(j)) = e_j .$$

Certainly $q \in \mathrm{RE}$, and we will establish the theorem if we can prove "continuity":

$$\mathrm{val}(p(n)) = \bigcup\{\mathrm{val}(p(\mathrm{fin}(j))) | e_j \subseteq \mathrm{val}(n)\}.$$

We proceed by contradiction. Suppose first we have a $k \in \mathrm{val}(p(n))$, where $k \notin \mathrm{val}(p(\mathrm{fin}(j)))$ whenever $e_j \subseteq \mathrm{val}(n)$. Pick r to be a primitive recursive function whose range is *not* recursive. Define s, primitive recursive, so that for all $m \in \omega$:

$$\mathrm{val}(s(m)) = \{j \in \mathrm{val}(n) | m \notin \{r(i) | i \leq j\}\}.$$

The set $\mathrm{val}(n)$ must be *infinite*, because p is extensional, and if $\mathrm{val}(n) = e_j = \mathrm{val}(\mathrm{fin}(j))$, then $k \notin \mathrm{val}(p(\mathrm{fin}(j))) = \mathrm{val}(p(n))$. Note that $\mathrm{val}(s(m))$, as a subset of the infinite set, is *finite* if m is in the range of r; otherwise it is equal to $\mathrm{val}(n)$. Again by the extensionality of p we see that $k \in \mathrm{val}(p(s(m)))$ if and only if m is *not* in the range of r. But this puts an r.e. condition on m equivalent to a non-r.e. condition, which shows there is no such k.

For the second case suppose we have a $k \notin \mathrm{val}(p(n))$, where for a suitable $e_j \subseteq \mathrm{val}(n)$ it is the case that $k \in \mathrm{val}(p(\mathrm{fin}(j)))$. Define:

$$t = \lambda m \in \omega . e_j \cup (\mathrm{val}(m) \supset \mathrm{val}(n), \mathrm{val}(n))$$

We have:

$$t(m) = e_j \quad , \text{ if } \mathrm{val}(m) = \bot;$$
$$= \mathrm{val}(n), \text{ if not.}$$

We choose u primitive recursive, where:

$$\mathrm{val}(u(m)) = t(m).$$

By the choice of k, and by the extensionality of p, and by the fact that $\mathrm{val}(n) \neq e_j$, we have:

$$k \in \mathrm{val}(p(u(m))) \text{ iff } t(m) = e_j$$
$$\text{iff } \mathrm{val}(m) = \bot .$$

But this is impossible, since one side is r.e. in m and the other is not by 3.4. As both cases lead to contradiction, continuity is established and the proof is complete.

Proof of 3.6. Consider a degree $\mathrm{Deg}(a)$. This set is closed under application, because:

$$u(a)(v(a)) = \mathrm{S}(u)(v)(a),$$

and $\mathrm{S}(u)(v)$ is r.e. if both u and v are. Note that it also contains the element G; hence, as a subalgebra, it is generated by a and G.

Let A be any finitely generated subalgebra with generators $a'_0, a'_1, \ldots, a'_{n-1}$. Consider the element $a = \mathrm{cond}(\langle a'_0, a'_1, \ldots, a'_{n-1}\rangle)(\mathrm{G})$. As in the proof of 3.1, a generates A under application. It is then easy to see why $A = \mathrm{Deg}(a)$.

Proof of 3.7. We first establish (16) and (17):

$$\begin{aligned} \mathrm{L} \circ \overline{u} \circ \mathrm{R} &= \lambda x.\mathrm{L}(\overline{u}(\langle 0, x\rangle)) \\ &= \lambda x.\mathrm{L}(\langle 1, u, x\rangle) \\ &= \lambda x.u(x). \end{aligned}$$

$$\begin{aligned} \overline{u} \circ \overline{v} \circ \mathrm{R} &= \lambda x.\overline{u}(\overline{v}(\langle 0, x\rangle)) \\ &= \lambda x.\overline{u}(\langle 1, v, x\rangle) \\ &= \lambda x.\overline{u(v)}(x) \\ &= \overline{u(v)}. \end{aligned}$$

Now starting with any $u \in \mathrm{RE}$, we write $u = \tau$, where τ is formed from G by application alone. By (17), we can write $\overline{u}$ in terms of $\overline{\mathrm{G}}$ and R using only $\circ$. That is, $\overline{u}$ belongs to a special subsemigroup. In view of (16), we find that $\lambda x.u(x)$ belongs to that generated by R, L and $\overline{\mathrm{G}}$. But

$$\mathrm{RE} \cap \mathrm{FUN} = \{\lambda x.u(x) \mid u \in \mathrm{RE}\},$$

and so the theorem is proved.

<u>*For Section 4.*</u> The notion of a *continuous lattice* is due to Scott (1970/7 and we shall not review all the facts here. One special feature of these lattices is that the lattice operations of meet and join ($\sqcap$ and $\sqcup$) are continuous (that is, commute with directed sups). As topological spaces, they can be characterized as those T_0-spaces satisfying the Extension Theorem (which we proved for $P\omega$ in 1.5).

Proof of 4.1. Consider a continuous function a, and let $A = \{x \mid x = a(x)\}$. By the Fixed-Point Theorem 1.4 we know that A is non-empty and that it has a least element under $\subseteq$. Certainly A is partially ordered by $\subseteq$; further, A is closed under *directed* unions but not under arbitrary unions. That is, A is not a complete sublattice of $P\omega$ with regard to the lattice operations of $P\omega$, but it could be a complete lattice on its own – if we can show sups exist. Thus, let $S \subseteq A$ be an arbitrary subset of A. By the Fixed-Point Theorem, find the least solution to the equation:

$$y = \bigcup\{x \mid x \in S\} \cup a(y)$$

Clearly $x \subseteq y$ for all $x \in S$; and so $x = a(x) \subseteq a(y)$, for all $x \in S$. This means that $y = a(y)$, and thus $y \in A$. By construction, then, y is an upper bound to the elements of S. Suppose $z \in A$ is another upper bound for S. It will also satisfy the above equation; thus $y \subseteq z$, and so y is the least upper bound. A partially ordered set with sups also has infs, as is well known, and is a complete lattice.

Suppose that a is a retract. We can easily show that the fixed-point set A (with the relative topology from $P\omega$) satisfies the Extension Theorem. For assume $f : X \to A$ is continuous, and $X \subseteq Y$ as a subspace. Now we can also regard $f : X \to P\omega$ as continuous because A is a subspace of $P\omega$. By 1.5 there is an extension to a continuous $\bar{f} : Y \to P\omega$. But then $a \circ \bar{f} : Y \to A$ is the continuous function we want for A, and the proof is complete.

The Space of Retracts. Let us define:

$$\mathsf{RET} = \{a \mid a = a \circ a\},$$

the set of all retracts, which is a complete lattice in view of 4.1. It will be proved to be not a retract itself by showing it is not a continuous lattice; in fact, the meet operation on RET is not continuous on RET.

The proof was kindly communicated by Y. L. Ershov and rests on distinguishing some extreme cases of retracts. Call a retract a *nonextensive* if for all nonempty finite sets x we have $x \not\subseteq a(x)$. Call a retract b *finite* if all its values are finite (*i.e.* $b(\top)$ is finite). If a is nonextensive and b is finite, then Ershov notes that they are "orthogonal" in RET in the sense that $c = a \sqcap b = \bot$. The reason is that, since $c \subseteq b$, it is finite; but $c \subseteq a$, too, so $c(x) \subseteq a(x)$ for all x. Because c is a retract, we have $c(x) = c(c(x)) \subseteq a(c(x))$. As $c(x)$ is finite and a is nonextensive, it follows that $c(x) = \bot$ for all x.

This orthogonality is unfortunate, because consider the finite retracts $b_n = \lambda x.e_n$. We have here a directed set of retracts where $\bigcup\{b_n \mid n \in \omega\} = \lambda x.\top = \top$. If $\sqcap$ were continuous, it would follow that for nonextensive a:

$$a = a \sqcap \top = a \sqcap \bigcup\{b_n \mid n \in \omega\} = \bigcup\{a \sqcap b_n \mid n \in \omega\} = \bot,$$

showing that there are no nontrivial such a. But this is not so.

Let $\ll$ be a strict linear ordering of ω in the order type of the rational numbers. Define:

$$a(x) = \{m \mid \exists n \in x.\, m \ll n\}.$$

We see at once that a is continuous; and, because $\ll$ is transitive and dense, a is a retract. Since $\ll$ is irreflexive, it is the case for finite nonempty sets x that $\max_{\ll}(x) \notin a(x)$; hence, a is nonextensive. As $a(\top) = \top$, we find $a \neq \bot$. The proof is complete.

Note that there are many transitive, dense, irreflexive relations on ω, so there are many nonextensive retracts. These retracts, like a above, are *distributive*. A nondistributive example is:

$$a' = \{m \mid \exists n, n' \in x.\, n \ll m \ll n'\}$$

Many other examples are possible.

Proof of 4.2. The relation $\leqslant$ is by the definition of retract reflexive on RET; it is also obviously antisymmetric. The transit goes like this: suppose $a \leqslant b \leqslant c$, then

$$a = a \circ b = a \circ b \circ c = a \circ c.$$

Similarly, $a = c \circ a$. Note, by the way, that $a \leqslant b$ implies that the range of a is included in that of b; but that the relationship $a \subseteq b$ does not imply this fact. The relationship $a \leqslant b$, however, is stronger than inclusion of ranges.

Proofs of 4.3 - 4.5. We will not give full details as all the parts of these theorems are direct calculations. Consider by way of example 4.3.(*i*). We find:

$$\begin{aligned}(a \hookrightarrow b) \circ (a \hookrightarrow b) &= \lambda u.\, b \circ (b \circ u \circ a) \circ a \\ &= \lambda u.\, b \circ u \circ a \\ &= a \hookrightarrow b,\end{aligned}$$

provided that a and b are retracts. A very similar computation would verify part (*iv*), if one writes out the composition:

$$(a \hookrightarrow b') \circ (f \hookrightarrow f') \circ (b \hookrightarrow a'),$$

and uses the equations:

$$f = b \circ f \circ a \quad \text{and} \quad f' = b' \circ f' \circ a'.$$

The main point of the proof of 4.6 has already been given in the text.

For Sections 5 - 7. Sufficient hints for proofs have been given in the text.

The author regrets that, due to unforeseen circumstances, it was impossible to include the bibliography in this version. The editors felt, however, that the loss would have been much greater had the paper itself been omitted from the volume.

Vol. 399: Functional Analysis and its Applications. Proceedings 1973. Edited by H. G. Garnir, K. R. Unni and J. H. Williamson. II, 584 pages. 1974.

Vol. 400: A Crash Course on Kleinian Groups. Proceedings 1974. Edited by L. Bers and I. Kra. VII, 130 pages. 1974.

Vol. 401: M. F. Atiyah, Elliptic Operators and Compact Groups. V, 93 pages. 1974.

Vol. 402: M. Waldschmidt, Nombres Transcendants. VIII, 277 pages. 1974.

Vol. 403: Combinatorial Mathematics. Proceedings 1972. Edited by D. A. Holton. VIII, 148 pages. 1974.

Vol. 404: Théorie du Potentiel et Analyse Harmonique. Edité par J. Faraut. V, 245 pages. 1974.

Vol. 405: K. J. Devlin and H. Johnsbråten, The Souslin Problem. VIII, 132 pages. 1974.

Vol. 406: Graphs and Combinatorics. Proceedings 1973. Edited by R. A. Bari and F. Harary. VIII, 355 pages. 1974.

Vol. 407: P. Berthelot, Cohomologie Cristalline des Schémas de Caracteristique p > o. II, 604 pages. 1974.

Vol. 408: J. Wermer, Potential Theory. VIII, 146 pages. 1974.

Vol. 409: Fonctions de Plusieurs Variables Complexes, Séminaire François Norguet 1970–1973. XIII, 612 pages. 1974.

Vol. 410: Séminaire Pierre Lelong (Analyse) Année 1972–1973. VI, 181 pages. 1974.

Vol. 411: Hypergraph Seminar. Ohio State University, 1972. Edited by C. Berge and D. Ray-Chaudhuri. IX, 287 pages. 1974.

Vol. 412: Classification of Algebraic Varieties and Compact Complex Manifolds. Proceedings 1974. Edited by H. Popp. V, 333 pages. 1974.

Vol. 413: M. Bruneau, Variation Totale d'une Fonction. XIV, 332 pages. 1974.

Vol. 414: T. Kambayashi, M. Miyanishi and M. Takeuchi, Unipotent Algebraic Groups. VI, 165 pages. 1974.

Vol. 415: Ordinary and Partial Differential Equations. Proceedings 1974. XVII, 447 pages. 1974.

Vol. 416: M. E. Taylor, Pseudo Differential Operators. IV, 155 pages. 1974.

Vol. 417: H. H. Keller, Differential Calculus in Locally Convex Spaces. XVI, 131 pages. 1974.

Vol. 418: Localization in Group Theory and Homotopy Theory and Related Topics. Battelle Seattle 1974 Seminar. Edited by P. J. Hilton. VI, 172 pages 1974.

Vol. 419: Topics in Analysis. Proceedings 1970. Edited by O. E. Lehto, I. S. Louhivaara, and R. H. Nevanlinna. XIII, 392 pages. 1974.

Vol. 420: Category Seminar. Proceedings 1972/73. Edited by G. M. Kelly. VI, 375 pages. 1974.

Vol. 421: V. Poénaru, Groupes Discrets. VI, 216 pages. 1974.

Vol. 422: J.-M. Lemaire, Algèbres Connexes et Homologie des Espaces de Lacets. XIV, 133 pages. 1974.

Vol. 423: S. S. Abhyankar and A. M. Sathaye, Geometric Theory of Algebraic Space Curves. XIV, 302 pages. 1974.

Vol. 424: L. Weiss and J. Wolfowitz, Maximum Probability Estimators and Related Topics. V, 106 pages. 1974.

Vol. 425: P. R. Chernoff and J. E. Marsden, Properties of Infinite Dimensional Hamiltonian Systems. IV, 160 pages. 1974.

Vol. 426: M. L. Silverstein, Symmetric Markov Processes. X, 287 pages. 1974.

Vol. 427: H. Omori, Infinite Dimensional Lie Transformation Groups. XII, 149 pages. 1974.

Vol. 428: Algebraic and Geometrical Methods in Topology, Proceedings 1973. Edited by L. F. McAuley. XI, 280 pages. 1974.

Vol. 429: L. Cohn, Analytic Theory of the Harish-Chandra C-Function. III, 154 pages. 1974.

Vol. 430: Constructive and Computational Methods for Differential and Integral Equations. Proceedings 1974. Edited by D. L. Colton and R. P. Gilbert. VII, 476 pages. 1974.

Vol. 431: Séminaire Bourbaki – vol. 1973/74. Exposés 436–452. IV, 347 pages. 1975.

Vol. 432: R. P. Pflug, Holomorphiegebiete, pseudokonvexe Gebiete und das Levi-Problem. VI, 210 Seiten. 1975.

Vol. 433: W. G. Faris, Self-Adjoint Operators. VII, 115 pages. 1975.

Vol. 434: P. Brenner, V. Thomée, and L. B. Wahlbin, Besov Spaces and Applications to Difference Methods for Initial Value Problems. II, 154 pages. 1975.

Vol. 435: C. F. Dunkl and D. E. Ramirez, Representations of Commutative Semitopological Semigroups. VI, 181 pages. 1975.

Vol. 436: L. Auslander and R. Tolimieri, Abelian Harmonic Analysis, Theta Functions and Function Algebras on a Nilmanifold. V, 99 pages. 1975.

Vol. 437: D. W. Masser, Elliptic Functions and Transcendence. XIV, 143 pages. 1975.

Vol. 438: Geometric Topology. Proceedings 1974. Edited by L. C. Glaser and T. B. Rushing. X, 459 pages. 1975.

Vol. 439: K. Ueno, Classification Theory of Algebraic Varieties and Compact Complex Spaces. XIX, 278 pages. 1975

Vol. 440: R. K. Getoor, Markov Processes: Ray Processes and Right Processes. V, 118 pages. 1975.

Vol. 441: N. Jacobson, PI-Algebras. An Introduction. V, 115 pages. 1975.

Vol. 442: C. H. Wilcox, Scattering Theory for the d'Alembert Equation in Exterior Domains. III, 184 pages. 1975.

Vol. 443: M. Lazard, Commutative Formal Groups. II, 236 pages. 1975.

Vol. 444: F. van Oystaeyen, Prime Spectra in Non-Commutative Algebra. V, 128 pages. 1975.

Vol. 445: Model Theory and Topoi. Edited by F. W. Lawvere, C. Maurer, and G. C. Wraith. III, 354 pages. 1975.

Vol. 446: Partial Differential Equations and Related Topics. Proceedings 1974. Edited by J. A. Goldstein. IV, 389 pages. 1975.

Vol. 447: S. Toledo, Tableau Systems for First Order Number Theory and Certain Higher Order Theories. III, 339 pages. 1975.

Vol. 448: Spectral Theory and Differential Equations. Proceedings 1974. Edited by W. N. Everitt. XII, 321 pages. 1975.

Vol. 449: Hyperfunctions and Theoretical Physics. Proceedings 1973. Edited by F. Pham. IV, 218 pages. 1975.

Vol. 450: Algebra and Logic. Proceedings 1974. Edited by J. N. Crossley. VIII, 307 pages. 1975.

Vol. 451: Probabilistic Methods in Differential Equations. Proceedings 1974. Edited by M. A. Pinsky. VII, 162 pages. 1975.

Vol. 452: Combinatorial Mathematics III. Proceedings 1974. Edited by Anne Penfold Street and W. D. Wallis. IX, 233 pages. 1975.

Vol. 453: Logic Colloquium. Symposium on Logic Held at Boston, 1972–73. Edited by R. Parikh. IV, 251 pages. 1975.

Vol. 454: J. Hirschfeld and W. H. Wheeler, Forcing, Arithmetic, Division Rings. VII, 266 pages. 1975.

Vol. 455: H. Kraft, Kommutative algebraische Gruppen und Ringe. III, 163 Seiten. 1975.

Vol. 456: R. M. Fossum, P. A. Griffith, and I. Reiten, Trivial Extensions of Abelian Categories. Homological Algebra of Trivial Extensions of Abelian Categories with Applications to Ring Theory. XI, 122 pages. 1975.